15th International Conference on Stability, Handling and Use of Liquid Fuels (IASH 2017)

Stability, Handling and Use of Liquid Fuels

AA009115

Rome, Italy
10-14 September 2017

Editor:

Robert E. Morris

ISBN: 978-1-5108-7066-6

Printed from e-media with permission by:

Curran Associates, Inc.
57 Morehouse Lane
Red Hook, NY 12571

Some format issues inherent in the e-media version may also appear in this print version.

Copyright© (2017) by International Association for Stability, Handling and Use of Liquid Fuels (IASH)
All rights reserved.

Printed by Curran Associates, Inc. (2018)

For permission requests, please contact International Association for Stability, Handling and Use of Liquid Fuels (IASH) at the address below.

International Association for Stability, Handling and Use of Liquid Fuels (IASH)
3525 Piedmont Road
5 Piedmont Center
Suite 300
Atlanta, GA 30305
USA

Phone: +1 404.760.2829

www.iash.net

Additional copies of this publication are available from:

Curran Associates, Inc.
57 Morehouse Lane
Red Hook, NY 12571 USA
Phone: 845-758-0400
Fax: 845-758-2633
Email: curran@proceedings.com
Web: www.proceedings.com

15th International Conference on Stability, Handling and Use of Liquid Fuels (IASH 2017)

Stability, Handling and Use of Liquid Fuels

Rome, Italy
10-14 September 2017

IASH - International Association for Stability, Handling and Use of Liquid Fuels, Inc.

TABLE OF CONTENTS

SESSION 3: FUEL CONTAMINATION
Rick Kamin, Session Chair

SESSION 4: BIODIESEL AND DIESEL
Melanie Thom, Session Chair

SESSION 5: AVGAS and GASOLINE

Anne Gandubert, Session Chair

SESSION 6: ALTERNATIVE AVIATION FUEL

Edwin Corporan, Session Chair

SESSION 7: MICROBIOLOGY

Dietmar Posselt, Session Chair

SESSION 8: FUEL TESTING AND TEST METHODS – Section 1

Steve Westbrook, Session Chair

SESSION 8: FUEL TESTING AND TEST METHODS – Section 2
Amy Carico , Session Chair

SESSION 8: FUEL TESTING AND TEST METHODS – Section 3
Anthony Kitson-Smith, Session Chair

SESSION 9: FUEL PROPERTIES AND EFFECTS – Section 1

David Abdallah, Session Chair

SESSION 9: FUEL PROPERTIES AND EFFECTS – Section 2

David Abdallah, Session Chair

SESSION 10: FUEL CHEMISRTY, RESEARCH AND DEVELOPMENT – Section 1

Matthew DeWitt, Session Chair

SESSION 10: FUEL CHEMISTRY, RESEARCH AND DEVELOPMENT – Section 2

Matthew DeWitt, Session Chair

SESSION 11: FUEL CHEMISTRY, RESEARCH AND DEVELOPMENT – Section 3

Robert Morris, Session Chair

POSTER SESSION

Steven Zabarnick, Session Chair

IASH ®, The International Association for

Stability, Handling and Use of Liquid Fuels

www.iash.net

Proceedings of the 15th International Conference on Stability Handling and Use of Liquid Fuels

Edited by

Dr. Robert E. Morris
IASH Secretary
Nova Research, Inc.
1900 Elkin Street, Suite 230
Alexandria, VA 22308

Rome, Italy
September 10-14, 2017

PREFACE

Theme: **Fuel Specifications and the Impact of Distribution and New Production Technologies**

Location: **Rome, Italy**
Dates: **September 10-14, 2017**

The Sheraton Parco De'Medici Rome Hotel in Rome, Italy was the venue for the 15th International Conference organized by the International Association for the Stability, Handling and Use of Liquid Fuels. The conference was moved from October to September and had one of our largest European conferences!

The conference began with an opening reception held at the Sheraton Parco De'Medici on Sunday, 10 September, and ended with a closing reception poolside at the Sheraton on Thursday, 14 September. The Awards Dinner was held on Wednesday evening on the terrace of the Villa Miani overlooking Rome.

There were 166 delegates and 40 spouses/guests in attendance from 23 countries. Delegates viewed 21 poster presentations and 9 company exhibits at the conference.

The Keynote address was presented by Ortwin Costenoble, CEN Fuels Standardizaation. The title of his presentation was *"How European Standards Guarantee Fuel Quality, Vehicle Function and Emissions."* It was an excellent presentation relating to the theme of the 15th International Conference.

The Chevron Research Award of Excellence in Honor of John Bacha was presented to Mariam Ajam from Sasol in South Africa, for her paper entitled: *"Phenols and Aromatic Methyl Ethers from Biomass Pyrolysis Oil: Implications for Jet Fuel."*

Three other young researchers who submitted abstracts for the award were also given the opportunity to present their work at the conference and also received gratis registration. They were: Matthew Dwyer, the University of Sheffield (UK); Moshe Rabaev, Ben-Gurion University of the Negev (Israel); and Chrysovalanti Tsesmeli, Technical University of Athens (Greece).

The Nahum Por Award was presented to Anthony Kitson-Smith, The Clouds Network, for his leadership, support, and promotion of the Association over the past several years. IASH Lifetime Achievement Awards were given to the following individuals for their dedicated efforts in the field of aviation fuels throughout their careers:

- Gary Bessee, Southwest Research Institute
- Joanna Bauldreay, Shell Global
- Alistair Clark, Air BP
- Yaakov Sufrin, Petroleum & Energy Infrastructure
- Randy Williams, Honeywell Aerospace

Melanie Thom, Baere Aerospac, received the second Chairman's Award for her dedicated service to IASH in preparing the IASH Newsletter for several years.

Fifty-One presentations were given in eleven sessions that included the following topics:

- Session 1: Fuel Production and Quality Control
- Session 2: Marine Diesel Fuels
- Session 3: Fuel Contamination
- Session 4: Biodiesel and Diesel
- Session 5: AVGAS and Gasoline
- Session 6: Alternative Aviation Fuel
- Session 7: Microbiology
- Session 8: Fuel Testing and Test Methods – Section 1
- Session 8: Fuel Testing and Test Methods – Section 2
- Session 8: Fuel Testing and Test Methods – Section 3
- Session 9: Fuel Properties and Effects – Section 1
- Session 9: Fuel Properties and Effects – Section 2
- Session 10: Fuel Chemistry, Research and Development – Section 1
- Session 10: Fuel Chemistry, Research and Development – Section 2
- Session 11: Fuel Chemistry, Research and Development – Section 3

Conference Proceedings are available to members on-line through the IASH Library at www.iash.net. Contact the IASH Administrator for futther information.

The Conference Chairman and the IASH Board of Directors would like to thank all of the presenters for their time and effort in producing excellent presentations and the delegates themselves for supporting the conference by their attendance and active participation.

Many thanks to the 12 sponsoring organizations and companies without whose financial support we would not have been able to stage such a memorable 15th International Conference.

Gold Sponsors: Parker Hannifin and Pratt & Whitney

Silver Sponsors: AD Systems, Air BP, Falex, Shell Aviation, Southwest Research Institute; and U.S. Defense Logistics Agency Energy.

Exhibitors: AD Systems, Air BP, CHIMEC, ECHA Microbiology, Falex, Stanhope-Seta, Parker Hannifin, Southwest Research Institute, University of Dayton Research Institute

Pamela Serino
Conference Chair / 1st Vice-Chair IASH

PREVIOUS CONFERENCES

1st **Internal Conference on Long-Term Storage Stabilities of Liquid Fuels**, Tel Aviv, Israel, July 11-14, 1983. .

2nd **International Conference on Long-Term Storage Stabilities of Liquid Fuels**, San Antonio, Texas, USA, July 29-August 1, 1986.

3rd **International Conference on Stability and Handling of Liquid Fuels**, London, England, September 13-16, 1988.

4th **International Conference on Stability and Handling of Liquid Fuels**, Orlando, Florida, USA, November 19-22, 1991.

5th **International Conference on Stability and Handling of Liquid Fuels**, Rotterdam, the Netherlands, October 3-7, 1994. .

6th **International Conference on Stability and Handling of Liquid Fuels**, Vancouver, BC, Canada, October 13-17, 1997. .

7th **International Conference on Stability and Handling of Liquid Fuels**, Graz, Austria, September 24-29, 2000.

8th **International Conference on Stability and Handling of Liquid Fuels**, Steamboat Springs, Colorado, USA, September 14-19, 2003.

9th **International Conference on Stability and Handling of Liquid Fuels**, Sitges, Spain, September 18-22, 2005.

10th **International Conference on Stability and Handling of Liquid Fuels**, Tucson, Arizona, USA. October 5 – 11, 2007.

11th **International Conference on Stability and Handling of Liquid Fuels**, Prague, Czech Republic. October 18 – 22, 2009.

12th **International Conference on Stability and Handling of Liquid Fuels**, Sarasota, Florida, USA. October 16 – 20, 2011.

13th **International Conference on Stability and Handling of Liquid Fuels**, Rhodes, Greece, October 6 – 10, 2013.

14th **International Symposium on Stability and Handling of Liquid Fuels**, Charleston, Couth Carolina, USA, October 4 – 8, 2013.

15th **International Conference on Stability and Handling of Liquid Fuels**, Rome, Italy, September 10-14, 2017.

CONFERENCE SPONSORS

Gold Sponsors

Parker Hannifin

Pratt & Whitney

Silver Sponsors

AD Systems

Air BP

FALEX

Shell Aviation

Southwest Research Institute

U.S. Defense Logistics Agency Energy

CONFERENCE EXHIBITORS

AD Systems

Air BP

Chimec

ECHA Microbiology

FALEX

Parker Hannifin

Southwest Research Institute

Stanhope-Seta

University of Dayton Research Institute

Analysis and Diagnosis Systems

AD Systems specializes in the design, manufacturing, selling and servicing of unique test equipment for the evaluation of physical properties and performances of fuels and lubricants.

All of the products offered by AD systems were developed in response to specific needs within the petroleum industry.

AD Systems has in the past and is currently working with governmental agencies and research departments of petroleum companies to develop prototypes and subsequently to manufacture various instruments and systems. Our customers include oil refineries, commercial laboratories, fuel distributors, terminals, blenders, lubricant plants, OCM labs, research labs, marine, aviation companies and military.

AD Systems cumulates 60 years of experience in designing test equipment for petroleum industry. Our commitment is to deliver prompt, friendly and attentive customer service and technical support.

AD Systems is an active member of the American Society for Test and Materials (ASTM®) as well as with Energy Institute, GFC and BNPé.

SOME OF OUR PRODUCTS

Jet Fuel Thermal Oxidation Deposit Rater - DR10 (ASTM D3241, IP 323)

The DR 10 accurately determines the thickness of the deposit, reducing test subjectivity.

The DR10 test report provides a 3-D profile of the deposit distribution on the tube surface. The software automatically detects and reports the Standard Spot value which is the mean deposit thickness of the thickest 2.5 mm2 area as defined in ASTM D3241. The average and the maximum thickness deposit values, as well as the calculated total deposit volume are reported.

Automated Smoke Point – SP10 (ASTM D1322, IP 598)

This unique automated instrument uses a patented system (License TOTAL RM) based on a video camera that measures the flame. The flame image is digitalized. The dedicated software determines the height of the flame when its shape corresponds to the one described in the test method. The SP10 is an Automated Smoke Point instrument that strictly follows the test method while both eliminating the subjectivity inherent to the manual test and significantly improving repeatability and reproducibility. Additionally, the SP10 eliminates all safety risks linked to the visual observation of an open flame. The test can now be run unattended which drastically reduces labor costs.

NACE Spindle Corrosion Test – CT10 (NACE TM0172)

The innovative CT10 instrument images the whole surface of the specimen. Operation is based on a homogeneous lighting source, CCD camera, specimen rotation system, and specially designed Windows CE® application software. The specimen is rotated and several images are taken. The software builds a flat image of the specimen surface, calculates the percentage of the corroded area and then translates it into a NACE rating. A detailed test report is ready in less than 5 minutes. The test can now be run unattended which reduces labor costs.

On-site Automated Fuel Stability and Compatibility Tester – ST10 (ASTM D4740)

AD Systems has developed a portable / on-site / on-board fully automated instrument for stability and compatibility testing of heavy fuel oil. This portable device is a complete automation of the ASTM D4740 method. This method is very simple and practical for the test but it had the serious handicap to rely on the know-how and subjectivity of the operator. We have grouped all the phases of the test in a suitcase, the conditioning of the samples, and the preparation of the spot, its drying and the automatic rating by camera and associated software. The only phases under the load of the operator are the sampling and the deposition of the drop with the aid of a micropipette supplied with the equipment. As a result, the apparatus can be used anywhere by untrained persons for this test whenever power is available. The apparatus is provided with a database for storing the results. The image of the spot is memorized at the time the spot is rated and we therefore ensure perfect traceability of the test.

AD SYSTEMS IN A FEW DATES

2008 – Creation of AD Systems

2009 – Launching of the DT100 - Dispersancy Tester for in service lubricant

2010 – Launching of the DR10 – JFTOT® Deposit Rater

Launching of the AutoREID - Vapor Pressure for crude oil

2011 – Launching of the SP10, approval of the DR 10 by Energy Institute (IP597)

2012 – Approval of the SP10 by Energy Institute and ASTM

2013 – AD Systems moves in a new 600 sq. meter facility
Launching of the CT10 – NACE Spindle Corrosion Test

2014 – Approval of the DR10 by ASTM

2017 – Launching of the ST10 - On-site Automated Fuel Stability and Compatibility Tester – ST10 (ASTM D4740)

SOME OF CUSTOMERS

ExxonMobil, BP, Shell, Chevron, Total, Petrobras, PDVSA, Intertek, SGS, US Army, Saudi Aramco, UOP, SWRI, Neste Oil, Etc. Instruments in use every day in over 100 countries.

AD systems
P.A. Portes de la Suisse Normande
Allée de Cindais
14320 Saint André sur Orne
France

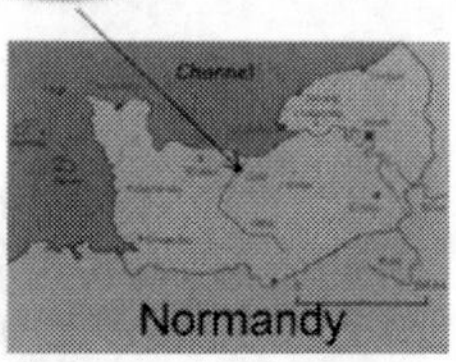

For more information, visit AD Systems website at www.adsystems-sa.com

air bp
A global leader in aviation fuel & services
As the aviation division of BP, Air BP is one of the world's largest suppliers of aviation fuel products and services. It currently supplies around seven billion gallons of jet kerosene and aviation gasoline each year. Air BP fuels around 6,000 flights everyday – that's four every minute – at more than 800 locations in over 50 countries. Customers include domestic and international airlines, the military, business and private aircraft owners, international airports and airfield operators.
Supporting its fuel offer, Air BP's team of experts provide a range of services for customers including technical expertise via a complete aviation fuel consultancy service, including the design, build and operation of fuelling facilities, to help customers protect their operations and manage risks.
Other services on offer include supporting customers to meet their environmental goals, the Air BP Sterling Card for efficient general aviation refuelling and RocketRoute MarketPlace enabling users to purchase fuel and connect with thousands of ground handlers, fixed base operators (FBOs) and aviation providers worldwide.
With over 90 years history and experience, Air BP is committed to continue providing expert support and tailored fuelling solutions that add value to our customers' businesses today and tomorrow.
www.airbp.com

CHIMEC COMPANY PROFILE

Active worldwide since over 45 years, CHIMEC has distinguished itself in developing and applying innovative solutions to the Oilfield, Refining and Petrochemical industries.
CHIMEC holding includes five different fully owned companies. The parent company is CHIMEC S.p.A., based in Rome, Italy, where our Commercial Offices, Development & Marketing, R&D, Technical Department and product plant are located.
CHIMEC worldwide activities are supported by four subsidiaries: CHIMEC GmbH (Germany), OOO KIMEK (Russian Federation), CHIMEC Asia Pacific Pte Ltd (Singapore), CHIMEC Gulf Fze (United Arab Emirates) and one sub subsidiary: Chimec Thailand (Thailand).
Several commercial offices are locates in the industrial areas where Chimec operates on a daily basis, Europe, Asia, Africa, Middle and Far East.
Our production includes five main business line: Oilfield, Process, Fuel Additives, Water and Decontamination. We are specialized in the development and application of performance chemicals and technologies for the E&P, Refining and Petrochemical industries as well as other industrial areas (metal working, pharmaceutical, etc.).
Our specialists work in very different areas, such as oil & gas fields, refineries, chemical and petrochemical plants, steel mills, paper mills, sugar refineries and pharmaceutical industry.
Our additives, technologies and consulting services are aimed at optimizing processes, maximizing profits and reducing the environmental impact.
Fuel Additives Development & Marketing is responsible for the technical and commercial development of performance chemicals applied to petroleum and petrochemical products as well as biofuels. Our first target is the development of additives which meet market requirements in terms of performance, quality and cost efficiency and to help our customers to solve their fuel's stability, performance, storage, handling and environmental issues, understanding their specific needs and finding the right treatment in terms of cost and performance.
To meet this aim, we work in strict collaboration with our clients formulating, when necessary, tailored products and guaranteeing on-going technical and analytical support, including specific monitoring and predicting tools. During the years we created a network with Universities, Research centers, Institutions and external laboratories, not to forget our customers, that are our most important collaborators.
Our portfolio includes all types of chemical additives normally used to improve the quality and performances of fuels (gasoline, kerosene, diesel, bunker) and of other petroleum products (intermediates, bitumen, residue). With our broad range of antioxidant and stabilizers we can support our clients in preventing changes in quality. Our flow improvers, conductivity improvers and H2S scavengers guarantee an easy and safe use and transport of any type of fuel. Lubricity improvers and corrosion inhibitor, together with our innovative DCA, allow to protect our final clients' engines.
A range of biocides and biodispersants developed to solve all the problems caused by the microbiological contamination of fuels, in particular gasoil/biodiesel blends.

For further details please visit Chimec website at *www.chimec.com* or contact us at *fueladditives@chimec.it*.

echa
Results that count

IASH - International Association for
Stability, Handling and Use of Liquid Fuels, Inc.

ECHA Microbiology has been helping industry to solve microbiological contamination and corrosion problems for more than 30 years.

By helping our customers understand and avoid the consequences and costs of microbiological contamination, we increase their safety, efficiency and predictability.

We offer consultancy, testing and training services and have globally renowned expertise in the aviation, marine, land transport and energy sectors. We also produce one of the world's most widely-used microbiological test kits for fuels; trusted by many of the world's leading producers and users.

As an expert team of industrial microbiologists, we spend all our time solving your problems or engaged in research and development.

We are regarded as leading authorities on the causes and consequences of microbial contamination – as well as the solutions.

Laboratory Analysis

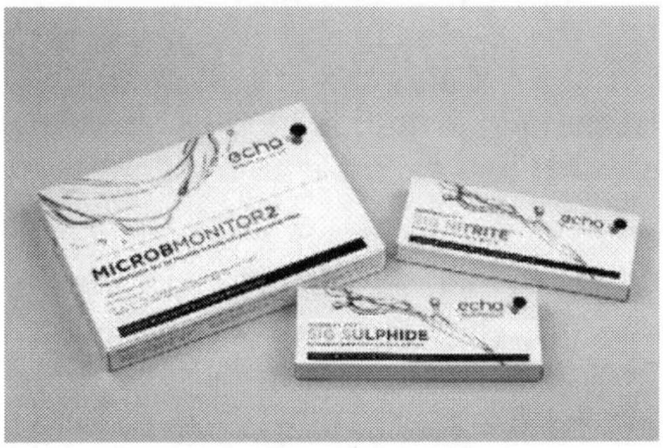

Microbiological Test Kits

Training Courses

Site Surveys

T: +44 (0)29 2036 5930 **E:** info@echamicrobiology.com **W:** www.echamicrobiology.com

© ECHA Microbiology Ltd, Units 22-23 Willowbrook Technology Park, Cardiff, CF3 0EF, United Kingdom

TRUSTED + INNOVATIVE + ACCURATE + DEPENDABLE

WWW.FALEX.COM

Falex Corporation provides high quality petroleum and tribology testing equipment and services for the evaluation of the physical properties of fuels, lubricants, and materials. Founded in 1927, Falex offers an unparalleled level of experience supplying standard and custom laboratory instruments, high precision test consumables and spares, laboratory testing services, and expert instrument calibration and repair.

Falex instruments stand the test of time, capable of providing precise test data for decades. The world's leading aerospace, petroleum, automotive, metalworking, and materials companies rely on Falex Testing Equipment as part of their daily production and development process.

The Falex Tests are well established standards for commercial and military product specifications, research and development programs, quality control, and technical support programs.

CONTACTS:

James Hepp
Tel: +1 (630) 556-4835
E-Mail: heppjp@compass-instruments.com

Mike Croudace
Tel: +1 (714) 697-0157
E-Mail: mcroudace@compass-instruments.com

Compass Instruments, Inc. is the North American Distributor for Falex Corporation.

Laboratory Testing Equipment for determining the physical properties and performance characteristics of Fuels, Lubricants, and Materials. North American Distributors of Waukesha Engine-GE CFR Products, Eralytics, Falex Corporation, Normalab, PCS Instruments, Parker Hannifin, PiloDist, Shinto, and Strama-MPS. Instrument Sales and Service. Oxidation ASTM D3241, Octane ASTM D2699 and D2700, Cetane (D613 and D7170), Vapor Pressure (D5188, D5191, D5492, D6377, D6378, D6897), Fuel Lubricity (D5001, D6078 and D6079), Particle Counters, Distillation (D86 and D1160), Flash Point (D56, D92, D93), Oxidation, Penetration, Cold Flow (D2500) and (D97), Tribology Testing. www.compass-instruments.com (630) 556 4835 Sugar Grove, IL USA

Parker Hannifin

Parker Hannifin Manufacturing (UK) Ltd.
Hydraulic Filter Division Europe
Condition Monitoring Centre
Tel: +44(0)1842 763299

The world's best filtration starts with the world's best engineering.

With annual sales of $14 billion in fiscal year 2017, Parker Hannifin is the world's leading diversified manufacturer of motion and control technologies and systems, providing precision-engineered solutions for a wide variety of mobile, industrial and aerospace markets.
Parker Hannifin's Filtration Group manufactures a comprehensive range of filtration and monitoring equipment that is proven in both civilian and military aviation industries. The range includes Parker Condition Monitoring, and Parker Velcon Aviation & Diesel Fuel Handling products. From the refinery to the injector, at the terminal and on the forecourt, Parker Filtration has a filter vessel and element solution to fuel your delivery needs.

Parker Kittiwake - Condition Monitoring Business Unit

A leading manufacturer of condition monitoring technologies including automatic laser particle counters, metallic wear debris sensors, oil testing and analysis instrumentation and sensors as well as Acoustic Emissions for use with bearings and rotating machinery.

These products are used alongside the Parker filtration technology in a number of industries, including commercial marine, oil and gas, power generation and defence markets. The Parker Kittiwake instruments and test kits enable fast on-site maintenance decisions to be made, providing laboratory grade oil condition results in minutes. Parker Kittiwake DIGI Test Kits give simple accurate results for TAN, TBN, Insolubles and Water in oil. For the measurement of ferrous wear debris, the Ferrous Wear Meter (FWM) is a highly accurate instrument designed to measure the contamination of an oil sample from a variety of types of machinery. Another unique offering from Parker Kittiwake is the MHC-Bearing Checker, a hand-held instrument which uses acoustic emissions technology to analyse bearing condition and lubrication state.

Parker's renowned particle counter has been re-engineered and calibrated for use in fuels and allows quick, easy economical fuel condition checks for aviation and diesel fuels. A fast, 2 minute test will allow you to check contamination levels, trends and integrity in a far more consistent reliable and repeatable way than other traditional methods.
DefStan 9191 issue 8, now includes particle counting as an alternative to the current subjective test methods such as Clear & Bright and Gravimetric Millipore, in accordance with an Energy Institute (EI) – IP 564 method statement.

Parker Hannifin, has enhanced its range of icount online particle detectors with the introduction of an innovative new self-contained and portable device for measuring the quality and cleanliness of hydraulic oils and hydrocarbon fuels.

The icountFS (Fuel Sampler), incorporates a laser detection particle counter, battery and pumps into a single integrated unit for the real time condition monitoring of fluids. The iFS provides a compact, simple to setup and easy to use solution for providing an immediate warning of particle and moisture contamination in fluids, allowing remedial action to be taken quickly before system degradation or failure occurs. Available in three versions, for monitoring the quality of hydrocarbon fuels, and either offline or inline oil sampling.

Parker Velcon Aviation & Diesel Fuel Handling

Parker Velcon is the global leader in the world able to meet the exacting requirements as specified by the Energy Institutes in order to be able to supply filtration equipment inside the Airport fence. Parker Velcon Aviation Fuel Filtration products can be found in the following applications:

- Fuel Storage Facilities
- Fuel Storage Equipment Manufacturers
- Aviation Fuel Test Equipment Suppliers
- Aircraft De-fuelling Equipment Suppliers
- Aircraft Services (Manufacture/Supplies)
- Fuel System Design/Manufacture
- Fuel Distributors
- Petroleum Equipment/Product Manufacturers
- Bulk Liquid Equipment/Service Providers

Latest EI Qualification Test

Parker Velcon vessels and elements demonstrate a breadth of experience in the fuel supply industry, producing the most advanced Aviation Fuel EI 1581 5th Edition qualified Filter Water Separators, EI 1590 microfilters and EI 1583 6th Edition Monitors, as well as a wide range of industry standard interchangeable products.
Parker Velcon Contaminant Analyzer (VCA) is the first full-flow, on-line fuel condition monitor with the capability of discerning between solid and water contaminants simultaneously.
The Flow Differential Pressure Monitor (FDPM) was pioneered by Parker Velcon to provide corrected differential pressure for aircraft refuelers.
The Differential Pressure Monitor (DPM) is the latest technology design to be fully compliant with JIG Bulletin No. 58 that is required by all aviation refuelers by 12/31/2013.

CLEANDiesel for Diesel Fuel Filtration

Today's heavy duty diesel engines are powered by high pressure common rail. To help engines run more efficiently, increase mean time between failures and comply with regulatory emission requirements, it starts with CLEAN DRY FUEL. By leveraging what Parker Velcon has learned over the past 60 years, CLEANDiesel solutions were developed to help users achieve fuel clean enough to fly. Parker Velcon CLEANDiesel solutions are designed for several key markets including: Mining, Power Generation, Fleet Transportation, Refineries & Terminals, and Pipelines.

Pratt & Whitney

Pratt & Whitney, a United Technologies Corp. company (NYSE:UTX), is a world leader in the design, manufacture and service of aircraft engines and auxiliary power units. Pratt & Whitney has approximately 35,000 employees who support more than 11,000 customers around the world.

Pratt & Whitney develops game-changing technologies, such as the PurePower® PW1000G engine, with patented Geared Turbofan™ technology, for the next generation of passenger aircraft.

Pratt & Whitney has built a long and distinguished record of providing top-of-the-line engines to 34 armed forces around the world. Our military engines power front line fighters, such as the F-15 Eagle, F-16 Fighting Falcon, F-22 Raptor and F-35 Lightning II, as well as the C-17 Globemaster III military transport and Boeing's KC-46, the U.S. Air Force's airlift tanker.

Pratt & Whitney Canada has produced more than 80,000 engines and currently there are more than 50,000 engines in service with more than 10,000 operators in over 200 countries and territories.

Pratt & Whitney works closely with customers to drive the highest level of performance throughout the service life of each engine through its EngineWise™ service offerings. By sharing technical expertise and business intelligence, and applying new developments in data analytics, Pratt & Whitney is helping operators improve engine predictability and reliability.

Learn more at www.pw.utc.com

Shell Aviation

Shell Aviation is a leading global supplier of aviation fuels and lubricants.

Every day, at over 800 airports in around 40 countries, Shell Aviation provides fuel for almost 7,000 aircraft, refuelling a plane every 12 seconds. Shell Aviation's customers range from private customers such as pilots, corporate jet operations and flying clubs to the largest global airlines.

Shell is active across the full value chain. We produce fuels and lubricants (in a range of grades); market, sell and distribute them; and offer further services that help our customers operate as efficiently as possible.

Supporting Aviation

Shell has been committed to aviation from the very beginning of flight, providing quality performance enhancing fuels and lubricants to the entire aviation community.

- In 1909, we supplied fuel for the first cross-channel flight between England and France.
- In 1919, we fuelled the first transatlantic flight.
- In 1976, we supplied fuel and lubricants for the British Airways Concorde aircraft on its first commercial flight from London to Bahrain.
- In 1989, Shell supplied fuel for the record breaking non-stop Qantas flight of a Boeing 747 from London to Australia.

- In October 2009, the world's first commercial passenger flight powered by fuel made from natural gas by Shell was flown by Qatar Airways from London Gatwick to Doha.

Enabling the Future

Shell has invested heavily in the future of aviation, particularly in research, technical advances and people – at local, regional and global levels.

Shell is one of the few energy companies with research and development facilities dedicated to the aviation sector that cover fuels, fuel handling and lubricants.

We are constantly working on fuel and lubricants research and development projects in our renowned aviation research laboratories at Houston Technology Centre in the US.

The company's operating methods focus on local strength, backed by global support. Shell Aviation places great emphasis on training and keeping the skills and professionalism of its people maintained at high levels. We understand that consistency, reliability and safety are important to our customers.

Partnership is at the heart of how Shell Aviation develops innovative products to help answer the needs of its customers. We participate in and co-operate with leading industry bodies and equipment manufacturers to refine fuels and lubricants specifications and improve test methodology to be ready for current and future generations of aircraft, engines and systems.

Find out more about Shell Aviation and how we can work together by visiting: www.shell.com/aviation

SOUTHWEST RESEARCH INSTITUTE

Southwest Research Institute® (SwRI®), headquartered in San Antonio, Texas, is one of the oldest and largest independent, nonprofit, applied research and development (R&D) organizations in the United States. Founded in 1947, SwRI provides contract research and development services to industrial and government clients in the United States and abroad. The Institute is governed by a board of directors, which is advised by approximately 100 trustees.

Southwest Research Institute, which has a long history of evaluating physical properties of conventional and unconventional fuels, is proud to be a sponsor of the 14th International Symposium on Stability, Handling and Use of Liquid Fuels. SwRI's capabilities include performing fit-for-purpose testing per ASTM D4054 for approval of new fuels and additives; material capability testing; development of new industry test methods; EI 1581, EI 1583, EI 1590, and EI 1535 aviation filtration testing; evaluation of electronic sensors for determining fuel cleanliness; SAE and ISO automotive filtration tests; and debris analysis for issues in the field.

Based on preliminary consultation with a client, SwRI prepares a proposal outlining the project's scope of work. Subject to client wishes, programs are kept confidential. As part of a long-held tradition, patent rights arising from sponsored research are often assigned to the client. SwRI generally retains the rights to Institute-funded advancements.

SwRI offers multidisciplinary, problem-solving services in a variety of areas in engineering and the physical sciences. Historically, more than 4,000 projects are open at the Institute at any one time. These projects are funded almost equally between the government and commercial sectors. SwRI's total revenue for fiscal year 2014 was $549 million. In 2014, SwRI managed more than 73 projects with expenditures of more than $6.9 million to its internally sponsored R&D program, which is designed to encourage new ideas and innovative technologies.

SwRI's headquarters occupies more than 2 million square feet of office and laboratory space on more than 1,200 acres in San Antonio. The Institute has technical offices and laboratories in Ann Arbor, Mich.; Beijing, China; Boulder, Colo.; Dayton, Ohio; Hill Air Force Base (Ogden), Utah; Hanover and Rockville, Md.; Minneapolis, Minn.; Oklahoma City, Okla.; Warner Robins, Ga.; and Durham, N.H.

At the close of fiscal year 2014, the staff numbered 2,771, including 294 professionals who hold doctorate degrees and 508 with master's degrees. In 2014, staff members published 620 papers in the technical literature; made 504 presentations at technical conferences, seminars and symposia around the world; submitted 54 invention disclosures; filed 42 patent applications; and received 73 U.S. patent awards. The Institute supports professional development of its staff through on-site technical and training courses and tuition reimbursement.

The Institute holds more than 1,100 patents awarded to its staff members, has earned 38 R&D 100 awards, and has been inducted in the U.S. Space Foundation's Space Technology Hall of Fame. The Institute has received three Department of Defense James S. Cogswell Outstanding Industrial Security Achievement Awards. The American Society of Mechanical Engineers has recognized our split-Hopkinson pressure bar apparatus (2006) and the Southern Gas Association analog (1990), developed by SwRI in 1955 for the natural gas industry, as ASME National Historic Engineering Landmarks. Several SwRI divisions have achieved ISO 9001 or ISO 14001 certification and ISO/IEC 17025 accreditation. The Ford Motor Company has designated the Institute a Tier 1 product development engineering services supplier and has awarded the Institute its Q1-2000 award.

SwRI offers a wide-ranging network of representatives to discuss Institute capabilities, business opportunities and projects. In addition to offices worldwide, the Institute maintains a comprehensive presence on the Internet. SwRI retains an extensive network of consultants in Asia and Europe and has established numerous technology alliances with internationally known corporations and organizations.

Pertinent to the 14th International Symposium on Stability, Handling, and Use of Liquid Fuels, SwRI has the capabilities of performing fit-for-purpose testing per ASTM D4054 for approval of new fuels and additives; material capability testing; development of new industry test methods; EI 1581, 1583, 1590, and 1535 aviation filtration testing; evaluation of electronic sensors for determining fuel cleanliness; SAE and ISO automotive filtration tests; and debris analysis for issues in the field.

The Institute has 9 technical divisions cooperating in multidisciplinary approaches to problem-solving.

For more information about SwRI and its capabilities, contact the Business Development Office, Southwest Research Institute, 6220 Culebra Road, P.O. Box 28510, San Antonio, Texas 78228-0510, Phone (210) 522-2122, Fax (210) 522-3496.

SOUTHWEST RESEARCH INSTITUTE

Southwest Research Institute® (SwRI®), headquartered in San Antonio, Texas, is one of the oldest and largest independent, nonprofit, applied research and development (R&D) organizations in the United States. Founded in 1947, SwRI provides contract research and development services to industrial and government clients in the United States and abroad. The Institute is governed by a board of directors, which is advised by approximately 100 trustees.

Southwest Research Institute, which has a long history of evaluating physical properties of conventional and unconventional fuels, is proud to be a sponsor of the 14th International Symposium on Stability, Handling and Use of Liquid Fuels. SwRI's capabilities include performing fit-for-purpose testing per ASTM D4054 for approval of new fuels and additives; material capability testing; development of new industry test methods; EI 1581, EI 1583, EI 1590, and EI 1535 aviation filtration testing; evaluation of electronic sensors for determining fuel cleanliness; SAE and ISO automotive filtration tests; and debris analysis for issues in the field.

Based on preliminary consultation with a client, SwRI prepares a proposal outlining the project's scope of work. Subject to client wishes, programs are kept confidential. As part of a long-held tradition, patent rights arising from sponsored research are often assigned to the client. SwRI generally retains the rights to Institute-funded advancements.

SwRI offers multidisciplinary, problem-solving services in a variety of areas in engineering and the physical sciences. Historically, more than 4,000 projects are open at the Institute at any one time. These projects are funded almost equally between the government and commercial sectors. SwRI's total revenue for fiscal year 2014 was $549 million. In 2014, SwRI managed more than 73 projects with expenditures of more than $6.9 million to its internally sponsored R&D program, which is designed to encourage new ideas and innovative technologies.

SwRI's headquarters occupies more than 2 million square feet of office and laboratory space on more than 1,200 acres in San Antonio. The Institute has technical offices and laboratories in Ann Arbor, Mich.; Beijing, China; Boulder, Colo.; Dayton, Ohio; Hill Air Force Base (Ogden), Utah; Hanover and Rockville, Md.; Minneapolis, Minn.; Oklahoma City, Okla.; Warner Robins, Ga.; and Durham, N.H.

At the close of fiscal year 2014, the staff numbered 2,771, including 294 professionals who hold doctorate degrees and 508 with master's degrees. In 2014, staff members published 620 papers in the technical literature; made 504 presentations at technical conferences, seminars and symposia around the world; submitted 54 invention disclosures; filed 42 patent applications; and received 73 U.S. patent awards. The Institute supports professional development of its staff through on-site technical and training courses and tuition reimbursement.

The Institute holds more than 1,100 patents awarded to its staff members, has earned 38 R&D 100 awards, and has been inducted in the U.S. Space Foundation's Space Technology Hall of Fame. The Institute has received three Department of Defense James S. Cogswell Outstanding Industrial Security Achievement Awards. The American Society of Mechanical Engineers has recognized our split-Hopkinson pressure bar apparatus (2006) and the Southern Gas Association analog (1990), developed by SwRI in 1955 for the natural gas industry, as ASME National Historic Engineering Landmarks. Several SwRI divisions have achieved ISO 9001 or ISO 14001 certification and ISO/IEC 17025 accreditation. The Ford Motor Company has designated the Institute a Tier 1 product development engineering services supplier and has awarded the Institute its Q1-2000 award.

SwRI offers a wide-ranging network of representatives to discuss Institute capabilities, business opportunities and projects. In addition to offices worldwide, the Institute maintains a comprehensive presence on the Internet. SwRI retains an extensive network of consultants in Asia and Europe and has established numerous technology alliances with internationally known corporations and organizations.

Pertinent to the 14th International Symposium on Stability, Handling, and Use of Liquid Fuels, SwRI has the capabilities of performing fit-for-purpose testing per ASTM D4054 for approval of new fuels and additives; material capability testing; development of new industry test methods; EI 1581, 1583, 1590, and 1535 aviation filtration testing; evaluation of electronic sensors for determining fuel cleanliness; SAE and ISO automotive filtration tests; and debris analysis for issues in the field.

The Institute has 9 technical divisions cooperating in multidisciplinary approaches to problem-solving.

For more information about SwRI and its capabilities, contact the Business Development Office, Southwest Research Institute, 6220 Culebra Road, P.O. Box 28510, San Antonio, Texas 78228-0510, Phone (210) 522-2122, Fax (210) 522-3496.

SETA STANHOPE-SETA

IASH 2017

Analytical & Physical Test Solutions for Hydrocarbon Petroleum Products

Stanhope-Seta design and manufacture innovative test equipment for hydrocarbon petroleum products, using the latest measurement technologies to provide enhanced analytical solutions for exploration, refining, petrochemical, distribution and marketing operations. Latest methods of test and product releases comprise:

Cold Filter Blocking (CFBT) – Middle Distillate Fuel Oils, IP 618

CFBT is a performance based diesel fuel test which is intended to highlight potential filtration issues such as poorly blended cold flow additive, low level polyethylene, saturated mono glycerides, sterol glucosides and water, or in some cases ineffective chemistry to deal with the challenging cold winter conditions, even for non-biodiesel distillate fuels.

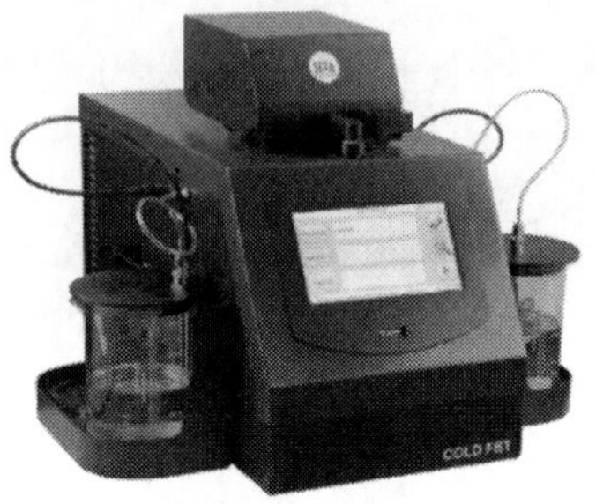

The UK Energy Institute has developed a new method, IP 618, for measuring the CFBT at +3 and -1 deg C, and IP 618 has now been included in the UK national annex of BS EN 590 as a reporting value during March 2018.

Seta-CFBT is a small scale laboratory analyser, fully compliant with IP 618, and uses a cold measurement cell and filter to automatically predict how diesel fuel may behave at low ambient temperatures by replicating vehicle/engine field cold conditions and testing the fuel's filter blocking tendency.

Water Separability – Aviation Turbine Fuel, ASTM D8073

Monitoring surfactants in jet fuel is a key specification test. In conjunction with D-2 Incorporated and industry partners a new Water Separability Test has been developed, WSI, with the objective to automate many of the fuel handling steps associated with the traditional method and achieve a step change in precision.

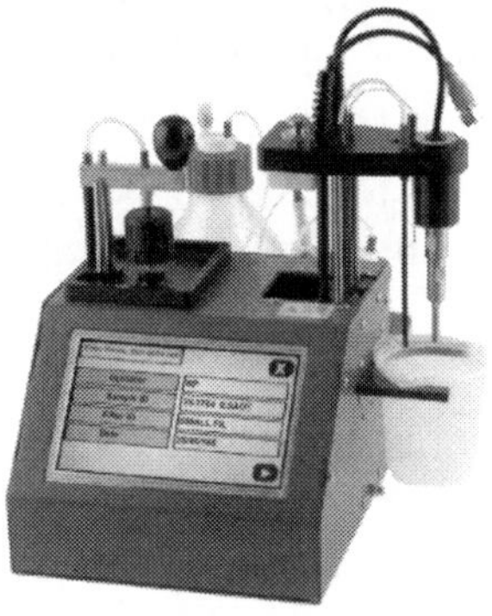

An ASTM task group successfully concluded the performance evaluation and method development resulting in a new standard, ASTM D8073, published in 2016.

The small scale tester is fully automatic with simple touch screen operation and uses patented Sonic Mixing Technology which eliminates any variables with water droplet size and ensures the same amount of energy is applied to each sample.

FAME in Jet & Diesel - ASTM D7797, D7963, IP 583

ASTM D1655, DEF STAN 091-91, EI 1530, JIG bulletin 75, ISO 8217
MIL-STD-3004D, MIL-DTL-83133J, MIL-DTL-5624W, MIL-DTL-16884

Following approval of the rapid screening methods into Jet Fuel specifications, the same technology has been successfully applied for the measurement of FAME in diesel at trace levels using a modified test procedure and new chemometrics. This technique is standardised as ASTM D7963 and is suitable for MIL-DTL-16884 (F76) and now adopted into the ISO 8217 Marine Fuel specification for DM,DF and RM grades.

FIJI provides the benchmark test which is fully approved, fast, and now a universal rapid screening tool for the measurement of FAME from 10 mg/kg to 20%. Tests are completed in 15 minutes with an easy 'load and go' operation.

Stanhope-Seta, London Street, Chertsey, Surrey, KT16 8AP, UK
t: +44 (0) 1932 564391 | e: sales@stanhope-seta.co.uk | www.stanhope-seta.co.uk

SETA STANHOPE-SETA

IASH 2017

Fuel Cleanliness, Particle Counting – ASTM D7619, IP 565

MIL-STD-3004D, MIL-DTL-83133J, MIL-DTL-5624W and DEF STAN 091-91

- Compact and portable
- Rapid testing, 4 minutes
- Six size channels
- ISO 11171, NIST 2806b
- Micron 'c' reporting
- ISO 4406 codes

Vapour Pressure - ASTM D6378, D5191, D5188, EN 13016-1, EN 13016-3, IP 394

- Rapid automatic vapour pressure testing
- Triple and single expansion test methods
- In the field temperature and pressure calibration
- Easy to maintain
- Wide format colour touch screen
- Intuitive operating software

Flash Point – ASTM D3278, D3828, D7236, BS EN ISO 3679, EPA 1020 A & B, IP 523

DEF STAN 091-91, ASTM D1655, MIL-STD-3004D, MIL-DTL-83133J, MIL-DTL-5624W

- Rapid, less than 2 minutes
- Search option
- Small sample size - 2 or 4ml
- Automatic flash detection
- Automatic barometric correction
- USB port and result storage
- Colour touch screen
- Cost effective
- Portable operation
- Simple calibration

About Stanhope-Seta

From headquarters in the UK, Stanhope-Seta manufactures and exports instruments to over 150 countries worldwide. A global network of distributors provides local customers with technical advice and comprehensive maintenance support.

An established reputation for instruments of the very highest quality and reliability is supported by an ISO 9001:2008 approved integrated material and production control system coupled with stringent test, inspection and calibration procedures.

The company received a coveted Queen's Award for International Trade in 2012 and Innovation in 2014.

THE QUEEN'S AWARDS FOR ENTERPRISE: INTERNATIONAL TRADE 2012

THE QUEEN'S AWARDS FOR ENTERPRISE: INNOVATION 2014

Enquiries

For further information please contact us or visit our website www.stanhope-seta.co.uk

Ian Mylrea, R&D Manager | DD: +44 (0)1932 574589 | E: im@stanhope-seta.co.uk

Stanhope-Seta, London Street, Chertsey, Surrey, KT16 8AP, UK

t: +44 (0) 1932 564391 | e: sales@stanhope-seta.co.uk | www.stanhope-seta.co.uk

The University of Dayton Research Institute (UDRI) is the research arm of the University of Dayton, located in Dayton, Ohio USA. UDRI is an international leader in scientific and engineering research, serving government, industry, and not-for-profit customers. Our full-time professional staff of engineers and scientists conducts research and provides support in a wide variety of technical areas.

The Research Institute's efforts in the area of advanced fuels include studies in:

- Development and evaluation of alternative fuels and fuel technologies
- Improved high temperature thermal stability fuels and additives
- Improved low temperature flowable fuels and additives
- Development of advanced fuels and additives for reduced combustion particulate and gaseous emissions
- Advanced chemical analyses of fuels and additives
- Chemical kinetics and fluid dynamics modeling of fuels and combustion
- Development and testing of advanced combustion concepts

UDRI is a pioneer in testing and evaluation of alternative jet fuels. Our researchers conducted many of the detailed laboratory evaluations and supported field trials of Fischer-Tropsch, HEFA, and ATJ fuels and fuel blends which resulted in ASTM and military certification. We also coordinate the FAA D4054 Clearinghouse for ASTM certification and qualification of alternative jet fuels.

UDRI has experience in identification and mitigation of biocontaminants for ground and aircraft fuel systems. We pioneered DNA sequencing methods for identification and quantification of microbial contamination. We have studied the role of fuel composition (e.g., alternative fuel species) on the growth of fuel microbes. We continue to study the development of non-toxic advanced fuel antimicrobial molecules and the molecular mechanisms of fuel species metabolism.

UDRI is a world leader in studies of jet fuel thermal stability in both oxidative and pyrolytic environments. Our researchers have many years expertise in thermal stability additive testing, evaluation, and formulation. We have pioneered the development and interpretation of thermal stability tests such as:

- Quartz crystal microbalance (QCM)
- Extended duration thermal stability test (EDTST)
- Isothermal corrosion oxidation test (ICOT)
- Near isothermal flowing test rig (NIFTR)
- Phoenix & ECAT thermal stability rigs
- Advanced reduced scale fuel system simulator (ARSFSS)

UDRI is a leader in the development of unique analysis techniques for the identification and quantification of important fuel species. We specialize in difficult analyses which require custom preseparation methods, as well as quantification of fuel species classes. Preseparation techniques pioneered by UDRI researchers for use for fuel include solid-phase extraction and high performance liquid chromatography. These techniques are then combined with gas chromatography with various detectors or with multi-dimensional gas chromatography (GCxGC). Our researchers also employ unique derivatization techniques which greatly improve difficult analyses, such as silylation for analysis of metal deactivator additive and reaction with triphenylphosphine for analysis of hydroperoxides.

UDRI is the primary U.S. Air Force technical contractor responsible for the development, testing, and evaluation of the advanced high temperature thermal stability fuel JP-8+100 and the improved low temperature flowable fuel JP-8+100LT. These improved fuels have saved the Air Force millions of dollars per year in maintenance and logistics costs.

UDRI has expertise in improving the low temperature flowability of fuels via development of flow improving additive packages. We have developed numerous techniques to study fuel crystallization and flowability including:

- Scanning Brookfield viscometry
- Cold-stage optical microscopy
- Differential scanning calorimetry
- Custom fuel system simulators
- Custom additive screening tests

In addition we have developed numerous computational fluid dynamic models of test devices and aircraft fuel system components to better understand fuel tank temperature profiles and the effects of viscosity, heat transfer, and solidification. These models have supported the conversion of the U.S. Air Force from JP-8 to Jet A fuels at CONUS locations.

UDRI researchers are actively involved in performing measurement of combustion particulates and gaseous emissions for test rigs as well as actual gas turbine engines. We employ numerous methods for characterization of emissions including particle number density, particle size distribution, particle mass concentration, and particle count using instruments such as:

- Condensation nuclei counter
- Scanning mobility particle sizer
- Differential mobility analyzer
- Tapered element oscillating microbalance
- Laser particle counter

In addition, we have pioneered the evaluation and development of emissions reducing additives.

UDRI researchers are pioneering techniques for prediction of thermal-oxidative fuel deposition. A methodology has been developed, which employs a combination of unique chemical kinetic mechanisms with computational fluid dynamics techniques which will allow the prediction of fuel deposition and oxidation for various fuels as a function of temperature and time in complex fuel system geometries. Such models can be used to improve fuel system design, make better use of the fuel heat sink, predict nozzle lifetime, and improve reliability and time between maintenance.

UDRI has a Cooperative Research and Development Agreement (CRADA) with the U.S. Air Force Research Laboratory Fuels & Energy Branch at Wright-Patterson Air Force Base which allows use of their one-of-a-kind facilities for third party research and testing. Your organization gains the dual benefits of research and testing using these unique facilities along with analysis and interpretation of results by UDRI's fuel experts.

For additional information please contact:

Steven Zabarnick, Ph.D., Division Head, Energy & Environmental Engineering
University of Dayton Research Institute
300 College Park, Dayton, OH USA 45469-0043
Telephone: (937) 255-3549
Email: Steven.Zabarnick@udri.udayton.edu

Please visit our web page: http://www.udri.udayton.edu

Defense Logistics Agency Energy
8725 John J. Kingman Road
Fort Belvoir, VA 22060-6220
Website: www.energy.dla.mil

Mission: To provide the Department of Defense and other government agencies with comprehensive energy support in the most effective and efficient manner possible.

The Defense Logistics Agency Energy provides energy to the military services, the Department of Defense, and other federal agencies at more than 4,000 locations worldwide. Although its beginnings date back to World War II, in 1962 DLA Energy became part of the Defense Logistics Agency (then known as the Defense Supply Agency). Since 1964, DLA Energy has operated independently to purchase and manage energy products and services for the Defense Department. Prior to 1998, DLA Energy was known as the Defense Fuel Supply Center. Between 1998 and 2010, DLA Energy was known as the Defense Energy Support Center.

DLA Energy manages petroleum products – jet fuels, aviation and automotive gasoline, heating oil, lubricants – in addition to coal, natural gas, electricity, alternative fuels and missile fuels. The major fuels DLA Energy procures are JP5, a high flash jet fuel used shipboard by the Navy; JP8, a kerosene-based jet fuel comparable to commercial jet fuel; Jet A1, a commercial jet fuel with additive; and F76, a naval diesel fuel.

DLA Energy is organized to work with customers and suppliers to accomplish all areas of its mission. Bulk Petroleum Products and Bulk Petroleum Supply Chain Services provide procurement and distribution of petroleum products, the majority of DLA Energy's business. Direct Delivery Fuels manages fuels that are delivered directly from the contractor to the customer. Defense Fuel Support Point Management manages fuel terminal and storage operations, including government-owned, contractor-operated and contractor-owned, contractor-operated facilities.

Outside of petroleum, Installation Energy provides support for natural gas, coal, renewable energy, electricity and energy savings performance contracts. Utility Services manages the utility services contracting mission supporting the utility privatization programs of service partners. The Aerospace Energy business unit provides propellants, cryogenics and gases in support of weapon systems and space programs.

DLA Energy launched several innovative programs to bring fuel to its customers economically and efficiently. The Aviation Into-Plane Reimbursement Card, Ships' bunkers Easy Acquisition Card and the DoD Fleet Card programs allows military service and federal civilian agency planes, ships and ground vehicles, respectively, to purchase fuel commercially. This replaces more complicated, paper-based methods used in the past.

Other initiatives to reduce paper processing and deliver product more efficiently to customers include the Bid Evaluation Model, a database that analyzes transportation rates, delivery modes and shipping points to determine the lowest laid-down cost for a contract award. Under the system, evaluation time has been reduced to 20 minutes – down from a previous 24-hour turnaround time. Also, the Enterprise Business System processes fuel transactions electronically through a Web-based program that allows for ordering, invoicing and receipting over the Internet.

As the supplier of fuel for DoD, DLA Energy plays a critical role overseas in support of U.S. military installations and operations. The agency powered world-class global capability through delivery of nearly 100 million barrels of fuel, 112 million cubic feet of helium and over 1.1 billion kilowatt hours of renewable power in fiscal year 2014. Expert support extended from the reliable in Afghanistan to the revolutionary and nearly instantaneous in Liberia combating Ebola, ensuring uninterrupted and dependable service to all our DoD customers around the globe.

DLA Energy fuel support also continues into humanitarian efforts around the globe. For Operation United Assistance Ebola relief operations, DLA Energy made a spot buy for 10,000 gallons of diesel fuel and 2,500 gallons of gasoline to have product ready to support any short notice requirements, and arranged to have 30 International Standards Organization containers to store the fuel where it was needed the most. Another 2,914,455 gallons of Jet A fuel was procured for Dakar, Senegal, along with 1,546,992 gallons of motor gasoline, diesel fuel and Jet A1 fuels in Monrovia, Liberia, to support Ebola relief operations.

Alternative fuels and renewable energy have also been major endeavors for DLA Energy. The agency provides expertise for technical and administrative actions for research and development projects and congressional earmarks relating to alternative mobility fuels and renewable energy products. The latest projects include procurement of direct sugar to hydrocarbon aviation fuel to support the Army's program to certify the fuel as a JP8 blend component, procurement of catalytic hydrothermal conversion synthetic aviation and marine diesel fuel to support the Navy's alternative fuels certification program, and support of Defense Production Act Title III efforts. Assisting with DPA work involves providing technical expertise and support in evaluating proposals for the Department of the Navy, Department of Agriculture and Department of Energy Biorefinery DPA Title III initiative, which resulted in the award of government funds to three enterprises for the construction of commercial facilities to produce military grade biofuels by 2017.

DLA Energy's objective this year is lowering long-term warfighter risk by ensuring reliable, capable and affordable capabilities. The agency aims to be all things energy for DoD – agile, effective and efficient.

IASH 2017, the 15th INTERNATIONAL SYMPOSIUM ON STABILITY, HANDLING AND USE OF LIQUID FUELS
Rome, Italy
10-14 September 2017

How European Standards Guarantee Fuel Quality, Vehicle Functioning and Emissions

Ortwin Costenoble, M.Sc., Netherlands Standardization Institute (NEN), Delft, the Netherlands

Abstract

Guaranteeing fuel quality in Europe and selling the product to customers, is not standard. Even though the European Standardization Committee (CEN) develops fuel specification and test method standards which become the national standard in 37 European countries. The fact that Directives align the legal requirements regarding emissions of vehicles, the biofuel blends allowed on the market and the alternative fuel infrastructure and marking across Europe, several peculiarities remain.

Mr Costenoble has been secretary of the CEN fuels standardization committee for 15 years. His lecture is an introduction to the European regulatory framework concerning fuels quality and vehicle emissions. He will explain the harmonizing concepts and the differences that occur in the continent due to national regulatory and climatic conditions. After this talk you won't be a EU fuels' expert nor know how the qualities demanded should be maintained throughout the logistic system, but you will be much more aware of what requirements apply in the chain from refinery production via blender tank to consumer outlet and filling station nozzle.

IASH 2017, the 15TH INTERNATIONAL SYMPOSIUM ON STABILITY, HANDLING AND USE OF LIQUID FUELS
Rome, Italy
10-14 September 2017

PHENOLS AND AROMATIC METHYL ETHERS FROM BIOMASS PYROLYSIS OIL: IMPLICATIONS FOR JET FUEL STABILITY

Mariam Ajam[1], Carl L Viljoen[1], Chris Woolard[2], and Eric van Steen[2]

[1] *Sasol Energy Technology, P.O. Box 1, Sasolburg, 1947, Republic of South Africa mariam.ajam@sasol.com and carl.viljoen@sasol.com*

[2] *Department of Chemical Engineering, University of Cape Town, Private Bag X3, Rondebosch, Cape Town, 7700, Republic of South Africa chris.woolard@uct.ac.za and eric.vansteen@uct.ac.za*

ABSTRACT

Jet fuel is used as a heat sink for aircraft systems and is thus required to be thermally stable. Jet fuel instability could cause deposits, which may foul aircraft components. Renewable jet fuel can be produced from biomass pyrolysis oil, although concerns may exist about its high phenolic content, since some phenols have a high propensity to form deposits. In this paper, the removal of detrimental phenolic species in jet fuel by conversion to aromatic methyl ethers was evaluated as a novel method of improving the thermal stability of jet fuels derived from biomass. The thermal and oxidative stability of a model fuel and a jet fuel derived from biomass pyrolysis oil, containing a variety of phenols and/or aromatic methyl ethers, was studied using a Quartz Crystal Microbalance (QCM) technique and the Jet Fuel Thermal Oxidation Test (JFTOT) method. The jet fuels containing phenols indicated strong depositing propensities, whereas the jet fuels containing the corresponding aromatic methyl ethers showed reduced depositing propensity. Small amounts of phenols present in the kerosene derived from biomass pyrolysis oil were converted to the corresponding methyl ether using an organic synthesis technique and this was shown to improve the thermal stability of the fuel significantly. This study demonstrated that aromatic methyl ethers are of a high relative thermal stability when compared to their phenolic counterparts. The reduced amount of deposits formed in fuels containing aromatic methyl ethers may be attributed to the strong ether bond linkage (O-CH_3) that is stable under the thermal oxidative stressing conditions experienced by jet fuel.

KEYWORDS: jet fuel thermal and oxidative stability, biomass pyrolysis oil, phenols, aromatic methyl ethers

1. INTRODUCTION

Jet fuel acts as a coolant in various aircraft systems by absorbing excess heat. Once the fuel temperature reaches about 140 °C, the fuel reacts with dissolved oxygen by an autoxidation chain mechanism (shown in Figure 1) to form oxidised species and deposits.[1] Deposit formation can lead to fouling of critical fuel system parts and may eventually result in catastrophic engine failure, if left unchecked.

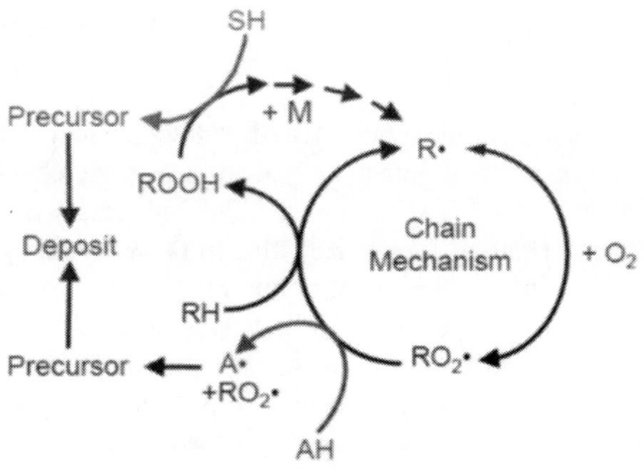

Figure 1. Jet fuel autoxidation chain mechanism.

Figure 1 shows the oxidation cycle which starts with the production of an initial hydrocarbon radical (R•). This hydrocarbon radical (R•) undergoes bimolecular addition of dissolved oxygen to form a peroxy radical (ROO•). The peroxy radical undergoes an intermolecular hydrogen abstraction reaction with a jet fuel hydrocarbon species (RH) (most likely an alkyl-substituted aromatic species) to form a hydroperoxide (ROOH) and a new hydrocarbon radical (R•), which propagates the chain. Hydroperoxides are the primary autoxidation product at temperatures below 120 °C, but they can undergo unimolecular degradation at higher temperatures to form alkoxy and hydroxyl radicals (RO• and •OH respectively, not shown in Figure 1), which accelerates the oxidation process. Although the rate of autoxidation is a critical factor when assessing jet fuel stability, the amount of deposits formed is also of paramount importance. The exact mechanism by which jet fuel deposition occurs is, however, poorly understood compared to that of autoxidation. It is believed that the peroxy radical (ROO•) may react with hetero-atomic species in the fuel, such as phenolic species, which are known to be antioxidants (AH), and reactive sulfur species (SH), to form deposit precursors. Phenols act as antioxidants due to the low bond dissociation energy of the hydrogen attached to the oxygen atom.[2] Phenols inhibit the autoxidation chain mechanism by donating a hydrogen atom to the chain-propagating peroxy radical (ROO•). Although phenols typically slow down oxidation by acting as antioxidants, further reactions eventually lead to deposit formation in jet fuel.[3]

Balster et al. investigated the relationship between the concentration of hetero-atomic (oxygen, sulfur and nitrogen-containing) species and deposit formation in twenty different petroleum-derived jet fuels. They showed a correlation between the total amount of hetero-atomic species and the amount of deposit formed. Sobkowiak et al. also found correlations between the trace hetero-atomic species in fuels (such as phenol, indole, and carbazole) and the deposits formed.[4] Abdallah et al. investigated the influence of phenol and alkyl-substituted phenols spiked at 400 ppm into petroleum-derived jet fuel on Jet Fuel Thermal Oxidation Test (JFTOT) breakpoint temperatures.[5] The results showed that unsubstituted phenol had the greatest negative impact, causing a 10 °C reduction in the JFTOT breakpoint temperature. In another study, Garver showed that, as the phenolic concentration increased in crude-derived jet fuels, the fuel had a greater tendency to form deposits.[6] It is believed that phenols can undergo oxidative phenolic coupling reactions to form dimers, trimers, tetramers and ultimately high molecular weight compounds.[7] Beaver et al. proposed a mechanism of jet fuel deposition involving the coupling of phenolic oxidation products, such as quinones, with electron-rich aromatic species, such as indoles and carbazoles, to produce high molecular weight compounds referred to as soluble macromolecular oxidatively reactive species (or SMORS).[8] Phenols have been found to be the predominant species of the total trace polar hetero-atomic species present in conventional jet fuel; with concentrations as high as 2000 ppm being reported.

To address growing climate change concerns, the development of renewable fuel sources, such as those from lignocellulosic biomass (or biomass pyrolysis oil), are of increasing interest to the aviation industry.[9] The conversion of biomass pyrolysis oil to renewable jet fuels has attracted some interest, although it is still in the very early stages of development.[10] It is anticipated that phenolic species may occur in biomass pyrolysis oil-derived jet fuels because phenolic structures occur naturally in biomass. Phenols are very resistant to catalytic hydroprocessing due to the high strength of the bonds between oxygen and aromatic carbons.[11] It is therefore likely that phenolic species may be present in the final jet fuel, depending on the severity of the catalytic hydroprocessing step. The presence of trace phenolic species and other hetero-atomic species may affect the implementation and widespread use of biomass pyrolysis oil-derived jet fuels due to instability concerns, unless physical removal or conversion techniques can produce stable fuels. The thermal stability of jet fuels can be improved by removing phenols by either severe hydroprocessing[12], solid-phase extraction (such as silica or clay treatment), the conversion of phenols to silyl ethers[13] or the etherification of phenols to alkyl aryl ethers. The latter two approaches are able to convert problematic species (phenols) into stable species (silyl ethers or alkyl aryl ethers). Alkyl aryl ethers, for example, aromatic methyl ethers can easily be formed from phenols using well-known organic synthetic techniques (such as the O-methylation reaction of phenols). To the best of the authors' knowledge, the etherification of phenols to alkyl aryl ethers as a means of improving jet fuel stability has not been explored (e.g. simplified O-methylation reaction of phenol to aromatic methyl ether is shown in Figure 2). Etherification of phenols could improve the stability of jet fuel by eliminating the oxidative coupling reactions typical of phenolic species.

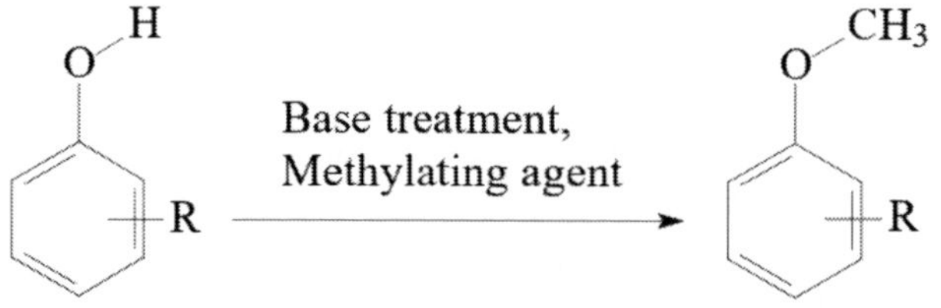

Figure 2. The etherification (or O-methylation) of phenols to aromatic methyl ethers.

Little is understood of the thermal stability of alkyl aryl ethers under the conditions to which jet fuels are exposed during application. Studies have shown that aryl and benzyl ethers in synthetic paraffinic kerosene, for instance, have excellent seal swelling characteristics.[14,15] These studies, however, did not address their impact on thermal stability. Aromatic methyl ethers contain a strong ether bond linkage ($O-CH_3$).[16] Consequently, aromatic methyl ethers are expected to be able to withstand jet fuel oxidative stressing conditions. Etherification of intrinsic phenolic constituents to aromatic methyl ethers could be a viable strategy for improving the thermal stability of jet fuels derived from biomass pyrolysis oil. The objective of this study was therefore to show that aromatic methyl ethers exhibit excellent thermal oxidative stability when used in jet fuel because they do not form deposits, unlike phenols. A second objective of this study was to show that the conversion of phenols present in jet fuel to aromatic methyl ethers yields a thermally stable jet fuel.

2. EXPERIMENTAL

2.1. Materials

a) Real and Model Fuels

The characteristics of the fuels used in this study are shown in Table 1.

Table 1. Characteristics of the fuels used in this study.

No.	Base fuel	Characteristics of base fuel
1	*Biomass-derived Kerosene*	A representative biomass pyrolysis oil-derived kerosene generated from the catalytic hydrotreatment of biomass pyrolysis oil using sulfided NiMo/Al_2O_3 catalyst in a continuous fixed-bed reactor, was received from BTG in the Netherlands. Based on two-dimensional GC with time of flight mass spectrometry, the Biomass-derived Kerosene had a carbon distribution from C_8-C_{16} and consisted predominantly of cycloparaffins (72.2 mass%) and aromatic hydrocarbons (17.3 mass%). In addition, n-paraffins (3.6 mass%) and iso-paraffins (6.6 mass%) were also present. Oxygenates were also found to be present at trace levels (±0.25 mass%).
2	*Model Fuel*	Based on the composition of the Biomass-derived Kerosene, a Model Fuel consisting of a highly cycloparaffinic stream (Nappar 10, a C_8-C_{11} stream obtained from Protea Chemicals) and an aromatic hydrocarbon stream (Fluidar 100, a C_8-C_{11} stream obtained from Engen) was blended in an 80/20 volumetric ratio. The Model Fuel was treated with dry silica before further use, to remove peroxides and other polar species that may interfere during the stability tests. In silica treatment, the Model Fuel (1000 mℓ) was eluted through a dry silica column (100 g).

b) Model Phenols and Aromatic Methyl Ethers Mixtures

Phenols and their corresponding aromatic methyl ether blends were investigated as dopants in the Model Fuel. Each individual phenol and aromatic methyl ether compound (≥99.0% purity) was obtained from Sigma-Aldrich, and used as received. These mixtures were blended together to mimic the species that would typically be present in petroleum-derived jet fuel (Blend 1) and biomass pyrolysis oil-derived jet fuel (Blend 2). Each of these blend mixtures was doped separately at 1 vol% into the Model Fuel.

Table 2. A summary of the phenols blends and aromatic methyl ether blends.

No.	Phenols	Corresponding aromatic methyl ethers
1	*Model Phenols Blend 1:* A blend of 56 vol% phenol, 34 vol% 4-methylphenol and 10 vol% 2,4-dimethylphenol	*Model Aromatic Methyl Ethers Blend 1:* A blend of 56 vol% anisole, 34 vol% 4-methylanisole and 10 vol% 3,4-dimethylanisole
2	*Model Phenols Blend 2:* A blend of 5 vol% phenol, 50 vol% 4-methylphenol, 15 vol% 2,4-dimethylphenol, 20 vol% 2-methoxyphenol and 10 vol% 4-ethylphenol	*Model Aromatic Methyl Ethers Blend 2:* A blend of 5 vol% anisole, 50 vol% 4-methylanisole, 15 vol% 3,4-dimethylanisole, 20 vol% 2-dimethoxybenzene and 10 vol% 4-propylanisole

2.2. Methods

a) O-Methylation Conversion of Phenols to Aromatic Methyl Ethers in Biomass-derived Kerosene

The following procedure was followed for the etherification of phenols present in the Biomass-derived Kerosene: the kerosene (100 mℓ, 83.51 g, containing 0.1% phenols) was added to anhydrous tetrahydrofuran (5% of kerosene volume) in a three-necked round bottom flask (1000 mℓ) equipped with a stirrer bar and reflux condenser, under a continuous stream of nitrogen. The mixture was cooled in an ice bath before the addition of portions of 60% NaH in a mineral oil dispersion (2 equivalents, 0.12 g, 0.002 moles). NaH was added to the cold liquid mixture in portions to limit the vigorous bubbling of the reaction mixture as hydrogen was formed. The reaction mixture was stirred at room temperature for 1 h to allow the complete formation of the phenolate ions. The methylating agent, $(CH_3)_2SO_4$ (2 equivalents, 0.20 g, 0.002 moles) was added to the reaction mixture via a syringe through a septum. The round bottom flask containing the reaction mixture was placed in an oil bath and stirred at 80 °C for 48 h. These reaction conditions were adapted from the method of Stoochnoff & Benoiton.[17]

The reaction mixture was cooled to room temperature. The reaction mixture was quenched with distilled water (100 mℓ) and stirred for 1 h to decompose all unreacted NaH. Aqueous NH_3 (5%, 50 mℓ) was added to the water mixture and stirred for 0.5 h to quench any unreacted $(CH_3)_2SO_4$. Aqueous HCl (1M, 150 mℓ) was added to the reaction mixture and stirred for 1 h to reduce the pH to approximately 4 (litmus paper turned pink). Two layers were visible, viz. the kerosene top layer and the aqueous bottom layer. The kerosene layer was washed with distilled water (5 x 50 mℓ) and subsequently dried with anhydrous $MgSO_4$. After filtration, the remaining tetrahydrofuran solvent was removed by rotary evaporation at 70 °C under vacuum. The kerosene layer was weighed and analysed by two-dimensional GC.

b) Techniques to Monitor Jet Fuel Thermal and Oxidative Stability

Three techniques were used to investigate the thermal stability of the jet fuels in this study.

The Quartz Crystal Microbalance (QCM) Technique

The QCM instrument (shown in Figure 3) has been extensively used to monitor the thermal and oxidative stability of jet fuels with good reproducibility and accuracy.[18,19] The experimental reproducibility of a QCM for measuring mass deposition has been reported to be ±20%.[20] In this study, the repeatability of the QCM test for the Model Fuel after four measurements was 2.41 $\mu g/cm^2$ ± 0.48. The QCM is capable of monitoring both the consumption of the headspace oxygen and the deposit formation during fuel degradation. In this way, the QCM is able to discriminate between the thermal and oxidative stability of various fuels under similar reaction conditions. The amount of deposits formed is based on the relationship between the change in surface mass and the resonance frequency of the quartz crystal.

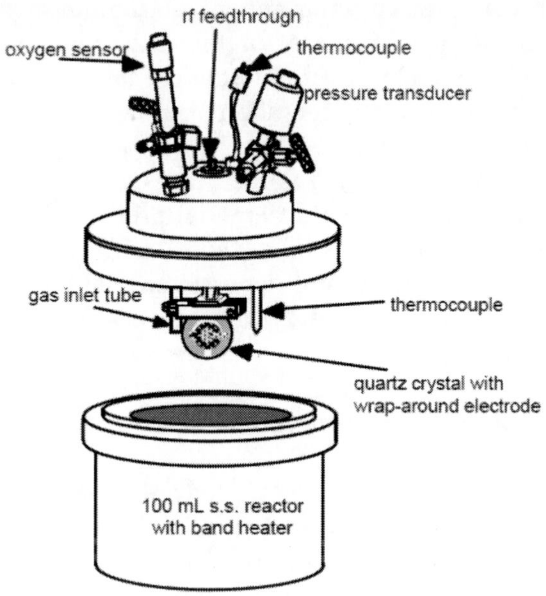

Figure 3. A schematic diagram of a QCM used to monitor jet fuel stability.[21]

In a typical experiment, 60 mℓ of fuel was saturated with air by sparging, prior to heating up to 140 °C for 15 h in a closed system (time zero was taken from the time the reactor started heating up to the required temperature). The time to reach a constant temperature of 140 °C was typically between 20 and 25 min. The total amount of deposit formed and the oxygen consumption parameters, at the end of 15 h were used to characterise the thermal and oxidative stability characteristics of the various fuel blends. In the current study, the term complete oxygen consumption is used to refer to the time taken for 99% of the oxygen in the headspace to be consumed during the reaction (i.e. 1% of the oxygen remaining in the headspace). The thermal oxidative stability characteristics of jet fuels obtained using a QCM can be classified according to their oxidation and deposit forming properties. Table 3 summarises the QCM results grouped into three classes, according to the oxidation properties and the amount of deposit formed.

Table 3. Typical deposition and oxidation values used to interpret QCM results.

Fuels that oxidise:	**Fuels that deposit in mass ranges from:**
<5 h Fast oxidisers	0-2 μg/cm^2 Light depositor
5-10 h Moderate oxidisers	2-5 μg/cm^2 Moderate depositor
>10 h Slow oxidisers	>5 μg/cm^2 Heavy depositor

The Jet Fuel Thermal Oxidation Test (JFTOT) Method

The JFTOT method (Alcor JFTOT Mark III), based on the ASTM D3241 standard method, is used and approved world-wide to determine if jet fuel meets the current thermal stability specification. It measures the deposit formation tendencies of jet fuel when heated at a specified temperature in the presence of air and an aluminium metal surface. JFTOT is used as a qualitative pass/fail rating instrument for thermal oxidation stability at 260 °C (for petroleum-derived jet fuels) and 325 °C (for alternative fuels, including renewable jet fuels). A higher test temperature is required for alternative and renewable fuels to provide an additional margin that ensures that reactive species are absent in these fuels. The degree of fuel degradation due to thermal oxidation is then measured by the deposits formed and the corresponding pressure build-up across a differential pressure filter. The JFTOT pass rating criteria for a jet fuel is a deposit rating of <3 and a differential pressure increase of <25 mmHg. Another technique for quantifying deposits on JFTOT tubes is laser ellipsometry. This technique was used in the current study to provide further information on the deposits formed for each fuel system tested.

JFTOT Tube Analysis using Laser Ellipsometry

An ellipsometry technique (Falex Ellipsometer, Model 430) was used to quantify the deposits on the JFTOT heater tubes and to provide more information into the fuel's thermal stability characteristics by enabling the investigation of the morphology of deposits, based on the depth and thickness of the deposits formed on the tube. The JFTOT heater tube was carefully placed inside the ellipsometer, without touching the centre portion of the tube, and the laser was switched on to start the measurement of deposit thickness on the tube. At the end of the measurement procedure, the maximum average deposit thickness was recorded in nanometers (nm). The deposit thickness measurements consist of 1200 data points collected along the axial direction of the heater tube (50 points) at equally spaced radial circumference positions (24 points). Extensive prior studies on this ellipsometry technique recognised that an average deposit thickness of 85 nm correlates to a JFTOT tube rating pass of 3.[22] A deposit thickness above 85 nm represents a fail of the stability test. The repeatability of the deposit thickness measurement of a new blank JFTOT tube after six measurements was 9.8 nm ± 1.9.

3. RESULTS AND DISCUSSION

3.1. The Effect of Model Phenols and Aromatic Methyl Ethers in a Model Fuel

As the effect of aromatic methyl ethers on the thermal and oxidative stability of jet fuel was not known, this study investigated the effect on surface deposition when aromatic methyl ethers were added to a Model Fuel. For comparison purposes, the corresponding phenols were also tested to investigate their effect on surface deposition.

a) Model Phenols and Aromatic Methyl Ether Blend 1

Model Phenols Blend 1 and Model Aromatic Methyl Ethers Blend 1 were separately evaluated in the Model Fuel at a 1 vol% blend level. A 1 vol% level of oxygenates in jet fuel is seen as a worst-case scenario that will typically not be seen in real commercial jet fuels. The presence of high quantities of oxygenates in jet fuel is currently limited because of their potential impact on the ability of a jet fuel to meet stringent specifications, such as thermal stability, minimum energy density, freezing point, volatility, water solubility and corrosion effects.[23] This blend level was chosen to aggravate any possible deposition that might be caused by these oxygenates in the fuel. The deposition and oxygen consumption profiles are shown in Figure 4.

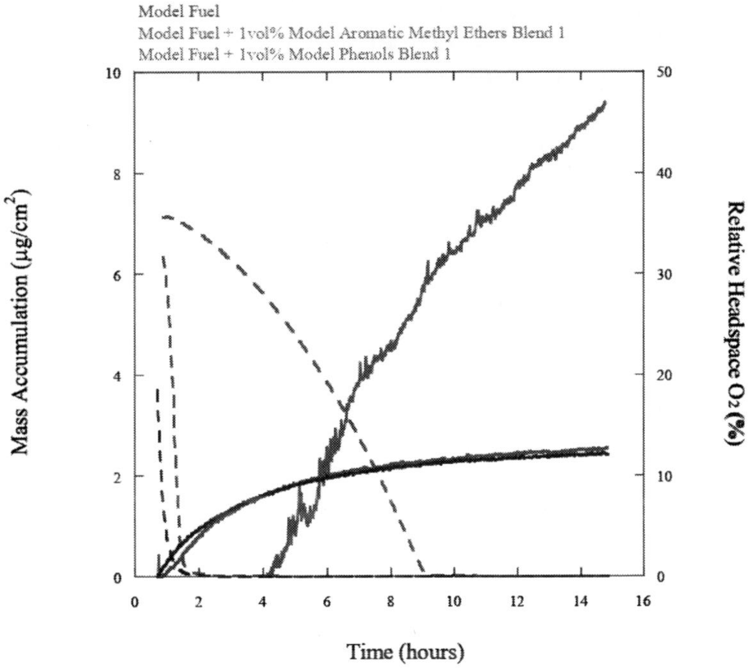

Figure 4. Deposition (solid lines) and oxygen consumption (dashed lines) profiles of the Model Fuel containing 1 vol% Model Phenols Blend 1 or Model Aromatic Methyl Ethers Blend 1.

The Model Fuel was a moderate depositor (2.41 μg/cm^2) and a fast oxidising fuel (complete oxygen consumption within 1.3 h). The moderate deposit formation of the Model Fuel is typical of fuels containing a high concentration of aromatics (20 vol% in this case).[24] Upon evaluating the deposit formation tendencies of the individual components that make up the Model Fuel (viz. a highly cycloparaffinic component, Nappar 10 and a highly aromatic component, Fluidar 100), it was found that the cycloparaffinic stream was a low depositor (0.72 μg/cm^2), whereas the aromatic stream was an extremely high depositor (18.70 μg/cm^2). These individual streams were also silica treated before stability testing.

From the deposition profiles in Figure 4, it was immediately evident that the Model Phenols Blend 1 produced a substantial amount of deposits after 15 h (9.37 μg/cm^2), whereas the

Model Aromatic Methyl Ethers Blend 1 produced similar quantities of deposits (2.55 μg/cm^2), compared to the Model Fuel (2.41 μg/cm^2). In the first 4 h of the experiment containing phenols, there was a delay in deposition, after which it started to increase significantly, not levelling off even after 15 h. It is known that non-hindered alkylated phenols are detrimental to jet fuel stability, which was also evident in these results. The aromatic methyl ethers were found not to have a detrimental effect on fuel stability. In the case of the fuel containing aromatic methyl ethers, there was a lower rate of oxygen consumption (complete oxygen consumption by 1.6 h) compared to the Model Fuel (complete oxygen consumption by 1.3 h). Phenolic impurities (0.03% 2,3,5,6-tetramethylphenol, measured by GC-MS) were found to be present in the pure aromatic methyl ethers blend. Phenolic species are known to delay oxidation by intercepting peroxy radicals formed in the autoxidation cycle. This was observed with the significant slowing of the oxygen consumed in the case of the fuel blend containing the 1 vol% Model Phenols Blend 1 (complete oxygen consumption by 8.9 h). Aromatic methyl ethers were not expected to delay oxidation, since there are no readily abstractable phenolic hydrogen atoms available to interfere with the autoxidation cycle.

The Model Fuel was doped with Model Aromatic Methyl Ethers Blend 1 at concentrations of 1, 3 and 5 vol%. The aromatic methyl ethers were added at a much higher concentration to determine their effect on stability. The deposition and oxygen consumption profiles of the Model Fuel and the fuel containing varying concentrations of Model Aromatic Methyl Ethers Blend 1 are shown in Figure 5.

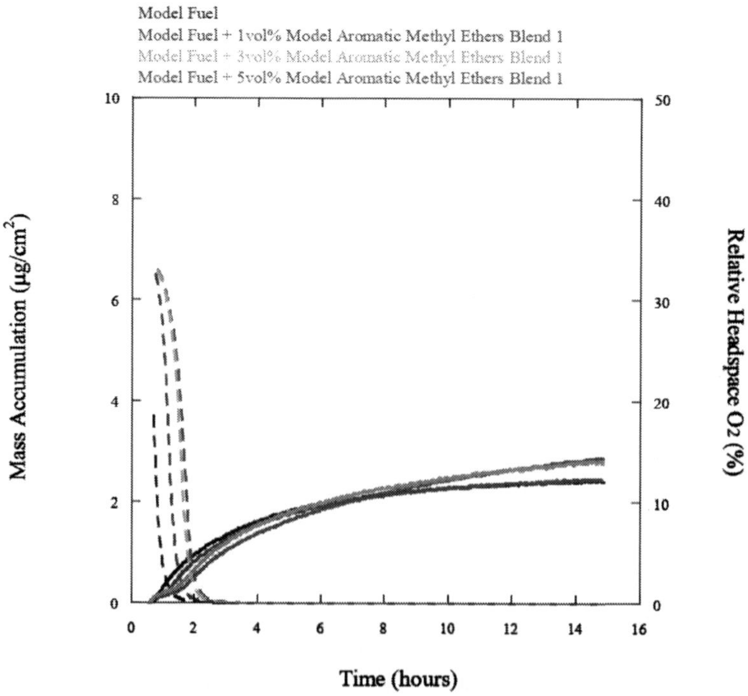

Figure 5. Deposition (solid lines) and oxygen consumption (dashed lines) profiles of the Model Fuel containing 1, 3 and 5 vol% of Model Aromatic Methyl Ethers Blend 1.

As the Model Aromatic Methyl Ethers Blend 1 concentration in the Model Fuel increased from 1 to 5 vol%, a further reduction occurred in the rate of oxygen consumption (complete oxygen consumption for the Model Fuel was 1.3 h compared to 2.2 h for fuel containing 5 vol% Model Aromatic Methyl Ethers Blend 1). This can be attributed to the fact that, as the aromatic methyl ethers concentration increased, the phenolic impurities present also increased, causing a slowing in the oxidation rate. At an aromatic methyl ethers concentration above 1 vol%, deposition was slightly inhibited (as measured by the onset of deposit formation), which was believed to be due to the phenolic impurities. After 15 h reaction time, the amount of deposits formed in these blends was greater than in the Model Fuel and its blend with 1 vol% Model Aromatic Methyl Ethers Blend 1 (2.87 μg/cm^2 at 5 vol% Model Aromatic Methyl Ethers Blend 1 vs. 2.41 μg/cm^2 for the Model Fuel).

b) Model Phenols and Aromatic Methyl Ether Blend 2

The deposition and oxygen consumption profiles of the Model Fuel and the fuel containing Model Phenols and Aromatic Methyl Ethers Blend 2 at a 1 vol% blend level are shown in Figure 6.

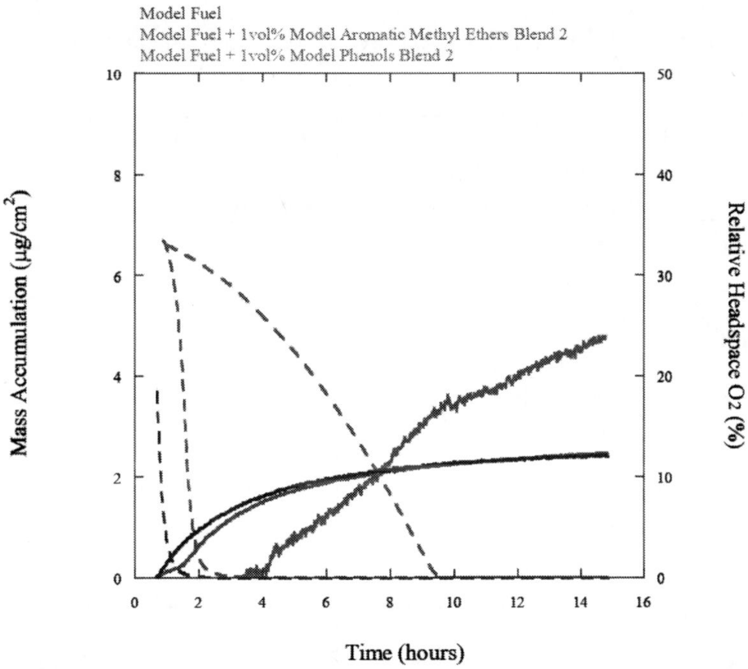

Figure 6. Deposition (solid lines) and oxygen consumption (dashed lines) profiles of the Model Fuel containing 1 vol% Model Phenols Blend 2 or Model Aromatic Methyl Ethers Blend 2.

Figure 6 showed that the Model Phenols Blend 2 produced a higher amount of deposits after 15 h (4.79 μg/cm^2) compared to the Model Aromatic Methyl Ethers Blend 2 (2.46 μg/cm^2). As with the Model Phenols Blend 1, in the first 4 h of the experiment, there was a delay in deposition, after which it started to increase significantly, not levelling off after 15 h. The deposition in the case of the Model Phenols Blend 2 (4.79 μg/cm^2) was however much less compared to the Model Phenols Blend 1 (9.37 μg/cm^2). This was believed to be due to the type of phenolic species present in the blends. Phenol, in particular, has been shown to be a high depositing species in studies by Abdallah et al. and Hui-Peng et al.[25] The Model Phenols Blend 1 contained a higher concentration of phenol (56 vol%) compared to the Model Phenols Blend 2 (5 vol% of phenol present). With the deposition profile of the fuel containing the Model Aromatic Methyl Ethers Blend 2, a slight inhibition was observed in the deposit formation. This can be attributed to the phenolic impurities (0.06% 2,3,5,6-tetramethylphenol, measured by GC-MS) present in the pure aromatic methyl ethers blend. The Model Aromatic Methyl Ethers Blend 2 contained a higher concentration of phenolic impurities compared to the Model Aromatic Methyl Ethers Blend 1 and this was clearly observed in the slight initial inhibition of deposit formation and the longer delay in oxygen consumed (complete oxygen consumption within 2.2 h) compared to the Model Fuel (complete oxygen consumption by 1.3 h).

c) JFTOT and Laser Ellipsometry Results

The deposition characteristics of the Model Fuel containing phenols and aromatic methyl ethers mixtures were measured by the JFTOT method and Laser Ellipsometry. An extreme JFTOT temperature of 325 °C (typically used for alternative fuels, including renewable fuels) was used to measure the deposit formation characteristics of these blends. The results of the blends are shown in Table 4. For comparison purposes, the deposition results for different Model Fuel blends by the QCM technique are also shown in Table 4.

Table 4. The JFTOT and QCM results of the Model Fuel containing phenols and aromatic methyl ethers mixtures.

Model Fuel Blend	QCM deposition (μg/cm²)	JFTOT temperature (°C)	Differential pressure (mmHg)	Tube rating	Tube colour
Base Model Fuel	2.41	325	0	1 PASS	No visible colour
With 1% Model Aromatic Methyl Ethers **Blend 1**	2.55	325	0	<1 PASS	No visible colour
With 1% Model Phenols **Blend 1**	9.37	325	0	>4P FAIL	Yellow-brown Peacock
With 1% Model Aromatic Methyl Ethers **Blend 2**	2.46	325	0	1 PASS	No visible colour
With 1% Model Phenols **Blend 2**	4.79	325	0	>4P FAIL	Yellow-brown Peacock

The Model Fuel containing the aromatic methyl ethers mixtures passed the JFTOT method at the stringent temperature of 325 °C. The Model Fuel containing the phenols mixtures failed the JFTOT method at 325 °C. This clearly showed the detrimental effect of phenolic species on the jet fuel thermal stability.

A laser ellipsometry technique was used to measure the deposit thickness on the aluminium surface of the JFTOT tube for each of the fuels tested. The results for the Model Fuel containing the Phenols and Aromatic Methyl Ethers Blend 1 are shown in Figure 7.

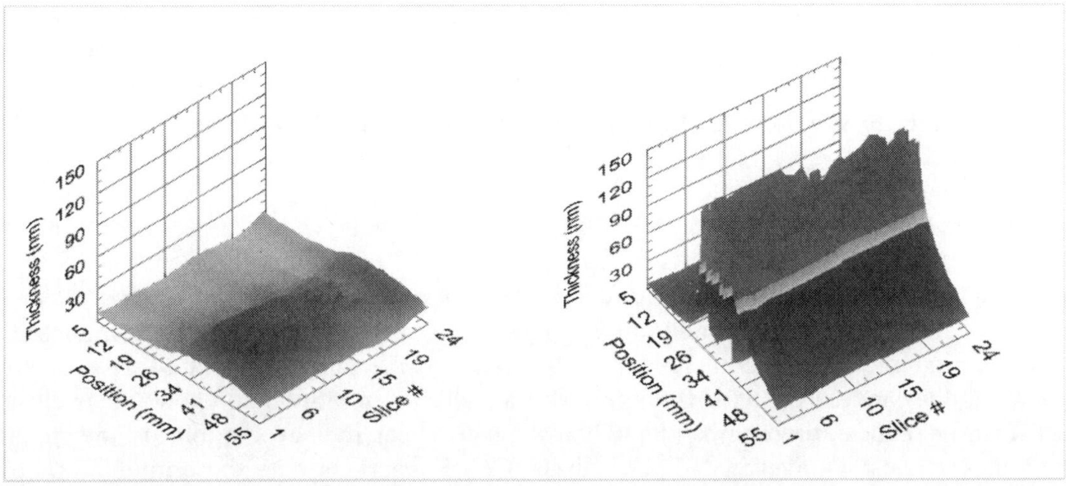

Figure 7. Contour plots of the deposit thickness on the JFTOT tubes for the Model Fuel containing the Aromatic Methyl Ethers Blend 1 (left) and the Phenols Blend 1 (right).

The average deposit thickness on the JFTOT tube for the Model Fuel containing the Aromatic Methyl Ethers Blend 1 was 26.8 nm. This provided evidence that the aromatic methyl ethers are not detrimental to jet fuel stability. For the Model Fuel containing the Phenols Blend 1, the deposit thickness on the JFTOT tube was 446.9 nm. This is a significant amount of deposit that formed on the tube, clearly showing the detrimental nature of phenols with regards to jet fuel's thermal stability. In comparison, the Model Fuel with no oxygenates present had a deposit thickness on the JFTOT tube of 47.0 nm, which was slightly more than the deposit thickness on the JFTOT tube for the Model Fuel containing the aromatic methyl ethers mixtures.

The results for the Model Fuel containing the Phenols and Aromatic Methyl Ethers Blend 2 are shown in Figure 8.

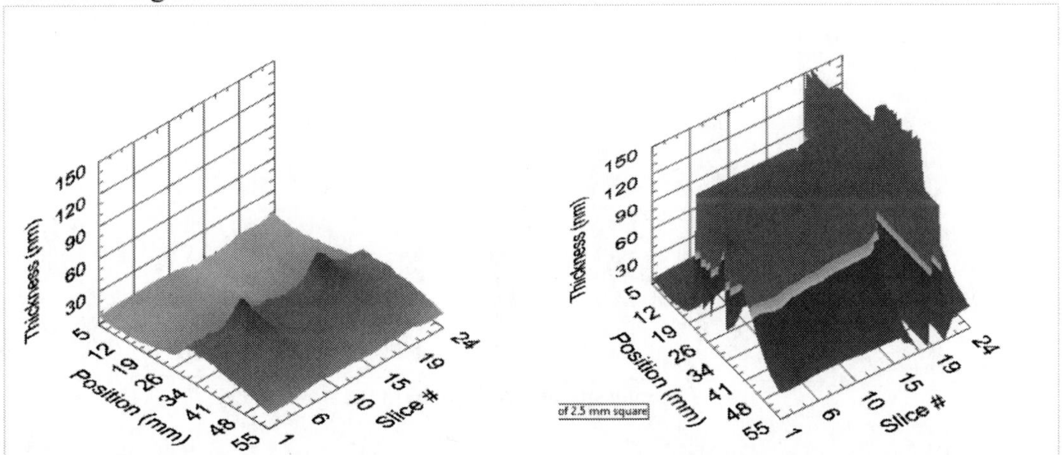

Figure 8. Contour plots of the deposit thickness on the JFTOT tubes for the Model Fuel containing Aromatic Methyl Ethers Blend 2 (left) and the Phenols Blend 2 (right).

The average deposit thickness on the JFTOT tube for the Model Fuel containing the Aromatic Methyl Ethers Blend 2 was 44.0 nm. This deposit thickness was greater than for the Model Fuel containing the Aromatic Methyl Ethers Blend 1 (26.8 nm), though less than the required 85 nm, which signifies a pass rating. For the Model Fuel containing the Phenols Blend 2, the deposit thickness on the JFTOT tube was 339.8 nm. This deposit thickness was less than for the Model Fuel containing the Phenols Blend 1 (446.9 nm). The reason for this difference in deposit thickness was likely due to there being more phenol present in the Phenols Blend 1 (56 vol%) compared to the Phenols Blend 2 (5 vol%), as also observed with the QCM results.

3.2. The Effect of the Type of Phenolic Species on Jet Fuel Deposition using the QCM technique

Few studies have thus far investigated the effects of different types of phenols on the thermal stability of the fuel.[25,26] The results of the current study suggest that different model phenolic mixtures had different effects on a fuel's deposition characteristics. For this reason, the influence and reactivity of individual phenolic compounds (various non-hindered and sterically hindered alkylated phenols) on deposition, as measured by a QCM, were evaluated by spiking the Model Fuel with 1000 ppm (by weight) of an individual phenolic species. This significantly reduces the complexity when measuring jet fuel deposition, as the impact of only one species is investigated. For these QCM reactions, the experimental time was extended beyond the typical 15 h period to 36 h to allow complete oxygen consumption and to allow the deposit level to plateau out. The results were separated according to the phenolic species that form greater amounts of deposits than the original Model Fuel (shown in Figure 9) and the phenolic species that form lower amounts of deposits than the original Model Fuel (shown in Figure 10). The values plotted in Figures 9 and 10 are the final mass per area accumulated using a QCM at the end of each 36 h run at 140 °C.

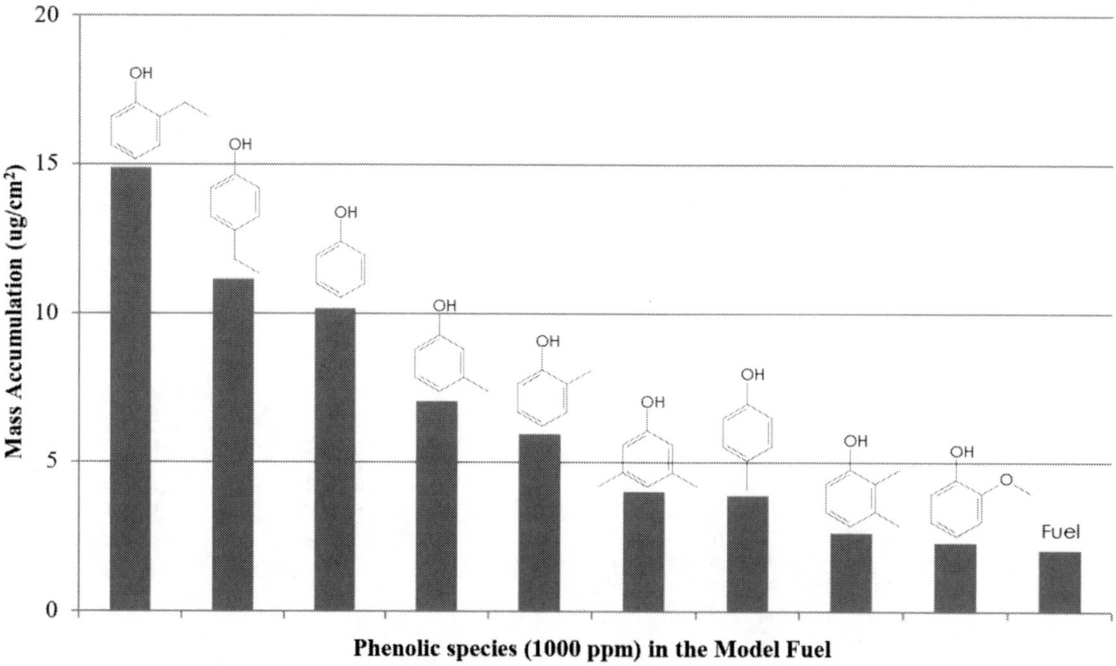

Figure 9. The mass accumulation results after 36 h of individual phenolic species that form higher amounts of deposits than the Model Fuel.

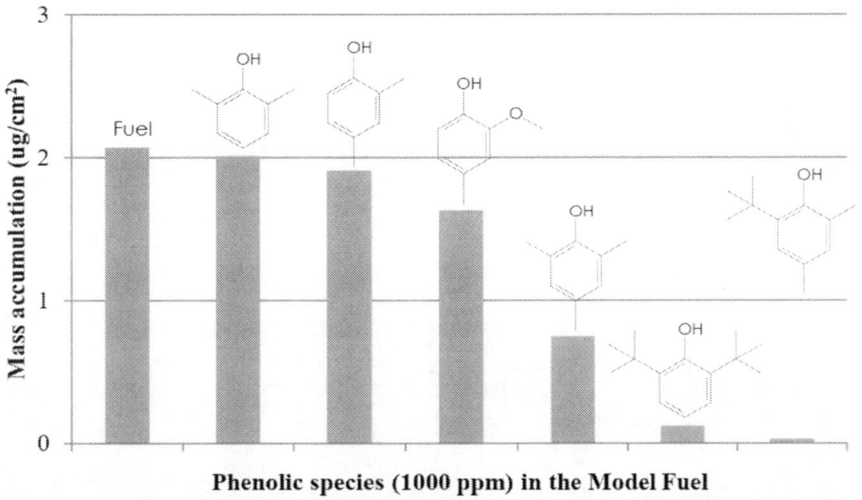

Figure 10. The mass accumulation results after 36 h of individual phenolic species that form lower amounts of deposits than the Model Fuel.

The type of phenolic species was shown to have a major impact on the fuel deposition. Deposition depends to a large extent on the substitution (both position and group) on the phenol molecule. The addition of more sterically hindered phenols to the Model Fuel resulted in a significant reduction in deposition. Conversely, non-hindered (single or doubly alkylated) phenols tend to form significant quantities of deposits. The amount of deposits formed, however, also depends on the type of alkyl group and the alkylation position on the phenol. For instance, when comparing 2-methylphenol, 3-methylphenol and 4-methylphenol, it was observed that, for the para-substitution on the phenol molecule, the amount of deposits formed was reduced. The ortho and para positions on the methylphenols seemed to be preferred for lower amounts of deposits. The same rule applied for 2-ethylphenol and 4-ethylphenol. The addition of two or three alkylated groups on the phenol molecule has a greater reducing effect on deposition compared to one alkyl group on the phenol molecule, although it also depended on where on the phenol molecule these alkyl groups were situated. Bulkier groups on the phenol molecule have the greatest effect on reducing deposition. Furthermore, phenolic species that contain benzylic hydrogens on the alkyl group (such as 2-ethylphenol and 4-ethylphenol) tend to form greater amounts of deposits. This is due to the weaker bond dissociation energy of the benzylic hydrogens (Φ-CH_2-H), which tends to be more reactive. These benzylic sites are thus more prone to oxidation.

3.3. The Thermal and Oxidative Stability of Biomass-derived Kerosene and Methods for Improving its Thermal Stability

The Biomass-derived Kerosene failed the JFTOT at 275 °C. For this reason, the Biomass-derived Kerosene would be classified as an unstable fuel. This was believed to be due to the presence of polar oxygenated species (±0.25 mass%) in the fuel, as indicated by two-dimensional gas chromatography (2D-GC with time of flight mass spectrometry) analysis. The trace levels of phenols and other oxygenated species (containing a reactive O-H bond) present in the Biomass-derived Kerosene were etherified by the O-methylation reaction. The 2D-GC analysis of the etherified Biomass-derived Kerosene showed that the phenolic species were successfully converted to aromatic methyl ethers. The concentration of the aromatic methyl ethers was found to be slightly higher than the amount of phenols present. This could be due to other aromatic oxygenates present in the fuel also undergoing the etherification reaction, thereby forming higher amounts of ethers. This method was evaluated to determine whether the deposition characteristics of an unstable fuel could be improved by etherification. It is also known that silica treating a fuel removes polar hetero-atomic species such as phenols.[27] For comparison purposes, the silica treatment of the Biomass-derived Kerosene and its subsequent effect on jet fuel stability was also evaluated. The etherification reaction resulted in a fuel loss (by mass) of 11%. This was much lower than the silica treatment, which resulted in a 16% fuel loss. The deposition and oxygen consumption profiles of the untreated, silica treated and etherified kerosene are shown in Figure 11.

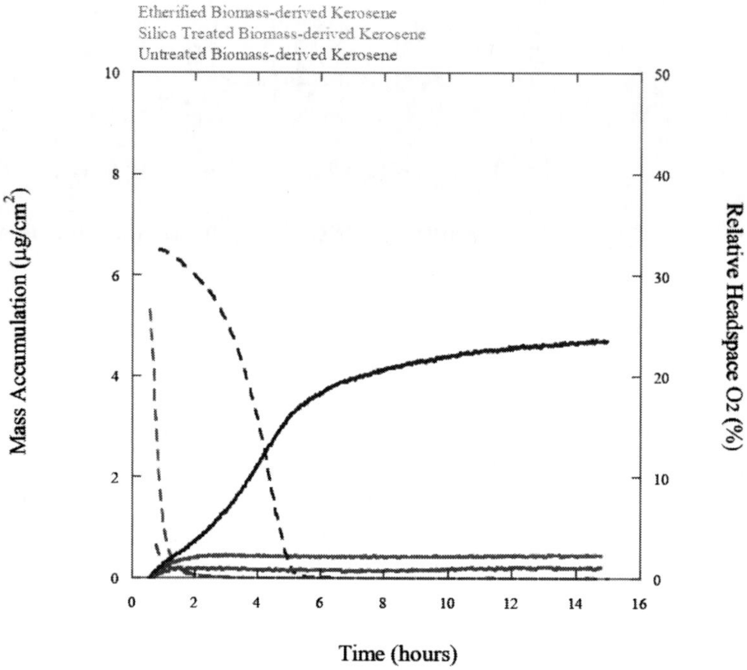

Figure 11. Deposition (solid lines) and oxygen consumption (dashed lines) profiles of the untreated Biomass-derived Kerosene, silica treated Biomass-derived Kerosene compared to the etherified Biomass-derived Kerosene.

The untreated Biomass-derived Kerosene was a moderate depositing fuel (4.69 μg/cm^2) and a moderate oxidising fuel (complete oxygen consumption in 5.0 h). The moderate oxygen uptake is typical of fuels containing trace hetero-atomic species.[28] The etherification reaction of phenols to aromatic methyl ethers resulted in an increase in oxidation rates and a significantly decreased deposition (0.22 μg/cm^2). The deposition profile of the silica treated fuel (0.44 μg/cm^2) is shown for comparison purposes. The extent of deposition in the silica treated and etherified kerosenes were reduced by more than 90%. The etherified and the silica treated Biomass-derived Kerosenes oxidised rapidly, clearly showing the absence of the hetero-atomic species (oxygen consumption was completed within 1.5 h). This suggests that in the etherified kerosene, the easily abstractable hydrogen atoms in the phenolic and other oxygenated species were converted to methyl groups.

4. CONCLUSIONS

Phenols have been implicated in jet fuel deposition reactions. This study shows that different phenolic species had different effects on jet fuel deposition. Non-hindered phenols formed significant amounts of deposits. Sterically hindered phenols, which are typically added as antioxidants to highly refined jet fuels, had no harmful effect on jet fuel deposition. The phenols typically present in petroleum- and biomass pyrolysis oil-derived jet fuels are the non-hindered types and are therefore more likely to be harmful to jet fuel stability. Hence these are the phenols that need to be removed or converted.

It was hypothesised that the stability of biomass pyrolysis oil-derived kerosene could be improved significantly by converting detrimental phenols to aromatic methyl ethers. This is consistent with the presence of the strong ether bond linkage ($O\text{-}CH_3$) of the aromatic methyl ether species. This study showed that aromatic methyl ethers are relatively stable species under fuel stressing conditions compared to their phenolic counterparts. Furthermore, it confirmed that the chemical transformation of phenols to aromatic methyl ethers indeed improved the deposition characteristics of the jet fuel. A novel pathway for improving the thermal stability of jet fuels derived from biomass, by in-situ etherification of the inherent phenolic constituents to the corresponding aromatic methyl ethers has been successfully demonstrated.

5. REFERENCES

1. Kuprowicz, N.J., Zabarnick, S., West, Z.J., and Ervin, J.S. (2007), "Use of measured species class concentrations with chemical kinetic modelling for the prediction of autoxidation and deposition of jet fuels", *Energy Fuels*, 21, 530-544.

2. Zabarnick, S., and Phelps, D.K. (2006), "Density functional theory calculations of the energetics and kinetics of jet fuel autoxidation reactions", *Energy Fuels*, 20, 488-497.

3. Balster, L.M., Zabarnick, S., Striebich, R.C., Shafer, L.M., and West, Z.J. (2006), "Analysis of polar species in jet fuel and determination of their role in autoxidative deposit formation", *Energy Fuels*, 20, 2564-2571.

4. Sobkowiak, M., Griffith, J.M., Wang, B., and Beaver, B. (2009), "Insight into the mechanisms of middle distillate fuel oxidative degradation Part 1: On the role of phenol, indole and carbazole derivatives in the thermal oxidative stability of Fischer-Tropsch/petroleum jet fuel blends", *Energy Fuels*, 23, 2041-2046.

5. Abdallah, D.J., Kreno, L., DiSanzo, F.P., Gaughan, R.G., Genowitz, M.W., Wells, P.P., Hoskin, D.H., Penn, K.M., Brown, B., and Wiseman, D. (2013), "Understanding phenol levels in jet fuel", *Proceedings of the 13th International Conference on Stability, Handling and Use of Liquid Fuels*; Rhodes, Greece, October 6-10.

6. Garver, J.M. (1995), "Phenolic measurements in jet fuels by UV absorption for the analysis of thermal oxidative stability", *Master of Science Thesis* at Wright State University, Dayton, Ohio, USA.

7. White, C.M., Jones, L., and Li, N.C. (1983), "Ageing of SRC-11 middle distillate from Illinois No. 6 coal II. Further evidence of phenolic oxidative coupling in coal-derived liquids", *Fuel*, 62, 1397-1403.

8. Beaver, B., Gao, L., Burgess-Clifford, C., and Sobkowiak, M. (2005), "On the mechanisms of formation of thermal oxidative deposits in jet fuels. Are unified mechanisms possible for

both storage and thermal oxidative deposit formation for middle distillate fuels?", *Energy Fuels*, 19, 1574-1579.

9.International Air Transport Association (IATA). (2014), IATA 2014 Report on alternative fuels, Retrieved from: http://www.iata.org/publications/documents/2014-report-alternative-fuels.pdf

10. International Air Transport Association (IATA). (2013), IATA 2013 Report on alternative fuels, Retrieved from: http://www.iata.org/publications/documents/2013-report-alternative-fuels.pdf.

11. Furimsky, E. (2000), "Catalytic hydrodeoxygenation", *Applied Catalysis A: General*, 199, 147-190.

12. Batts, B.D., and Fathoni, A.Z. (1991), "A literature review on fuel stability studies with particular emphasis on diesel oil", *Energy Fuels*, 5, 2-21.

13. Zabarnick, S., Mick, M.S., Striebich, R.C., and Grinstead, R.R. (1999), "Model studies of silylation agents as thermal-oxidative jet fuel additives", *Energy Fuels*, 13, 154-159.

14. Burnham, R. (2012), "The effect of different chemical classes on the swelling of NBR O-rings in blends with synthetic paraffinic kerosene", MSc Dissertation; University of Cape Town, Cape Town, South Africa.

15. Scully, C. (2013), "The effect of hydrocarbon exposure and temperature on the behaviour of elastomeric seal", MSc Dissertation; University of Cape Town, Cape Town, South Africa.

16. Musgrave, O.C. (1969), "The oxidation of alkyl aryl ethers", *Chem. Rev.*, *69(4)*, 499-531.

17. Stoochnoff, B.A., and Benoiton, N.L. (1973), "The methylation of some phenols and alcohols with sodium hydride/methyl iodide in tetrahydrofuran", *Tetrahedron Letters*, 1, 21-24.

18. Klavetter, E.A., Martin, S.J., and Wessendorf, K.O. (1993), "Monitoring jet fuel thermal stability using a quartz crystal microbalance", *Energy Fuels*, 7, 582-588.

19. Zabarnick, S. (1994), "Studies of jet fuel thermal stability and oxidation using a quartz crystal microbalance and pressure measurements", *Ind. Eng. Chem. Res.*, 33, 1348-1354.

20. Zabarnick, S., Zelesnik, P., and Grinstead, R.R. (1996), "Jet fuel deposition and oxidation: Dilution, materials, oxygen and temperature effects", *J. Eng. Gas Turbines Power*, 118, 271-277.

21. Mick, M. S. (1998), "Fundamental studies of the oxidation and deposition of jet fuels", Master Thesis, University of Dayton, Dayton, OH, USA.

22. Schlepp, E. E., Yu, S., and Croudace, M. (2013), "Overview of ellipsometer as an absolute way to measure D3241 deposit thickness", *Proceedings of the 13th International Conference on Stability, Handling and Use of Liquid Fuels*; Rhodes, Greece, October 6-10.

23. SWAFEA formal report D.3.2 (SW_WP3_Aero Engine Suitability Synthesis_C Lewis_Mar_2011) (2011), Released on 01.10.2011, Submitted to the European Commission under contract Tren/F2/408.2008/SI2.518403/SI2.59012.

24. DeWitt, M.J., West, Z., Zabarnick, S., Shafer, L., Striebich, R., Higgins, A., and Edwards, T. (2014), "Effect of aromatics on the thermal-oxidative stability of synthetic paraffinic kerosene", *Energy Fuels*, 28, 3696-3703.

25. Hui-Peng, L., Jian, S., Hua, Z., Li, Z., and Yu-tai, Q. (2009), "Effects of phenols on the stability of FCC diesel fuel", *Petroleum Science and Technology*, 27, 486-497.

26. Epping, R., Kerkering, S., and Andersson, J.T. (2014), "Influence of different compound classes on the formation of sediments in fossil fuels during aging", *Energy Fuels*, 28, 5649-5656.

27. Zabarnick, S., and Grinstead, R.R. (1994), "Studies of jet fuel additives using the quartz crystal microbalance and pressure monitoring at 140 °C", *Ind. Eng. Chem. Res.*, 33, 2771-2777.

28. Jones, E.G., and Balster, L.M. (1997), "Impact of additives on the autoxidation of a thermally stable aviation fuel", *Energy Fuels*, 11, 610-614.

IASH 2017, the 15th INTERNATIONAL SYMPOSIUM ON STABILITY, HANDLING AND USE OF LIQUID FUELS
Rome, Italy
10-14 September 2017

Stability characterization of resid hydrocracking effluents by the S-Value method (ASTM D7157): Effect of the analytical parameters

Jérémie Barbier[1,*], André Diot[2], Matthieu Drellard[1], João Marques[1]

[1]*IFP Energies nouvelles, Rond-point de l'échangeur de Solaize, BP 3, 69360 Solaize, FRANCE*
[2]*ROFA France, 6 Rue Raymond Poincaré, 25300 Les Allies, France*

[*]*Corresponding author e-mail: jeremie.barbier@ifpen.fr*

Ebullated-bed resid hydrocracking (RHC) process is used to upgrade vacuum residues feedstock. This is a catalytic process that allows to convert heavy oils into gasoline and gasoil and to remove contaminants such as sulphur and metals from heavy feeds. By 2020 the sulphur content of marine fuel oils should not exceed 0.5% has decided by the International Maritime Organization in order to significantly reduce sulphur emissions from ships. RHC process may contribute to the production of low sulfur marine fuels.

The stability of RHC heavy effluents must be controlled in order to limit sediments and sediments precursors formation. Therefore, the S-value method (ASTM D7157) was developed to quantify the intrinsic stability of asphaltenes in residual products from thermal and hydrocracking processes. This method consists on a titration using a solvent/antisolvent couple (toluene/n-heptane) and an optical detection of the asphaltenes precipitation. 3 parameters are derived: *S* parameter which is representative of the intrinsic stability of oil, *Sa* parameter that quantifies the peptizability of asphaltenes and *S0* that represents the peptizing power of oil.

The objective of this study is to investigate the relevance of the S-Value method to characterize the stability of RHC effluents. Different RHC pilot plant effluents, obtained from an Ural straight run vacuum residue treated under different operating conditions of residence times and temperatures, were used in this study. Results show that for RHC effluents with weak intrinsic stability, the S-value parameters have poor repeatability and the confidence intervals of parameters are significantly larger than the ones described in the standard method. Therefore, two parameters of S-value method were studied: the anti-solvent addition rate flow and the polarity differential of solvent/anti-solvent couple. Results show that the kinetics of asphaltenes precipitation is a key phenomenon, and titration conditions must be adapted in order to improve the characterization of high conversion RHC effluents. A new software was developed allowing to adapt these conditions.

IASH 2017, the 15th INTERNATIONAL SYMPOSIUM ON STABILITY, HANDLING AND USE OF LIQUID FUELS
Rome, Italy
10-14 September 2017

FACTORS AFFECTING FILTERABILITY OF MIDDLE DISTILLATE FUELS

David A. Daniels, Alex Belly, Andrew McKnight, Eric Payne, Arthur Fogiel, Christopher LeMieux

Innospec Fuel Specialties, 200 Executive Drive, Newark, Delaware, U.S.A. 19702
David.Daniels@innospecinc.com

ABSTRACT

Modern diesel engines continue to evolve to meet ever more stringent emissions standards. This evolution includes fuel injection equipment (FIE) design features that are increasingly sensitive to fuel-borne contaminants. Subsequently engine performance and durability are linked to fuel cleanliness.

The industry has turned to ultrafine filtration, both on-board the vehicle as well as in the upstream diesel distribution system, in order to meet today's cleanliness challenges. The introduction of ultrafine filtration has generated a new set of concerns and problems for the industry as a whole.

This paper is the first in a series that highlights what may be important differences between various high efficiency bulk filtration media. The differences could pose serious operational problems and expense to the end user.

A climate-controlled filtration rig was used to evaluate the flow rate through various filter media under real world temperature conditions. The filtration rig clearly demonstrated dramatic flow rate variation with different bulk fuel filters of similar porosity and efficiency. This paper further highlights that different filters have disparate sensitivity to fuel temperature, fuel properties, heritage additives, porosity, and filter media.

KEYWORDS: Filtration, Cold Flow Improver, soft particulate

INTRODUCTION

Challenging emissions regulations have lead OEMs to push the boundaries of engine technologies. A primary strategy to reduce engine emissions is the use of fuel injection equipment (FIE) with ever-tightening tolerances between moving components. Fuel containing extremely low levels of hard particulate is critical to maintaining the longevity of the FIE. The industry has increasingly turned to ultra-fine filtration to achieve particulate levels that are compliant with OEM requirements[1].

Problems

With the introduction of ultra-fine filtration, numerous field issues have arisen surrounding the topic. It should be no surprise that the introduction of ultra-fine filtration has shown an increase in the amount of complaints of premature filter plugging. There have been numerous publications about filters plugging due to materials that are not classified as "hard particulate". A wide variety of "soft particulate" sources are thought to be a root cause of the increasing field issues.

Soft particulate encompasses a large category of substances that are commonly found within middle distillate fuels. A number of publications have pointed to different soft particulates such as glycerin[2], carboxylate salts[3], and heritage fuel additives[4] for prematurely plugging bulk fuel filters.

Current methods for evaluating the efficiency of fuel filters such as ISO 16889:2017[5], ISO 19438:2003[6], and SAE J1985[7] do not examine soft particulates. Similarly fuels are evaluated according to various specifications depending on region such as ASTM D975-17[8], EN 590:2013[9] and CGSB 3.517:2017[10] which do not include any filtration requirements. End users currently do not have the necessary information to make educated decisions about their filtration systems.

The issues of soft particulates in ultra-fine filtration systems have not been adequately studied which therefore places an undue burden on end users to determine what types of filters are "fit for purpose" with their operations, fuel supply and heritage additives.

Search for Solutions

Without any current industry standard laboratory methods or data examining the true effects of soft particulate on ultra-fine filters, the end user is unable to properly evaluate their filtration systems to prevent issues that may occur. It is clear that there are not any current methods that directly examine bulk filtration of middle distillate fuels. An effort was made to closely examine ultra-fine filtration under conditions that would better estimate actual field conditions.

An in-house filtration rig was specifically designed to examine the low temperature operability of middle distillate fuels using commercial bulk filter media in order to fill the current methodology gap. The low temperature operability rig (LTOR) is housed within a climate-controlled chamber that allows for no-harms testing under common field conditions.

EXPERIMENTAL

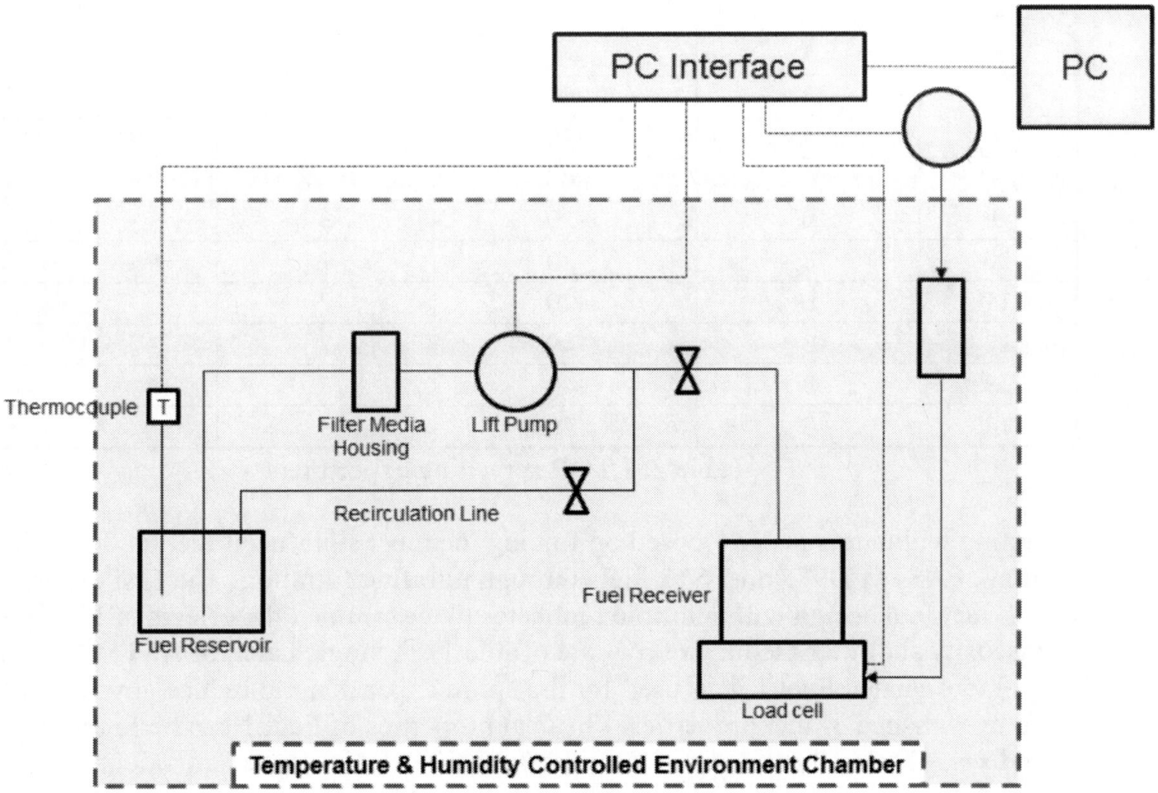

Figure 1: Simplified schematic of the LTOR

The LTOR is an in-house filtration rig that is contained within an environmental chamber with programmable temperature and humidity control. Testing conditions are controlled through a computer interface to adapt to a number of different test conditions. Each test involves a conditioning period prior to pulling a sample of fuel through a commercial filter patch using a vehicle lift pump. The sample that is passed through the filter patch is then collected in a collection vessel. The mass change within the collection vessel is monitored during the entire test (Figure 1).

These initial experiments used samples of ASTM D975 no. 1 S15 fuel to remove any potential effects caused by wax crystallization (particulate) at the test temperatures. The samples were clay filtered to remove any additives that could have been added prior to receipt. The fuels were then filtered at 0.8 µm using nitrocellulose paper filters (Millipore no. AAWP04700) to remove any particulates that would interfere with the selected filters.

Filters were selected from two commercial manufacturers designated as manufacturer A (A) and manufacturer B (B). These filters are rated as $\beta_{\mu m(c)}$ 2000 according to ISO 16889 for their respective porosity by the manufacturer. After the original projects were complete an additional filter was added which will be designated manufacturer C (C). C is rated as $\beta_{\mu m(c)}$ 500 according to SAE J1985 at 4µm by the manufacturer.

Standard	Manufacturer	Temperature (°C)	Porosity (μm)	DFN
1	A	-20	4	
2	A	-20	7	
3	A	-20	4	
4	A	-40	7	
5	A	-40	4	
6	A	-40	7	
7	A	-40	4	
8	A	-40	7	
9	B	-20	3	
10	B	-20	7	
11	B	-20	7	
12	B	-40	7	
13	B	-40	3	
14	B	-40	3	

Table 1: LTOR screening experiment

The testing within this paper focused on finding the most significant factors that affected filtration of ASTM D975 no.1 S15 fuels through ultrafine filtration. The first experiment was a 2^3 factorial design with multiple replicates that examined the effects of filter porosity, fuel viscosity and filter media on flow rate (Table 1). A single batch of ASTM D975 no.1 S15 fuel, designated Fuel 1, was used for the entire experiment to reduce any variability caused by a change in fuel properties. Physical properties of Fuel 1 can be found in Appendix A. The analysis assumes that the 3μm filters from B are equivalent to the 4μm filters from A.

STD	Manufacturer	WASA (ppm AI)	EVA (ppm AI)	DFN
1	A	0	0	
2	A	0	0	
3	A	0	0	
4	B	0	0	
5	B	0	0	
6	A	320	0	
7	B	320	0	
8	A	0	500	
9	B	0	500	
10	A	320	500	
11	B	320	500	
12	A	160	250	
13	A	160	250	
14	B	160	250	
15	B	160	250	
16	A	160	250	
17	B	160	250	
18	A	160	250	
19	B	160	250	

Table 2: LTOR additized filtration experiment

The second experiment was also a 2^3 factorial design with multiple replicates that examined the effects of common cold flow improver (CFI) chemistries on flow rate. The chemistries used were ethylene vinyl-acetate copolymer (EVA) and a common wax anti settling additive (WASA) (Table 2). A new batch of ASTM D975 no.1 S15 fuel, designated Fuel 2, was used for this entire experiment to reduce any variability caused by a change in fuel properties. Physical properties of Fuel 2 can be found in Appendix A.

Standard	Manufacturer	Temperature (°C)	Porosity (μm)	DFN
1	A	-40	4	
2	A	-40	4	
3	A	-40	4	
4	A	-20	4	
5	A	-20	4	
6	B	-40	3	
7	B	-40	3	
8	B	-20	3	
9	B	-20	3	
10	C	-40	4	
11	C	-40	4	
12	C	-20	4	
13	C	-20	4	

Table 3: LTOR examination of a third filter

A third filter manufacturer was added after the completion of experiment 1 and 2 to compare against manufacturers A and B. A new batch of ASTM D975 no.1 S15 fuel, designated Fuel 3, was used for this entire experiment to reduce any variability caused by a change in fuel properties. Physical properties of Fuel 3 can be found in Appendix A. The experiment also examined repeatability of the first two experiments using a new fuel.

Testing procedure

A standardized method was made to ensure results between the individual runs and the projects were consistent. The two projects did use different fuel samples, the physical properties of each fuel can be found in Appendix A.

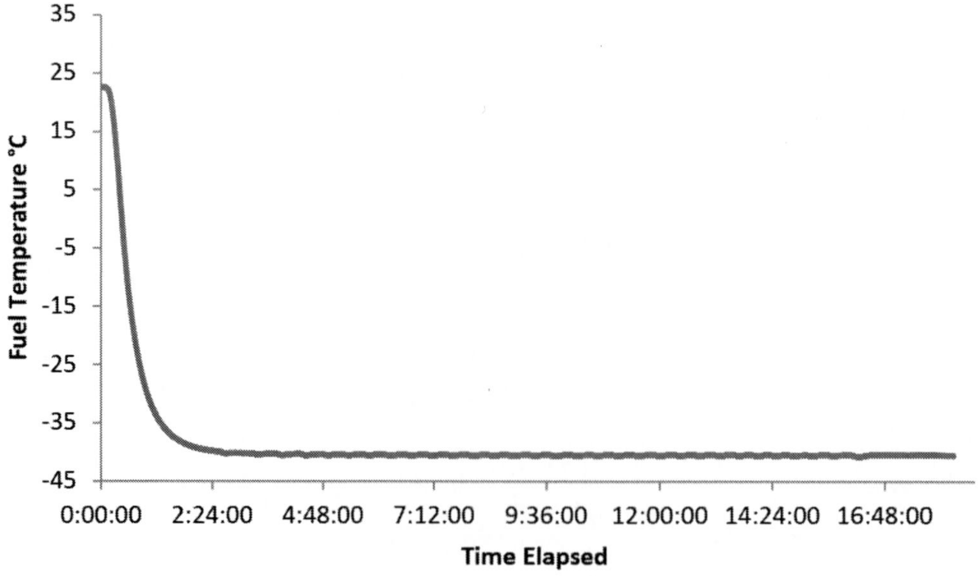

Figure 2: Fuel temperature profile during LTOR testing

Prior to starting the test, the environmental chamber was purged of water to ensure a dry testing environment. The fuel sample is then loaded into the fuel reservoir. The fuel is then recirculated for 10-20 seconds to prime the filter. After recirculation the fuel is crash cooled to the testing temperature and held at temperature for 18 hours (figure 2). After the cooling period the pump is initiated until the test is completed. The Test is considered completed after passing 600g of test fuel or if there is not a significant mass change for one minute.

Evaluation of filter testing

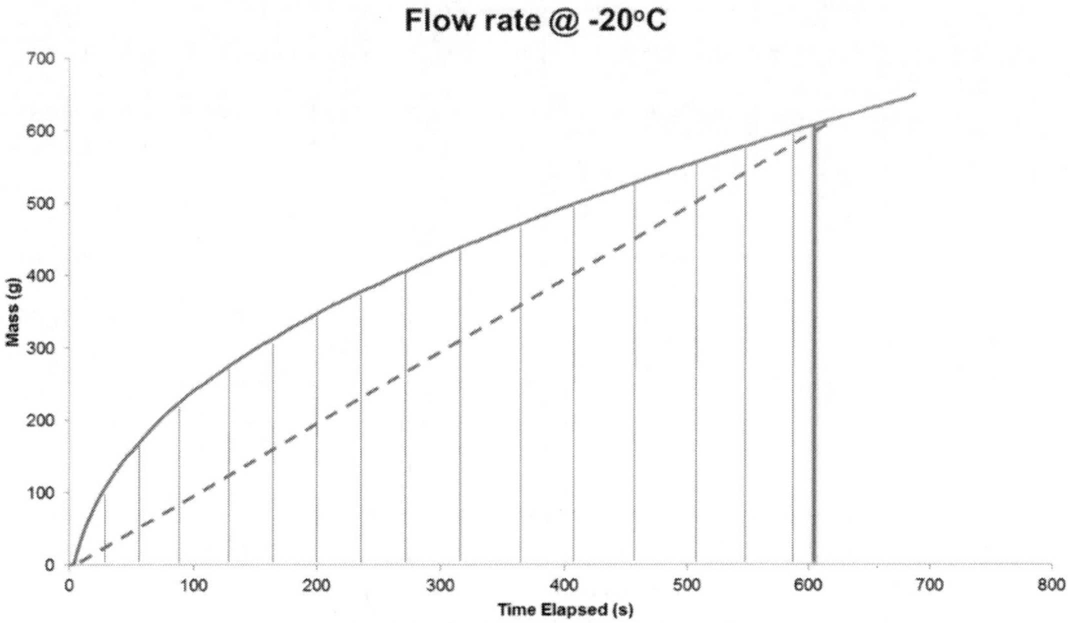

Figure 3: Visualization of the DFN

Assuming no particulate is present, flow through a filter should be approximately linear; however when a plugging occurs this is often not the case. A new metric, labeled the derived flow number (DFN), was created to ensure uniform analysis of samples if there was not an approximately linear flow rate. By integrating under the mass collection curve a singular value can be obtained to fairly analyze every sample. Because the number is an integral of the collection graph, a high DFN indicates a sample that shows a greater resistance to flow (Figure 3).

RESULTS AND DISCUSSION

Untreated fuel testing

Standard	Run #	Manufacturer	Temperature (°C)	Porosity (μm)	DFN
10	1	B	-40	7	6789
2	2	B	-40	4	47048
4	3	A	-40	7	13062
13	4	A	-40	7	17304
12	5	A	-20	4	25368
8	6	A	-20	7	6292
7	7	A	-40	4	84858
11	8	A	-40	7	37665
5	9	A	-20	4	19202
1	10	B	-20	7	5222
9	11	B	-20	4	18847
14	12	A	-40	4	78498
3	13	B	-40	4	20943
6	14	B	-20	7	5746

Table 4: Results for the LTOR Screening Experiment* (Fuel 1)

The analysis of these samples was performed using Design Expert 10 by Stat-ease. Overall the analysis showed only 2 outliers over the 14 run experiment. Table 4 contains the collected results from the LTOR screen experiment, with the yellow rows showing the outliers which were removed prior to the ANOVA analysis. Finally the data was transformed using the natural log function with no added constant. The natural log transformation functions to reduce the range of DFNs obtained throughout the experiment.

Results from the ANOVA analysis showed five significant effects within the model. The three main effects (viscosity, porosity, and filtration media) were all shown to be significant for the model. In addition to the main effects an interaction was seen between filtration media and temperature as well as an interaction between temperature and porosity.

* Outliers were traced to power issues with the lift pump

Effect	Coefficient
Porosity	-0.75
Temperature	-0.41
Manufacturer	0.21
Manufacturer * Temperature	-0.13
Temperature * Porosity	0.13

Table 5: LTOR screening experiment coefficients

Table 5 gives the coefficients used to model the regression for the LTOR screening project. The positive value for the manufacturer coefficient indicates higher DFNs for filters from A. Each numeric value is only applicable in this particular fuel, within these specific conditions and filters. With the level of variability seen throughout middle distillates, even within the confines of ASTM D975, a single predictive model would be too general to determine the DFN for all fuels. These coefficients do give insight to the magnitude of change they will exhibit in future studies. Table 5 is ordered by magnitude of effect on DFN to give more insight to which factors will have the greatest impact for future studies.

Porosity changes seem to have much larger sway in the DFN than the other factors (Table 5). This can be linked with the relatively small change in filter size (from 7μm to 4μm). The temperature coefficient is approximately half of the magnitude of the porosity, yet is over a much larger range (from -20 °C to -40 °C). The effect from filter media is rather small but still significant. The positive value in this case indicates that manufacturer A filters will cause a higher DFN, indicating worse flow rates when compared with manufacturer B filters.

This is particularly surprising as an assumption was made at the outset of this work that the 3μm filter from manufacturer B was equivalent to the 4μm from manufacturer A. One might ordinarily expect to find that the manufacturer B's 3 μm filter would filter more slowly due to decreased porosity size, especially with porosity having the primary effect.

The two factor interactions in this model were found to have the same coefficient. These should have only minor impacts when compared to the main factors. Additionally the assigned values matched with the standard deviation found in the model, the complete regression statistics can be found in Appendix B.

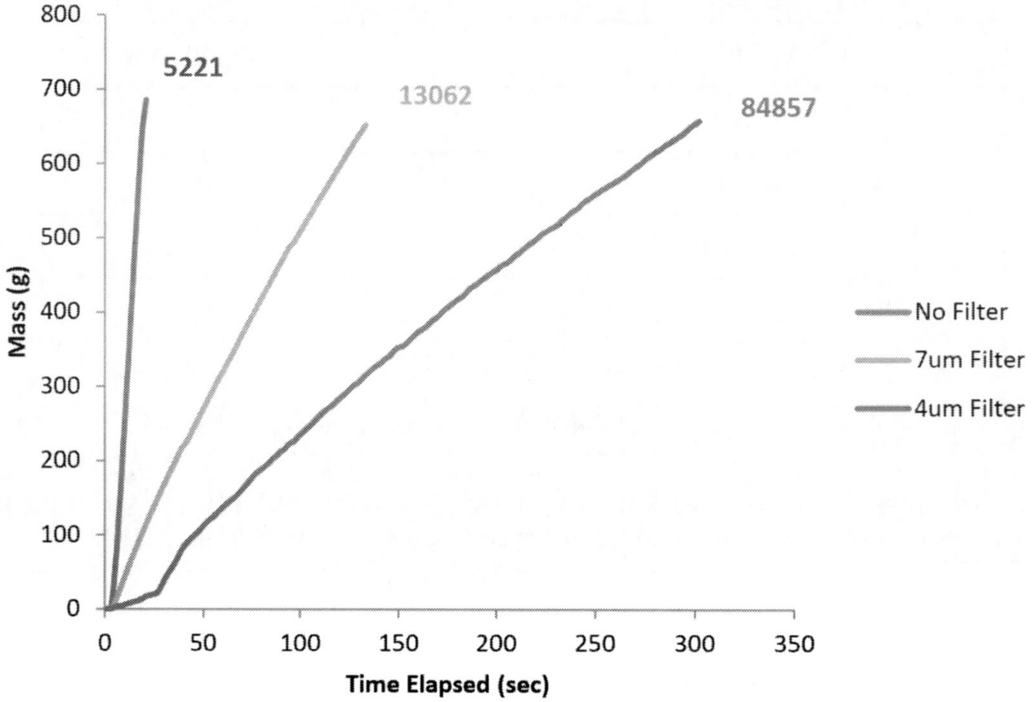

Figure 4: DFN with decreasing porosity using A at -40°C (Fuel 1)

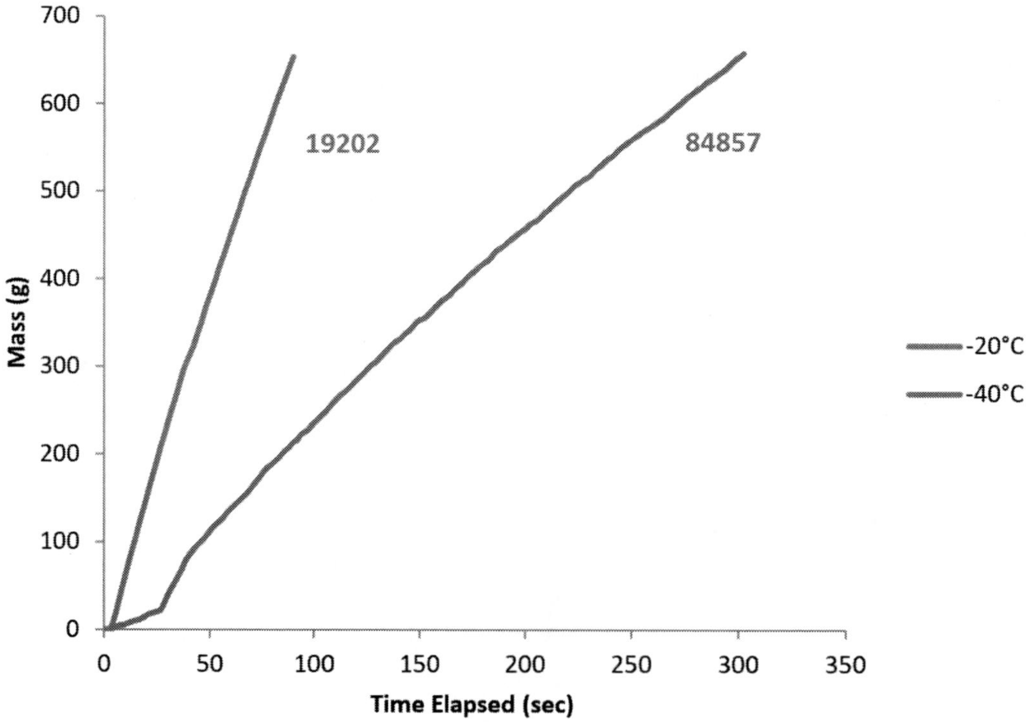

Figure 5: DFN with increasing viscosity using A's 4μm filters (Fuel 1)

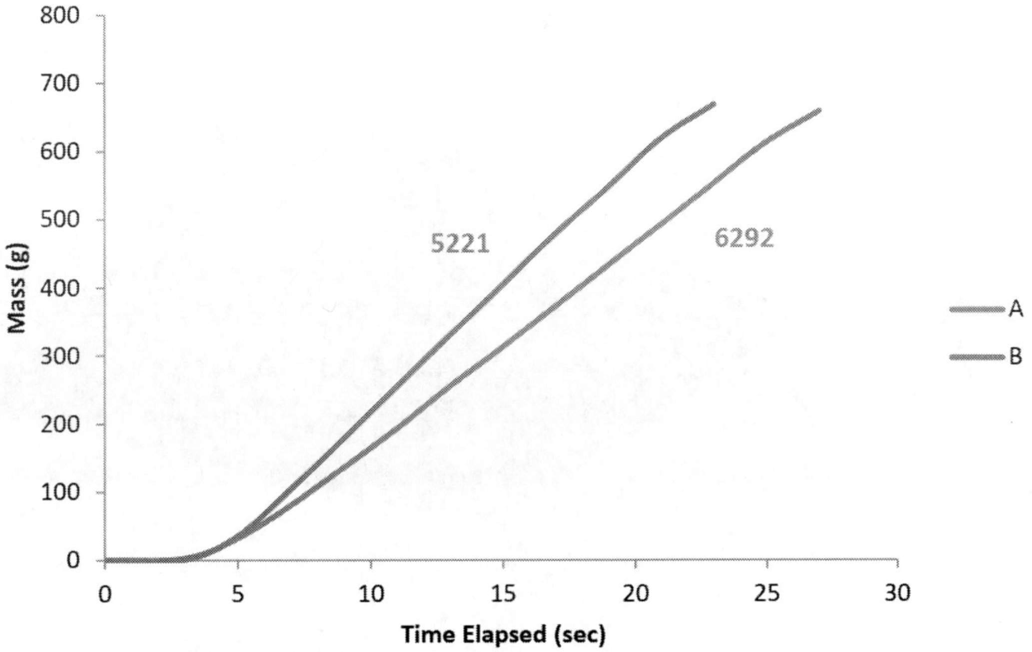

Figure 6: DFN comparison between A and B 7μm filters at -20°C (Fuel 1)

Figures 4-6 examine how each independent factor influences the DFN in Fuel 1. The changes in the DFN follow the trends set by the regression model. It also confirms the variables of interest for untreated fuels will be the filter porosity and sample viscosity. Note that the numbers indicated on the graphs of Figures 4-6 are the DFN.

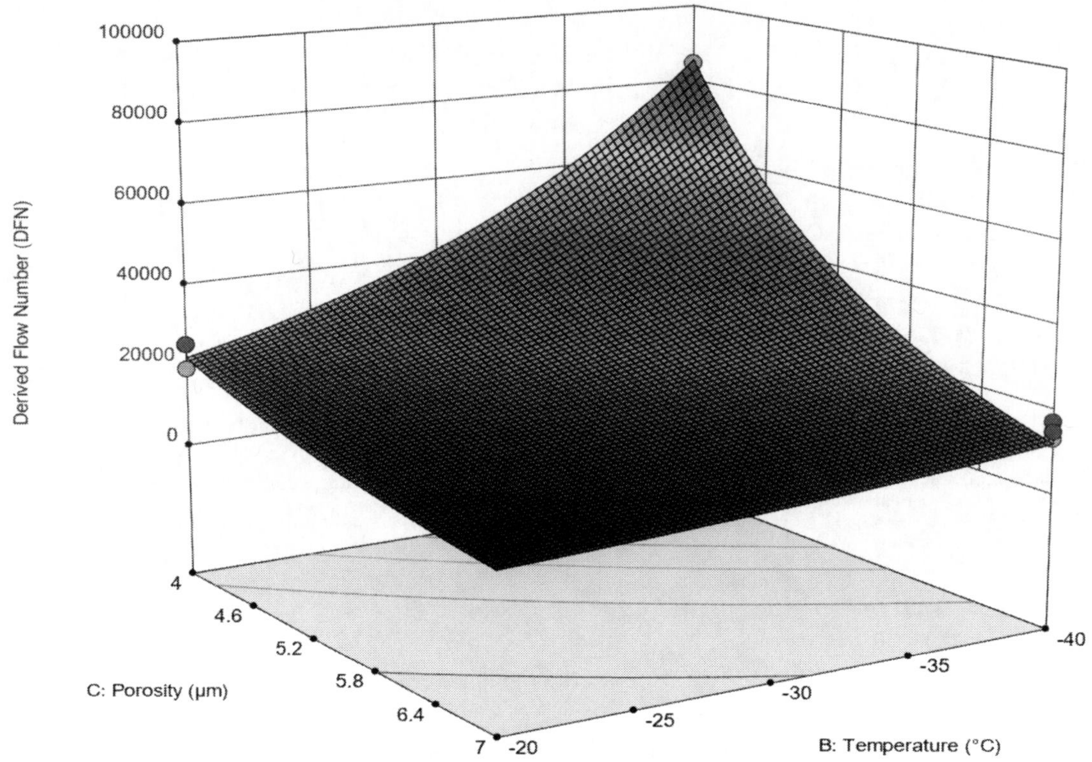

Figure 7a: DFN response curve for A in Fuel 1

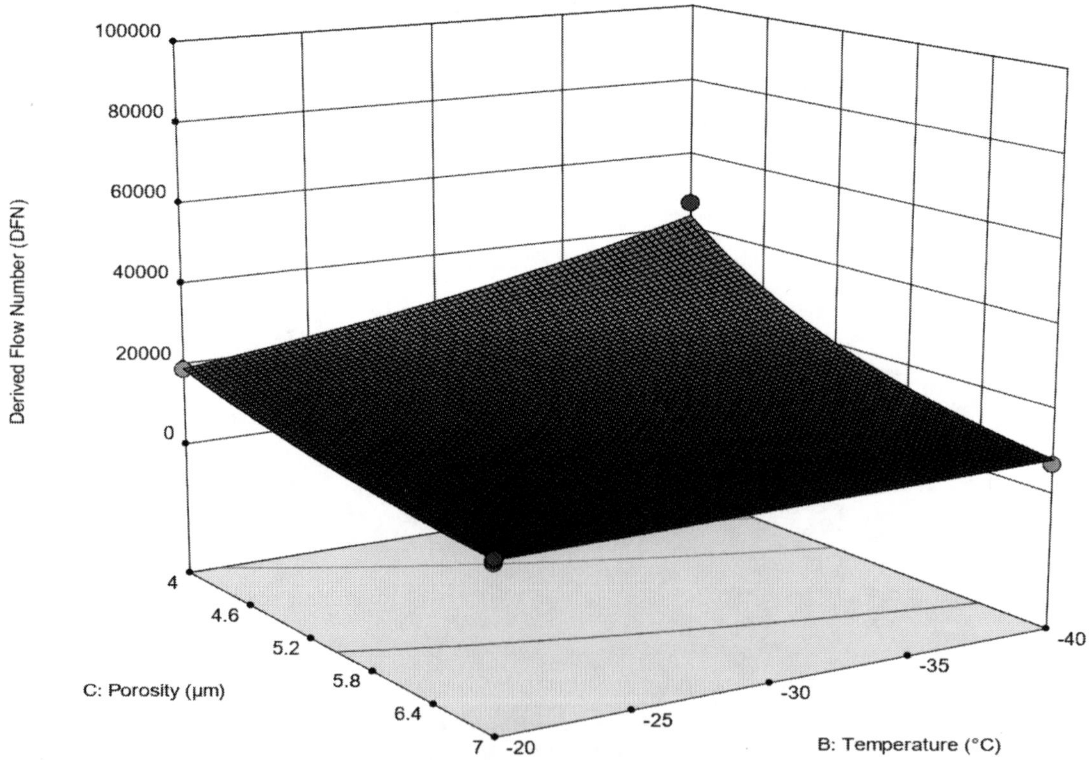

Figure 7b: DFN response curve for B in Fuel 1

Figure 7a and 7b give a visual representation of the response curves for A and B filters respectively. When the graphs are compared side by side a few important observations can be made in order to distinguish the filter media. First, A's filters show a large increase in DFN under the most strenuous conditions (-40 °C with the 4μm filter). The B filters do experience a similar increase under these conditions, yet the DFN is approximately half of A's filter under the same conditions. Additionally the -40 °C value using the equivalent 7um filters showed significant difference between the filtration media. The results found at -20 °C were very similar between the two manufacturers at both porosity sizes. This seems to indicate that A's filters are more sensitive to changes in viscosity than the B's filters.

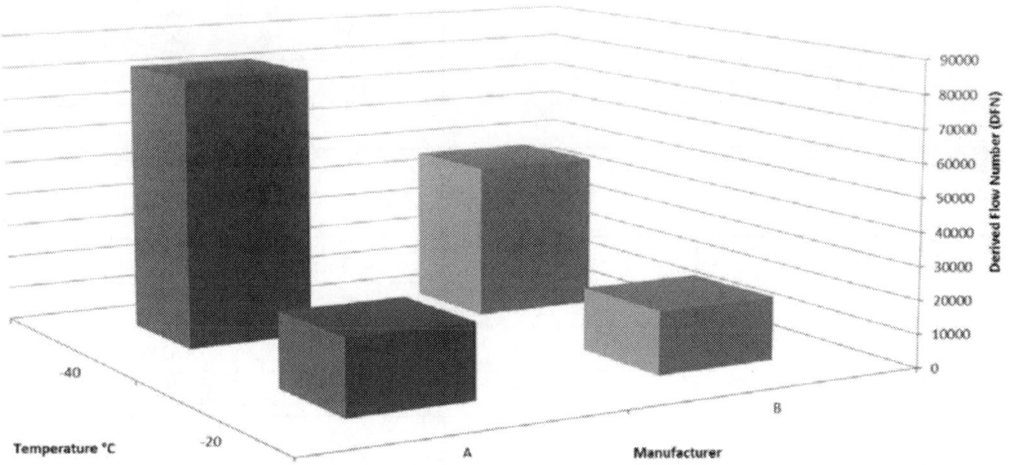

Figure 8a: DFN response for 4μm filters in Fuel 1

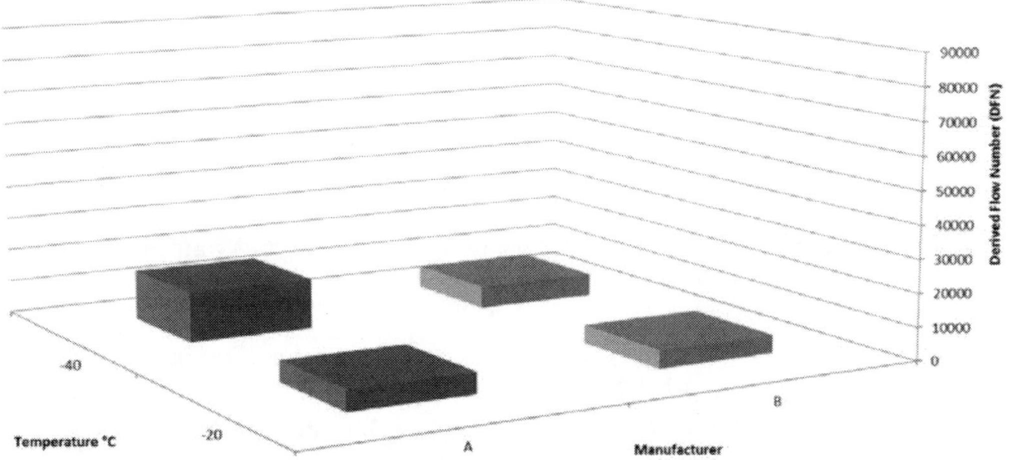

Figure 8: DFN response for 7μm filters in Fuel 1

Figure 8a and 8b give a side by side comparison of the filter manufacturers for 4μm and 7μm filters respectively. Again a large discrepancy can be seen between the most strenuous conditions (-40 °C with the 4μm filter). Using 7μm filters resulted in very little change regardless of the testing conditions in Fuel 1 during this experiment.

Additized fuel testing

STD	Run #	Manufacturer	WASA (ppm Al)*	EVA (ppm Al)	DFN
1	10	A	0	0	49786
2	16	A	0	0	52798
3	17	A	0	0	36070
4	1	B	0	0	13294
5	19	B	0	0	10427
6	12	A	320	0	81433
7	9	B	320	0	13232
8	13	A	0	500	515038
9	2	B	0	500	10051
10	11	A	320	500	484686
11	6	B	320	500	15429
12	3	A	160	250	233071
13	4	B	160	250	10355
14	5	B	160	250	27706
15	14	A	160	250	410773
16	18	A	160	250	529372
17	7	B	160	250	10629
18	8	A	160	250	262883
19	15	B	160	250	25148

Table 6: Results for the LTOR additized filtration experiment† (Fuel 2)

The analysis of these samples was performed using Design Expert 10 by Stat-ease. Overall the analysis showed 4 outliers over the 19 run experiment. Table 6 contains the collected results from the LTOR screen experiment, with the yellow rows showing the outliers which were removed prior to the ANOVA analysis. Finally the data was transformed using the natural log function with no added constant. The natural log transformation functions to reduce the range of DFN values obtained throughout the experiment.

An additional note to consider is the drastic difference between manufacturer A and B's untreated sample runs in Fuel 2. The LTOR screening experiment observed that the DFN differences between these exact two filters was relatively low in the first sample. While the fuel used to determine additive effects on filtration was different (but with very similar physical properties) it is very surprising to see such a change.

* ppm is parts per million by volume (µl/L)

† Outliers were traced to power issues with the lift pump

Results from the ANOVA analysis showed four significant effects within the model. The three main effects (manufacturer, concentration of EVA, and concentration of WASA) were all shown to be significant for the model. In addition to the main effects an interaction was seen between filtration media and EVA content.

Effect	Coefficient Estimate
Manufacturer	-1.29
Manufacturer * EVA	-0.55
EVA	0.54
WASA	0.14

Table 7: LTOR additized filtration experiment coefficients

EVA concentration will have a moderately antagonistic effect on the DFN of a fuel prior to wax formation. Alternatively the interaction between manufacturer and EVA concentration is almost equivalent. From the set definitions in the design this interaction states that B's filters will have an approximately net zero effect from EVA concentration. On the other hand A's filters will have a doubling effect in relation to the EVA concentration. This is also paired with the large manufacturer coefficient (Table 7). The WASA samples had a small negative effect on DFN, however the coefficient is below the standard deviation for the model. Additional regression statistics can be found in Appendix C.

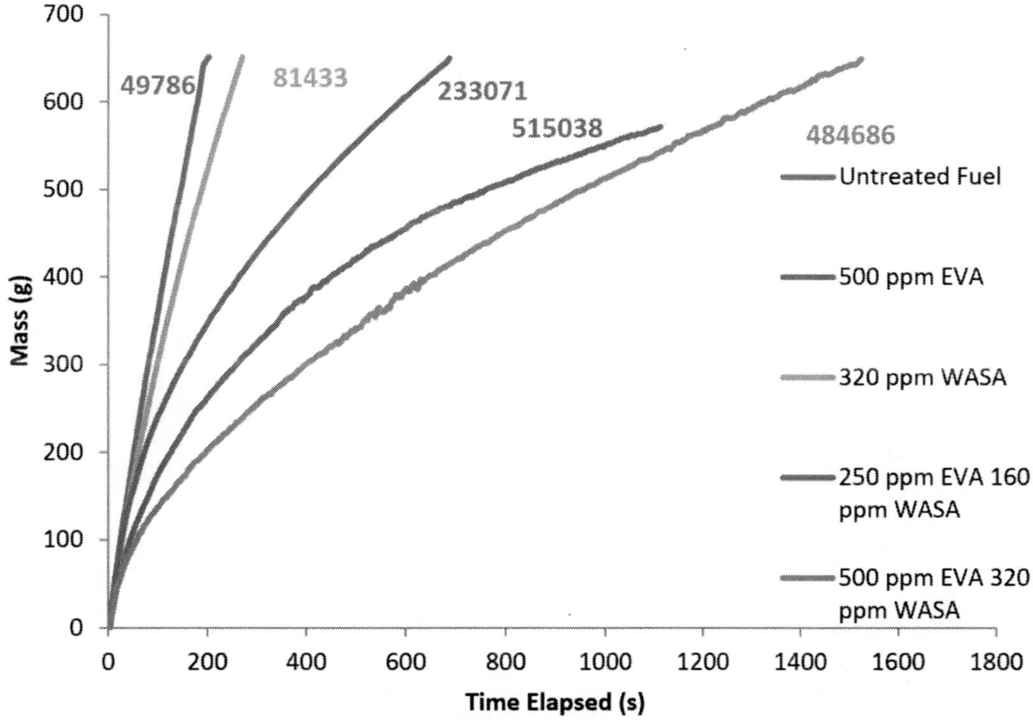

Figure 9: Comparison of DFN after being additized using A's 4μm filters at -20°C (Fuel 2)

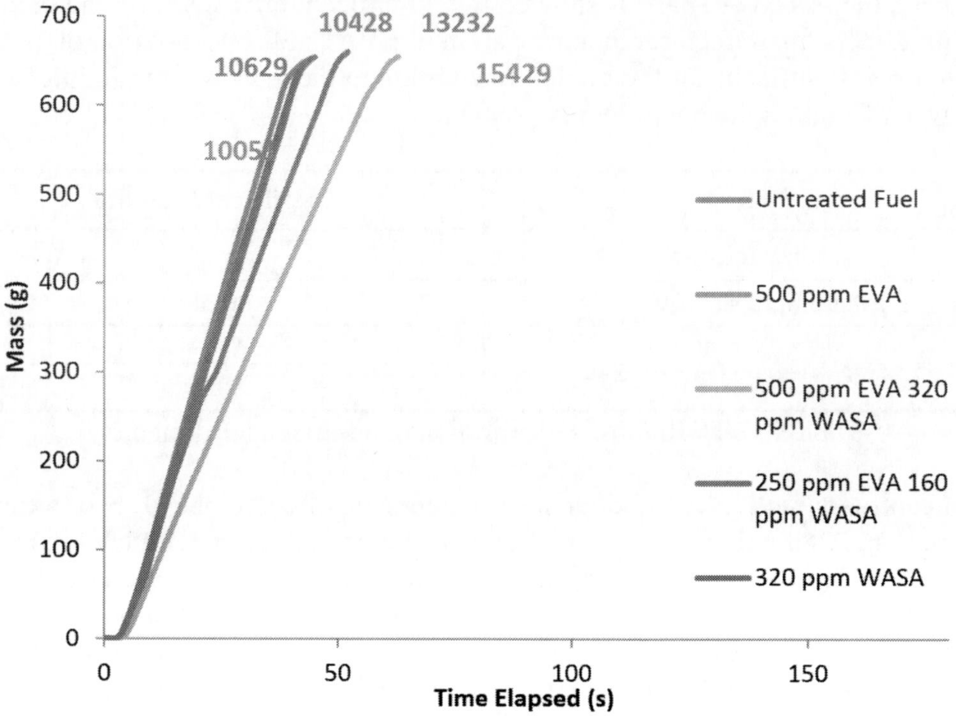

Figure 10: Comparison of DFN after being additized using B's 4μm filters at -20°C using a 1/10th time scale (Fuel 2)

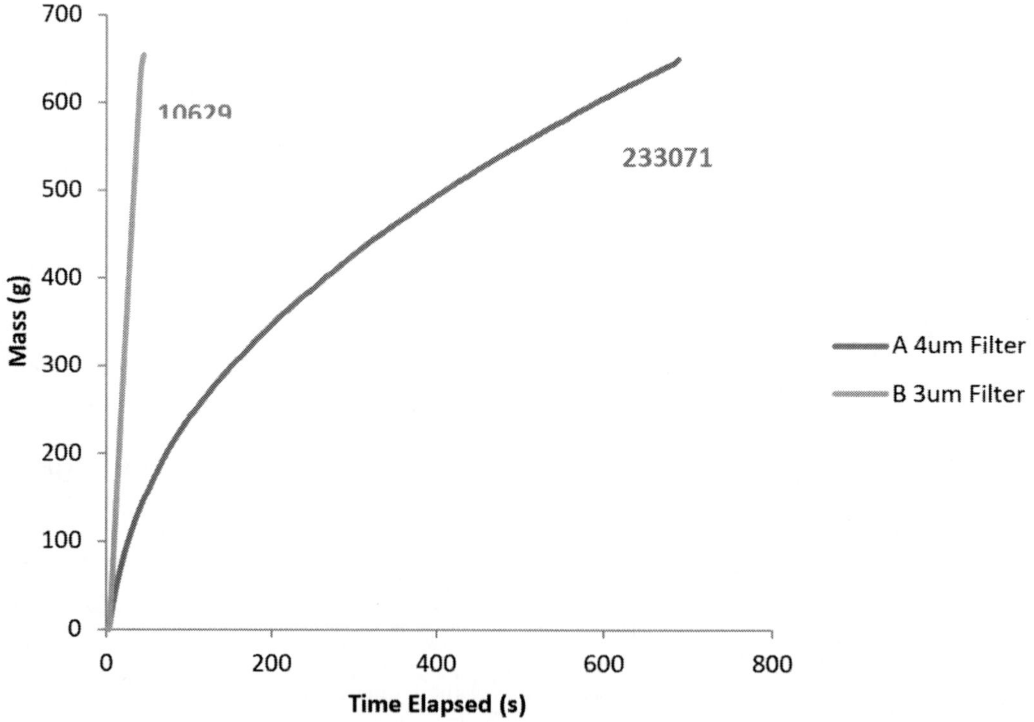

Figure 11: Comparing DFN of A and B filters at -20°C additized with 250 ppm EVA and 160 ppm WASA (Fuel 2)

Figure 9-11 show a drastic difference between the two filter manufacturers across the entire experiment. A direct comparison between manufacturer A and B filters shows drastically different results even at a moderate treatment (Figure 11). The x-axis on Figure 9 is a full order of magnitude larger than Figure 10's x-axis to accommodate the large filtration times. Using B's filters, DFN was slightly improved over the untreated sample when the sample was additized with EVA. However, the samples treated with 320ppm of WASA had worse DFN with B's filters (Figure 10). DFN of A's filters had a large negative impact with the use of additives. The response to EVA was especially bad in A's filters. The sample treated with 500 ppm of EVA caused A's 4μm filter to plug after ~570g (Figure 9).

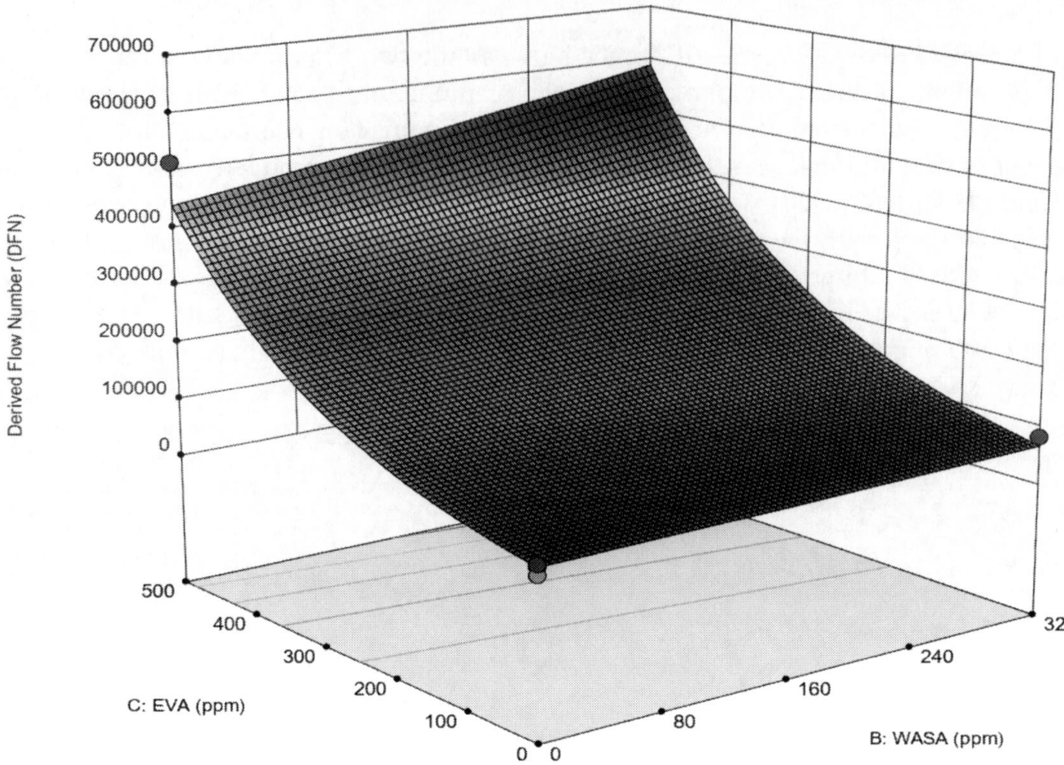

Figure 12a: DFN response curve for A's 4μm filters in Fuel 2 with CFI additives

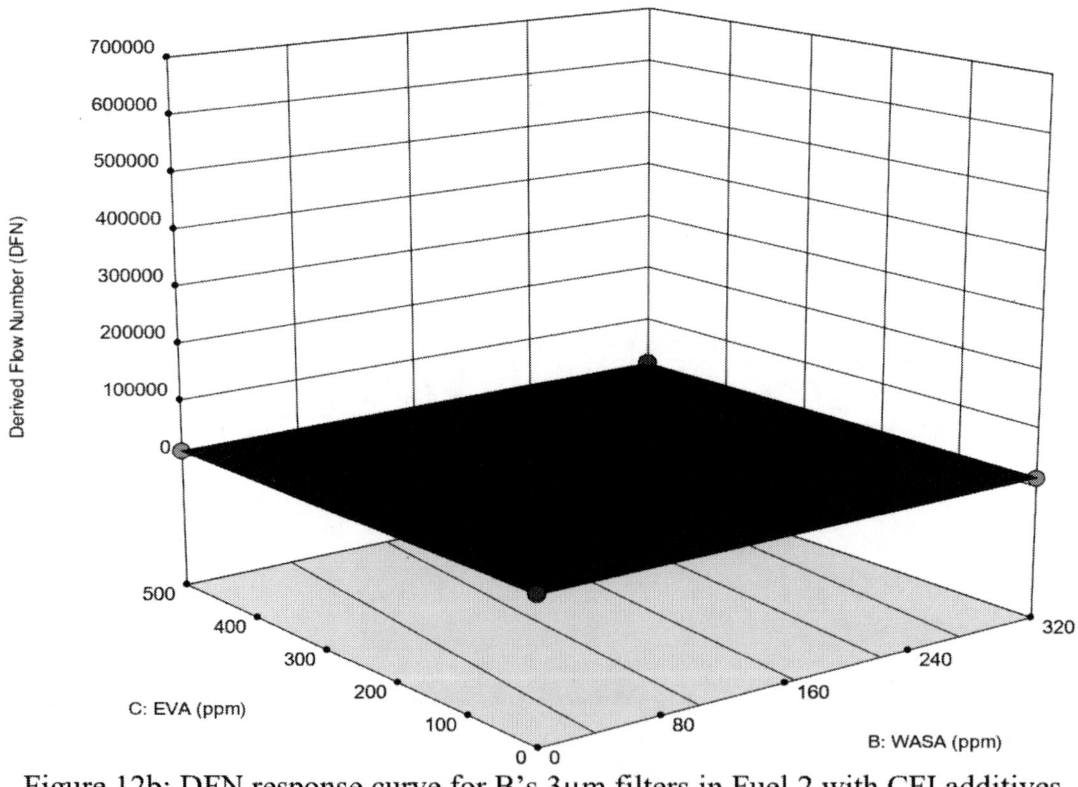

Figure 12b: DFN response curve for B's 3μm filters in Fuel 2 with CFI additives

The response curves reiterate the significant difference between the filtration media and their interactions with CFI additives. The B filters had almost no response to additive concentration resulting in a flat plane (Figure 12b). A's filters saw a very large response to EVA concentration, with only a minor impact from added WASA (Figure 12a).

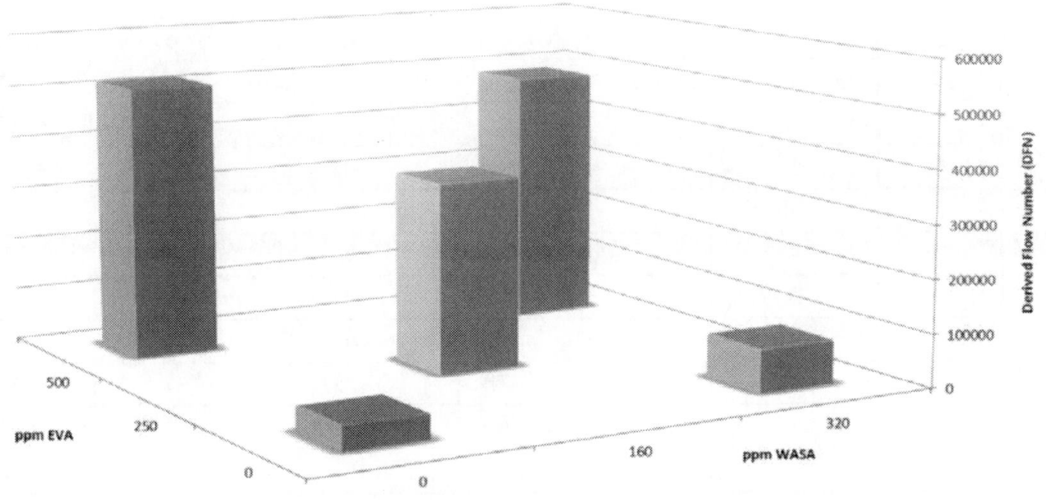

Figure 13a: DFN response in Fuel 2 using A's 3 μm filters with CFI additives

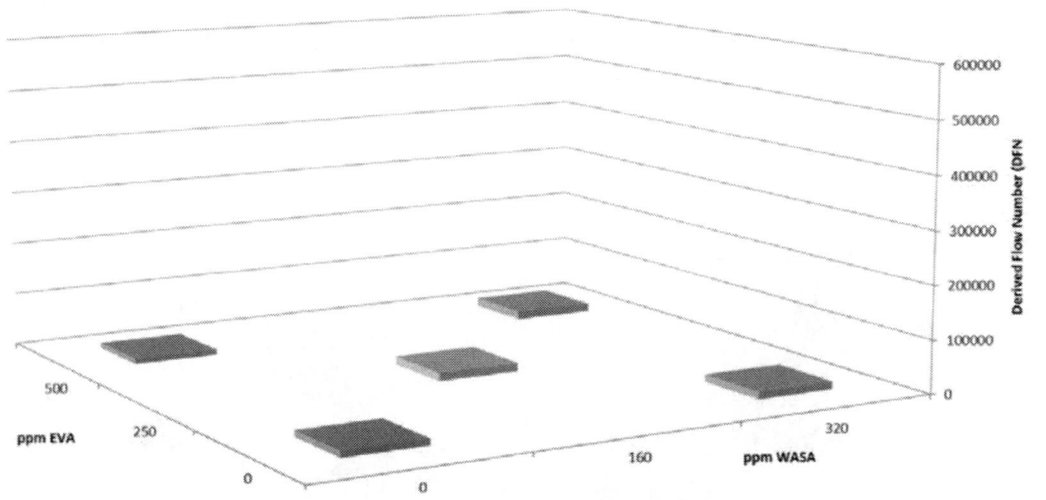

Figure 13: DFN response in Fuel 2 using B's 3 μm filters with CFI additives

Figure 13a and 13b also show the large difference between the A and B. Figure 13a also helps to show that elevated WASA content could slightly lower the DFN in Fuel 2 for A's filters.

Examining manufacturer C

STD	Run #	Manufacturer	Temperature (°C)	Porosity (µm)	DFN
1	8	A	-40	4	59561
2	10	A	-40	4	60851
3	12	A	-40	4	68846
4	1	A	-20	4	18130
5	11	A	-20	4	17741
6	3	B	-40	3	5590
7	14	B	-40	3	7372
8	4	B	-20	3	3339
9	9	B	-20	3	3694
10	6	C	-40	4	19282
11	13	C	-40	4	26740
12	5	C	-20	4	10199
13	7	C	-20	4	5927

Table 7: Results from the LTOR third filter examination (Fuel 3)

A new filter was sourced from manufacturer C to better investigate the differences seen between filters from A and B. Use of a two level factorial design was no longer appropriate after the introduction of a third filter media. Only the most restricted porosity was utilized because the tighter clearances saw the greatest differentiation between manufacturers.

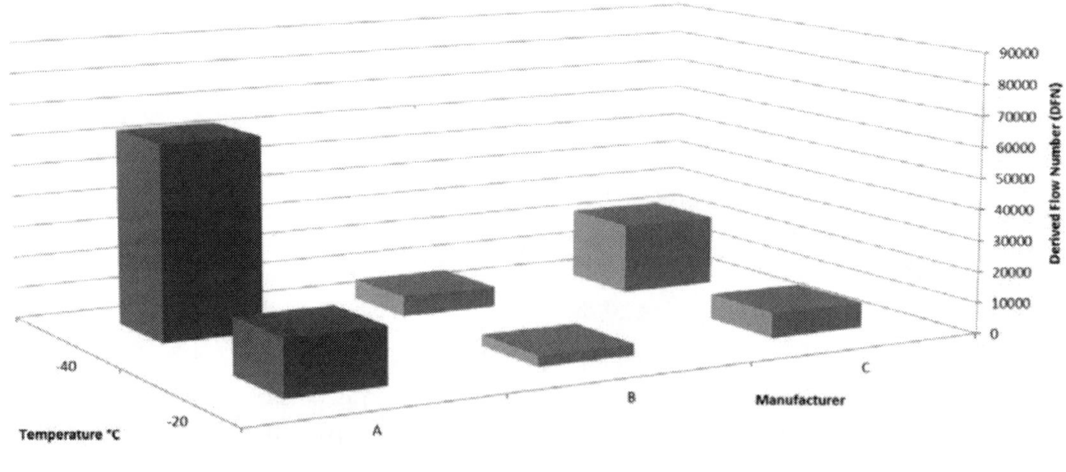

Figure 14: DFN response in 4 µm filters

The new experiment utilized Fuel 3 and three separate filter manufacturers to determine if the earlier projects were repeatable. Viscosity was still a major factor in determining the DFN values. Fuel 3 also saw significant differentiation between the three filter media (Figure 14). C's filters were much lower than manufacturer A's filter, however manufacturer B's filter still had the lowest DFN.

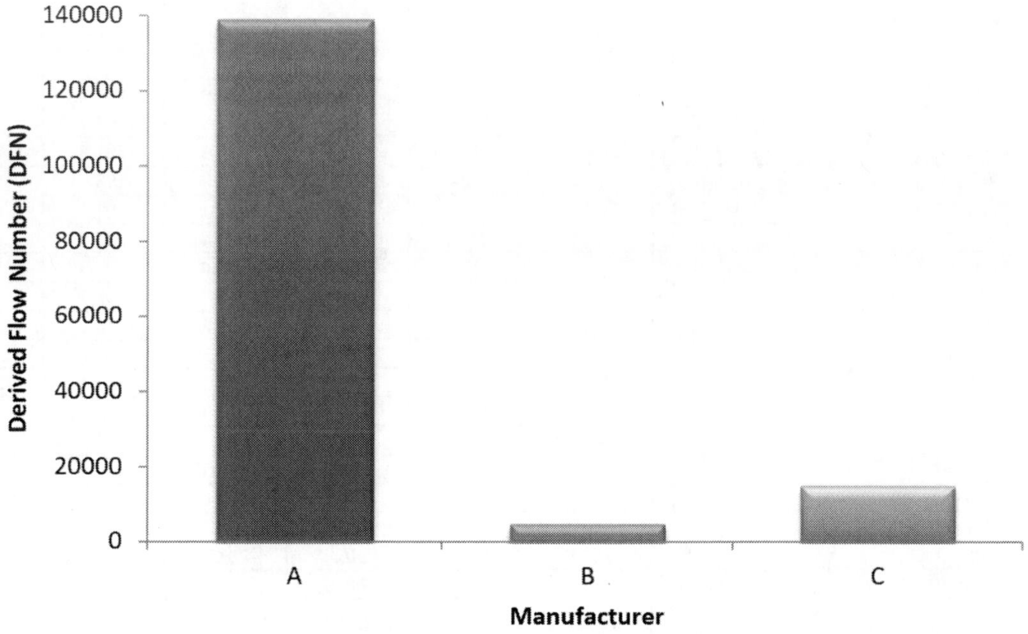

Figure 15: DFN response of additized fuels in filters ≤ 4μm

The three filters were also tested using additized fuel. C's filters were still found to have a DFN between A's filters and B's filters. C's filters were also found to have a slight improvement when CFI componentry was used. CFI componentry was only found to have an antagonistic effect on DFN when A's filters were used (Figure 15).

CONCLUSION

Current ultra-fine filtration has a number of methods to evaluate the removal of "hard particulate" from liquids however field filtration issues are often attributed to "soft particulate". The current industry standard methods used to evaluate bulk filtration appear inadequate to determine all potential particulate problems that may arise in the field. A significant difference could be seen in the flow rate of an untreated ASTM D975 no. 1 S15 fuel through different commercial filter media. This effect was exaggerated in another ASTM D975 no. 1 S15 fuel treated with heritage CFI componentry.

The filtration methods which are currently used to distinguish various bulk fuel filters are not appropriate to determine how the filters will interact with real middle distillate fuels and soft particulates. End users need a new method to better understand their requirements for a "fit for purpose" filtration system.

Appendix A- Physical Property Data for Fuel Samples

		Fuel 1		Fuel 2		Fuel 3	
Recovered Vol %		Boiling Point °C		Boiling Point °C		Boiling Point °C	
		ASTM D86 Correlation	ASTM D2887	ASTM D86 Correlation	ASTM D2887	ASTM D86 Correlation	ASTM D2887
IBP		162.4	106.8	163.3	116.1	157.5	115.5
5% vol		182.8	158.5	178.7	154.5	169.1	143.5
10% vol		190.8	172.6	187.5	168.5	177.3	159.0
15% vol		195.7	181.6	193.4	178.9	183.1	168.1
20% vol		200.6	189.6	198.5	187.5	187.7	175.2
25% vol		204.5	196.9	202.7	195.5	191.8	182.6
30% vol		207.6	202.0	206.4	200.8	195.6	189.9
35% vol		210.7	208.0	209.9	207.1	199.5	197.1
40% vol		213.9	214.0	213.5	213.4	203.4	201.8
45% vol		217.4	218.8	217.1	218.4	207.5	208.1
50% vol		221.3	224.4	220.9	224.0	211.7	214.6
55% vol		225.6	230.3	224.9	229.5	216.1	219.5
60% vol		230.2	236.2	229.0	235.3	220.7	225.8
65% vol		234.8	241.2	232.9	239.9	225.4	232.2
70% vol		239.0	248.6	236.7	246.7	230.2	237.8
75% vol		243.0	255.2	240.2	253.2	235.1	246.1
80% vol		247.3	261.9	244.0	258.9	240.4	254.2
81% vol		248.4	263.5	244.9	260.4	241.6	255.4
82% vol		249.4	265.2	245.8	262.2	242.8	256.2
83% vol		250.6	266.6	246.8	263.7	244.0	258.0
84% vol		251.9	268.2	247.9	265.2	245.5	260.2
85% vol		253.4	270.2	249.1	266.5	247.0	262.5
86% vol		255.0	271.7	250.5	268.1	248.8	264.7
87% vol		256.8	273.0	252.1	270.0	250.8	266.6
88% vol		258.9	274.9	253.9	271.6	253.0	269.1
89% vol		261.1	277.4	255.9	272.6	255.6	271.4
90% vol		263.7	279.7	258.1	274.5	258.4	272.9
95% vol		280.0	294.7	272.8	287.4	277.3	291.4
FBP		299.7	342.9	290.8	332.1	300.5	343.7
Physical Properties							
Biodiesel Content		0		0		0	
Density, kg/L @ 15°C		0.82133		0.82096		0.80796	
API Gravity °		40.61		40.69		43.46	
Viscosity @ -20°C		5.8380		6.6630		5.1252	
Viscosity @ -30°C		8.4690				7.2442	
Viscosity @ -40°C		13.0360				8.4213	
ASTM D7039 Sulfur, ppm		≤ 15		≤ 15		7.8	
ASTM D5773 CP, °C		≤ -45		-47		-48	
ASTM D5949 PP, °C		≤ -55		-54		-57	
ASTM D6304C Water via KF, ppm		≤ 30		≤ 30		15.2	
Sim Dist		ASTM D86	ASTM D2887	ASTM D86	ASTM D2887	ASTM D86	ASTM D2887
IBP, °C		162.4	106.8	163.3	116.1	157.5	115.5
FBP, °C		299.7	342.9	290.8	332.1	300.5	343.7
T90-T20°C		63.1	90.1	59.6	87.0	70.7	97.7
FBP-T90°C		36.0	63.2	32.7	57.6	42.1	70.8
T95-T90°C		16.3	15.0	14.7	12.9	18.9	18.5
FBP-T95°C		19.7	48.2	18.0	44.7	23.2	52.3

Table 8: Physical property data for Fuel 1, 2 and 3

APPENDIX B- LTOR Screening Experiment Regression Statistics

Source	Sum of Squares	Degrees of Freedom	Mean Square	F Value	Probability > F Value
Model	10.50	5	2.10	118.96	6.46E-06
Manufacturer	0.45	1	0.45	25.46	2.34E-03
Temperature	1.92	1	1.92	108.98	4.53E-05
Manufacturer* Temperature	0.18	1	0.18	10.40	1.80E-02
Temperature * Porosity	0.20	1	0.20	11.14	1.57E-02
Residual	0.11	6	0.02		
Lack of Fit	0.02	2	0.01	0.47	6.58E-01
Pure Error	0.09	4	0.02		
Cor Total	10.61	11			

Table 9: LTOR screening experiment ANOVA results

Std. Dev.	0.13	R-Squared	0.99
Mean	9.77	Adj R-Squared	0.98
C.V. %	1.36	Pred R-Squared	0.96
PRESS	0.39	Adeq Precision	29.22
-2 Log Likelihood	-22.70	BIC	-7.79
		AICc	6.10

Table 10: LTOR screening experiment model statistics

Factor	Coefficient Estimate	Degrees of Freedom	Standard Error	Lower 95% Confidence	Upper 95% Confidence	VIF
Intercept	9.71	1	0.040	9.62	9.81	
Manufacturer	0.21	1	0.041	0.11	0.30	1.09
Temperature	-0.41	1	0.040	-0.51	-0.32	1.06
Porosity	-0.75	1	0.040	-0.85	-0.66	1.06
Manufacturer *Temperature	-0.13	1	0.041	-0.23	-0.03	1.09
Temperature * Porosity	0.13	1	0.040	0.04	0.23	1.06

Table 11: LTOR screening project coefficient analysis

APPENDIX C- LTOR Additized Filtration Experiment Regression Statistics

Source	Sum of Squares	Degrees of Freedom	Mean Square	F Value	Probability > F Value
Model	20.18	4	5.04	113.03	8.90E-06
Manufacturer	16.89	1	16.89	378.42	1.20E-06
WASA	0.20	1	0.20	4.50	7.81E-02
EVA	2.78	1	2.78	62.34	2.19E-04
Manufacturer * EVA	3.07	1	3.07	68.85	1.66E-04
Residual	0.27	6	0.04		
Lack of Fit	0.15	3	0.05	1.36	4.04E-01
Pure Error	0.11	3	0.04		
Cor Total	20.45	10			

Table 12: LTOR additized filtration experiment ANOVA results

Std. Dev.	0.21	R-Squared	0.99
Mean	10.62	Adj R-Squared	0.98
C.V. %	1.99	Pred R-Squared	0.95
PRESS	0.97	Adeq Precision	27.94
-2 Log Likelihood	-9.65	BIC	2.34
		AICc	12.35

Table 13: LTOR additized filtration experiment model statistics

Factor	Coefficient Estimate	Degrees of Freedom	Standard Error	Lower 95% Confidence	Upper 95% Confidence	VIF
Intercept	10.74	1	0.068	10.57	10.91	
Manufacturer	-1.29	1	0.067	-1.46	-1.13	1.08
WASA	0.14	1	0.068	-0.02	0.31	1.05
EVA	0.54	1	0.068	0.37	0.70	1.05
Manufacturer * EVA	-0.55	1	0.067	-0.71	-0.39	1.08

Table 14: LTOR additized filtration experiment coefficient analysis

REFERENCES

1. Blizard, N. C. "Case Studies - Operational Benefits of Improved Diesel Fuel Filtration on Injection Equipment Durability" American Filtration and Separations Society Presentation, April 2017.

2. Goertz, M.; Moravec, D.; Dallas, A. "Glycerin Related Premature Plugging of Fuel Filters: Revisiting Glycerin Solubility Limits" 14th International Symposium on Stability, Handling and Use of Liquid Fuels, October 2015.

3. Doyle, J.; Goertz, M.; Grossbauer, S. "Summary of HPCR Diesel Injector Wear Research and the Implications of Adequately Filtering Fuel" 14th International Symposium on Stability, Handling, and Use of Liquid Fuels, October 2015.

4. Goertz, Matthew. Potential Problems of High Efficiency Fuel Filtration and Cold Flow Improvers, 14th International Symposium on Stability, Handling, and Use of Liquid Fuels October 2015.

5. ISO16889, Multi-pass Method for Evaluating Filtration Performance of a Filter Element, June 2008.

6. ISO19438, Filtration Efficiency Using Particle Counting and Contaminant Retention Capacity, June 2006.

7. SAE J1985, Initial Single-Pass Efficiency Test Method, October 2013.

8. ASTM D975-17, Standard Specification for Diesel Fuel Oils, ASTM International, West Conshohocken, PA, 2017.

9. EN590:2013+A1:2017, Automotive Fuels – Diesel – Requirements and Test Methods, British Standards Institution, September 2017.

10. CAN/CGSB 3.517, National Standard of Canada - Diesel Fuel, Canadian General Standards Board, June 2017.

IASH 2017, THE 15th INTERNATIONAL CONFERENCE ON STABILITY, HANDLING AND USE OF LIQUID FUELS
Rome, Italy
10-14 September 2017

A THEORETICAL AND FULL SCALE INVESTIGATION INTO THE USE OF TANK SETTLING TO REMOVE WATER AND SOLID CONTAMINANTS FROM AVIATION FUEL

Steven D Anderson, Alisdair Q Clark, Andrew Glendinning, Gary Norris, Air BP

Air BP, Sunbury Business Park, Chertsey Road, Sunbury-on-Thames, Middlesex, TW16 7LN UK.
alisdair.clark@bp.com

ABSTRACT

Aviation represents a unique and important part of the global transport infrastructure with more than 100,000 flights each day linking countries and continents across the world carrying 3.3 billion passengers per annum [1]. To achieve this goal the Aviation Industry have developed technically advanced engines and aircraft with stringent specifications for fuel quality [2]. In this paper a detailed study of fuel cleanliness is undertaken with focus on tank settling time to remove trace contaminants. Traditionally tank settling has been seen as a primary route to eliminate particulate and water matter with guidelines of 3 hours per meter tank height [3]. A theoretical and full scale investigation challenges the validity of this approach in light of soluble/free water equilibrium, common contaminants and use of modern filtration equipment. As part of the study a typical aviation terminal tank of 60 feet height is examined over 48 hours on receipt of a cargo which has experienced a 20°C temperature change since load-port. Analytical techniques include Karl-Fischer, particle counting/co-solvent, scanning electron microscopy to determine the true value of tank settling/floating suction on fuel cleanliness.

1. INTRODUCTION

Aviation represents a significant and unique aspect of the global transport network. Each day over 100,000 flights carry more than 9 million passengers in speed and comfort across the world [1], safely passing through climatic conditions of freezing temperature and thin air. This has only been made possible by the combined development of aircraft, engine and fuel technology. Expectations for fleet reliability are high with modern turbine engines seeking in excess of 10,000 hours operation before major overhaul and aircraft carrying more than 240,000 litres of product in advanced fuel systems for inter-continental transport[4].

To meet the performance requirements fuel of excellent cleanliness is required as even small amounts of contaminant can quickly accumulate given the large volumes utilised. Industry specifications and guidelines [5] have stringent limits, for example less than 0.20 mg/litre dirt and less than 30 ppm of water at the point of delivery into plane. To meet this challenge, the Industry have sought to develop analytical methods to measure, and procedures/equipment to remove, trace dirt and water from aviation fuel, monitoring quality throughout the distribution chain, for example Table 1.

Table 1 - Examples of Fuel Cleanliness Methods, Filtration Performance Criteria and Guidelines

Method/Guide	Description
IP423[6] / ASTM D5452[7]	Trace dirt - Gravimetric 0.8 micron Millipore.
IP 564[8], 565[9], 577[10]	Trace dirt / free water - Particle counting.
ASTM D3240[11]	Trace free water - Detection of undissolved water.
EI 1590[12]	Filter performance – Aviation microfilters.
EI 1581[13]	Filter performance - Aviation filter-water separators.
EI 1583[14]	Filter performance - Aviation filter monitors
EI/JIG 1530[15]	Fuel Handling
ASTM Manual 5[16].	Fuel Handling

Within this paper, a particular aspect of fuel handling procedure is examined in detail – that of tank settling to promote the removal of trace dirt / free water from aviation fuel.

2. TANK SETTLING

Tank settling, where product remains static in a tank to allow trace dirt / water to settle, has been used in fuel supply for many years. Industry guidelines [3, 15] cite 3 hours for each meter height of tank with 24 hours total. From an operational perspective, this is a significant undertaking as market demand for product necessitates consistent supply. Provision must be made for such a time delay within a distribution chain which is constantly flowing, for example Table 2.

Table 2 - Example Flow Rates for Aviation Jet Fuel In the Distribution System

Location	Flow Rate litres / hr
Refinery Production Unit Run-down	130,000
Ship to Shore Transfer	2 million
Major US Pipeline	4 million
Terminal to Pre Airport Terminal	300,000
Airport hydrant to Inter-continental aircraft	240,000

An understanding of the value of tank settling is therefore important, combined with detailed knowledge of tank design, use of floating suction (Figure 1), and down-stream filtration.

Figure 1 - Floating Suction Schematic

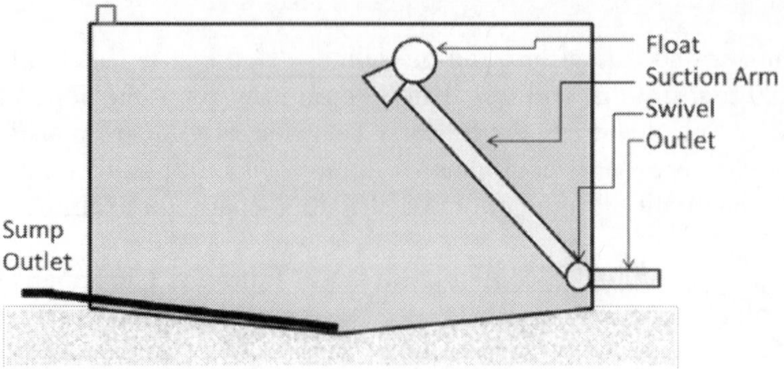

3. TANK SETTLING STUDIES

Two full scale studies have been undertaken at an aviation terminal to determine the influence of tank settling time on Jet fuel cleanliness. The first study examined removal of trace dirt and the second trace free water. The use and value of floating suction was also considered as part of the project. Typical receiving tank dimensions were as follows:

- Height 18.3 metres.
- Volume 16,000 m^3.
- Side sampling at ground level.
- Sump for dirt / water removal and sight glass to ensure detection, Figure 2.

Figure 2
Sump In-Line Sight Glass to Detect Water/Dirt – Fuel Interface

4. SETTLING OF TRACE DIRT

4.1 Samples

Two import cargos were selected to determine the effectiveness of tank settling on trace dirt removal, Table 3:

Table 3 - Import Cargos Analysed

Cargo	Volume m^3
1	16642
2	51466

The following samples were taken during discharge, initial and final tank settling, for analysis of trace dirt distribution across the cargos and tank, Table 4:

Table 4 - Cargo Sampling Regime

Sample Time	Discharge	After 2 hours settling	After 24 hours settling
Tank Layer	Start Middle End	Upper Middle Lower Bottom	Upper Middle Lower Bottom

4.2 Analysis

Analysis of dirt in aviation fuels is analytically challenging due to:

- Only trace quantities generally being present.
- The precision of available laboratory test methods.
- Possible interference from trace free water in techniques such as particle counting.

Efforts to minimize the impact on data quality were sought by:

- Sampling 2 cargos for comparative analysis.
- Selecting a more severe 0.4 Micron Millipore for Gravimetric analysis to:
  - Capture more dirt/increase gravimetric result.
  - Be more representative of performance microfilter elements (0.5 micron, nominal).
- Using co-solvent/Annex B in particle counting methods to eliminate interference from any trace free water.
- Utilizing Scanning Electron Microscopy (SEM) for particle size/composition analysis in support of the Millipore measurement.

The analytical methods selected are summarized in Table 5.

Table 5 - Summary of Analytical Methods

Method	Comment
Gravimetric Millipore Colorimetric Millipore	0.4 Micron, 2 litres product filtered "
Particle Counting	IP 565 with Annex B Co-solvent, counts and ISO code
SEM with X-Ray	Selected Millipores, resolution to <1μm with counting groups of <1, 1-2, 2-4, 4-10 and ≥10 μm

4.3 Results

Results for the two cargos are presented in Tables 6 and 7, together with Figures 3 to 10.

Table 6 - Results for Cargo 1

Analysis	Vessel Discharge Samples			After 2 Hours Settling				After 24 Hours Settling			
	Start	Middle	End	Upper	Middle	Lower	Bottom	Upper	Middle	Lower	Bottom
IP423 Gravimetric mg/l	0.4	0.5	0.3	0.25	0.25	0.2	0.1	0.25	0.2	0.35	0.3
IP423 Colour Rating	B3	B2	B2	B3	B2	B2	B2	B3	B3	B2	B3
IP 565 Particle Counting Counts/ml											
≥4µm(c)	715.2	839.7	774.4	627.2	1072.1	759.9	723.1	753.5	899	748.3	832.7
≥6µm(c)	279.2	324.6	318.5	243.6	390	290.2	286.1	264.7	317.2	306.1	274.4
≥14µm(c)	31.8	58.5	41.2	16.1	25.7	20.3	21	14.1	20.7	25.2	16.8
≥21µm(c)	6.8	18.2	9.7	2.9	3.8	3	2.9	2.4	2.7	3.4	2.9
≥25µm(c)	2.6	9.3	4.8	1.2	2	1.4	1.2	1	1	1.6	1.3
≥30µm(c)	0.8	3.7	2	0.5	1	0.7	0.5	0.4	0.3	0.7	0.5
ISO Code (4/6/14/30 µm)	17/15/12/7	17/16/13/9	17/15/13/8	16/15/11/6	17/16/12/7	17/15/12/7	17/15/12/6	17/15/11/6	17/15/12/5	17/15/12/7	17/15/11/6
SEM % of Particles Counted											
<1.00 µm				37.5		55.3	69.1	48.0		49.3	53.6
1.00-2.00 µm				35.4		28.6	21.5	31.9		29.1	28.5
2.00-4.00 µm				20.0		12.9	7.4	15.2		15.7	13.3
4.00-10.00 µm				6.1		3.0	1.8	4.7		5.3	4.2
≥10.00 µm				1.1		0.1	0.2	0.3		0.5	0.4

Table 7 - Results for Cargo 2

Analysis	Vessel Discharge Samples			After 2 Hours Settling				After 24 Hours Settling			
	Start	Middle	End	Upper	Middle	Lower	Bottom	Upper	Middle	Lower	Bottom
IP423 Gravimetric mg/l	0.15	0.15	0.25	0.15	0.25	0.15	0.25	0.10	0.30	0.30	0.25
IP423 Colour Rating	B1	B1	B2	B2	B2	B2	B2	B2	B2	B2	B2
IP 565 Particle Counting Counts Per ml											
≥4μm(c)	350.9	111.8	204.6	502.6	618.5	559.2	1014.5	612.9	595	741.2	592.5
≥6μm(c)	119.9	36.1	80.5	222.8	259.8	214.6	364	208.6	184.9	252.5	214.1
≥14μm(c)	9.7	3.6	12.7	24.2	35.6	20.1	31.6	15.7	14.8	20.8	16.2
≥21μm(c)	2.2	0.9	2.6	5.3	8.1	3.9	6.2	3.6	2.8	3.6	3.8
≥25μm(c)	1.1	0.5	1.1	2.5	3.7	1.6	2.7	1.7	1.4	1.9	1.9
≥30μm(c)	0.5	0.2	0.4	1.1	1.3	0.7	0.9	0.6	0.5	1	0.6
ISO Code (4/6/14/30 μm)	16/14/10/6	14/12/09/6	15/14/11/6	16/15/12/7	16/15/12/7	16/15/12/7	17/16/12/7	16/15/11/6	16/15/11/6	17/15/12/7	16/15/11/6
SEM % of Particles Counted											
<1.00 μm	-	58	-								
1.00-2.00 μm	-	24	-								
2.00-4.00 μm	-	12	-								
4.00-10.00 μm	-	5	-								
≥10.00 μm	-	1	-								

Figure 3
Cargo 1 Middle Discharge 0.4μm Millipore

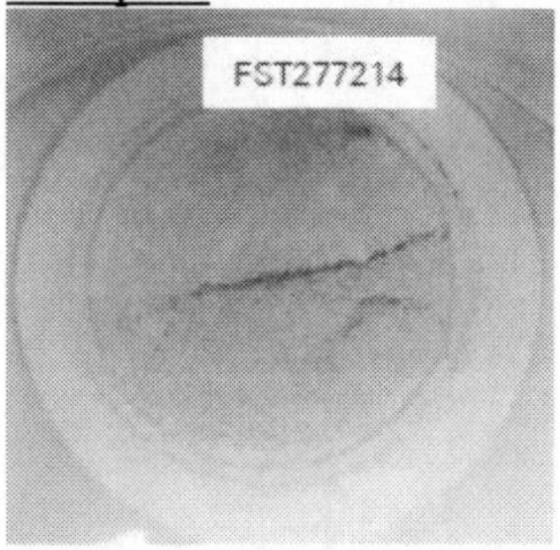

Figure 4
Cargo 2 Middle Discharge 0.4μm Millipore

Figure 5
Cargo 1 0.4μm Millipore Upper 2 hrs

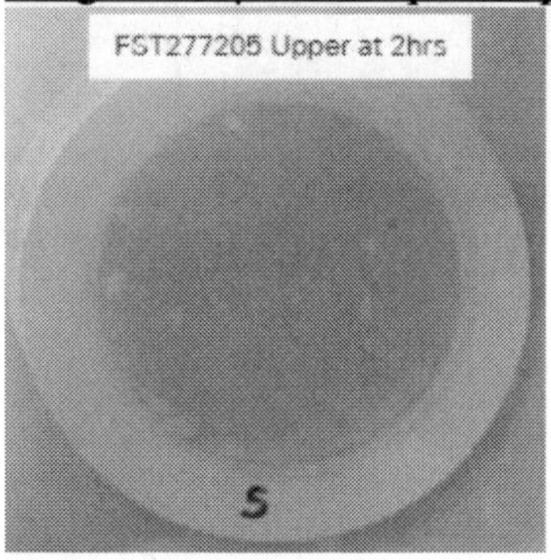

Figure 6
Cargo 1 0.4μm Millipore Upper 24 hrs

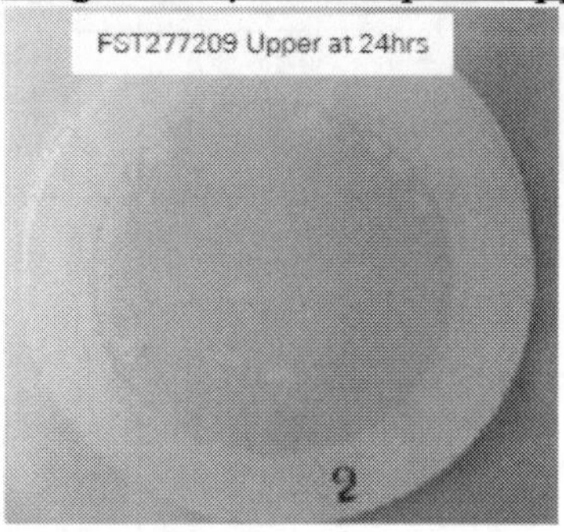

Figure 7
Cargo 1 0.4μm Millipore Lower 2 hrs

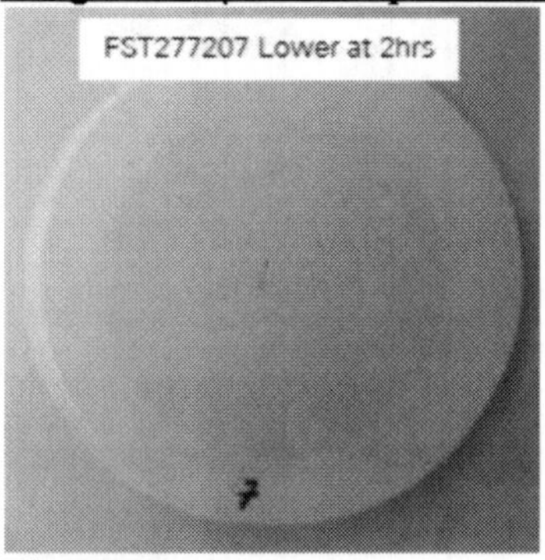

Figure 8
Cargo 1 0.4μm Millipore Lower 24 hrs

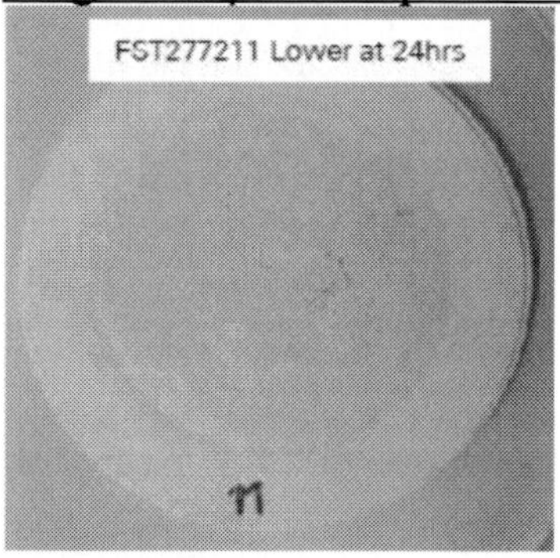

Figure 9
Cargo 1 0.4μm Millipore Bottom 2 hrs

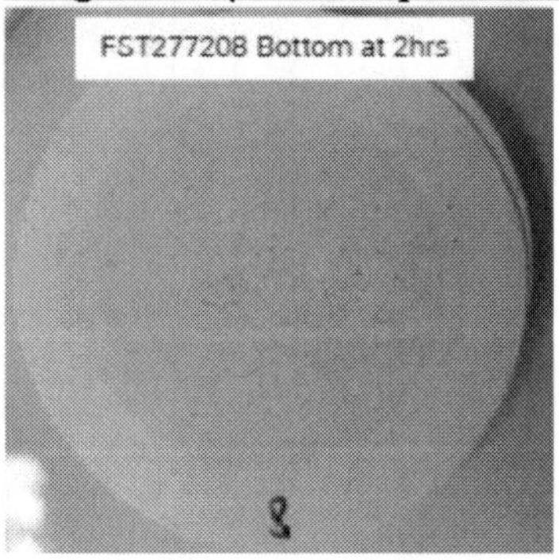

Figure 10
Cargo 1 0.4μm Millipore Bottom 24 hrs

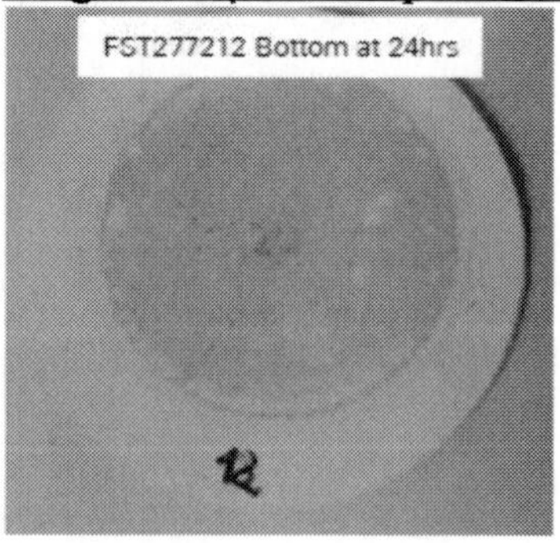

4.4 Discussion

Weight of Dirt Delivered into Tank

The weight of dirt delivered to the storage tanks as part of each cargo can be estimated using the:

- Average Gravimetric result.
- Volume of cargo.
- Assumption that this is largely trace rust, Fe_2O_3, density 5242 kg/m^3.

Results are given in Table 8.

Table 8 - Estimate of Dirt Delivered to Receiving Terminal

Cargo	kg Dirt
1	9.4
2	6.7

These results may appear low but the size distribution of the particles has a significant influence on the effect down-stream which is discussed in the following section.

Tank Settling Time

Gravimetric and IP565 particle size data may be utilised to given an indication of the change in product cleanliness based on tank settling time. Gravimetric data are presented for the two cargos in Tables 9 and 10 respectively, with particle counting data in Tables 11 and 12 respectively.

Table 9 - Cargo 1 Gravimetric Data

Vessel Discharge Time:	Start	Middle	End		Average
Discharge Samples mg/l	0.4	0.5	0.3		0.40
Tank Layer	Upper	Middle	Lower	Bottom	Average
After 2 Hours Settling mg/l	0.25	0.25	0.2	0.1	0.20
After 24 Hours Settling mg/l	0.25	0.2	0.35	0.3	0.28

Table 10 - Cargo 2 Gravimetric Data

Vessel Discharge Time:	Start	Middle	End		Average
Discharge Samples mg/l	0.15	0.15	0.25		0.18
Tank Layer	Upper	Middle	Lower	Bottom	Average
After 2 Hours Settling mg/l	0.15	0.25	0.15	0.25	0.20
After 24 Hours Settling mg/l	0.1	0.3	0.3	0.25	0.24

Table 11 - Cargo 1 Particulate Data – All Particles ≥4μm

Vessel Discharge	Start	Middle	End		Average
Discharge Samples counts/ml	715.2	839.7	774.4		776.4
Receiving Tank	Upper	Middle	Lower	Bottom	
After 2 Hours Settling counts/ml	627.2	1072.1	759.9	723.1	795.6
After 24 Hours Settling counts/ml	753.5	899	748.3	832.7	808.4

Table 12 - Cargo 2 Particulate Data – All Particles ≥4μm

Vessel Discharge	Start	Middle	End		Average
Discharge Samples counts/ml	350.9	111.8	204.6		222.4
Receiving Tank	Upper	Middle	Lower	Bottom	
After 2 Hours Settling counts/ml	502.6	618.5	559.2	1014.5	673.7
After 24 Hours Settling counts/ml	612.9	595	741.2	592.5	635.4

- Both cargos were well within Defence Standard 91-091 cleanliness limits of 1.0 mg/l dirt maximum.
- All particle counting results met MIL-DTL-83133J incidental contaminants Table III limits of ISO 19/17/14/13 (≥4μm, ≥6μm, ≥14μm, ≥30μm)
- Cargo 1 contained a higher level of dirt. On delivery further particles were entrained into the product but after 24 hours settling, the cargo was similar to point of discharge. Some settling was also apparent from the change in tank Bottom sample Millipore colour, Figures 9 and 10.
- Cargo 2 contained a low level of dirt. On delivery the cargo mixed with particles which were already present in the tank giving a higher gravimetric/particle count even after 24 hours settling.

These results indicate:

- When particulates have contaminated a tank, product cleanliness will remain an issue until overhaul.
- After 2 hours settling the upper layer cleanliness is similar to after 24 hours, offering some support to use of floating suction for early discharge for cargos containing significant trace dirt.
- After 24 hours particulates appear dispersed throughout the tank, an indication that further benefits from floating suction/settling are limited.

As a general observation, the ability to fill a tank with minimum disturbance to settled particulate may be more useful than floating suction based on the data, emphasising the importance of good engineering design. In addition, poor design of floating suction systems can result in settled dirt entering the distribution system as the tank approaches empty. This is illustrated in Figure 11 where the floating suction may have contributed to removal of dirt from the settled layer leaving a clean area of tank floor.

Figure 11 - Clear Area of Settled Dirt around Floating Suction Head at Lower Level

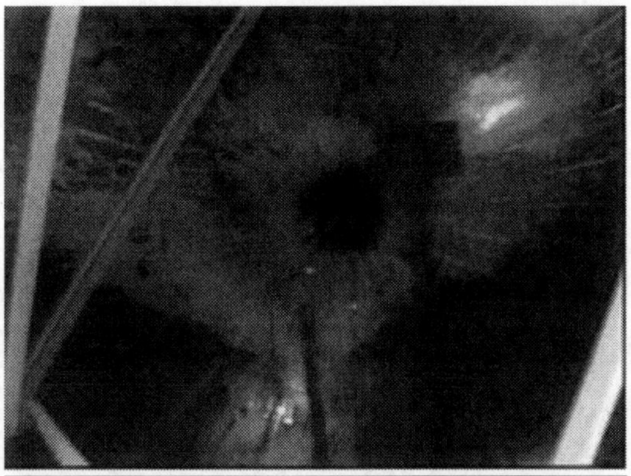

4.5 Theoretical Studies

Particle Settling Rate
Industry work [3] offers an estimate of settling rate according to fuel type, Table 13:

Table 13 - Equations to Estimate Rust Settling Rate

Fuel	Equation
Avgas	$y = 3.280\times10^{-3}\,x^3 - 8.651\times10^{-3}\,x^2 + 3.508\,x - 3.352$
JP-4, Jet B	$y = 4.346\times10^{-2}\,x^2 + 1.639\,x - 1.847$
Jet A, Jet A-1, JP-5, JP-8	$y = -1.492\times10^{-4}\,x^3 + 5.028\times10^{-2}\,x^2 + 9.785\times10^{-1}\,x - 1.979$

Where
x = Particle Size in μm
y = Settling Rate at 15.5°C, m/s x 10^{-5}

Results for various particle sizes are presented for Jet A/A-1 in Table 14. General guidance for settling is 3 hours per meter tank height, 24 hours total, the latter being representative of a 12 meter high tank containing particles ≥10μm. Particles of ≤2μ are noted to not separate/remain entrained in the fuel due to Brownian Motion.

Table 14 - Rust Particle Size versus Settling Rate

Rust Particle Size μm	Settling Rate meters per hour
20	1.3
17	1.0
15	0.8
10	0.5
5	0.1
2	0

Particle Size Distribution
Normally no data are available for particles of <4μm due to the limitations of standard light obscuration technology. However, by utilizing electron wavelengths/SEM data an estimate

has been possible for Cargo 1. This is based on a 'pro-rata' calculation using the counts for all particles ≥4μm which are measured by both techniques and re-allocating to counts/ml. Results are presented in Table 15.

Table 15 - Estimate of Particle Size Distribution for Cargo 1 Based on Combined SEM/Light Obscuration Data

Time	After 2 Hours Settling			After 24 Hours Settling		
Counts per ml	Upper	Lower	Bottom	Upper	Lower	Bottom
<1.00 μm	3261.8	13465.9	24120.6	7271.3	6292.5	9847.0
1.00-2.00 μm	3077.8	6962.0	7489.3	4831.8	3707.5	5228.9
2.00-4.00 μm	1737.0	3151.4	2569.6	2302.9	2006.8	2450.8
≥4μm(c)	627.2	759.9	723.1	753.5	748.3	832.7
≥6μm(c)	243.6	290.2	286.1	264.7	306.1	274.4
≥14μm(c)	16.1	20.3	21.0	14.1	25.2	16.8
≥21μm(c)	2.9	3.0	2.9	2.4	3.4	2.9
≥25μm(c)	1.2	1.4	1.2	1.0	1.6	1.3
≥30μm(c)	0.5	0.7	0.5	0.4	0.7	0.5

Based on the average result after 24 hours, 79% of particles are <2μm /entrained in the cargo with little/no chance of separation by settling. As such, these particles will enter the distribution system and impact filtration units downstream. Current light obscuration methods were unable to detect approximately 93% of particles which may be captured by EI 1590 1μm nominal aviation micro-filters.

Particle Mass and Surface Area

Both cargos contain a large number of particles beyond the limit of normal detection. However, an indication of mass/surface area is useful to assess impact down-stream:

- Large numbers of small particles can have little combined mass but high projected surface area and distribution system filter blocking tendency.
- Acceptable into-plane fuel quality is based on mass, mg/litre.

To gain an insight, the average mass/surface area for Cargo 1 was calculated by combining the SEM/IP565 data after 24 hours settling with the particle size allocations in Table 16.

Table 16 - Assumptions for Particle Size for Mass/Area Calculations

SEM/IP565 Size Range	Allocate Particle Diameter μm	After 24 Hours Settling, Upper, Lower, Bottom Tank Sample Average. particles/ml
<1.00 μm	0.5	7803.6
1.00-2.00 μm	1.5	4589.4
2.00-4.00 μm	3	2253.5
≥4μm(c)	4	778.2
≥6μm(c)	6	281.7
≥14μm(c)	14	18.7
≥21μm(c)	21	2.9
≥25μm(c)	25	1.3
≥30μm(c)	30	0.5

Using the counts, size range and density of 3352 kg/m^3, calculated from the x-ray determined composition of the contaminants, the projected area of the dirt within Cargo 1 and mass of particles can be deduced, Tables 17 and 18, Figure 12.

Table 17 - Calculated Projected Area and Mass of Particles in Cargo 1

Analysis	Result		
Calculated Projected Area of Particulates in 16642 m^3	802 m^2		
Calculated Mass of Particles in 16642 m^3	9.5 kg		
Calculated Gravimetric Result	0.57 mg/l		
Measured Gravimetric Results Upper, Middle, Lower, Bottom Samples for Tank after 24 hours settling	U: 0.25 mg/l M: 0.20 mg/l L: 0.35 mg/l B: 0.30 mg/l	Repeatability 0.22 mg/l	Reproducibility 0.40 mg/l

Table 18 - Weight Distribution of Dirt by Particle Size

Assumed Particle Diameter µm	Impact on Cargo Cleanliness mg/l
0.5	0.00185937
1.5	0.02952511
3	0.11598
4	0.09493238
6	0.11599887
14	0.09781095
21	0.05119383
25	0.03871909
30	0.02744886

Figure 12 - Graph of Weight Distribution of Dirt by Particle Size

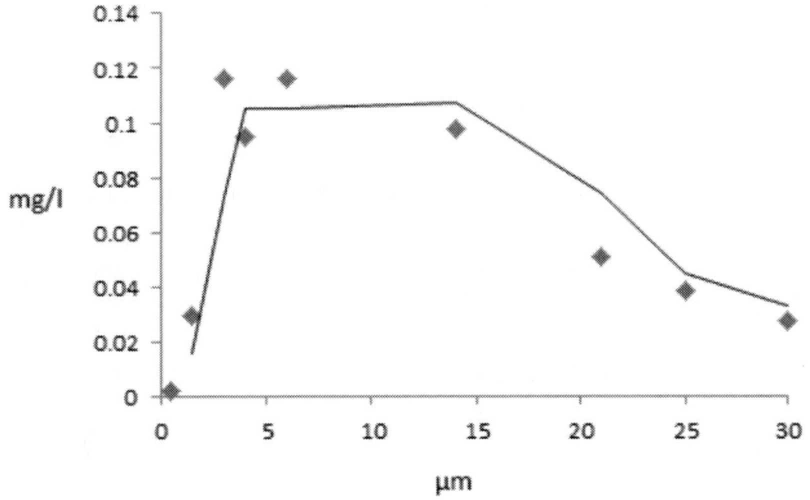

Given the broad assumptions the calculated Gravimetric mass is of the correct order of magnitude when compared to experimental analysis and within test Reproducibility. The projected surface area of the particles, at 802 m^2, is significant, particularly when compared against the 111 m^2 projected area of a microfilter vessel containing 25 elements. This illustrates the advantage of particle size analysis in the lower size range for determining fuel cleanliness/impact on downstream filtration.

Considering the particle size/weight distribution data, Figure 12, Aviation Industry targets of 0.2 mg/l dirt into-plane suggest focus should be placed on removing particles in the size range 2 – 20 μm. This is on the basis of:

- Large numbers of very small particles and small numbers of large particles contribute little to overall Gravimetric weight.
- Mid-range particles are sufficient in number/mass to have a major influence.
- Such particles are not removed by tank settling and rely on filtration media for capture.

EI 1590 specifications and qualification procedures for aviation fuel microfilters are therefore well placed with nominal ratings of 1 – 10 μm, larger dirt having more potential to settle on standing.

5. SETTLING OF TRACE FREE WATER

5.1 Samples

An 82589 m^3 Jet cargo was identified for the trace free water settling study. The cargo had the advantage of:

- Being loaded in hot/humid environment of +30 °C maximising water solubility due to temperature.
- Containing 15% aromatics, typical of global production.
- Arriving into the receiving terminal under ambient conditions of +8 °C, 94% humidity, 1024 hPa pressure [17], promoting free water separation within the cargo and water adsorption from humidity / condensation within the receiving tank.
- Having typical Jet density, 795.5 kg/m^3.
- Containing 0.9 mg/litre static dissipater additive which can assist in water entrainment.

A series of samples were taken from the receiving tank upper and lower levels to determine the rate of water settling, Table 19. To ensure these were representative, lines were flushed with over 4x internal volume prior to sampling.

Table 19 - Trace Free Water Samples

Sample	Upper Tank Level 5 litre Drum and Bottle for Particle Counting	Lower Tank Level 5 litre Drum and Bottle for Particle Counting
On Receipt	✓	✓
After 3 hrs settling	✓	✓
After 6 hrs settling	✓	✓
After 12 hrs settling	✓	✓
After 24 hrs settling	✓	✓
After 48 hrs settling	✓	✓

5.2 Analysis

Two analysis methods were utilised to determine the change in water content of the tank over time, Karl Fischer and Particle Counting.

Karl Fischer ASTM D6304 [18]

The Karl Fischer method was used to determine the total (dissolved + free) water content of the samples. To ensure every trace of water was detected, including droplets which may have attached to the sample container wall, a modified version of SAE ARP 1401[19] was used featuring propan-2-ol:

- 295 mls of propan-2-ol of pre-determined trace water content was added to the jet fuel sample container.
- The container was tumbled to mix the propan-2-ol into the jet, acting as a co-solvent and dissolving all traces of free water present.
- The total water content was determined by Karl Fischer.
- The weight of jet fuel analysed was determined by weighting the drum + sample then empty drum.

This method provided an accurate analysis of the total amount of water present in the tank upper and lower layers during the 48 hour settling period.

Particle Counting IP 565

On-site particle counting was used to determine the settling rate of total contaminant (dirt + water) of size >4 µm over time. In addition, earlier work [20, 21] had demonstrated that water content might be identified by the reduction in count through use of a co-solvent or propriety chemical as detailed in Annex B of IP 565. This approach was also evaluated using an external laboratory, accepting some variance may occur due to sample shipment/handling.

5.3 Results

Karl Fischer

Results for the Karl Fischer analysis are provided in Table 20.

Table 20 - Karl Fischer Analysis of Total (Soluble + Free) Water in Receipt Tank Over Time

Sample	Upper Layer		Lower Layer	
	Jet Water Content mg/kg	Visual Confirmation Sample Free of Water / Haze	Jet Water Content mg/kg	Visual Confirmation Sample Free of Water / Haze
When tank filled	91	✓	87	✓
After 3 hrs settling	89	✓	79	✓
After 6 hrs settling	83	✓	86	✓
After 12 hrs settling	89	✓	87	✓
After 24 hrs settling	87	✓	82	✓
After 48 hrs settling	86	✓	77	✓

The repeatability of ASTM D6304 is provided in Table 21.

Table 21 - Repeatability of ASTM D6304

Average Result, mg/kg	Repeatability mg/kg
80	0.5
90	0.6
100	0.6

Particle Counting
Results for the particle counting analysis are provided in Tables 22 to 24.

Table 22 - Ship Tank Sample for Comparison with Terminal Tank Data

Analysis and Time	>4 µm / ml	>6 µm / ml	>14 µm / ml	>21 µm / ml	>25 µm / ml	>30 µm / ml
Ship Tank 3P	1874.2	627.0	67.2	20.3	10.6	5.0

Table 23 - Upper Tank Layer Particle Counts versus Settling Time

Analysis and Time	>4 µm / ml	>6 µm / ml	>14 µm / ml	>21 µm / ml	>25 µm / ml	>30 µm / ml
0 hrs settling	13802.3	4417.1	88.7	23.1	10.9	3.8
0 hrs Annex B	595.0	205.6	23.9	7.5	4.1	2.1
3 hrs Settling	1652.6	735.4	35.0	4.2	1.9	0.6
3 hrs Annex B	278.3	111.4	16.6	4.9	3.0	1.8
6 hrs settling	234.0	89.8	11.7	3.2	1.7	0.8
6 hrs Annex B	319.6	125.1	14.9	3.4	1.7	0.6
12 hrs settling	246.6	91.8	10.3	2.5	1.2	0.4
12 hrs Annex B	139.2	60.2	9.6	2.7	1.3	0.6
24 hrs settling	340.4	114.2	10.6	3.0	1.7	0.6
24 hrs Annex B	212.1	80.1	12.9	4.6	3.0	1.6
48 hrs settling	307.5	100.5	10.5	3.1	1.3	0.6
48 hrs Annex B	109.4	48.7	7.5	2.5	1.5	0.8

Table 24 - Lower Tank Layer Particle Counts versus Settling Time

Analysis and Time	>4 µm / ml	>6 µm / ml	>14 µm / ml	>21 µm / ml	>25 µm / ml	>30 µm / ml
0 hrs settling	1675.1	857.5	86.9	22.8	10.8	3.9
0 hrs Annex B	627.5	209.1	18.7	4.5	1.7	0.6
3 hrs Settling	468.9	176.8	22.7	4.8	2.4	1.0
3 hrs Annex B	369.5	133.4	16.8	5.5	3.5	2.0
6 hrs settling	446.2	288.9	96.3	45.1	26.5	12.2
6 hrs Annex B	202.6	82.0	13.3	4.6	2.8	1.5
12 hrs settling	194.1	73.5	9.8	3.3	1.5	0.6
12 hrs Annex B	234.4	93.3	13.6	4.4	2.5	1.1
24 hrs settling	261.1	89.6	9.5	2.7	1.5	0.7
24 hrs Annex B	194.1	85.7	16.3	5.7	3.0	1.7
48 hrs settling	342.5	118.6	10.7	2.6	1.2	0.3
48 hrs Annex B	103.5	46.8	7.9	2.5	1.6	0.7

The precision of IP 565 is provided in Table 25.

Table 25 - Precision of IP 565 from Test Method, Section 13.

Parameter	Range of Applicability Counts / ml	Repeatability Counts / ml	Reproducibility Counts / ml
>4 µm	216 – 45061	0.07861 * (x + 2000)	0.09959 * (x + 2000)
>6 µm	74 – 22483	0.1545 * (x + 500)	0.1993 * (x + 500)
>14 µm	9 – 4105	0.3485 * (x + 50)	0.4567 * (x + 50)
>21 µm	2 - 1444	0.5202 * (x + 5)	0.7272 * (x + 5)
>25 µm	1 – 776	0.6023 * (x + 2)	0.9096 * (x + 2)
>30 µm	1 - 365	0.7937 * (x + 0.5)	1.2781 * (x + 0.5)

5.4 Discussion

Karl Fischer

Based on a tank volume 16000 m^3 Karl Fischer results indicate an average of 1133 kg of water (free + dissolved) was entrained with the Jet fuel hydrocarbon layer, rather than rapidly separated to tank sump, in the receiving tank. Assuming a similar distribution across the 82589 m^3 cargo, a total of 5847 kg water was delivered from the ship into storage.

Interestingly, the highest concentration of water, 91 mg/kg, was in the upper tank layer. This may be due to one or a combination of factors:

- Initial tank fill coming from the lowest point in the ships tanks with the greatest risk of free water.
- Initial tank fill mixing with tank heel / residual free water trapped on the base of the tank and rising on filling.
- Upper fuel layer entraining condensation water from tank walls during filling.
- Fuel temperature difference.

A graph for the water content of the tank upper / lower layers is provided in Figure 13.

Figure 13 - Trend in Tank Water Content Over Time

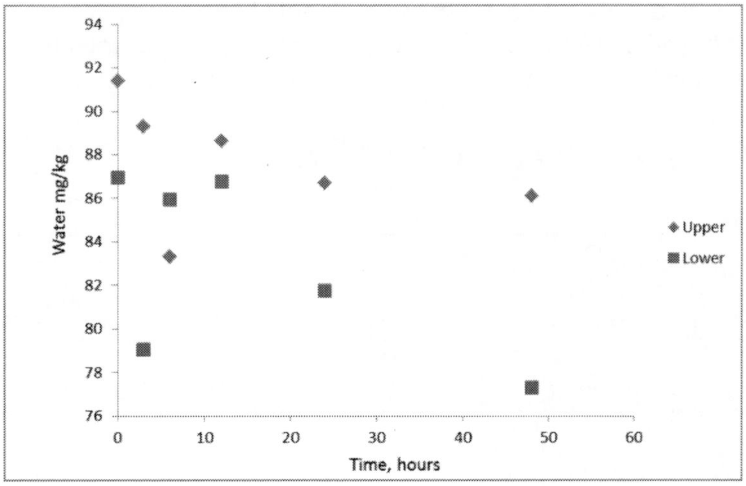

While Karl Fischer gives a measurement of the total water content present in the fuel, an estimate of the dissolved / free water content is possible from CRC work [3]. An equation for the volume of dissolved water present in a typical Jet A-1 is given by Equation 1.

Equation 1 - CRC Water Solubility in Jet A-1

$$Y = 0.00269 * \exp(0.02962 * X)$$

Where
X = Temperature, °C
Y = Water Solubility, volume %

Using temperature data at time of tank receipt and after 3 hours settling, the split between soluble and free water can be calculated, Table 26.

Table 26 - On Receipt Soluble / Free Water Concentrations Derived Using CRC Data

Tank Layer	Temp °C	Soluble Water % v/v	Soluble water mg/kg	Free water mg/kg	Total Water mg/kg
Receipt Upper	11.9	0.0038	48	43	91
Receipt Lower	10.5	0.0037	47	40	87
3 Hr Settling Upper	13.3	0.0040	50	38	89
3 Hr Settling Lower	10.9	0.0037	47	32	79

Data indicate:

- Soluble water is at a greater concentration than free water in the product.
- The higher water content of the upper layer is influenced by temperature promoting solubility.
- Free water can transition to soluble water as noted for the upper layer.

Assuming the upper and lower layers had reached an average temperature of 12.1 °C the soluble / free water content for the tank after 48 hours settling were calculated, Table 27.

Table 27 - After 48 Hours Settling Soluble/Free Water Concentrations Derived using CRC Data

Tank Layer	Temp °C	Soluble Water % v/v	Soluble water mg/kg	Free water mg/kg	Total Water mg/kg
Upper	12.1	0.0038	48	38	86
Lower	12.1	0.0038	48	29	77

Based on the average result for the upper and lower layers and 16000 m^3 volume the kg of soluble / free water on receipt and after 48 settling was calculated, Table 28. This indicated 96 kg of the free water had settled to the tank sump after 48 hours.

Table 28 - Kg Water Versus Settling Time

Time of Tank Release	Soluble Water kg	Free Water kg
Immediately After Receipt	605	528
After 48 hours settling	611	426

Particle Counting

IP 565 particle counting provides a measure of both dirt and free water droplets in Jet fuel. Using the >4 µm channel a general view of cargo cleanliness / settling can be obtained.

The ship tank sample >4 µm counts/ml, 1874.2, are similar to the receiving tank lower layer counts on cargo receipt 1675.1 but significantly less than the upper layer counts of 13802.3. This might be due to:

- Sampling protocol
- Variation in ship tank cleanliness
- The upper layer of the tank being composed of the lower (more contaminated) layer from the ship tanks.
- Incoming product mixing with settled trace dirt/water and condensation in the receiving tank.

An indication of the rate of settling was obtained, using the >4 µm channel over time, Figure 14.

Figure 14 - Particles and Water Droplets >4 µm

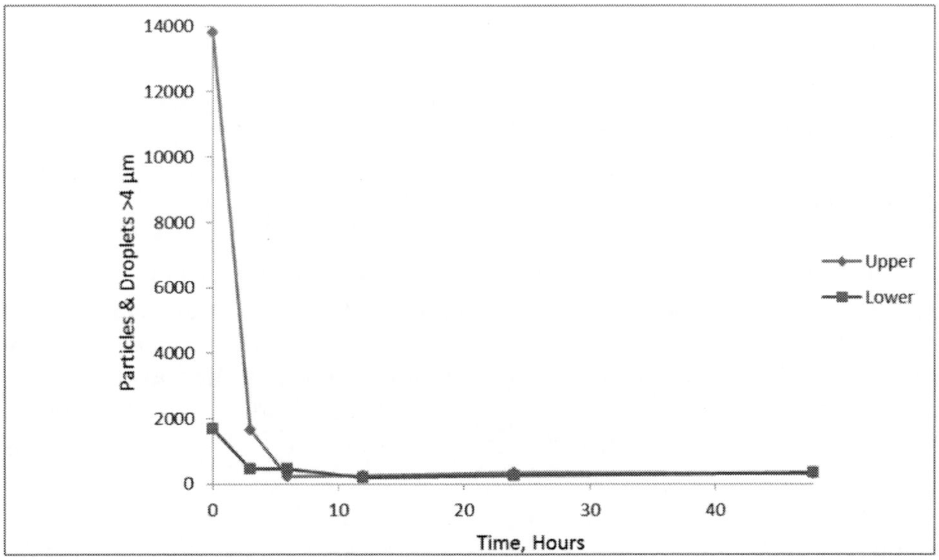

Results indicate that after 6 hours the majority of dirt particles / free water droplets in the >4 µm size range capable of settling had partitioned from the fuel.

An estimate of the contribution due to water can be made by the difference between:

- IP 565 Total Particle Count
- IP 565 Annex B Particle Count – Water Droplets Eliminated by Co-Solvent

With the total particle count determined on site and the Annex B results in an external laboratory, some accounting for test precision was necessary. This was achieved by:

- Calculating the average particles + water and particles only (Annex B) result.
- Applying IP 565 Section 13 precision criteria.
- Determining if the difference between results was >2R.

Results determined significant water content in the >4 and >6 µm channels for the upper and lower layers on receipt and after 3 hours for the upper layer, Table 29. This may be due to the high pump shear rate for rapid ship to shore transfer. All other data showed no significant difference.

Table 29 - Significant Particle Counting Differences Due to Trace Water

Sample / Counts	Upper Layer		Lower Layer	
	>4 µm / ml	>6 µm / ml	>4 µm / ml	>6 µm / ml
0 hrs settling	13802.3	4417.1	1675.0	857.5
Annex B Water Droplets	595.0	205.6	627.5	209.1
3 hrs settling	1652.6	735.4		
Annex B Water Droplets	278.3	111.4		
Remaining Settling Time	All counts similar in range 109 to 341	All counts similar in range 48 to 126	All counts similar in range 103 to 469	All counts similar in range 46 to 289

Results indicate if a significant amount of water is present, Annex B offers a viable method to eliminate the impact on particle count / return data to trace dirt content only. This is in alignment with current MoD / Energy Institute SC-B-11 work to evaluate Annex B as possibly method for refinery Test Certificate analysis.

Data indicate that on filling the upper tank layer contained the highest concentration of free water and confirm that after 6 hours settling this has largely reached equilibrium levels in agreement with the Karl Fischer analysis. In addition, the raw data, suggests the possibility of larger dirt particles / coalesced water droplets passing down through the lower layer in a similar time-frame, Figure 15. Data also confirm earlier published work[20] regarding a broad range of water droplet size in line with natural distributions.

Figure 15 - Lower Layer >14, >21, >25, >30 μm Channel Counts Per ml with Settling Time

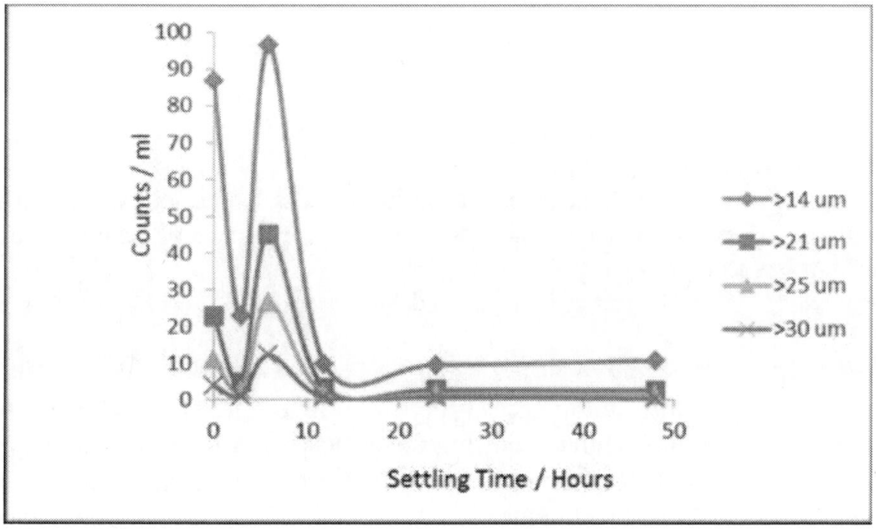

5.5 Theoretical Studies

CRC data [3] offers a route to calculate water settling rate based on droplet size, Equation 1.

Equation 1- Water Settling Rate for Jet A-1

$$y = -1.015\times10^{-4}\, x^3 + 1.229\times10^{-2}\, x^2 - 8.977\times10^{-2}\, x + 0.8069$$

Where
x = Water droplet size in μm
y = Settling Rate at 15.5°C, m/s x 10^{-5}

This indicates water droplets of <4 μm no longer settle / are entrained in product, Table 30.

Table 30 - CRC Calculated Settling Rate for a Range of Water Droplet Sizes

Water Droplet Size μm	Settling Rate meters per hour
60	1.0
40	0.6
20	0.2
10	0.1
4	0

A recently CRC study [22] into fuel-water behaviour offers a more detailed analysis of free water settling, combining both experimental and theoretical work. This highlights the complexity of the area with effects due to the log-normal water droplet size distribution, exponential decay in water concentration with time and the application of Stoke's law to investigate the influence of droplet size diameter. Typical terminal velocities for a 10 μm droplet were estimated in the range approximately 5 x 10^{-6} to 1.5 x 10^{-5} m/s. As such, 10 μm droplets would not be expected to settle from the upper tank layer after 48 hours. However, the potential influence of diffusion is also noted together with data for the temperature-solubility relationship for water in Jet fuel. Combining the CRC and present study suggests that the significant reduction in tank water droplet content for the first 3 to 6 hours settling is not driven by droplet size alone but:

- Diffusion / the probability of collision and coalescence with other dirt particles / water droplets to form larger, heavier drops with an exponential increase in terminal settling velocity.
- Equilibrium between water droplet solubility/condensation from the bulk jet fuel.

This would be in agreement with published work on cloud/rain formation [23] and the peak in size distribution for the lower tank level after 6 hours. Such a conclusion fits well with filter water coalescer operation where small droplets are proactively brought into contact to achieve the same effect.

6. CONCLUSIONS

- Two experimental studies into tank settling to remove trace dirt and water from aviation fuel have been completed with supporting theoretical work based on Industry data.
- The first study assessed cargo cleanliness on discharge, after 2 hours and 24 hours settling and featured the analytical techniques:

Method	Comment
Gravimetric Millipore	0.4 Micron, 2 litres product filtered
Colorimetric Millipore	0.4 Micron, 2 litres product filtered
Particle Counting	IP 565 with Annex B Co-solvent, counts and ISO code.
Scanning Electron Microscope (SEM)	Selected Millipores, resolution to <1µm with counting groups of <1, 1-2, 2-4, 4-10 and ≥10 µm

- Results indicated:
    - Incoming product fully met Defence Standard 91-091 cleanliness criteria and MIL-DTL-83133J incidental contaminants Table III limits of ISO 19/17/14/13.
- Based on the first cargo, when in storage:
    - 2 hours after discharge use of floating suction offered some benefits to access upper layer cleanliness.
    - After 24 hours settling 79% of particles were <2 µm with no chance of settling and no benefit from floating suction. 93% were also <4 µm and undetectable by IP565 light obscuration technology.
- A combination of composition, particle counting and SEM data allowed the cleanliness of product to be calculated, the result of 0.57 mg/l being of a similar order of magnitude to the average Gravimetric result, 0.28 mg/l and within test Reproducibility.
- The projected area of the unsettled particles was estimated at 802 m^2, a significant figure versus projected downstream microfilter pleat area.
- Calculations included size-weight-number distribution which indicated particles in the range 2 – 20 µm would have most influence of Gravimetric result, supporting EI selection of microfilter qualification criteria in the 1 – 10 micron nominal size range.
- The second study featured water settling where a cargo loaded in a warm/humid environment of +30 °C was received into terminal tankage at +8 °C and 94% humidity.
- Water content was assessed in the tank layers on receipt and after 3, 6, 12, 24, 48 hours settling time using Karl-Fischer and IP565 analysis.

- Karl-Fischer / CRC data gave the following average results for soluble/free water in the tank on receipt and after 48 hours settling:

Time	Total Water	Soluble Water[3] kg	Free Water kg
Immediately After Receipt	1133	605	528
After 48 hours settling	1037	611	426

- A greater average water concentration in the upper layer, 87.5 mg/kg compared to the lower layer 83 mg/kg, was confirmed by particle counting and possibly due to:
    - Initial tank fill coming from the lowest point in the ships tanks with the greatest risk of free water.
    - Initial tank fill mixing with the receiving tank heel / residual free water and rising on tank filling.
    - Upper fuel layer entraining condensation water from tank walls during filling.
- A higher average temperature of the upper tank layer compared to the lower layer was noted 13.3 versus 10.9 °C respectively after 3 hours settling. This promoted water solubility in the upper layer.
- IP 565 particle counting data indicated a significant amount of combined trace water / dirt separated after 6 hours with theoretical work suggesting diffusion and coalescence of water drops/dirt promoting settling and accelerating terminal velocity.

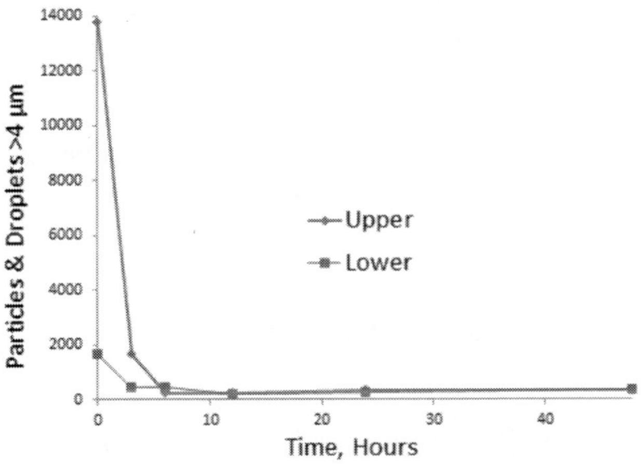

- Use of Annex B to eliminate interference from water in the particle counting method was found successful and in alignment with MoD / Energy Institute work.
- Water was present across a range of size ranges with most droplets of small diameter. This may be due to the high pump shear rate of ship to shore transfers.
- Accepting the precision of the method / scatter in the data, results indicated for tank settling:
    - The mechanism of droplet collision / coalescence appeared significant in eliminating the small droplets.
    - 90% of >4 µm water droplets left the upper layer after 3 hours.
    - Particle count in the lower layer increased after 6 hours then fell to base-line level, presumably as the trace coalesced water/dirt moved down through this layer to tank base.

7. ACKNOWLEDGEMENT

The expertise, help and enthusiasm of the BP Terminal, Air BP Product Quality, BP Bochum and Intertek Sunbury Technology Centre teams is acknowledge in the successful completion of this project. Thank You.

8. REFERENCES

[1] Aviation Benefits Beyond Boarders, Air Transport Action Group, 33 Route de L'Aeroport, 1215 Geneva 15, Switzerland. July 2016.

[2] EI 1550 Handbook on Equipment used for the Maintenance and Delivery of Clean Aviation Fuel. 2nd Edition, 2014. The Energy Institute, 61 New Cavendish St, Marylebone, London W1G 7AR, UK.

[3] CRC Aviation Fuel Properties Handbook. 4th Edition, 2014. The Coordinating Research Council, 5755 North Point Parkway, Suite 265, Alpharetta, GA 30022 USA.

[4] M Frost, Aircraft Fuel & Inerting Systems - Impacts of Fuel Properties. Aviation Technical Committee Meetings of the Coordinating Research Council Inc. Nashville, Tenesse, USA, 4 – 7th May 2015.

[5] Guidelines for Aviation Fuel Quality Control and Operating Procedures for Joint Into-Plane Fuelling Services, Joint Inspection Group Issue 10 July 2008.

[6] IP 423 Determination of Particulate Contaminant in Aviation Turbine Fuels By Laboratory Filtration. Energy Institute 61 New Cavendish St, London W1G 7AR.

[7] ASTM D5452 Standard Test Method for Particulate Contamination in Aviation Fuels by Laboratory Filtration. ASTM International, 100 Barr Harbor Drive, PO Box C700, West Conshohocken, PA, 19428-2959 USA.

[8] IP 564 Determination of the Level Of Cleanliness Of Aviation Turbine Fuel – Laboratory Automatic Particle Counter Method. Energy Institute 61 New Cavendish St, London W1G 7AR.

[9] IP 565 Determination of the Level of Cleanliness of Aviation Turbine Fuels - Portable Automatic Particle Counter Method. Energy Institute 61 New Cavendish St, London W1G 7AR.

[10] IP 577 Determination of the Level of Cleanliness of Aviation Turbine Fuel – Automatic Particle Counter Method Using Light Extinction. Energy Institute 61 New Cavendish St, London W1G 7AR.

[11] ASTM D3240 Standard Test Method for Undissolved Water In Aviation Turbine Fuels. ASTM International, 100 Barr Harbor Drive, PO Box C700, West Conshohocken, PA, 19428-2959 USA.

[12] EI 1590 Specifications and Qualification Procedures for Aviation Fuel Microfilters. Energy Institute 61 New Cavendish St, London W1G 7AR.

[13] EI 1581 Specifications and Laboratory Qualification Procedures for Aviation Fuel Filter/Water Separators. Energy Institute 61 New Cavendish St, London W1G 7AR.

[14] EI 1583 Laboratory Tests and Minimum Performance Levels for Aviation Fuel Filter Monitors. Energy Institute 61 New Cavendish St, London W1G 7AR.

[15] EI/JIG Standard 1530 Quality Assurance Requirements for the Manufacture, storage and distribution of Aviation Fuels to Airports. Energy Institute 61 New Cavendish St, London W1G 7AR.

[16] ASTM Manual 5, Aviation Fuel Quality Control Procedures. ASTM International, 100 Barr Harbor Drive, PO Box C700, West Conshohocken, PA, 19428-2959 USA.

[17] UK Meteorological Office Data.

[18] ASTM D6304 Standard Test Method for Determination of Water in Petroleum Products, Lubricating Oils, and Additives by Coulometric Karl Fischer Titration. ASTM International, 100 Barr Harbor Drive, PO Box C700, West Conshohocken, PA, 19428-2959 USA.

[19] SAE ARP 1401 SAE International Warrendale, Pennsylvania; Troy, Michigan, USA.

[20] Dispersed Water and Particulates in Jet Fuel: Size Analysis under Operational Conditions and Application to Coalescer Disarming, Industrial and Engineering Chemistry Research, 2011.

[21] Report on the Use of the Chemicals Resolver and Isopropyl Alcohol for Removing the Contribution of Water Droplets from Energy Institute Particle Counting Assay Methods. IP564, 565, 577-RR 2013. Energy Institute 2012.

[22] CRC Project AV-19-14, Alternative Aviation Fuels – Water Solubility and Demulsibility Impact. University of Dayton Research Institute 2017. The Coordinating Research Council, 5755 North Point Parkway, Suite 265, Alpharetta, GA 30022 USA.

[23] The Formation of Rain by Coalescence, Bowen E. G. Australian Journal of Scientific Research, Series A: Physical Sciences, vol. 3, p. 193 – 212. 1950.

EFFECTS OF AROMATIC TYPE AND CONCENTRATION ON PROPERTIES AND STABILITY OF ALTERNATIVE MARINE DIESEL

Jinxia Fu[1], Scott Q. Turn[2]

[1] *University of Hawaii, Honolulu, HI USA, jinxiafu@hawaii.edu*
[2] *University of Hawaii, Honolulu, HI USA, sturn@hawaii.edu*

The use of biofuels has been a part of the U.S. Navy's approach to deploy the "Great Green Fleet". Hydroprocessed renewable diesel (HRD-76), and direct sugar to hydrocarbon (DSH-76), also known as synthesized isoparaffin (SIP-76), have been produced in sufficient quantity for testing by the US Navy, and demonstrated as compatible replacements or blend stock with marine diesel (NATO F-76). Operational limitations with respect to the lubricity, seal swell and cold-flow properties, however, need to be improved before using 100% HRD-76 or SIP-76. Aromatics are known to improve seal swell and low-temperature characteristics, and a minimum of 8 wt% aromatics is required in marine diesel or diesel blends. In this study, three commercial aromatic blends and 15 pure alkyl aromatics which are present in petroleum diesel were added (individually and as a blend) to HRD-76 and SIP-76 at various concentrations to investigate their influence on material compatibility, storage and oxidation stability, physicochemical properties, and low-temperature qualities. Both experimental and computational methods were used to study the propensity of different aromatic species to enhance qualities of HRD-76 and SIP-76. In addition, the impacts of aromatic additives on biofuel water content and vapor pressure were also investigated.

IASH 2017, the 15th INTERNATIONAL SYMPOSIUM ON STABILITY, HANDLING AND USE OF LIQUID FUELS
Rome, Italy
10-14 September 2017

THE IMPACT OF CHANGING REGULATIONS ON MARINE FUEL QUALITY AND THE ROLE AND EFFECTIVENESS OF FUEL ADDITIVES

Simon Mulqueen, Michael Banning

Innospec Limited

In the late 1980's the International Maritime Organisation (IMO) started work on reducing air pollution from ships. Since then in a parallel movement to automotive application, the introduction of further legislation and emission regulations have seen the introduction of low sulfur fuels. Further controls have also been implemented with the introduction of emissions control areas (ECA) in coastal waters and ships trading in designated emission control areas now have to use on board fuel oil with a sulphur content of no more than 0.10% from 1 January 2015, against a limit of 1.00% sulphur in effect up until 31 December 2014

As a consequence from 2015 there has been a change in the middle distillate versus residual fuel demand, directly as a result of the ECA legislation. The higher middle distillate fuel consumption has inevitably suppressed residual fuel quality, and therefore value. Subsequently refiners are looking to adopt deeper conversion processes to meet the extra demand of the market, resulting in a further deterioration of the ignition, combustion and stability properties of the provided residual fuel.

The 2015 ECA legislation further challenges fuel stability since future fuel changeover will be between middle distillate and residual fuels and at a slower change over rate, maximizing the incompatibility risk. Middle distillate fuel can also represent its own risks, since increased consumption requires additional storage space onboard, along with longer storage periods. This means risks of inadequate lubricity, stability and cold flow properties must be considered.

The market has also seen the emergence of several hybrid "ECA Compliant" fuels, which themselves present specific challenges. In all of the above fuel additives can play a key role in mitigating many of the quality and operability concerns.

This paper discusses and demonstrates how fuel additives are a reliable and cost effective solution focused directly on solving fuel quality issues, along with the potential operational challenges. The introduction of new legislation has marked a step change in the grades of bunker fuels available and in service. As a result a step change in fuel quality has occurred, which may act to further exacerbate existing challenges. Challenges associated with residual and distillate fuels will be examined, plus those associated with changeover between the two. A summary of new ECA Compliant fuels, their properties, and their advantages and disadvantages versus conventional fuels will also be considered.

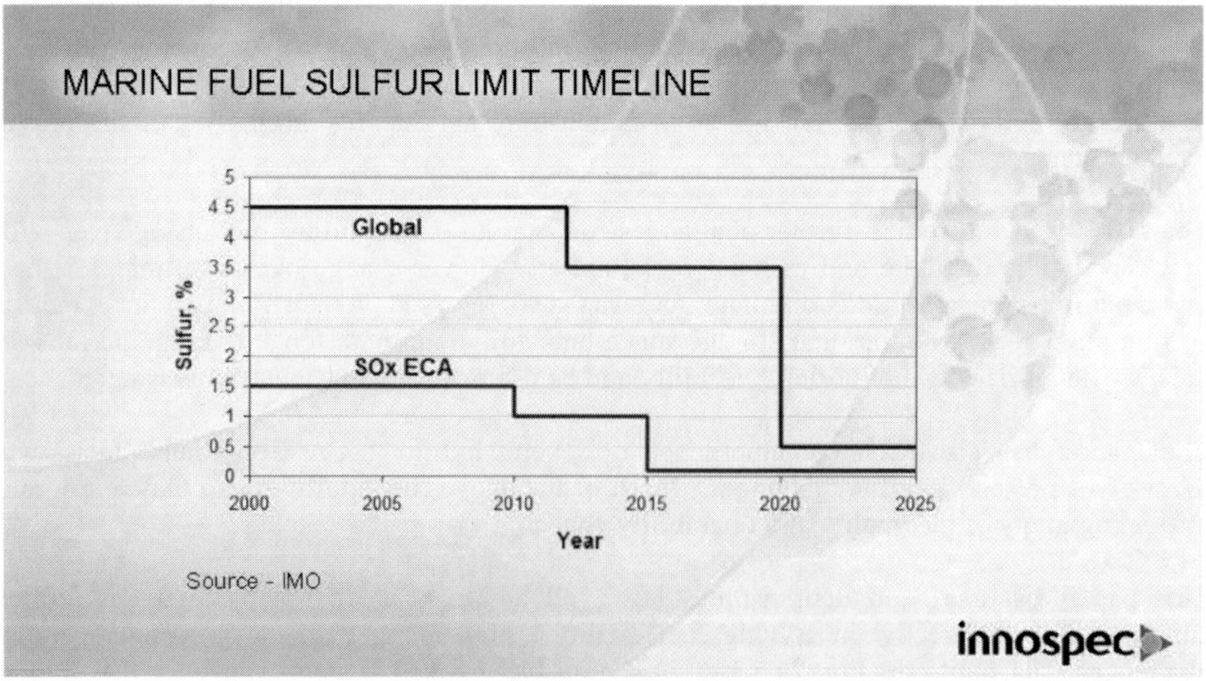

This slide provides an overview of the sulfur limits for marine fuel imposed by the International Maritime Organisation on commercial shipping. There are two significant changes to note here, the first of which happened in 2015, where the sulfur limit was reduced from 1.0% m/m to 0.1%

within the Emission Control Area(s) (ECA). Whilst this is a large reduction, its impact on the wider fuel supply chain was not really felt due to the relatively small geographical area of the ECAs (which will be detailed in a later slide). What is perhaps the most substantial change, and the subject of this presentation, is the blue line which is the global sulfur cap. On 1st January 2020 the global sulfur cap will reduce from 3.5 to 0.5%.

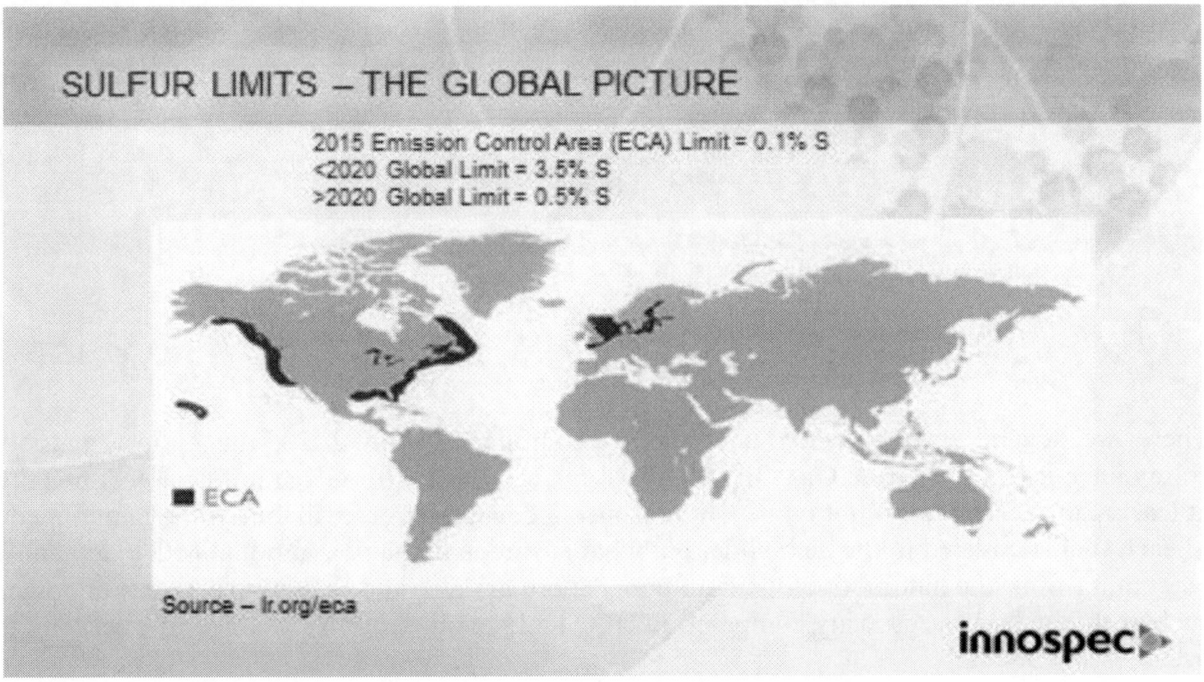

Here is a map detailing the ECAs throughout the world (highlighted in blue), these ECAs limit the sulfur content of fuel in use to 0.1%. It is important to note that there are other ECAs currently in operation, most notably in China, however these areas impose a sulfur limit of 0.5%. What is most significant is the light blue area, where the global sulfur cap will apply.

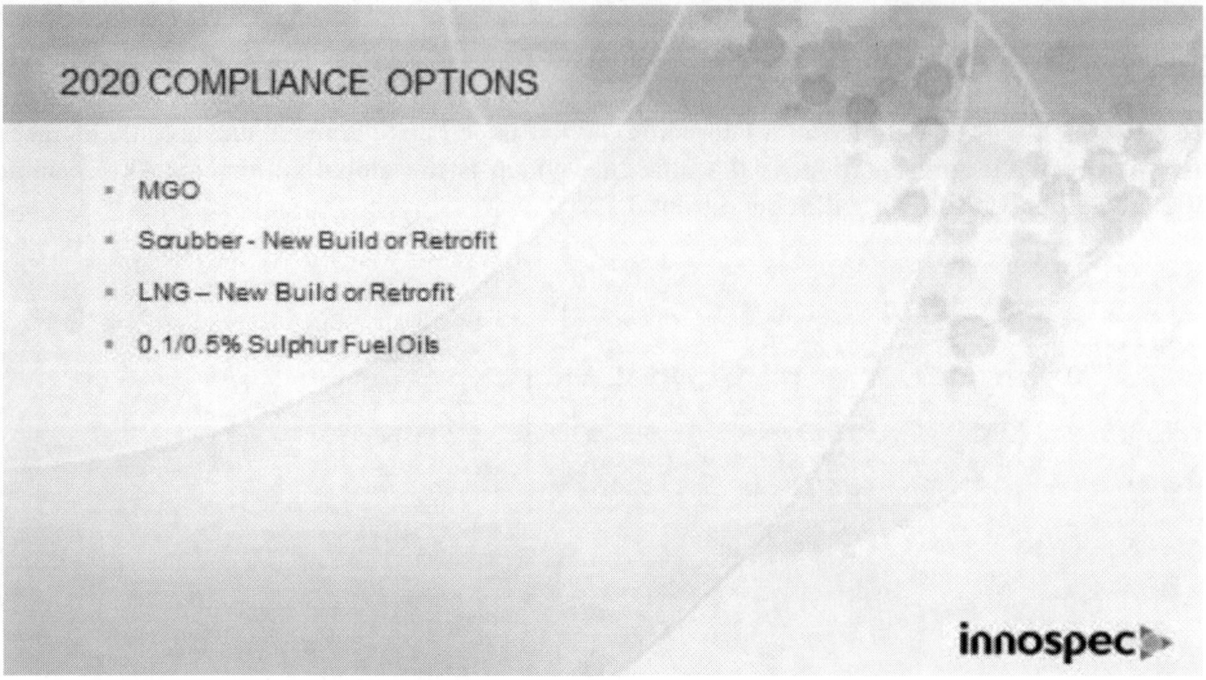

There are several options when it comes to compliance with 2020, the easiest method of compliance is to use Marine Gas Oil (MGO), however there will be a significant cost penalty of at least $200 US per ton if not more. There is also a concern over availability, the anticipated use of each fuel is covered in the next slide, but what is sure is that any additional MGO demand will be coming from the automotive distillate pool. There are also the operational issues to consider, such as the fuel's low viscosity compared to HFO and potential lubricity concerns over long term use.

Using a complaint fuel is not the only option; it is possible to use an exhaust gas cleaning system or scrubber. These systems essentially use Sea Water to scrub the sulfur out of the exhaust, this allows the vessel to continue to burn HFO, they are however a significant piece of equipment in terms of capital and physical size and weight. Whilst it may seem attractive to use a scrubber and benefit from the continued use of HFO or any other high sulfur fuel, installation costs vary drastically dependant on vessel type and the Return on Investment/payback period is a difficult number to quantify. When retrofitting a scrubber, it is not only the physical volume of the equipment that must be considered, but the mass of it and the potential change in vessel stability, not to mention considerations such as additional power requirements in terms of electricity and crew to operate the plant. It is anticipated that scrubbers will be an attractive option for new build vessels, where the installation costs are negligible compared to refit costs. A recent trend for many new builds is to given the notation of 'Scrubber Ready', which means that sufficient space (for the scrubber and piping) and generating capacity have been included in the vessel design. This allows the vessel flexibility in the future, as most of the cost of a retrofit has been mitigated.

The same can be said for LNG, as a retrofit this is a significant investment and additional storage tanks must be fitted onboard. The main concern for LNG is its availability; at present the infrastructure is improving which may be viable for ferries, cruise or container ships on a fixed route. However the infrastructure will not be at a sufficient level to support the spot market for some time.

This leaves us with compliant fuels, 0.1% and 0.5%S 'hybrids'. In the run up to 2015 we saw the introduction of 0.1% Ultra Low Sulfur Fuel Oils (ULSFO), which are a hybrid fuel predominantly consisting of VGO, MGO and a certain percentage of residual fuel. These fuels are now commonplace with the ECAs mentioned before, and provide a cost effective and technical solution to MGO. It is widely anticipated that a similar approach will be adopted for 0.5% Very Low Sulfur Fuel Oils (VLSFO).

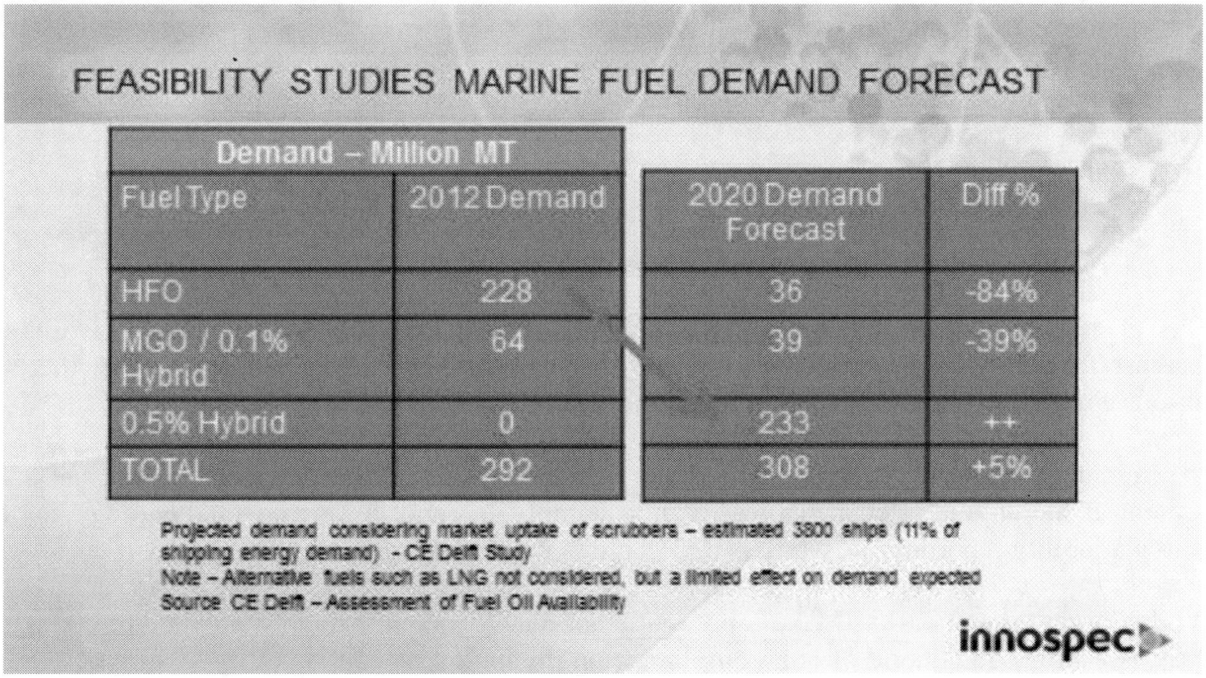
FEASIBILITY STUDIES MARINE FUEL DEMAND FORECAST

Demand – Million MT			
Fuel Type	2012 Demand	2020 Demand Forecast	Diff %
HFO	228	36	-84%
MGO / 0.1% Hybrid	64	39	-39%
0.5% Hybrid	0	233	++
TOTAL	292	308	+5%

Projected demand considering market uptake of scrubbers – estimated 3800 ships (11% of shipping energy demand) - CE Delft Study
Note – Alternative fuels such as LNG not considered, but a limited effect on demand expected
Source CE Delft – Assessment of Fuel Oil Availability

innospec

The above chart shows the total demand for marine fuel in 2012; these figures have come from the CE Delft study commissioned by the IMO to make the 2020 decision. In 2012 the demand for HFO was 228 Million MT, however looking at CE Delft's projections for 2020, it appears that this has dropped by over 80% to 36 million MT. This demand has shifted to a 0.5% Hybrid fuel (or VLSFO), which is significant.

The remaining HFO will be consumed by ships equipped with scrubbers, which CE Delft has estimated to be 3800 ships, however the Exhaust Gas Cleaning System Association puts their own estimations at around the 2000 installed and operational systems by 2020.

It is important to note that for this study, alternative fuels have not been considered, but their impact is anticipated to be negligible. Also interesting to note, is that several industry sources currently expect there to be a surplus of approximately 120 Million MT of HFO, this is taking into account additional uptake of HFO from other sources such as power generation and residual content used for 0.1% and 0.5% blends.

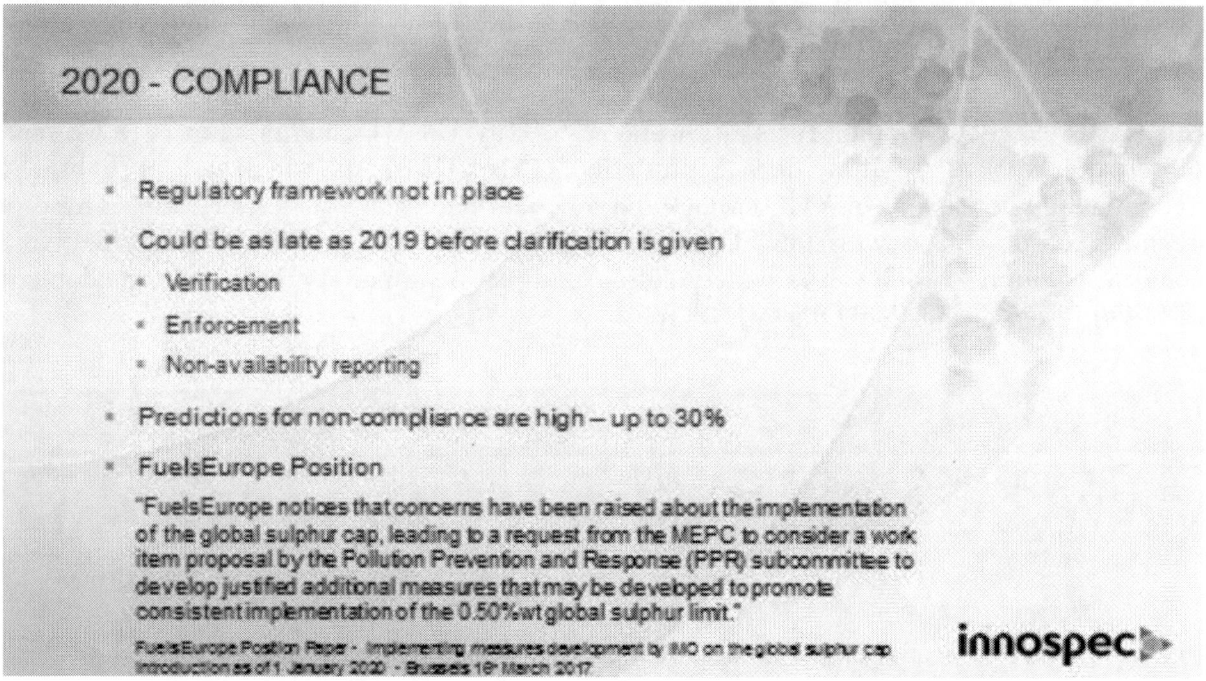

This change in 2020 is unprecedented within the marine industry, looking at the availability forecast from the CE Delft study, which is made using hypothetical blend models, there are serious concerns over compliant fuel availability.

At present the regulatory framework is not in place, this includes verification of fuel sulfur content at the engine inlet, penalties for non compliance and no agreed methodology for fuel non-availability reporting. It is anticipated that this will not come until 2019.

Whilst larger companies will take a proactive approach to compliance, with smaller companies there is a strong likelihood of non-compliance on the high seas as the ability to enforce the new rules is incredibly difficult. Some sources are predicting that non-compliance may be as high as 30%.

In March this year FuelsEurpoe published their position on this topic, stating that "FuelsEurope notices that concerns have been raised about the implementation of the global sulphur cap, leading to a request from the MEPC to consider a work item proposal by the Pollution Prevention and Response (PPR) subcommittee to develop justified additional measures that may be developed to promote consistent implementation of the 0.50%wt global sulphur limit."

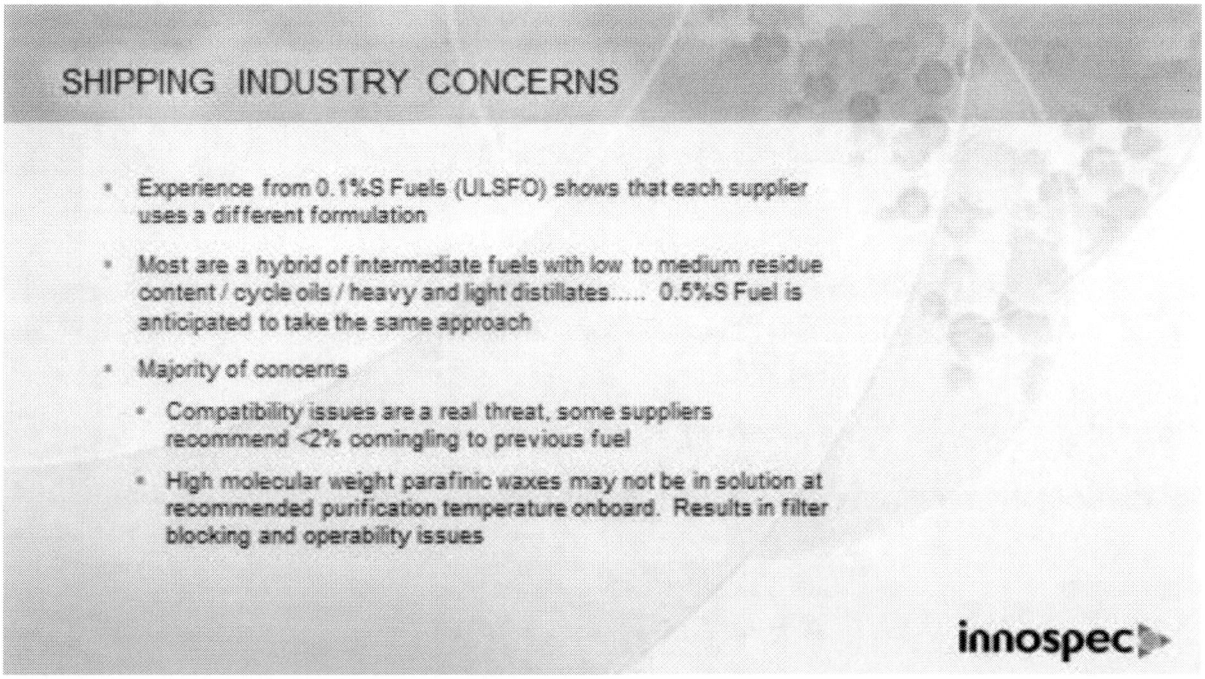

Not only are there concerns over compliance, there are concerns from the shipping industry regarding stability and handling of an unknown fuel.

Experience gained since 2015 with the 0.1%S ULSFO shows that each supplier has a different formulation; most are a hybrid of intermediates and have relatively low residue content. It is anticipated that many 0.5%S fuels will take a similar approach, with potentially more residual content.

The majority of concerns come from experience with incompatibility between two batches of fuel onboard. Most suppliers recommending less than 2% comingling occurs, which is not always practical onboard a vessel, particularly for example for a vessel on the spot market.
Another concern that has come to light is waxing of a 0.1% ULSFO fuel, where higher weight waxes may not be in solution at the recommended purification temperatures. With a ULSFO the purification temperature is largely dictated by the required injection viscosity at the engine.

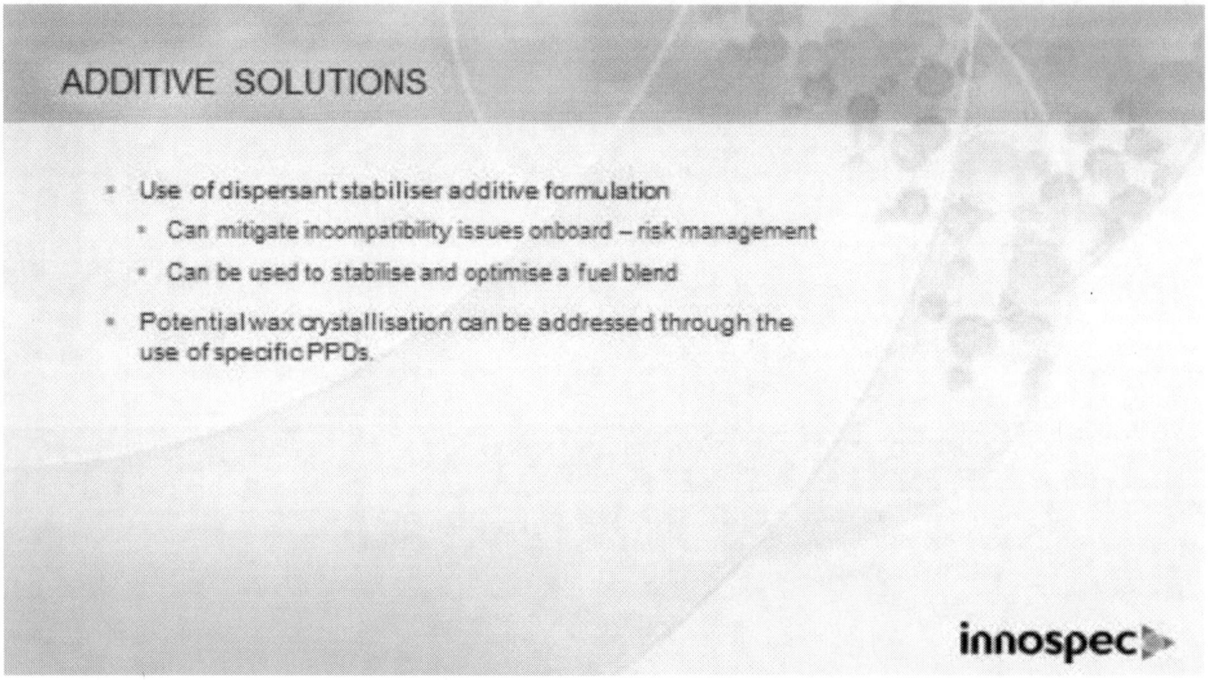

The issues highlighted can be mitigated by the use of additives, a dispersant stabiliser can be used for risk management onboard and reduce incompatibility between different parcels of fuel. Or on the supply side can be used to stabilise and optimise a potential fuel blend. Wax crystallisation can be addressed through the use of specific additives for that specific blend stream.

Here are a few examples of operability issues onboard vessels that I have seen since the introduction of ULSFOs, most issues such as waxes and instability are seen in the fuel handling side of the system, such as the purifier. However in some instances, this instability can lead to supply lines becoming blocked and fuel starvation occurring. The image on the bottom left is a supply line to a set of three generators, which tripped due to fuel starvation.

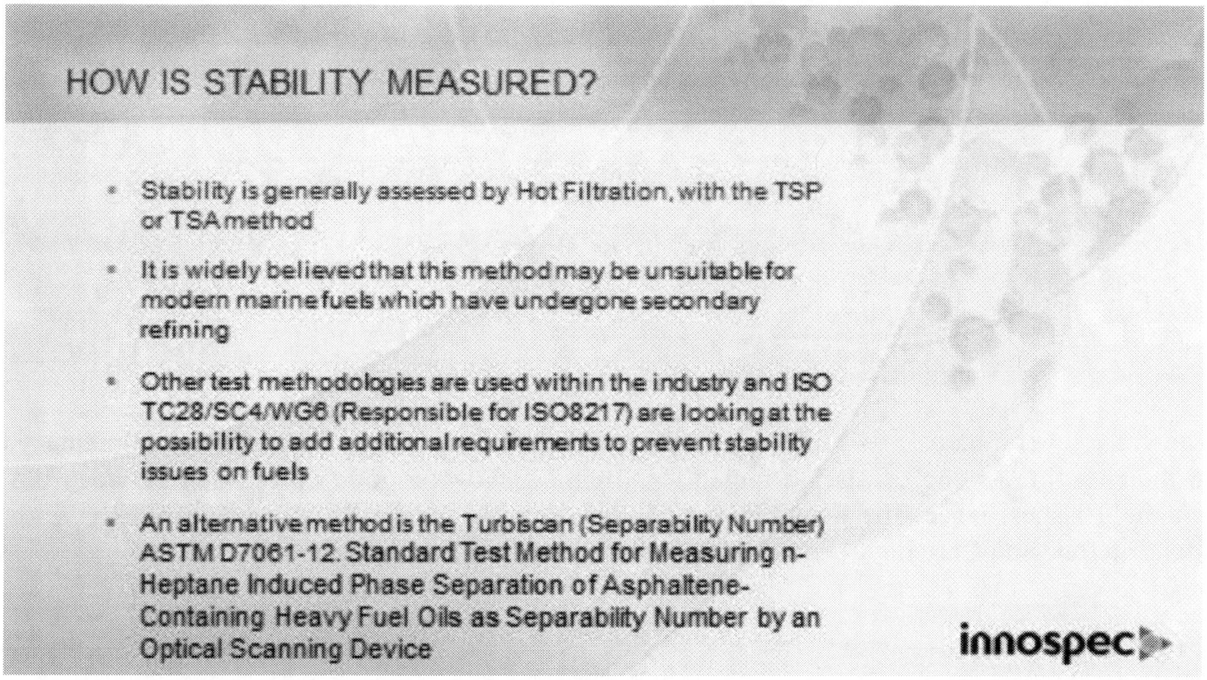

Stability of Marine HFO is measured within ISO 8217 fuel standard by Holt Filtration, however over recent years it has become apparent that this method may not be the most suitable for today's fuels. Following the IMO's 2020 decision, they requested that ISO look at ensuring that there is standardisation in place for the new fuels expected to reach the market in 2020.

The ISO working group has set up a sub group whom are looking at other testing methodologies for stability, there is a wide array of methodologies under review and one such method is ASTM D7061 or Turbiscan. This method stresses the fuel by adding toluene and n-Heptane which causes the asphaltenes to flocculate. The rate of precipitation is converted into a Separability Number (SN).

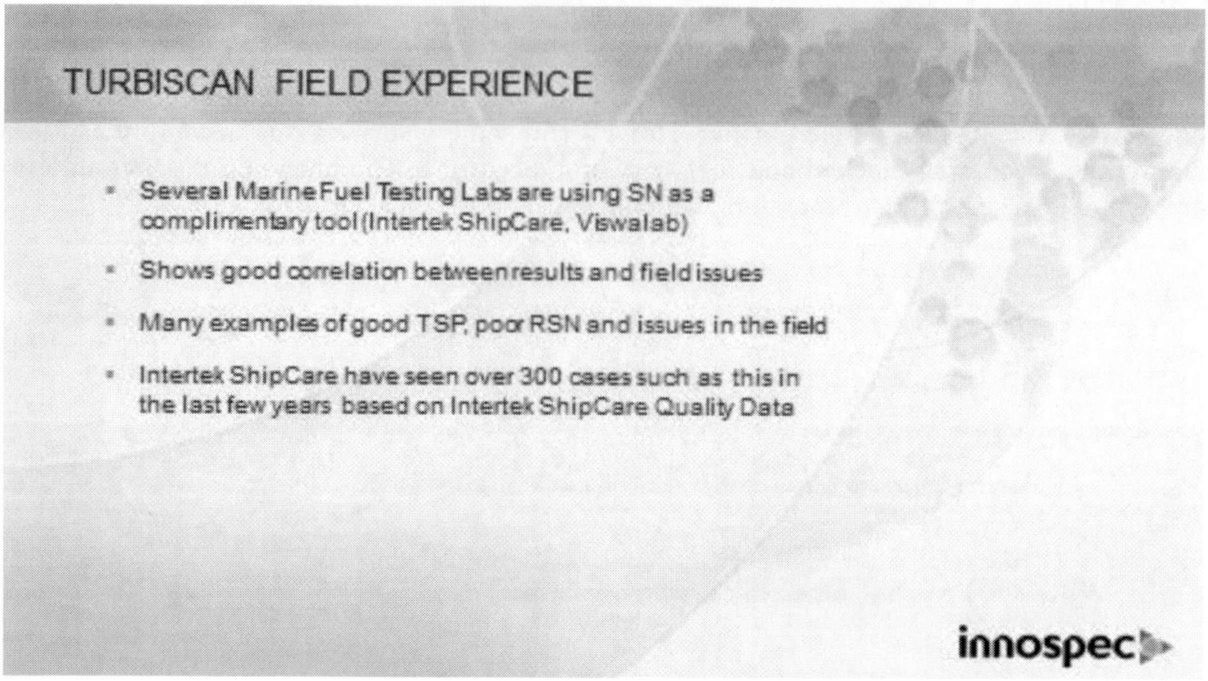

There are several marine fuel testing labs using the Turbiscan equipment as a complimentary tool to their routine testing of marine fuels. This method shows a good correlation between the SN and field issues, where the total sediment is on spec but the vessel still has problems with the fuel. Intertek ShipCare have seen over 300 cases like this over the last few years.

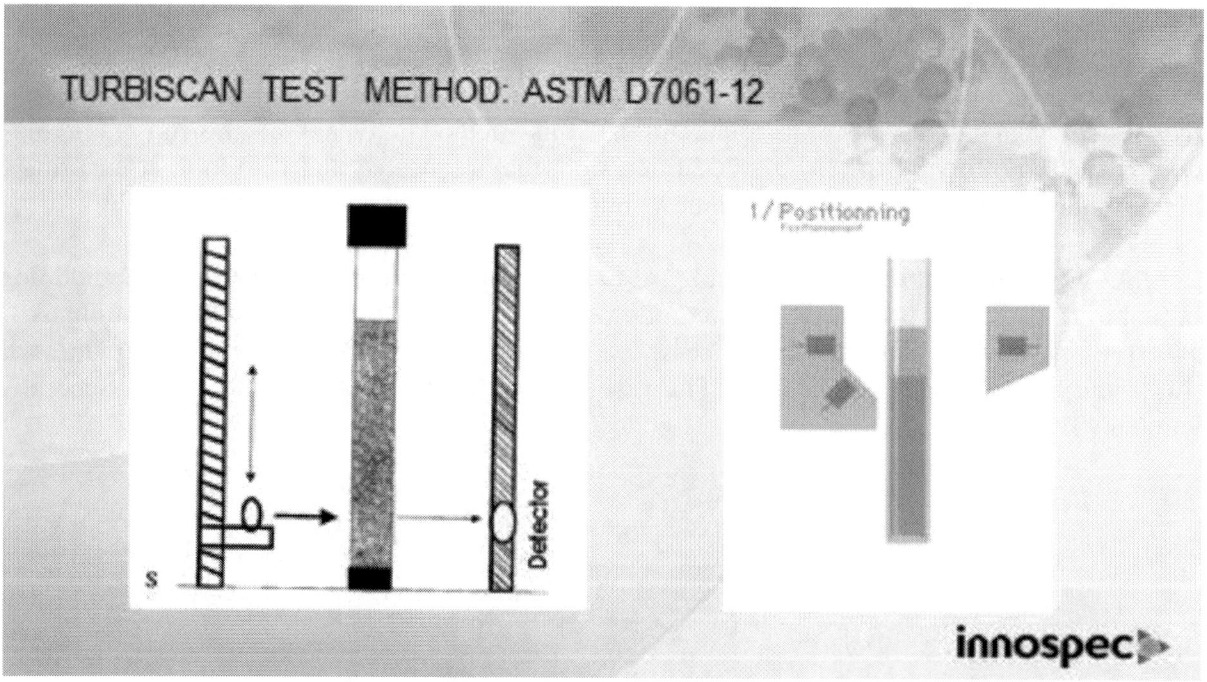

Turbiscan uses an optical measurement device, where light is directed at the test tube. When the sample is homogeneous the light will not pass through the sample, this light is not picked up by the detector and therefore is not measured (known as backscatter), however when the Asphaltenes begin to precipitate, light will pass through the sample and be picked up by the detector, this is known as transmission.

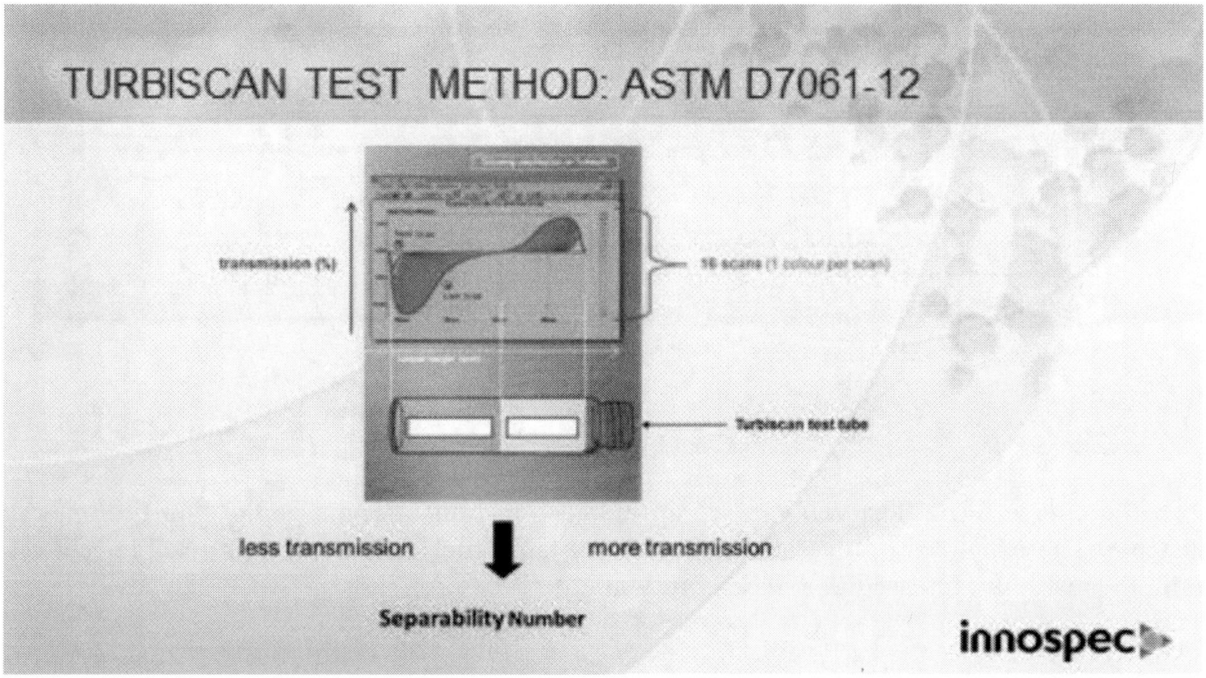

The piece of equipment records the data as transmission percentage, with backscatter shown below the reference line and transmission above. The sample is scanned once per minute of the test duration of 15 minutes.

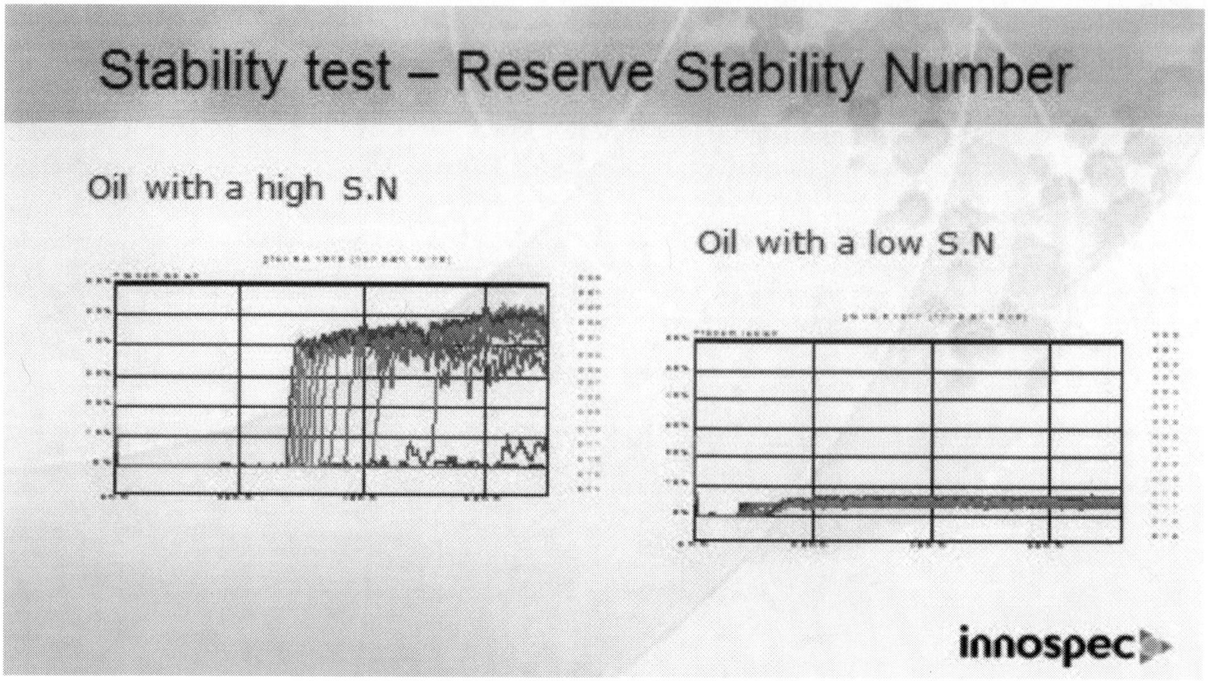

Over the course of the test you can see an increase in transmission, as the fuel becomes more unstable. The examples on this slide show an unstable fuel on the left and a stable fuel on the right, the result is expressed as a Separability number.

INTERPRETATION OF RESULTS

S.N Between 0 and 5	Good Asphaltene peptisation state, fuel is considered stable
S.N Between 5 and 10	Indicates a risk of fast drift towards higher S.N values. Care must be taken that the storage conditions do not lead to a further agglomeration process.
S.N Greater than 10	Indicates a highly degraded Asphaltene peptisation state, potential for operability problems due to the size of Asphaltene agglomerates.

innospec

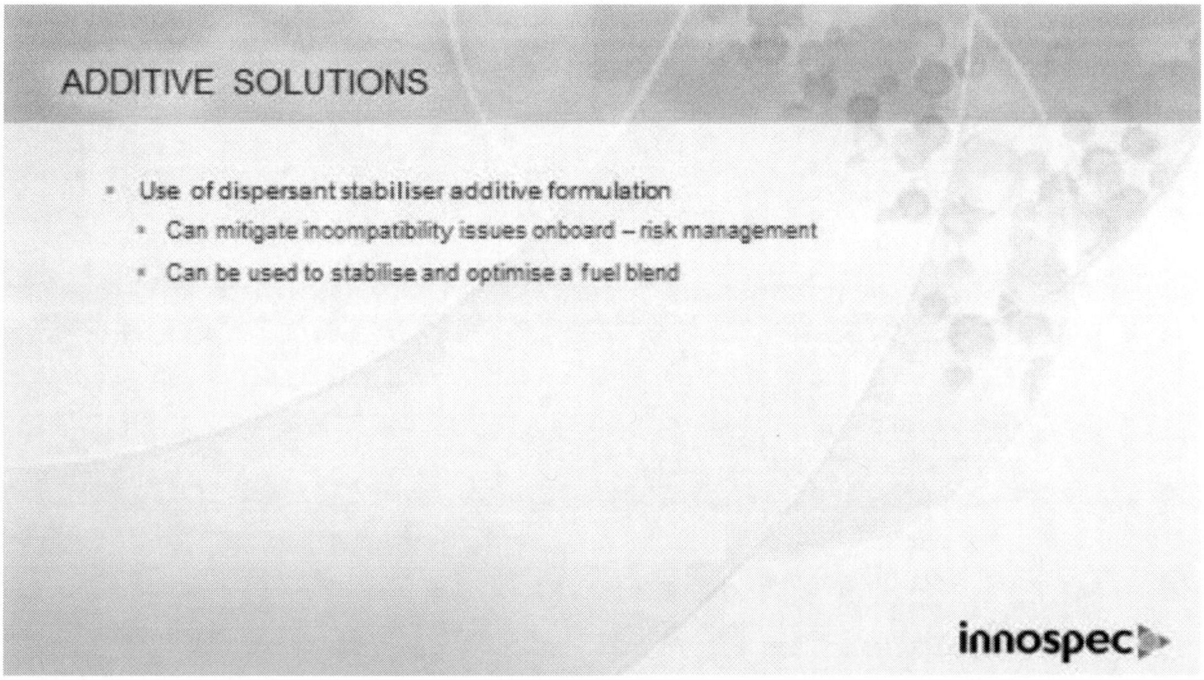

Additives can be used to mitigate stability/incompatibility issues onboard, or to optimise fuel blends.

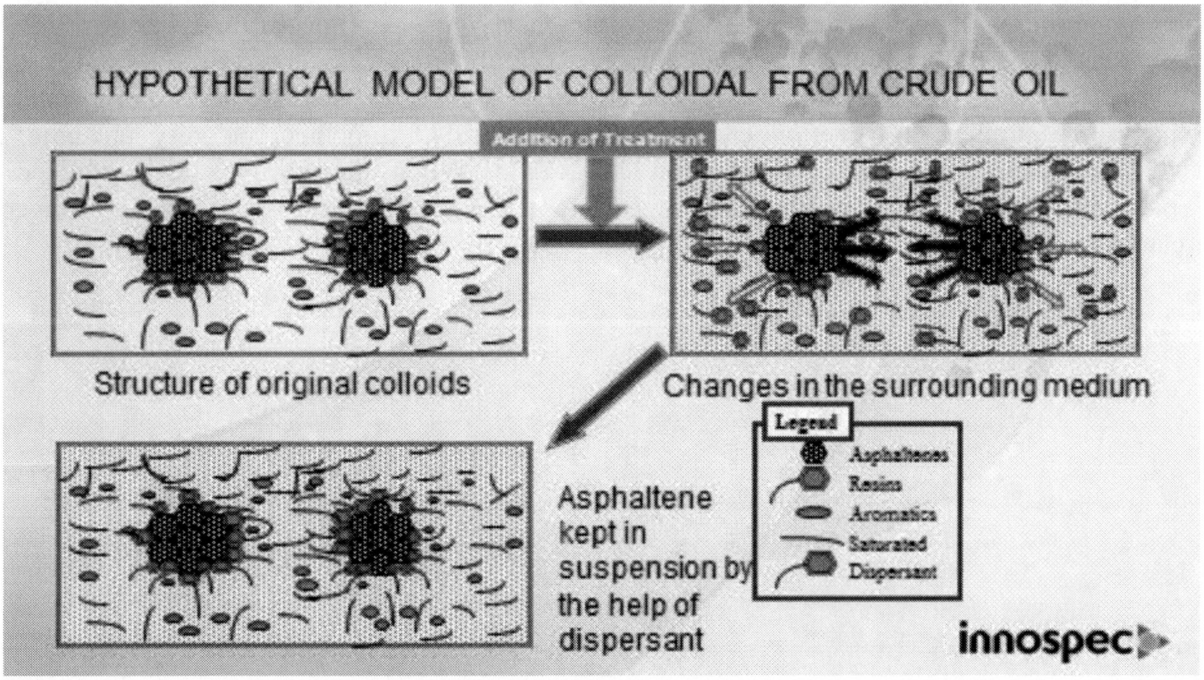

In this Colloidal model, Asphaltenes are held in suspension and stable by the resins surrounding them. When changes in the surrounding medium occur (such as in the refining process), some of these resins are removed, causing the asphaltenes to agglomerate and fall out of suspension.

When dispersants are added, they simulate the natural resins that have been lost, keeping the asphaltenes in suspension.

ADDITIVE SOLUTION - STABILITY

Blend	Base RSN	Additised RSN
100% HFO 1	9.02	0.31
50% HFO 1 / 50% ULSFO 1	11.23	0.35
50% HFO 1 / 50% ULSFO 2	11.34	0.71
100% HFO 2	7.70	0.10
50% HFO 2 / 50% ULSFO 3	8.40	0.10
50% HFO 2 / 50% ULSFO 4	9.0	1.00

innospec

Innospec has conducted many simulations, looking to replicate onboard conditions and review the compatibility of fuels onboard. The examples here look at High Sulfur HFO and ULSFO, these tests are designed to simulate fuel changeover onboard a vessel, which will happen when entering or exiting an ECA, here we can see that the HFO used are in the mid-range of Separability number, however when blended with ULSFO they become very unstable, and asphaltene crash out and fuel starvation can occur. When the samples are additised the Separability number comes right down into the stable range, where no issues during changeover are anticipated.

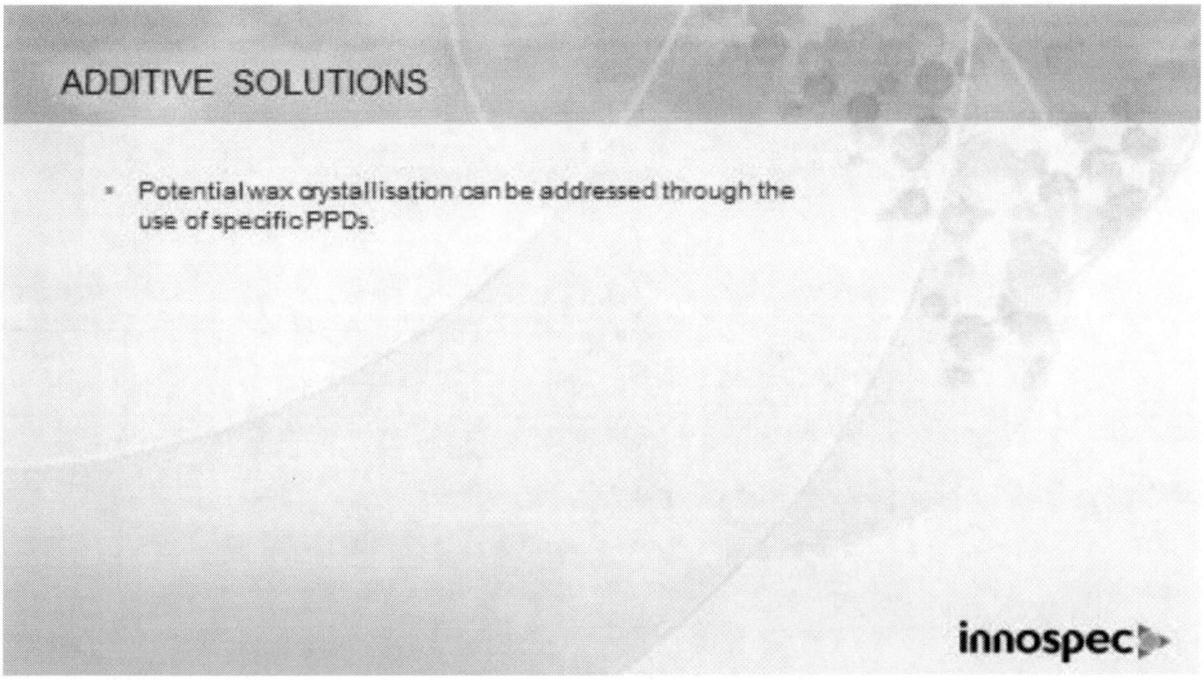

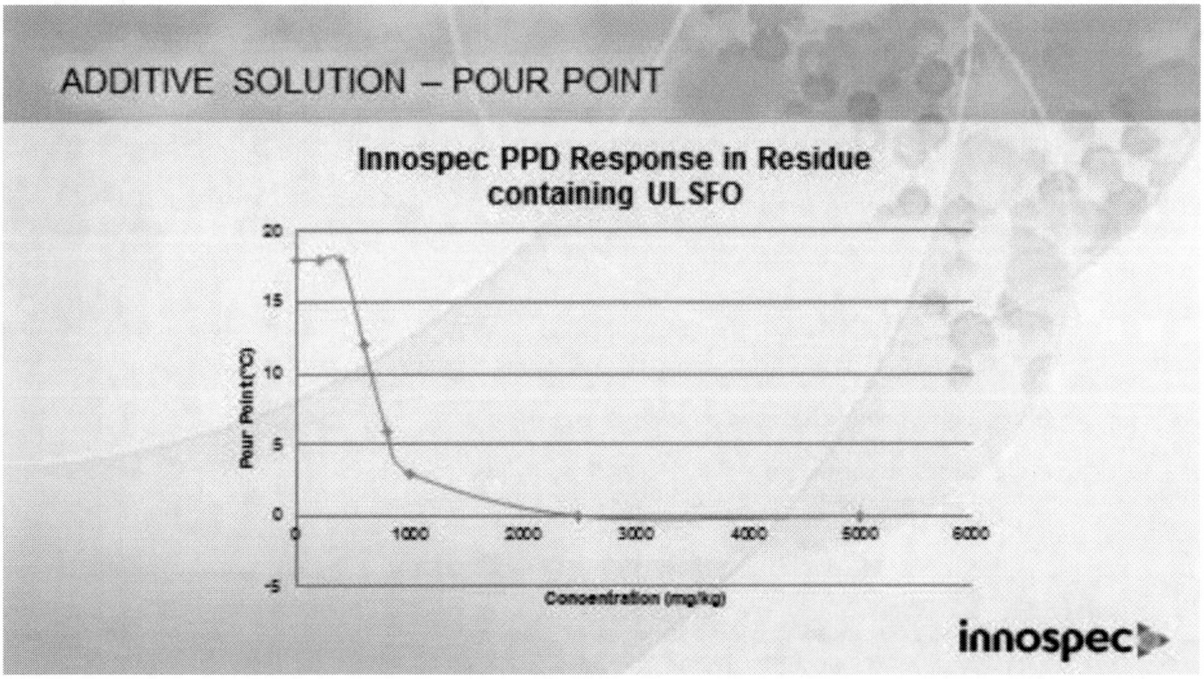

Here is an example of the response seen in a ULSFO which contains approximately 20% residue, the main purpose of this exercise was to eliminate the need for tank heating onboard the ship, which is a significant cost.

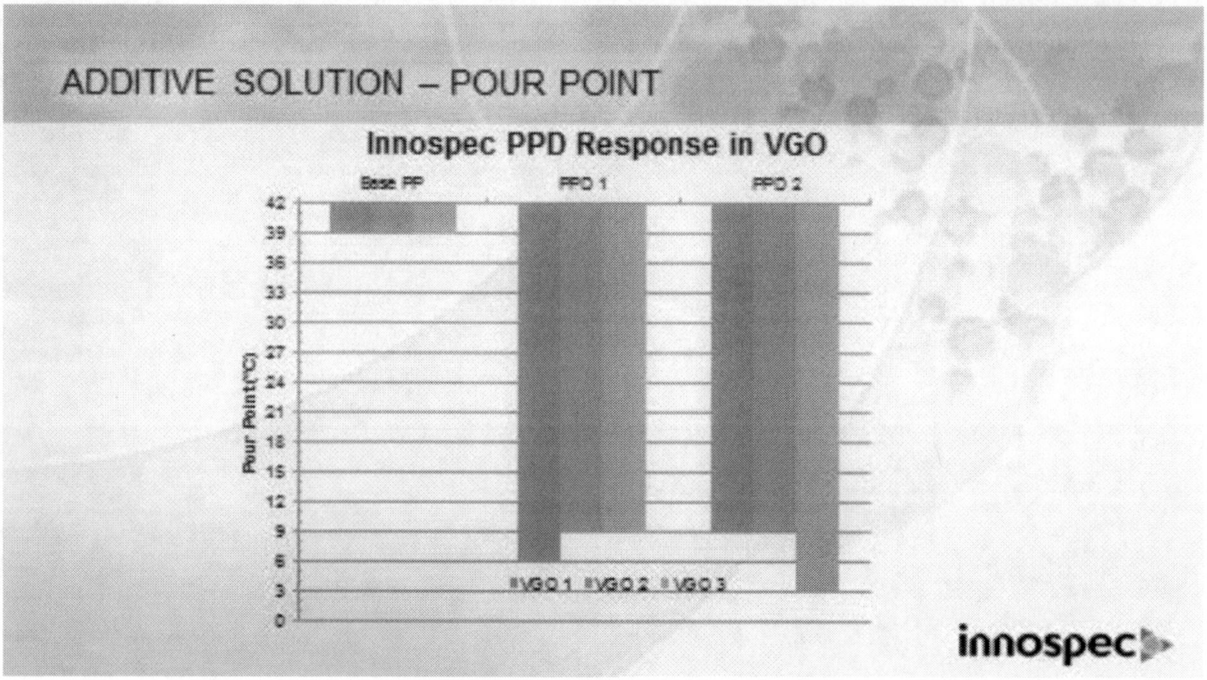

Here is an example of responses in several neat VGOs, the reason we have looked at this is that VGO is a principal component in most ULSFO, and widely anticipated to feature heavily in VLSFO. This reduction in PP eliminates the need for tank heating onboard, as well as improving handling characteristics in shore tanks.

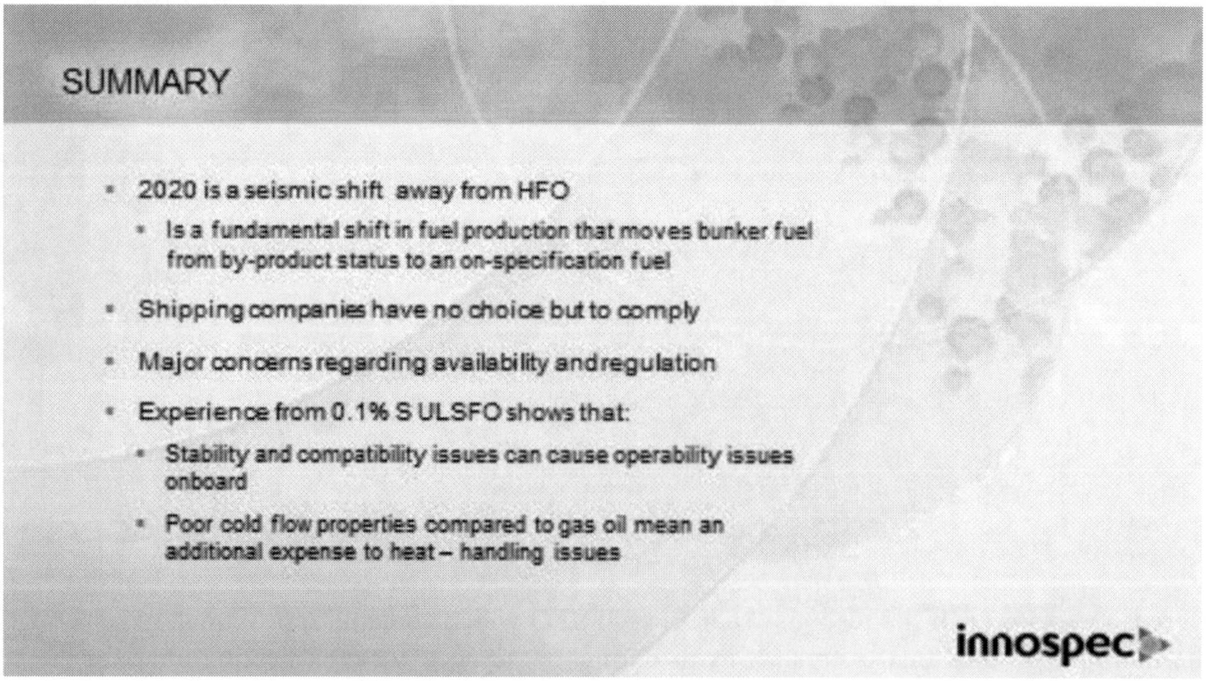

To summarise, 2020 is a seismic shift away from HFO, which is a fundamental shift in fuel production that moves bunker fuel from by-product status to an on-specification fuel.

Shipping companies have no choice but to comply, although there are major concerns around fuel availability, regulation and non-compliance at sea.

Experience from ULSFO shows that the main concerns with regards to 2020 fuel use is stability and compatibility cold flow properties.

IASH 2017, the 15th INTERNATIONAL CONFERENCE ON STABILITY, HANDLING AND USE OF LIQUID FUELS
Rome, Italy
10-14 September 2017

IMPACT OF WATER BOTTOMS ON AVIATION TURBINE FUEL CHLORIDE CONTENT

Paul P. Wells[1]

[1] ExxonMobil Research & Engineering, 600 Billingsport Road, Paulsboro, NJ 08066, USA, paul.p.wells@exxonmobil.com

ABSTRACT

Elevated chloride levels in aviation turbine fuel can cause corrosive and abrasive wear of aircraft fuel control systems. Chloride contamination can be the result of carryover from a refinery salt dryer or seawater contamination during marine transport. A series of high profile incidents were caused by chloride contamination in 1974. In the most serious incident, two of four Boeing 747 main engines were shutdown in flight. Analysis revealed that aircraft fuel tank water sumps contained 11,900 to 21,900 ppm sodium chloride. An ensuing report by the International Air Transport Association (IATA)[1] noted that there were no detrimental effects on aircraft hardware when water sumps contained up to 500 ppm sodium chloride.

Issued in April 1990 (but since withdrawn), the Australian Defence Standard 5240B for aviation turbine fuel contained a requirement for ≤0.15 mg/L sodium chloride. With the passage of time, it is unknown why this concentration was selected or if there is any relationship to the IATA guidance above.

ASTM D7959, a standard test method to measure chloride levels in aviation turbine fuel has been developed and a reference to the methodology has been added to the D1655 appendix. The details of the test will be reviewed.

The impact of water bottoms on aviation turbine fuel chloride content via D7959 will be investigated. The key variables investigated will be chloride content of the water bottoms and additive content of the fuel. The ability of current D1655 requirements for water separation and conductivity to function as a substitute control of fuel chloride content will also be investigated.

KEYWORDS

Aviation turbine fuel, jet fuel, chloride, salt, contamination, water bottoms, heel

INTRODUCTION

The potential consequences of gross chloride contamination are illustrated in the following passages of an International Air Transport Association (IATA) document.[1]

"In May, 1974, sodium chloride contamination in aviation turbine fuel occurred to three separate B747 aircraft in passenger service on the Australia-United Kingdom route. Malfunction of fuel metering to an engine resulted in a single engine shutdown on one aircraft, while on the other two aircraft, serious, almost catastrophic, corrosion damage to the airframe fuel supply system and engine fuel and vane control components caused a double engine shutdown in flight. ... Experience has shown from many water samples from different aircraft in airline service that saline concentration of less than 100 ppm in the sump water is normal, and up to 500 ppm appears to have no detrimental operating effects to the systems. On the [aircraft mentioned above], samples of water removed from the tanks in London varied from tank to tank, ranging from 11,900 to 21,900 ppm of sodium chloride (1.19% to 2.19%). ... Salt does not mix chemically with the hydrocarbon of fuel, but may be carried in discrete crystals in the fuel and as brine in the undissolved water in the fuel."

No exact cause of the contamination was identified in the IATA paper, however detection of sodium chloride and no other salt would seem to indicate that the contamination did not occur during marine transport of the fuel. Seawater contains a mixture of salts, not sodium chloride exclusively.

Since sump water is not always available or readily accessible in the aviation fuel distribution system, the IATA guidance regarding maximum sump water saline concentration cannot always be investigated. Aviation turbine fuel is filtered numerous times as it progresses from the refinery to the wingtip of the aircraft to remove dirt and free water. Storage tanks and filter vessels must be dewatered regularly to prevent microbial contamination. An added complication is that the water required for analysis is that which evolves from the fuel over time as it cools. Therefore, simply pouring water through an aviation turbine fuel storage tank and collecting a sample from the sump would most likely not be sufficient to extract chloride from the fuel or provide an accurate assessment. Establishment of an appropriate chloride limit for the aviation turbine fuel would aid assessment.

Published by the Australian Department of Defence in April 1990, Def (Aust) 5240B established a requirement of ≤0.15 mg/L sodium chloride in aviation turbine fuel. This equates to ≤0.091 mg/L chloride. Unfortunately with the passage of time, it is unknown why this concentration was selected or if there is any relationship to the IATA guidance above. Def (Aust) 5240B has since been withdrawn.

Chloride concentration of aviation turbine fuel, measured as part of a 1960 survey of trace contaminants, has been reported as high as 0.82 ppm.[2] Chloride data is reported as "In solution" and "In solution and suspension", but unfortunately the methodology used to quantify the chloride is not listed. Where determined on the same sample, the "In solution and suspension" values tend to be greater (ranging from 0.02 to 0.82 ppm in the survey) compared to the "In solution" values (ranging from <0.01 to 0.61 ppm in the survey). Copper chloride or hypochlorite sweetening processes to treat aviation turbine fuel were apparently in use during this time period and may have provided a source of chloride not present (or at least much less prevalent) in modern refineries.

Chloride contamination of aviation turbine fuel can be the result of either carryover from a refinery salt dryer or seawater contamination during marine transport. Salt dryers can be present in a refinery as part of an aviation turbine fuel polishing process. The salt dryers are typically placed upstream of clay treatment vessels to remove water which would hinder the performance of the clay. Sodium chloride, calcium chloride or a mixture of the two is typically used in the salt dryer. Water in the aviation turbine fuel will be attracted to the salt and subsequently form a brine which is removed from the bottom of the vessel. Failure to routinely remove this brine, operation at higher than design flowrates or high operating temperatures[3] could lead to brine carryover. Salt grade selection is also important as too high a fines concentration could lead to an unacceptable level of salt in the aviation turbine fuel. An important industry document related to the marine transport of petroleum products is HM 50.[4] Table 4 within that document provides vessel compartment cleaning recommendations after discharging one product and before loading the next. For a number of scenarios, a seawater wash of the compartment(s) is recommended prior to loading aviation turbine fuel. HM 50 Section 2.12.7 states "In all cases salt contamination needs to be minimized after washing with sea water. A fresh water rinse is recommended following sea water wash but thorough mopping may suffice." It is unclear how much seawater would remain after thorough mopping. Energy Institute/Joint Inspection Group (EI/JIG) 1530[5] Section 10.1.4 concerns the assessment of a vessel prior to loading and acknowledges that "...due to modern environmental controls the ability to make any form of physical inspection is rare...". EI/JIG 1530 Section 8.3.5.1 concerns procedures to be followed before discharge of a vessel and states that all compartments shall be checked for the presence of water and that key personnel should be contacted if "...significant levels of water are observed...". Significant, however, is not defined. The chloride content of ocean water is extremely high, slightly less than 19,000 ppmw.[6]

EXPERIMENTAL

The ASTM D7959 Standard Test Method for Chloride Content Determination methodology was developed as a rapid means of quantifying chloride in aviation turbine fuel without sophisticated laboratory equipment. Unfortunately a relatively large sample (i.e. 3 L) of fuel is needed for analysis in order to determine the low chloride concentrations of interest. Samples less than 3 L can be analyzed as described in the D7959 Appendix, but no precision information is available for those results. Per the D7959 method, the sample container is shaken and then a 3 L sample is placed in a 4 L beaker for analysis. Sample in excess of 3 L is either discarded or transferred to separate container. A 5 mL sample of pH 7 buffer solution is then added to the now empty sample container to interact with any residual sample clinging to the inside surface, the container is capped and shaken. The buffer solution is then carefully collected and transferred to the 4 L beaker. A 20 mL sample of iso-octane, serving as a rinse to help collect any residual buffer solution, is then added to the sample container, the container is again capped and shaken. The iso-octane and any residual buffer solution is also added to the 4 L beaker. The 3 L test sample is extracted with the buffer solution via contact on a magnetic stir plate for 10 minutes. Mixing energy is standardized by increasing the stir bar speed until the horizontal bars of an ASTM distillate fuel bar chart taped onto the back of the 4 L beaker are no longer visible. Once mixing is complete the sample is allowed to settle and the buffer solution is collected in the corner of the beaker. The buffer solution is then obtained with a pipette and placed into a small vial with a Hach Quantab® Low Range (30-600 mg/L Cl-) test strip. Silver ions within the test strip combine with chloride ions in the extract as it is drawn up by capillary action to form a white column of silver chloride.[7] The test strip reading is then converted to chloride concentration via the scale on the Hach Quantab bottle. This concentration (for the 5 mL aqueous sample) is then converted to that of the original 3 L test sample by multiplying by a correction factor of 0.0017.

Hach Quantab test strip readings below the initial conversion scale value provided on the bottle were used to determine chloride concentration for this study. For example, the conversion scale for Hach Quantab lot A7016 is provided in Table 1 and shown graphically in Figure 1. Conversion is not provided for test strip readings below 1.4.

Table 1: Conversion Scale for Hach Quantab Bottle Lot A7016

Quantab Strip	Cl-, mg/L	Quantab Strip	Cl-, mg/L	Quantab Strip	Cl-, mg/L
1.4	33	3.6	125	5.8	298
1.6	39	3.8	136	6.0	321
1.8	45	4.0	148	6.2	346
2.0	52	4.2	161	6.4	372
2.2	60	4.4	175	6.6	400
2.4	68	4.6	190	6.8	431
2.6	76	4.8	205	7.0	464
2.8	85	5.0	221	7.2	500
3.0	94	5.2	239	7.4	539
3.2	104	5.4	257	7.6	581
3.4	114	5.6	277	7.8	628

Figure 1: Conversion Scale for Hach Quantab Bottle Lot A7016

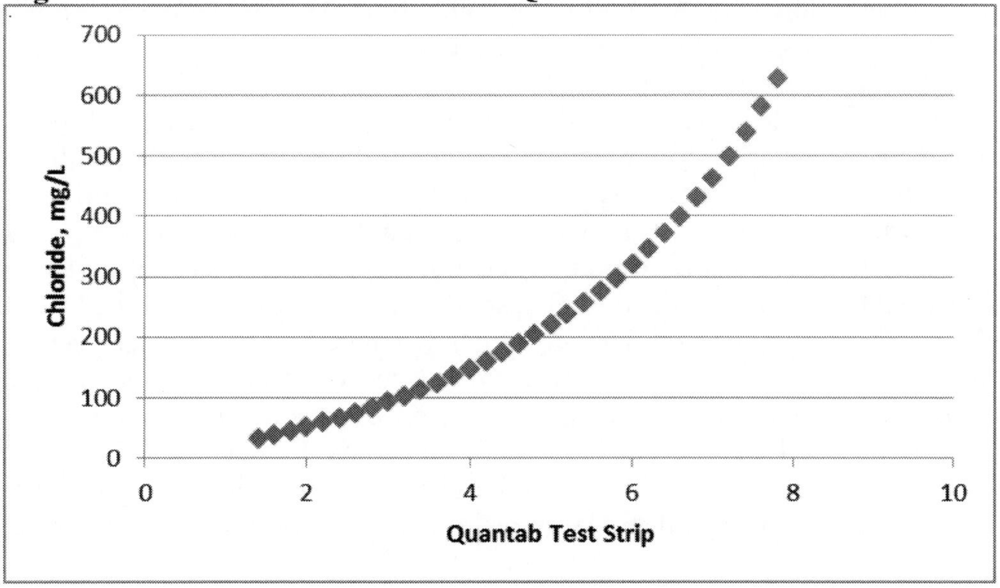

In order to convert test strip readings below 1.4 for the A7016 lot, the first four scale values were plotted along with the origin (i.e. a test strip reading of 0 was assumed to signify 0 mg/L chloride) as shown in Figure 2. A polynomial regression line provides the best fit through these points. The resulting equation is then used to convert test strip readings below 1.4 into chloride concentration.

Figure 2: Conversion Scale for Hach Quantab Bottle Lot A7016 Near Origin

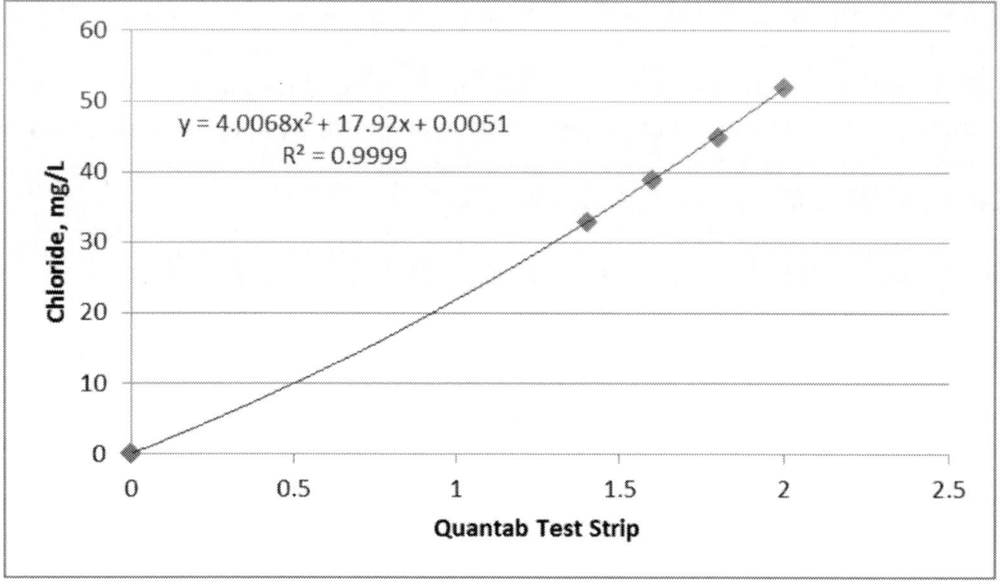

For this study, a large, 4 L graduated cylinder and magnetic stir plate were chosen to contact aviation turbine fuel with an aqueous heel. Immediately after contacting, the magnetic stir bar was removed and the graduated cylinder was set aside for settling. After the pre-determined setting period, aviation turbine fuel was siphoned from the top of the graduated cylinder, leaving a 250 mL or ~3 cm buffer of fuel on top of the aqueous heel. The fuel was not decanted since it was felt that decanting would cause too much disturbance and too great a possibility for carryover of the aqueous heel. Siphoning the aviation turbine fuel was also felt to more closely mimic floating suction within a storage tank. Any fuel siphoned from the cylinder in excess of the 3 L required for D7959 analysis could be used for bench testing.

The main variables investigated in this study were additive content of the aviation turbine fuel and chloride content of the aqueous heel. With regard to aviation turbine fuel additives, both approved (Stadis® 450, DCI-4A, Turboline® FS100) and unapproved (Aerosol OT, Infineum R655) additives were investigated. Stadis 450 and DCI-4A are approved for use in aviation turbine fuel according to ASTM D1655 at maximum concentrations of 5 mg/L and 23 mg/L, respectively. Turboline FS100 is a high temperature fuel stabilizer and detergent additive approved for use in JP-8+100 military jet fuel. The recommended dosage of this material is 625 ppmv. At a dosage of 100 ppm, Turboline FS100 was found to cause rapid loss of coalescence in small-scale aviation turbine fuel rig testing.[8] Aerosol OT is a strong surfactant used to prepare reference fluids utilized in the D3948, D7224, and D8073 test methods. Infineum R655 is an ester-based lubricity improver for diesel fuel which was found to cause rapid loss of coalescence in aviation turbine fuel rig testing (both small and large scale) at a gross miss-additization dosage of 200 ppm.[9] "Gross miss-additization" signifies that the aviation turbine fuel was mistaken for diesel fuel and the full treat rate (recommended for diesel fuel) was added. The aqueous heels evaluated were ASTM D1193 Type II water (5 µg/L maximum chlorides), ASTM D1141 synthetic seawater without heavy metals, a 50/50 vol% mixture of Type II water and synthetic seawater, and Type II water containing 500 mg/L NaCl. Type II water containing 500 mg/L NaCl was chosen due to its reference in the aforementioned IATA report.

Settling times up to and including 24 hours were investigated. According to EI/JIG 1530[10], aviation turbine fuel must be settled three hours per meter depth or 24 hours, whichever is less. Given the ever increasing demand for aviation turbine fuel, terminals/depots do not always have the ability to settle for longer periods of time and still meet supply commitments. Therefore, settling times in excess of 24 hours were not investigated in this study.

Aviation turbine fuel import retain samples were analyzed in a limited commercial survey. Samples were obtained from a refinery that has received up to three aviation turbine fuel cargoes each month for the past several years. Prior to discharging the vessel, composite samples are taken in three 5 L epoxy lined cans. One can is used to recertify the aviation turbine fuel, another can is used for filtration testing, and the third can is kept as a retain sample. Only after acceptable recertification and filtration test results are obtained is the vessel discharged. Retain samples are normally kept for 90 days and then discarded. For this survey, at the end of 90 day period these samples were shipped to the ExxonMobil Research Laboratory in Paulsboro, NJ for analysis. No attempt is made to determine the source of chloride as this would require cation determination which is outside the scope of this study.

RESULTS AND DISCUSSION

The initial test procedure utilized follows.

Initial Test Procedure

1. Add 500 mL of aqueous heel to a 4 L Nalgene® graduated cylinder. The cylinder is approximately 24" high with an approximate 4" internal diameter.
2. Fill to the 4 L mark with the aviation turbine fuel to be evaluated.
3. Add a coated magnetic stir bar (see Figure 3) and mix at a setting of 8.5 for 10 minutes. A Corning® model PC-11 magnetic stirplate was used for this study. The magnet may not remain centered throughout the 10 minutes. If this happens, pause the timer, turn off the stirplate and allow magnetic stir bar to center. Increase stir speed to 8.5 over approximately 10 seconds. Continue the timer countdown.
4. After 10 minutes, turn off the stirplate, remove the graduated cylinder from the stirplate platform and remove the stir bar using a plastic coated, metal tipped wand.
5. Cover the top of the graduated cylinder with Parafilm® and leave undisturbed for a prescribed period of time (default being 24 hours).
6. After settling, siphon the test fuel leaving a 250 mL (approximately 3 cm) buffer of fuel above the heel. (Do not decant as this causes too much disturbance of the heel.) A straight length of ¼", 316 stainless steel (SS) tubing was used to siphon the fuel.

7. Conduct D7959 analysis, retaining fuel in excess of 3 L (~250 mL) in a glass bottle.

Figure 3: Magnetic Stir Bar Utilized for Testing

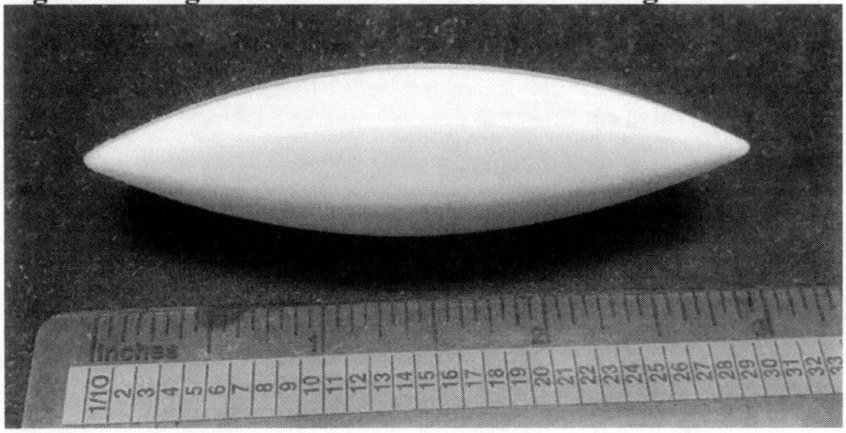

Phase I Testing

The impact of select additives on aviation turbine fuel chloride content was investigated in testing Phase I. Results, generated via the initial test procedure, are shown in Table 2. All testing was conducted with a synthetic seawater without heavy metals heel and a subsequent settling time of 24 hours. Results in Table 2 are color-coded. Any result which does not comply with either ASTM D1655 or CGSB-3.23 is colored red; any chloride result which exceeds the Def (Aust) 5240B limit of ≤0.091 mg/L is colored blue. The base aviation turbine fuel produced a low chloride result of 0.01 mg/L, indicating that the fuel is effectively shedding the synthetic seawater droplets after 24 hours. The addition of AOT to the fuel increases post-contact chloride content. Addition of 1.2 and 3.0 mg/L AOT increases chloride content to 0.07 and 0.25 mg/L, respectively. It is hypothesized that the presence of AOT causes smaller synthetic seawater droplets to form which settle at a lower rate per Stokes' Law. Although outside the scope of this study, particle counting would help to confirm this hypothesis and will be considered for future testing. The presence of AOT also had an impact on D3948 micro-separometer test results post synthetic seawater contact. Addition of 1.2 and 3.0 mg/L AOT caused D3948 results to decease to 80 and 54, respectively. The D1655 limit for aviation turbine fuel without electrical conductivity additive is 85 minimum.

Synthetic seawater contact appeared to be extracting some amount of AOT from the aviation turbine fuel sample. Table 3 of the D3948 method provides expected performance with various AOT concentrations (dispersing agent in mL/L is equivalent to AOT concentration on a mg/L basis). The maximum AOT concentration listed in D3948 is 0.8 mg/L and the maximum expected result at this concentration is 77. The post synthetic seawater contact D3948 result of 80 obtained in Phase I with 1.2 mg/L AOT exceeded the maximum expected result listed in the D3948 table for 0.8 mg/L AOT.

Stadis 450 and DCI-4A were evaluated both separately and in combination in Phase I. Stadis 450 was evaluated at a typical concentration of 1.5 mg/L as well as the maximum permitted concentration of 5.0 mg/L. The chloride content of the fuel post-contact appeared to be impacted by Stadis 450 concentration. At a Stadis 450 concentration of 5.0 mg/L, the chloride content of the fuel post synthetic seawater contact was 0.14 mg/L, exceeding the Def (Aust) 5240B limit.

The test fuel, however, is not commercially viable since it exceeded ASTM D1655 and CGSB-3.23 conductivity limits of 600 pS/m maximum and the D3948 limit of 70 specified for fuel with electrical conductivity additive in D1655. By contrast, DCI-4A did not appear to impact the chloride content of the fuel. At the maximum permitted DCI-4A concentration of 23 mg/L, chloride content of the test fuel post synthetic seawater contact was just 0.01 mg/L. The combination of Stadis 450 and DCI-4A however appeared to be antagonistic with regard to fuel chloride content post synthetic seawater contact. Combining the maximum permitted concentration of Stadis 450 with the maximum permitted concentration of DCI-4A produced a fuel chloride content of 0.50 mg/L. This is the highest chloride result obtained in all phases of testing following a 24 hour settling period (higher results were obtained, but with a shorter settling period). This particular test fuel again is not commercially viable since it produced an unacceptably high conductivity and an unacceptably low D3948 result. Finally, two different intermediate combinations of Stadis 450 and DCI-4A were evaluated in an attempt to produce a test fuel that met industry requirements for D2624 and D3948 while exceeding the Def (Aust) 5240B specification for fuel chloride content. Neither combination met this requirement. A combination of 1.0 mg/L Stadis 450 with 8.0 mg/L DCI-4A produced a borderline D3948 result of 68 (i.e. borderline with regard to a 70 minimum limit) but a chloride result of only 0.07 mg/L. The combination of 1.5 mg/L Stadis 450 with 16.0 mg/L DCI-4A produced a chloride result of 0.23 mg/L (exceeding the Def (Aust) 5240B limit) and an acceptable conductivity but the D3948 result was quite low at 53.

Conductivity does not appear to be viable as a substitute control for elevated chloride levels. AOT concentration was found to impact chloride level but AOT had little impact on conductivity. In addition, test fuels at the high end of the conductivity scale produced highly variable chloride results. The ASTM D1655 and CGSB-3.23 conductivity limits of 600 pS/m maximum offered no protection from high fuel chloride levels. Examples can be found in Table 2 in which test fuels either met or greatly exceeded the ASTM D1655 and CGSB-3.23 conductivity limit of 600 pS/m maximum and produced chloride concentrations in excess of 0.091 mg/L.

Table 2: Phase I Results Using Initial Test Procedure

						Chloride, mg/L (D7959)
						syn. seawater heel
				Cond., pS/m	Alumicel	24 hr settling
Midas #	AOT, mg/L	Stadis 450, mg/L	DCI-4A, mg/L	D2624	D3948	8.5 mixer for 10 min
16-94075	- -	- -	- -	14 @ 22.7°C	98	0.01
16-51555	1.2	- -	- -	8 @ 22.3°C	80	0.07
16-58621	3.0	- -	- -	9 @ 23.7°C	54	0.25
16-51562	- -	1.5	- -	405 @ 21.8°C	56	0.05
16-58625	- -	5.0	- -	1,669 @ 22.7°C	56	0.14
16-58624	- -	- -	23.0	2 @ 23.2°C	99	0.01
16-58626	- -	5.0	23.0	1,752 @ 23.7°C	0	0.50
16-94073	- -	1.0	8.0	380 @ 22.7°C	68	0.07
16-94065	- -	1.5	16.0	596 @ 23.2°C	53	0.23
Notes:						
Red text signifies non-compliance with D1655 or CGSB-3.23						
Blue test signifies non-compliance with Def Aust 5240 B						

In Phase I, no test fuel produced a chloride content which exceeded the Def (Aust) 5240B limit while meeting the applicable D3948 limit of either 70 or 85 depending upon the presence of electrical conductivity additive. Results are shown in Figure 4. The upper right-hand quadrant formed by the intersection of the dashed red lines is empty. D3948 is one of three water separation bench tests applicable to aviation turbine fuel. D7224 is similar to D3948 but utilizes a different test coalescer cartridge (i.e. MCell vs. Alumicel). D7224 is specified in CGSB-3.23 and is reportedly less impacted by the presence of electrical conductivity additive. D8073 is a relatively new methodology which determines water separation characteristics of aviation turbine fuel via the detection of a water soluble dye downstream of a coalescing cartridge. The relationship of D7224 and D8073 to chloride content is also of interest, but in order to evaluate all three water separation bench tests a larger sample size following aqueous heel contacting was required.

Figure 4: Relationship between Aviation Turbine Fuel MSEP and Chloride Content

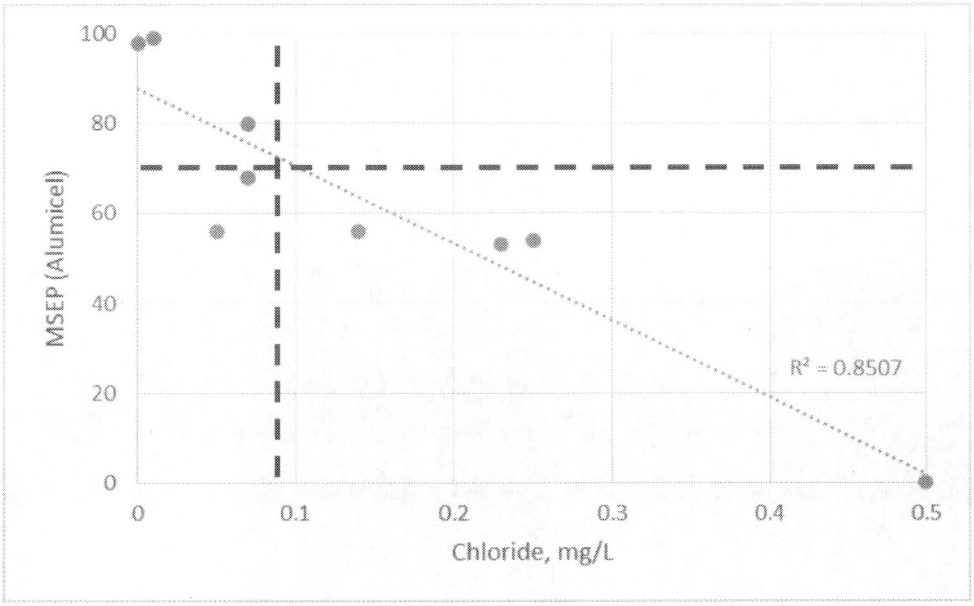

In order to generate a larger sample after aqueous heel contact, the following modified test procedure was utilized.

Modified Test Procedure

1. Add 250 mL of aqueous heel to a 4 L Nalgene graduated cylinder.
2. Fill to ~4.25 L with the aviation turbine fuel to be evaluated. A line, placed 3 cm above the 4 L mark on the graduated cylinder with a marker, served as the ~4.25 L reference point.
3. Stabilize the graduated cylinder on the magnetic stir plate platform as shown in Figure 5.
4. Add a coated magnetic stir bar and mix at the maximum setting for 10 minutes. The magnet may not remain centered throughout the 10 minutes. If this happens, pause the timer, turn off the stirplate and allow magnetic stir bar to center. Increase stir speed to maximum over approximately 10 seconds. Continue the timer countdown.
5. After 10 minutes, turn off the stirplate, remove the graduated cylinder from the stirplate platform and remove the stir bar using a plastic coated, metal tipped wand.
6. Cover the top of the graduated cylinder with Parafilm and leave undisturbed for a prescribed period of time (default being 24 hours).
7. After settling, siphon the test fuel leaving a 250 mL (approximately 3 cm) buffer of fuel above the heel. (Do not decant as this causes too much disturbance of the heel.) Siphon directly into a 1-gal epoxy lined can (to minimize transfer steps) using a "fishhook" wand with vacuum controlled at 8 in Hg. The wand was prepared from ¼", 316 SS tubing. The length of the wand was approximately 26 ½" from the top of the nut union to the bottom of the 180° bend. The length from the bottom of the 180° bend to the suction point was approximately 2 ¼". The 180° bend was intended to further reduce any disturbance between the fuel and aqueous heel. The vacuum set-up and "fishhook" wand are shown in Figures 6 and 7, respectively.
8. Conduct D7959 analysis, retaining fuel in excess of 3 L (~750 mL) in a glass bottle.

Figure 5: Graduated Cylinder Secured to Magnetic Stirplate Platform

Figure 6: Vacuum Set-Up

Figure 7: "Fishhook" Siphon

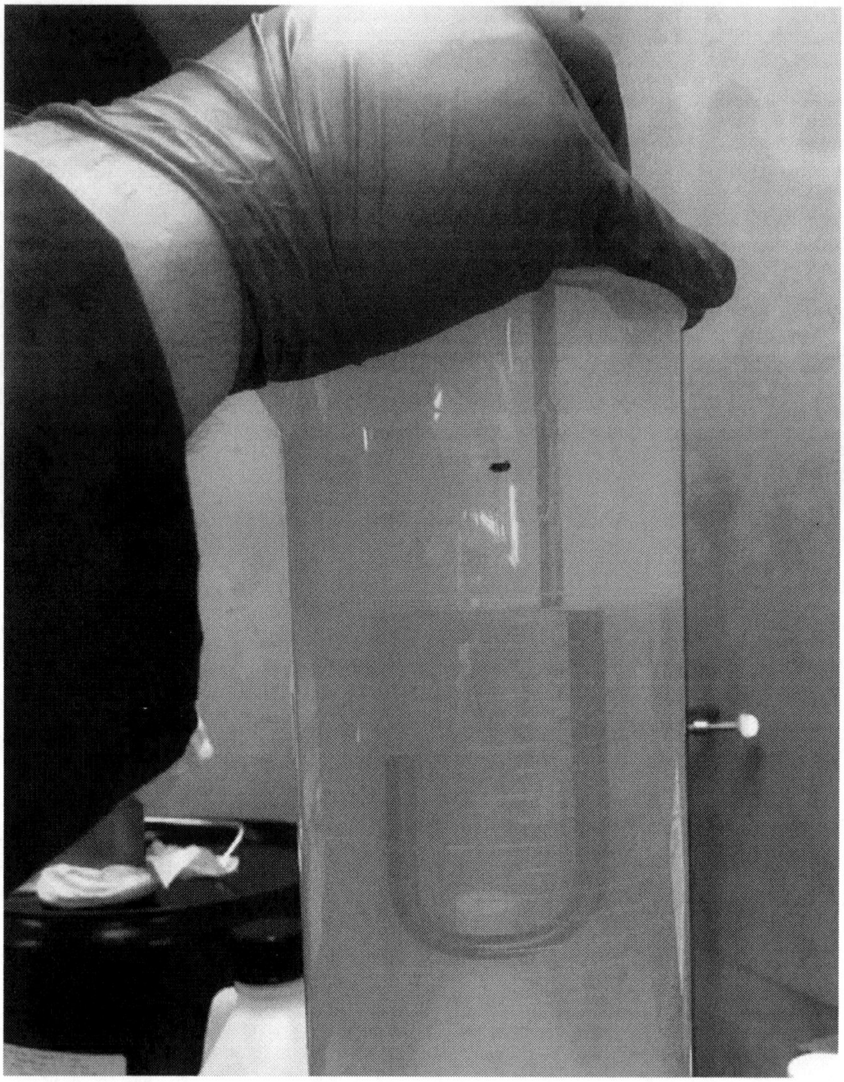

Comparison of Initial and Modified Test Procedures

A comparison of chloride test results obtained using the initial and modified test procedures is provided below in Table 3. The comparison was felt necessary to determine whether or not observations from Phase I testing would still be germane or capable of being read-across to subsequent phases of testing. In all cases, the test fuel was additized with 3.0 mg/L AOT and settling time was 24 hours. A description of "500 mL heel, fill to 4 L" in Table 3 represents data obtained via the initial test procedure; likewise, a description of "250 mL heel, fill to 4.25 L" in Table 3 represents data obtained via the modified test procedure.

There was good agreement in the chloride results obtained from both the initial and modified test procedures. Chloride results from the original test procedure ranged from 0.10 to 0.35 mg/L and chloride results from the modified test procedure ranged from 0.19 to 0.35 mg/L. Results are plotted in Figure 8. The modified test procedure actually produced a tighter grouping of chloride results. Based on this data, the test procedures are considered roughly equivalent. The average chloride result for all runs (i.e. both original and modified procedures) was 0.23 mg/L.

Water separation bench test results from all three water separation methodologies were poor as expected. The D8073 method has not yet been adopted for either the D1655 or CBSB specifications, so those results are presented without color-coding.

Table 3: Comparison of Chloride Results from Initial and Modified Test Procedures

		Chloride, mg/L	Alumicel	MCell	WSI	Cond., pS/m
Midas #	Description	D7959	D3948	D7224	D8073	D2624
16-58621	Jet plus 3.0 mg/L AOT post syn. seawater (500 mL heel, fill to 4 L)	0.25	54	- -	- -	9 @ 23.7°C
17-49335	Jet plus 3.0 mg/L AOT	- -	54	44	58.1	31 @ 22.3°C
17-49337	... post synthetic seawater contact (500 mL heel, fill to 4 L)	0.35	61	- -	- -	32 @ 22.3°C
17-49338	... post synthetic seawater contact (250 mL heel, fill to 4.25 L)	0.19	55	52	59.8	33 @ 21.8°C
17-49339	... post synthetic seawater contact (500 mL heel, fill to 4 L)	0.13	72	- -	- -	24 @ 22.3°C
17-49340	... post synthetic seawater contact (250 mL heel, fill to 4.25 L)	0.23	52	52	57.4 HW	33 @ 22.3°C
17-51134	Jet plus 3.0 mg/L AOT	- -	52	0	55.8	27 @ 24.1°C
17-51155	... post synthetic seawater contact (500 mL heel, fill to 4 L)	0.32	63	- -	- -	17 @ 24.1°C
17-51164	... post synthetic seawater contact (250 mL heel, fill to 4.25 L)	0.35	56	47	53.5 HW	30 @ 23.7°C
17-65786	Jet plus 3.0 mg/L AOT	- -	55	0	54.7	31 @ 22.7°C
17-65792	... post synthetic seawater contact (500 mL heel, fill to 4 L)	0.10	- -	53	- -	33 @ 22.7°C
17-65793	... post synthetic seawater contact (250 mL heel, fill to 4.25 L)	0.19	50	48	57.8	40 @ 22.7°C
17-65795	... post synthetic seawater contact (500 mL heel, fill to 4 L)	0.23	59	- -	- -	33 @ 22.7°C
17-65796	... post synthetic seawater contact (250 mL heel, fill to 4.25 L)	0.23	48	49	56.6 HW	40 @ 22.7°C
	Avg	0.23				
Notes:						
HW = high water alarm						
Red text signifies non-compliance with D1655 or CGSB-3.23						
Blue test signifies non-compliance with Def Aust 5240 B						

Figure 8: Comparison of Chloride Results from Initial and Modified Test Procedures

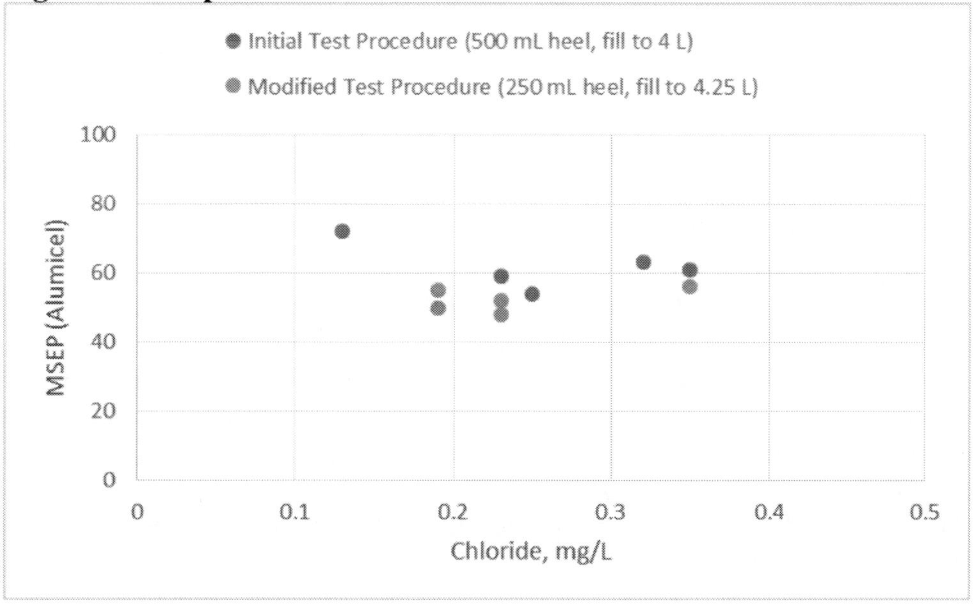

Impact of Settling Time

Settling time was found to impact chloride results. That is, as settling time increased, chloride results decreased. Results (obtained using the modified test procedure) are provided in Table 4.

For this testing, a larger 5 gallon batch of test fuel containing 3.0 mg/L AOT was prepared in an epoxy-lined can to eliminate variation in AOT dosage as a variable. From this 5 gallon batch of test fuel, four Nalgene graduated cylinders were prepared for evaluation, with sufficient remaining test fuel for baseline water separation and conductivity bench testing (i.e. prior to seawater contact). The highest chloride result obtained was 0.92 mg/L from the sample that was allowed to settle for 4 hours after synthetic seawater contact. Samples which settled for both 2 and 8 hours produced identical chloride results of 0.68 mg/L. Chloride concentration is not believed to peak after 4 hours of settling; the difference between chloride concentrations of 0.92 and 0.68 mg/L is believed to be within test variability. At 16 hours settling the chloride concentration decreased to 0.41 mg/L. For reference, the average chloride result obtained after 24 hours settling (from Table 3) was 0.23 mg/L. As mentioned above, settling time was limited to a maximum of 24 hours due to the ever increasing demand for aviation turbine fuel and inability of terminals/depots to retain the fuel for longer periods of time while still meeting supply commitments.

Water separation bench test results from all three methodologies were poor as expected with one exception. A relatively high D8073 rating of 84.3 was obtained on the sample which was settled for 8 hours. This appears to be an anomalous result.

Table 4: Impact of Settling Time on Chloride Results (Modified Test Procedure)

		Chloride, mg/L	Alumicel	MCell	WSI	Cond., pS/m
Midas #	**Description**	**D7959**	**D3948**	**D7224**	**D8073**	**D2624**
17-49336	Jet plus 3.0 mg/L AOT	- -	51	44	53.1	30 @ 24.1°C
17-50606	... post synthetic seawater contact (2 h settling)	0.68	55	50	51.9 HW	31 @ 24.1°C
17-50607	... post synthetic seawater contact (4 h settling)	0.92	63	51	55.3 HW	36 @ 24.1°C
17-50608	... post synthetic seawater contact (8 h settling)	0.68	54	51	84.3	37 @ 24.1°C
17-50609	... post synthetic seawater contact (16 h settling)	0.41	56	46	59.1	40 @ 24.1°C
	Average of 24 h settling runs	0.23				
Notes:						
HW = high water alarm						
Red text signifies non-compliance with D1655 or CGSB-3.23						
Blue test signifies non-compliance with Def Aust 5240 B						

Impact of Fuel Additives and Aqueous Heels

In the next series of experiments, the impact of various fuel additives and aqueous heels was explored. As in the previous phase of testing, larger 5 gallon batches of test fuel were prepared in epoxy-lined cans to eliminate variation in additive dosage as a variable. From each 5 gallon batch of test fuel, four Nalgene graduated cylinders were prepared for evaluation, with sufficient remaining test fuel for baseline water separation and conductivity bench testing (i.e. prior to seawater contact). A different aqueous heel was placed in the bottom of each graduated cylinder (i.e. ASTM D1193 Type II water, ASTM D1141 synthetic seawater without heavy metals, a 50/50 vol% mixture of Type II water and synthetic seawater, and Type II water containing 500 mg/L NaCl). Results obtained using the modified test procedure and 24 hour settling are provided in Table 5.

The 500 mg/L NaCl aqueous solution produced a chloride result close to its calculated value of 303 mg/L. Since the aqueous solution was within the detection range of the Quantab Low Range test strips (i.e. 30 – 600 mg/L), an analysis was conducted to provide added confidence that the solution was correctly prepared. The test strip from that analysis is shown in Figure 9. A test strip reading of 5.6 corresponds to a chloride concentration of 277 mg/L (see Table 1), reasonably close to the calculated value.

Figure 9: Quantab Chloride Result for ASTM Type II Water Plus 500 mg/L NaCl

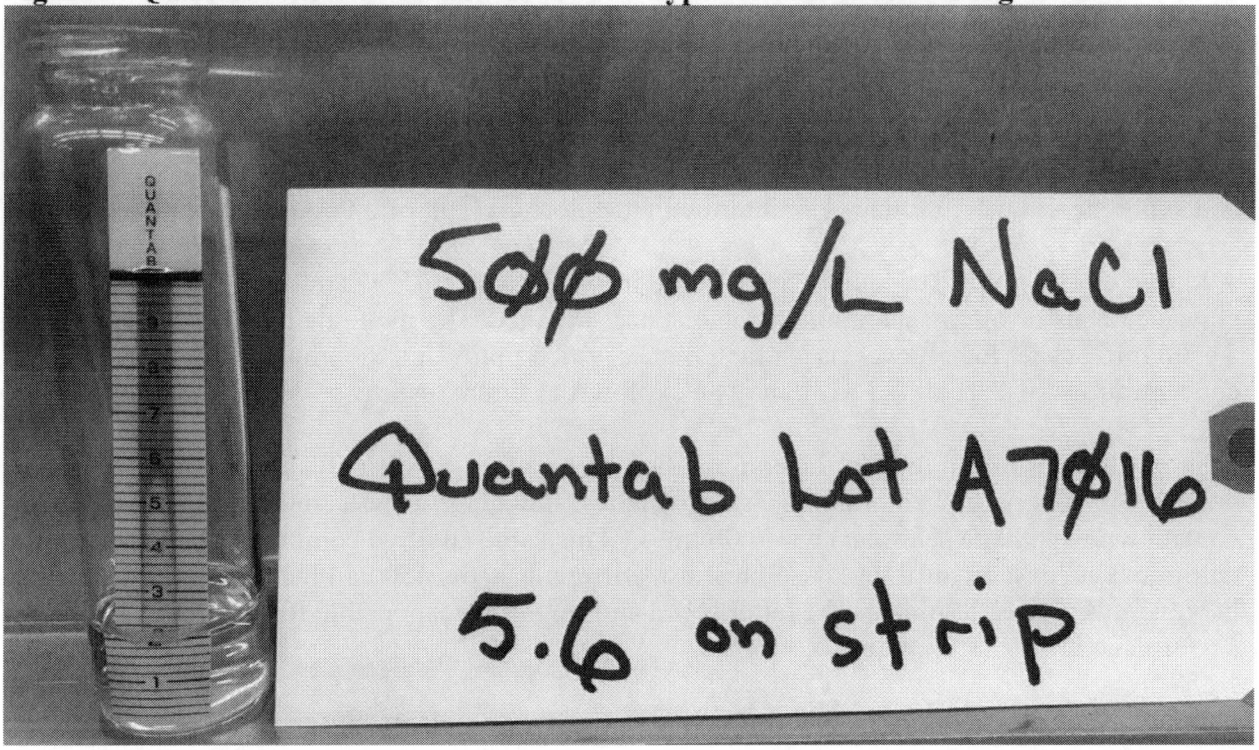

Base aviation turbine fuel (without additives) produced low chloride and conductivity levels and very high and acceptable water separation bench test results as expected. Although the chloride results were low, the salinity of the aqueous heel did appear to impact results. No chloride was detected in the fuel following contact with either Type II water or Type II water containing 500 mg/L NaCl. A chloride content of 0.06 mg/L was measured following contact with synthetic seawater; a chloride content of 0.03 mg/L was measured following contact with the 50/50 vol% mixture of Type II water and synthetic seawater.

The impact of aqueous heel salinity on fuel chloride content was most noticeable with fuel containing 3.0 mg/L AOT. Very low chloride levels were detected in the fuel following contact with either Type II water or Type II water containing 500 mg/L NaCl. A chloride content of 0.30 mg/L was measured following contact with synthetic seawater. Slightly less than half this chloride concentration (i.e. 0.12 mg/L) was measured following contact with the 50/50 vol% mixture of Type II water and synthetic seawater.

Based on the water separation bench test results, aqueous heel contacting appeared to be extracting AOT from the test fuel. This was most noticeable when contacting with either Type II water or Type II water containing 500 mg/L NaCl (i.e. lower salinity levels). The test fuel prior to aqueous heel contacting produced D3948, D7224 and D8073 results of 0, 43, and 55.3, respectively. The test fuel after contacting with Type II water produced elevated D3948, D7224 and D8073 results of 74, 76, and 67.7, respectively, indicative of a fuel containing a lower concentration of AOT. Surprisingly, the water separation bench test results were even higher after contacting with Type II water containing 500 mg/L NaCl. Water with a low (but greater than nil) salinity content appeared to be aiding extraction of AOT from the fuel; there is not currently enough data however to determine definitively.

Addition of 200 ppmv Infineum R655 to the fuel decreased D3948 water separation bench test results but had little impact on fuel chloride levels. Surprisingly, the additive had little impact on either D7224 or D8073 performance. The highest fuel chloride level of 0.08 mg/L was obtained following contact with synthetic seawater, slightly higher than the comparable result produced with base fuel (i.e. 0.06 mg/L). Here again, contact with the 50/50 vol% mixture of Type II water and synthetic seawater produced an intermediate chloride result of 0.06 mg/L.

Addition of 100 ppmv Turboline FS100 (similar to Infineum R655) had little impact on fuel chloride levels or water separation bench test performance. The treat rate approved for use in JP-8+100 military jet fuel, however, is much higher at 625 ppmv and the impact of higher concentrations of Turboline FS100 may be explored in future testing.

Contrary to Phase I test results, the combination of 1.5 mg/L Stadis 450 and 16 mg/L DCI-4A had little impact on fuel chloride levels. The highest fuel chloride level measured (i.e. following contact with synthetic seawater) was 0.06 mg/L. This same additive combination produced a chloride level of 0.23 mg/L in Phase I testing. Different Stadis 450 and DCI-4A additive batches, however, were used to prepare the latter blend and might be responsible for the difference in performance.

Unusually low conductivity results were obtained on the sample of fuel containing 1.5 mg/L Stadis 450 and 16 mg/L DCI-4A after contact with the 50/50 vol% mixture of Type II water and synthetic seawater. Two conductivity measurements were made and both were low compared to the other samples in this group. The cause of these low conductivity results is not apparent.

Table 5: Impact of Aviation Turbine Fuel Additives and Heel on Chloride Results (Modified Test Procedure)

		Chloride, mg/L	Alumicel	MCell	WSI	Cond., pS/m
Midas #	**Description**	**D7959**	**D3948**	**D7224**	**D8073**	**D2624**
17-63808	Base Jet	- -	100	99	93.1	4 @ 21.8°C
17-63850	... post ASTM Type II water contact	0.00	100	99	93.4	14 @ 22.7°C
17-63812	... post synthetic seawater contact	0.06	100	100	97.4	5 @ 21.3°C
17-63842	... post 500 mg/L NaCl aqueous contact	0.00	98	99	95.8	13 @ 22.7°C
17-63860	... post 50/50 mix of Type II water and syn. seawater	0.03	100	99	97.0	12 @ 23.4°C
17-51847	Jet plus 3.0 mg/L AOT	- -	0	43	55.3	40 @ 24.1°C
17-51855	... post ASTM Type II water contact	0.02	74	76	67.7	18 @ 25.1°C
17-51859	... post synthetic seawater contact	0.30	46	45	56.8	29 @ 25.1°C
17-51864	... post 500 mg/L NaCl aqueous contact	0.03	82	88	75.7	17 @ 25.1°C
17-51865	... post 50/50 mix of Type II water and syn. seawater	0.12	63	65	62.7	21 @ 24.1°C
17-61573	Jet plus 200 ppmv Infineum R655	- -	81	97	92.6	12 @ 22.7°C
17-63898	... post ASTM Type II water contact	0.01	68	96	87.8	13 @ 22.3°C
17-63927	... post synthetic seawater contact	0.08	76	93	94.9	15 @ 22.3°C
17-63922	... post 500 mg/L NaCl aqueous contact	0.01	82	95	93.5	15 @ 22.3°C
17-63946	... post 50/50 mix of Type II water and syn. seawater	0.06	79	96	91.5	5 @ 21.3°C
17-63971	Jet plus 100 ppmv Turboline FS100	- -	98	99	92.6	12 @ 21.8°C
17-64010	... post ASTM Type II water contact	0.00	98	98	93.3	15 @ 21.8°C
17-64004	... post synthetic seawater contact	0.07	97	96	94.5	15 @ 21.3°C
17-64013	... post 500 mg/L NaCl aqueous contact	0.00	95	98	93.3	15 @ 21.3°C
17-64014	... post 50/50 mix of Type II water and syn. seawater	0.05	96	96	93.2	15 @ 21.3°C
17-65451	Jet plus 1.5 mg/L Stadis 450 & 16 mg/L DCI-4A	- -	65	93	95.7	654 @ 22.7°C
17-65453	... post ASTM Type II water contact	0.01	60	86	98.1	749 @ 22.7°C
17-65455	... post synthetic seawater contact	0.06	58	88	97.2	658 @ 22.7°C
17-65454	... post 500 mg/L NaCl aqueous contact	0.00	75	89	96.8	761 @ 22.7°C
17-65452	... post 50/50 mix of Type II water and syn. seawater	0.04	65	90	98.2	262 @ 21.8°C, 488 @ 22.3°C
	Notes:					
	Red text signifies non-compliance with D1655 or CGSB-3.23					
	Blue test signifies non-compliance with Def Aust 5240 B					

Survey of Import Retain Samples

Twenty percent of the aviation turbine fuel import retain samples (or 13 samples) from the limited commercial survey exceeded a chloride concentration of 0.09 mg/L. A total of 65 samples have been analyzed to date. The four highest results obtained were 1.10, 0.54, 0.51 and 0.51 mg/L chloride.

A number of the import retain samples which produced elevated chloride results also produced very high and acceptable water separation bench test results. Bench testing was conducted on most of the import retain samples which exceeded the Def (Aust) 5240B chloride limit. A few samples, however, were either discarded or not available in sufficient quantity for additional bench testing. Results are provided in Table 6. One sample (from Vessel #8) failed to meet the D1655 or CGSB requirement for water separation performance. (Based on conductivity, it is assumed that this sample did not contain static dissipator additive.). This result is surprising since, in theory, this sample met recertification test requirements upon arrival at the refinery which includes water separation performance. It is also curious that the D8073 WSI result was much higher than either the D3948 or D7224 result for this sample.

For the import retain samples, water separation bench test performance did not appear to be a viable substitute control for elevated fuel chloride content. Results are plotted in Figure 10. Unlike the earlier laboratory testing, the majority of the points from these commercial samples appear in the upper right-hand quadrant formed by the intersection of the dashed red lines.

Table 6: Results from Refinery Aviation Turbine Fuel Import Retain Samples

		Chloride, mg/L	Alumicel	MCell	WSI	Cond., pS/m
Midas #	Vessel	D7959	D3948	D7224	D8073	D2624
17-8084	#1	0.51	99	97	97.4	8 @ 22.3C
17-8076	#2	0.51	- -	- -	- -	47 @ 22.7C
17-8082	#3	0.47	99	98	88.4	4 @ 22.7C
15-64506	#4	0.25	101	100	96.5	4 @ 21.8C
17-8060	#5	0.23	97	97	96.7	30 @ 22.7C
15-64546	#6	0.15	97	93	97.9	3 @ 21.8C
17-8087	#7	0.13	- -	- -	95.8	10 @ 21.8C
15-64592	#8	0.10	58	76	89.6	43 @ 21.8C
15-64576	#9	0.10	100	99	97.1	20 @ 22.7C
Notes:						
Red text signifies non-compliance with D1655 or CGSB-3.23						

Figure 10: Relationship Between Aviation Turbine Fuel MSEP and Chloride Content for Refinery Import Retain Samples

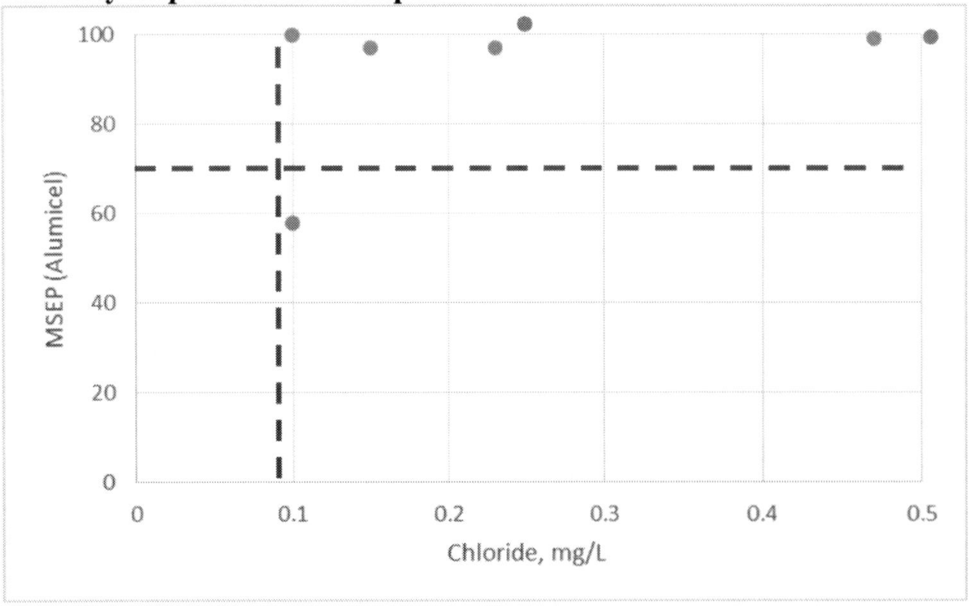

CONCLUSIONS

In Phase I testing consisting of synthetic seawater contact and subsequent 24 hour settling, AOT and Stadis 450 concentration of the fuel correlated to increased fuel chloride concentration. Although DCI-4A alone had little impact, the combination of Stadis 450 and DCI-4A appeared to be antagonistic with regard to fuel chloride content.

The chloride content of the aqueous heel was found to correlate to the chloride content of the fuel after contacting and settling. Contact with synthetic seawater (containing approximately 19,000 ppmw chloride) produced the highest fuel chloride results in this study. There was little difference in fuel chloride concentration after contacting with either Type II water or Type II water containing 500 mg/L NaCl.

As expected, the chloride content of fuel was found to decrease with increased settling time. Due to the lack of commercial practicality, settling times in excess of 24 hours were not investigated.

Neither conductivity nor water separation bench test performance appeared to be viable as a substitute control of aviation turbine fuel chloride content. In laboratory testing, both AOT (which has little impact on conductivity) and Stadis 450 (designed to increase conductivity) were found to impact fuel chloride levels. A number of samples from a limited import retain survey, produced both high chloride results (i.e. in excess of 0.091 mg/L) and excellent water separation bench test performance, effectively eliminating this property as a possible control for fuel chloride content.

REFERENCES

[1] "Guidelines for Sodium Chloride Contamination Troubleshooting and Decontamination of Airframe and Engine Fuel Systems", International Air Transport Association, 2nd Ed., February 1998, pg. 1

[2] Aviation Fuels, Maxwell Smith, 1970, pgs. 463-465

[3] "The Design and Operation of Salt Towers for Drying Petroleum Products", E. I. Du Pont De Nemours & Company, File No. 5509, December 1960, pg. 6

[4] HM 50 "Guidelines for the Cleaning of Tanks and Lines for Marine Tank Vessels Carrying Petroleum and Refined Products", 3rd edition, March 2014

[5] EI/JIG Standard 1530, "Quality Assurance Requirements for the Manufacture, Storage and Distribution of Aviation Fuels to Airports", 1st edition, October 2013

[6] The Oceans Their Physics, Chemistry, and General Biology, Sverdrup, Johnson, and Fleming, 1942, Table 33

[7] ASTM D7959-15 Standard Test Method for Chloride Content Determination, Section 3

[8] SwRI 10th International Filtration Conference, "Response of Micro-Separometer (MSEP) Cartridges to Different Jet Fuel Surfactants", Paul P. Wells, Sept 2010

[9] Institute of Petroleum Research Report, "Investigation Into the Effects of Lubricity Additives on the Performance of Filter/Water Separators", Daniel E. Kadlecek, March 2003

[10] EI/JIG Standard 1530, "Quality Assurance Requirements for the Manufacture, Storage and Distribution of Aviation Fuels to Airports", 1st edition, October 2013, Section 8.4.3

IASH 2017, the 15TH INTERNATIONAL SYMPOSIUM ON STABILITY, HANDLING AND USE OF LIQUID FUELS
Rome, Italy
10-14 September 2017

APPLICATION OF MATRIX-ASSISTED LASER DESORPTION/IONIZATION MASS SPECTROMETRY TO THE CHARACTERIZATION OF HIGH MOLECULAR WEIGHT FUEL CONTAMINANTS

Thomas N. Loegel, Jeffrey A. Cramer, and Iwona A. Leska*

Naval Research Laboratory, 4555 Overlook Ave. SW, Washington, D.C., USA 20375
** NOVA Research Inc., 1900 Elkin Street, Suite 230, Alexandria, VA 22308*

ABSTRACT

During even routine fuel handling operations, it is possible to introduce high molecular weight (HMW) contaminants, such as lubricants and greases, to fuels, thus contaminating them and negatively impacting their stability and performance. To address this fuel handling challenge, analytical methodologies based on high-performance liquid chromatography combined with an evaporative light scattering detector (HPLC-ELSD) have previously been developed in our laboratory to detect and quantify HMW contaminants in fuels. However, both the high molecular masses that are required to pass through the HPLC columns and the relatively limited selectivity of the ELSD data can interfere with attempts to identify the sources of HMW contamination using HPLC-ELSD methodologies. This presentation will cover the use of matrix-assisted laser desorption/ionization (MALDI) mass spectrometry, combined with a sampling technique adapted from the forensics literature, to characterize the high molecular masses collected from contaminated fuel samples more thoroughly.

INTRODUCTION

Previous Work. Nonpolar high molecular weight (NP-HMW) substances such as lubricants, greases, and hydraulic fluids can potentially appear in jet and diesel fuel supplies due to various supply chain and point-of-use failures. However, petrochemical fuels are already complex mixtures of organic compounds, thus detecting the existence of problematic organic compounds within them, let alone surmising potential points of origin, can be challenging. Significant progress in detecting NP-HMW contaminants has already been achieved using ultra-high performance liquid chromatography combined with evaporative light scattering detection (UHPLC-ELSD).[1] Specifically, the methodology developed utilizes a reverse phase liquid chromatography separation over oven-heated graphite columns, and was explicitly designed to, whenever possible, make use of commonly available, minimally-modified instrumentation and detection schemes as well as minimally-hazardous solvents. Chemometric techniques, such as partial least squares (PLS) and principal component analysis (PCA), can also be incorporated into the overall methodology to assess the underlying linear mathematical factors within data sets, allowing for at least some NP-HMW characterization capabilities, including quantification.[2]

Despite these successes, however, analysis challenges can arise due to the fundamental lack of chemical-specific information in ELSD spectra. An example of such a challenge is shown in Figure 1. Two polyol ester-based lubricants, each blended separately and together with two different jet fuels, are assessed via PCA modeling. In the resultant model, the samples possessing larger amounts of contamination fall further away from the control fuel sample in directions consistent with which of the two lubricants are present, as would be expected, and these trends remain consistent across many replicates, indicating their underlying robustness. However, this model also interprets a third set of replicates, in which a different polyol ester-based lubricant contaminates a different jet fuel, as being very nearly identical to one of the other lubricants' sets of replicates, both quantitatively and with respect to which lubricant is predicted as being. Although the former comparison may be appropriate, as both sets of samples possess 0.1% lubricant, the latter conflation of the two lubricants would, in scenarios involving unknown samples, complicate definitive assessments of potential sources of contamination.

Matrix-Assisted Laser Desorption/Ionization (MALDI) Mass Spectrometry. MALDI involves the co-crystallization of a neat analyte, such as a contaminated fuel, and a laser-absorbing matrix into a single spot on a target plate. When this spot is desorbed by an appropriate laser, the matrix is actually desorbed, as opposed to the analyte itself, conveying a "soft" ionization to the analyte that can subsequently be used for time-of-flight measurements of mass/charge (m/z) ratios. This technique is typically used to characterize much larger molecules, such as proteins and peptides, but heavy lubricant and grease molecules can easily be accommodated as well.[3]

MALDI provides interesting advantages in the present context. Because the analyte molecules are not necessarily decomposed, the spectrometric data can convey straightforward and definitive characterization information of immediate chemical relevance. Also, because the molecules typically present in petrochemical fuels are, almost by definition, much smaller than the molecules found in NP-HMW contaminants, MALDI inherently minmizes the fuel itself as a potential interferent. The hypothesis thus put forth upon considering MALDI as an NP-HMW characterization strategy is that it can provide identification capabilities through an assessment of high mass ranges. Provided that different NP-HMW contaminants possess different chemical information at these high mass ranges, than this technique can function as an effective characterization complement to the UHPLC-ELSD technique indicated previously.

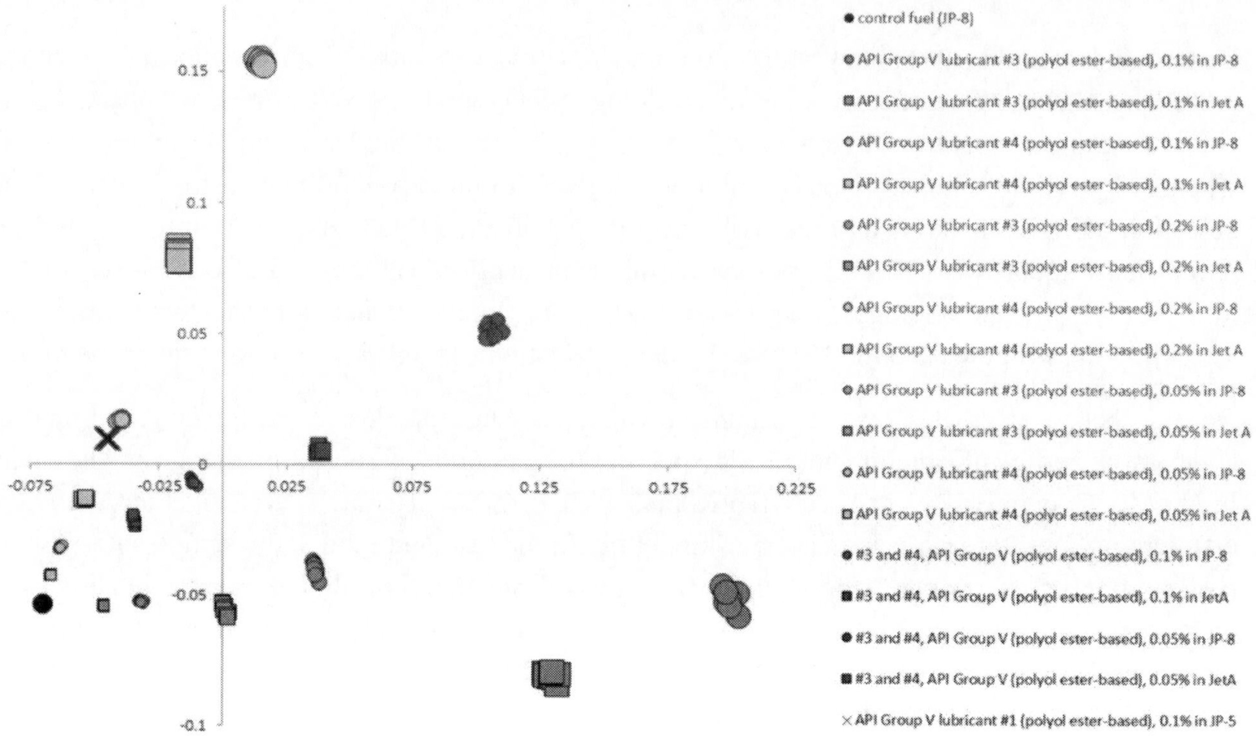

Figure 1. PCA score plot (PC2 vs. PC1) produced from the blends of three polyol ester-based lubricants with various jet fuels. Note that 0.1% samples of lubricant #1 and lubricant #3 might be mistaken for each other if one set of samples were of unknown origin.

Development of Matrix Application Strategy. Despite MALDI's theoretical potential, however, a strategy is required to allow for the crystallization of a usable sampling matrix around nonpolar fuel molecules that would resist a more standard polar solvent usage. The strategy thus developed makes use of a firm brushing of matrix into a small amount of a liquid sample, followed by the addition of the polar solvent into the resulting paste-like material. This forces a novel type of fuel/matrix co-crystallization suitable for MALDI analysis. The concept of this technique was originally derived from the analysis of fingerprints,[4] but was significantly streamlined because no spatial features need to be preserved in the present application.

EXPERIMENTAL

Lubricants, Greases, and Contaminated Fuel Samples. Generally speaking, manufacturers are reluctant to provide complete breakdowns of proprietary lubricant and grease compositions, but sample labeling has been sufficient to provide enough compositional information for the purposes of the present work. Two different API Group V polyol ester lubricants were used to initially determine if characteristic data features could distinctly separate superficially similar lubricants and also to determine which of two MALDI matrices might be more effective in the present context, all within the context of contaminated fuel samples. Also, a grease composed of fatty acid salts and/or esters and a hydrotreated mineral oil lubricant (either API Group II or III) were both used to further evaluate the effectiveness of the two MALDI matrices. Finally, a graphite-based grease was used to determine how the developed technique would respond to a contaminant that, at least superficially, would appear to lack the NP-HMW compositional components necessary for an effective analysis. To produce contaminated fuel samples when necessary, the indicated lubricants were blended with JP-5 at a concentration of 0.1%.

MALDI Data Collections. In order to determine which would be more suitable for the intended analysis, two matrices were used in the present work. The matrix α-cyano-4-hydroxycinnamic acid (CHCA) was used because it was known to be well-suited for use with analytes of less than 10,000 Daltons and available MALDI instrumentation. The matrix 7,7,8,8,-tetracyanoquinodimethane (TCNQ), meanwhile, was used because of previous results in the literature indicating its utility with a class of materials considered to be similar to lubricants and greases, i.e. insoluble carbonaceous pitches.[5] The solvent utilized for both matrices was a mixture of 30 ml acetonitrile, 10 ml H_2O , and 40 microliters trifluoroacetic acid. Multiple replicates, representing different spots on the overall plate, were produced to account for variances in spot preparation.

MALDI Instrument. The MALDI instrument used was an Applied Biosystems SCIEX 4700 Proteomics Analyzer. While the instrument's controlling computer has many post-collection analysis options geared towards the identification of specific peptides and proteins, data collected from the instrument were simply exported to an external computer for the present visual comparisons. Spectrometric data was collected from 200-1200 Daltons with a focus mass of 400 Daltons. To account for variances in sample spot preparation, each replicate's data is the sum of the time-of-flight spectra collected from 2,000 separate laser shots, with the laser's target within each sample spot moving after every 100 laser shots.

Software. Data were initially collected using the MALDI instrument's built-in 4700 Explorer software (version 2.0, Applied Biosystems) and converted to an exportable ASCII format using the similarly built-in Data Explorer software (version 4.5, Applied Biosystems). Data visualization algorithms were developed using functionalities provided by MATLAB (MathWorks, Inc., Natick, MA), the PLS_Toolbox for MATLAB (Eigenvector Research, Inc.), and the MATLAB-based Calibration and Standard Toolboxes (FABI / ChemoAC Consortium).

RESULTS AND DISCUSSION

Initial Proof Of Concept and Matrix Comparison, Polyol Ester-Based Lubricants in JP-5 Jet Fuel. Although the two lubricants are ostensibly of the same API grouping, the exact composition of a "polyol ester" is sufficiently broad that many types of lubricant compositions are possible. When the two lubricants, after being blended at a concentration of 0.1% in JP-5 jet fuel, were initially analyzed using CHCA matrix, mass spectral data were obtained that clearly indicated differences between them, as can be seen in Figure 2. The fact that these differences were seen across all collected replicates would seem to indicate how robust the detection of such differences can be in real-world scenarios.

Furthermore, when one replaces CHCA with TCNQ, these same features become more pronounced, as can be seen in Figure 3. This not only provides evidence that the features are due to the sample and are not artifacts of the matrix itself, but also indicates that TCNQ is a more reliably matrix choice in the present application.

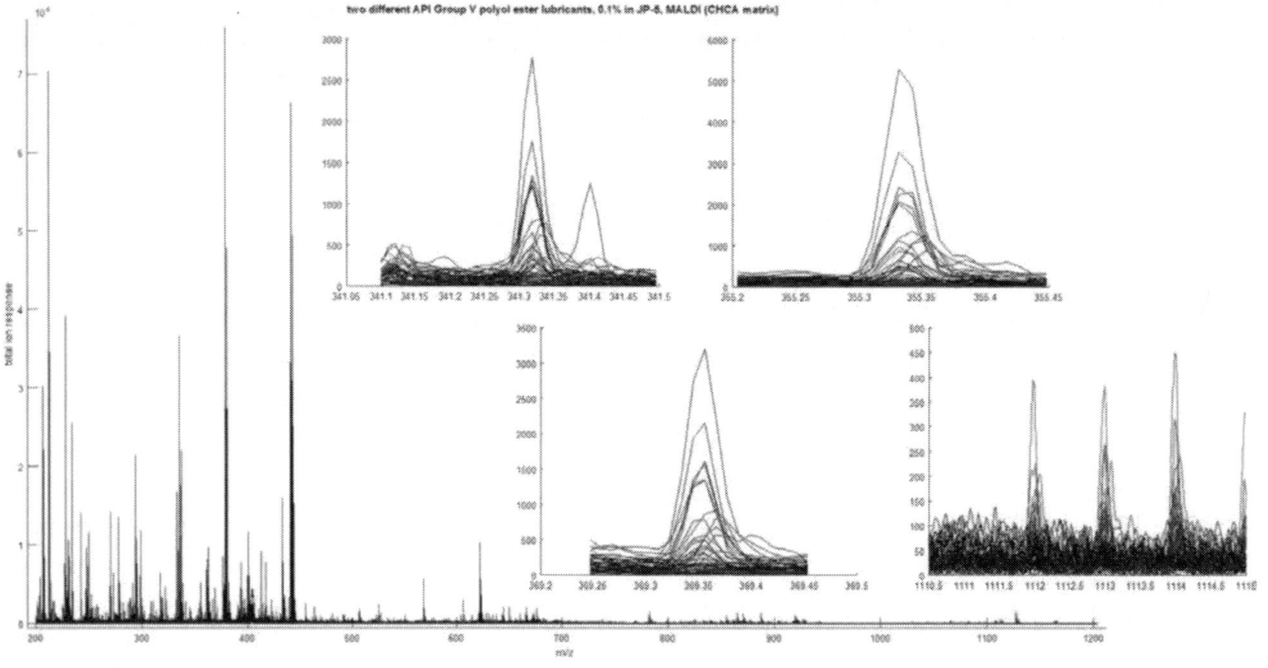

Figure 2. MALDI mass spectra collected from two different polyol ester-based lubricants using a CHCA matrix. One set of replicates is plotted in blue, the other in red. The insets indicate several mass ranges over which the replicates from one of the lubricants displays features that are not possessed by the other lubricant.

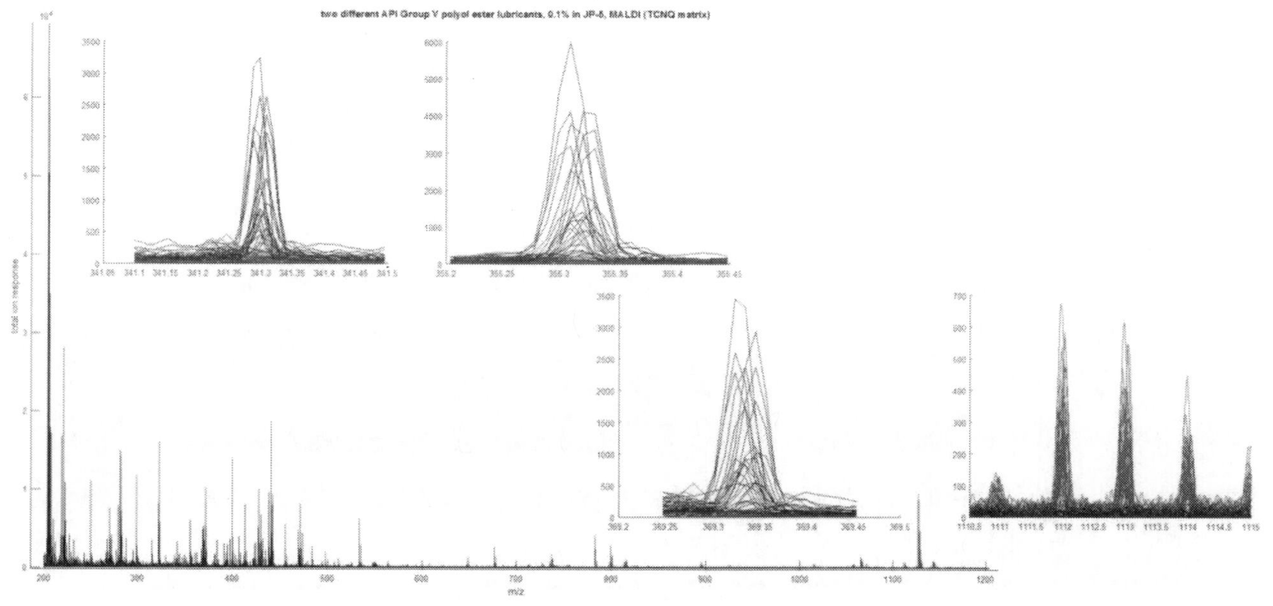

Figure 3. MALDI mass spectra collected from two different polyol ester-based lubricants using a TCNQ matrix. One set of replicates is plotted in blue, the other in red. The insets indicate several mass ranges over which the replicates from one of the lubricants displays features that are not possessed by the other lubricant.

Continued MALDI Matrix Comparisons With Fatty Acid Salts/Esters Grease And Hydrotreated Mineral Oil Lubricant. In an abstract sense, the most straightforward analysis strategy for NP-HMW contamination characterization would be to find the most prominent characteristic MALDI data features in any given neat HMW candidate, then attempt to find these features in the data collected from a contaminated fuel sample. This operation was thus first carried out using a fatty acid salts/esters grease and CHCA matrix, and the results can be found in Figures 4a and 4b. Figure 4a shows that a few characteristic data features do indeed make themselves apparent in the case of the grease. However, these same features seem to disappear entirely when considering this same grease when it is diluted in jet fuel. This could indicate that the grease simply isn't soluble in the fuel, but when TCNQ is used as the matrix instead, not only are the same data features present in the neat sample (Figure 5a), but they remain apparent in the fuel-diluted sample as well (Figure 5b). As was indicated previously, TCNQ appears to be a more effective matrix in the present context. Similar results can be found for, and similar conclusions can be drawn from, a hydrotreated mineral oil lubricant in Figures 6a, 6b, 7a, and 7b, with the primary difference being that CHCA shows at least an exceedingly minor capacity as a viable MALDI matrix in this case.

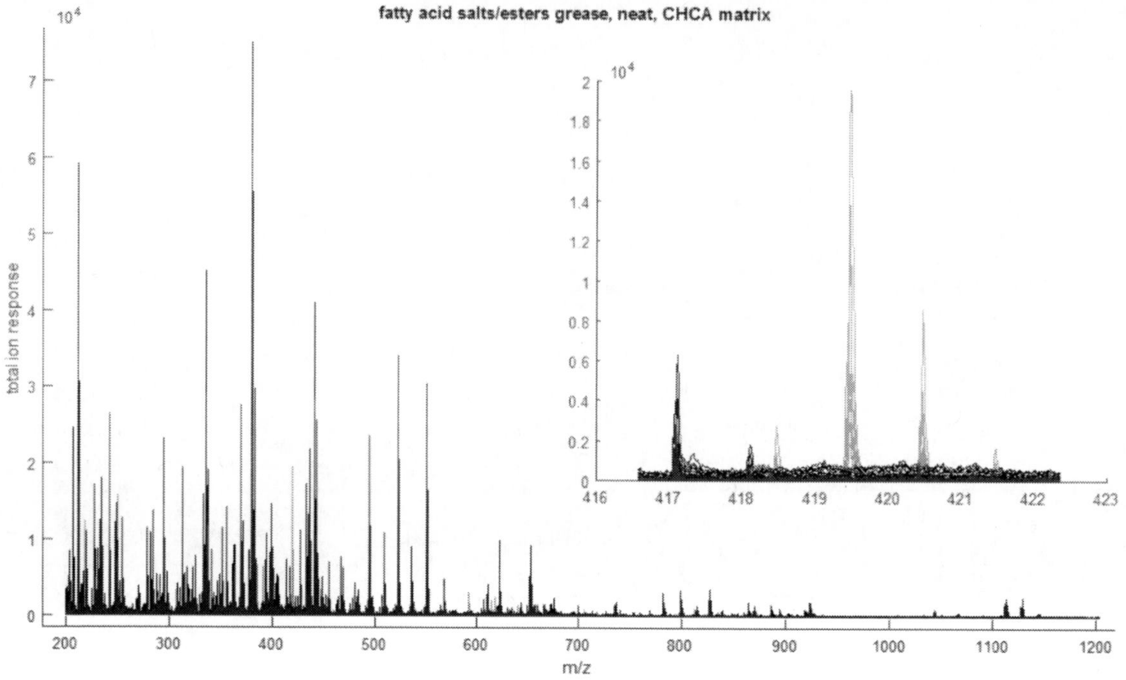

Figure 4a. MALDI data collected from a fatty acid salts/esters grease prior to fuel dilution using CHCA matrix. Neat control fuel replicates are plotted in black, the results from other HMW samples are plotted in blue, and the characteristic features being assessed are plotted in green.

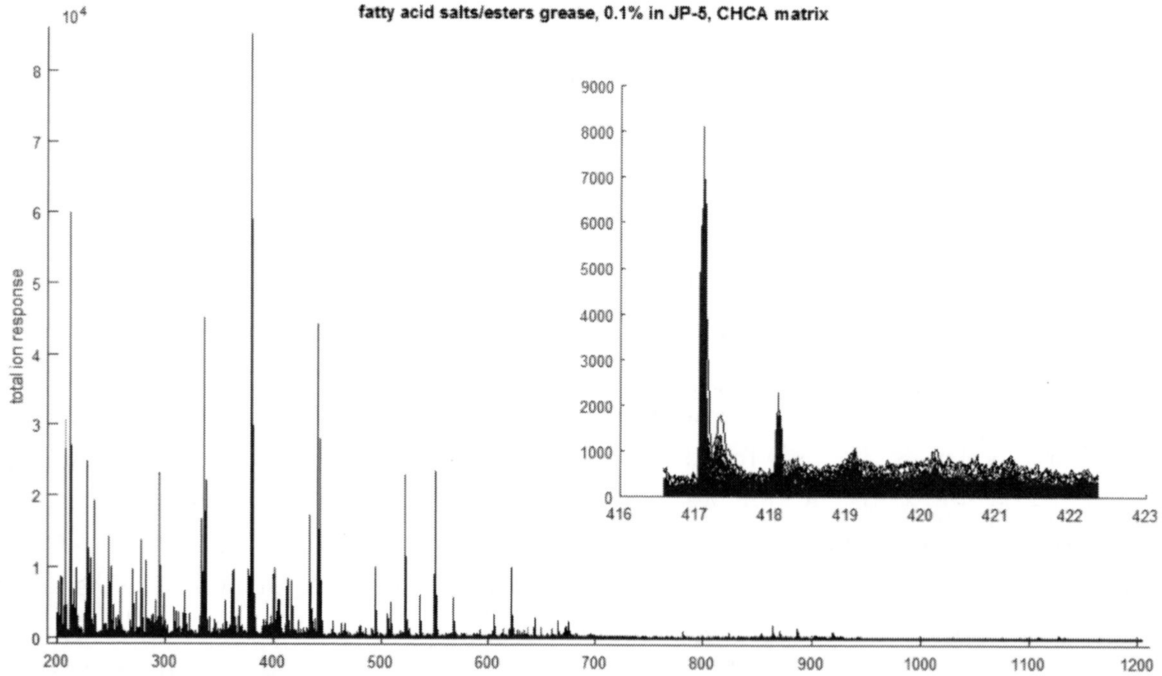

Figure 4b. MALDI data collected from a fatty acid salts/esters grease after fuel dilution using CHCA matrix. Neat control fuel replicates are plotted in black, the results from other HMW samples are plotted in blue, and the characteristic features being assessed are plotted in green.

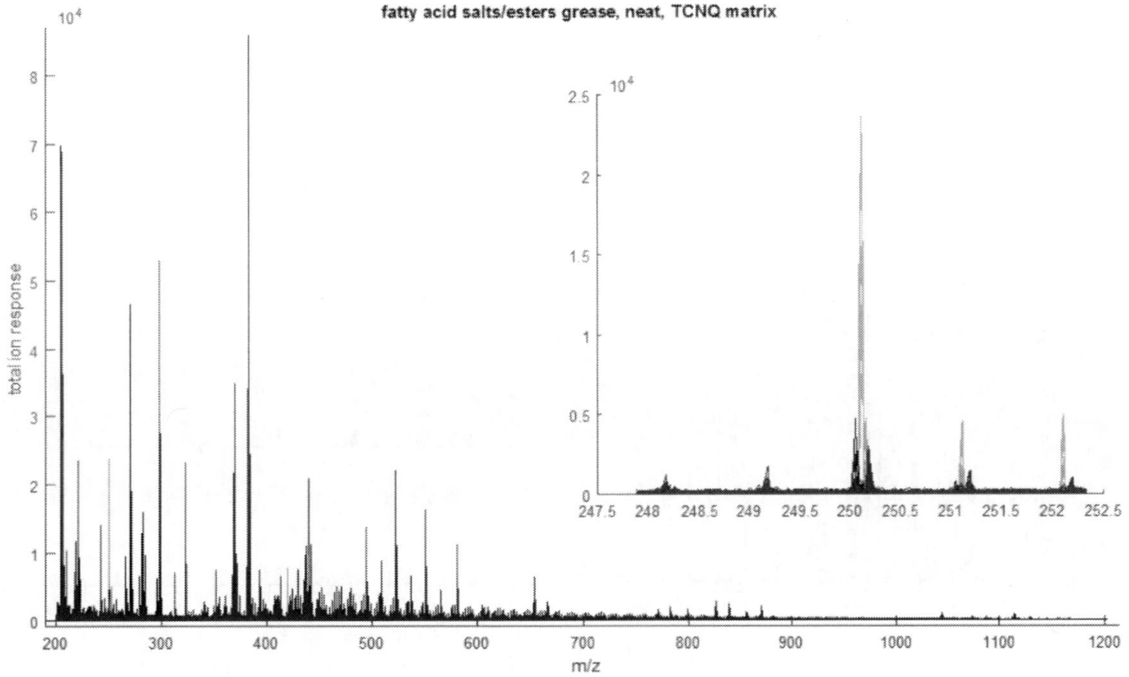

Figure 5a. MALDI data collected from a fatty acid salts/esters grease prior to fuel dilution using TCNQ matrix. Neat control fuel replicates are plotted in black, the results from other HMW samples are plotted in blue, and the characteristic features being assessed are plotted in green.

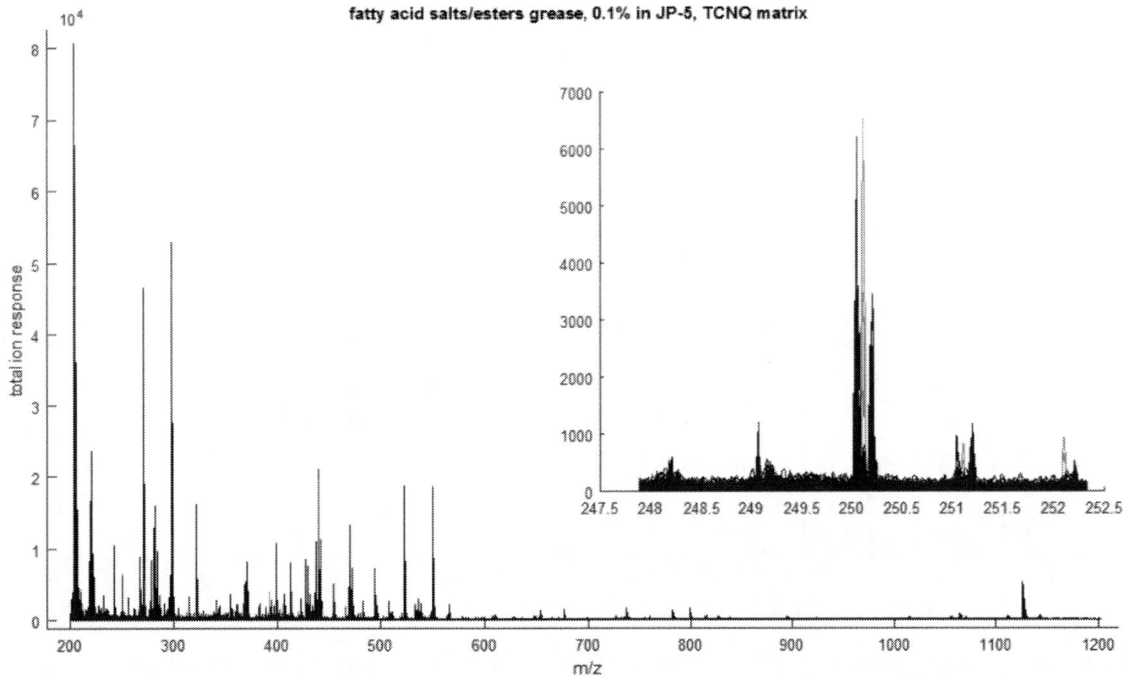

Figure 5b. MALDI data collected from a fatty acid salts/esters grease after fuel dilution using TCNQ matrix. Neat control fuel replicates are plotted in black, the results from other HMW samples are plotted in blue, and the characteristic features being assessed are plotted in green.

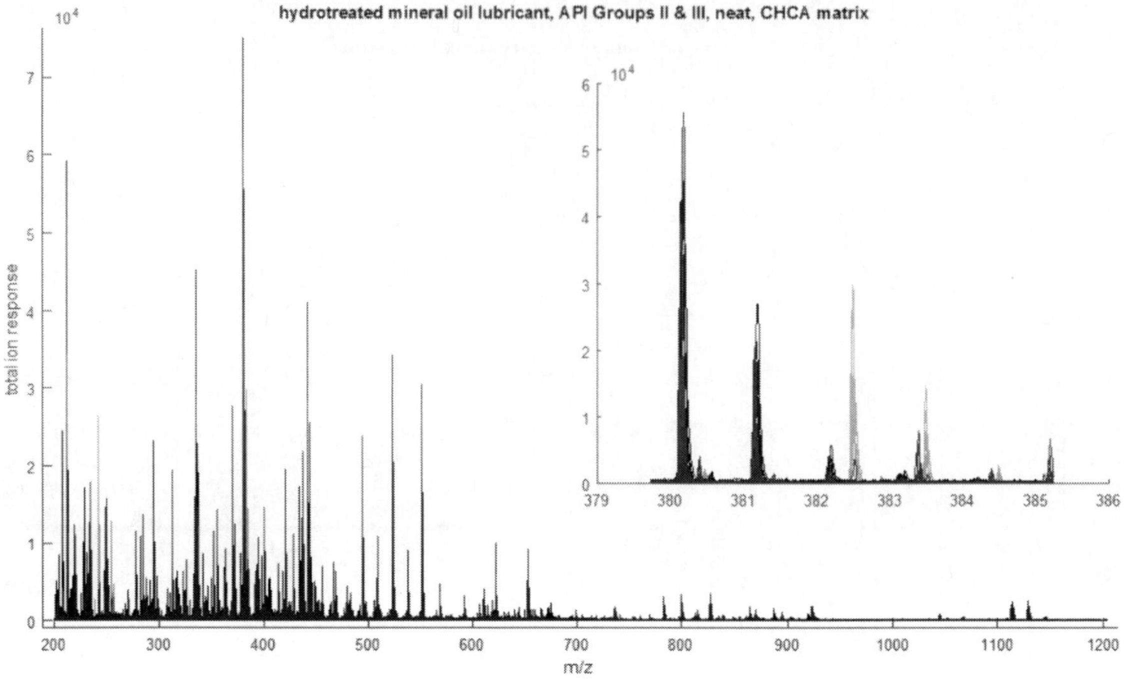

Figure 6a. MALDI data collected from a hydrotreated mineral oil lubricant prior to fuel dilution using CHCA matrix. Neat control fuel replicates are plotted in black, the results from other HMW samples are plotted in blue, and the characteristic features being assessed are plotted in green.

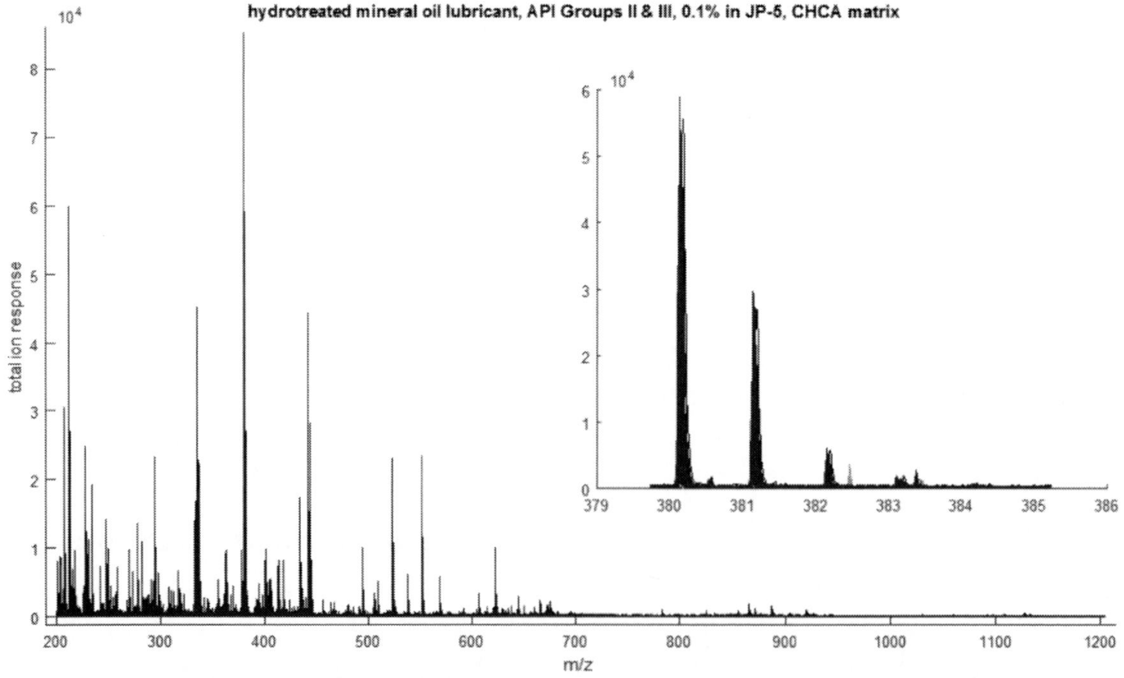

Figure 6b. MALDI data collected from a hydrotreated mineral oil lubricant after fuel dilution using CHCA matrix. Neat control fuel replicates are plotted in black, the results from other HMW samples are plotted in blue, and the characteristic features being assessed are plotted in green.

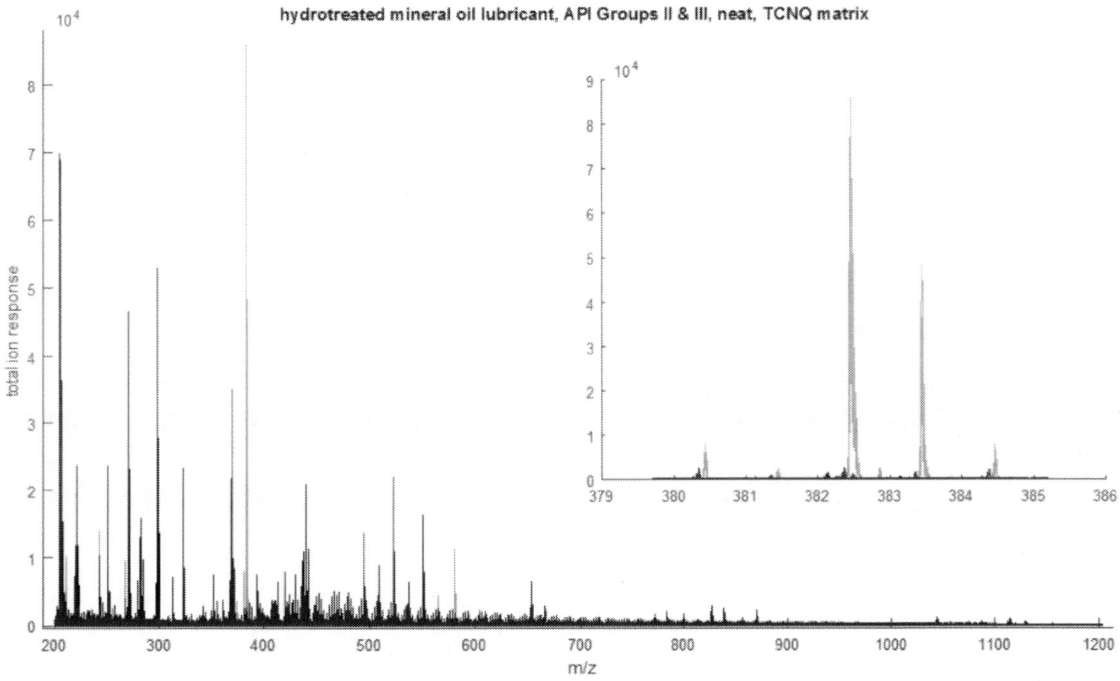

Figure 7a. MALDI data collected from a hydrotreated mineral oil lubricant prior to fuel dilution using TCNQ matrix. Neat control fuel replicates are plotted in black, the results from other HMW samples are plotted in blue, and the characteristic features being assessed are plotted in green.

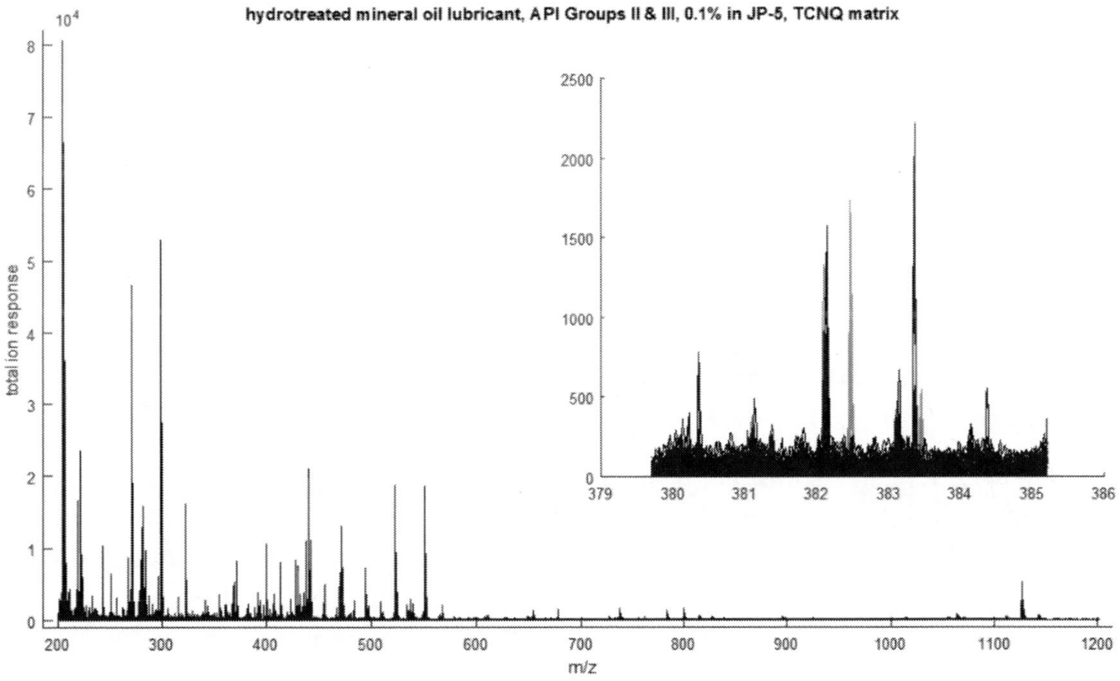

Figure 7b. MALDI data collected from a hydrotreated mineral oil lubricant after fuel dilution using TCNQ matrix. Neat control fuel replicates are plotted in black, the results from other HMW samples are plotted in blue, and the characteristic features being assessed are plotted in green.

Graphite-Based Grease Results. Given the evidence accumulated up to this point, it was deemed appropriate to examine a final grease using only TCNQ matrix. This graphite-based grease is considered an interesting target for analysis, not only because graphite as a compound should not be able to present itself as distinct features in MALDI mass spectra collected in the range of 200-1200 Daltons, but also because graphite itself can be used as a MALDI matrix,[6] a fact that might interfere with consistent detection capabilities across multiple samples. However, as can be seen in Figures 8a and 8b, this grease does show characteristic peaks in both its neat and fuel-diluted samples. These distinctive features are most likely representative of the non-primary, non-graphite "Proprietary amine" component indicated in the grease's labeling. This shows that the MALDI-based analysis strategy thus developed will have applicability even if the actual compounds to be detected are not the primary components of the original source of contamination.

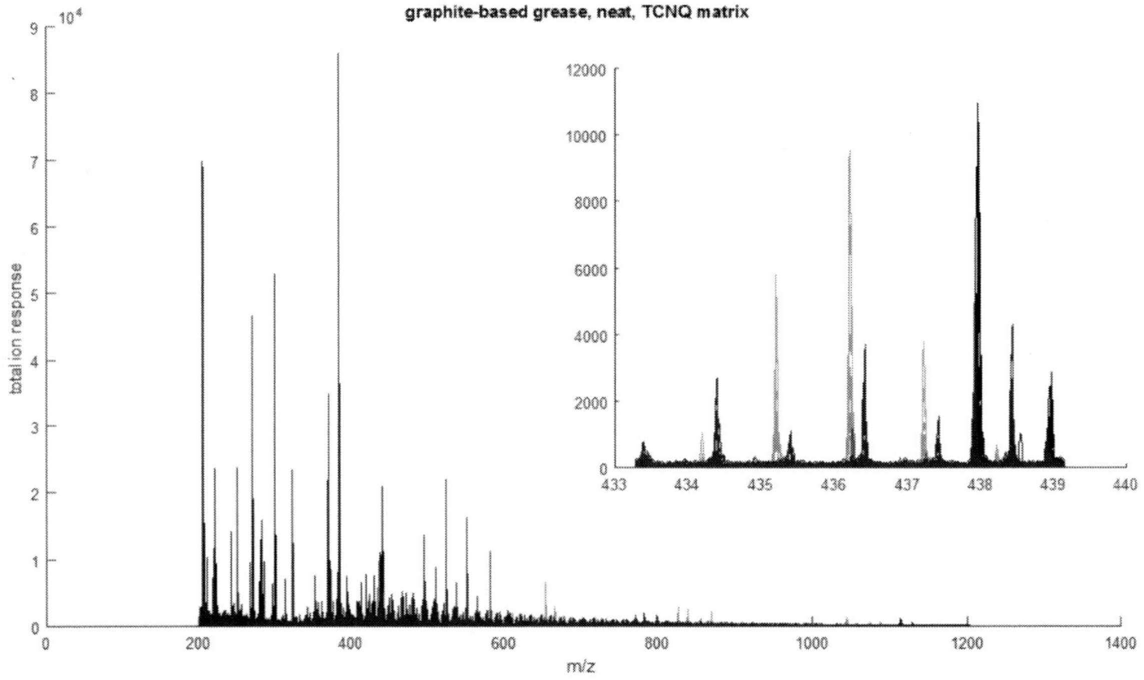

Figure 8a. MALDI data collected from a graphite-based grease prior to fuel dilution using TCNQ matrix. Neat control fuel replicates are plotted in black, the results from other HMW samples are plotted in blue, and the characteristic features being assessed are plotted in green.

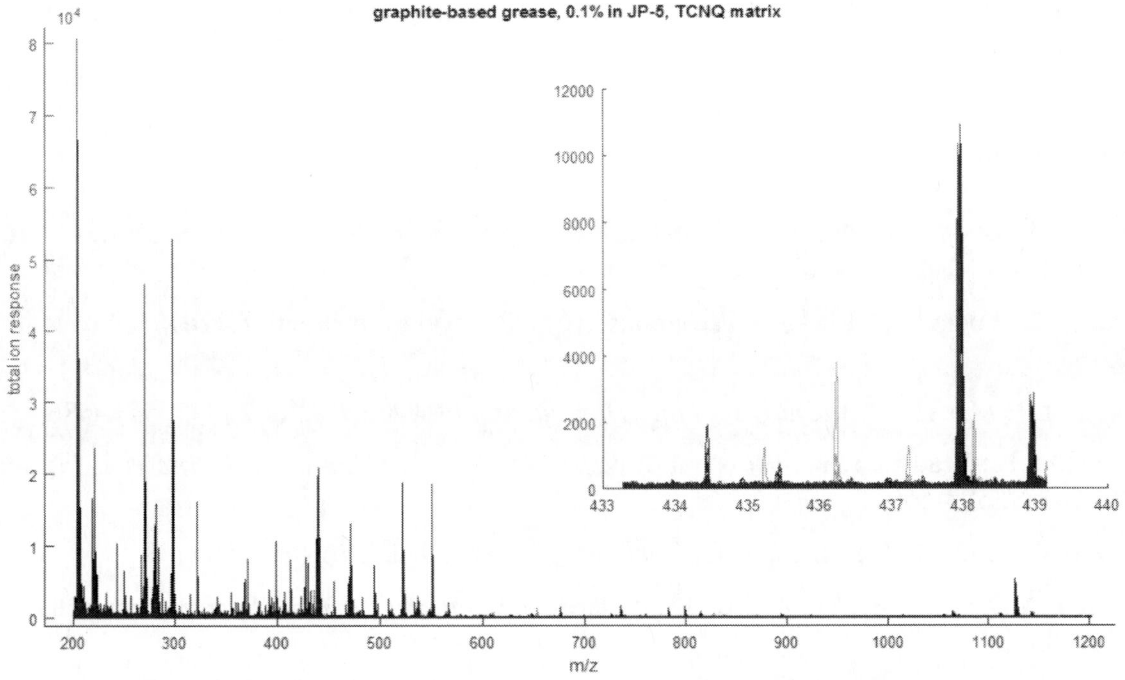

Figure 8b. MALDI data collected from a graphite-based grease after fuel dilution using TCNQ matrix. Neat control fuel replicates are plotted in black, the results from other HMW samples are plotted in blue, and the characteristic features being assessed are plotted in green.

CONCLUSIONS

A MALDI matrix application strategy initially developed for the analysis of fingerprints has been adapted for fuel samples for the purposes of NP-HMW contaminant characterization. While most fuel molecules are too small to be detected using MALDI, the larger polymeric molecules of at least some NP-HMW contaminants can be, allowing MALDI to function as a complementary technique to UHPLC-ELSD for the identification of specific sources of fuel contamination.

TCNQ appears to be a more effective matrix than CHCA. Further work will explore not only the use of alternative MALDI matrices, but also whether or not the use of solvent can be eliminated completely, a technique modification that has been applied effectively in the literature in the case of TCNQ.[7]

ACKNOWLEDGEMENTS

This work was funded by the Defense Logistics Agency Energy (DLA-E, program manager Philip Chang).

LITERATURE CITED

[1] Cramer, J.; Loegel, T.; Morris, R.; Leska, I. Presentation at CRC Aviation Meetings **2016**, May 2 – 5, 2016.

[2] Cramer, J.; Loegel, T.; Leska, I. Presentation at CRC Aviation Meetings **2017**, May 1 – 4, 2017.

[3] Cohen, L.H.; Gusev, A.I. *Analytical and Bioanalytical Chemistry* **2002**, *373*, 571–586.

[4] Ferguson, L.; Bradshaw, R.; Wolstenholme, R.; Clench, M.; Francese, S. *Analytical Chemistry* **2011**, *83*, 5585-5591.

[5] Edwards, W.F.; Jin, L.; Thies, M.C. *Carbon* **2003**, *41*, 2761–2768.

[6] Warren, A.D.; Conway, U.; Arthur, C.J.; Gates, P.J. *Journal of Mass Spectrometry* **2016**, *51*, 491-503.

[7] Trimpin, S.; Grimsdale, A.C.; Räder, H.J.; Müllen, K. *Analytical Chemistry* **2002**, *74*, 3777-3782.

IASH 2017, the 15th INTERNATIONAL SYMPOSIUM ON STABILITY, HANDLING AND USE OF LIQUID FUELS
Rome, Italy
10-14 September 2017

IDENTIFICATION AND QUANTIFICATION OF PHOSPHATE ESTER-BASED HYDRAULIC FLUID IN JET FUEL

Diana Gertopski Stamker[1], Konstantin Tartakovsky[1], Moshe Rabaev[1,*]

[1]*Israeli Air Force, Depot 22, Fuel and Chemistry Department, POB 02538*

ABSTRACT

Phosphate ester-based hydraulic fluids are commonly used in aviation, among others due to their fire resistance properties. However, they also cause corrosion of metals and swelling of elastomer and polymer seals. Contamination of jet fuel with hydraulic fluid may cause serious engine failure, and due to the corrosive effect of phosphate esters on cobalt, the level of phosphate ester must not exceed 1 ppm in jet fuel. The jet fuel should be tested for traces of phosphate compounds. This article describes methods for identification and quantification of Exxon and Skydrol hydraulic fluid contamination in jet fuel. Both of the fluids are composed of tri-butyl phosphate. Jet fuel chromatogram peaks interfere with identifying this compound's main peak, thus two techniques to separate the jet fuel peaks from the tri-butyl phosphate peaks were developed. One is based on a solid phase extraction procedure followed by identification and quantification which is carried out using a gas chromatograph equipped with a mass spectrometer or by a gas chromatograph equipped with a flame ionization detector. The other separation technique is single-staged, where the separation and analysis are carried out by a gas chromatograph equipped with a DB-WAX column and MS. Lowest level of detection reached was 0.1 ppm of tri-butyl phosphate in jet fuel; approximately 10 times lower than the limit. Linearity was achieved in all analysis methods up to contamination of 100 ppm.

***Keywords**:* Jet fuel, Organophosphate, Skydrol, Exxon, solid phase extraction, GC-FID, GC-MS.

* Corresponding author. Tel.: +972 88685967; fax: +972 8 8685298. E-mail address: rabaevm@gmail.com (M. Rabaev)

1. INTRODUCTION

The hydraulic pumping system of the aircraft is immersed in the jet fuel tanks which are operated by a hydraulic fluid. Any slight leaking failure in the system, which results in higher pressure of the hydraulic fluid than that of the surrounding fuel, can cause serious contamination of the fuel in the tanks.

Hydraulic fluids used in aviation hydraulic systems contain a variety of compounds based on phosphate esters. These compounds play a significant role in fire resistance but may also cause corrosion of metals and swelling of polymer seals. Contamination of jet fuel with hydraulic fluid may endanger operating seals and engine parts, which are made of materials that are incompatible with phosphate compounds, in turn causing serious engine failure. Due to the corrosive effect of phosphate esters on cobalt, the level of phosphate ester contamination must not exceed 1 ppm of hydraulic fluid in engines manufactured using cobalt-containing alloys and must be less than 10 ppm in engines manufactured without cobalt alloys.

Several methods for the identification and quantification of hydraulic fluids[1,2] and organophosphate[3,4] contaminations in turbine engine oil, occupational air[5], human plasma[6], soil[7] and water[8] were reported in recent years. Despite the serious danger of hydraulic fluid contamination in jet fuel, only scarce methods of identification and quantification of the contamination have been reported. One such method was described by E. Spila et al[9] and another in a patent of Werner et al[10]. The two methods refer to a contamination caused by Skydrol hydraulic fluid.

The Israeli Air Force (IAF) uses both brands of hydraulic fluid - Skydrol and Exxon. Quantification and identification of Skydrol can performed according to its di-butyl phenyl phosphate component, as described by E. Spila et al.. Identification and quantification of the second oil - Exxon cannot be performed in the same way since it does not contain di-butyl phenyl phosphate. The identification and quantification of both Skydrol and Exxon can be carried out through analysis of their tri-butyl phosphate compound, which exists in both fluids. The ion chromatogram of tri-butyl phosphate shows a main peak at 99 m/z that was considered non-significant for analysis because of the interference of peaks originating from the matrix of hydrocarbons. The failures of the hydraulic system in air refueling jets that have led to contamination in the jet fuel, together with the use of hydraulic fluid mixtures in the IAF, have revealed a need for a universal method for identification and quantification of both hydraulic fluids in jet fuel.

In the present work we managed to separate the tri-butyl phosphate compound that is common in both Skydrol and Exxon from the matrix of jet fuel hydrocarbons by using two methods presented in the article. One separation method was developed by using solid phase extraction (SPE) columns. The use of SPE technique resulted in isolation of the 99 m/z signal originating from both hydraulic fluids, so that identification and quantification could be carried out using a gas chromatograph equipped with a mass spectrometer (GC-MS) or by a gas chromatograph equipped with a flame ionization detector (GC-FID). The second method is based on gas chromatographic separation and is carried out by a GC equipped with a DB-WAX column and MS analysis for identification and quantification.

2. EXPERIMENTAL

One method is comprised of a separation stage that is carried out by SPE technique followed by injection into a gas chromatograph where the analysis is performed by an MS or FID detector (SPE-GC-MS/FID). The gas chromatographic separation method (GC-MS) is performed in one stage by injecting the sample without any prior procedure. The gas chromatographic separation is performed by a DB-WAX column and analysis is carried out by an MS detector.

2.1. Reagents

Commercial products were used in preparation of standard solutions: Exxon (ExxonMobil), Skydrol LD-4 (Solutia Inc.), Hexane (anhydrous, 95%, Sigma-Aldrich), Methanol (HPLC grade 99.9%, Sigma Aldrich), Jet fuel JP8, Florisil solid phase extraction columns (500 mg/6 mL, Starata).

2.2. Instruments

2.2.1. GC-MS method:

A Varian 450-GC gas chromatograph was equipped with a DB-WAX (30 meter, 0.25 mm ID, 0.25 µm df) column. The MS analysis was carried out by a Varian 300 MS system.

2.2.2. SPE-GC-MS/FID method:

The apparatus used for GC-MS analysis was comprised of a Varian 450-GC gas chromatograph equipped with an Rxi®-5ms (crossbond® 5% diphenyl 95% dimethyl polysiloxane, 30 meter, 0.25 mm ID, 0.25 µm df) column. The MS analysis was carried out in a Varian 300 MS system. The apparatus used for GC-FID analysis was composed of a gas chromatograph and an Rxi®-5 ms column as described above. The analysis was carried out by an FID detector.

2.3. Procedure

Stock solution was prepared by addition of 0.08 mL of hydraulic fluid (Exxon/Skydrol) to 80 mL of jet fuel. Preparation of standard solutions was done by dilution of stock solution in jet fuel. Dilution was done according to the desired concentrations of standard solutions.

2.3.1. GC-MS method:

The sample and the standard solution were diluted in hexane; 2 mL of the sample/standard solution were added to 20 mL of hexane. After the dilution 1 µl was injected to the GC-MS.

2.3.2. SPE-GC-MS/FID method:

Each standard solution was treated using the SPE technique:

The SPE column was loaded with 20 mL of standard solution; 15 mL of hexane loaded in the column in order to extract jet fuel leftovers and 5 mL of methanol used to extract the polar phosphate fraction from the SPE column.
The methanol fraction containing the phosphate ester was analyzed by 1 µl that was injected into the GS-MS or into the GC-FID system.

2.4. System conditions

2.4.1. GC-MS method:

GC-MS analysis was carried out under the following conditions: injection volume: 1µl, Injector: 117, injector temperature: 200°C, split ratio: 1:5, Carrier gas: Helium, column flow: 1.5 mL/min.

2.4.2. SPE-GC-MS/FID method:

GC-MS analysis was carried out under the following conditions: injection volume: 1µl, Injector: 117, injector temperature: 240°C, split ratio: 1:10, Carrier gas: Helium, column flow: 1.5 mL/min. GC-FID analysis was carried out under the conditions: injection volume: 1µl, Injector: 117, injector temperature: 240°C, splitless, make up flow: 28 mL/min, hydrogen flow: 45 mL/min, air flow: 300 mL/min, column flow: 1.5 mL/min, detector temperature: 230°C.

3. RESULTS AND DISCUSSION

Both hydraulic fluids and jet fuel were separately injected to the GC-MS in order to examine their composition. It can be seen that ion 99 m/z is common to both hydraulic fluids and originates from tri-butyl phosphate (Figure 1). Ion 99 m/z is inseparable from the saturated aliphatic ion fraction originating from the non-contaminated jet fuel (Figure 2).

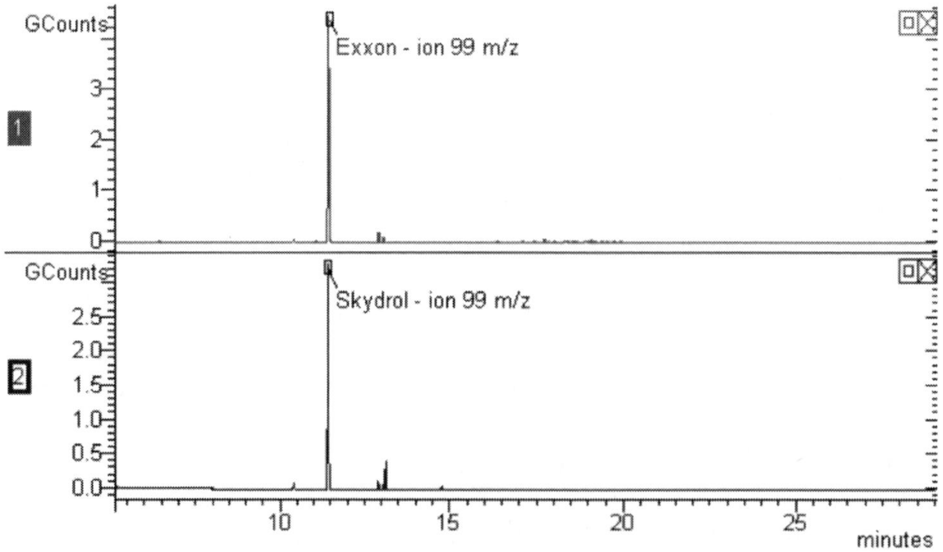

Figure 1: Chromatogram of Exxon and Skydrol. Ion 99 m/z indicated.

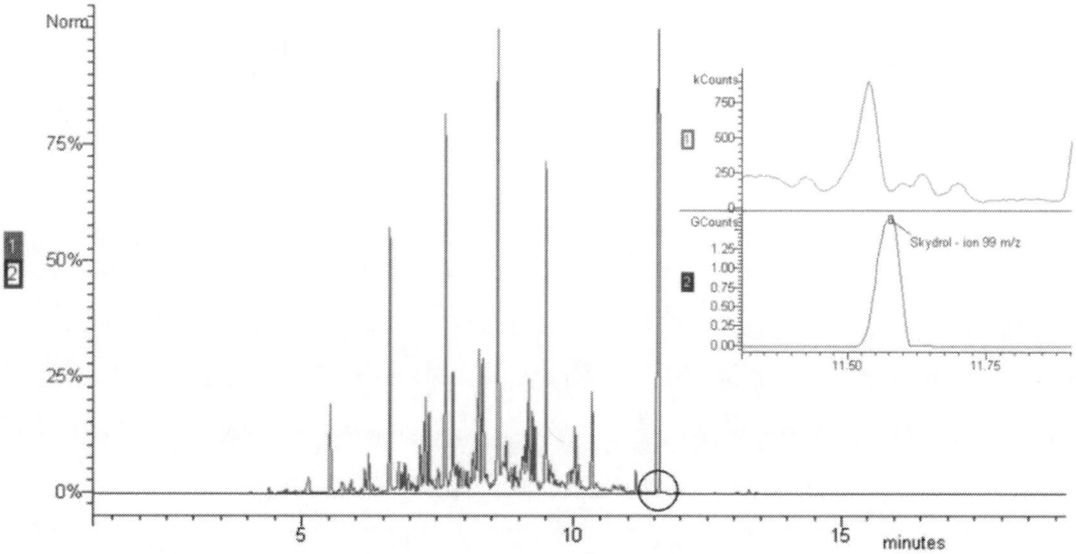

Figure 2: Chromatogram of ion 99 m/z in Skydrol and in uncontaminated jet fuel.

3.1. GC-MS method

The separation of ion 99 m/z originating from tri-butyl phosphate from the matrix of hydrocarbon peaks was achieved by using a DB-WAX column. It is evident that ion 99 m/z is not present in the chromatogram of uncontaminated jet fuel, but it can be seen in the standard solutions beginning with 10 ppm of hydraulic fluid in jet fuel (Figure 3). The calibration curve in ranges of 10 ppm – 100 ppm of hydraulic fluid shows a linear relationship R^2=0.9998 (Figure 4). Quantification and identification of the contamination reached a lowest detection limit of 10 ppm of contamination in jet fuel.

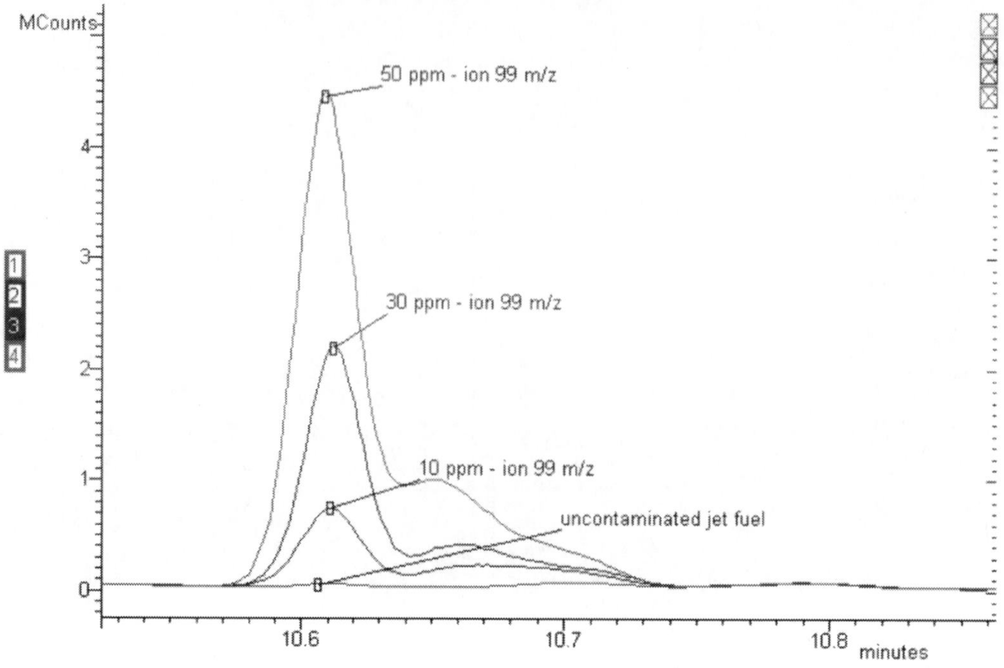

Figure 3: Chromatogram (10.5 min – 10.9 min) of uncontaminated jet fuel and standard solutions of hydraulic fluid in jet fuel. Ion 99 m/z was indicated only in standard solutions and not in uncontaminated jet fuel.

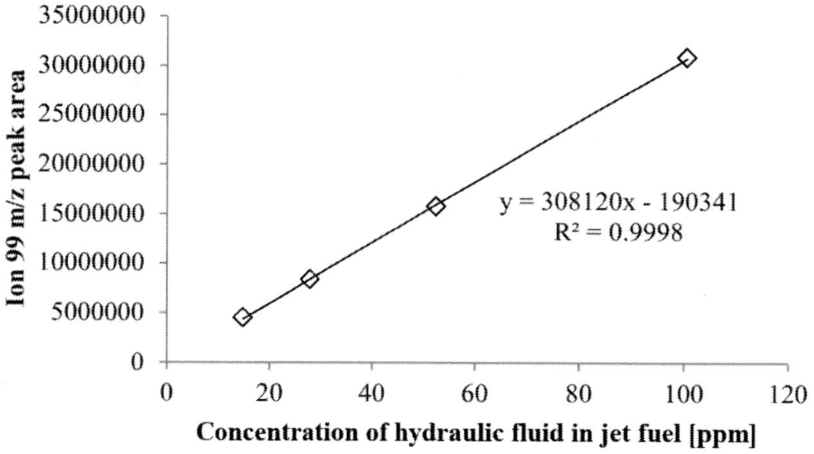

Figure 4: Contamination concentration (10 ppm -100 ppm) versus ion 99 m/z peak area calibration curve.

3.2. SPE-GC-MS/FID method

In this separation method the tri-butyl phosphate compound was separated from the jet fuel before the chromatographic analysis was performed. The separation was carried out by SPE that was performed according to the procedure described (section 2.3.2). Solid extraction concentrates the sample and separates the jet fuel from hydraulic fluid. After separation, the methanol fraction containing the phosphate was analyzed by GC-MS. The chromatogram of the methanol shows a peak of ion 99 m/z. In the chromatogram of the jet fuel fraction no peaks of ion 99 m/z are seen (Figure 5), indicating no traces of tri-butyl phosphate are left in the jet fuel after solid phase extraction. Quantification and identification of the contamination showed a lowest detection limit of 0.5 ppm of contamination in jet fuel. The calibration curve at a range of 0.5 ppm – 100 ppm of hydraulic fluid shows a linear relationship R^2=0.999.

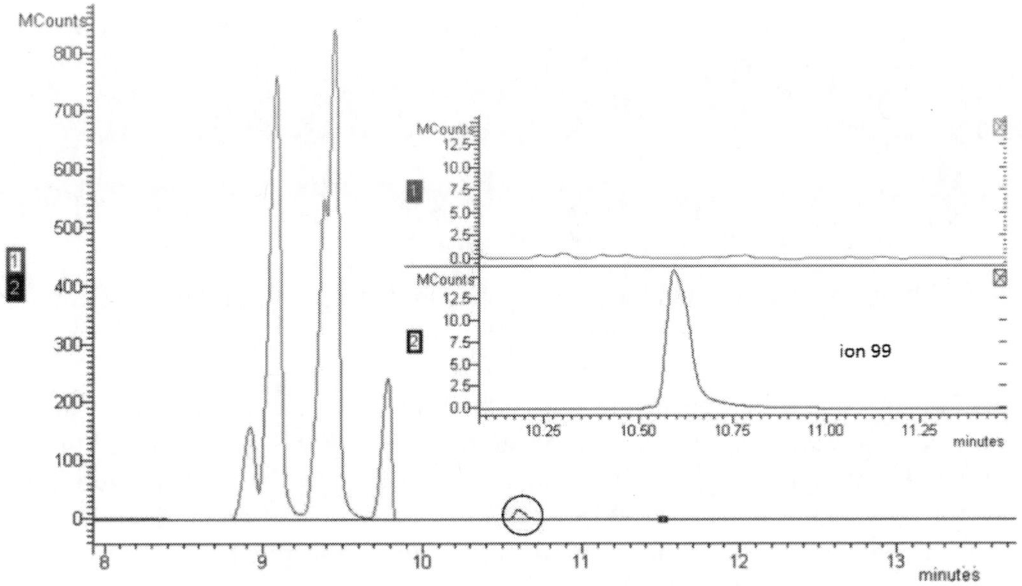

Figure 5: Chromatograms (8 min – 14 min) of jet fuel fraction (red chromatogram) and methanol fraction (blue chromatogram) after solid phase extraction of 11 ppm of hydraulic fluid in jet fuel. Ion 99 m/z was indicated only in methanol fraction.

The other analysis method was based on GC-FID. After SPE the methanol fraction containing the phosphate was analyzed by GC-FID. The chromatogram of the methanol shows a peak in retention time of 5.3 minutes while the chromatogram of the jet fuel that passed the SPE column does not contain this peak. Quantification and identification of the contamination showed a lowest detection limit of 0.1 ppm of contamination in jet fuel. The calibration curve at a range of 0.1 ppm – 100 ppm of hydraulic fluid shows a linear relationship R^2=0.9996.

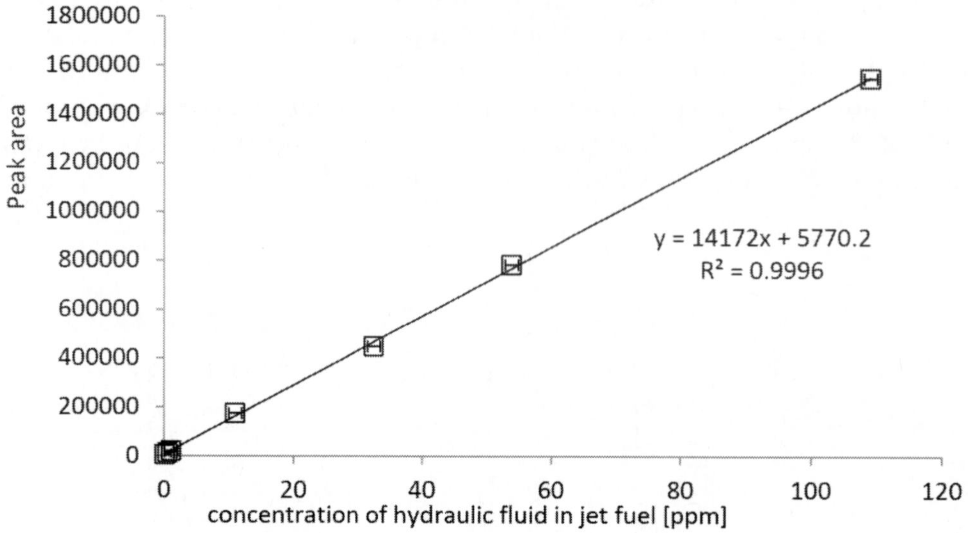

Figure 6: Calibration curve of standard solutions from 0.1 ppm to 100 ppm of contamination by the GC-FID method.

The quantification of contamination in the methods mentioned above was according to the tri-butyl phosphate reagent that was separated from the jet fuel matrix. This quantification method is not specific for only one type of hydraulic fluid contamination, unlike the method described by E. Spila et al. for only Skydrol. The GC-MS method is a single step method so there is no need for any sample preparation, unlike the method described in the patent of Werner et al. This method may be used when determination of massive concentration of contamination is needed, or for contamination detection in jet fuel used in engines not manufactured using cobalt metals. The SPE-GC-FID method followed by identification and quantification by GC-FID provides a better limit of detection of contamination, lower than that of the method described by E. Spila et al. This method requires the use of much more accessible equipment, since GC-FID is more common than GC-MS or GC-FDP, which were used in the article and in the patent mentioned above.

4. CONCLUSIONS

Methods for identification and quantification of jet fuel contamination by hydraulic fluids containing tri-butyl phosphate reagent were developed. Analysis was performed both by a gas chromatograph equipped with a mass spectrometer and a gas chromatograph equipped with a flame ionization detector. Lowest level of detection reached was 0.1 ppm of tri-butyl phosphate in jet fuel; approximately 10 times lower than the limit. Linearity was achieved in all analysis methods up to contamination of 100 ppm.

Utilizing this analysis during proactive maintenance can be successful in a number of failure cases, in which leakage detection is made possible before the problem becomes severe. These early checks cannot be performed through other means of maintenance which apply existing detection methods.

5. REFERENCES

[1] Webster R, A method for the identification and quantitation of hydraulic fluid contamination of turbine engine oils by gas chromatography-chemical ionisation mass spectrometry, Lubrication Science. (2012) 373–381. doi:10.1002/ls.

[2] A.E. Habboush, S.M. Farroha, H.I. Khalaf, Extraction-gas chromatographic method for the determination of organophosphorus compounds as lubricating oil additives, Journal of Chromatography A. 696 (1995) 257–263. doi:10.1016/0021-9673(94)01008-3.

[3] G. De Nola, J. Kibby, W. Mazurek, Determination of ortho-cresyl phosphate isomers of tricresyl phosphate used in aircraft turbine engine oils by gas chromatography and mass spectrometry, Journal of Chromatography A. 1200 (2008) 211–216. doi:10.1016/j.chroma.2008.05.035.

[4] M. Bernabei, R. Seclı, G. Bocchinfuso, Determination of Additives In Synthetic Base Oils For Gas Turbine Engines, J. Microcolumn Separations. (2000) 585–592. doi:10.1002/1520-667.

[5] K. Solbu, S. Thorud, M. Hersson, S. Øvrebø, D.G. Ellingsen, E. Lundanes, P. Molander, Determination of airborne trialkyl and triaryl organophosphates originating from hydraulic fluids by gas chromatography-mass spectrometry. Development of methodology for combined aerosol and vapor sampling, Journal of Chromatography A. 1161 (2007) 275–283. doi:10.1016/j.chroma.2007.05.087.

[6] L.M. Schopfer, P. Masson, P. Lamourette, S. Simon, O. Lockridge, Detection of cresyl phosphate-modified butyrylcholinesterase in human plasma for chemical exposure associated with aerotoxic syndrome, Analytical Biochemistry. 461 (2014) 17–26. doi:10.1016/j.ab.2014.05.021.

[7] M.D. David, J.N. Seiber, Analysis of organophosphate hydraulic fluids in U.S. Air Force base soils, Archives of Environmental Contamination and Toxicology. 36 (1999) 235–241. doi:10.1007/s002449900466.

[8] I. Rodríguez, F. Calvo, J.B. Quintana, E. Rubí, R. Rodil, R. Cela, Suitability of solid-phase microextraction for the determination of organophosphate flame retardants and plasticizers in water samples, Journal of Chromatography A. 1108 (2006) 158–165. doi:10.1016/j.chroma.2006.01.008.

[9] E. Spila, G. Sechi, M. Bernabei, Determination of organophosphate contaminants in jet fuel, Journal of Chromatography A. 847 (1999) 331–337. doi:10.1016/S0021-9673(99)00033-3.

[10] J. Werner, J.R. Tamas, Method and system for monitoring for the presence of phosphate esters in jet fuel, 8,426,212., 2013

IASH 2017, the 15th INTERNATIONAL CONFERENCE ON STABILITY, HANDLING AND USE OF LIQUID FUELS
Rome, Italy
10-14 September 2017

COLD FILTER BLOCKING TENDENCY AS A PREDICTOR FOR LOW TEMPERATURE OPERABILITY OF DIESEL FUELS

Ian. P. Mylrea[1], Jerry Burton[2], David Swan[3]

[1]*Stanhope-Seta, London Street, Chertsey, KT16 8AP, UK, im@stanhope-seta.co.uk*
[2]*Tesco Stores Ltd, Checkout Building (2nd floor, pillar N), Shire Park, Welwyn Garden City, Herts, AL7 1GA, jerry.burton@uk.tesco.com*
[3]*Greenergy Biofuels, Teesside Limited, Seal Sands Terminal, South Site, Seal Sands Road, Middlesbrough, TS2 1UB, david.swan@greenergy.com*

ABSTRACT

A new Energy Institute test method, IP 618 (Cold Filter Blocking Tendency) has been developed in response to diesel fuel filterability issues during winter months. Over several winters the number of diesel vehicle breakdowns in the UK due to blocked fuel filters is considerably above the norm, when compared to the warmer months. As was stated in the introductory remarks at the June 2015 CEN/TC19/WG24 Filter Blocking Workshop, in London, "a wide range of vehicles is being affected in several European countries, and there is a possible causal link with base fuel quality, FAME composition, cold flow additives and oxidation stability affects"
[1]. It is considered by many experts to be a complex issue, not only related to the fuel, but also the design of the vehicle filtration systems. A correlation has been seen between cold days in the UK and the number of vehicle breakdowns, with overnight minimum temperatures below 3 °C causing a significant rise. Due to the complexity of the issue experts were motivated to develop a laboratory scale performance test to help identify those fuels likely to cause field issues. Following extensive industry participation, IP 618 was published with full precision in November 2016 and was used during the winter 2016/17 to gain experience on field fuels. This paper describes the background and development of the test method, the work to-date and available data.

KEYWORDS

Diesel; Filter blocking tendency; FBT; cold filter blocking tendency; CFBT

INTRODUCTION

In the UK during successive winters from 2012/13 the number of small and light duty diesel vehicles breaking down due to fuel filter blocking has increased above the norm. In some months more than five times the expected level has been observed. A single root-cause has so far not been identified, although a correlation with temperatures below 3°C has been evident. This paper reviews historical diesel fuel filter blocking issues in the UK and New Zealand, as well as the current problem in the UK. It goes on to discuss the supply of diesel fuel in the UK and the difficulty in pin-pointing a single root-cause for the issue. Finally it describes the development of a new test method, IP 618 Determination of Cold Filter Blocking Tendency, which has been developed by the Energy Institute to assign a cold performance parameter to diesel fuels in the hope that this can be correlated to field performance such that a limit value can be established to protect vehicles from breakdown due to filter blocking. The new cold performance parameter is included in the UK National Annex of BS EN 590 along with an ambient filter blocking test, IP 387.

HISTORY

British Rail

During the winters of 1987/88 & 1988/89 fuel starvation and loss of power were reported for diesel locomotives by British Rail[2]. Filter blockage was cited as the principal reason for power failure. At the time fuel was supplied predominantly from UK based refineries. The summary indicated that the reasons for failure were not identical for each year. In 1987/88 incorrect blending of a fuel additive was the main reason for filter blocking, whereas in 1988/89 oil formulation coupled with storage at low temperatures were cited as the main reason for filter blocking. New test procedures were developed and it was recommended to develop a quality control test for fuel at low temperatures. The report also stated that there were grounds for concern about fuel quality from a number of suppliers to the Railway industry.

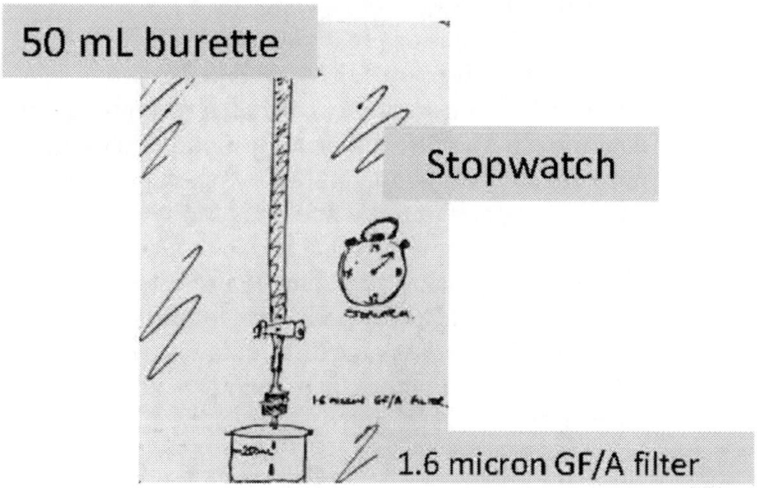

Figure 1

The fuel additive in question was poly(ethylene-vinyl acetate) or EVA which is a type of cold flow improver. Ineffective blending by one refinery enabled fuels to block filters at room temperature. Heating the fuel to 40 °C was stipulated as effective in solving the issue. A filter blocking test carried out at room temperature (approximately 25°C) was developed using a simple burette and a filter paper holder. A drawing of the apparatus is shown in Figure 1. A known volume (3 x 50 mL) of fuel was passed through a 1.6 micron filter paper. If the time to filter was less than 10 minutes then the fuel was deemed to be suitable for use. During testing only 80% of fuels supplied met the required standards.

Given the successful outcome of the investigation in 1987/88, it was a surprise to British Rail that during December 1988 and January 1989 similar problems occurred. This time though fuels generally performed well at room temperature and long chain hydrocarbons (waxes) were detected on blocked filters. Some storage issues were identified with microbial growth and iron oxide contamination present. A parallel study indicated that the majority of failures were due to waxes coming out of solution at low temperature. Refinery practices to optimise economic performance was discussed, with 5 % of long chain hydrocarbon typically being added.

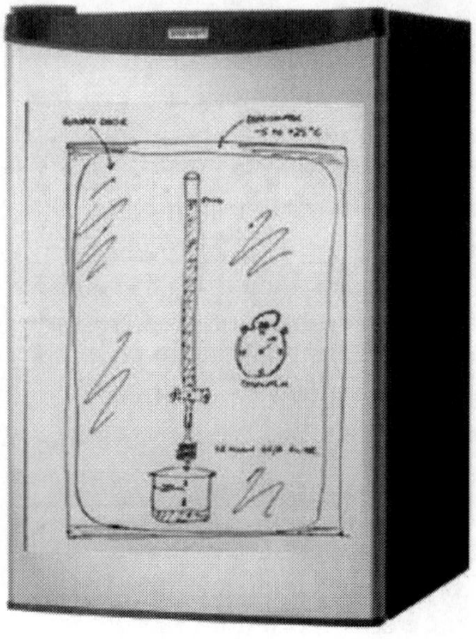

Figure 2

To research the problem a wide range of winter grade fuels from major refineries were analysed to see whether the storage at low temperatures affected the fuel and if these effects were reversible. Samples were divided and stored in a refrigerator for 24, 48 and 72 hours at 3°C. Subsequently cold filter plugging point (CFPP) and cloud point (CP) were determined. The filter blocking test outlined earlier was further developed to measure the property at lower temperatures, by placing the apparatus and samples into a refrigerator fitted with a transparent plastic door, as per Figure 2. Samples were stored for 24, 48 and 72 hours, with recovery periods of 0, 6 and 24 hours. A recovery period being defined as the time allowed for the sample to come to room temperature. The 10 minute test time limit was no longer acceptable as the fuels took longer to run at lower temperature due to an increase in viscosity. A new "average differential time" (ADT) parameter was developed, whereby an increase in time of more than 10 % for subsequent 50 mL filter volumes to filter was deemed unacceptable.

Overall findings were as follows: storing samples at low temperatures increased ADT; longer low temperature storage also caused the ADT to increase; most fuels were significantly affected in the first 24 hours of cold storage; in terms of recovery times, some fuels recovered and some did not; monitoring either CFPP and CP did not allow prediction of filter blocking tendency even after long term cooling; a low temperature performance test should be introduced into specifications.

New Zealand
During March to May 2001, diesel vessels and vehicles reported filter blocking[3]. Issues coincided with introduction of a new additive, which was later reported by the media as Clariant Dodiflow ®[4]. The additive was reported as "undissolved" in diesel. Following investigations IP 387 was introduced into the diesel specification in January 2002 with a maximum limit of 2.0. This was later relaxed to 2.5[5]. In addition the specification wording was changed to state that the fuel must have acceptable filterability and be fit for common purposes.

UK, 2012 to 2017
Since the winter of 2012/13 an increased number of small and light duty vehicles were reported breaking down, due to diesel fuel filter blocking, as shown in Figure 3. December 2012 saw three times the usual number of breakdowns.

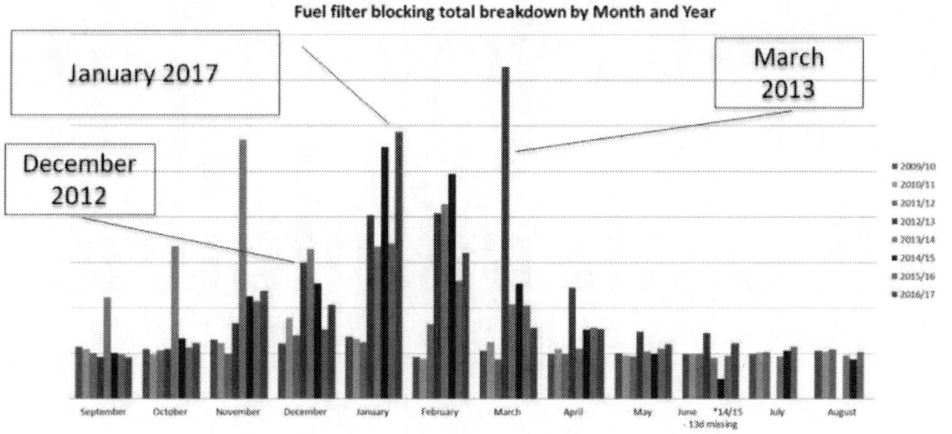

Figure 3

A small number of breakdowns due to filter blocking are accepted, and are typically at a relatively static monthly level probably caused by poor vehicle maintenance. March 2013 saw the highest number of breakdowns with more than seven times the norm. Several investigations have been carried out including vehicle filter analysis. During the winter 2012/13 the main material found was saturated monoglycerides (SMG). Problems continued, as did further investigations, and in winter 2013/14 polyethylene and polyamide were detected on blocked fuel filters and from fuel samples taken from failed vehicles. In 2014/15 high levels of EVA were recovered from filters. Experiments by Afton Chemicals demonstrated that the ambient FBT test would pick up polyethylene at very low levels.

Data collected indicates that there is a correlation[6] between low temperature and the number of breakdowns as shown by Figure 4.

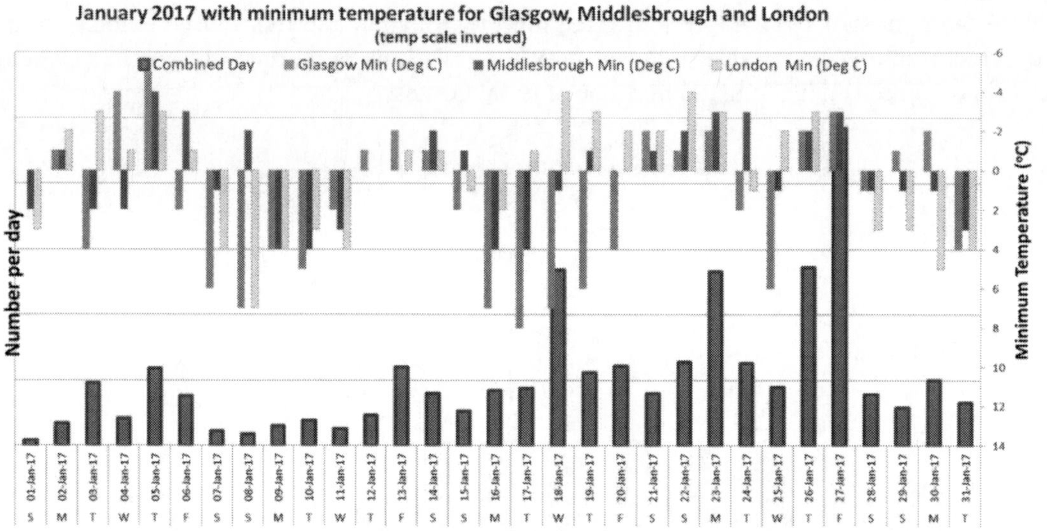

Figure 4

The data in Figure 4 is for January 2017, which was the second highest number of breakdowns at almost six times the norm. Along the x-axis are plotted a bar proportional to the number of breakdowns per day. Above are plotted the minimum temperature on that day for Glasgow, Middlesbrough and London. The temperature is plotted so that bars rise up from a central axis to show lower temperatures. Looking at the left side of the graph, for the first week, where temperatures tended to drop there was a corresponding rise in breakdowns. The following week was milder and there was a corresponding drop in breakdowns. Towards the right as the month concluded there were several cold days in London and Middlesbrough starting from 18th January and continuing until 27th January, whereby the temperature stayed below zero, which culminated in January being the second worse month for breakdowns.

A model was developed which was able to predict increases in breakdowns based on expected temperatures. (Figure 5)

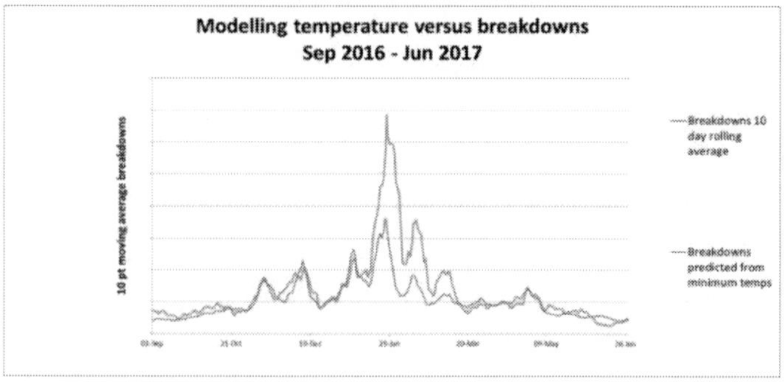

Figure 5

Further information gathered over several winters highlighted the North East & South East/ East

Anglia England regions and Scotland as worst affected when normalised for the vehicle parc (that is the number of vehicles registered in each area).

The UK Department for Transport fuel quality monitoring scheme for diesel fuel samples obtained from retail stations collected data for FBT (IP 387), CP (IP 219 / EN 23015), CFPP (IP 309 / EN 116), biodiesel content (IP 579 / EN 14078), and particle counting (IP 565). All samples were found to conform to the BS EN 590 requirements. Over several winters FBT improved significantly, with the samples obtained in the winter of 2016/17 all meeting the specification requirement of <2.52 for FBT. Biodiesel content, CP and CFPP were as expected. Particle counts were higher than expected, although no specification exists for particle counts the worldwide fuel charter proposed 18/16/13, and counts were in excess of these ISO 4406 codes. Some data was obtained with and without water removal, and water did not solely account for the higher readings.

COMPLEXITY OF THE PROBLEM

Diesel fuel as supplied is a mixture of mineral derived diesel, biodiesel and additives. Each component itself is a mixture, and comes from a unique refining or process set-up. Whereas in the 80's it may have been possible to trace batches of diesel from a retail site back to a specific refinery production the picture is now far more complex.

Diesel from crude oil

Diesel is manufactured by refining crude oil, and crude oil varies depending on its source. Ultimately diesel is a complex mixture of hydrocarbons, with varying overall aromatic content, molecular weight distribution, and traces of other chemicals which are in the same boiling point range, all dependent on the feedstock and processes. No two refineries are identical. Refinery optimisation occurs in order to meet economic requirements.

The number of UK based refineries has reduced from nineteen in 1975 to six by 2017. UK road diesel consumption[7] has gone up by more than 25% over the last decade alone and stands at 24m tonnes, compared to refining output of 14m tonnes. Some diesel is actually exported, so in the end the amount of imported diesel sits at around 46%. The amount of imported diesel has increased by more than 50% since 2010.

Biodiesel

To meet environmental renewable energy targets up to 7% biodiesel is blended in to refined mineral diesel. In the UK approximately 0.6m tonnes was blended in during 2015. Biodiesel is manufactured by trans-esterification of plant oils and animal fats. The source of the raw material affects the physical properties significantly due to changes in the molecular weight, composition and chemistry. Coconut methyl ester (CME) is very saturated, can have an Iodine value of typically[8] 6 to 12, and contains predominantly C10 and C12 molecules, whereas rapeseed methyl ester (RME) is more unsaturated, the Iodine value can be typically 96 to 117, and contains predominantly C16 and C18 molecules (Iodine value is a measure of the unsaturation in fatty acids. As such the cloud point and cold filter plugging point can be different depending on the raw material. Over several years the types of FAME being blended into diesel in the UK has shifted[9] such that used cooking oil methyl ester (UCOME) now accounts for 83% of biodiesel and soy (SME) & rapeseed (RME) have reduced to effectively zero. Used cooking oil comes from all over the world to be processed and turned into biodiesel and itself is a complex mixture.

Additives

There are many types of diesel additives, and unlike the aviation industry where individual additives are approved by name in the jet specifications, they are not formally approved in the diesel specification. Additives are used in biodiesel and diesel separately, and each fuel is required to meet its own specification, before being blended together. Additives might be added to make a premium product, or simply to enable the specification to be met, for example cold filter plugging point (CFPP). Some additives are for other purposes altogether, such as pipeline drag reducer, which is used to speed up flow in a pipeline or corrosion inhibitor which is added to protect the pipeline. Going back to the British Rail situation in the late 80's, fuel was supplied from a limited number of refineries, and it would have been feasible to list all the additives and the treat rates with some cooperation. In the present day with 46% imported fuel, and biodiesel it would be an impossible task to try to list the components added together to make the final fuel.

Supply chain
Imported fuels are supplied to the UK by ship which are loaded at foreign terminals supplied by pipelines or other ships. Pipelines stretch into Europe and the Russian Federation. For almost 50% of diesel fuel in the UK the origin of fuel is now difficult to trace; the crude oil it came from, the original plant oil or animal fat from which biodiesel was produced, the additives, and the refining process are all variables. The supply chain is far more complex when compared to the time of the British Rail investigation.

Vehicles and user behaviour
Diesel cars have evolved significantly over the last twenty years with the advent of direct injection common rail engines, where extremely high pressures now put the fuel and injection systems under new stresses. The need for clean, dry fuel is now more significant than ever. Fuel filters spend many thousands of miles in the car before being replaced and see many different batches of fuel. The vehicle parc consists of old and new vehicles, and a large variety of makes and models. Some users frequently refill the tank, whilst others might take several weeks before they refill. All of this adds to a complex picture, whereby measurement of the fuel in the tank at the moment of failure, may not give such a good insight into a single reason why the filter was blocked.

TEST METHOD DEVELOPMENT

With such a complex picture, and in the knowledge that temperatures below 3°C are causal, a cold performance test was required which can predict the use of the fuel in a vehicle at low temperature.

A full size rig test was developed by Infineum[10] through the BSI PTI/2 Task force on diesel filter blocking. The rig contained a car diesel tank (approximately 55L), a car diesel filter and a pump downstream of the filter i.e. the fuel pump sucked fuel through the filter. Sensors monitored the flow and pressure before and after the filter. The whole apparatus was contained in a large chest freezer style refrigerator. This enabled fuel to be cooled, cycled and filtered to explore the effects on the base mineral diesel fuel, biodiesel and additives. Effects were demonstrated with pressure increases signifying filter blocking seen for some test fuels. Although useful as a research tool this rig is not suitable to form the basis of a simple laboratory test.

The Stanhope-Seta multi-filtration (MFT) apparatus used for IP 387 is similar on a laboratory scale, but did not have the ability to cool the fuel. Stanhope-Seta developed a 'proof of concept' prototype by immersing the key parts into a low temperature liquid bath. The prototype enabled fuels to be tested at different temperatures as shown in Table 1.

Table 1

	CFBT (°C)			3rd Party IP 387	
Sample ID	-1	3	10	20	FAME (%)
SA1761	2.25	1.64	1.14	1.05	3.6
SA1756	2.36	1.34	1.13	1.03	3.7
SA1757	3.48	2.25	1.26	1.29	2.4
SA1762	4.4	2.52	1.34	1.35	2
SA1759	5.1	3.48		1.87	0.9
SA1760	6.08	3.88	1.64	2.24	3.3
SA1763	7.57	3.88	2.69	1.87	0.7
SA1758	10.05	4.4	1.69	2.36	2.3

The fuels were obtained from service stations and the initial work showed that fuels could be differentiated by an FBT test at different temperatures. In addition there was no apparent relationship between the biodiesel content and the FBT, with the lowest concentration sample being one of the worst performers. Having demonstrated the principle authorisation was given to start a task group to develop the test method. At the time it was unclear what temperature the test would be carried out and so a set-up which enabled two test temperatures was preferred and was useful from a data gathering perspective. The two temperatures proposed were +3 °C and -1 °C. Three prototypes were assembled which contained all the key elements; pump, pulse damper, heat exchanger and filter. The proposed method was drafted, with the principle being very similar to IP 387, where the fuel is pumped through a pulse damper to an encapsulated filter at a constant flow rate. A pressure transducer measures the pressure increase across the filter. As the filter blocks the pressure goes up. The test finishes when either 300 mL of fuel has gone through the filter, or when the pressure reaches 105 kPa. A calculation is used to transfer the results to an index called filter blocking tendency. The new test method had two key differences in that the fuel passed through a heat exchanger to condition it at the test temperature and the filter was encapsulated tightly in aluminium which itself formed part of the heat exchanger. This enabled the fuel and filter to be at the required test temperature. The test was called Cold Filter Blocking Tendency and an acronym was developed which also indicated the temperature of the test; CFBT(-1) means the cold filter blocking tendency at -1 °C. A pilot study was proposed to make a limited estimate of precision in order to validate and publish the proposed method. The pilot study contained ten fuels as shown in Table 2.

Table 2

Sample	Sample	FAME Added (%)	FAME (%)	CP (°C)	CFPP (°C)
1	BASE FUEL BSI	0	0	-5	-16
2	BSI TF Fuel 2 (Sample 3)	0	7	-4	-10
3	BSI TF Fuel 3 (Sample 9)	0	7	-4	-11
4	BASE FUEL BSI	0			
5	Terminal Fuel 1 + FAME	5	<0.05*	-6*	-11*
6	Terminal Fuel 2 + FAME	5	<0.05*	-6*	-16*
7	EI ICPMS April 2015	0	4.98	-5.8	-25.5
8	EI ICPMS June 2015	0	0	-4	-14.5
9	HIGH CLOUD POINT (-3)	0	0	-3	-14
10	B0 + FAME	7			
n/a	Market FAME (added to 5,6 and 10), sampled August 2015	n/a	n/a	n/a	n/a
	*properties of the base fuel and not the resulting blend				

Each sample was divided into bottles and labelled with a random number. Each operator had two bottles of each sample. Three sets of apparatus were used each by a different operator. Each operator had a unique running order. The CFBT(-1) and CFBT (+3) were determined for each sample. The overall outcome of the pilot study was that similar precision to IP 387 was demonstrated. An appendix was included to describe a cold soak procedure in case it was required in the future and the proposed method was published.

During the development a small number of samples were tested using the proposed method and the BSI/Infineum rig, and results from the new bench test made sense when compared to the rig and expectations about the samples.

An Energy Institute funded inter-laboratory precision study was planned and a workshop to disseminate information was hosted by Essar and SGS at the Stanlow Manufacturing Complex. This provided an excellent opportunity to critique the test method and give operators hands on experience of the test. The inter-laboratory study was performed shortly afterwards and again contained ten samples, in duplicate which were blind coded. The samples were despatched to nine individual labs for measurement within a fixed timeframe to avoid ageing affects. The cold soak procedure was also included. The samples were winter grades obtained from four UK refineries meeting BS EN 590; two service station samples from East Anglia and the Northeast meeting BS EN 590; one USA sourced sample meeting ASTM D975 and three samples blended from an imported European fuel. Biodiesel was added to the non-service station samples. Biodiesel concentration ranged from 0 % to 20 % and was derived from used cooking oil, rapeseed, tallow and soy. 720 tests were carried out in total during the study.

The EI statistician calculated precision[11] and equations were proposed for repeatability (r) and reproducibility (R) as shown below in Table 3.

Table 3

Type	Range of results	Repeatability	Reproducibility
CFBT(+3)	1.0 to 4.4	$0.07374x^{2.2}$ (70)	$0.09827x^{2.2}$ (59)
CFBT(-1)	1.0 to 8.4	$0.08953x^{1.7}$ (79)	$0.1181x^{1.7}$ (91)
CSCFBT(+3)	1.0 to 4.1	$0.1032x^{2}$ (73)	$0.1740x^{2}$ (63)
CSCFBT(-1)	1.0 to 8.4	$0.1143x^{1.7}$ (80)	$0.2279x^{1.7}$ (74)

The numbers in brackets in the r and R columns are the degrees of freedom. Precision for CFBT(-1) was seen as the likely specification target and Table 4 below shows the precision at various levels.

Table 4

Range of results: 1.0 to 8.4

		CFBT(-1)	
FBT	Volume (mL)	r	**R**
1	300	0.1	**0.1**
1.5	268	0.2	**0.2**
2	173	0.3	**0.4**
2.5	131	0.4	**0.6**
3	106	0.6	**0.8**
3.5	89	0.8	**1.0**
4	77	0.9	**1.2**
4.5	68	1.2	**1.5**
5	61	1.4	**1.8**
5.5	55	1.6	**2.1**
6	51	1.9	**2.5**
6.5	47	2.2	**2.8**
7	43	2.4	**3.2**
7.5	40	2.8	**3.6**
7.8	39	2.9	**3.9**
8	38	3.1	**4.1**

The precision compared well against the pilot study and IP 387, and the conclusion was drawn that the precision estimate was robust. The test method was updated to include the precision equations and was published as IP 618.

EXPERIENCE OF IP 618 TODATE

During the winter 2016/17 the test method was used by several labs to assess retail fuel samples, as shown by Figure 6, Figure 7 and Figure 8. The figures show 3 columns per sample, the blue column for the ambient FBT test (IP 387), the red column for the new CFBT(+3), cold filter blocking tendency at +3°C and the green column the new CFBT(-1), cold filter blocking tendency at -1°C.

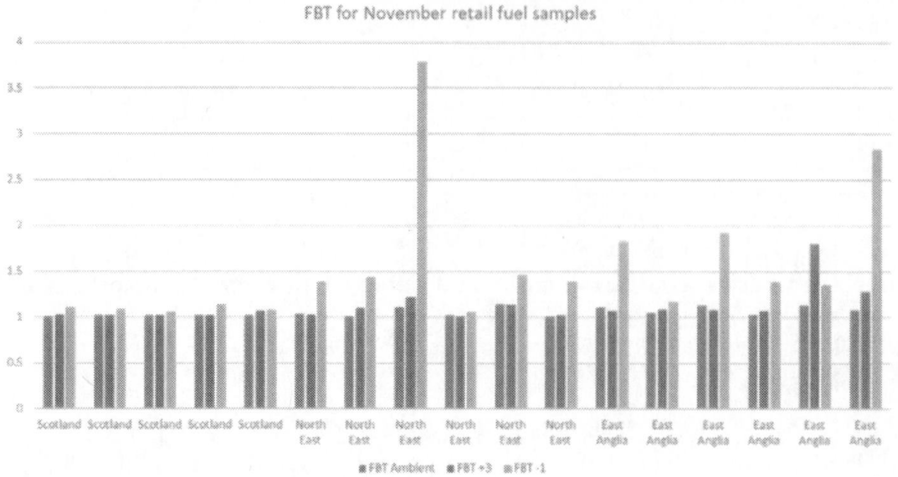

Figure 6

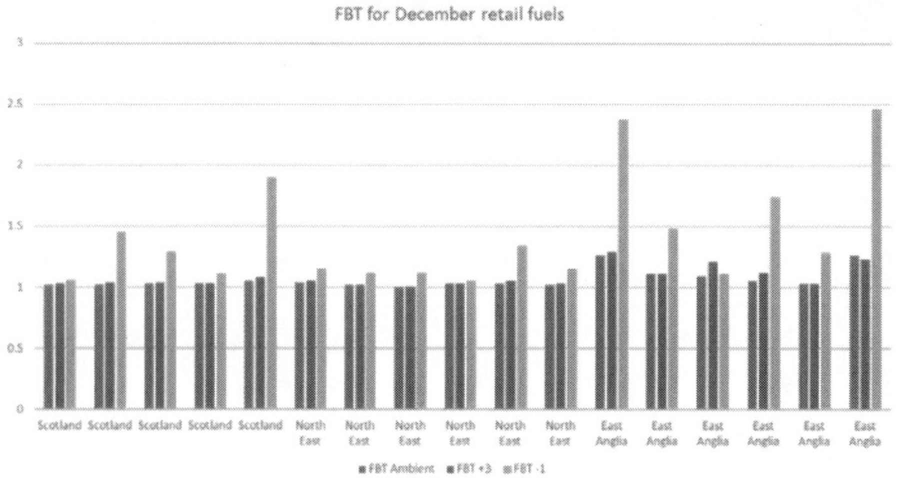

Figure 7

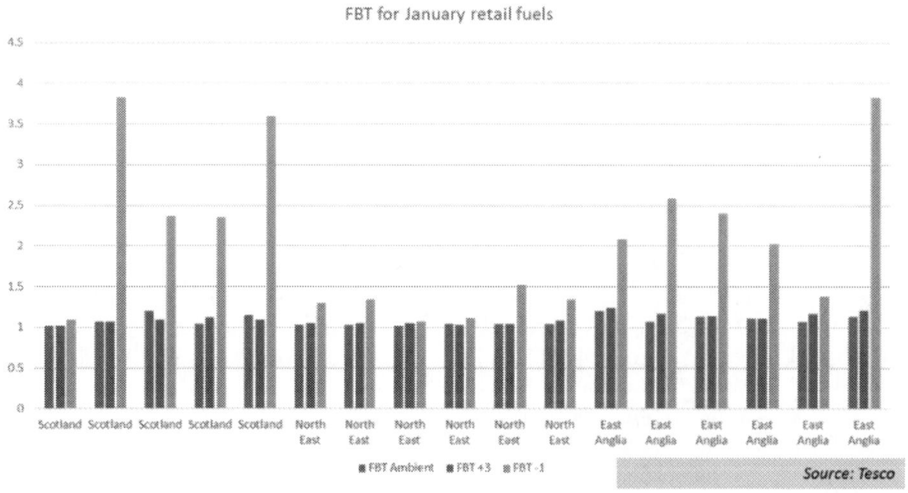

Figure 8

The data is interesting as it shows by CFBT(-1), that -1 °C is a key temperature to differentiate fuels. The ambient FBT and CFBT(+3) tests generally agree with each other, within precision. Samples were all well within the 2.52 limit for ambient FBT. Some samples showed elevated readings for CFBT(-1), especially in East Anglia, one area with a higher frequency of vehicle breakdowns. Scotland and the North East gave a mixed picture and it is difficult to draw a conclusion.

Some data, shown in Figure 9, provided anonymously from terminals, shows that different base fuels' cold properties vary considerably. CFBT(-1) varies independently of CFPP and CP Some values are significantly higher than 2.52.

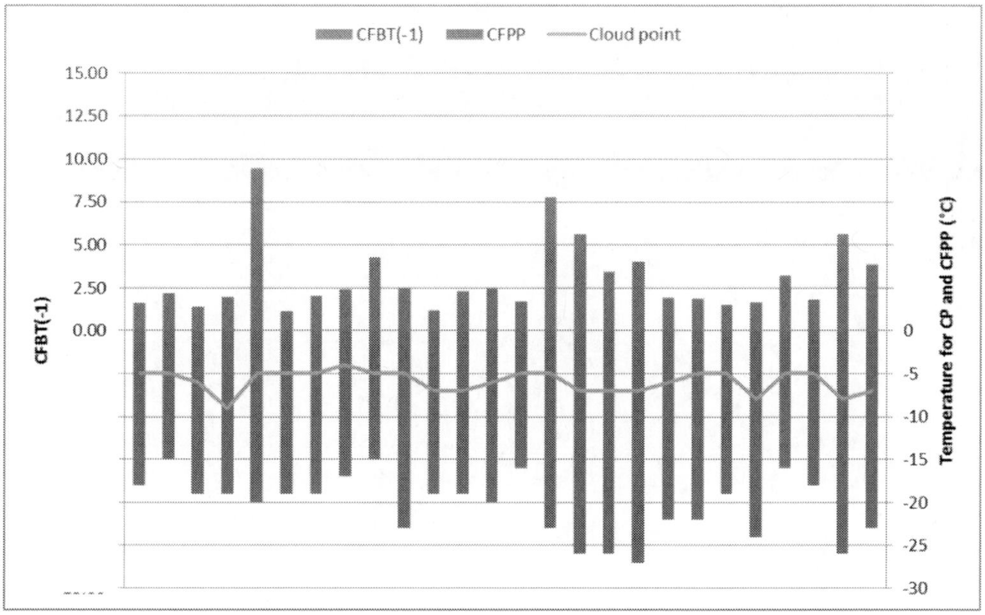

Figure 9

Data gathered during the winter 2016/17 further enhanced the view that the new IP 618 CFBT test gives more insight into the low temperature filtration characteristics of the fuel and is able to differentiate fuels which otherwise would appear similar by the ambient FBT test IP 387.

SUMMARY

Current and historical diesel quality issues have been discussed along with the complexity of investigating the current issue in the UK, which all lead to the development and publication of a new cold performance based test method for diesel, IP 618 by the Energy Institute, as described by this paper.

The new test method, IP 618 (report) and IP 387 (maximum limit 2.52) are now incorporated in the National Annex of BS EN 590 for the winters of 2017/18 and 2018/19.

ACKNOWLEDGEMENTS

Afton Chemicals,
British Rail
BSI PTI/2 Task Force on Filter Blocking
EI TMS B-5 CFBT Task Group Members
Essar Oil,
ExxonMobil,
Greenergy,
P66,
Infineum,
Innospec,
Saybolt,
SGS
Stanhope-Seta
UK Energy Institute
UK Department for Transport

[1] CEN/TR 16982:2016 (E) Diesel blends and fuels – Cold Filterability Issues
[2] "Coolflow" an investigation into the filterability of cooled fuel oils, P Middleton & A J Field, Reference TM-AS-006, January 1990
[3] www.scoop.co.nz/stories/BU0105/S00233.htm
[4] tvnz.co nz/content/42318/2591764.xhtml 25/
[5] Engine Fuel Specifications Regulations 2011 SR 2011/352, Schedule 3
[6] BSI PTI/2 Task force on diesel filter blocking
[7] DUKES 2017 Chapter 3: Petroleum
[8] Biodiesel The Comprehensive Handbook; M Mittelbach, C Remschmidt
[9] Renewable Transport Fuel Obligation Statistics: Period 8 2015/16, report 6, UK Department for Transport
[10] www.infineuminsight.com/insight/jan-2016/improving-diesel-filterability
[11] CFBT/CSCFBT Research Report – IP PM ES Determination of cold filter blocking tendency, I Mylrea, 22/9/2016

IASH 2017, the 15TH INTERNATIONAL CONFERENCE ON STABILITY, HANDLING AND USE OF LIQUID FUELS
Rome, Italy
10-14 September 2017

A USER'S PERSPECTIVE AND EXPERIENCE WITH PARTICLE COUNTING IN LIQUID FUELS

Joel A. Schmitigal[1]

[1]United States Army Tank Automotive Research Development and Engineering Center, Mail Stop 110, 6501 E. 11 Mile Road, Warren, Michigan, United States of America, joel.a.schmtigal.civ@mail.mil

ABSTRACT

Following several years of research within the Department of Defense, limits for light obscuration particle counting were published in MIL-DTL-83133J in December of 2015 followed by inclusion in MIL-STD-3004D change 1 in March 2016. Over 2500 field samples have been analyzed by both particle counter technologies and gravimetric analysis for comparison purposes, yielding a 92% go/no-go agreement between the established 19/17/14/13 ISO code limits and 1.0 mg/L gravimetric contamination limit.

INTRODUCTION

The U.S. Army maintains the mission of providing quality fuel to U.S. and Allied troops in tactical environments. Operation in extreme environments with tactical equipment can be particularly challenging for soldiers tasked with provisioning fuel meeting the requirement needs of both fix winged and rotary aircraft as well as ground systems. Presently, requirements as outlined require a dedicated group of specifically trained fuels personnel to perform several tests monthly or daily at each filter separator in a fuel distribution system looking for traces of sediment and water in the fuel (1) (2) (3).

Particulate and water are the two most common types of fuel contamination and are similar in that both can be removed from fuel by settling, or by a filter/separator. Particulate contamination is of particular concern as hard particulates can cause wear within the fuel distribution and injection system leading to long term mechanical damage. Filter plugging leading to reduced filter life or engine fuel starvation, and fuel system blockage caused by ultrafine particulates building up deposits that restrict fuel flow in fine tolerance components. Water contamination presents its own problems in that water can freeze at low temperatures leading to filter plugging or fuel system blockage resulting in engine starvation. Water can also contribute to fuel system corrosion and feed biological activity while in storage.

The Army historically utilized several techniques to ensure that aviation fuels are clean and dry. Fuel filter effectiveness testing is utilized to determine if filters are operating effectively though conducting periodic fuel sampling for gravimetric analysis. The Army utilizes two methods for measuring particulate contamination by gravimetric analysis: ASTM D2276 - Standard Test Method for Particulate Contaminant in Aviation Fuel by Line Sampling, and ASTM D5452 - Standard Test Method for Particulate Contamination in Aviation Fuels by Laboratory Filtration. Additionally the Army employs ASTM D2276 Filter Membrane Color Ratings found in Appendix 1 as a quantitative metric for determining particulate contamination present in the fuel. Undissolved water content is determined by performing ASTM D3240 – Standard Test Method for Undissolved Water in Aviation Turbine Fuels, commonly termed AquaGlo testing. Additionally the Army utilizes ASTM D4176 – Standard Test Method for Free Water and Particulate Contamination in Distillate Fuels (Visual Inspection Procedures), as a final check of fuel to ensure aviation fuel is clear and bright before flight operations. These test methods have several drawbacks including: sample size, timeliness of data due to the turn-around time needed to get the test results, operator subjectivity, lack of detailed analysis, poor repeatability and reproducibility and limitations in providing reliable data.

The U.S. Army developed particle counting limits based on ISO 4406 of 19/17/14/13 that closely correspond to their receipt and issue limit of 1.0 mg/L for particulate contamination and 10ppm undissolved water limit based on particle counts fuels developed in a laboratory setting with standard test dusts and controlled water injections (4). The 14 and 13 of the 19/17/14/13 ISO code is associated with approximately 5ppm free water, as the 10ppm free water has a tendency to overwhelm the particle counting channels. In December 2015 MIL-DTL-83133J was published including incidental contaminant limits for particle counting of ISO 4406 limits of 19/17/14/13 which was followed by the update of MIL-STD-3004 to revision D with Change 1 to include the same limits.

APPROACH

The Army sampled fuels for over 4 years comprising of over 2500 unique samples at TARDEC's Army Petroleum Laboratory (APL) in New Cumberland, Pennsylvania. To evaluate the light obscuration particle counter limits the fuels were tested via particle counter as specified by IP 564 or IP 565 and ASTM D5452 to measure the particulate contamination content by gravimetric determination. Additionally free water content was determined by performing ASTM D3240 – Standard Test Method for Undissolved Water in Aviation Turbine Fuels when particle counts exceeded the 19/17/14/13 limits, but the gravimetric measurements were below 1.0 mg/L. The USAF provided 475 additional data points to the Army in an effort to establish into plane limits that are commensurate with the 0.5 mg/L limit employed for USAF aircraft.

ANALYSIS

A total of 2509 samples were tested for particulate contamination during the evaluation period. 2137 of the tested samples passed both the 1.0 mg/L gravimetric limit and the 19/17/14/13 particle count limit, with 372 samples failing either the particle count or gravimetric limit, or both test limits. 294 of the 372 samples tested failed the particle count limit and 246 of the 372 samples failed the gravimetric limit, 168 of the samples failed both the particle count and the gravimetric limit. 78 samples gave a false negative response to the proposed particle count limits by failing the gravimetric limit but passing the particle count limit. 126 samples gave a false positive response to the particle count limit by failing the particle count limit, but passing the gravimetric limit of 1.0 mg/L.

False Negatives

An analysis of the 78 false negatives which had a gravimetric measurement over 1.0 mg/L but a particle count under 19/17/14/13, showed that 58 of these samples may have a gravimetric contaminate load lower than the 1.0 mg/L limit based on the precision statement provided in ASTM D5452. While the published repeatability of ASTM D5452, Equation 1, only spans from 0 to 0.6 mg/L extrapolating the formulas to the 1.0 mg/L contaminate level provides a repeatability of 0.42 mg/L. Also the sample volume used to develop this formula was 5 liters, the testing performed by the Army only utilized 1 liter compounding to potential error. Analysis of the poor repeatability of ASTM D5452 does not address all false negatives as the particle counting methods also introduce repeatability and reproducibility errors aspects associated with their own methods as displayed in Figure 1.

$$r = 0.415x^{0.5}$$

Equation 1. ASTM D5452 repeatability

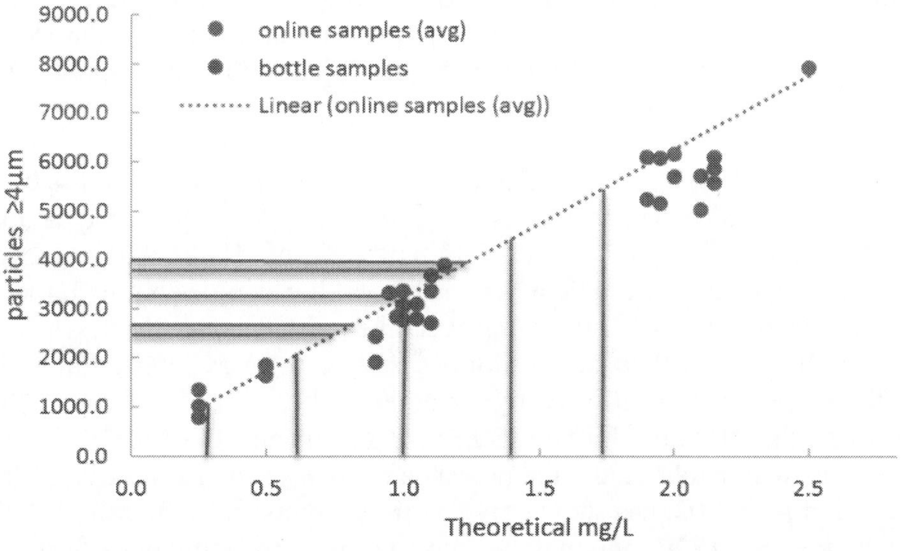

Figure 1. Reproducibility (red) and repeatability (blue) of ASTM D5452 extrapolated at 1 mg/L and IP 564 at 3242 particles ≥4 µm (c) for ISO 12103-1 A2 fine test dust

20 samples that have a high gravimetric measurement cannot be accounted for with particle counter readings. There is the potential that particles are present on the gravimetric filter pad

that has a nominal pore size of 0.8μm that could be either smaller than 4μm (c) or greater than 200μm (c) in the case of the IP 564 instrumentation or 70μm (c) in the case of the IP 565 specified instrumentation. Visual analysis of filter pads with a greater than 1.0 mg/L particle loading, as particles greater than 70μm or 200μm (c) would be visible to the naked eye and heavily contribute to the mass load of the tested filter, yet not be measured by the particle counters. Running the gravimetric samples in duplicate could also establish the actual repeatability of ASTM D5452 within the laboratory potentially identifying the actual source of error.

False Positives

An analysis of the 126 false positives which had a gravimetric measurement under 1.0 mg/L but a particle count over 19/17/14/13, showed that 56 of these samples may have a gravimetric contaminate load higher than the 1.0 mg/L limit based published repeatability of ASTM D5452. 118 of these samples where high in the 6μm (c),14μm (c), and/or 30μm (c) channels indicating the potential for free water contamination. 16 of these samples were tested via ASTM D3240 and confirmed to contain greater than 5.0 ppm free water, 15 samples contained 1-5 ppm, while 41 fuel samples were absent of free water, and 46 samples were not tested for free water. The utilization of glassware to transfer the fuel through the ASTM D3240 filter pad may have adversely impacted the free water measurements.

Into Plane Limit Evaluation

The United States Air Force requested the Army to propose into plane limits for particle counting based on the Air Force's 0.5 mg/L gravimetric limit as found in MIL-STD-3004D Change 1, and also found in ATA 103 (5). To support this request the Air Force provided particle count performed via ASTM D7619 and gravimetric data via ASTM D5452. Additionally the 2509 samples collected by the Army laboratory in New Cumberland were analyzed. Preliminary limits were established from the laboratory analysis of ISO-12103-1 A2 fine test dust at 0.5 mg/L giving an ISO 4406 code of 18/16/14.

The particle counts for the 475 samples from the USAF were compared to the 0.5 mg/L into plane gravimetric limit to determine a level of agreement. 435 of the tested samples passed both the 0.5 mg/L gravimetric limit and the 18/16/14 particle count limit, while 5 samples failed both limits, giving an agreement level of 92.6%. Thirteen of the samples not in agreement are false negatives, failing the gravimetric limit but passing the particle count limit, while the number of false positives, failing the particle count limit but passing the gravimetric limit is 22. The false negative readings may again be affected by the repeatability of ASTM D5452 which is 0.293mg/L at the 0.5mg/L limit, 8 of the 13 false negatives and 10 of the 22 false positives may also be affected by this. Additionally the presence of free water at levels as low as 5 ppm, 15ppm specified in ATA 103, has shown give particle counts exceeding the 19/17/14/13 limits implemented by the U.S. DOD, and may be contributing to the number of false positives identified at the 18/16/14 levels.

Expanding on the 0.5 mg/L gravimetric limit agreement with the 18/16/14 ISO code limit with the entirety of the Army's 2509 field samples detailed above 1603 samples passed both the gravimetric and particle count limits, while 348 samples failed both the particle count and gravimetric limits for an agreement level of 77.8%. The Army samples that are do not fall into agreement of with one another, give 347 false negatives, and 211 false positives. There is potential that the false negative readings are affected by the poor repeatability of ASTM D5452 with 287 of the 347 false negatives and 123 of the 211 false positives falling within the published repeatability range. Additionally the presence of free water may affect the number of false positives identified.

CONCLUSIONS AND RECOMENDATIONS

The laboratory data collected supports the U.S. Department of Defense ISO code limits of 19/17/14/13, developed from 1.0 mg/L concentration levels for the A1 and A2 test dusts, and down to a 5 ppm free water presence. The field data also supports the 18/16/14 ISO code condemnation limits when compared to the USAF 0.5 mg/L into plane limit. Additionally the technologies have demonstrated their advantages in field operations.

In addition to requirements to ensure that fuel is safe for flight operations, as DoD components bring more high pressure common rail engines into their respective inventories fuel injector wear should be evaluated to ensure that vehicles will meet the stated mean miles between hardware mission failure requirements, when operating on contaminated fuels.

The Department of Defense intends to move forward with the establishment of limits for fuel contamination that is allowable in aircraft and ground systems based on particle counter technologies.

REFERENCES

1. **Department of Defense Standard Practice.** Quality Assurance/Surveillance for Fuels, Lubricants and Related Products. *MIL-STD-3004D w/change 1.* December 7, 2016.

2. **Headquarters Department of the Army.** Petroleum Laboratory Testing and Operations. *Technical Manual No. 4-43.31.* Washington DC : s.n., May 30, 2012.

3. —. Concepts and Equipment of Petroleum Operations. *Field Manual No. 10-67-1.* Washington DC : s.n., April 2, 1998.

4. **Schmitigal, J.** Laboratory Evaluation of Light Obscuration Particle Counters used to Establish use Limits for Aviation Fuel. *TARDEC Technical Report 27480.* Warren, Michigan : s.n., December 2015.

5. **American Transport Association.** ATA Specification 103: Standards for Jet Fuel Quality Control. 2009. Vol. 2009.1.

6. **Schmitigal, J., Cox, D., Boenker, M., Krizovensky, J.0.** Update on the US Army's Fuel Contamination Detection Efforts. Seattle, Washington : Coordinating Research Council, Inc. 2011 Aviation Technical Committee Meetings, May 2011.

7. **Besse, G., Schmitigal, J.** Army's Evaluation of Aviation Fuel Contaminants Using Electronic Sensors. Alexandria VA : Coordinating Research Council, Inc. Vols. 2013 Aviation Technical Committee Meetings, May 2012.

8. **Schmitigal, J. and Bramer, J.** Field Evaluation of Particle Counter Technology for Aviation Fuel Contamination Detection – Fort Rucker. *TARDEC Technical Report 23966.* 2013.

9. **Schmitigal, J.** Evaluation of Particle Counter Technology for Detection of Fuel Contamination Detection utilizing Advanced Aviation Forward Area Refueling System. *TARDEC technical report.* Warren : s.n., 2014.

10. —. Evaluation of On-line Light Obscuration Particle Counters for Detection of Fuel Contamination. Warren, MI : s.n., August 2014. Vol. TARDEC Technical Report 25587.

11. **Dickerson, T.** Navy Field Evaluation of Particle Counter Technology for Aviation Fuel Contamination Detection. *NF&LCFT REPORT 441/14-003.* Patuxent River, MD : Naval Air Systems Command, February 6, 2014.

12. **Bigus, J. and Krizovensky, J.** Evaluation of On-line Light Obscuration Particle Counters for Detection of Fuel Contamination. Patuxent River, MD : Evaluation of On-line Light Obscuration Particle Counters for Detection of Fuel Contamination, February 9, 2012.

13. **Bessee, G.** Procurement of Instruments and Limited Comparison Study for Field Evaluation of Automatic Particle Counters. 2014. Vol. SwRI® Project No. 08.15954.08.001.

14. **Fielder, M.** The use of Light Obscuration Particle Counting to Study Dispersed Solid Contamination in Aviation Turbine Fuel at Kandahar Airbase, Afghanistan. [ed.] Parker. Sarasota, FL : IASH 2011 12th International Conference on Stability, Handling and Use of Liquid Fuels, October 2011.

15. **Bauer, C.** Bulk Diesel Fuel Filtration –A Sensible Investment. [ed.] Pall Corporation. Sarasota, FL : IASH 2011 12th International Conference on Stability, Handling and Use of Liquid Fuels, 2011.

16. **Dallas, A.,Block, J., Klick, P., Grove, B., Zastera, D., Doyle, J., Johnson, P., Elsayed, Y.** Contamination Found on Diesel Fuel Storage Tank Filters. [ed.] Donaldson Filtration Solutions. Sarasota, FL : IASH 2011 12th International Conference on Stability, Handling, and Use of Liquid Fuels, October 2011.

17. **Cummins Inc., BOSCH.** Joint Position on Fuel Quality for Common Rail Diesel Engines. December 7, 2011.

18. **Worldwide Fuel Charter, Fifth Edition. September 2013.**

19. Commonwealth of Australia. Petroleum, Oils and Lubricants Manual DEF(AUST)5695B. ***Australian Defense Standard.***

20. Pamas, Parker, Particle Solutions. A Proposal for quantitative Particle Counting Limits in Def-Stan 91-91. ***AFC Meeting.*** **February 28-29, 2012.**

IASH 2017, the 15TH INTERNATIONAL SYMPOSIUM ON STABILITY, HANDLING AND USE OF LIQUID FUELS
Rome, Italy
10-14 September 2017

STORAGE STABILITY STUDIES AND SHELF LIFE DETERMINATIONS OF COMMERCIAL BRAZILIAN BIODIESELS STOCKED IN SUBTROPICAL CONDITIONS IN CARBON STEEL CONTAINERS

Eduardo H. de S. Cavalcanti[1*], Adriane Zimmer[2], Marco Flôres Ferrão[3], Fatima Menezes Bento[4]

[1]Instituto Nacional de Tecnologia, eduardo.cavalcanti@int.gov.br
[2]Universidade Federal do Rio Grande do Sul, bio_didi@yahoo.com.br
[3]Universidade Federal do Rio Grande do Sul, mfferrao@gmail.com
[4]Universidade Federal do Rio Grande do Sul, fatimabento@ufrgs.br

*Correspondingand presenting author

ABSTRACT

In the present work we have studied the long term storage stability of three different commercial Brazilian methyl esters made from pure soy oil (SoyB100) containing TBHQ commercial antioxidant, 65% soybean 35% beef tallow (SoyBTBlend6535) and pure beef tallow (BTB100pure). They initially remained stored in HDPE60 L containers sent completely full bythe biodiesel producers and kept stored at thelaboratory premises for 60 days. This preliminary period was meant as most of commercial biodiesel stay in average 60 days stored or travelling before being pumped to the diesel, notably at far end regions of the country.Subsequently 1L carbon steel canisters with a small top hole giving access to the atmosphere were 80%filled with these biodiesels and kept stored for 90 days under sheltered but exposed to air, local heat and humidity conditions.Temperature (T) and relative humidity (UR%) at the open corridor of the laboratory- where the carbon steel canisters were left aside, were on-line monitored during the 90 days testing period. Monthly analyses of aspect, water content, specific gravity, kinematic viscosity, acid number and oxidation stability of the stored biodiesels were also conducted.The main conclusions that were reached are the following:

In terms of the water content, all of them went out of specification (above 200 ppm) after 30 days of storage in the carbon steel canisters. As far theoxidative stability is concerned only beef tallow biodieselremained above the limit of 6 hours valid on those days (summer of 2013) established by that time by the Brazilian Oil, Gas and Biofuels Regulatory Agency (ANP) specification for biodiesel, over the 90 days of storage in the carbon steel canisters. The soya methyl ester went out of specification after 30 days of storage. Regarding the monitoring of the acid index, only beef tallow biodiesel showed an index above the limited accepted value of 0.5 mg KOH/g going outside the specification after 60 days of storage under the experimental conditions. With regards to the aspect, specific gravity and kinematic viscosity, these properties remained within the acceptable range by the standards all over the 90 days storage period. In the present work limited shelf life of commercial soya biodiesel notably with respect to the water content and oxidation stability index was found. As aresult, special care should be addressed by the time that this sort of biodiesel is added to the diesel unless more robust antioxidant and multipack additives are used and special handling practices are adopted. This is particularly truein situations of: a) low turnover distribution chains where fuel remains stored for longer than usual; b) off-road applications notably at far distant areas of the manufacturing biodiesel industries and refinery regions and c) intermittent use situationsand first filling applications where the fuel stay under atmospheric storageconditions for long periods.

Keywords: storage stability, biodiesel, shelf life, fuel stability

1. INTRODUCTION

The storage stability of biodiesel is a matter of great concern notably in continental countries that possess complex distribution and retail fuels chains.This is particularly true in Brazil also due to the long distances from the production biodiesel regions to the refineries where it is added to the diesel.On top of thatthere areat least six biomes in Brazil ranging from cold high land regions in the South up to hot wet tropicaland semi-arid climates in the in the Northand Northeast, respectively. Subtropical weather however prevails is most of the economically active andurban regions of the country where most diesel consumers arelocated.This diversity added to the fact that most of the biodiesel plants are concentrated in the Midwest and Southern parts of the country poises an extra logistic challenge to keep the quality under full control all over the distribution, commercialization and retail chains. The biodiesel should be 100% in line with the specifications, so that total quality should be guaranteed up to the moment of fuel ignition in the diesel engines. At present almost 85% of the Brazilian Biodiesel is made of soya oil. It is well known the high susceptibility to oxidation of soya derived biodiesel products. This sort of aging is aggravated in tropical humid countries due to the high hygroscopic nature of biodiesel. Whilst air humidity tends to induce the incorporation of dissolved water and the establishment of hydrolytic degradation processes in the mass of stored biodiesel, the action of the oxygen of the air promotes its oxidation of biodiesel, with serious consequences for the end user, such as : Incorporation of dissolved water, cloudiness, release of free radicals, acidification, formation of abioticsludge, and in the case of the presence of water free of microbial sludge, besides attacks to polymeric and metallic materials, corrosion, turbidity and varnishing of surfaces, clogging and leaks. As a result the tendencyof biodiesel to suffer significant changes over atmospheric storagetime is always a matter of concern, asphysical and chemical keys properties tend to goout of specification with serious risks to the end users of diesel-biodiesel fuel, particularly for pumps, parts, tanks and injection systems components of vehicle and industrial engines employed in urban, off-road and in-land applications.

Unlike hydrocarbons of fossil origin that are relatively inert and retain their physicochemical characteristics practically unchanged for long periods of storage, biodiesel degrades over time, notably in aerated environments and hot and humid climates such as those prevailing in Brazil. At the same time, we also have temperate and cold climates, which are responsible for reducing the fluidity of certain types of biodiesel, such as beef tallow and palm oil biodiesel, which in extreme conditions such as winter, especially in the Southern Region tend to promote nucleation of crystals and solidify in the storage tanks.

The study of the capacity of water absorption by biodiesel over time is of great importance due to its high hygroscopicity. While diesel is extremely stable biodiesel incorporates water to your molecule with ease. In fact, due to its high hygroscopicity, biodiesel can become a product out of specification during storage [1]. It should be noted that the maximum limit of water content accepted the National Agency of Oil, Gas and Biofuels specifications [2] at the time of the present study was 350 ppm.

Studies conducted by CENPES / PETROBRAS [3] show that methyl biodiesel presents a higher capacity to absorb moisture than ethyl biodiesel and that biodiesel from castor oil usually incorporates more water than soybean. They recommended that a good monitoring of the evolution of the water content should be conducted until moments before adding biodiesel to the diesel. Studies conducted by INT also signal the reduced shelf life of soybean biodiesel.

It is also emphasized that as the dissolved water content in biodiesel increases, a slight increase in its acidity is observed. This may raise the corrosiveness in relation to metallic materials, such as steel, and also the detergency of biodiesel, in relation to polymeric materials. Studies conducted by INT in partnership with UFRGS highlight this aspect for the case of soybean methyl biodiesel stored in steel containers [4,5]. That is, hydrolysis of biodiesel tends to interfere not only with its own physical and physicochemical characteristics, but may also affect the integrity and stability of materials that are in contact with this biofuel, such as metallic and polymeric materials present in parts car and storage structures. It should be also mentioned that when water contents raises to above the dissolution limits of water in biodiesel, small water droplets appear and coalesced ones tend to settle in the bottom of the tanks. This free water layer presence tends to lead to biofouling formation and biocorrosion establishment in tank bottoms [6].

A second, equally important kind of degradation process concerns oxidation. We can define it as the interaction of oxygen molecules with the fatty acid chains present in biodiesel [1]. Like butter that oxidizesatambient temperature (turning it rancid), biodiesel tends to oxidize to a greater or lesser degree, as the buritic acid,the oxidation of the triglycerides present in the butter leads to the formation of compounds with unpleasant smell and flavor. In the case of biodiesel, oxidation has greater consequences, as it not only results in the release of free radicals, the formation of relatively unstable hydro peroxides, which decompose, promoting the increase of acidity and polymerization, as well as promoting the formation of gums and sludge.With major implications for product quality assurance. Due to its differentiated chemical composition, despite its similarities in terms of performance with the diesel molecule, which accredits it as an advantageous substitute for diesel, biodiesel as a rule is more susceptible to oxidation. Take, for example, the case of soybean oil, which contains eleven types of fatty acids in its composition. Among these, four types of carbon chains (> 50%) predominate with 18 carbon atoms, such as stearic (18: 0), oleic (18: 1), linoleic (18: 2) and linolenic fatty acids (18: 3), based on about 85% of the methyl biodiesel produced in Brazil today. The digits 0, 1, 2 and 3 denote the number of unstaurations (double bonds) in the chain. It should be noted that the transesterification process that converts this oil into biodiesel carries these carbon chains into the final product. These double bonds, in fact, act with sites of destabilization of the molecule, that is, as an entrance gate for the devastating action of oxygen. While stearic acid is immune to oxidation (which predominates in sebum biodiesel, for example), the unsaturated acids that predominate in the soybean oil composition exhibit very high rates of oxidation [7]. It should be noted that such fatty acids as linoleic and linolenic respectively exhibit oxidation rates 41 and 98 times more intense than that of oleic acid. Therefore, as already mentioned, which make up more than 50% as both methyl and soybean biodiesel, such as cotton, make such products extremely susceptible to oxidation.

It is interesting to note that almost all the scientific publications on stability observed in the literature deal with varieties of overseas biodiesel, such as rapeseed oil derived ones [8]. They are also restricted to the storage conditions of the northern hemisphere. There are few studies that portray the oxidative profile and the stability to the storage of the national biodiesel and under national climatic conditions.

Indeed, storage stability becomes worrying when biodiesel is stored for a long time. We refer to large volumes remain stored for long periods indoors, semi-closed and atmospheric - with relief valves only, or eventually opened for supply operations. These conditions are characterized by the presence of air, moisture and temperature variations in the range of - 5 ° C to 49 ° C. They are also characterized by the absence of thermal shocks - such as those found in the vicinity of the vehicular and stationary engines, a process of instabilities that will be discussed in the continuation of this communication. Change in composition and loss of adherence to specifications is probably one of the first concerns of these associated problems arising from the extent of exposure or contact with atmospheric air. In addition, storage stability can also be influenced by the effect of temperature, exposure to sunlight, as well as when biodiesel is placed in contact with metal alloy surfaces rich in transition metals such as copper and zinc [10,11]. Operational changes in storage containers, loading operations under adverse weather conditions, such as under high relative humidity conditions, and the nature of the flow in the supply can also influence the behavior of biodiesel [12].

Such concerns afflict not only the scientific and technological community, producers, large buyers, distributors, as well as ANP, SINDICOM and FECOMBUSTÍVEIS, who also expressed their concerns as soon as the B5 blend was made mandatory in 2010. In this regard, it should be highlighted developed by INT, published in a workshop promoted by EMBRAPA, which revealed that about 25% of biodiesel shipments reached their destination in distributors, outside the specification [12].

International reports highlight changes in some properties of biodiesel and its composition over the time of storage. The main properties that change are: dissolved water content (DWC), acid value (AV), peroxide value (PV), induction time (IT) or oxidative stability index (OSI), kinematic viscosity (KV), among others [8,11,13,14].
The main biodiesel properties sensitive to the incidence of degenerative processes of abiotic origin (hydrolytic and oxidative) that may suffer significant changes during storage, as well as the limits reported for the storage of biodiesel and the effect of contact with metals, based on the literature consulted [5]. In the Brazilian case soybean biodiesel, soybean / tallow and beef tallow blends dominate our market. We therefore sought to examine this issue in the present work and to define the shelf life of these three most common types of domestic methane biodiesel under actual storage conditions in steel containers. In steel containers.

2. EXPERIMENTAL

2.1 Analytical Methods

During the 90 days of storage, 800 mL of biodiesel were kept in carbon steel containers under sheltered conditions exposed to heat and humidity. Each biodiesel was added in two sets of four 1L containers made of AISI 1020 carbon steel (thus observing a head-space of 80%). One inch diameter threaded covers containing 0.8 mm mesh stainless steel shield screens were used in order to allow only moisture ingress and not external impurities. We attempted to portray the most critical storage situation in which atmospheric tanks are commonly used. Containers were used in triplicate, for each monthly periodic removal. Due to the susceptibility to atmospheric corrosion on the inner walls of these containers, a process of pre-blasting the internal parts of the containers was carried out according to the LACOR-INT procedure. During the test period the recipients remained in a corridor of the LACOR-INT premises, close to the window to be exposed to external ambient conditions of heat and humidity. The ambient temperature and relative humidity outside the containers were monitored with the aid of atemperature sensor and external humidity withthe help of a digital thermo hygrometer three times a day. During the 90 day period, monthly monitoring of six physical-chemical characteristics was carried out, according to the methodology applied by the laboratory in previous studies [14, 5].

2.2 Biodiesel samples

Commercial biodiesel volumes were sent by the manufacturers in 100% filled high-density polyethylene bottles (HDPE) up to the mouth and sealed. They were kept for 60 days in these hermetic conditionsstored in the laboratory. After this preliminary periodthey were opened and filling of steel containers started. In parallel, samples were collected and sent for characterization analyzes, called time 0 samples, at LACOR-INT laboratories, in order to verify if the biodiesel still complied with the Parameters established by the ANP, 60 days after being receipt at INT. These 60 days prior to the steel storage were established based on the information of the distribution market. Cases were reported in which the addition of biodiesel to the diesel was given after a maximum of 60 days of manufacture by the producers. This was especially noted in regions where the receiving and distributing bases were at great distances from the producing centers (e.g. north and northeast regions).

During the test period the tanks remained in a corridor of the LACOR-INT premises, close to the window to be exposed to external ambient conditions of heat and humidity. During the 90-day period, the monthly monitoring of six physical-chemical characteristic described in Table 1 was carried out, according to the methodology applied by the laboratory in previous studies [4,5].

The appearance of the biodiesel samples was determined following the method described in ABNT NBR 16048 - Biodiesel: Determination of appearance, whose evaluation consisted in observing the level of turbidity and the presence of impurities in the same at a temperature of 26°C ± 1°C. A standard turbidity classification chart in accordance with ASTM D 4176 - Standard test method for free water and particulate contamination in distillate fuels (visual inspection procedures) was used. The density was determined at 20°C using the method according to ASTM D 4052 - Density and relative density of liquids by digital density meter. The equipment used was the ANTON PAAR Digital

Density Meter, model DMA 5000. Kinematic viscosity at 40°C was also monitored monthly. The method adopted was the one recommended by ABNT NBR 10441 - Petroleum products - Transparent and opaque liquids - Determination of kinematic viscosity and calculation of dynamic viscosity. It was used as a set of Cannon Fenske viscometers, calibrated chronometers with 0.01s resolution, thermostatic bath, and a thermometer calibrated with ± 0.05 ° C subdivision and a manual vacuum pump. The monthly monitoring of the water content was done using the Karl Fischer Coulometric method. The standard method ASTM D 6304 - Standard Test Method for Determination of Water in Petroleum Products, Lubricating Oils, and Additives by Coulometric Karl Fischer Titration was adopted. The equipment used was the Karl Fischer Titrator METROHM (model 756 KF) coupled to a Mettler XP-205 analytical balance of 0.1 mg of resolution.

Table 1 - Properties of biodiesel time 0 samples

Property	Test method	Units	ANP 2012 spec limits	SoyB100	SoyBTBlend6535	BTB100pure
Appearance	ABNTNBR16048-12	-	-	limpid	limpid	limpid
Water content	ASTM D 6304-07	ppm	max.350(1)	302.0	349.4	156.0
Oxidation stability	EN 14112-03	hours	min 6(2)	8.7	7.8	>24
Density (20°C)	ASTM D 4052-09	$kg.m^{-3}$	850- 900	881.2	879.5	871.5
Viscosity (40°C)	ASTM D 445-11	$mm.s^{-2}$	3 - 6	4.01	4.53	4.88
Acid value	ASTM D 664-11	mg KOH/g	max.0.50	0.39	0.20	0.46

(1) max 200 ppm as established by the ANP 2015 specification limit
(2) min 8hs as established by the ANP 2015 specification limit

The acidityvalue assessmentwas performed according to ASTM D 664 - Acid number of petroleum products by potentiometric titration. The equipment used was aMetrohm, model 905 Titrando, with the aid of the software TIAMO 2.2. The standard method ABNT NBR 14248 (ABNT, 2009) was also used. For the determination of oxidative stability, the standard EN 14112 - Fat and oil derivatives - Determination of oxidation stability (accelerated oxidation test) was followed and the equipment used was the Rancimat 873 METROHM.

Multivariate analysis of the data through principal component analysis (PCA) and hierarchical cluster analysis (HCA) was performed using the Chemostat program[15] for water content (WC), oxidation stability index (OSI), density (D), viscosity (V) and acid value (AV).

3. RESULTS AND DISCUSSION

The variability of temperatureand relative humidityof the external environment wasmonitored during the 90 days storage period, between October and December 2013. According to the treatment of the data, the relative air humidity remained within a range of 30 to 80% (mean

RH% = 55%), while the temperature outside the recipients varied between 24 and 34°C (mean T = 29°C).

The methodology used to evaluate the biodiesel's physical appearance aimed to observe the degree of turbidity and the presence of impurities, compared to an ASTM standard letter of turbidity ranging from 1 (limpid) to 6 (high level of turbidity). The classification identified for the three biodiesels studied was degree 1, proving to be clear and free of impurities as specified in the ANP 2012 resolution [2].

Figure 1shows the appearance of the three biodiesels at the start of the90 days of storage.

SoyB100Soy **BTBlend6535** **BTB100pure**

Figure 1. Initial appearance (time 0- October 2013) of the biodiesels tested showing clear and impurities-free condition.

Of interest to note that during the 90 days of the tests, the original appearance was maintained and the presence of impurities was not verified, which gave them the LII classification (clear and free of impurities) according to the ANP 2012 resolution [2].It should be noted that this appearance was kept for all the three biodiesels (grade 1 turbidity), that is, they remained completely translucent throughout the storage.

Figure 2 shows the evolution of the water content with storage time of the three biodiesels samples.

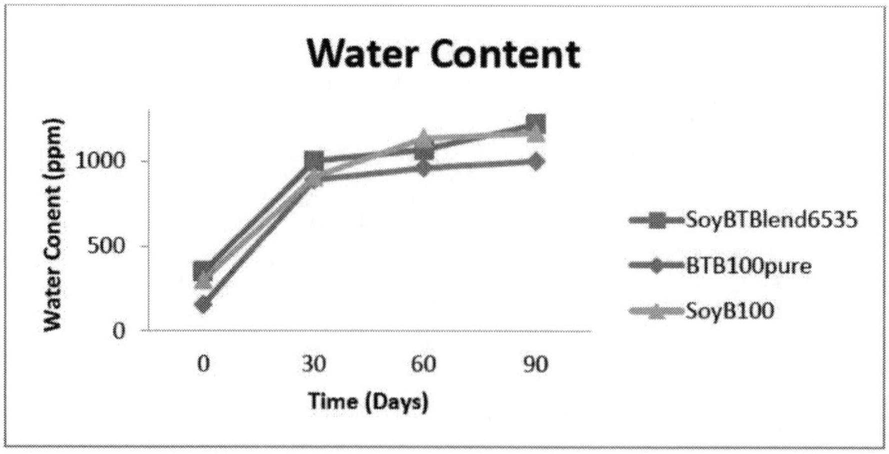

Figure 2. Water content evolution of the three biodiesels samples over storage time.

An increase was observed in the water content duringthe storage period, exceeding the maximum water content limit of 350 ppm, established by the 2012 ANP Resolution [2], after 30 days of storage. This is due to the hygroscopicbehavior of the biodiesel. Moisture is an essentialparameter in biodiesel quality, since water can cause areaction of unwanted hydrolysis, producing free fattyacids, which may cause corrosion and engine problems [5].

Figure 3 illustrates the evolution of the oxidation stability with storage time of the three biodiesels samples.

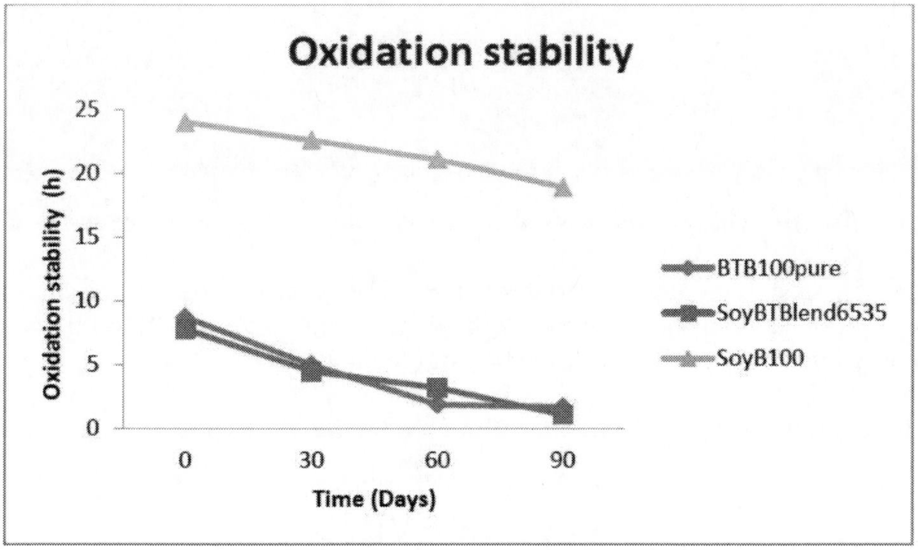

Figure 3. Oxidation stability index evolution of the three biodiesels samples over storage time.

As it can be seen the oxidation stability index (OSI) decay of the three biodiesels, which reflects the antioxidant reserve still present in biodiesel. We can also observe in Figure 3 a large decrease of soybean biodiesel was observed despite the presence of antioxidant according to the manufacturer's declaration. The blend 65% soybean and 35% beef tallow biodiesel also showed OSI reduction. However, with regards to pure beef tallow biodiesel, the induction period remained high, which is justified by its composition in which relatively stable saturated compounds predominate. It is also worth mentioning that after the 30 days of storage, both biodiesel 100% soybean and biodiesel 65% soybean and 35% tallow, started to show values of the induction period below the limit specified by ANP Resolution [2], which gave them a limited shelf life less than 30 days, under the storage conditions of the present study.

Figure 4 shows the evolution of the density values at 20 °C of the samples during the evaluation period.

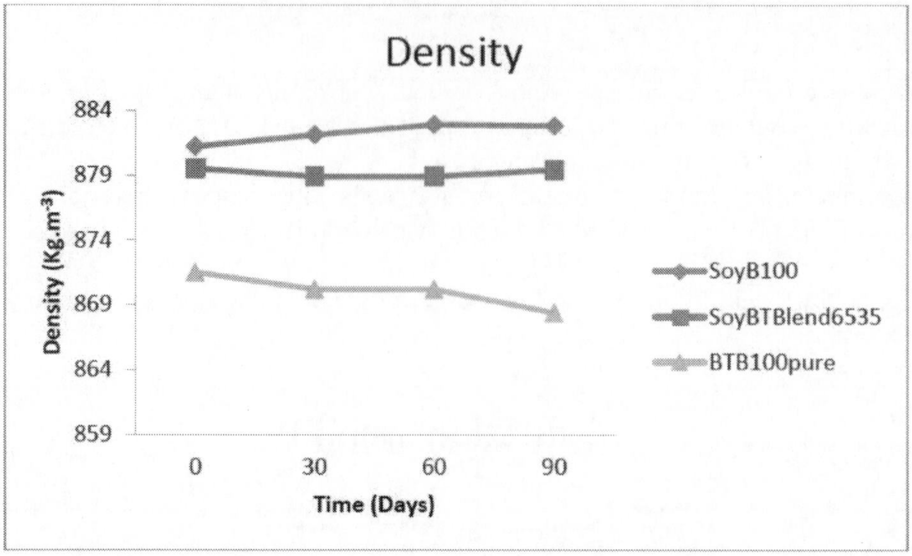

Figure 4. Density values evolution of the three biodiesels samples over storage time.

Initially all the methyl biodiesel samples exhibited the density values within the specification range and were thus maintained over the ninety days. The small variation found between the analyzes was found to be within the variation of the method.

The evolution of the viscosity values of the methyl biodiesel samples are shown in Figure 5.

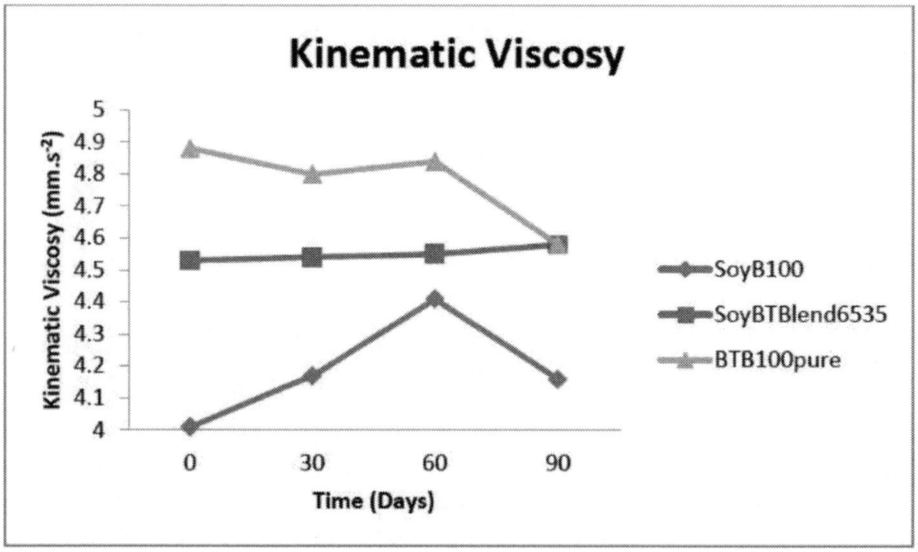

Figure 5.Influence of storage time on the kinematic viscosity of biodiesels samples.

With regards to the viscosity values they were kept within the limits of the specification during the ninety days of storage for the three samples studied. Some oscillations observed were also considered as marginal.

Figure 6 shows the changes in acid value over the storage period of ninety days. As expected, acid value increased with storage time for all biodiesels.

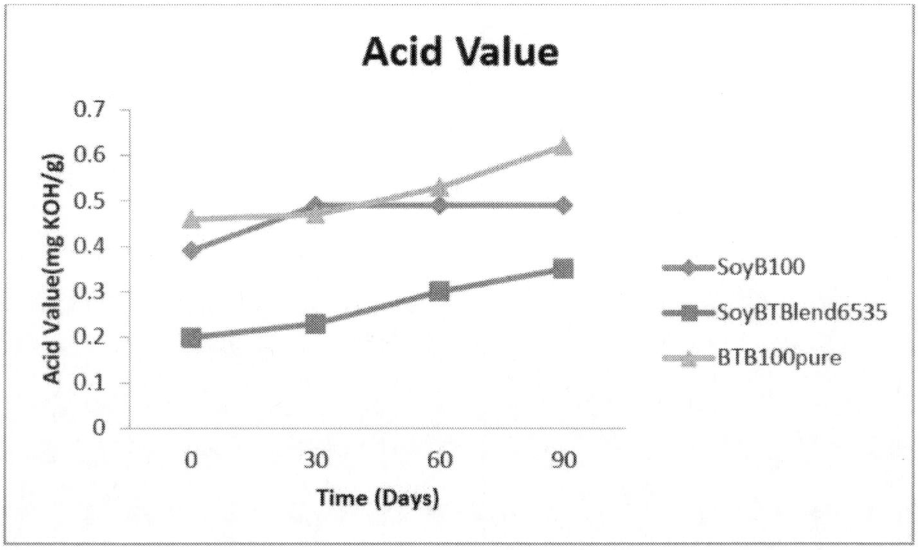

Figure 6. Influence of storage time on the acid value of biodiesels samples.

With respect to the 100% pure soy biodiesel after an initial increase, there was no preponderant increase in the acidity value, so that they remained withinthe limits of the specification of 0.5. In relation to the sample of biodiesel composition 65% soybean and 35% beef tallow, the initial acidity values were relatively smaller. Afterwards a slight increase was observed butwasnot significant to supersede the specification limit. It should be highlighted however that the pure tallow biodiesel, presented the highest initial acidic value, but as the storage time passed, a value above the requirement specified started to appear after 60 days of storage.

In the dendrogram of Figure 7 we can observe the formation of 2 groups, one being formed by the experiments that used only beef tallow biodiesel and another one formed by the samples that contained only biodiesel of soybean or mixture of this one with beef tallow biodiesel.

A more complete evaluation can be obtained by the Principal Component Analysis (PCA) (Figure 8), in which we can attribute the separation of the experiments with beef tallow biodiesel (negative scores in PC1) to the higher values of viscosity (V) and oxidative stability (OS), while the Other experiments (with positive PC1 scores) are characterized by higher density (D). In addition, PC2 is related to the storage time of the experiments, indicating that as it progresses, water content (W) andacid value (AV) increase independently of the composition of the biodiesel used.

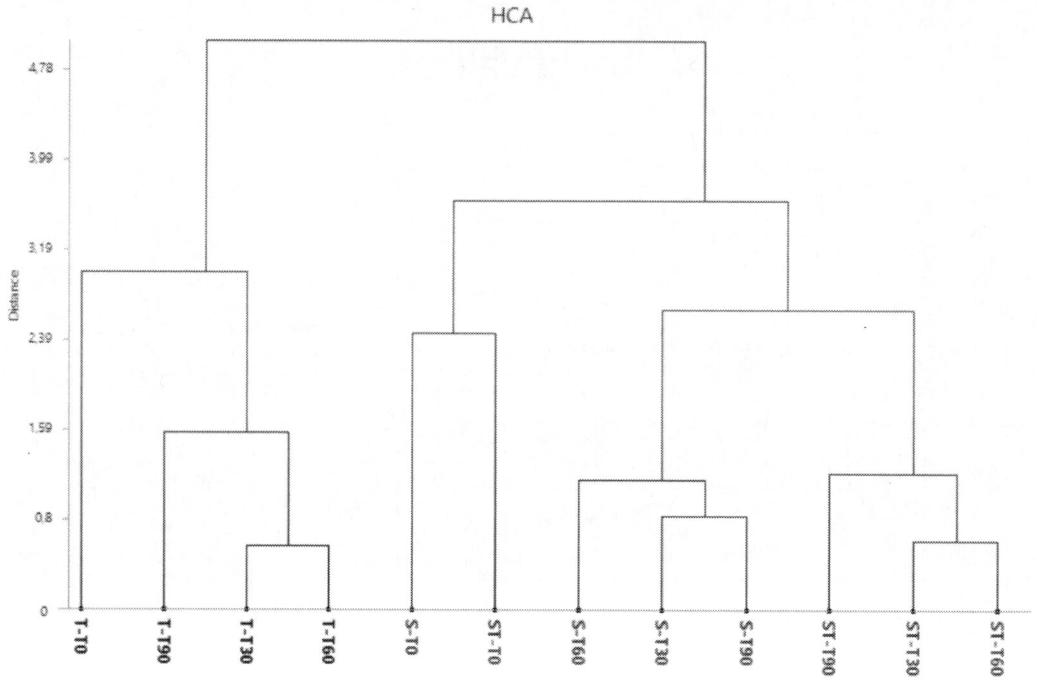

Figure 7.Dendrogram for soybean biodiesel – GREEN, soybean/beef tallow (65%/35%) biodiesel – RED and beef tallow biodiesel - BLUE.

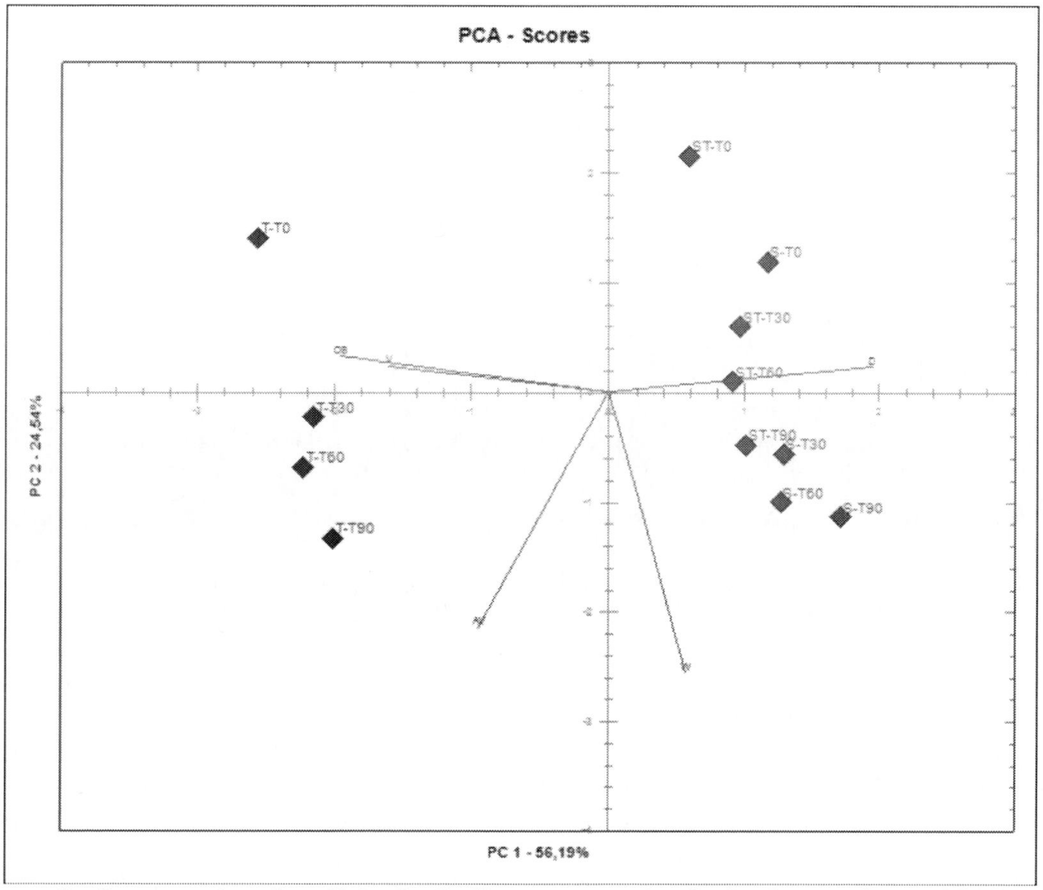

Figure 8. Scores plot for Principal Component Analysis for soybean biodiesel – GREEN, soybean/beef tallow (65%/35%) biodiesel – RED and beef tallow biodiesel – BLUE.water content (WC), oxidation stability index (OSI), density (D), viscosity (V) and acid value (AV)

4. CONCLUSIONS

The main conclusions that were reached are the following:

In terms of the water content, all of them went out of specification (above 200 ppm) after 30 days of storage in the carbon steel canisters. As far the oxidative stability is concerned only beef tallow biodiesel remained above the limit of 6 hours valid on those days (summer of 2013) established by that time by the Brazilian Oil, Gas and Biofuels Regulatory Agency (ANP) specification for biodiesel, over the 90 days of storage in the carbon steel canisters. The soya methyl ester OSI went out of specification after 30 days of storage. Regarding the monitoring of the acid index, only beef tallow biodiesel showed an index above the limited accepted value of 0.5 mg KOH/g going outside the specification after 60 days of storage under the experimental conditions. With regards to the aspect, specific gravity and kinematic viscosity, these properties remained within the acceptable range by the standards all over the 90 days storage period. In the present work limited shelf life of commercial soya biodiesel notably with respect to the water content and oxidation stability index was found. As a result, special care should be addressed by the time that this sort of biodiesel is added to the diesel unless more robust antioxidant additives are used and special handling practices are adopted.

5. ACKNOWLEDGMENTS

The authors are grateful to National Council for Scientific and Technological Development (CNPq), Agency for P&D Studies and Projects (FINEP), National Technology Institute (INT), Ministry of Science, Technology and Innovation (MCTI), Federal University of Rio Grande do Sul (UFRGS) as well as to the RBIOCOMB Network (www.rbiocomb.com.br) for their support and financial grants.

6. BIBLIOGRAPHIC REFERENCES

1. HOSHINO V., IWATA V., KOSEKI H. Oxidation, Stability and Risk Evaluation of Biodiesel. Thermal Science, v. 11, No. 2, p..87-100, 2007.

2. ANP, Resolução ANP N° 14, de 11.5.2012 - DOU 18.5.2012 – Access: http://nxt.anp.gov.br/nxt/gateway.dll/leg/resolucoes_anp/2012/maio/ranp%2014%20-%202012.xml.

3. VIERA J. A. V., LIMA M. S. S., DIAS, B. S. Estudo de Higroscopicidade do Biodiesel. Petro & Química, p. 64-68, November 2006.

4. CAVALCANTI, E. H. S. et al .Vida de prateleira e avaliação da contaminação microbiana de biodiesel e de misturas B5 e B20 armazenados em recipientes de aço de 1L - resultados preliminares (30 dias de estocagem). In: 4° Congresso da Rede Brasileira de Tecnologia de Biodiesel - 7° Congresso Brasileiro de Plantas Oleaginosas, Óleo, Gorduras e Biodiesel, 2010, Belo Horizonte. 4° Congresso da Rede Brasileira de Tecnologia de Biodiesel - 7° Congresso Brasileiro de Plantas Oleaginosas, Óleo, Gorduras e Biodiesel,. v. 03. p. 1353-1354, 2010.

5. ZIMMER A., CAZAROLLI J. , TEIXEIRA, R.M. , VISCARDI, S.L.C. , CAVALCANTI E.H.S., GERBASE A.E., FERRÃO M.F. , PIATNICKI C.M.S., BENTO, F.M. Monitoring of efficacy of antimicrobial products during 60days storage simulation of diesel (B0), biodiesel (B100) and blends (B7 and B10). Fuel (Guildford), v. 112, p. 153-162, 2013.

6. LEE J.S., RAY R.I., LITTLE B.J., An assessment of alternative diesel fuels: microbiological contamination and corrosion under storage conditions. Biofouling Vol. 26, No. 6, 623–635, 201o.

7. KNOTHE, G. et al, The biodiesel handbook, AOCS Press, 2005.

8. MITTELBACH M, GANGL S. Long storage stability of biodiesel made from rapeseed and used frying oil. Journal of the American Oil Chemists' Society, v. 78, p. 573–7, 2001.

9. YANG Z, HOLLEBONE B.P., WANG Z., YANG C., LANDRIAULT M. Effect of storage period on the dominant weathering processes of biodiesel and its blends with diesel in ambient conditions. Fuel,v. 104, p.342–50, 2013.

10. BONDIOLI P, GASPAROLI A, DELLA BELLA L, TAGLIABUE S, TOSO G. Biodiesel stability under commercial storage conditions over one year. European Journal of Lipid Science and Technology, v.105, p.735–41, 2003.

11. LEUNG D. Y. C, KOO B.C.P, GUO Y. Degradation of biodiesel under different storage conditions. Bioresource Technology, v. 97, p. 250–6, 2006.

12. CAVALCANTI, E. H.S., RESENDE, V., PACHECO, H.G.J., Monitoramento de não conformidades de natureza abiótica e de origem operacional observadas no biodiesel e na mistura biodiesel/diesel. Anais: Workshop Controle Pleno da Qualidade do Biodiesel e Mistura com Diesel, Brasília, DF, p.19-32 (Access on-line link: https://www.embrapa.br/agroenergia/busca-de-publicacoes/-/publicacao/990144/anais), 2012.

13. SCHOBER S., M. MITTELBACH M., The impact of antioxidants on biodiesel oxidation stability. European Journal of Lipid Science and Technology, v. 106, p. 382–389, 2004.

14. ASHRAFUL, A. M., MASJUKI, H. H., KALAM M. A., ASHRAFUR RAHMAN, S. M. HABIBULLAH M., and SYAZWAN, M. Study of the Effect of Storage Time on the Oxidation and Thermal Stability of Various Biodiesels and Their Blends. Energy Fuels. v. 28 (2), p. 1081–1089, 2014.

15. HELFER G. A., BOCK, F., MARDER L., FURTADO J. C., COSTA A., FERRÃO M. F.. CHEMOSTAT: Exploratory multivariate data analysis software. Química Nova, 38(4), 575-579, 2015.

IASH 2017, the 15TH INTERNATIONAL CONFERENCE ON STABILITY, HANDLING AND USE OF LIQUID FUELS
Rome, Italy
10-14 September 2017

THERMAL AND OXIDATIVE INSTABILITY IN BIODIESEL BLENDS DURING VEHICLE USE AND ONBOARD FUEL STORAGE

Steven R. Westbrook[1]

[1] US Army TARDEC Fuels & Lubricants Research Facility (SwRI), Southwest Research Institute, 6220 Culebra Road, San Antonio, Texas 78238 USA. swestbrook@swri.org

ABSTRACT

This paper covers the results of Coordinating Research Council (CRC) Project AVFL-17c to study the thermal and oxidative stability of biodiesel blends during vehicle use and onboard storage. The CRC considers the work reported herein as Phase I of a potentially larger-scope project. The objective of this first phase was to examine the impacts of various aspects of fuel stability (temperature, pressure, oxidation, free radical formation, acid formation, etc.) on biodiesel blends during onboard vehicle use and storage conditions

The project was divided into 6 different tasks: 1) conduct a literature review of oxidation and thermal fuel stability parameters as pertains to vehicle use and storage, 2) make selections of pertinent vehicle technology pertaining to possible stability parameters, 3) identify and obtain test fuels, 4) vehicle testing to collect fuel system condition data and effects on biodiesel degradation, 5) develop bench-scale, fuel-stressing methodology to mimic the conditions that fuel experiences in HPCR fuel systems, and 6) provide recommendations for Phase II (fuel effects testing) to CRC.

Biodiesel blends (B0, B5, and B20) were tested in vehicles using on-road and dynamometer testing. Fuel samples were collected from vehicle fuel tanks during test cycles and analyzed using tests selected to indicate changes in thermal and oxidative stability. Test fuels were also stressed using a bench-scale fuel stressing apparatus. Samples of the stressed fuel, taken at pre-selected intervals, were also tested using the same protocol as the vehicle fuel tank samples. Test results from all samples were analyzed for correlations between bench-scale and vehicle stressing.

INTRODUCTION

This paper is a summary overview of a more detailed report published by the Coordinating Research Council, Inc. Readers should obtain the report for a more detailed presentation of the research program and the results.[1]

As diesel engines for light-duty vehicles have evolved, the fuel burned in the engine has been exposed to increasingly severe levels of heat, pressure, and oxidation. These conditions can cause chemical changes (degradation) in the fuel resulting in undesirable, even damaging, consequences to the fuel system. The resistance of fuel to these changes is the stability of the fuel.

The commercial diesel fuel specification, ASTM D975[2], contains no requirement, only guidance, regarding the stability of diesel fuels. Much of the guidance, as it relates to vehicles and fuel systems, was developed for older technology engines. More modern, common-rail fuel systems expose fuel to significantly higher temperatures and pressures compared to these older technologies. Additionally, the D975 guidance was prepared prior to the allowance for up to 5% biodiesel (B5) and so may not adequately represent that potential concern.

As such, existing stability test methods and specification guidance do not adequately address stability of diesel fuel for modern diesel-powered vehicles. The Coordinating Research Council (CRC) solicited proposals for an investigation of diesel fuel stability to address these concerns.

As mentioned above, fuel stability is typically defined as "the resistance of the fuel to physical and chemical changes brought about by the interaction of the fuel with its environment.[3]" Most research has divided the general topic of fuel stability into three, broad categories: thermal, oxidation, and storage. Thermal stability is the resistance of fuel to change caused by thermal stress. Oxidation stability is the resistance to change under oxidizing conditions. Storage stability is the resistance to change during the typical storage of fuel, prior to burning it in an engine. For most of the years that diesel fuel has been used, the need to measure, predict, and control the stability of the fuel has resulted in many studies. Some of those studies resulted in the development of standardized stability test methods. Within the United States, the most commonly used test methods are published by ASTM International. A partial listing of diesel fuel stability test methods is given below:

- Standard Test Method for High Temperature Stability of Distillate Fuels, ASTM D6468
- Standard Test Method for Oxidation Stability of Distillate Fuel Oil (Accelerated Method), ASTM D2274
- Standard Test Method for Distillate Fuel Storage Stability at 43 °C, ASTM D4625
- Standard Test Method for Assessing Distillate Fuel Storage Stability by Oxygen Overpressure, ASTM D5304
- Standard Test Method for Oxidation Stability of Middle Distillate Fuels – Rapid Small Scale Oxidation Test (RSSOT), ASTM D7545

Some test methods are published by other organizations, such as ISO:

- EN 15751, Fatty acid methyl esters (FAME) fuel and blends with diesel fuel – Determination of oxidation stability by accelerated oxidation method. [This test method is more commonly referred to as the Rancimat test.]

Since the introduction of the high-pressure, common-rail (HPCR) fuel system for diesel engines, there has been a growing acknowledgement within the industry that the stability tests listed above do not adequately account for the conditions experienced by the fuel in the fuel system. The temperatures and pressures within the HPCR system are far greater than those in the older technology fuel systems. For the above reasons, it has become necessary to develop new stability test methodology. And, that new methodology must be based on an accurate understanding of the conditions to which diesel fuel is exposed in a modern, diesel-powered, vehicle.

The CRC considers the work reported herein as Phase I of a potentially larger-scope project. The objective of this first phase is to examine the impacts of various aspects of fuel stability (temperature, pressure, oxidation, free radical formation, acid formation, etc.) on biodiesel blends during onboard vehicle use and storage conditions. More specific goals include:

- Identify the environmental conditions and chemical processes which influence thermal and oxidative instability.
- Investigate the relationship between such processes and a measurable property of the fuel such as the onset/presence of free radicals, peroxide formation, acid number, oxidation induction period, etc.
- Develop a surrogate process or performance test to simulate those impacts similarly to the effect of fuel stored in a light duty vehicle operated on a daily basis.

Create oxidized biodiesel blends according to a given oxidation stability requirement.

APPROACH

The project was conducted in several tasks, as detailed below.

- Task 1 – Conduct a literature review of oxidation and thermal fuel stability parameters as pertains to vehicle use and storage. The review will also include looking for possible fuel stability screening tools or test methods.
- Task 2 – Based on the results of Task 1, input from technical contacts, and SwRI experience, make selections of pertinent vehicle technology pertaining to possible stability parameters. This task will also include gathering available information regarding fuel residence time, fuel temperatures, and system pressures in actual light duty vehicles operating in hot, summer-time conditions.
- Task 3 – Identify and obtain test fuels.
- Task 4 – Vehicle Testing to collect fuel system condition data and effects on biodiesel degradation.
- Task 5 – Develop bench-scale, fuel-stressing methodology (mimic the conditions that fuel experiences in HPCR fuel systems)
- Task 6 – Provide recommendations for Phase II (fuel effects testing). *[This task is not discussed in this paper. Refer to the original report for this information.]*

RESULTS AND DISCUSSION

The results of the literature review are not presented here. Readers are encouraged to obtain the full report from the Coordinating Research Council.

Task 2 Test Vehicles

For this task, two different, light-duty, diesel-powered vehicles were selected:

- 2013 Chevrolet Silverado 3/4-ton pickup truck with the 6.6L diesel engine
- 2012 Mercedes ML350 Blue Tec

These two vehicles were used in order to have a light duty truck and a passenger automobile. It was thought that these vehicle types would offer some differences in the fuel systems. Which would, in turn, have different effects on the fuel. SwRI initially proposed using a Volkswagen Jetta as the passenger automobile; but, it was replaced with the ML350 when the manufacturer offered the Blue Tec as a loan to the project. Appendices B and C contain information regarding the instrumentation of each vehicle for the testing that was conducted.

Task 3 Test Fuels

For this task, four potential blend components were obtained:

- Two hydrocarbon-only (petroleum) ULSD
- Two B100 biodiesel
    - 1 soy derived
    - 1 mixed fats derived

Table 1 contains a description of each component; and, the complete analysis results are given in the original report. After consultation with AVFL panel members, it was decided to use diesel fuel CL14-6132 and biodiesel CL14-6152 to blend the various test fuels. The ULSD was thought to be more widely used than the TX-LED fuel and the soy biodiesel was chosen because it was more commonly used than mixed-fats biodiesel.

Table 1. Candidate Test Fuel Blend Components

Sample ID	Description
CL14-6132	ULSD B-0, DF-8486.
CL14-6133	#2 TX-LED 15ppm ULSD
CL14-6152	Soy Methyl Ester Biodiesel
CL14-6173	Biodiesel-Mixed Fats

Task 4 Vehicle Testing

The objectives of the vehicle testing were:

1. Mimic the degradation of specification-quality biodiesel blends seen by CRC AVFL members in other vehicle studies.
2. Identify analytical methods capable of detecting degradation of the test fuels during vehicle operation.
3. Determine the maximum temperatures and pressures that occur in the vehicle fuel systems. These parameters could then be used in designing non-vehicle, fuel-aging rigs.

4. Measure the conditions of fuel temperature and pressure in a light duty vehicle operating in summer.
5. Expose fuel(s) to these conditions and evaluate the effect on the fuel as measured by numerous candidate fuel stability test metrics.

On-Road Fuel Evaluation

An on-road "aging" of each test fuel was conducted in each vehicle. Approximately ½ tank of test fuel was put in the vehicle's tank. Mileage accumulation was conducted on the SwRI campus and the local public roads. The road course for the evaluation was the modified ASTM D5500 Intake Valve Deposit route used in CRC Project E-90-2b[4], which includes both city and highway driving as well as a soak and extended idle. Each cycle is approximately 23.5 miles and is followed by a 20-min soak and a 15-min idle.

1. During the first day of operation of each vehicle, driving cycles were performed until the fuel gauge was reading approximately ¼ of a tank. The parameters were continuously measured and recorded at a data acquisition rate of 1 Hertz while the engine is running, during the 20-min soak, and for 30 min after the engine was shut off at the end of the 15-min idle.
2. Once the vehicle's engine was shut off after the 15-min idle at the end of the first day of driving, a one-liter fuel sample (#1) was taken from the tank sample tube.
3. On the next morning, a 0.5 liter sample (#2) was taken from the fuel tank.
4. During the second day the road course was again driven until approximately 1 gallon of fuel was left in the tank.
5. A 0.5-gallon sample (#3) was collected at the end of driving on the second day.
6. A sample (#4) of the remaining 0.5-gallon fuel in the tank was collected on the next morning.
7. After each fuel was completed, the fuel tank of the vehicle was rinsed with the next fuel, the rinse fuel was drained, and the next test fuel was added to the fuel tank.
8. The process was repeated with each test fuel.

Chassis Dynamometer Fuel Evaluation

The chassis dynamometer portion followed the same basic test procedure used in the on-road tests. Each vehicle utilized the same instrumentation, fuels, and drive cycle. However, for these tests, the ambient temperature was raised to 95 °F and a supplemental heat source (heating blankets) was used to directly heat to the underside of the vehicle. The supplemental heat source will simulate radiant heat emitted from the road surface during extreme temperature conditions. The goal of this portion of the vehicle testing was to allow comparison with the on-road tests and see how the fuel system and fuel stability respond to extreme heat.

The dynamometer drive cycle was a simulation of the on-road route. This cycle was also developed during CRC Project E 90 2b, "Effects of Ethanol Blends on OBDII Systems of In-Use Vehicles," and includes the 20-min soak and 15-min idle period. The data acquisition system collected the same parameters used during the on-road tests.

For each vehicle, the chassis dynamometer was set for the inertia and road load coefficients listed in the EPA's Test Car Database. The vehicle's cold tire pressure was set to the manufacturer's specification and a speed-proportional fan was directed at the vehicle's radiator. The following steps give a more detailed test sequence for each vehicle:

1. Setup vehicle for chassis dynamometer operation including coast down regressions.
2. Drain fuel and add the test fuel to tank until gauge reads ½ full.
3. Conduct repeat drive cycles, at 95 °F, (including soak and idle portions) until fuel gauge reads ¼ full.
4. Take one-liter fuel sample directly after 15-min idle.
5. Soak vehicle overnight at standard ambient temperature.
6. Take one-liter fuel sample after overnight soak.
7. Repeat steps 3-6 with remaining ¼ tank of fuel. (total of 4 samples)
8. Repeat steps 2-7 with each test fuel.

After completion of the vehicle portion of this program, the instrumentation was removed and the vehicles were returned.

For the vehicle testing each vehicle was tested with B0 and B5 fuels. Each of these fuel/vehicle combinations was tested on an on-road course and a heated dynamometer.

The two test vehicles were each evaluated on 8 different test days. Representative temperature and pressure plots of the data collected during the vehicle testing are given in the full CRC report.

Table 2 is a summary of the highest temperatures recorded during the vehicle testing (both on-road and dynamometer). The maximum fuel rail pressures are also given in the table. There are some trends to the temperature and pressure data. Figure 5 is a plot of the temperature data and illustrates some of the trends:

1. The temperatures recorded during the on-road testing were consistently lower than the analogous temperatures during dyno-testing. Presumably, this was due to the lower ambient temperature, additional air flow cooling, and absence of heating mats.
2. In every instance, the highest temperature was either the Return Line Skin temperature or the Fuel Rail Skin temperature.
3. The temperatures recorded in the Silverado were consistently lower than the analogous temperatures recorded in the ML350 (with the exception of the fuel rail skin temperature on the dynamometer.
4. In general, there was very little difference in measured temperatures between B0 and B5 at any given measurement point under any given test condition.
5. The highest temperatures for both vehicles were recorded at the Fuel Rail Skin during dyno-testing.
6. Although fuel tank temperatures varied, both the temperature change at any given point and the overall temperature increase followed a generally consistent pattern throughout the testing.
7. As the temperature of the fuel in the fuel tank went up, the temperatures at the other measurement points increased accordingly.

Regarding the maximum recorded fuel rail pressures, the pressures were higher during on-road testing compared to dyno-testing; and, the Silverado pressures were higher than the ML350 pressures.

Reviewing the temperature data leads to the following observations:

1. Overall, the temperatures were higher for the dyno-testing compared with the on-road testing. This can be seen by comparing the plots in Figure 6. All measured temperatures, not just the recorded maxima, were consistently lower during on-road testing.
2. The under-carriage heating used in the dyno-testing resulted in higher and more sustained heating of the fuel in the fuel tank, compared with the on-road testing.
3. Throughout the dyno-testing, the fuel temperature in the fuel tank had the longest time at temperature. The fuel rail skin temperature had the next-longest time. Throughout the on-road heating, the fuel rail skin temperature had the longest time at temperature. The supply line skin temperature was the next-longest time. The effects of increased vehicle cooling during on-road testing are confirmed by this.
4. The reduced vehicle cooling that occurred during dyno-testing is also seen in that the individual temperature plots are all generally closer to the higher temperatures. In contrast, the fuel rail temperatures during on-road testing are significantly higher compared to the other recorded temperatures in each case.

In addition to the B0 and B5 blend testing reported above, a B20 blend was tested in the Silverado in the heated dynamometer facility. This B20 testing was conducted in order to obtain additional data on the effects of vehicle operation on a more concentrated biodiesel blend. The B20 used in this testing was blended from the same B0 and B100 used above. The blend was analyzed to confirm the blend concentration and the measured oxidation stability of the blend. Two separate vehicle tests were conducted with the B20. In the first, the vehicle went through the same dynamometer testing as described above but without air sparging of the fuel during the test. In the second, the fuel in the vehicle fuel tank was sparged with air 4 times during the test cycle. The sparging was conducted to evaluate the possible effects of increased air/fuel mixing that seemed to occur with the on-road testing. The fuel was sparged for a total of 1 min each time, through a fuel sampling line that went to the bottom of the fuel tank. The timing for the sparging and sampling was as follows:

1. Setup vehicle for chassis dynamometer operation.
2. Add B20 test fuel to tank until gauge reads ½ full.
3. Conduct repeat drive cycles at 95°F (including soak and idle portions) until fuel gauge reads ¼ full.
4. Take one-liter fuel sample directly after 15-minute idle.
5. Soak vehicle overnight at standard ambient temperature.
6. Take one-liter fuel sample after overnight soak.
7. Repeat Step 3 until approximately one gallon remains in tank.
8. Repeat Steps 4-6. (total of 4 samples collected)

Table 2. Summary of Data Maxima Recorded During Vehicle Testing

		Fuel Tank Bottom (FTB)	Supply Line Skin (SLS)	Return Line Skin (RLS)	Fuel in Line at Filter (FLF)	Fuel Rail Skin (FRS)	Fuel Rail Pressure
		°C					psi
ML350							
B0	Onroad	68	71	88	78	86	26341
	Dyno	82	87	96	92	98	23844
B5	Onroad	69	72	86	79	85	26409
	Dyno	82	87	93	92	99	25022
Silverado							
B0	Onroad	54	55	73	68	80	29407
	Dyno	82	83	91	89	104	27822
B5	Onroad	55	54	72	68	81	29799
	Dyno	79	82	89	87	103	26967

In addition to the B0 and B5 blend testing reported above, a B20 blend was tested in the Silverado in the heated dynamometer facility. This B20 testing was conducted in order to obtain additional data on the effects of vehicle operation on a more concentrated biodiesel blend. The B20 used in this testing was blended from the same B0 and B100 used above. The blend was analyzed to confirm the blend concentration and the measured oxidation stability of the blend. Two separate vehicle tests were conducted with the B20. In the first, the vehicle went through the same dynamometer testing as described above but without air sparging of the fuel during the test. In the second, the fuel in the vehicle fuel tank was sparged with air 4 times during the test cycle. The sparging was conducted to evaluate the possible effects of increased air/fuel mixing that seemed to occur with the on-road testing. The fuel was sparged for a total of 1 min each time, through a fuel sampling line that went to the bottom of the fuel tank. The timing for the sparging and sampling was as follows:

9. Setup vehicle for chassis dynamometer operation.
10. Add B20 test fuel to tank until gauge reads ½ full.
11. Conduct repeat drive cycles at 95°F (including soak and idle portions) until fuel gauge reads ¼ full.
12. Take one-liter fuel sample directly after 15-minute idle.
13. Soak vehicle overnight at standard ambient temperature.
14. Take one-liter fuel sample after overnight soak.
15. Repeat Step 3 until approximately one gallon remains in tank.
16. Repeat Steps 4-6. (total of 4 samples collected)

Vehicle-Aged Fuel Analysis

The fuel samples taken from the vehicles during operation were analyzed for thermal stability, Rancimat, RSSOT, and total acid number. These analyses were selected as they are the fuel properties commonly believed to be indicators of degrading fuel quality. The results of the analysis of B0 and B5 fuel samples taken during vehicle testing are given in the full report.

Examination of the vehicle-aged fuel test results leads to the following observations:

1. The fuels experienced very little change in thermal stability as measured by D6468; however, the additional heating that the fuels experienced in the dynamometer testing seems to have resulted in slightly more degraded thermal stability compared to on-road testing in the Silverado.
2. The B0 samples showed no significant degradation in oxidation stability as measured by the Rancimat test (EN15751). This is to be expected since the Rancimat is not a reliable test of hydrocarbon fuel oxidation stability.
3. The B5 on-road samples from the Silverado had slightly poorer oxidation stability (by Rancimat) than did the dynamometer samples. We assume this is because of the fuel sloshing/air mixing that occurs during on-road testing. The B5 results from the ML350 were somewhat mixed.
4. The results also indicate that a strictly thermal stability test, such as D6468, will likely not be useful to measure changes to the stability characteristics of biodiesel blends during vehicle use and onboard storage.
5. For the B20 samples, the RSSOT and Rancimat results were worse than for the B0 and B5. Presumably this is due to the higher amount of biodiesel in the B20. The total acid number results were higher for the B20, compared with the other fuels. Again, it is assumed this is due primarily to the higher biodiesel content in the B20. Taken together, these results could indicate that higher biodiesel content blends are more susceptible to oxidation/degradation in vehicle operation. Additional testing with other fuels and longer vehicle test times is needed.
6. Based on the fuel sample analyses, the extent of fuel degradation during vehicle operation was comparatively slight. [Anecdotal reports from CRC AVFL members have described seeing significantly greater fuel degradation, in the field, with biodiesel blends that were specification quality when introduced to the vehicle.]

Considering all the information collected during the vehicle testing leads to the following points regarding a fuel-stressing apparatus.

1. The two scenarios utilized for the vehicle testing (on-road and heated dyno) resulted in measureable differences in the fuel temperatures throughout the vehicles. As such, a decision regarding fuel-stressing temperatures and/or fuel test method temperatures should be related to the vehicle-operation scenario that one hopes to replicate. Temperatures around 100 °C to 110 °C represent the highest temperatures encountered by the fuel in the current testing.
2. Heating of any fuel sump during processing or testing should be considered as this is similar to actual driving conditions. The maximum fuel tank temperature observed during vehicle testing was 82 °C (180 °F).
3. It is assumed that increasing the amount of oxygen available to the fuel during on-road operation (caused by sloshing) should lead to increased oxidation of the fuel. This assumption, however, was not consistently substantiated by the fuel testing results. It is worth noting that the test fuels had relatively short residence times in the vehicles prior to sampling and laboratory testing. Had the fuels been stored in the vehicles for longer

periods, the effect of the increased oxygen availability might have been more clearly demonstrated.

4. The fuels tended to undergo relatively little measurable degradation during the vehicle testing. Factors that influenced this include the use of specification quality fuel and the fact that most of the fuel spent little time in the hottest parts of the vehicle. As the fuel was exposed to higher temperatures and more oxygen, there were measurable changes in the stability of the fuel. Using lower-quality test fuel in the vehicles tests would likely have resulted in larger measureable changes in the fuel.
5. Since adding heat and oxygen accelerates the degradation of the fuel, selection of conditions for stressing/testing will require decisions about the type of environment (vehicle conditions) one wishes to replicate.
6. Accelerating the conditions above/beyond those encountered in the vehicles is possible and would accelerate the degradation of the fuel. But, some decision is required regarding the degree to which acceleration is representative of vehicle operation/storage. A single fuel stressing/testing regime to represent all scenarios is not very likely. Perhaps using a small number of stressing/testing regimes, similar to vehicle test cycles, is most appropriate.
7. Conducting some bench-scale testing with fuels of poorer initial stability would help answer some of these questions.
8. It seems that a combination of Rancimat, RSSOT, and acid number testing were the most reliable method of assessing the stability change of a fuel during stressing. However, these test methods do not adequately assess the tendency of a fuel to form deposits on metal surfaces. A test method, such as a modified ASTM D3241 D3241 provides a more robust assessment of the change in fuel stability and deposit-forming characteristics.
9. It should again be noted that these vehicle tests were very short term; less than 7 total hours in the vehicle. Therefore, this testing did not accurately represent the effects of on-board vehicle storage for longer periods of time. Recall that the primary goals of the vehicle testing were to collect accurate temperature and pressure data to assist with the design of a fuel-aging apparatus.

Task 5 Develop bench-scale, fuel-stressing methodology (mimic the conditions that fuel experiences in HPCR fuel systems)

The emphasis of Task 5 was to develop bench-scale fuel-stressing capability. The fuel-stressing capability was designed to provide a means to realistically stress test fuels in quantities up to about 50 gallons. The stressed fuels would then be available for other testing in the laboratory or in vehicles.

The fuel aging apparatus known as the SwHiReTS is a newly-developed (by SwRI) test technique based on the High Reynolds Number Thermal Stability (HiReTS) tester (See Figure 1). The HiReTS is the equipment used in ASTM D6811, Standard Test Method for Measurement of Thermal Stability of Aviation Turbine Fuels under Turbulent Flow Conditions. In the current configuration of the SwHiReTS, fuel is pumped through stainless steel tubing at up to 100 milliliters per minute. Test temperature can be set as high as 540 °C with an internal pressure of up to 6.9 megapascals (1000 psi). The fuel can be run under either laminar or turbulent flow conditions. The test tubing can be analyzed for deposits as a whole or sectioned for a per-length basis.

The fuel aging and analysis for this task were conducted according to the protocol given in Appendix E of the full report. One slight deviation from this protocol was approved by the project technical monitors, namely that when the fuel was not in the aging apparatus it was stored in an oven at 95-100 °F (not 90 °F).

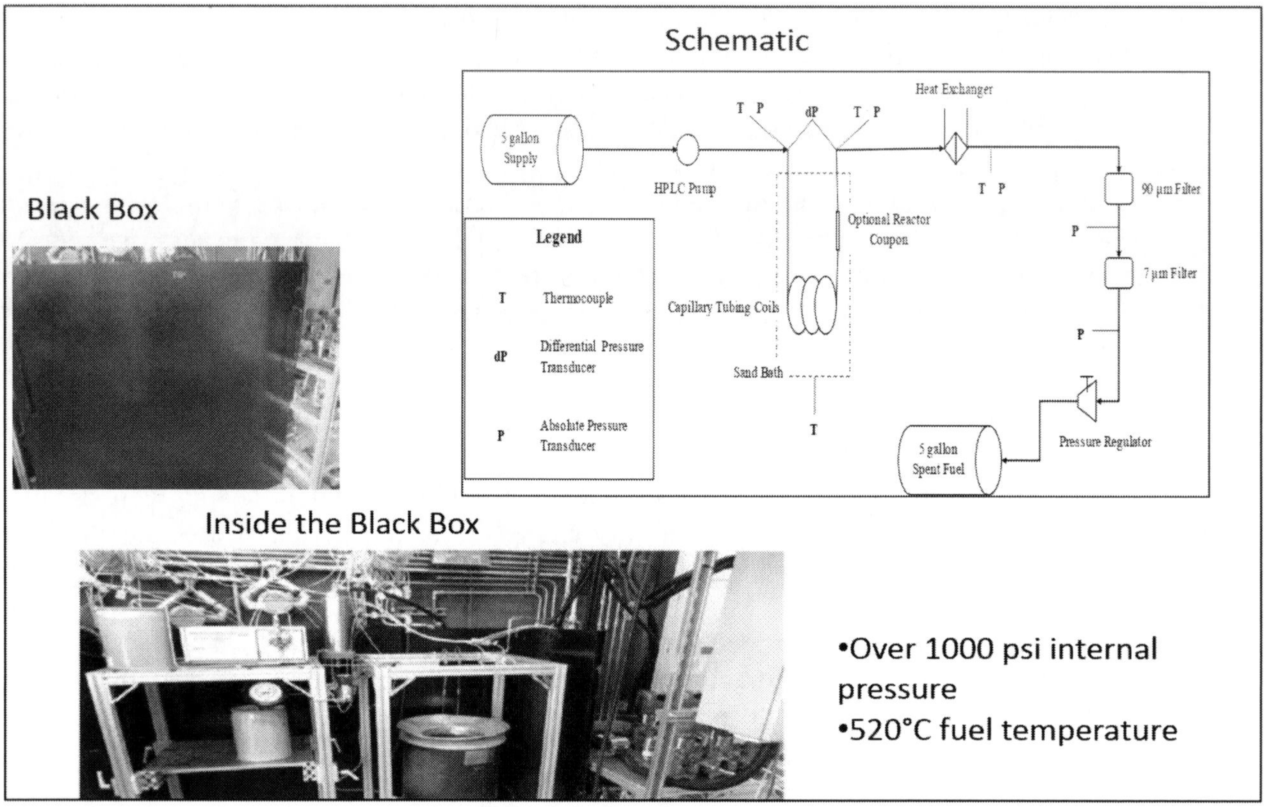

Figure 1. Fuel Stressing Apparatus

Two test fuels were used in this task. Both were B20 blends made from the B0 and B100 used for all the test fuels in this project. Test fuel 1 was the B20, as prepared. Test fuel 2 was the same B20 with added water (deionized water, 200 ppm, total) and added vegetable oil (food grade canola oil, 3 wt%). The Day 0 measured water levels for the Test 1 and Test 2 fuels were 147 and 155 ppm, respectively. Table 3 contains a description of the two test fuel blends used for this task. Each of the fuel samples taken during the fuel stressing process was analyzed by the tests listed in Table 4. Of note in this list is the inclusion of ASTM D3241, "Standard Test Method for Thermal Oxidation of Aviation Turbine Fuels D3241." This analysis was added to the list because the test has shown some ability to evaluate the deposit forming tendencies of diesel fuels.[5] According to the method:

> "This test method for measuring the high temperature stability of gas turbine fuels uses an instrument that subjects the test fuel to conditions that can be related to those occurring in gas turbine engine fuel systems. The fuel is pumped at a fixed volumetric flow rate through a heater, after which it enters a precision stainless steel filter where fuel

degradation products may become trapped. The apparatus uses 450 mL of test fuel ideally during a 2.5-h test. The essential data derived are the amount of deposits on an aluminum heater tube, and the rate of plugging of a 17 μm nominal porosity precision filter located just downstream of the heater tube."

For this project, the D3241 test temperature was set at 240 °C, because that is the temperature most often reported in the literature when the D3241 apparatus is used to test diesel fuel. The 240 °C set temperature results in a temperature profile across the heater tube. The deposits on each D3241 heater tube were analyzed using an ellipsometer. The following information, from Appendix A3 of ASTM D3241, describes ellipsometry and various terms related to D3241 heater tube deposits and ellipsometry:

A3.2.1.1 deposit profile—three-dimensional representation in terms of deposit thickness along and around the length of the heater tube test section.

A3.2.1.2 deposit thickness—the thickness of deposit present on the heater tube substrate surface expressed in nanometers (nm).

A3.2.1.3 ellipsometry—a technique used for measuring the optical properties of surfaces (refractive index and absorption coefficient) based on changes in the polarized state of light upon reflection from the surface.

A3.2.1.3.1 *Discussion*—In the presence of a thin transparent layer, with a known refractive index and absorption coefficient, ellipsometry can also be used to provide film thickness information.

A3.2.1.4 maximum deposit thickness—the maximum thickness of an average 2.5 mm^2 deposit present on the D3241 heater tube surface, expressed in nanometers, nm.

The fuel sample test results are given in the full report. We also ran D3241 on the two components of the B20 and the results are in Table 5. The diesel fuel failed. The failure mode in this case is that the total volume of deposit on the tube exceeded the ability of the ellipsometer to measure. Note that the instrument is able to measure a total volume of 3.99 x 10-5 cm^3. Since the deposit volume from the B0 is above this volume it can be seen that the amount of deposit from the B0 is at least one order of magnitude more than the deposits left by the test fuels. The B100 did run completely and we got a much higher deposit volume than the B20 tests. Additional testing is needed with other fuels to explore this further.

Figure 2 presents the ellipsometer plots for the D3241 tests results. The red areas are places where the deposit depth has exceed 85 nanometer deposit depth (this is the maximum deposit depth that the instrument is able to measure). Figure 3 is a plot of the RSSOT results for the samples from the fuel-stressing apparatus. Figure 4 is a plot of the Rancimat results for the samples from the fuel-stressing apparatus. Figure 5 is the plot of the D3241 total deposit volume for the samples from the fuel-stressing apparatus. Figure 6 is the plot of D3241 maximum deposit depth for the samples from the fuel-stressing apparatus. This number is an average of the highest values in a 2.5 mm^2 area. The RSSOT and Rancimat results tend to follow a similar pattern of decreasing induction period with increased aging, as would be expected. The Rancimat results demonstrated slightly more variation of induction period at the beginning of the aging; and, slightly more spread in the Fuel 1 vs. Fuel 2 results throughout the aging. This may be an indication of tighter precision for the RSSOT. Both the total deposit volume and the maximum deposit depth data show a trend of decreasing volume with Fuel 1 and increasing volume with Fuel 2, as the fuel is aged. The primary difference between these two test fuels is the 3% vegetable oil in Fuel 2. The decreasing volume with Fuel 1 may be an indication of deposits forming in the heating tubing of the fuel-stressing apparatus. Material deposited in the heating tubing would not be available for deposition on the D3241 heater tube. The increasing volumes with Fuel 2 may indicate the tendency of vegetable oil to form deposits on heated metal surfaces. In this case, the vegetable oil is in sufficiently high concentration to leave deposits in the heating tube and on the D3241heater tube as well. Additional work with other fuels and blends is needed to investigate this finding further.

Table 3. Fuel Stressing Apparatus Test Fuels and Target Contaminants Levels

Number	Blend	Total Glycerin	Total Water
1	B20 Soy	On-Spec	Low (~100 ppm)
2	B20 Soy	High (3.0 wt%)	High (200 ppm)

Table 4. Fuel Stressing Apparatus Fuel Sample Analyses

Test Method	Description
D664	Total Acid Number (TAN) by Potentiometric Titration
D3241	Jet Fuel Thermal Oxidation Test
D3703	Hydroperoxide Number by Reaction with KI and Titration
D6304	Water by Karl Fischer Titration
D6468	High Temperature Stability of Middle Distillate
D2887	Simulated Distillation by High Temperature Method
D7545	Oxidation Stability by Induction Period, a.k.a. Rapid Small Scale Oxidation Test (RSSOT)
EN15751	Oxidation Stability by Rancimat

Sampling Day	Fuel 1	Fuel 2
0	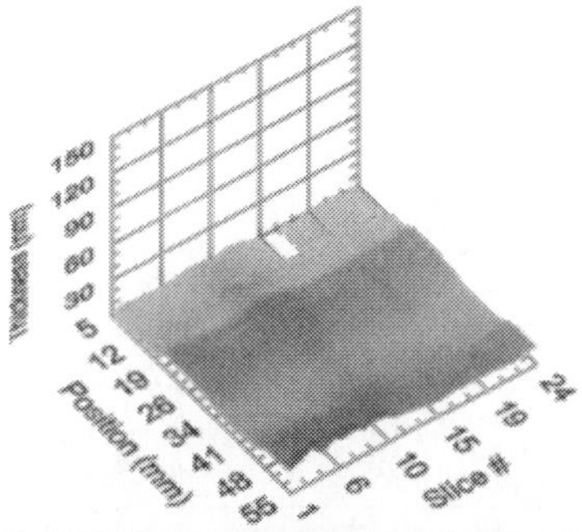	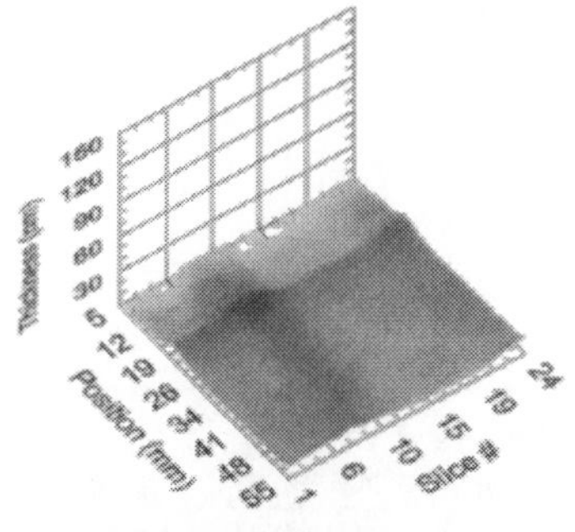
1	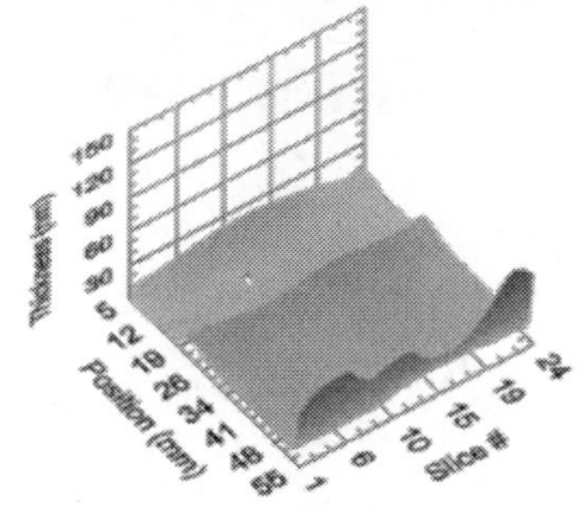	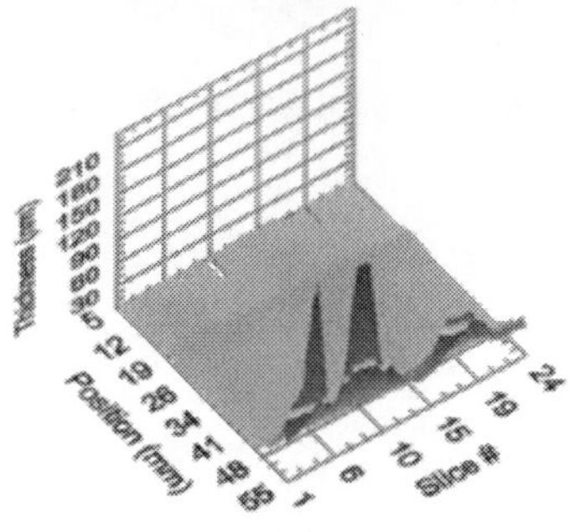
8	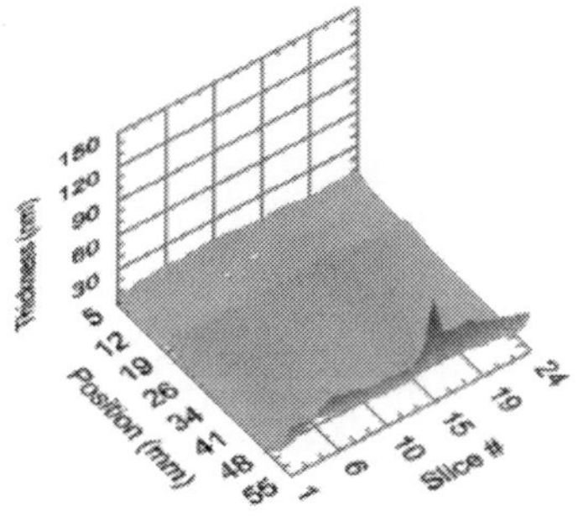	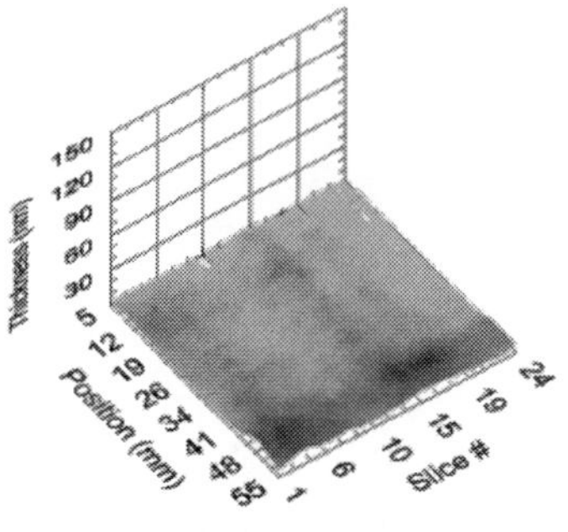

Figure 2. Plots of D3241 Heater Tube Deposits as Measured by Ellipsometry (Fuel samples from stressing-apparatus fuel aging. See Table 14 for additional sample information.)

(Figure Continued on Next Page)

<table>
<tr><th>Sampling Day</th><th>Fuel 1</th><th>Fuel 2</th></tr>
<tr><td>15</td><td>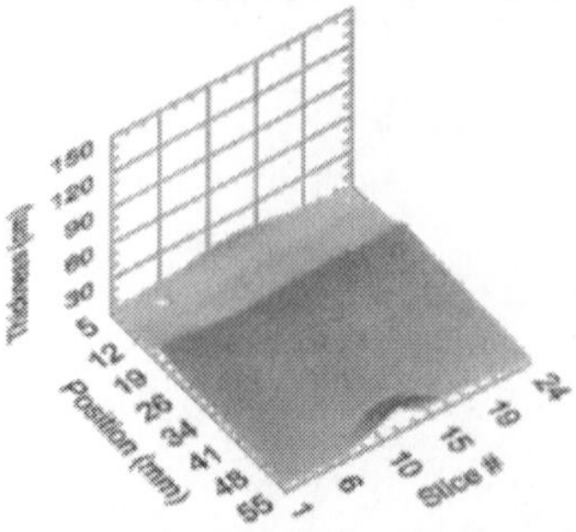
</td><td>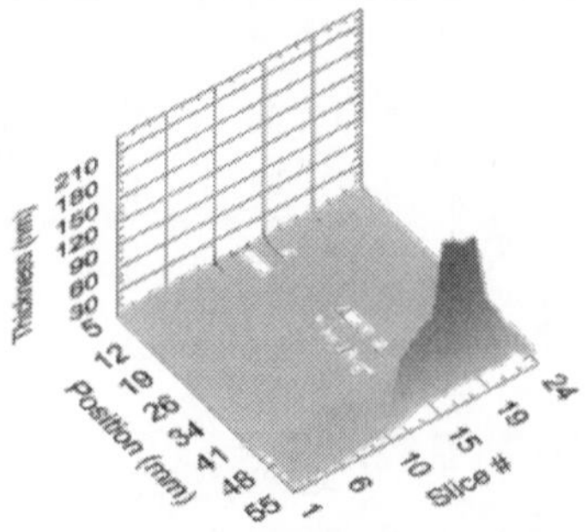
</td></tr>
<tr><td>22</td><td>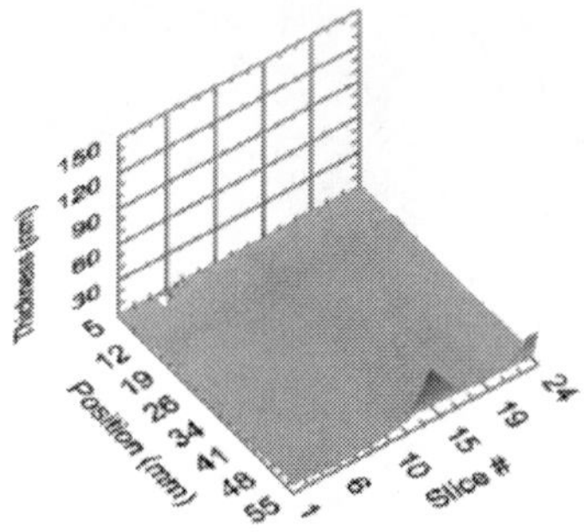
</td><td>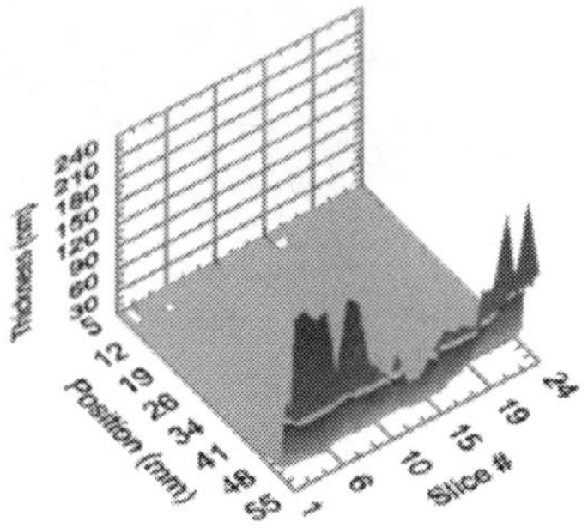
</td></tr>
<tr><td>29</td><td>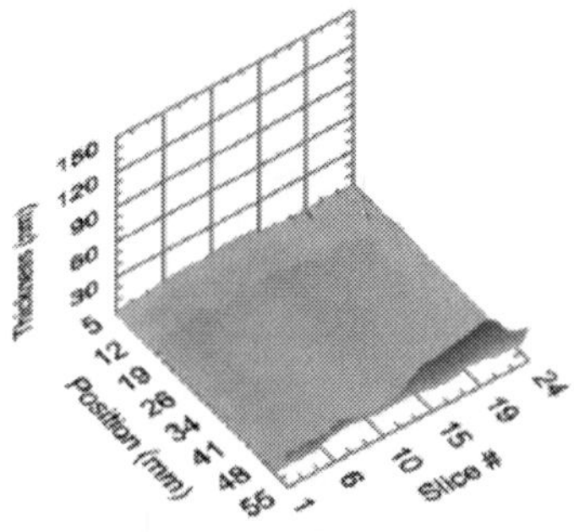
</td><td>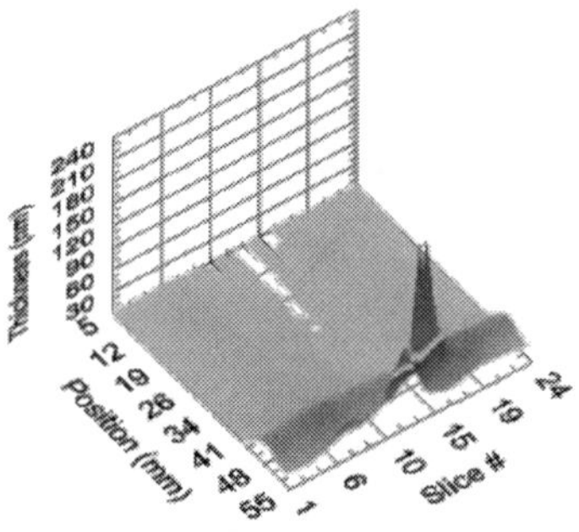
</td></tr>
</table>

Figure 2. (Cont'd). Plots of D3241 Heater Tube Deposits as Measured by Ellipsometry (Fuel samples from stressing-apparatus fuel aging. See Table 14 for additional sample information.)

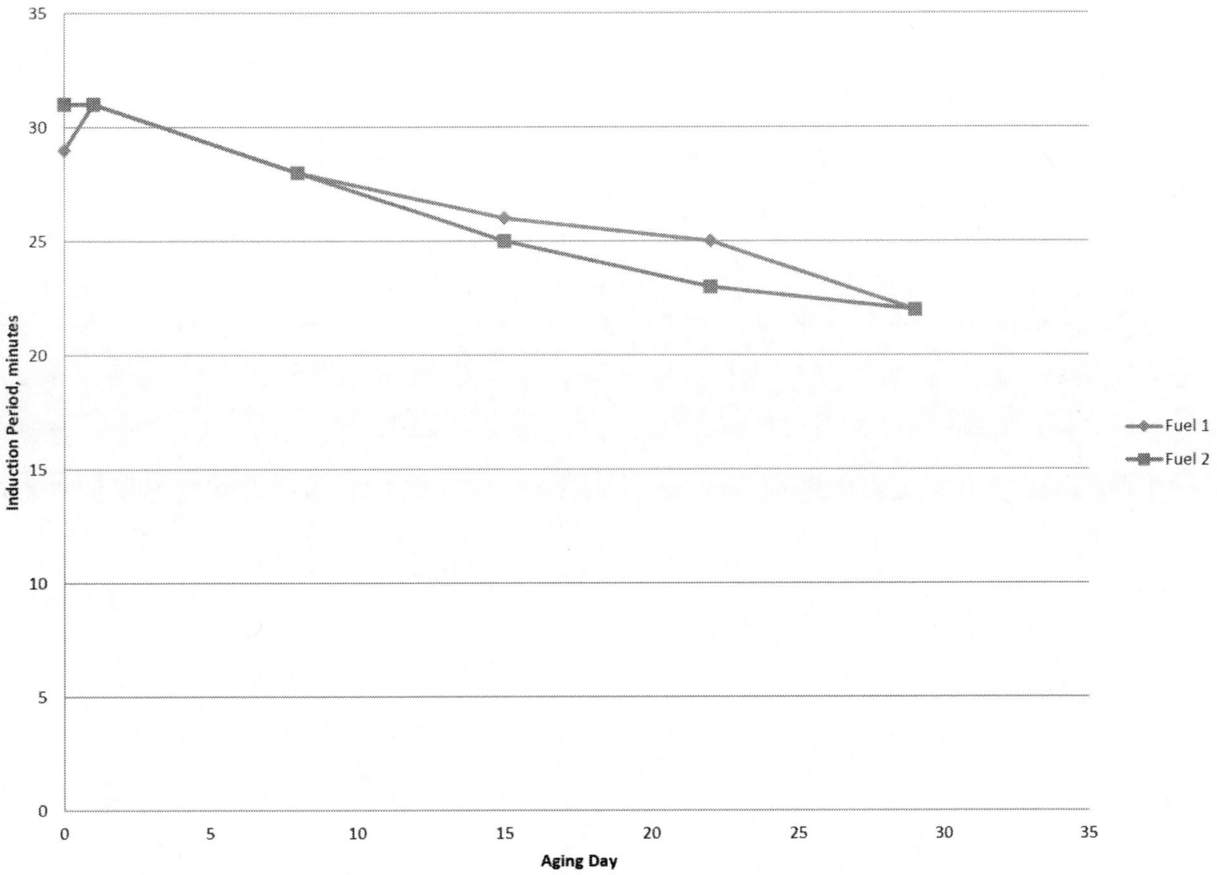

Figure 3. Stressing-Apparatus Fuel Samples RSSOT Test Results

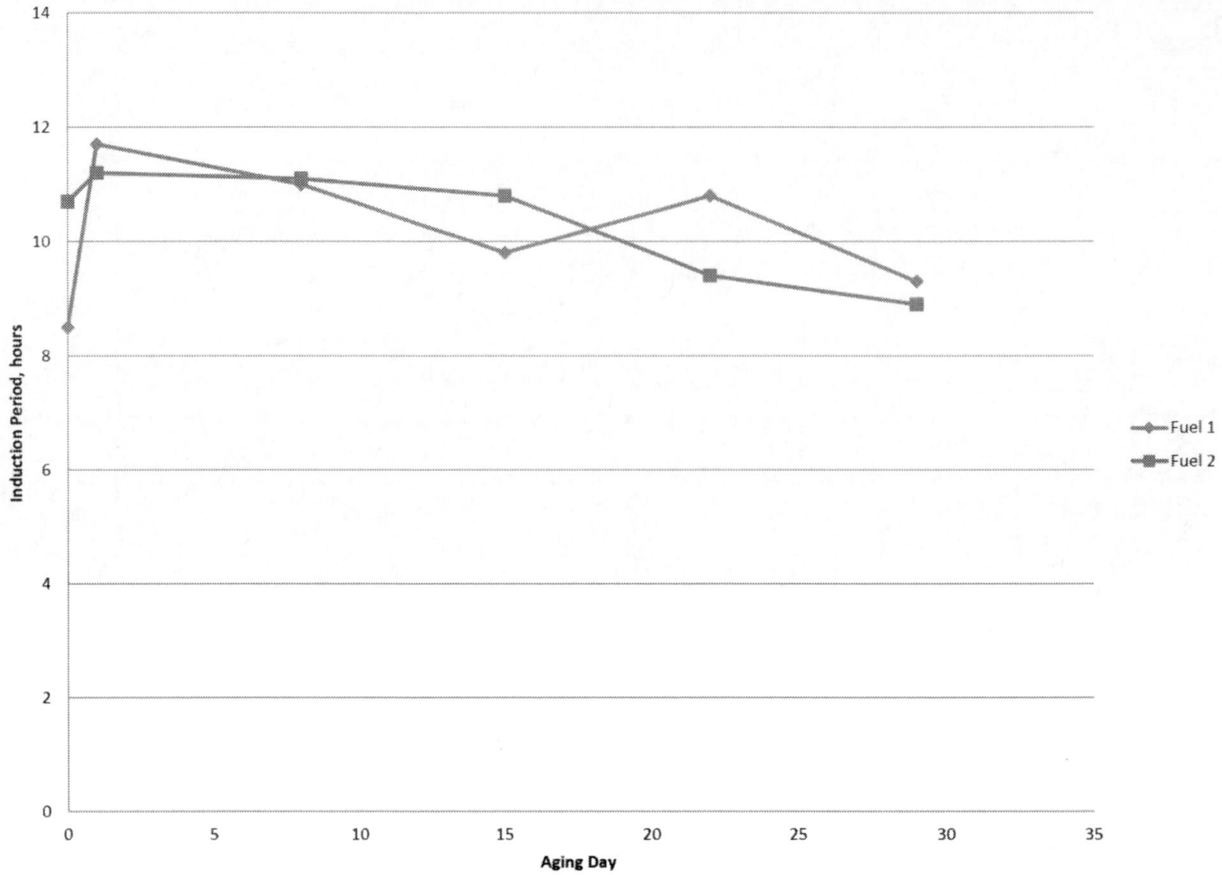

Figure 4. Stressing-Apparatus Fuel Samples Rancimat Results

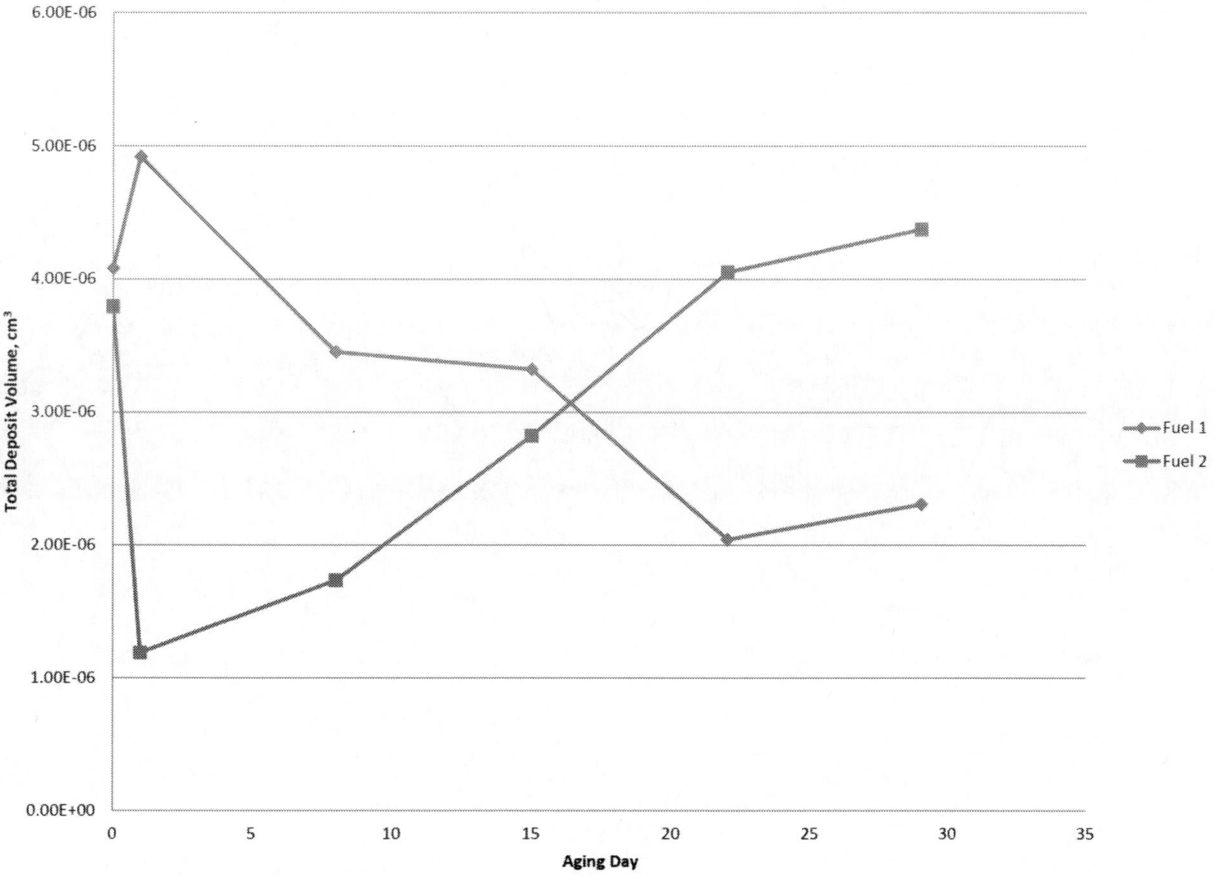

Figure 5. Stressing-Apparatus Fuel Samples D3241 Total Deposit Volume Results

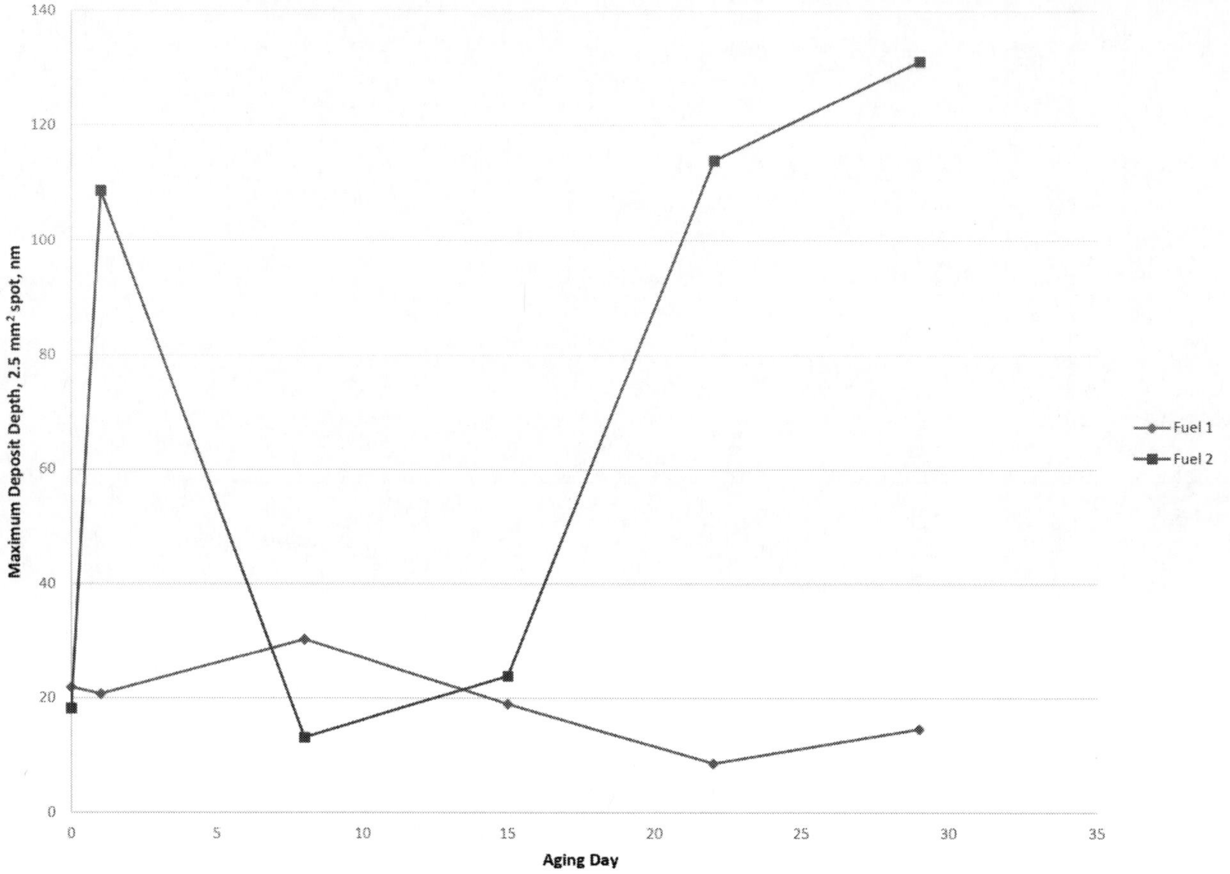

Figure 6. Stressing-Apparatus Fuel Samples D3241 Maximum Deposit Depth Results

Review of the data from the fuel aged in the stressing apparatus leads to the following observations and conclusions from the bench scale fuel stressing apparatus testing:

- The total water test results indicate that essentially none of the added water in Test Fuel 2 was dissolved by the fuel. The Day 0 water levels for the Test 1 and Test 2 fuels were 147 and 155 pm, respectively. The variability of the results is likely due to small amounts of water dropping out of solution during the period between sampling and analysis.
- The RSSOT still seems to be a reliable indicator of change in the oxidative stability of the aged fuel. This was the case for both test fuels.
- The Rancimat also provides a reliable indication of changes in oxidative stability so long as there is biodiesel in the fuel. The test method is not applicable to B0 fuels.
- For Fuel 1 there seems to be a general trend of decreasing D3241 deposit volume with aging time. For Test Fuel 2, there is a trend of increasing deposit volume with aging time.
- The max deposit depth seemed to be generally higher for fuel 2. Additional testing is needed to further investigate D3241 testing.
- The RSSOT and Rancimat results from the fuel aging tests correlated well with the analogous results from the B20 vehicle testing.

- The simulated distillation final boiling point for sample 9313 (Test Fuel 2, Day 29) is significantly higher than the final boiling point of the other samples. The chromatogram for this sample was closely examined for evidence of material in that boiling range. A very slight peak is visible but there is no way to determine if it is from contamination in the sample or the formation of material during aging of the fuel. It should be noted that the added vegetable oil would not typically be detected by D2887; however, if the vegetable oil degraded during aging, some of the breakdown byproducts might account for the higher boiling point material.

Table 5. Results of D3241 Testing of Base Blend Components

Fuel		B0	B100
Test Temperature	°C	240	240
ASTM Code	rating	3P*	1
Max Pressure Drop	mmHg	0^+	0
Max Deposit Depth	nm	208.075	227.203
Total Volume	cm^3	>3.99 E-5	9.65E-06

* *peacock*—A multicolor, rainbow-like tube deposit
+ 0 pressure drop indicates no filterable formed during testing

Figure 7 is a plot of RSSOT and Rancimat induction periods for both the vehicle (Silverado) and aging apparatus (Test Fuel 1) B20 samples. All the B20 Silverado results (both with and without sparging) were used. This plot shows that both the Silverado and the aging apparatus caused some oxidative degradation of the test fuel as indicated by changes in the measured induction periods. The results further indicate that the aging apparatus has the potential to age fuel similarly to aging in a vehicle. Additional work is needed to refine the aging apparatus conditions.

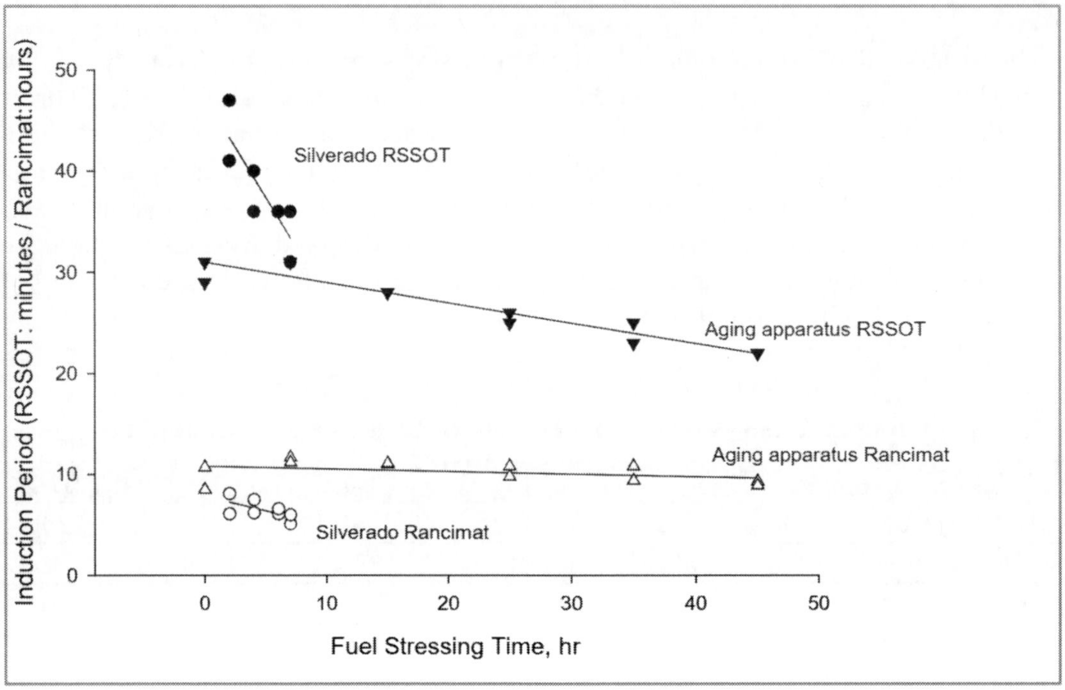

Figure 7. Plots of Induction Period Test Results for B20 Fuel

CONCLUSIONS

Two fuel blends, B0 and B5, were aged in two different vehicles (Chevrolet Silverado and Mercedes Blue Tec). The vehicle testing was conducted on a road course and in a heated dynamometer chamber. A third blend, B20, was aged in one vehicle in the heated dynamometer chamber.

Temperatures recorded from several fuel-system locations during the vehicle testing confirmed that the fuels got hotter as they moved through the system. The highest temperatures recorded in this study were in the range of 100-110 °C.

All three blends were made from the same B0 (commercially available ULSD, hydrocarbon only) and B100 (soy-derived biodiesel) to allow for property comparisons. The B20 and B20 with added vegetable oil and water were also aged in a fuel stressing apparatus. Fuel samples were collected throughout each aging regime and analyzed for indications of fuel aging.

Conclusions from the vehicle testing include:

1. The temperatures recorded during the on-road testing were consistently lower than the analogous temperatures during dyno-testing. Presumably, this was due to the lower ambient temperature, additional air flow cooling, and absence of heating mats.
2. In every instance, the highest temperature was either the Return Line Skin temperature or the Fuel Rail Skin temperature.

3. The temperatures recorded in the Silverado were consistently lower than the analogous temperatures recorded in the ML350 (with the exception of the fuel rail skin temperature on the dynamometer.
4. In general, there was very little difference in measured temperatures between B0 and B5 at any given measurement point under any given test condition.
5. The highest temperatures for both vehicles were recorded at the Fuel Rail Skin during dyno-testing.
6. Although fuel tank temperatures varied, both the temperature change at any given point and the overall temperature increase followed a generally consistent pattern throughout the testing.
7. As the temperature of the fuel in the fuel tank went up, the temperatures at the other measurement points increased accordingly.
8. Regarding the maximum recorded fuel rail pressures, the pressures were higher during on-road testing compared to dyno-testing; and, the Silverado pressures were higher than the ML350 pressures.
9. Overall, the temperatures were higher for the dyno-testing compared with the on-road testing. All measured temperatures, not just the recorded maxima, were consistently lower during on-road testing.
10. The under-carriage heating used in the dyno-testing resulted in higher and more sustained heating of the fuel in the fuel tank, compared with the on-road testing.
11. Throughout the dyno-testing, the fuel temperature in the fuel tank had the longest time at temperature. The fuel rail skin temperature had the next-longest time. Throughout the on-road heating, the fuel rail skin temperature had the longest time at temperature. The supply line skin temperature was the next-longest time. The effects of increased vehicle cooling during on-road testing are confirmed by this.
12. The reduced vehicle cooling that occurred during dyno-testing is also seen in that the individual temperature plots are all generally closer to the higher temperatures. In contrast, the fuel rail temperatures during on-road testing are significantly higher compared to the other recorded temperatures in each case.

Examination of the vehicle-aged fuel test results leads to the following observations:

1. The fuels experienced very little change in thermal stability as measured by D6468; however, the additional heating that the fuels experienced in the dynamometer testing seems to have resulted in slightly more degraded thermal stability compared to on-road testing in the Silverado.
2. The B0 samples showed no significant degradation in oxidation stability as measured by the Rancimat test (EN15751). This is to be expected since the Rancimat is not a reliable test of hydrocarbon fuel oxidation stability.
3. The B5 on-road samples from the Silverado had slightly poorer oxidation stability (by Rancimat) than did the dynamometer samples. We assume this is because of the fuel sloshing/air mixing that occurs during on-road testing. The B5 results from the ML350 were somewhat mixed.
4. The results also indicate that a strictly thermal stability test, such as D6468, will likely not be useful to measure changes to the stability characteristics of biodiesel blends during vehicle use and onboard storage.
5. For the B20 samples, the RSSOT and Rancimat results were worse than for the B0 and B5. Presumably this is due to the higher amount of biodiesel in the B20. The total acid

number results were higher for the B20, compared with the other fuels. Again, it is assumed this is due primarily to the higher biodiesel content in the B20. Taken together, these results could indicate that higher biodiesel content blends are more susceptible to oxidation/degradation in vehicle operation. Additional testing with other fuels and longer vehicle test times is needed.

6. Based on the fuel sample analyses, the extent of fuel degradation during vehicle operation was comparatively slight. [Anecdotal reports from CRC AVFL members have described seeing significantly greater fuel degradation, in the field, with biodiesel blends that were specification quality when introduced to the vehicle.]

Considering all the information collected during the vehicle testing leads to the following points regarding a fuel-stressing apparatus.

1. The two scenarios utilized for the vehicle testing (on-road and heated dyno) resulted in measureable differences in the fuel temperatures throughout the vehicles. As such, a decision regarding fuel-stressing temperatures and/or fuel test method temperatures should be related to the vehicle-operation scenario that one hopes to replicate. Temperatures around 100 °C to 110 °C represent the highest temperatures encountered by the fuel in the current testing. The maximum fuel tank temperature observed during vehicle testing was about 82 °C (180 °F).
2. Heating of any fuel sump during processing or testing should be considered as this is similar to actual driving conditions.
3. It is assumed that increasing the amount of oxygen available to the fuel during on-road operation (caused by sloshing) should lead to increased oxidation of the fuel. This assumption, however, was not consistently substantiated by the fuel testing results. It is worth noting that the test fuels had relatively short residence times in the vehicles prior to sampling and laboratory testing. Had the fuels been stored in the vehicles for longer periods, the effect of the increased oxygen availability might have been more clearly demonstrated.
4. The fuels tended to undergo relatively little measurable degradation during the vehicle testing. Factors that influenced this include the use of specification quality fuel and the fact that most of the fuel spent little time in the hottest parts of the vehicle. As the fuel was exposed to higher temperatures and more oxygen, there were measurable changes in the stability of the fuel. Using lower-quality test fuel in the vehicles tests would likely have resulted in larger measureable changes in the fuel.
5. Since adding heat and oxygen accelerates the degradation of the fuel, selection of conditions for stressing/testing will require decisions about the type of environment (vehicle conditions) one wishes to replicate.
6. Accelerating the conditions above/beyond those encountered in the vehicles is possible and would accelerate the degradation of the fuel. But, some decision is required regarding the degree to which acceleration is representative of vehicle operation/storage. A single fuel stressing/testing regime to represent all scenarios is not very likely. Perhaps using a small number of stressing/testing regimes, similar to vehicle test cycles, is most appropriate.
7. Conducting some bench-scale testing with fuels of poorer initial stability would help answer some of these questions.
8. It seems that a combination of Rancimat, RSSOT, and acid number testing were the most reliable method of assessing the stability change of a fuel during stressing. However,

these test methods do not adequately assess the tendency of a fuel to form deposits on metal surfaces. A test method, such as a modified ASTM D3241 provides a more robust assessment of the change in fuel stability and deposit-forming characteristics.

9. It should again be noted that these vehicle tests were very short term; less than 7 total hours in the vehicle. Therefore, this testing did not accurately represent the effects of on-board vehicle storage for longer periods of time. Recall that the primary goals of the vehicle testing were to collect accurate temperature and pressure data to assist with the design of a fuel-aging apparatus.

Review of the stressing apparatus-aged fuel data leads to the following observations and conclusions from the bench scale fuel stressing apparatus testing:

- The total water test results indicate that essentially none of the added water in Test Fuel 2 was dissolved by the fuel. The variability of the results is likely due to small amounts of water dropping out of solution during the period between sampling and analysis.
- The RSSOT still seems to be a reliable indicator of change in the oxidative stability of the aged fuel. This was the case for both test fuels.
- The Rancimat also provides a reliable indication of changes in oxidative stability so long as there is biodiesel in the fuel. The test method is not applicable to B0 fuels.
- For Fuel 1 there seems to be a general trend of decreasing D3241 deposit volume with aging time. For Test Fuel 2, there is a trend of increasing deposit volume with aging time.
- The max deposit depth seemed to be generally higher for fuel 2. Additional testing is needed to further investigate D3241 testing.
- The RSSOT and Rancimat results from the fuel aging tests and B20 vehicle testing demonstrated that both aging techniques caused some degradation of oxidation stability. Further study would be needed to develop a correlation between the two methods of fuel aging.
- The simulated distillation final boiling point for sample 9313 (Test Fuel 2, Day 29) is significantly higher than the final boiling point of the other samples. The chromatogram for this sample was closely examined for evidence of material in that boiling range. A very slight peak is visible but there is no way to determine if it is from contamination in the sample or the formation of material during aging of the fuel. It should be noted that the added vegetable oil would not typically be detected by D2887; however, if the vegetable oil degraded during aging, some of the breakdown byproducts might account for the higher boiling point material.

ACKNOWLEDGMENTS

The author wishes to thank the Coordinating Research Council for funding this study. Co-authors on the original study were Brent Shoffner, Matthew Blanks, Kevin Brunner, George Wilson, and Gregory Hansen, all of SwRI.

REFERENCES

[1] "Thermal and Oxidative Instability in Biodiesel Blends During Vehicle Use and Onboard Fuel Storage," CRC Report No. AVFL-17c, Coordinating Research Council, Inc., August 2016. www.crcao.com

[2] Standard Specification for Diesel Fuel Oils, ASTM Designation D975, ASTM International, West Conshohocken, PA.

[3] *Fuels and Lubricants Handbook: Technology, Properties, Performance, and Testing,* G.E. Totten, S.R. Westbrook, and R.J. Shah, eds. Manual 37, p. 125, ASTM International, West Conshohocken, PA. 2003.

[4] Shoffner, B, et al., "Effects of Ethanol Blends on OBDII Systems of In-Use Vehicles," CRC Project E-90-2b, Coordinating Research Council, 2012.

[5] Reid, J., et al., "Internal Injector Deposits From Sodium Sources," SAE No. 2014-01-1388.

IASH 2017, the 15th INTERNATIONAL SYMPOSIUM ON STABILITY, HANDLING AND USE OF LIQUID FUELS
Rome, Italy
10-14 September 2017

DEMYSTIFYING THE IMPACT DIESEL FUEL BULK COMPOSTION PLAYS IN INTERNAL INJECTOR DEPOSITS WITH VASE

David J Abdallah[1], Scott K Berkhous[1], Krystal B Wrigley[1], Matt I Watkins[1], Paul Lacey[2], Christian Daveau[2]

[1]*ExxonMobil Research and Engineering, 600 Billingsport Road, NJ 08066, USA, david.j.abdallah@exxonmobil.com*
[2]*Delphi Diesel Systems, Courteney Road, Gillingham, Kent, ME8 0RU, England*

ABSTRACT

Diesel thermal stability is becoming a critical property as common rail fuel systems increase in pressure, resulting in the fuel being exposed to extreme temperature environments prior to injection. If the temperature exceeds the thermal oxidative stability breakpoint of the fuel and its additives, it can lead to deposit formation on the hot internal injector surfaces causing sticking of close fitting parts, fuel flow restriction and reduced heat exchange efficiency. Owing to the important role deposit thickness plays on fuel-wetted hot metal surfaces, precise quantitative measurement is of great importance. Here we report the use of Variable Angle Spectroscopic Ellipsometer (VASE) to afford precise deposit thickness profiles on injector parts, which provides many benefits over a visual rating of injector parts.

A series of rig tests were conducted using commercially available fuel injectors with diesel fuels of varying compositions. Mapping of deposit thicknesses across the surface of the injector needles revealed thicker deposits toward areas nearer the "simulated" combustion chamber highlighting a greater level of heat soaking of the fuel as it approaches the tip of the hot nozzle (note, the injector was located in an electrically heated mounting block to simulate the combustion chamber in the rig tests). The magnitude, profile and optical characterization of deposit on the injector's parts is used to clarify the role of fuel components. In the limited set of fuels investigated the results suggest that conventional petroleum based fuels blended with biodiesel had relatively more deposits than the fuel without. Moreover, biodiesel blends containing cetane improver additive had relatively more deposit than the fuel without.

In a number of tests with fuels containing biodiesel, the deposits on parts were found to be more translucent than the deposit produced with conventional petroleum base fuel. This illustrates that differences in bulk fuel composition can lead to differences in the optical properties of the deposit. Plots of the imaginary part of the index of refraction of the deposit, derived from VASE analysis, substantiate these differences. Such differences in deposit optical properties reveals why visual comparisons are not directly comparable between different deposit chemistries. The ability of VASE analysis to discern these differences circumvents any issues associated with translucent deposits going un-noticed by visual evaluations and invalidate the long held expression that visibly apparent or strong color infers more deposit.

The VASE results provide a quantitative basis for understanding fuel composition trends in producing internal injector deposits.

INTRODUCTION

Advancements in common rail fuel injection systems continue to intensify the extreme temperature and pressure diesel fuel encounters prior to combustion. If the temperature in the fuel system exceeds the fuel and additive thermal oxidative stability breakpoint, autoxidation and subsequent reactions within the fuel can lead to deposits on the hot internal injector surfaces causing sticking of close fitting parts, fuel flow constraints and potentially reduced heat exchange efficiency. There have been numerous reports of internal injector deposits causing problems in diesel engines over the last decade commensurate with the introduction of high pressure common rail injector systems.[1] Investigations on internal injector parts often rely on deposit appearance to draw inferences on deposit thickness and to a lesser extent deposit chemistry. The intrinsic value of the visual appearance approach lies on the fact that greater opacity of carbonaceous deposits correlate to greater deposit thickness. Although this qualitative rating approach is true for relative comparisons between deposits with subtle compositional differences, it does not capture thickness variations when comparisons between deposits of different bulk compositions are made. Variation in fuel composition will often lead to differences in bulk deposits and the deposit appearance will look different. Over the last decade diesel fuel composition has changed in response to lower sulfur requirements, use of additives to improve performance and the introduction of biofuels.

Precise quantitative measurement are needed when comparing deposits from fuels that are substantially different. Variable Angle Spectroscopic Ellipsometer (VASE) was identified as a suitable technique for accurately measuring internal injector diesel fuel deposit thicknesses. Ellipsometry is a well-established technique for measuring the depth of thin films. Its main application is in the semi-conductor industry. VASE measures the change in polarization upon reflection of polarized light from a surface. Because VASE measures two quantities at each wavelength, namely the amplitude ratio and phase change, the technique obtains more information than available with standard laser ellipsometers and optical reflection interference approaches such as interferometry.

Here we use VASE measurements to provide a quantitative basis for understanding fuel composition trends in producing internal injector deposits.

EXPERIMENTAL

<u>Qualitative description of ellipsometric deposit measurements</u>

The comprehensive discussion of the ellipsometric measurement of deposit thickness has been described elsewhere.[2] An oversimplified discussion follows to give a qualitative, rudimentary understanding by considering a single wavelength and beam path. In ellipsometry, electromagnetic waves (beams) impinging on the deposit from a light source at a non-perpendicular angle can be described by two components with their individual vibration at right angles to each other. To understand how ellipsometry works, we need to consider the orthogonal beams separately because they behave independently. The combination of the 2 gives the overall ellipicity form of the light, as shown in Figure 1.

A plane can be defined by the area encompassed by the incident and reflected beam trajectories. One of the beams has its electric field in the plane (p-polarized), the other perpendicular to it (s-polarized). The s and p beams have a dissimilar propensity for reflection and transmission and the disparity is the largest near Brewster's angle. Brewster's angle is the angle used in ellipsometry where all of the p beam transmits and most of the s beam reflects.
For the s beam we will only consider the case where the entire beam is reflected and none transmits into the deposit. For the p beam we will only consider the case where the entire beam is transmitted into the deposit and none is reflected. The transmitted p beam traverses the deposit, fully reflects from the metal surface, and fully transmits through the top surface of the deposit back into the air.

Although the 2 beam simplification is not far from the reality, both beams have a reflected and transmitted component at every interface encountered (including the oxide layers on the metal) leading to multiple beams that contribute to the measurement result. Careful accounting of these beams is the basis of ellipsometry. To calculate the thickness of the layer or its refractive index from the intensities of the reflected light involves an extensive amount of straightforward mathematics and therefore is done by special ellipsometer software.

In our simple 2 beam account, the p beam will lag behind the s beam considering it had to traverse the deposit as both travel toward the detector. This lag is captured as a relative beam phase difference by the detector compared to their expected phase difference in the incident beam. The phase difference is known as delta (Δ) in ellipsometry.

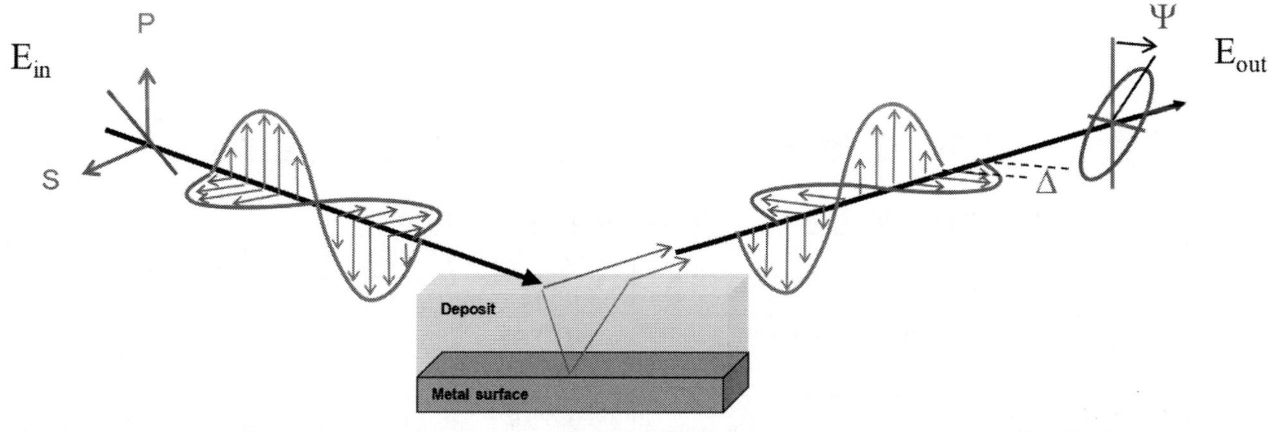

Figure 1. Depiction of a beam of light encountering a single light absorbing deposit layer on an opaque metal surface. Beam with the electric field perpendicular to the plane of incidence (in the direction of the blue arrow) tends to reflect more rather than penetrate into the deposit, in this highly simplified scheme it's all reflected. Conversely, a beam parallel to the plane of incidence (in the direction of the green arrow) tends to penetrate into the deposit rather than reflect, in the

simplified scheme it's all transmitted. The total reflected beam from the top of the deposit surface is a combination of the two independent beams that are transmitted from the deposit. The additional traveling through the deposit for the p beam creates a lag in the phase.

Seeing the lag, or phase difference, between the two beams is directly related to the path length in the deposit (or thickness of the deposit) it becomes apparent that the magnitude of the phase difference captures most of the information necessary to calculate the deposit thickness.

Light propagation in deposits can be described using a complex-valued refractive index. The imaginary part handles the attenuation, while the real part accounts for refraction. When the Δ is combined with the complex-valued refractive index (optical indices) of the deposit and information related to the wavelength and angle of incidence a model can be constructed to determine the deposit thickness. There is also a difference in intensity between the reflected and incident s and p beams due to absorption while traversing the deposit, but this is not as significant to the determination of deposit thickness. Absorption is taken into consideration in the amplitude ratio (Ψ) which is also measured by ellipsometry.

Thus, there are 2 overall changes that describe the combined s and p reflected beam, the amplitude ratio, Ψ, and the phase difference, Δ.[2a] Changing either Ψ or Δ is associated with changes in the ellipticity of the reflected beam which is the reason for the name ellipsometry.

Fuels and injector rig testing

A set of fuels spanning a range of commercial diesel products were used in the injector rig tests. The two types of B0 ULSD base fuels representing extremes in aromatics levels and one European B7 are described in Table 1. The biodiesel blend components, derived from a variety of feedstocks available in the market, are summarized in Table 2.

Table 1. Diesel fuels used in rig testing

	Mild ULSD referred to as US(M)	Severe ULSD referred to as US(S)	B7 EN590
PAH %	~6	17	4.7
Total aromatics %	~27	38	26.8
T95 °C	330 (321 for ID1)	340	349

Table 2. Biodiesel used for blending B20 fuels used in rig testing

B100 origin	Abbreviation for derived methyl ester
Yellow grease	YG(1)
Yellow grease (Hot room degraded)	YG(2)
Soy	SME
Used cooking oil	UCOME

The B7 (mixture of RME and UCOME) fuel represents a typical European diesel fuel meeting EN590 specification including lubricity and red dye additives. B20 fuels were formulated by blending 20% of the B100 with each US ULSD's. The base fuels were then dosed with 2-ethylhexyl nitrate cetane improver (CI) additive, where appropriate.

Details and procedure of the diesel fuel injector rig have been previously reported.[3] The rig is designed to accelerate formation of internal injector deposits. The conditions used in these tests simulates severe engine operating conditions, "thermal soak back", for high pressure common rail systems by employing a high injection pressure and temperature and low injected volume. Operating conditions for the rig are tabulated in Table 3.

Table 3. Diesel fuel injector rig conditions

	Standard	Severe	Units
Heater Set Point	200	210	°C
Rail Pressure	1800	2000	bar
Test Duration	7		hrs
Pump Speed	1750		rpm
Injection Frequency	12.5		Hz
Injected Fuel Flow Rate	5		g/min

A series of rig tests were conducted using two almost identical designs of commercially available injectors, and two rig conditions described in Table 3. The combination of fuel, injector type and rig conditions are categorized into 3 sets of rig tests and are summarized in Table 4 with ID's for each test used throughout the text. A new injector was used for each test.

Table 4. Combination of fuel, injector type and rig conditions in the 3 sets of tests

	Set 1							Set 2						Set 3				
ID	1	2	3	4	5	6	7	8	9	10	11	12	13	14	15	16	17	18
Fuel	US(M)	US(S)	Eur.	US(M)	US(S)	US(M)	US(S)	US(S)						US(S)		Eur.	US(S)	
Biodiesel	None		7%	20%				None		20%				None		7%	20%	
Type			Mix*	Soy		YG(1)				UCOME		YG(2)				Mix*	UCOME	
CI									Yes		Yes		Yes		Yes			Yes
Injector type	A							B						B				
Condition	standard							standard						severe				

*Purchased with a mix of RME and UCOME

Injectors were disassembled and rinsed using a squeeze bottle filled with isooctane after rig operation and allowed to dry before analyzing the deposits. Pictures of the parts were taken under diffuse light white conditions for visual rating.

<u>Metrology instrumentation</u>

Ellipsometry measurements were conducted using a J.A. Woollam Company, Model V M-2000® ellipsometer with CompleteEASE® software. Essential features of the VASE include; focusing probes to reduce the light beam from 2 mm to 250 μm diameter, near IR detector light source and detector to span wavelengths from 370 nm to 1690 nm. The application of near-infrared wavelengths allowed for a wider range of deposit thickness measurements considering deposits are more transparent at these longer wavelengths. Measurements were conducted at an angle of 70 degrees that corresponds to Brewster's angle affording the highest sensitivity. Data acquisition times were 2 seconds. A custom optical breadboard mounted on stage robotics allowed scanning of multiple measurements across the parts surface.

Analysis of the ellipsometric data

Modeling assumed a single light absorbing deposit layer on an opaque metal surface. Optical characterization of the metal substrate was done using clean unused parts and optical properties modeled using a B-Spline.[4] Deposit characterization was done using a Gen-Osc model from a position where the deposit was around ~100 nm thick, if not available the thickest deposit position was used. The Gen-Osc models the absorbing features of the deposit by describing the shape of the absorptions. After parameter fitting of the deposit optical model the optical parameters were frozen and only the thickness was allowed to fit for the remaining points in the scan.

In cases where the deposits are highly uneven, rough, blotchy, or of a heterogeneous morphology (which scatters lights) within the beam spot the reflected light was either insufficient or the ellipsometric data too complex to be modeled using the simple single layer approach described above and the thickness is not measured. In general, deposit measurement with the ordinary lacquer appearance is the most straightforward and reliable considering they best represent a homogeneous thin film whereas deposit areas that are dull and give a diffuse light appearance tend to be more problematic.

RESULTS AND DISCUSSION

Visual inspection

The established procedure for evaluating internal injector deposits is based on subjective visual examination, where the relative difference in color and/or opacity is used to rank the level of deposit. Prior work has demonstrated the ability to discern trace contamination and additive effects using this approach.[5] However, the current investigation looks at fuels with significant bulk compositional differences and multiple visual characteristics are exhibited between deposits on injector parts in this analysis that inhibit the ranking according to the fuel composition. As an example the tips of the injector needles are shown in Figure 2. Multiple visual characteristics are varying such as shininess/dullness, homogeneity, coloration, and opacity for the tests utilizing B20 fuels depending biodiesel type, injector type, rig conditions and the presence of CI. Ranking would require a seriatim of importance be established between the different visual characteristic types.

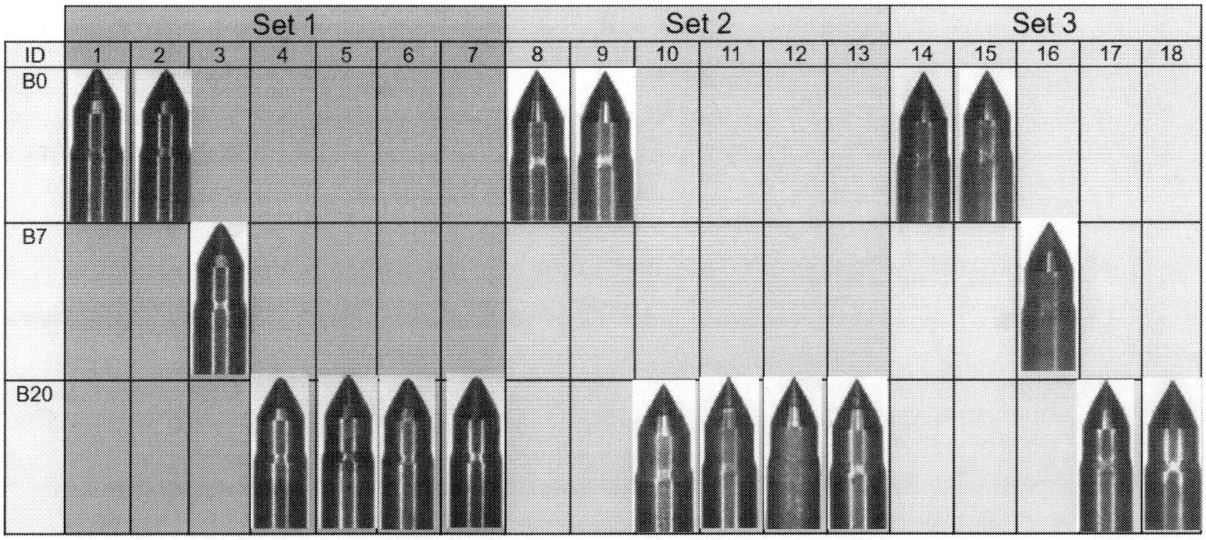

Figure 2. Optical images of the tips of injector needle from the injector rig tests

<u>Metrology</u>

Deposit thickness measured on an injector needle using VASE is shown in Figure 3. Each data point shows the average thickness of 2 measurements at the same axial position separated by 120 degrees rotation. For the most part the deposits were found to be radial symmetric around the circumference of each needle. The absolute thickness varied along the needle length and generally increased toward the tip as shown in Figure 3.

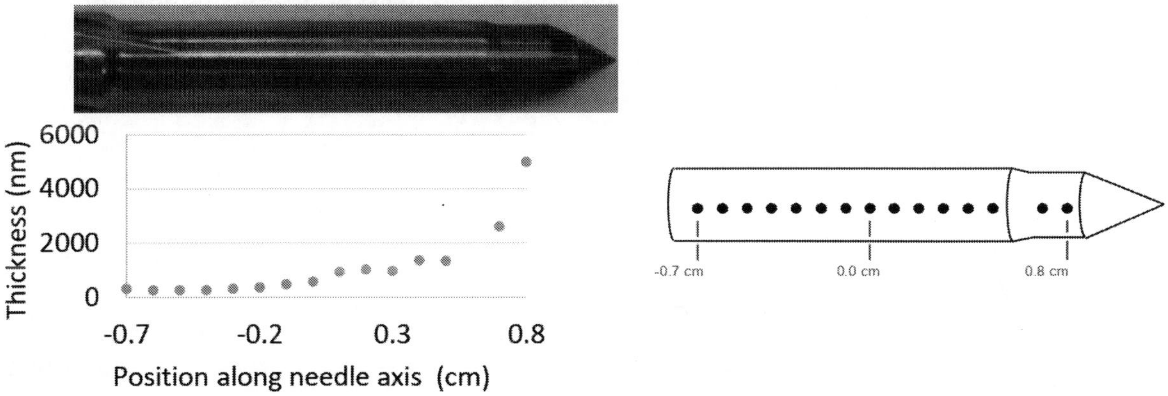

Figure 3. Thickness mapping using VASE along needle from ID #11 test.

Measurements on the injector needles exhibit a deposit thickness difference between petroleum base fuels and fuels containing biodiesel. Figure 4 plots the average injector needle deposit thickness from 6 injector tests in Set 2 with cetane value on the y-axis. In these tests a base fuel, base fuel + 20 UCOME and base fuel with 20% YG(2) were used. Each test was repeated with cetane improver in the fuel.

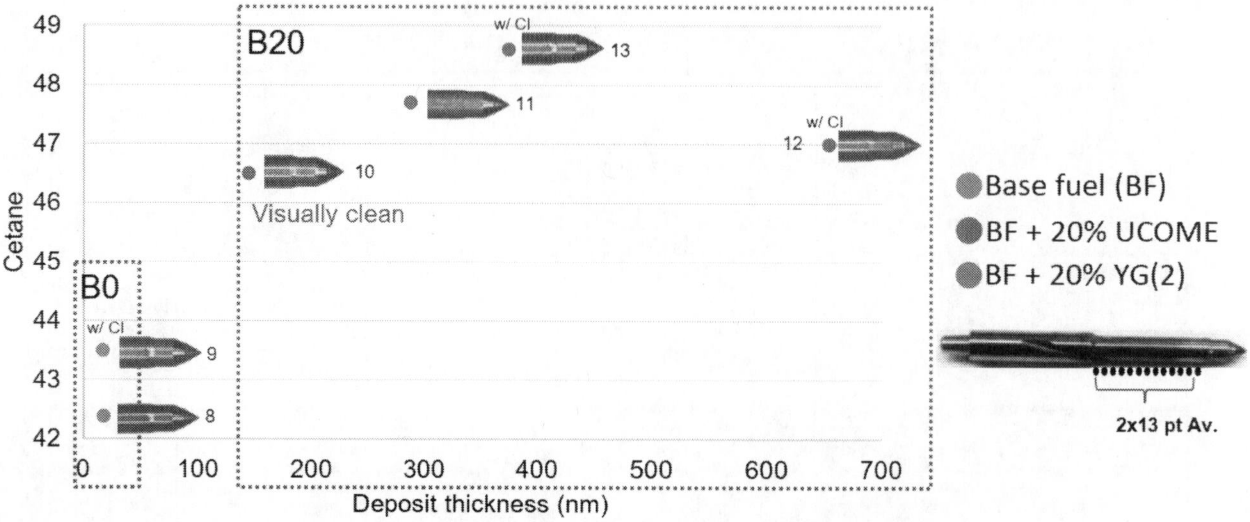

Figure 4. Average injector needle deposit thickness on fuel chemistry in Set 2. Average thickness is calculated from 13 measurement points every 1 mm along the axis over the wide diameter portion of the needle downstream of the spool and repeated after 120 degrees rotation of the needle for a total of 26 points. Data labels correspond to test ID #.

The deposit thickness data is plotted according to the fuels cetane value in Figure 4 and, as expected, fuels with CI have cetane values higher than their base fuel counterpart. Biodiesels inherently have higher cetane values relative to the high aromatic severe base fuel and leads to higher cetane values of the biodiesel blends over the base fuel. In both cases there is an increase in deposits for the B20 blends when CI is added. This is not entirely unexpected seeing CI is added to help lower stability of the fuel for better combustion. In addition the plots reveal that the B20 fuels have thicker deposit than the base fuels. To illustrate this further the data is reported with biodiesel content on the y-axis in Figure 5 with the additional data from Sets 1 and 3. The general trend is thicker deposit when biodiesel is present.

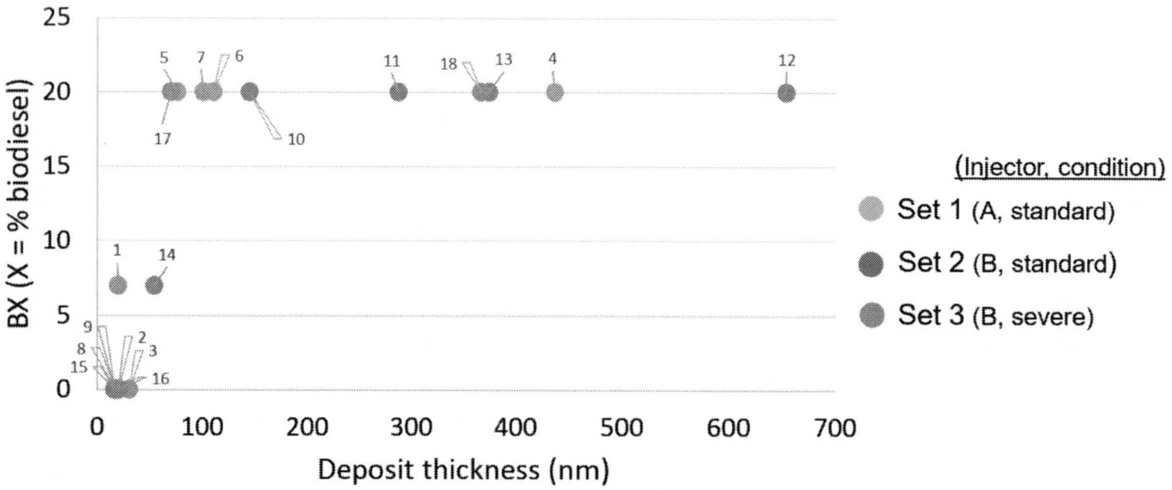

Figure 5. Average deposit thickness on injector needles with respect to level of biodiesel. Labels correspond to test ID# from table 4. Thickness averaging as described in Figure 4.

In many cases the deposit from fuels containing B20 exhibit a bright surface appearance that resemble a clean injector. This observation contradicts the VASE measurements and highlights the unreliability of visual rating.

Figure 6 highlights the contradiction between measurements and visual observation more clearly. The deposit produced by base fuel with 20% UCOME is almost invisible, despite being over an order of magnitude thicker than the dark brown deposit from base fuel alone. The appearance difference results from different optical properties of the 2 deposits. The bottom of the figure illustrates the imaginary part of the index of refraction which is related to the absorption characteristic of the deposit. The base fuel formed a dark brownish film and exhibits visual light absorption with a pronounced increase toward shorter wavelengths whereas the deposit from the B20 fuels is highly transparent giving it a bright appearance. The deposit from the B20 fuel exhibits subtle interference effects (polychromatic coloring) which exemplifies the thick nature of the deposit. Thus, translucent deposits are more likely to give a false positive appearance rating.

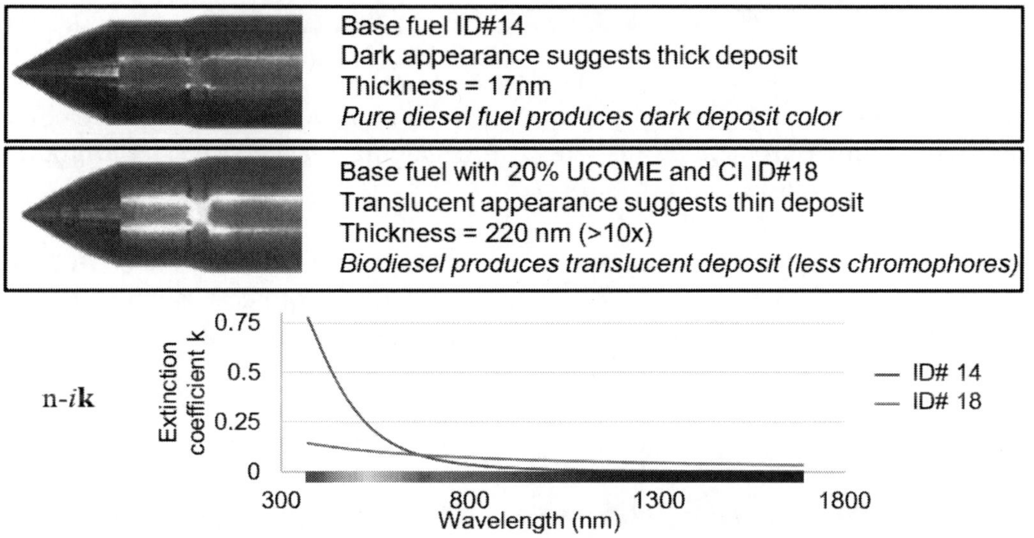

Figure 6. Imaginary part of the index of refraction for the two different types of deposit helps to explain how a thick deposits could appear thin and thin deposit could appear thick.

Deposit thickness measurements on the top surface of the nozzle body were also investigated. The deposit cross pattern exhibited in the contour plot (Figure 7) and visually in the image is formed by contact with an adjacent cross pattern shaped injector piece when assembled. Fuel fills the recessed volume in the outside area of the cross pattern and there is no physical contact between the two metal parts.

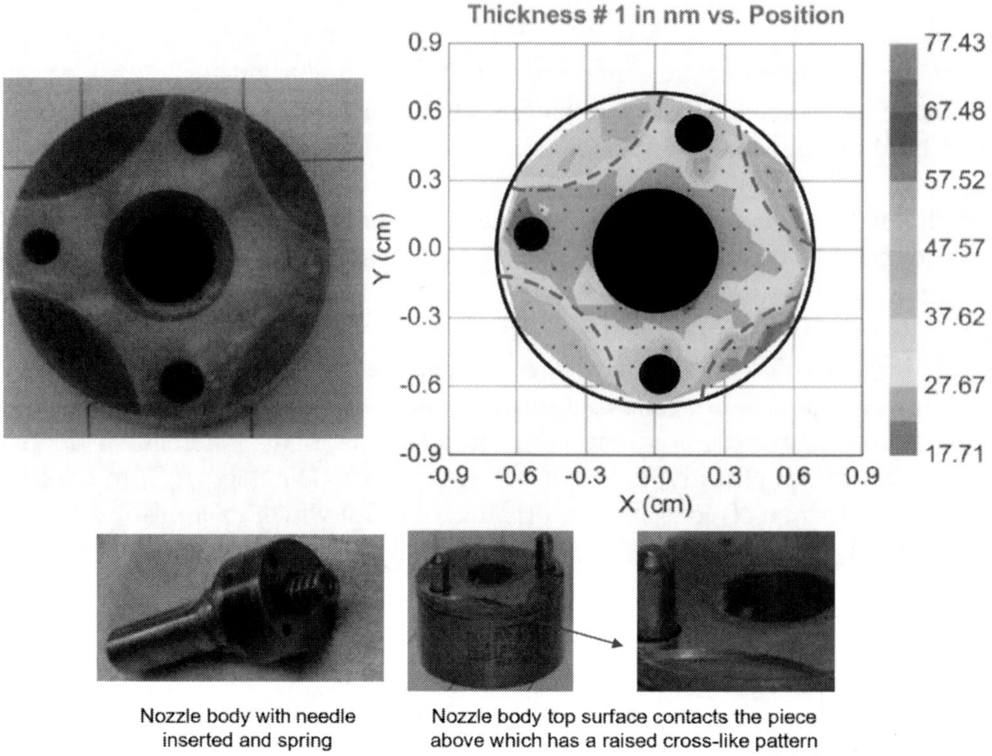

Figure 7. Example of a deposit thickness contour plot on the nozzle body top surface comprised of a hexagonal array of measurement points with 1 mm spacing. Dashed lines highlight the edge of the contacted surface of the adjacent injector part.

Deposit thickness measurements in the recessed (non-contacted) area are tabulated in Figure 8. For the most part, deposit thickness trends follow those described for needles where thick deposits from B20 fuels often appear less dark than thinner deposit from fuels with no biodiesel.

	Set 1							Set 2						Set 3				
ID	1	2	3	4	5	6	7	8	9	10	11	12	13	14	15	16	17	18
B0																		
B7																		
B20																		
Pos. 1	50	47	97	209	412	131	393	105	96	93	257	108	197	missed	34	68	144	179
Pos. 2	65	40	100	missed	380	337	433	106	120	120	301	104	196	116	32	69	119	200
Pos. 3	60	42	97	171	446	missed	missed	114	130	108	272	91	150	121	35	70	132	82
Pos. 4	54	40	96	389	349	363	1130	104	124	116	209	93	226	123	33	68	121	100
Average	57	42	98	256	397	277	652	107	118	109	260	99	192	120	34	69	129	140

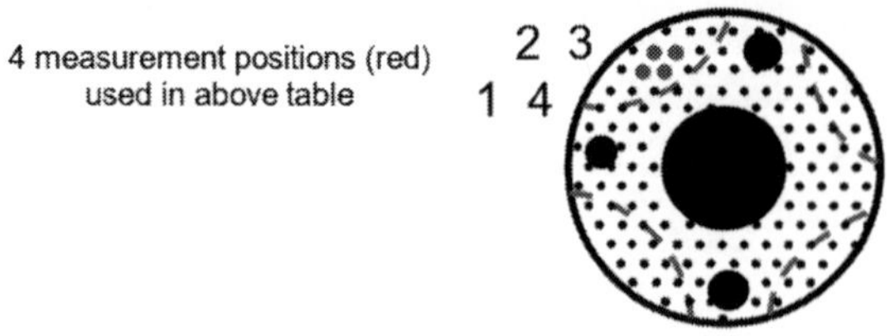

Figure 8. Nozzle body top surface images and deposit thicknesses in non-contact area.

CONCLUSIONS

Engines do not respond to deposit colors but rather deposit thickness and density (mass). Therefore, quantitative deposit rating approaches should measure these deposit characteristics. VASE has been identified as one such instrument. Mapping of deposit thicknesses across the surface of the injector needles revealed thicker deposits toward areas nearer the simulated combustion chamber, highlighting a greater level of heat soaking of the fuel as it approaches the hot nozzle tip.

The magnitude, profile and optical characterization of deposits on injector parts was used to clarify the role of fuel components in internal injector deposits. In the limited set of fuels investigated, the results revealed that conventional petroleum diesel blended with biodiesels had relatively more deposits than the fuel without. Moreover, biodiesel blends containing cetane improver additive had relatively more deposit than the fuel without. A few examples were given where fuels containing biofuels lead to thick internal injector deposits that could not have been observed by visual evaluation seeing they were highly transparent. Plots of the imaginary part of the index of refraction of the deposit, derived from VASE analysis, substantiate these differences. Such differences in deposit optical properties reveals why visual comparisons are not directly comparable. The ability of VASE analysis to discern these differences provides a more robust approach to qualitative understanding fuel composition trends in producing internal injector deposits.

1 Trobaugh, C., Burbrink, C., Zha, Y., Whitacre, S. et al., "Internal Diesel Injector Deposits: Theory and Investigations into Organic and Inorganic Based Deposits," SAE Int. J. Fuels Lubr. 6(3):772-784, 2013, doi:10.4271/2013-01-2670.

2 a.) Harland G. Tompkins and James N. Hilfiker, Spectroscopic Ellipsometry: Practical Application to Thin Film Characterization, Momentum Press, 2016. b.) Alan R. Kramer and Mark Ferrell 9th International Conference On Stability, Handling and Use of Liquid Fuels, Sitges, Spain September 18-22, 2005 page 468

3 a.) Lacey, P., Gail, S., Kientz, J., Milovanovic, N. et al., "Internal Fuel Injector Deposits," SAE Int. J. Fuels Lubr. 5(1):132-145, 2012, doi:10.4271/2011-01-1925., b.) Lacey, P., Gail, S., Kientz, J., Benoist, G. et al., "Fuel Quality and Diesel Injector Deposits," SAE Int. J. Fuels Lubr. 5(3):1187- 1198, 2012, doi:10.4271/2012-01-1693.

4 B. Johs and J.S. Hale, "Dielectric Function Representation by B-Splines", Phys. Stat. Sol. (a), 2008

5 https://crcao.org/publications/performance/index.html

IASH 2017, the 15TH INTERNATIONAL CONFERENCE ON
STABILITY, HANDLING AND USE OF LIQUID FUELS
Rome, Italy
10-14 September 2017

DEGRADATION OF DIESEL IN A MODERN FIE SYSTEM

Christopher R Smith[1], Kesavan Gopalan[2,3], Christopher J Chuck[3] and Christopher D Bannister[2]

[1]BP Formulated Products Technology, Technology Centre, Whitchurch Hill, Pangbourne, UK. christopher.smith3@uk.bp.com
[2]University of Bath, Department of Mechanical Engineering, University of Bath, UK
[3]University of Bath, Department of Chemical Engineering, University of Bath, UK

ABSTRACT

Trends in the automotive industry in recent years to reduce emissions and fuel consumption have led to diesel vehicles utilising injection systems capable of increasingly higher pressures, with typical peak rail pressures above 2000 bar now common. At the same time fuel composition has been changing around the world to meet legislation, most notably with the increased use of biodiesel. Since the introduction of biofuel regulations, FAME has been implicated in fuel degradation and filter blocking, which can cause fuel starvation and vehicle breakdown. However, more recently there has been growing concern that the increasing pressures and concomitant temperatures the fuel is exposed to in modern vehicles can themselves be the cause diesel degradation, even with the absence of bio-components.

A rig which simulates the pressures and temperatures that a fuel is subjected to in a modern vehicle was used to investigate the issue of fuel filter blocking. It was found that, at a rail pressure of 2000 bar, both B0 EN590 and B10 (SME) diesel caused the measured pressure differential across the fuel filter to raise at the same rate. This implies filter blocking caused by on board fuel degradation has the potential to occur broadly in a wide range of different fuel blends independent of bio-content. Further testing on the B10 fuel with a rail pressure of 2000 bar resulted in a pressure difference across the fuel filter of 0.5 bar within 200 hours, whilst the corresponding experiment at 1000 bar showed no filter pressure increase. This work suggests that the increases in rail pressures of modern diesel engines have a significant effect on the propensity of vehicle diesel filters to block due to fuel degradation. An initial study indicates that addition of specifically targeted additives into diesel would delay this filter blocking.

INTRODUCTION

Worldwide emissions standards are trending to progressively lower limits. For example, with regards the European diesel passenger car emissions legislation, from Euro 1 to Euro 6 there has been a fivefold reduction in the carbon monoxide (CO) and hydrocarbon (HC) and nitrogen oxides (NOx) limits and over a thirty times reduction in the particulate mass (PM) limit, see Figure 1.[1] The test cycles over which vehicles have to conform to these emissions limits have also recently become more severe, with the World Light-duty Test Cycle (all components) and Real Driving Emissions (for NOx and particle number) testing being introduced in September 2017 for new vehicle models and in September 2019 for all new vehicles sold in the EU.

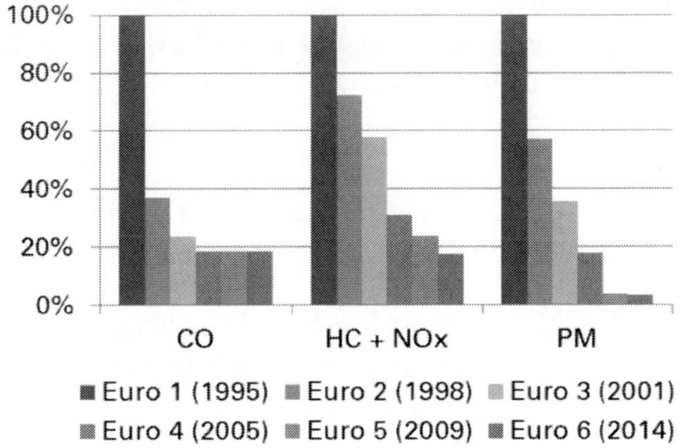

FIGURE 1 - REDUCTION IN DIESEL PASSENGER CAR EXHAUST EMISSIONS STANDARDS, DATA SOURCE DELPHI AUTOMOTIVE[1]

This increasingly stringent legislation has been met by diesel vehicles not only applying improved after treatment technologies, but by engine manufacturers improving the control of the combustion process within the cylinder. Common rail systems which allow high injection pressures have been a key enabler for the control of particulate formation through improved spray atomisation which reduces the occurrence of fuel rich combustion regions.[2,3] In addition, high maximum fuel pressures can be used to increase of the dynamic range of the injector. At low engine loads pressure can be reduced to allow very small quantities of fuel to be injected during pilot events thus avoiding very high rates of pressure rise which can increase NOx formation, while at full load, pressure can be increased to achieve the high fuel flow rates required for high specific power output.[4,5] Figure 2 illustrates the typical trend of increasing pressures that solenoid and piezo injectors have been capable of since 2010 and gives an estimate of the level they will reach in production by 2018.[6]

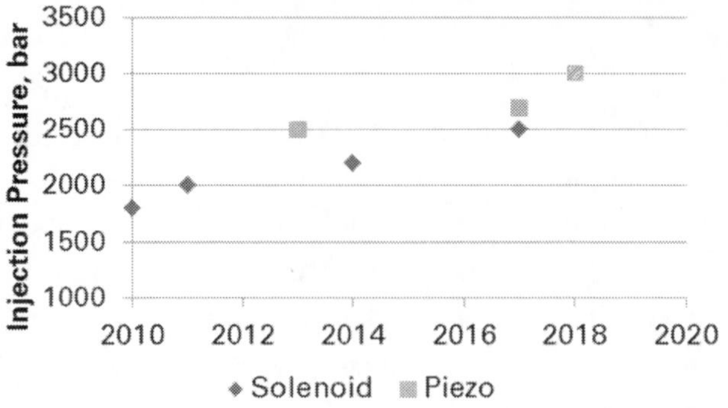

FIGURE 2 - MAXIMUM INJECTION PRESSURES OF BOSCH INJECTORS THROUGH THE YEARS, 2018 DATA POINT ESTIMATED BY BOSCH[6]

In parallel with vehicle technology adapting to meet emissions legalisation, a range of factors has led to the allowable composition of diesel fuel to evolve. In Europe the Renewable Energy Directive and Fuel Quality Directive have led to the introduction of Fatty Acid Methyl Ester (FAME) in diesel with a current limit of 7% in fuel and discussions in the industry on whether this can be safely increased to 10%.[7,8] Similarly in the US the biodiesel target is driven by the Renewable Fuel Standard with a current limit of 5% under ASTM D975 and under ASTM D7467 B6 to B20 diesel is permitted.[9,10]

With the introduction of FAME into diesel technical challenges have been well documented relating to feedstocks and related processes. For example, biofuel feedstocks tend to have a wide range of performance relating to wax crystallisation which in turn is related to the fatty acid structure, with parameters such as: levels of saturation, chain length and degree of branching all being important factors.[11] The production of biofuel can also lead to the introduction of traces of alternative reaction products, glycerols and sterol glucosides, in diesel. Both glycerols and sterol glucosides are relatively insoluble in diesel and these particles can act as seed crystals for wax formation.[12]

Over recent years there have been a spate of reports of vehicle breakdown due to blocking of diesel fuel filters, and injectors, which could not be ascribed to the typical bad actors: contamination in the fuel chain, incorrect application of additives or fouling due to biological growth. Fuels with and without bio-components have seen problems and on-road and off-road sectors have reported incidents. These facts seem to indicate that fuel degradation occurred during vehicle use at ambient atmospheric temperatures. A common theme relating reported incidents is the utilisation of high pressure common rail fuel injection systems. It has been suggested that the increases in injection pressures and concomitant high temperatures contribute to the filter blocking problem.[13,14,15]

This paper describes preliminary work carried out to replicate in a test rig the stress which fuel is exposed to in a running engine. This work aims to disentangle the long held belief of FAME as a bad actor in fuel filter blocking and supports the emerging theory that the use of higher pressure injection equipment (FIE) is a potential "catalyst" for on-board degradation of diesel fuel. Initial work to investigate possible approaches to prevent vehicle failure due to on board diesel degradation is underway.

EXPERIMENTAL WORK

TEST RIG

The test rig shown in Figure 3 was designed to mimic as closely as possible the real-world conditions that diesel fuel would be exposed to in a vehicle, and provide the ability to accelerate fuel degradation. To this end, the reaction parameters included variable injection pressure, an ability to heat the source fuel, control of the injection event and the ability to heat the injectors.

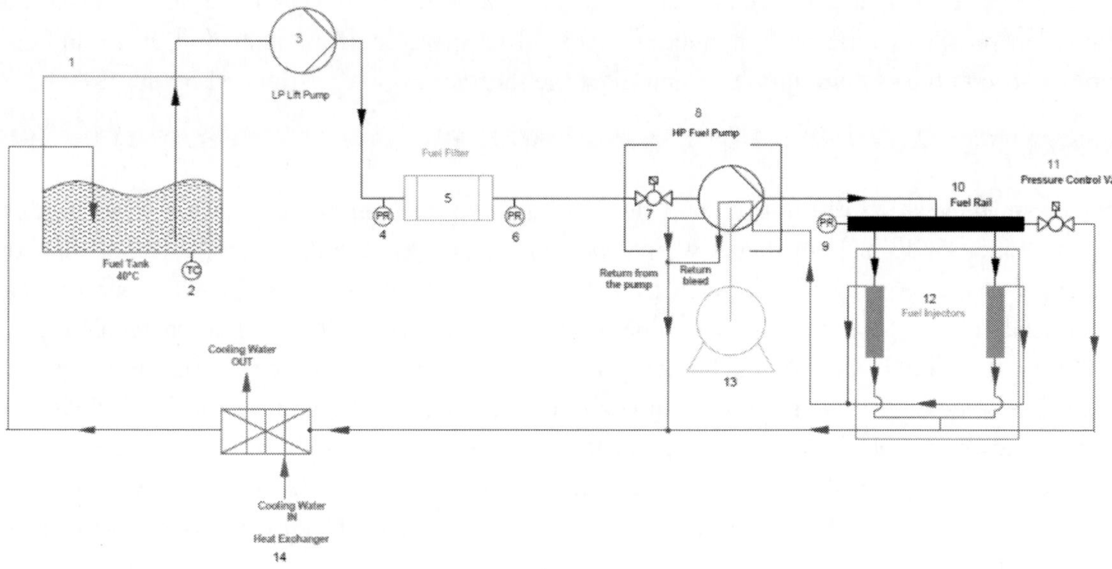

FIGURE 3 - SIMPLIFIED SCHEMATIC OF THE TEST RIG

TABLE 1 - KEY RELATING TO FIGURE 3.

Label	Description	Label	Description
1	Fuel tank	9	Rail pressure sensor
2	Fuel tank temperature	10	Fuel Rail
3	Injection fuel lift pump	11	Rail pressure control valve (rail assembly)
4	Pre-Filter fuel pressure	12	Fuel injectors
5	Fuel filter	13	FIE pump drive motor (integrated in rig)
6	Post-Filter fuel pressure	14	Heat exchanger
7	Flow control valve (FIE pump assembly)		
8	HP fuel Common Rail pump		

The fuel rig was built incorporating a commercially available fuel injection system consisting of two solenoid diesel injectors, a common rail (CR), high pressure pump (HP), a lift pump (LP), a fuel filter, a fuel cooler and a fuel tank. The layout of the recirculation test rig designed for this study is shown in Figure 3. The FIE pump was driven by an electric motor with an in-line torque flange recording the motor speed and torque. To accelerate the aging process, injected fuel had the option to be returned to the tank and recirculated. Spill returns from the common rail, high pressure pump and injectors returned fuel to the tank via a heat exchanger to maintain the fuel at 40 ℃ in the tank.

The injectors were mounted in a custom-made aluminium block with integrated cartridge heater elements and internal temperature measurement allowing the mounting temperatures to be set and controlled to simulate mounting within the cylinder head of a running engine. The common rail pressure and injector actuation was controlled by a Stardex® common rail testing system which allowed selection of rail pressure up to 2000 bar and user-defined injection profiles (duration and frequency). Fuel temperature was measured at a number of points around the system in order to monitor the thermal cycle experienced by the fuel.

TEST FUEL

As diesel fuel which did not contain biofuel had also been implicated in the filter blocking phenomenon a B0 EN590 diesel which had not been treated with additives other than that supplied at the refinery was selected as a test fuel. In Western Europe, FAME is commonly produced from rapeseed oil (RME) and soybean oil (SME). SME is known to have lower stability than RME due to its higher proportion of polyunsaturated compounds and was therefore selected for its increased propensity to degrade.[16] Although the current EN590 limit for biofuel in diesel is 7% the European Renewable Energy Directive[7] sets the objective of minimum 10% share of renewable energy in the transport sector by 2020. Therefore a fuel blend of 10% v/v EN14214 compliant SME in the B0 EN590 diesel was made to represent a possible future fuel.

In order to simplify the chemical analysis and minimise the test time, a B10 model fuel, incorporating the general chemistries found in diesel, was used. This model fuel was made from mixing individual compounds: an aromatic hydrocarbon, a FAME compound, a nitrogen containing heterocycle, a cyclic hydrocarbon and an aliphatic hydrocarbon.

TEST CONDITIONS

Prior to all tests a robust flushing procedure was utilised. The fuel temperature inside the tank was regulated to 40 °C, chosen as a representative temperature of a standard vehicle fuel tank. The tank was filled to two thirds of its capacity to allow air breathing while minimising system volume. The heating temperature of 190 °C of the two injector bodies was chosen based on a worst-case temperature of the cylinder head taking into account additional heat transfer generated by the combustion events. Each test was run continuously until a defined increase in the pressure drop (Δp) across the filter was observed. The filter pressure drop was used as a measure of rate of solid deposition on the filter, with a higher Δp indicating a lower pressure post filter that pre filter. Rail pressure and therefore injection pressure was varied between 2000 bar and 1000 bar to mimic high and mid load conditions: 2000 bar was used unless otherwise stated. The fuel injection events could be switched on, allowing fuel to be recirculated after passing through the injectors, accelerating the diesel degradation process. When the injectors were switched off, the fuel was to recirculated through the spill line. In the results presented, the injection events are switched on unless otherwise stated. In contrast to rigs used in other work[17], here the injectors are operated periodically, to mimic their mode of operation in an engine.

FUEL SAMPLING AND ANALYSIS

To give further information on the fuel degradation, fuel samples were taken periodically (every 24 h) and the solid matter separated from the filter at the end of the test. Analytical techniques such as elemental analysis (CHN), FT-IR and 1H NMR (see later for details) were used to examine fuel degradation throughout the testing.

The observed solid deposits were insoluble in most commonly used organic solvents such as hexane, ethanol, acetonitrile and diethyl ether. However, the majority could be dissolved in acetone. The extraction process involved the following steps;

(1) The filter containing solids was soaked for 48 hours in acetone in a beaker to allow solids from the cellulose filter to move in to the solvent phase.

(2) The acetone was removed in vaccuo from the liquor leaving behind a brown solid.

(3) The brown solid was subjected to hexane washes (5 x 20 ml) to ensure no trace fuel was present.

(4) The resulting solid was washed with a small quantity of acetone (1ml), dried at 60 °C overnight and subsequently analysed.

ANALYTICAL TECHNIQUES

Elemental analysis (CHN)

Samples were tested by Medac Ltd.® using a FlashEA® 1112 Elemental Analyser configured as a CHN-O analyser for the determination of carbon, hydrogen, nitrogen and oxygen contained in the organic and inorganic chemical compounds and substances. The determination of CHN was performed in a single sample analysis, whereas oxygen was determined separately. The technique used by Medac® for the determination of CHN was based on the quantitative "dynamic flash combustion" method. The technique used to analyse oxygen was based on the Unterzaucher modified method.

Nuclear Magnetic Resonance Spectroscopy (NMR)

Proton NMR spectra were recorded using a Bruker® Advance III NMR spectrometer operating at 500.13 MHz for ^{1}H. Samples were analysed using $((CH_3)_2CO)$ at 25 °C using standard Bruker pulse sequences (TopSpin® 2.1). Typically ^{1}H spectra were acquired with a spectral width of 20 ppm, and 16 transients. Spectra were referenced using the residual solvent signal, at 2.05 ppm for ^{1}H.

RESULTS AND DISCUSSION

TEST RIG REPEATABILITY

Evaluation of the experimental repeatability of the rig was undertaken by conducting a second test using the B0 fuel under identical test conditions: the rail pressure was held at 2000 bar and the fuel was recirculated through the injectors. The results are shown in Figure 4. The resolution of the pressure transducers in the pre- and post- fuel filter location was 0.01 bar giving the appearance of discrete test points. At each Δp the median time taken to achieve that pressure was recorded, with error bars representing the maximum and minimum times at which that pressure was recorded (this is true for all figures that show error bars unless stated). As can be seen from Figure 4 both runs of B0 diesel give very similar results: an initial offset of the filter blocking until 2700 min and a steady increase in pressure drop across the filter. Previous experiments showed this pressure profile was an expected one and that once a filter started to block, as in Figure 4, the pressure drop would continue to increase steadily. At a Δp of 0.5 bar injectors would fail due to deposits. Therefore to preserve equipment tests were generally stopped when the Δp reached 0.1 bar. Figure 6 illustrates the typical colour change in the periodically taken fuel samples as the filters blocked.

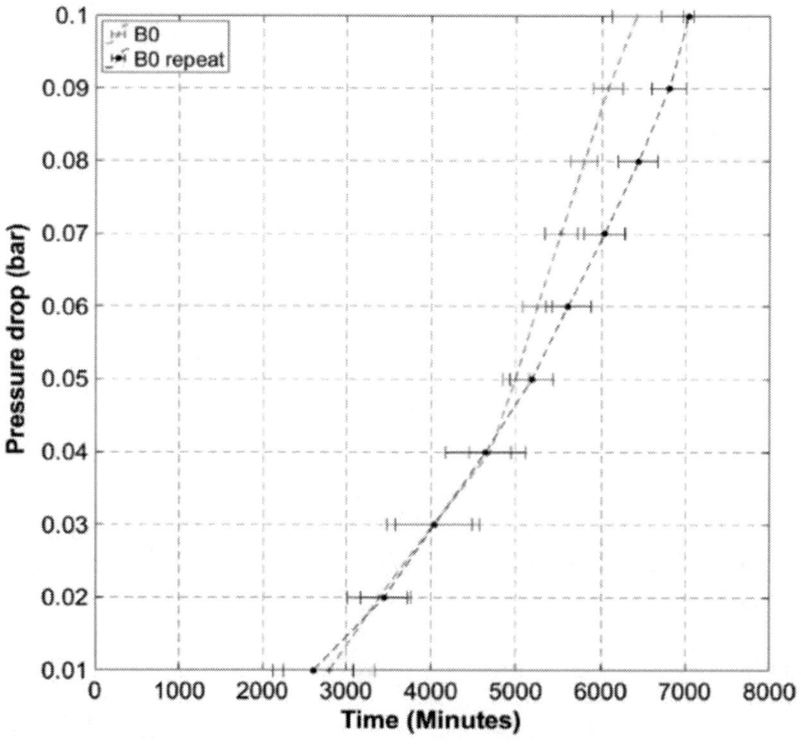

FIGURE 4 - TEST REPEATABILITY. TEST CONDITIONS: B0 FUEL; CRP 2000 BAR; INJECTION EVENT ON

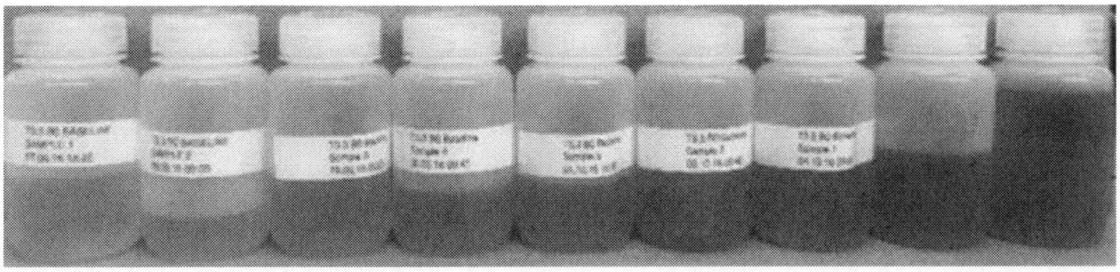

FIGURE 5 - TYPICAL COLOUR CHANGE OF FUEL SAMPLES TAKEN ON A 24 H BASIS, FIRST SAMPLE ON LEFT

EFFECT OF FUEL RECIRCULATION THROUGH THE INJECTORS

In order to accelerate the degradation process and reduce test times, the fuel degradation rig presented here allows the recirculation of fuel through the injector nozzles. Clearly, in a real engine, fuel which passes through the injector nozzles is burned in the combustion chamber but fuel may be recirculated around the rest of the system via the various spill outlets described above. It is possible that the cavitation and shear effects experienced by the fuel during injection into the cylinder are unique, compared to elsewhere in the vehicle fuel system and could give rise to different degradation processes than would be observed if the injected fuel was combusted and removed from the fuel system. For this reason, it was necessary to determine the impact of recirculating injected fuel through the injector nozzles rather than the spill line. To assess this, tests were undertaken using the B10 fuel at 2000 bar rail pressure, with and without the injection events switched on. Other than this the system setup remained unchanged for both tests, with fuel supplied to the injectors from the rail and returned via the spill line and pressure control valve or via injector nozzles, spill line and pressure control valve though in the latter case the majority of the fuel flows through injector nozzles. The results for this experiment are shown in Figure 6.

As in Figure 4, the B10 fuel begins to undergo degradation after approximately 2900 minutes with the injection event active and slightly later, at approximately 3900 minutes with the injector event deactivated. After the onset of the Δp rise, the rate at which deposition on the filter occurs is significantly higher with the active injection process resulting in a 42% reduction in the time taken for Δp to reach 0.1 bar. The magnitude of the variation in Δp between the two conditions is substantially greater than variations observed in the repeatability tests (Figure 4) and is thus considered to be significant.

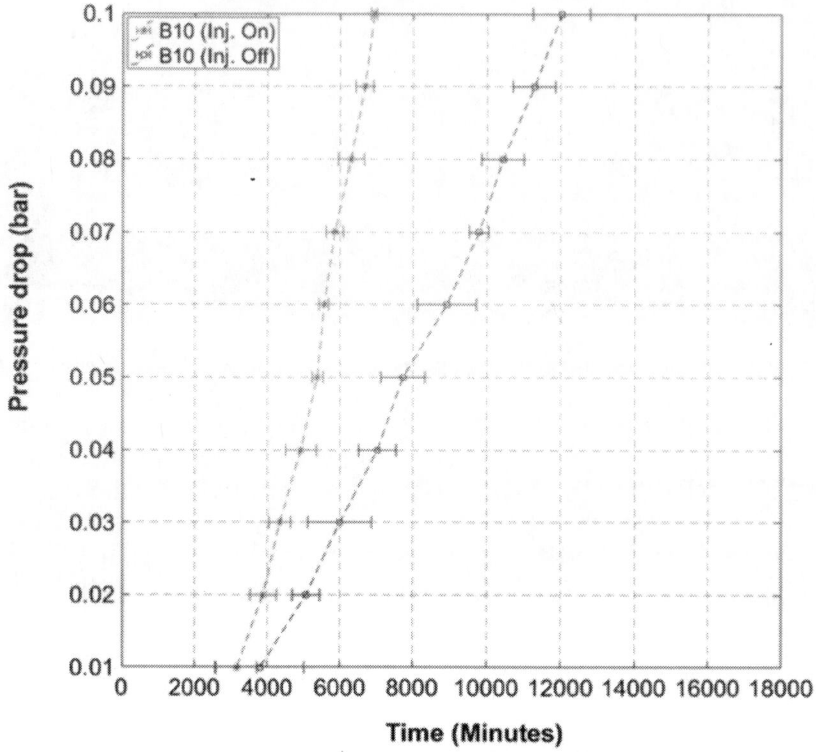

FIGURE 6 - EFFECT OF RECIRCULATION THROUGH THE INJECTOR NOZZLES. TEST CONDITIONS: B10 DIESEL; CRP 2000 BAR; INJECTION EVENT ON AND OFF

^{1}H NMR analysis of the acetone soluble solid residue was compared for both tests, and shown in Figure 7, to determine any structural differences in the residue resulting from the injection event. It should be noted that the resin of the filter is also soluble in acetone, and is observed in both spectra. However, after extraction from the filter material, the ^{1}H NMR spectra are highly similar to one another, showing no significant differences in the organic material between each run. The elemental analysis of the solids was also assessed, after being manually removed from the filter, and demonstrate the same relative ratio of carbon, hydrogen, oxygen and nitrogen within each of the solids (Table 2). Although both the NMR and elemental analysis is heavily influenced by the filter resin, and work is ongoing to improve the removal of the deposits from the filter resin, as the results of the analysis are so similar for both tests that the injection event therefore does not seem to substantively alter the composition of the solids formed, but successfully reduced the test duration by approximately 42%.

Either the shear and cavitation from the injection event does not play a large role in this mechanism or it is possible that similar shear and cavitation events are being experienced elsewhere within the rig, albeit to a lesser degree. Possible locations for these events would be the rail pressure regulator valve (which maintains the desired fuel pressure within the common rail system) and the valve chamber within the injectors. In both locations the fuel is discharged through a small orifice while the pressure is rapidly reduced from 2000 bar to approximately atmospheric in a similar way to the injection process. For the purposes of this investigation it was deemed that as the degradation mechanism was not significantly altered via recirculation of fuel through the injector nozzles and that test duration was significantly shortened, the tests would continue with the injection events switched on.

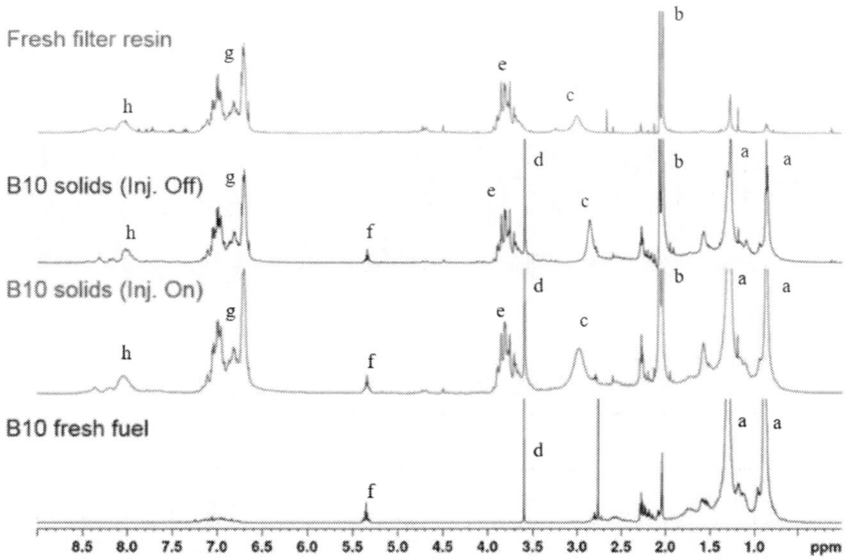

FIGURE 7 - ^{1}H NMR SPECTRA OF THE FRESH B10 DIESEL, FRESH FILTER RESIN, FILTER DEPOSITS FROM THE INJECTION EVEN OFF AND ON TESTS. PEAK ASSIGNMENT – A: (0.7-1.5 PPM), $-CH_2$, $-CH_3$; B: (2.04 PPM), SOLVENT PEAK (DEUTERATED ACETONE); C: (3 PPM, BROAD SINGLET) -OH PEAK; D: (3.6 PPM, SINGLET)-CH_3 ESTER PEAK CORRESPONDING FAME; E: (3.6-3.9PPM, MULTIPLET) ALPHA PROTONS; F: (5.3-5.4 PPM, QUARTET) OLEFIN PEAKS CORRESPONDING TO FAME; G: (6.5-7.2, MULTIPLET) AROMATICS; H: (8+ PPM) HETEROAROMATICS (NITROGEN/OXYGEN CONTAINING AROMATICS).

TABLE 2- ELEMENTAL ANALYSIS OF THE FRESH B10 DIESEL, FRESH FILTER RESIN, FILTER DEPOSITS FROM THE INJECTION EVENT OFF AND ON TESTS

	% C	% H	% N	% O
B10 fresh fuel	85.94	13.57	<0.1	1.61
B10 – Injection event ON	72.9	8.38	0.65	18.09
B10 – Injection event OFF	73.4	8.23	0.60	17.12
Fresh filter	70.89	6.92	1.47	20.32

EFFECT OF FAME ON DIESEL DEGRADATION

This paper has already outlined some of the associated problems with FAME that cause filter blocking, however these on the whole are related to climatic changes and storage of diesel. This investigation aimed to determine whether FAME containing fuel was any more susceptible, than a non-biofuel containing diesel, to degradation during vehicle use.

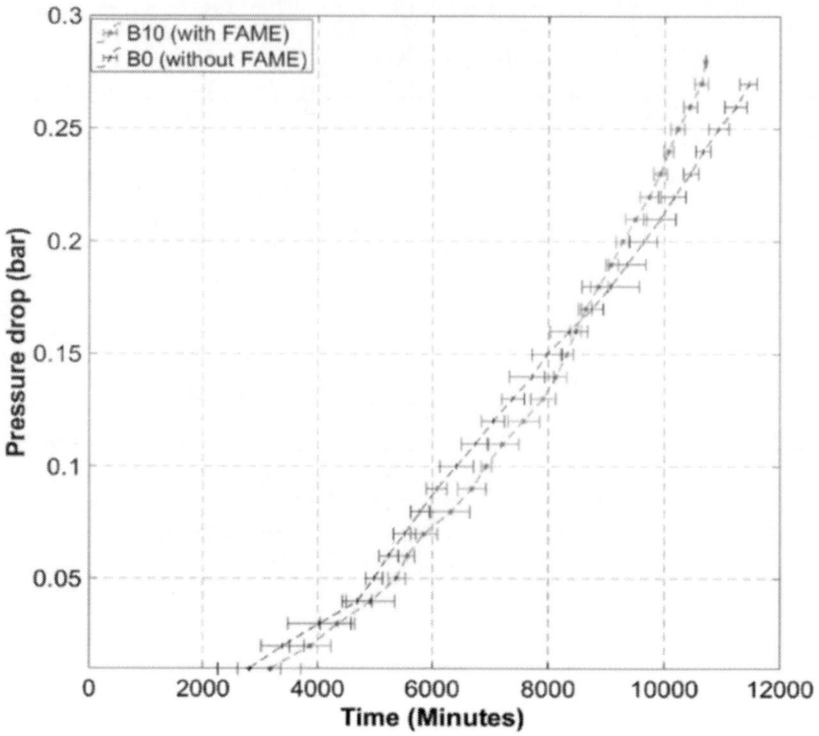

FIGURE 8 - EFFECT OF FAME ON FUEL DEGRADATION. TEST CONDITIONS: B0 AND B10 DIESEL; CRP 2000 BAR; INJECTION EVENT ON

Both the B0 and B10 diesel where tested with a rail pressure of 2000 bar and the injection event switched on. The experiment was carried out until a Δp was reached which exceeded 0.25 bar, the results are shown in Figure 8. Rather surprisingly the pressure traces for both fuels are almost identical, with only a small variation in the rate of Δp increase between 6000 and 8000 minutes. The magnitude of the difference is small and within the bounds of experimental accuracy. This illustrates that both FAME containing and non-FAME containing fuel seem to be equally susceptible to on-board degradation.

EFFECT OF INJECTION PRESSURE ON DIESEL DEGRADATION

As discussed in the introduction of this paper there seems to be growing concern that recent increases of operating pressures in engine fuel injection systems may be contributing to diesel degradation and consequent fuel filter and injector blocking. The relationship between rail pressure and deposit formation rate was therefore examined.

Further investigation into the chemistry of the fuel degradation initiated the development of a model fuel, consisting of components chosen to mimic B10 diesel (described in the experimental setup and procedures above). A comparison between filter pressure drop at the rail pressure of 1000 bar and 2000 bar was undertaken with the tests run for a fixed duration. The results are shown in Figure 9.

Due to the change in fluid properties there was an initial pressure drop across the filter of 0.04 bar for this test. The induction period at 2000 bar is reduced to approximately 1500 minutes (70% reduction compared to B10 fuel at this pressure) with the induction period for the 1000 bar test apparent with a rise in Δp observed after approximately the same duration. After this period it is evident in Figure 9 that the rise in Δp across the filter for the 1000 bar test continues at a much slower rate than for the 2000 bar test. It should be noted that the injection pulse timings for both 2000 bar and 1000 bar tests were kept the same. This of course would reduce the flow rate for the 1000 bar test around the spill line and through the injector nozzles. It is believed that this can account for some if not all of the slower rate of filter blocking shown in Figure 9. However the flow through the spill line and injectors would also be lower in the cases where the injection event was switched off.

The same experiment was carried out with the B10 fuel produced from EN590 diesel and SME, the results are shown in Figure 10. Unlike the previous graphs Figure 10 shows the output of the difference between the raw pressure transducer signals rather than a median result and range at a specific pressure. No degradation was observed in the B10 fuel at a rail pressure of 1000 bar within the test duration. The step in pressure at 10,000 minutes was attributed to a stop and restart of the rig and not degradation as the Δp did not continue to rise after this change, this was significantly different to the testing at 2000 bar, which showed a rapid increase in the measured filter pressure drop after approximately 5000 minutes. It would be expected that if fuel flow rate through the spill line and injectors were the dominant factor with respect to the rate of degradation, that the 1000 bar experiment would show a similar increase in pressure drop across the filter but at a lower rate than the 2000 bar test, as in Figure 9. With the 1000 bar test showing no sign of filter blocking in the time period shown it does seem that increasing rail pressure can have a detrimental impact on stability of the actual diesel fuel. It is postulated that the B10 fuel has a higher inherent solvency than the model fuel which allows any deposits formed at the lower flow rate to be solvated whereas the effect of pressure increase in the common rail causes degradation and deposit formation at a much higher rate which cannot be kept in solution.

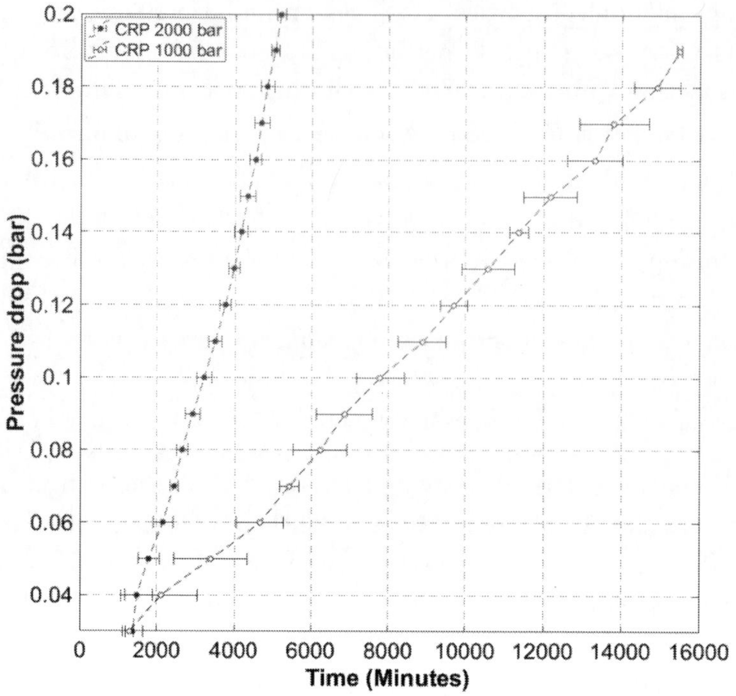

FIGURE 9 - EFFECT OF INJECTION PRESSURE ON FUEL DEGRADATION. TEST CONDITIONS: B10 MODEL DIESEL; CRP 1000 AND 2000 BAR; INJECTION EVENT ON

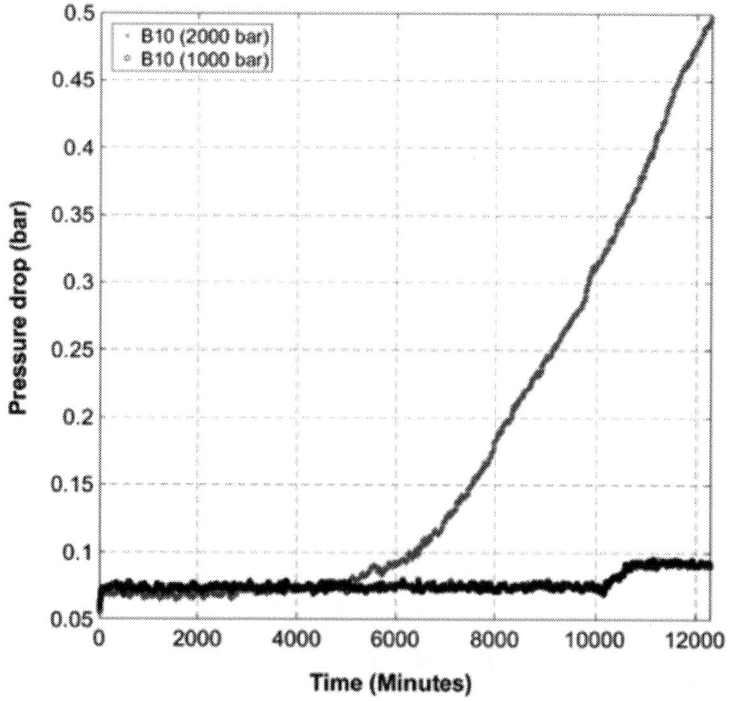

FIGURE 10 - EFFECT OF INJECTION PRESSURE ON FUEL DEGRADATION. TEST CONDITIONS: B10 DIESEL; CRP 1000 AND 2000 BAR; INJECTION EVENT ON

EFFECT OF ADDITIVES ON DIESEL DEGRADATION

This investigation suggests that the increasing rail pressure of modern diesel engines has a significant effect on the propensity of vehicle fuel filters to block and that fuels with and without FAME are likely to be susceptible to degradation leading to such issues. If this is the case what can be done to combat a possible global problem?

A preliminary study indicates that addition of specifically targeted additives into diesel would delay this filter blocking. Figure 11 illustrates an experiment where the B0 diesel used in the previous tests was blended with two concentrations, A and B, of an additive component (Conc. B > Conc. A). For this work the rail pressure was set to 2000 bar and the fuel recirculated through the injectors. As can be seen in Figure 11 not only does the inclusion of a specifically targeted additive significantly increase the time required for the Δp to reach 0.1 bar, but subsequent increase in concentration of the same additive component increases that time significantly again. Note that the error bars shown in Figure 11, unlike those shown in the rest of the paper, are the average of multiple tests with the bars indicating ±2 standard deviations.

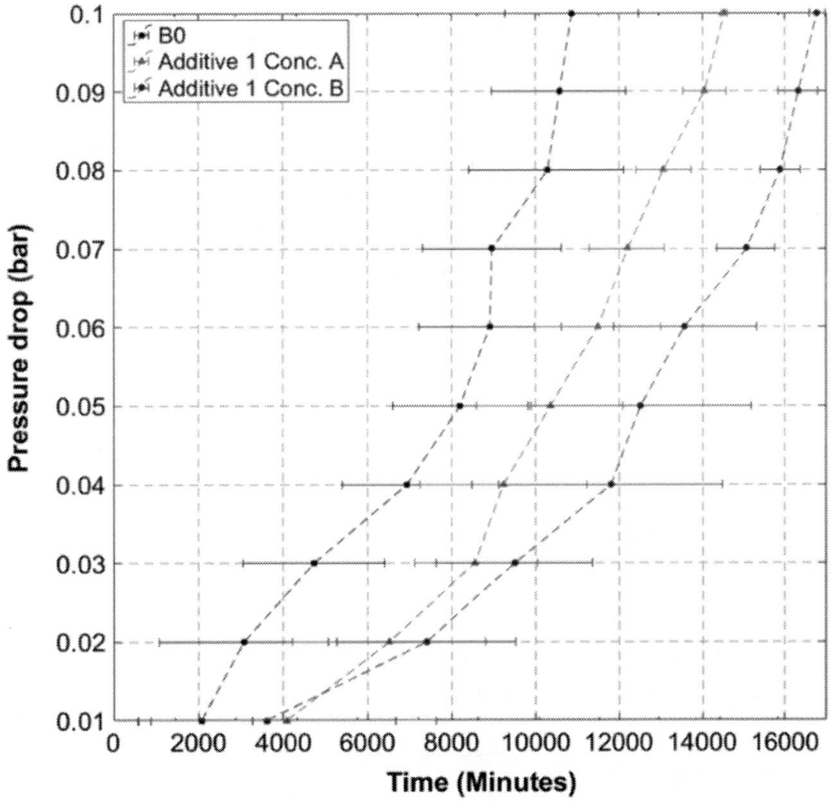

FIGURE 11 - EFFECT OF ADDITIVES ON FUEL DEGRADATION. TEST CONDITIONS: B0 DIESEL PLUS TWO CONCENTRATIONS OF AN ADDITIVE COMPONENT; CRP 2000 BAR; INJECTION EVENT ON

CONCLUSIONS

A rig which simulates the pressures and temperatures that a fuel is subjected to in a modern diesel engine has been developed and was used to investigate the issue of vehicle fuel filter blocking. It was found that at a rail pressure of 2000 bar both B0 EN590 and B10 (SME) diesel caused filter blocking to the same degree. This implies filter blocking caused by on-board fuel degradation has the potential to occur in a wide range of different fuel compositions independent of bio-component content. Further testing on the B10 fuel with a rail pressure of 2000 bar resulted in a pressure difference across the fuel filter of 0.5 bar within 200 hours, whilst the corresponding experiment at 1000 bar showed no filter pressure increase. This work suggests that the increases in rail pressures of modern diesel engines has a significant effect on the propensity of vehicle fuel filters to block. An initial study indicates that addition of specifically targeted additives into diesel would delay this filter blocking.

ACKNOWLEDGEMENTS

The authors would like to thank the Powertrain and Vehicle Research Centre at the University of Bath, UK, where the work for this paper was undertaken and BP International Plc for permission to publish this paper.

ABBREVIATIONS

Δp	Delta Pressure (Pressure drop between the pre and post fuel filter pressure transducers)
Bx	Biodiesel blend ratio. x=0 represents no biodiesel content, x=10 means 10 % v/v biodiesel in diesel.
CHNO	Carbon, Hydrogen, Nitrogen, Oxygen
CO	Carbon monoxide
CR	Common Rail
CRP	Common Rail Pressure
FIE	Fuel Injection Equipment
FAME	Fatty Acid Methyl Ester
FT - IR	Fourier Transform Infrared Spectroscopy
HC	Hydrocarbon
^{1}H NMR	Proton Nuclear Magnetic Resonance
HP	High Pressure
HPCR	High Pressure Common Rail
NOx	Nitrogen Oxides
PM	Particulate Mass
ppm	parts per million
RME	Rapeseed Methyl Ester
SME	Soybean Methyl Ester

REFERENCES

1. Delphi Automotive LLP, "2017/2018 Worldwide Emissions Standards, Passenger Cars and Light Duty Vehicles," https://www.delphi.com/docs/default-source/worldwide-emissions-standards/delphi-worldwide-emissions-standards-passenger-cars-light-duty-2017-2018.pdf, accessed Aug. 2017.
2. Shundoh, S., Kakegawe, T., Tsujimura, K., and Kobayashi, S., "The Effect of Injection Parameters and Swirl on Diesel Combustion with High Pressure Fuel Injection", SAE paper 910489, 1991.
3. Xu, Z., Li, X., Guan, C., and Huang, Z., "Effects of Injection Pressure on Diesel Engine Particle Physico-Chemical Properties", *Aerosol Science and Technology,* 2014, 48, 128-138
4. Eastwood, P., Morris, T., Tufail, K., Winstanley, T., Hardalupas, Y., and Taylor, A., "The Effects of Fuel-Injection Schedules on Emissions of NOx and Smoke in a Diesel Engine during Partial-Premix Combustion", SAE paper 2007-24-001, 2007.
5. Aoyama, T., Mizuta, J., and Oshima, Y., "NOx Reduction by Injection Control", SAE paper 900637, 1990.
6. Zeh, D., Hammer, J., Uhr, C., Rückle, M., Rettich, A., Grota, B., Stöcklein, W., Gerhardt, J., Naber, D., and Raff, M., "Bosch Diesel Injection Technology – Response for Every Vehicle Class" in 23rd Aachen Colloquium Automobile and Engine Technology, 2014.
7. Directive 2009/28/EC of the European Parliament and the Council of 23 April 2009 on the promotion of the use of energy from renewable sources and amending and subsequently repealing Directives 2001/77/EC and 2003/30/EC
8. Directive 2009/30/EC of the European Parliament and the Council of 23 April 2009 amending Directive 98/70/EC as regards the specification of petrol, diesel and gas-oil and introducing a mechanism to monitor and reduce greenhouse gas emissions and amending Council Directive 1999/32/EC as regards the specification of fuel used by inland waterway vessels and repealing Directive 93/12/EEC
9. United States Environmental Protection Agency, "Renewable Fuel Standard Program," https://www.epa.gov/renewable-fuel-standard-program, accessed Aug. 2017.
10. ASTM Standard D7467, 2017, "Standard Specification for Diesel Fuel Oil, Biodiesel Blend (B6 to B20)", ASTM International, West Conshohocken, PA, 2017, DOI: 10.1520/D7467-17, www.astm.org
11. Echim, C., Maes, J., and De Greyt, W., "Improvement of cold filter plugging point of biodiesel from alternative feedstocks", *Fuel,* 2012, 93, 642
12. Jolly, L., Kitano, K., Sakata, I., Strojek, W., and Bunting, W., "A Study of Mixed-FAME and Trace Component Effects on the Filter Blocking Propensity of FAME and FAME Blends", SAE paper 2010-01-2116, 2010.
13. Barker, J., Richards, P., Goodwin, M., and Wooler, J., "Influence of High Injection Pressure on Diesel Fuel Stability: A Study of Resultant Deposits", SAE paper 2009-01-1877, 2009.
14. Cook, S., and Richards, P., "Possible Influence of High Injection Pressure on Diesel Fuel Stability: A Review and Preliminary Study", SAE paper 2009-01-1878, 2009.

15. Gopalan, K., Smith C., Pickering, S., Chuck, C., Bannister, C., "Factors affecting diesel fuel degradation using a bespoke high-pressure fuel system rig", *Proc. IMechE Part D: Journal of Automobile Engineering*, 2017, Special Issue, DOI: 10.1177/0954407017723796

16. Yamane, K., Kawasaki, K., Sone, K., Hara, T., and Prakoso, T., "Oxidation stability of biodiesel and its effects on diesel combustion and emission characteristics", *International Journal of Engine Research,* 2007, 8 (3), 307-319

17. Lockett, R., and Jeshani, M., "An experimental investigation into the effect of hydrodynamic cavitation on diesel", *International Journal of Engine Research,* 2013, 14 (6), 606-621

IASH 2017, the 15TH INTERNATIONAL CONFERENCE ON STABILITY, HANDLING AND USE OF LIQUID FUELS
Rome, Italy
10-14 September 2017

ADVENTURES IN DIESEL STABILITY: STANDARDISING THE QUARTZ CRYSTAL MICROBALANCE METHOD

David J. Evans, Renée L Webster, Paul M. Rawson, Nathan Matheson and Christy-Anne Stansfield

Defence Science and Technology Group, 506 Lorimer St, Fishermans Bend, Victoria, Australia, 3207. David,Evans@dsto.defence.gov.au

Diesel stability requirements in the fuel specifications have a sole focus around trying to predict the "use by date" of the fuel. This requirement has greater relevance for organisations that tend to hold fuel for extended durations versus the retail sector. Additionally, changes in fuel system design such as high pressure common rail injection fuel systems have placed greater emphasis on the ability of diesel fuels to resist thermal oxidative processes. Current diesel fuel specifications do not include a thermal stability requirement. Current storage stability tests such as ASTM D2274 and D5304 are well established however these were validated solely on high sulfur fuels. These tests have not been good at identifying problematic fuels versus newer methods such as the rapid small scale oxidation test (RSSOT) and quartz crystal microbalance (QCM).

QCM testing of a range of ultra-low sulfur diesel fuels showed that the fuels are low sediment formers producing generally less than 2 $\mu g/cm^2$ of deposit in 15 hours. The fuels were all observed to be fast oxidisers with the 90% of available oxygen being consumed at various rates. In order to assess the stability of the fuels a method of standardising the QCM results was investigated. This paper will discuss the methods used and relationships between stability and oxidation products formed.

IASH 2017, the 15th INTERNATIONAL SYMPOSIUM ON STABILITY, HANDLING AND USE OF LIQUID FUELS
Rome, Italy
10-14 September 2017

CARBONACEOUS INTERNAL DIESEL INJECTOR DEPOSITS: MECHANISMS, CHARACTERIZATION, AND REMEDIATION

Andrew McKnight[1], Jim Barker[2], and Jacqueline Reid[3]

1 Innospec Fuel Specialties LLC, 200 Executive Drive, Newark DE 19702, U.S.A., andrew.mcknight@innospecinc.com
2 Innospec Limited, Manufacturing Park/Oil Sites Rd, Ellesmere Port CH65 4EY, United Kingdom, jim.barker@innospecinc.com
3 Innospec Limited, Manufacturing Park/Oil Sites Rd, Ellesmere Port CH65 4EY, United Kingdom, jacqueline.reid@innospecinc.com

ABSTRACT

As the global diesel vehicle base continues to shift to high pressure common rail engine technology to meet ever stricter emissions legislation, the result is a heightened sensitivity to deposit formation. A significant industry focus on internal diesel injector deposits over the last decade has generated significant knowledge on contributing factors. Several primary deposit types have been identified. One such type, carbonaceous deposits, is inherently a concern because it is derived from thermo-oxidative breakdown of fuel inside the vehicle instead of fuel-borne contaminants. This paper will provide examples where these deposits have been found, explanation of the formation mechanism and characterization techniques, fuel effects, and effectiveness of deposit control additives to remediate these deposits.

KEYWORDS: High pressure common rail, IDID, carbonaceous, hydropyrolysis, GC-MS, TPO, FIB-SEM, TOF-SIMS, DCA

INTRODUCTION

Ever-tightening emissions legislation globally has been a driving force behind the development and implementation of advanced diesel engine technology. Most notably emission regulations as specified in Euro 6 [1] in Europe, EPA and CARB in USA [2] and others worldwide [3] have been the key driver in the introduction of even higher pressure common rail HPCR injector equipment in on- and off-road applications. Common rail engines not only offer improvement in emissions but are also more efficient and allow more operational flexibility than their predecessors. But as fuel injection pressures have increased in order to meet these regulations there has been a concomitant increase in temperature and shear within the fuel injectors and hence an increase in distressing of the fuel within the injector. As such, the use of such systems in diesel powertrains and problems experienced since 2005 has led to intense industry activity on high pressure common rail systems.

The evolution in engine technology has occurred in parallel with the introduction of Ultra-Low Sulfur Diesel (ULSD), resulting changes in fuel solubilizing properties and the blending of biodiesel to meet the renewable fuels mandate has seen a number of reports worldwide of injector problems in the field because of deposit formation [4]. The injector malfunction has manifested itself in:

- No cold start performance
- Higher emissions
- Stalling and misfiring
- Rough idling
- Increased smoke production
- Reduced fuel economy

As a result there is significant interest in diesel injector deposits with CRC (Central Research Council, Diesel Performance Group, Deposit Panel, Bench/Rig Investigation sub panel) in the United States and CEN (Committee European de Normalisation TC19/WG24 Injector Deposit Task Force, Engine Test) and the CEC (Coordinating European Council TDG F-110) as well as numerous research groups worldwide.

THE HIGH PRESSURE COMMON RAIL

It is not uncommon for injector temperatures to exceed 300 °C, and injection pressures are increasingly as high as 3000-4000 bar. Moreover, there is considerable recirculation of fuel within the various return lines of a typical HPCR system. This fuel "spill-over", or return to the vehicle tank, can be in the range of 180-600 liters/hour for a Class 8 vehicle, compared to a typical fuel consumption of roughly 50 liters per hour during normal highway operation.

Figure 1: Typical HPCR system

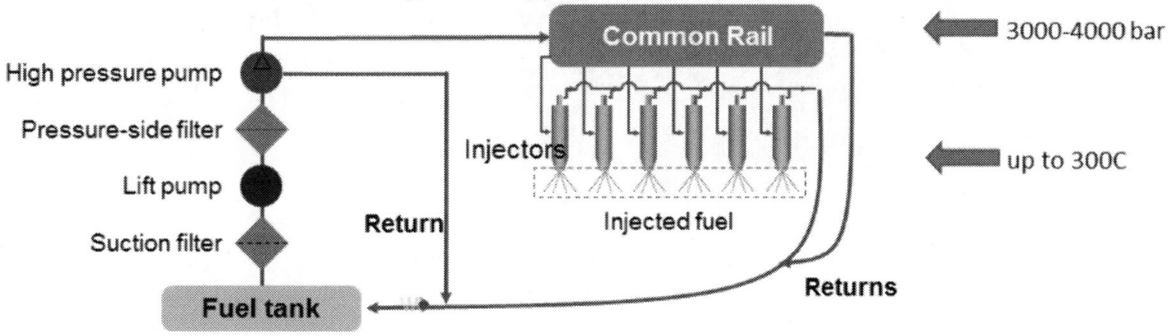

DIESEL DEPOSITS

Originally [5] deposits from the field were seen on the filters of fuel injector equipment FIEs, but then on the injector nozzles and more recently on the internal parts of the injector. In some cases deposits were found in all three locations [6]. The current focus is on Internal Diesel Injector Deposits (IDID). There are five primary industry-recognized types of IDID [7]:

Table 1: Types of IDID

Type	Description	
Carbonaceous	black in color, carbon-based	
Metal Soaps	white in color, sodium-carboxylate based	
Amide	Brown-colored, polymeric	
Lacquer	may be a carbonaceous based precursor? Described by fringe patterns.	
"Sticky"	often seen in conjunction with aged fuels	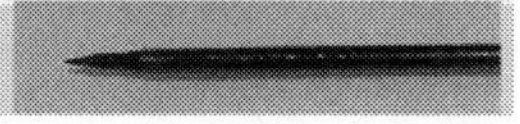

IDID deposits may be found in various locations throughout the injector body. All of the deposits can be detrimental to the operation of fuel filters and injectors (Figures 2,3).

Figure 2: Examples of "Blackened" Fuel Filters from Carbonaceous Deposits

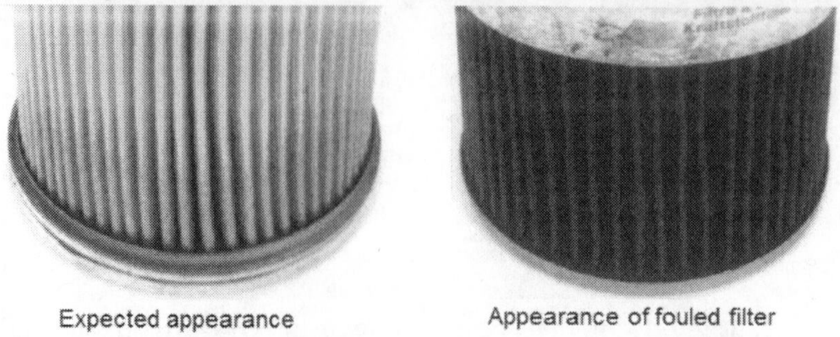

Figure 3: Examples of IDID in Injectors

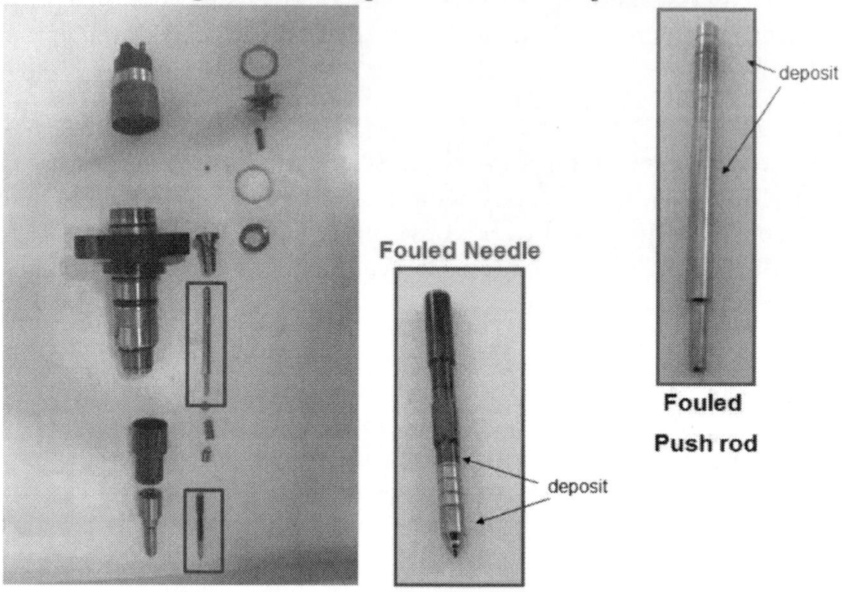

These and other FIE deposits may be attributed to a number of sources which are depicted in Figure 4. The sources described can of course act both singly and in combination to produce the deposit. But it is notable that very careful housekeeping, high quality biodiesel, appropriate use of additive, and lack of adulteration by lubricant oil alone do not obviate the possibility thermal-oxidative breakdown of fuel to form carbonaceous deposits and deposit precursors.

Figure 4: Possible Causes for Diesel System Deposits

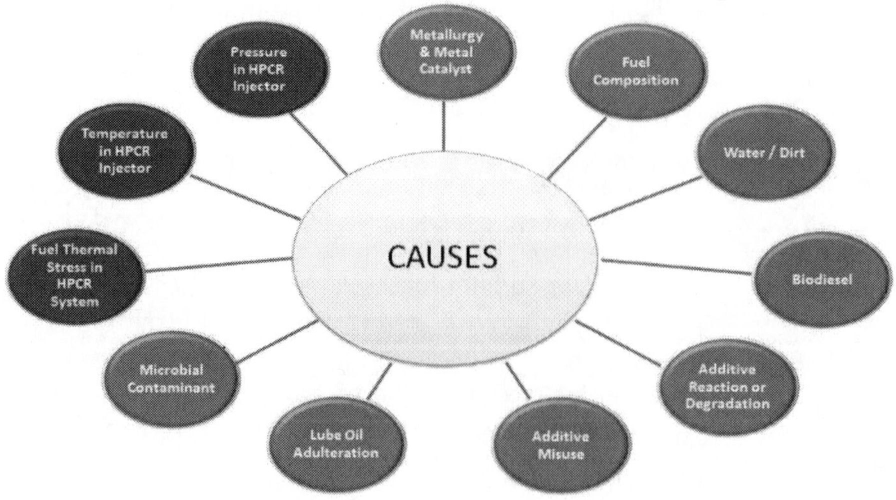

CARBONACEOUS DEPOSIT FORMATION MECHANISM

The chemistry involved in thermal-oxidative breakdown of linear and cyclic hydrocarbons is extremely complex and the subject of considerable study. For example, comprehensive analyses of structure, kinetics and mechanism have been investigated and account for the oxidation of hexadecane at temperatures between 120 and 190°C; intramolecular reactions leading to, among other things, keto-hydroperoxides [8], fragmentation to yield aldehydes, olefins and carboxylic acids [9] as well as formation of cyclic ether products [10].

To illustrate the rate of decomposition reactions possible with ULSD systems, three model fuels were blended from pure components (Table 2). Each of the components is a surrogate for the various general hydrocarbon classes found in fuel, such as alkanes, cycloalkanes, aromatics, polyaromatics, and olefins.

Table 2: Model Fuel Compositions

Hydrocarbon Class	Component	Model 1	Model 2	Model 3	
Alkanes	Hexadecane	50	50	49	% vol
Cycloalkanes	Decalin	30	30	30	% vol
Aromatics	n-Butylbenzene	20	15	15	% vol
Polycyclic aromatics	Naphthalene	0	5	5	% vol
Olefins	Decene	0	0	1	% vol

Each of the model fuel blends were refluxed at 150 °C with flowing air, and aliquots were removed every hour for determination of peroxide values (ASTM 7447-03) and total acid number (ASTM D 974-08, but at 0.01 M KOH). The results, shown in Figure 5, show that all of the model fuels uniformly yielded a relatively rapid rise in peroxide levels starting at 2 hours, then leveling off at hour 5 and then dropping off at hour 7. In parallel, an induction period of roughly 4 hours was observed for acid number at which point the fuel acid content began to steadily rise (as the peroxide value began to saturate).

It is also interesting to note that the level of aromatics, polyaromatics, and olefins in this study did not appear to have a significant impact on the peroxide levels or acid numbers. Only model fuel 1, with higher alkyl benzene content, gave a sharper rise in acid number presumably due to the reactivity of the benzylic hydrogens.

Figure 5: Effect of Reflux Time on Peroxide Formation and Acid Number in Model Fuels

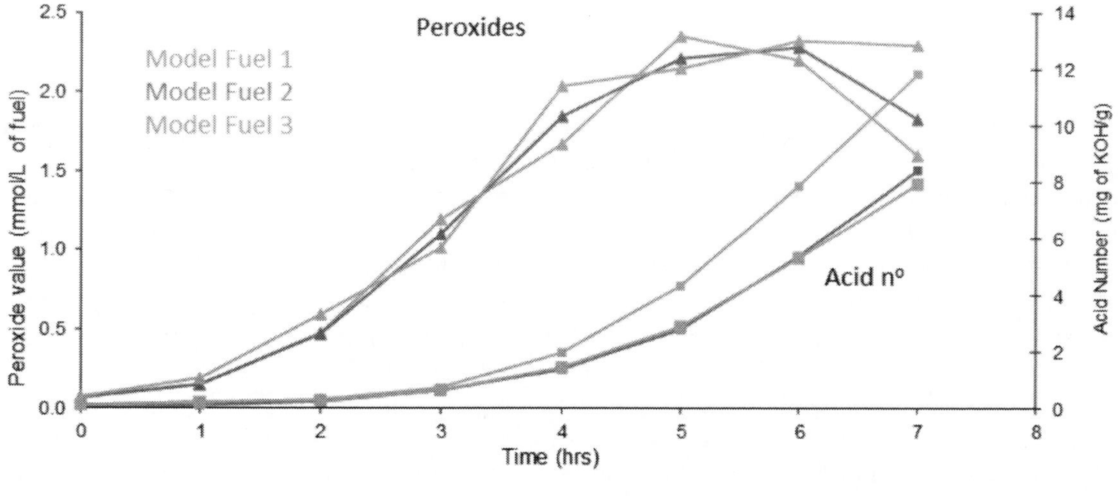

Nonetheless, this simple study demonstrates that even in far less severe conditions than the high pressure common rail – much lower temperature, 1 atmosphere pressure, and very short timescale – it is possible to generate highly reactive species by thermal oxidation processes. These reactive species, in turn are able to undergo further reaction to form deposit precursors.

Following on this and other work, a formation mechanism for carbonaceous deposits has been previously proposed [4] which likely mirrors the slow process for formation of graphite and diamonds in the natural world (Figure 6), whereby key features include:

- all hydrocarbon fuels are capable of providing carbonaceous deposits.
- the inter-fuel variation lies in the intensity and duration of thermal, oxidative and mechanical stresses required to bring this about.
- The initiation stages are most likely the formation of hydroperoxides from n-alkanes (or n-alkane-like fragments of other molecules) via hydrogen abstraction.
- these initial stages may be metal-catalysed or due to exceptionally reactive fuel components.
- dependent on the levels of the relevant materials in a particular sample, O- and/or N-containing aromatic species act as radical traps, become oxidized and undergo coupling and/or acid/base reactions.
- molecular weight and polarity of oxidized products increases and/or fuel cools such that a solubility limit is reached and phase separation occurs.

Figure 6: "The Journey From Fuel to Graphite"

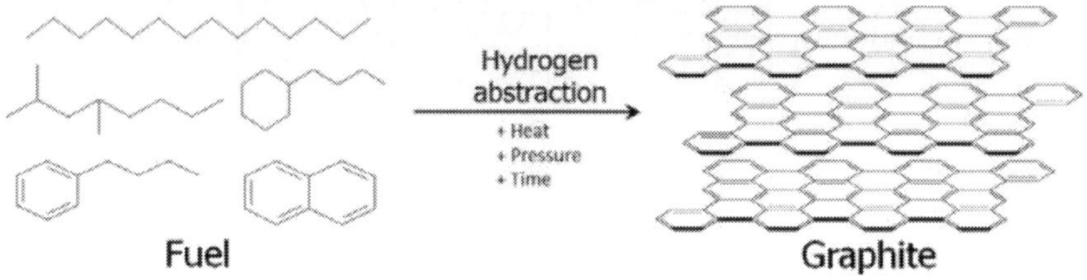

CARBON DEPOSIT CHARACTERIZATION

A number of analytical techniques have been used to determine the nature of IDID deposits which has then led to the development of deposit control additives (DCA) to mitigate these deposit effects. Examples of these techniques are as follows.

Temperature Programmed Oxidation (TPO) [11,12]

This method involves raising the temperature of in-situ carbon in an oxygen atmosphere and measurement of the evolved carbon dioxide. TPO can differentiate carbon structures, whereas amorphous carbon is more reactive than structured carbon, oxidizing at lower temperatures and more highly "coked" deposits show higher burn-off temperature. More highly coked deposits also tend to be enriched in their elemental carbon content and deficient in hydrogen and polar elements such as oxygen, nitrogen, and sulphur. Comparison of field deposits is therefore possible against known coke standards as well as to highly ordered diamond-like coatings (DLC) found the internal surfaces of HPCR fuel injectors. Figure 6 below illustrates that a typical fuel filter carbonaceous deposit burns off at a much lower temperature than deposits found inside the fuel injector, which can have peak combustion temperatures ranging from 440 to 750 °C. Interestingly, deposits from the upper section of the needle and the control piston have burn-off temperatures that are very similar to that of the DLC, suggesting that considerable coking can occur even on internal parts that are not adjacent to the combustion chamber (eg the injector nozzle).

Figure 7: TPO

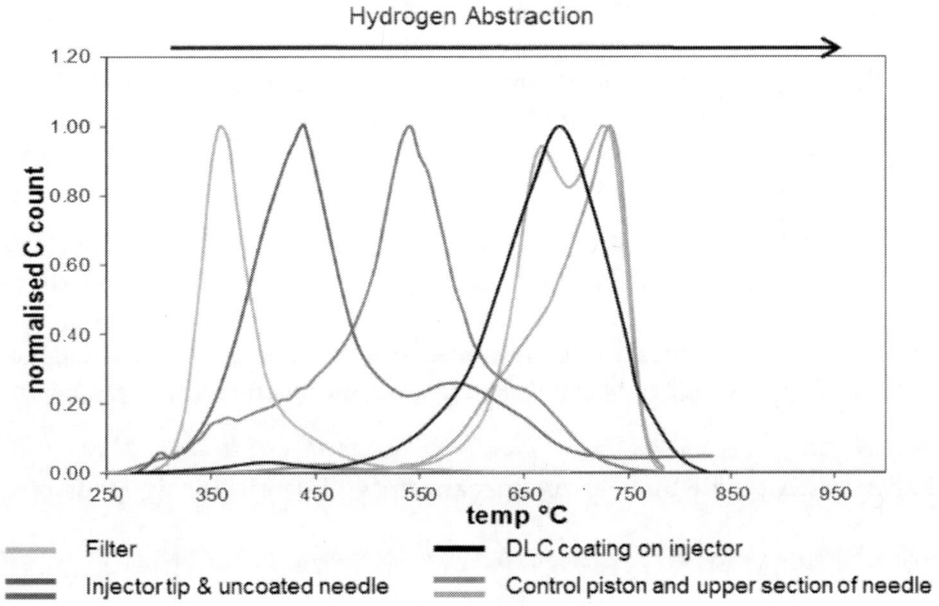
Hydrogen Abstraction
1.20
1.00
0.80
0.60
0.40
0.20
0.00
normalised C count
250
350
450
550
650
750
850
950
temp °C
Filter
DLC coating on injector
Injector tip & uncoated needle
Control piston and upper section of needle

Hydropyrolysis (Hypy)

This technique is worth mentioning as it has been used to characterize deposits from the tip of the injector although data for IDID is limited with only a study showing a long chain n-alkane series profile for an IDID [13]. Nonetheless, the technique has provided information concerning deposits from the injector orifices. It is a catalytic technique which delicately strips molecules of their functional groups but retains their carbon skeletons and stereo chemistries intact. The resulting products from the hydropyrolysis are then subjected to Gas Chromatography/Mass Spectrometry (GC/MS) analysis.

Hypy-GC/MS results showed the presence of alkylbenzenes, alkylnapthalenes and alkylphenathranes (Figure 7) indicative of species on the path to graphitic type structures. Moreover, use of this technique on filter and injector deposits has shown that filter deposits are a complex mixture of graphitic carbon, polyaromatics, cycloalkanes, aromatics, straight chain & substituted alkanes, acids, and inorganics. And the identification of n-C16 & n-C18 carbon backbones in the MS spectra are indicative of carboxylic acid residues.

Figure 8: Characteristic Single Ion Chromatograms (SIC) for injector tip Hydropyrolisis product.

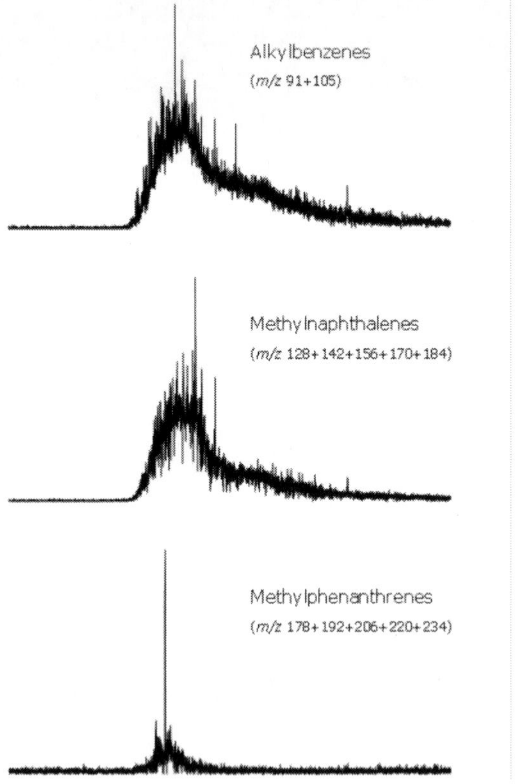

Raman Spectroscopy (RS) *[7]*

Raman spectroscopy unlike FTIR has the ability to observe ordered and disordered carbonaceous deposits and has been used to map an area of an injector needle. This has shown the variation across it (Figure 10) using the change in the intensity of the I_G and I_D bands. The ratio I_G/I_D demonstrates how the ratio of ordered to disordered carbon (graphitic and non-graphitic) changes across the surface.

Figure 9: Raman map across injector surface

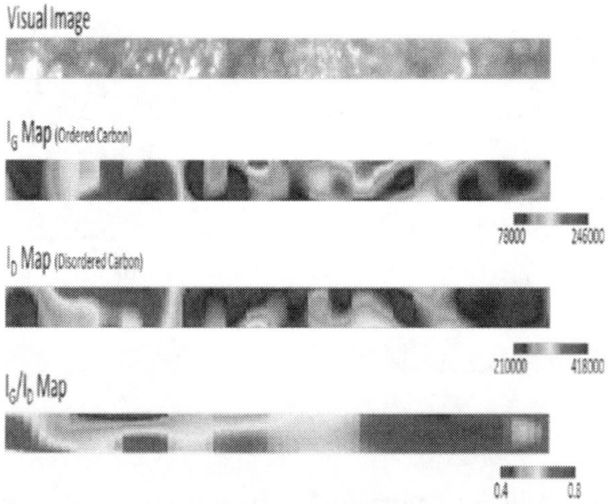

Time of Flight Secondary Ion Mass Spectrometry (Tof-SIMS) *[14].*

The previous techniques are surface based. The true nature of IDIDs in layers was only noted with the use of the Tof-SIMS technique as this allowed "drilling" below the surface with an ion source. The technique has also shown the presence of aromatic structures within the carbonaceous layers, consistent with a move toward graphitic type products [15]. More specifically, it was noted that poly-aromatic hydrocarbons (PAHs) constituting carbonaceous deposit are more concentrated closer to the surface in relation to iron oxide. The data is further evidence towards highly structured ring systems being found in carbonaceous internal injector deposits.

Figure 10: Schematic of depth profile of injector deposit for Tof-SIMS

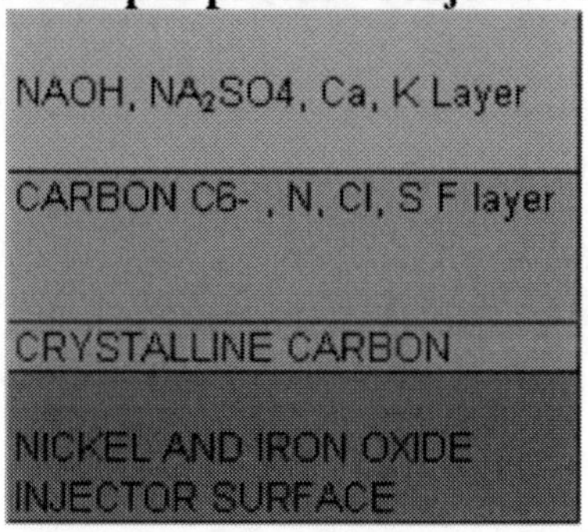

Focused Ion Beam Scanning Electron Microscopy (FiB-SEM) *[16, 7]*

Due to the multi-layered nature of internal diesel injector deposits, it is difficult to analyse these in situ. The previous techniques discussed would involve simply analysing the surface in a non-destructive or destructive fashion. Even less sophisticated deposit analysis would arise from physically scraping or solvent dissolution of the surface deposit. However, use of FIB-SEM, which is used extensively in the semiconductor industry, allows the careful removal of a trench of deposit material from the internal surface using the ion beam. This deposit sub-section can be subjected to other types of analysis techniques to determine the provenance or compositional history of each layer.

In one illustrative example (Figure 11), a sample was obtained from a deposit from a jammed field injector push rod from the United States. The injector had failed while using ULSD fuel, which resulted in a no-start issue. The injector was dismantled with minimal disruption to the deposit. Application of the ion beam to a specific portion of the deposit by varying the FiB current allows for a material slice of the deposit of desired width and depth to be carefully removed for further analysis for progressive structural assessment.

Figure 11: SEM Images of Injector Deposits during FIB-SEM process

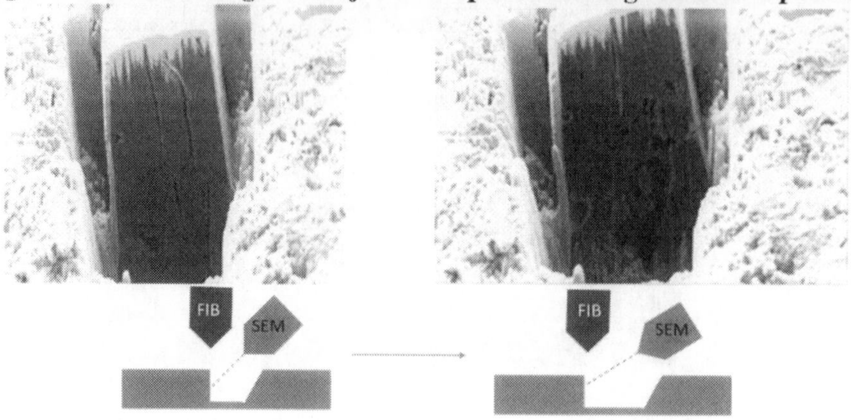

Transmission Electron Microscopy (TEM)

TEM describes a sample's topography and morphology on a nanometer scale thus allowing the graphitic nature of deposits to be seen. This can be used to characterize bulk injector deposits removed by standard mechanical scraping in a lab environment. One example, shown in Figure 13, is from an injector needle deposit scraping and shows pseudo-graphitic content that is similar to that previously found on diesel filters although with much higher crystallinity [17]. However, without proper mapping of where in the injector the graphitic content comes from, or what layer, it is difficult to make conclusions about when the deposit may have been formed in the injector plugging history.

Figure 12: TEM micrograph of injector needle deposit

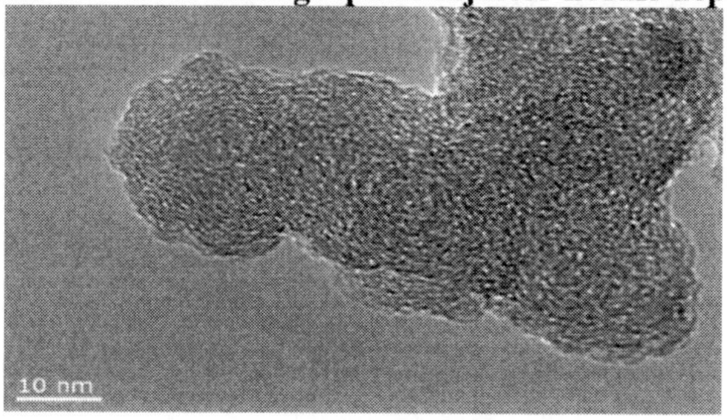

TEM analysis can also be combined with the aforementioned FiB-SEM technique. The push rod deposit analyzed above (Figure 11), once removed, showed a distinct multi-layer structure when viewed side-on by TEM, with one platinum layer from the FiB technique and the other three layers pertaining to the push rod metal surface and two other layers from the field deposit. The middle layer nearest to the surface (Layer 1) resembled a more amorphous liquid state by further TEM analysis. The layer closest to the original push rod surface (Layer 2) exhibited a nanostructure with a mixture of crystalline and amorphous carbon (Figure 14), providing further evidence of how varied temperature regimes on the injector internal surfaces can give varying levels of "coking" of diesel deposit precursors.

Figure 13: Schematic of Deposit Layers and TEM Showing Deposit Layers

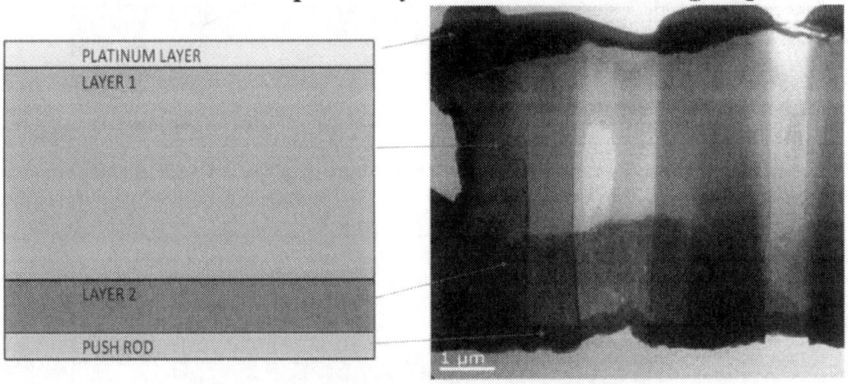

Figure 14: TEM Image of Middle Layer from FiB-SEM Tranch

The above work is not a comprehensive account of the knowledge on carbonaceous based IDIDs in diesel vehicles. There is also much work available from test rigs and other techniques which have led to a much better understanding of IDIDs.

IMPACT OF DEPOSIT CONTROL ADDITIVES

One ongoing area of significant activity in industry groups (CEC, CRC, etc.) is to develop test methods and/or rigs which can demonstrate the effectiveness of deposit control additives (DCA) at remediating or mitigating IDI deposits. But much of this work has focused on the role of external contaminants, such as zinc and sodium carboxylates, as accelerating the formation of deposits. Currently there are no adequate HPCR surrogate fouling tests developed by the industry to simulate the phenomenon of in situ coking of fuel to form carbonaceous deposits without the presence of a catalytic accelerant.

One such keep clean test methodology we developed is described here wherein an injector rig fitted with four new HPCR medium duty injectors is used for each test along with two new fuel filters. All other components (high pressure pump, common rail) were from the same manufacturer as the injector. The rig test parameters are indicated in Table 3. Careful cleaning of the rig was performed before each test to remove traces of deposits, additives, etc.from the previous run. The sequence of methanol (MeOH) and then the glycol ether methoxypropoxypropanol (MEPP) was chosen as MEPP is more easily removed by diesel, while the more aggressive cleaning solvent MeOH is not.

Table 3: Injector Rig Test Conditions

Variable	Setting
Injectors	4 new Bosch CRIN1 injectors per run
Filters	2 new Fleetguard FF167 filters per run
Pre Rig Clean	Repeat rinses with MeOH, MEPP and test diesel
Run time	100 hours
Cycle	Standard Rig test cycle
Injector Tip Temperature	180° C
Fuel Tank Temperature	35° C
Fuel Volume	32 litres
RPM flywheel (engine equiv.)	1200 (2400)

A keep clean run length 100 hours was chosen as part of the original optimization process to maximize deposit formation. Fuel passes through the rig in a 'single pass' (no recirculation) and fuel is not injected into a combustion chamber assembly as would normally occur in an engine. The rig runs through a cycle which last for 24 hour before repeating (Table 4). At times injector heaters and injectors are switched off to allow for periods of 'soaking' to enhance deposit dissolution from the fuel and building up on the internal metal surfaces. The injector tip and tank temperatures were derived as part of a previous cycle optimization process, while the flywheel speed was set at 1200 rpm, equivalent to an engine speed of 2400rpm.

Table 4: Rig Test Cycle

Stage	Duration (hours)	Cumulative Run time (hours)	Pressure (bar)	Injector Firing?	Injector Heater?
1	2.0	2.0	1200	Y	Y
2	2.0	4.0	1350	Y	Y
3	0.5	4.5	1350	N	N
4	2.0	6.5	400	Y	Y
5	1.0	7.5	1000	Y	Y
6	12.5	20.0	1000	Y	N
7	4.0	24.0	1000	N	N

Two separate runs were performed on the rig using a market S50 diesel fuel with 5% FAME (B5): one with just the base fuel and another with a keep clean level of a deposit control additive. The injectors were then removed at the end of each test and disassembled for subsequent analysis by TPO. Figures 15-17 show both the appearance of each critical injector part along with the TPO combustion profile of the bulk deposits on each part of the injector.

For the injector tip (Figure 15), only the part below the dotted line was analyzed. This is because the upper section was in direct contact with the injector tip heater and as such could give a falsely high result. The peaks on the TPO trace show that in the base fuel a more stable carbon deposit with a much higher peak burn off temperature (570 °C) than for the comparable TPO scan for the treated fuel injector tip. This higher peak temperature will correspond to a more graphitic deposit which should be more difficult to remove. In addition, the overall injector tip carbon level for the treated fuel test run was significantly lower than for the base fuel run, demonstrating the keep clean effect of the deposit control additive. This is also easily seen from the visual assessment of each injector tip.

Figure 15: Injector Tip for Base and Treated Fuel, with TPO Scan

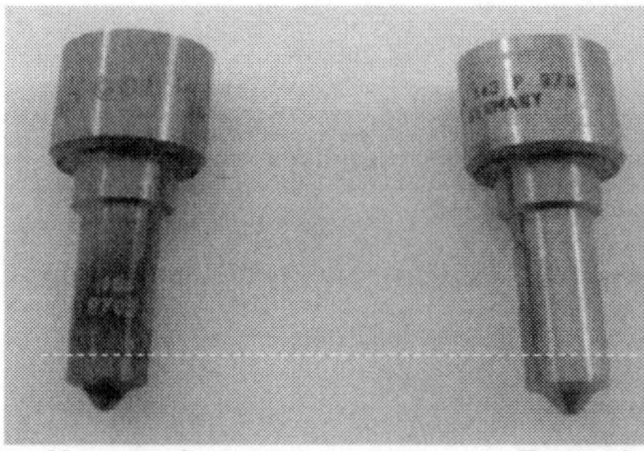

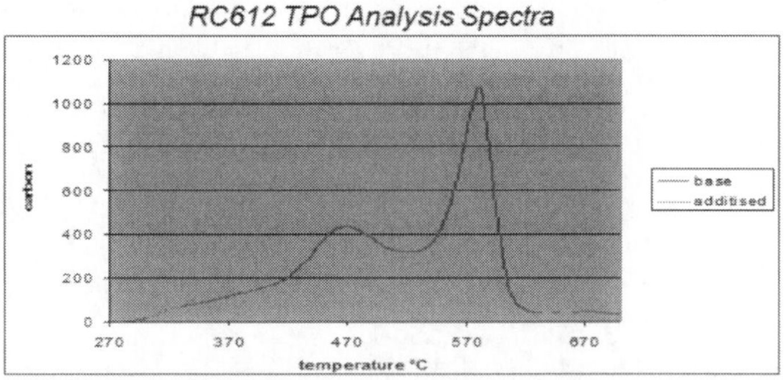

Untreated Treated

The injector needle (Figure 16) also yielded different appearance for the treated fuel run, with this needle appearing clean while the base fuel needle was noticeably dirtier. This level of fouling is heavy enough to lead to injector sticking and loss of performance, although no sticking was observed during the 100 hour test run. The needles were then sectioned using a diamond saw to allow TPO testing of lower needle section to avoid the influence of the DLC coating on the upper section of the needle. Both lower needle sections showed similar TPO combustion profiles but the treated fuel carbon deposit level was lower than for the base fuel. Also, the untreated B5 gave higher levels of increasingly stable deposits above 600°C on the needle.

Figure 16; Injector Needle for Base and Treated Fuel, with TPO Scan

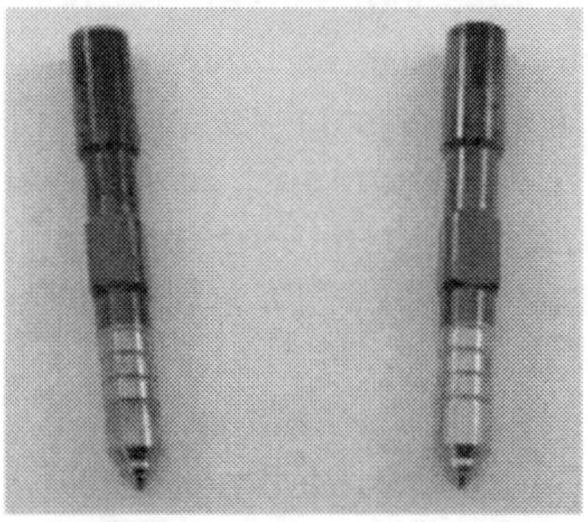

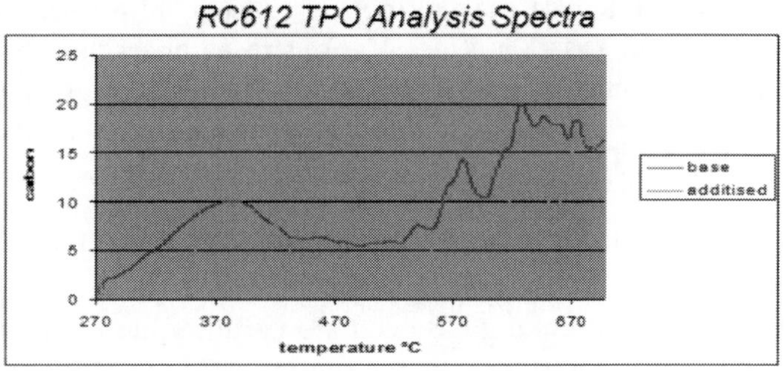

Untreated Treated

The injector push rod (Figure 17), like the other parts, also was visibly dirtier for the base fuel run compared to the treated fuel run, And much like for the needle, both the treated and untreated fuel runs showed similar TPO combustion profiles. Nonetheless, the treated fuel carbon deposit level was lower than for the base fuel and the untreated B5 gave higher levels of increasingly stable deposits above 600°C on the push rod. It is worth noting that deposits on the push rod are more common but the level seen in this test is quite high. Such deposits can eventually cause the firing mechanism to stick or block resulting in loss of performance.

Figure 17: Injector Push Rod for Base and Treated Fuel, with TPO Scan

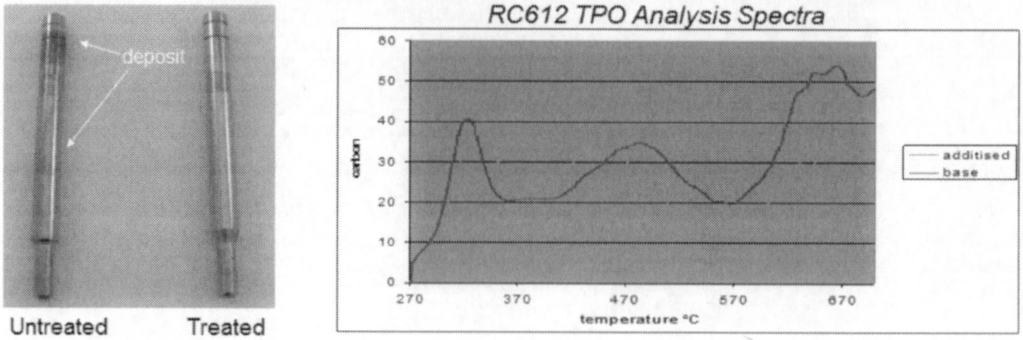

Overall the rig study demonstrated that it is possible to simulate coking of fuel in a relatively short time scale in a laboratory rig, with commensurate build-up of deposits on the internal diesel injector deposits. The carbon deposits generated by this rig test yielded peak burn-off temperatures that are qualitatively similar to ones observed from problem field injectors (Figure 7). And the efficacy of a deposit control additive to lower the overall level of deposit (by both visual and TPO analysis) was demonstrated.

CONCLUSIONS

High pressure common rail IDID injector problems have existed for a number of years and have proven to be complicated. In particular, even in the absence of fuel contaminants such as hard particles and metal carboxylates, the problem of fuel coking and subsequent formation of carbonaceous deposits on injector internals will continue to be an issue. Various researchers have proposed chemical degradation pathways that can lead to deposits and deposit precursors, led by hydrogen abstraction reactions and oxidation reactions to form peroxide and acid intermediates. An array of analytical techniques have been deployed to try and give insight into the fuel decomposition chemistry as well as the makeup of the carbon deposits. The deposit analytical techniques range from bulk compositional assessment (TPO, HyPy) to surface-based (Raman), to sub-surface (ToF-SIMs, FiB-SEM/TEM). All the methods used together help to give insight into the bulk structure of the carbonaceous deposits, especially as regards their "history", and also into the degree to which extreme coking has occurred (ie amorphous vs crystalline morphologies), likely due to the thermal and pressure conditions existing for that particular field issue. Lastly, the use of deposit control additives has been an effective market response to this issue of carbonaceous deposits since the HPCR engine and ULSD fuels converged to give rise to this problem. Here in this paper was detailed a surrogate rig test which successfully simulates this coking phenomenon on a relatively short scale and also shows the effectiveness in using deposit control additives to lower the severity of the problem ("keep clean").

CONTACT

Andrew McKnight
Tel: 302.451.1388

REFERENCES

[1] SMMT European engine emission standards.http://www.smmt.co.uk/industry-topics/environment/intro/european-engine-emission-standards/ (accessed 13/01/2016).

[2] MILLER, P.; SOLOMON, M., A Brief History of Technology-Forcing Motor Vehicle Regulations. 2009.

[3].United States Emission Standards. https://en.wikipedia.org/wiki/United_States_emission_standards (accessed 13/01/2016).

[4] COOK, S.; RICHARDS, P. Possible Influence of High Injection Pressure on Diesel Fuel Stability: Review and Preliminary Study; SAE Technical Paper: 200, 2009-01-1878

[5] BARKER, J.; RICHARDS, P.; GOODWIN, M.; WOOLER, J. Influence of High Injection Pressure on Diesel Fuel Stability: A Study of Resultant Deposits; SAE Technical Paper: 2009. 2009-01-1877

[6] BARKER, J.; SNAPE, C.; SCURR, D. Diesel Deposits, 9th International Colloquium Fuels Conventional and Future Energy for Automobiles, 2013 TAE Esslingen, 149.

[7] BARKER, J.; REID, J.; PIGGOT, M.; FAY, M.DAVIES A,PARMENTER C,WESTON N, SNAPE C, SCURR D, WESTON N ANGEL-SMITH S; The Characterisation of Diesel Internal Injector Deposits by Focused Ion-Beam Scanning Electron Microscopy (FIB-SEM), Transmission Electron Microscopy (TEM), Atomic Force Microscopy and Raman Spectroscopy; SAE Technical Paper 2015; 2015-01-1826.

[8] JENSEN, R.; KORCEK, S.; MAHONEY, L.; ZINBO, M. Liquid-Phase Autoxidation of Organic Compounds at Elevated Temperatures. 1. The Stirred Flow Reactor Technique and Analysis of Primary Products from n-Hexadecane Autoxidation at 120-180°C. J. Amer. Chem. Soc. 1979, 101, 7574-7584

[9] JENSEN, R.; KORCEK, S.; MAHONEY, L.; ZINBO, M. Liquid-Phase Autoxidation of Organic Compounds at Elevated Temperatures. 2 Kinetics and Mechanisms of the Formation of Cleavage Products in n-Hexadecane Autoxidation. J. Amer. Chem. Soc. 1981, 103, 1742-1749

[10] JENSEN, R.; KORCEK, S.; ZINBO, M. Formation, Isomerization, and Cyclization Reactions of Hydroperoxyalkyl Radicals in Hexadecane Autoxidation at 160-190°C. J. Amer. Chem. Soc. 1992, 114, 7742-7748

[11] BARKER, J.; CHEESEMAN, B.; PINCH, D.; RICHARDS, P. Temperature Programmed Oxidation as a Technique for Understanding Diesel Fuel System Deposits; SAE Technical Paper: 2010, 2010-01-1475.

[**12**] BARKER, J.; LANGLEY, G.; RICHARDS, P. Insights into Deposit Formation in High Pressure Diesel Fuel Injection Equipment; SAE Technical Paper: 2010, 2010-01-2243.

[13] BARKER J, RICHARDS P, SNAPE S, MERADITH W, Diesel Injector Deposits-An Issue That Has Evolved with Engine Technology SAE Technical Paper 2011. 2011-10-1923.

[14] BARKER J SNAPE C, SCURR D Information on the Aromatic Structure of Internal Diesel Injector Deposits from Time of Flight Secondary Ion Mass Spectrometry (Tof-SIMS) SAE Technical paper 2014, 2014-01-1387.

[15] BARKER J, SNAPE C, SCURR D, A Novel Technique for Investigating the Characteristics and History of deposits Formed Within High Pressure Fuel Injection Equipment SAE Int. J. Fuels Lubr. 5(3) 2012, doi: 104271/2012-01-1685

[16] BARKER, J.; REID, J.; SMITH, S.; SNAPE, C.; SCURR, D.;PIGGOT, M.; FAY, M.; DAVIES, A.; PARMENTER, C.; WESTON, N. Internal Diesel Injector Deposits (IDID). 11th International Colloquium Fuels Conventional and Future Energy for Automobiles, 2017 TAE Esslingen.

[17] BARKER, J.; REID, J. Injector and Fuel System Deposits. 10th International Colloquium Fuels Conventional and Future Energy for Automobiles, 2015 TAE Esslingen.

IASH 2017, the 15TH INTERNATIONAL CONFERENCE ON STABILITY, HANDLING AND USE OF LIQUID FUELS
Rome, Italy
10-14 September 2017

A NOVEL APPROACH FOR THE EVALUATION OF THE DEPOSIT-FORMING TENDENCIES OF SPARK-IGNITION FUEL

Steven R. Westbrook[1] and George R. Wilson, III[2]

[1]US Army TARDEC Fuels & Lubricants Research Facility (SwRI), Southwest Research Institute, 6220 Culebra Road, San Antonio, Texas 78238 USA. swestbrook@swri.org

[2]US Army TARDEC Fuels & Lubricants Research Facility (SwRI), Southwest Research Institute, 6220 Culebra Road, San Antonio, Texas 78238 USA. gwilson@swri.org

ABSTRACT

As plug in, hybrid automobiles with range extending onboard generators increase in popularity, so does the concern over the storage stability of motor gasoline. The concern is because gasoline will oxidize to form gums and other compounds that can cause deposits in the engine. A range extended hybrid vehicle is more vulnerable to this problem since gasoline could remain in the fuel tank for several months if the engine is not often used. This project used a variety of test gasoline fuels to investigate a proposed new method to assess the deposit-forming tendencies of gasoline. Test fuels included high and low oxidation stability fuels, gasolines that had been stored in vehicle fuel tanks for long periods, and fuels treated with various additives to modify their stability characteristics. One test fuel was also aged under controlled, accelerated conditions to further represent stored fuel. Stability-related properties of the fuel were tested using ASTM methods. An existing method for thermal-oxidative stability of aviation fuel was modified to test the gasoline samples as well.

INTRODUCTION

As plug in, hybrid automobiles with range extending onboard generators increase in popularity, so does the concern over the storage stability of motor gasoline. The concern is because gasoline will oxidize to form gums and other compounds that can cause deposits in the engine. A range extended hybrid vehicle is more vulnerable to this problem since gasoline could remain in the fuel tank for several months if the engine is not often used.

Motor gasoline, technically known as automotive spark-ignition fuel, remains the primary transportation fuel in the United States. As a transportation fuel, gasoline is produced with the understanding that it will be used within a relatively short time period, typically 1-2 months. Because of this understanding, commercial gasoline is not produced with the requirement to remain stable during longer periods of storage. The specification used for the production and sale of gasoline in the United States is ASTM D4814[1] and contains the following statement regarding storage stability of gasoline:

> "The induction period as measured in the oxidation stability test is used as an indication of the resistance of fuel to gum formation in storage. Experience indicates that fuels with an induction period equal to or greater than [240 minutes] generally have acceptable short-term storage stability. However, correlation of the induction period with the formation of gum in storage can vary markedly under different storage conditions and with different fuels."

The "induction period," referenced in the above quote, is the test result from the ASTM D525[2] test method. The specification requirement is that the fuel have a minimum induction period of 240 minutes. According to the ASTM D525 method:

> "...the induction period may be used as an indication of the tendency of motor gasoline to form gum in storage. It should be recognized, however, that its correlation with the formation of gum in storage may vary markedly under different storage conditions and with different gasolines."

Also, the ASTM D525 method contains no procedure to assess chemical changes in the fuel that may have occurred during the testing. A new testing methodology and protocol is required in order to properly evaluate both the risk of unwanted deposits and approaches to mitigate the problem. This study investigated a new application of existing technology to evaluate the stability and deposit forming tendencies of spark ignition fuels.

APPROACH

A combination of conventional tests for gasoline stability, along with a test method designed for aviation fuel, was used to analyze test fuels that covered a range of fuel stability characteristics. Each of the test methods is summarized below, followed by a description of the test fuels used in the study.

Oxidation Stability Test

It is known that as gasoline ages it undergoes some oxidation. In other words, it experiences a decrease in oxidation stability with time. For some fuels, this decrease may be very small. For other fuels the decrease can be quite dramatic. For this study, we used ASTM D7525[3] as the method to assess changes in the oxidation stability of the aged fuel. This test is also known as the Rapid Small Scale Oxidation Test, RSSOT. Under this test method, a 5 mL sample is introduced into a pressure vessel which is then charged with oxygen to 500 kPa at a temperature of 15 to 25 °C. The test is initiated by heating the pressure vessel to a temperature of 140 °C. The pressure is recorded continuously until the breakpoint is reached (i.e. the induction period).

The break point is recorded when the pressure in the test apparatus, is 10 % below the maximum pressure of the actual test run. The induction period is the time elapsed between starting the heating of the sample vessel and the breakpoint, measured in minutes. Changes in the induction period of aged fuels are a reliable measure of changes to the oxidation stability of the fuel.

Gum Content

Standard test method ASTM D381[4], covers the determination of the existent gum content of aviation fuels, and the gum content of motor gasolines, or other volatile distillates, in their finished form, including those containing alcohol, ether-type oxygenates, and deposit control additives. Under this method, a 50 ± 0.5 mL quantity of fuel is evaporated under controlled conditions of temperature and air flow. For motor gasoline, the residue is weighed before and after extracting with heptane and the results reported as milligrams per 100 mL. According to the method:

> "The true significance of this test method for determining gum in motor gasoline is not firmly established. It has been proved that high gum can cause induction-system deposits and sticking of intake valves, and in most instances, it can be assumed that low gum will ensure absence of induction-system difficulties. The user should, however, realize that the test method is not of itself correlative to induction-system deposits. The primary purpose of the test method, as applied to motor gasoline, is the measurement of the oxidation products formed in the sample prior to or during the comparatively mild conditions of the test procedure. Since many motor gasolines are purposely blended with nonvolatile oils or additives, the heptane extraction step is necessary to remove these from the evaporation residue so that the deleterious material, gum, may be determined. With respect to aviation turbine fuels, large quantities of gum are indicative of contamination of fuel by higher boiling oils or particulate matter and generally reflect poor handling practices in distribution downstream of the refinery."

Accelerated Stability and Fuel Aging

The purpose of accelerated aging is to estimate or predict the quality stored fuel at some time in the future. The typical method used to accelerate the aging of fuels is to heat the fuel for a specific time period at a pre-selected temperature. Since many of the reactions involved in the degradation of the fuel are oxidation reactions, an overpressure of oxygen is also common. For any given reaction, there is a unique rate constant, *k*. If the increase in temperature is not too great, the dependence of the rate constant on temperature fits the empirical equation shown in eq. 1. This equation is named for Arrhenius, who first proposed it in 1889.[5]

$$k = Ae^{-E_a/RT} \qquad \textit{(eq. 1)}$$

A = pre-exponential factor or Arrhenius constant
E_a = activation energy
R = gas constant
T = temperature, Kelvin

However, since each gasoline is a mixture of hundreds of different compounds, there are an immense number of possible, individual reactions that one would have to consider in order to calculate which fuel-degradation reactions would occur during a given period of fuel storage. And, each gasoline would potentially present a different mixture of compounds, requiring a different set of calculations. This is why test methods to evaluate the stability of fuel use some combination of heat and oxygen to artificially age the fuel. After aging, the fuel is analyzed for some combination of properties such as gum and particulate formation as measures of the degradation of the fuel during aging.

As mentioned above, the current test methods for gasoline stability are either ASTM D525 or ASTM D7525. But, these test methods measure only the consumption of oxygen by the reduction of pressure in the test vessel as an indication of the oxidation stability of the fuel. There is no provision to evaluate the chemical changes that may have occurred in the fuel. Specifically, whether the aged fuel has a greater tendency to form deposits on hot, metal, fuel-system components. For this reason, we used ASTM D4625[6], for the accelerated aging of the fuel. Although ASTM D4625 is primarily used for fuels other than gasoline, it can be adapted for testing gasoline and has been widely used to evaluate stability of gasoline in many past studies.[7,8,9]

Under ASTM D4625, the test fuel is aged in glass containers at 43 °C (110 °F) for pre-selected time periods. After aging for the selected time period, a sample is removed from storage, cooled, and analyzed for particles and gums that have formed during storage. For most practical purposes, it has been shown that aging fuel at 43 °C results in an approximately four-fold acceleration of the degradation compared to storage at an ambient temperature of 21 °C. That is, a week of storage at 43 °C is roughly equivalent to a month of storage at approximately 15 °C.[10] We used ASTM D4625 to artificially age samples of our stable gasoline for 4 weeks and 8 weeks. The aged fuel was then tested with other test methods as an evaluation of the change in the fuel during storage.

Thermal-Oxidation Stability Testing and Deposit Measurement

While measureable changes in the oxidation stability of the fuel are important, they are only of significant concern to the degree that they lead to a greater propensity of the fuel to form deposits in the engine. As stated above, oxidation stability is not necessarily a good indicator of deposit-forming tendency. We used ASTM D4625 to age the fuel so we could have some measure of the fuel-insoluble material that forms in the aged fuel. However, ASTM D4625 does not provide a measure of deposit-forming tendencies on hot, metal surfaces. To evaluate this property of the fuel, we used a modified version of ASTM D3241.[11] The test method uses 450 mL of test fuel during a 2.5-h test. The fuel is pumped at a fixed volumetric flow rate through a tube-in-shell heat exchanger that uses a resistively heated tube, usually aluminum, as a test coupon. Subsequently, it passes through a precision stainless steel filter where fuel degradation products may become trapped. For this testing, we used a specific type of ASTM D3241 device, the Alcor JFTOT II®[12], because it allows the complete containment of the volatile and extremely flammable test fluids. The essential data derived are the amount of deposits on the heater tube, and the rate of plugging of a 17 μm nominal porosity precision filter located just downstream of the heater tube. The amount of deposits left on the heater tube are quantified using both visual and ellipsometric[13] techniques. The ellipsometric technique provides an absolute measurement of the deposit thickness on the heater tube. The ellipsometric method also provides a 3-dimensional map of the deposits (similar to the one shown in Figure 1) and a calculation of the total deposit volume. These maps give a quantitative assessment of any changes in the deposit forming tendencies of the fuels. The ellipsometer used for this study was a FALEX Model 430 ETR (Ellipsometric Tube Rater).

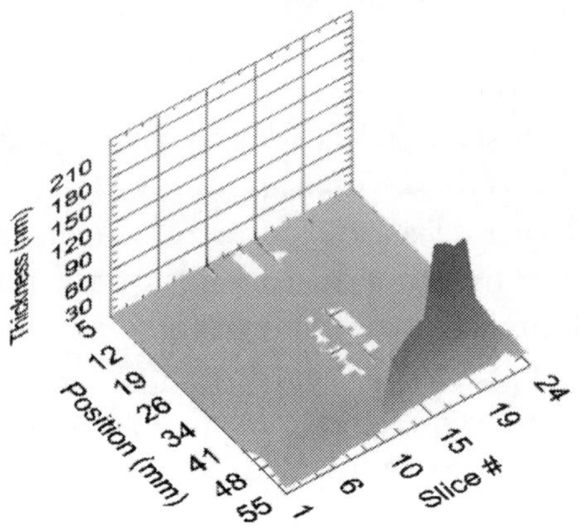

Figure 1. Plot of Deposit Position and Thickness – Ellipsometric Technique

Test Fuels and Additives

Since the objective of this study was to evaluate a potential new method for assessing the deposit forming tendencies of gasoline, there was no need to assemble a large variety of test fuels. For this work we started with a Pump Gasoline (stable), samples from two different vehicle storage tanks (unknown storage times and stability), and a Research Gasoline (unstable). We analyzed these four samples and then used the Pump Gasoline and Research Gasoline, (blended with selected additives or aged), to vary the stability of the fuels for further testing. The unstable, Research Gasoline was treated with three different additives that have been used to improve the stability of gasoline. The stable, Pump Gasoline was aged by ASTM D4625 and tested. This fuel was also treated with 2 different levels of m-toluidine (see Figure 2) which has previously been shown to increase deposits in fuels. The fuels are summarized in Table 1.

Table 1. Test Fuels

Laboratory Number	Description
CL17-0556	Pump Gasoline
CL17-0668	Gasoline from vehicle "A" fuel tank; unknown storage time
CL17-0669	Gasoline from vehicle "B" fuel tank; unknown storage time
CL17-0708	Research Gasoline
CL17-0871	CL17-0708 + 12 ppm hindered phenol antioxidant
CL17-0872	CL17-0708 + 1.95 mL/L of commercial gasoline treatment "A" (CL17-0863)
CL17-0873	CL17-0708 + 3.15 mL/L of commercial gasoline treatment "B" (CL17-0864)
CL17-0957	CL17-0556 + 50 ppm m-toluidine
CL17-0971	CL17-0556 + 100 ppm m-toluidine

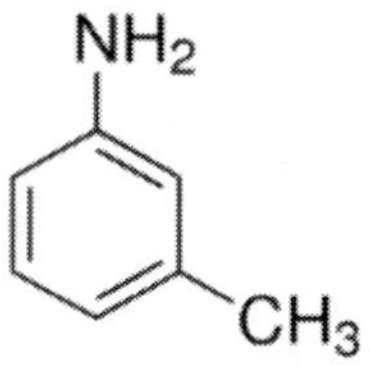

Figure 2. ***m*-toluidine**

RESULTS AND DISCUSSION

The test results are presented and discussed in two parts. The first part covers the RSSOT, gum content, and accelerated storage testing. The second part covers the thermal stability and deposit measurement results (ASTM D3241).

Oxidation Stability – RSSOT

The ASTM gasoline specification sets a minimum of 240 minutes for the measured induction period by ASTM D525 or ASTM D7525. The four base fuels covered a wide range of results for this test (see Table 2). The stable fuel induction period was very long at 5,719 minutes. The Research Gasoline had a very short induction period of 23 minutes. Based on the test results for the two samples from vehicle fuel tanks, it is apparent that one fuel had been in the tank for a long time (26 minute induction period) and the other fuel had likely been in the vehicle fuel tank for a much shorter period (1,109 minute induction period).

Table 3 presents the results for the testing of the Research Gasoline blended with selected stabilizer additives. All three additives have been used in gasoline; but, they are all intended to be put in new, fresh fuel. This Research Gasoline is not the same quality as refinery fresh fuel so it was expected that the additives might not provide much improvement in the gasoline stability. In fact, only commercial additive "B" resulted in any measureable improvement in the RSSOT results, and the improvement was not significant.

Table 4 contains the results of treating the Pump Gasoline with *m*-toluidine, a compound that has shown a tendency to increase thermal deposits[14]. Due to the reactive nature of the toluidine, we anticipated that the RSSOT results would get worse, but we were unsure of the degree of degradation. The testing showed a very significant decrease in the measured induction period of the fuel.

Table 5 presents the results for the accelerated aging (ASTM D4625) of the Pump Fuel. The accelerated aging was conducted as a means to age the fuel under controlled conditions. The goal was to age the fuel and determine if any of the test methods could quantify a change in the fuel. The RSSOT induction periods for the aged fuel showed very significant decrease, compared with the fresh fuel. This aging caused induction period reductions that were equivalent to the change caused by the addition of the toluidine.

In summary, the RSSOT did measure differences in the oxidation stability of the test fuels.

Gum Content

The gum content analysis, ASTM D381, typically provides two pieces of information regarding the gasoline that was analyzed; an indication of detergent additive content (unwashed gum) and a measure of the gum content of the fuel itself (washed gum). Higher solvent-washed gum values are a concern. The ASTM specification limit for solvent-washed gum content is 5 mg/L. Also, according to the specification, solvent-washed gum can plug filters and cause deposits on fuel system surfaces. According to Section X1.14.2 of the gasoline specification:[15]

> "Solvent-washed gum can contribute to deposits on the surfaces of carburetors, fuel injectors, and intake manifolds, ports, valves, and valve guides. The impact of solvent-washed gum on malfunctions of modern engines is not well established and the current specification limit is historic rather than the result of recent correlative studies. It depends on where the deposits form; the presence of other deposit precursors, such as airborne debris, blowby and exhaust gas recirculation gases, and oxidized engine oil; and the amount of deposits."

Only the two fuel samples taken from vehicle fuel tanks had solvent-washed gum results that were at or near the specification limit. Given that the evaporation step of the gum content test uses high temperature air, it could be expected that a fuel with poor oxidation stability might form gum during the test. Yet, such was not the case with the Research Gasoline. These results support previous findings that the gum test is not well-correlated with other stability properties of modern gasoline.

Table 2. Test Results for Base Fuels

Test	ASTM Method	Units	CL17-0556	CL17-0708	CL17-0668	CL17-0669
RSSOT	**D7525**	min	5719	23	26	1190
Gum Content	**D381**					
Unwashed		mg/100 mL	1.5	0.50	35.00	28.50
Washed		mg/100 mL	<0.5	<0.5	5.00	4.00
Jet Fuel Thermal Oxidation Test	**D3241**					
Test Temperature		°C	240	240	240	240
ASTM Code		--	--	--	--	--
Maximum Pressure Drop		mm/hg	1	1	14	0
Ellipsometer		nm	48.193	28.968	158.174	56.509
Total Volume		cm^3	7.89E-06	2.65E-06	1.90E-05	1.30E-05

Table 3. Test Results for Unstable Gasoline Treated with Selected Stabilizer Additives

Test	ASTM Method	Units	CL17-0708	CL17-0871	CL17-0872	CL17-0873
RSSOT-Gasoline	**D7525**	min	23	27	23	40
Gum Content	**D381**					
Unwashed		mg/100 mL	0.50	1.50	5.50	5.50
Washed		mg/100 mL	<0.5	0.50	1.50	1.00
Jet Fuel Thermal Oxidation Test	**D3241**					
Test Temperature		°C	240	240	240	240
ASTM Code		--	--	--	--	--
Maximum Pressure Drop		mm/hg	1	0	0	1
Ellipsometer		nm	28.968	194.531	45.098	35.931
Total Volume		cm^3	2.65E-06	1.30E-05	3.00E-06	1.40E-05

Table 4. Test Results for Stable Gasoline Treated with *m*-toluidine at 50 ppm and 100 ppm

Test	ASTM Method	Units	CL17-0957	CL17-0971
RSSOT - Gasoline	**D7525**	min	119	138
Gum Content	**D381**			
Unwashed		mg/100	6.50	6.50

		mL		
Washed		mg/100 mL	2.00	1.00
Jet Fuel Thermal Oxidation Test	**D3241**			
Test Temperature		°C	240	240
ASTM Code		--	N/A	N/A
Maximum Pressure Drop		mm/hg	16	30
Ellipsometer		nm	87.735	125.881
Total Volume		cm³	1.20E-05	1.9E-05

Table 5. Results of Accelerated Storage Stability Testing (ASTM D4625) of Stable (Pump) Gasoline (CL17-0556)

Test	**ASTM Method**	**Units**	**4 weeks**	**8 weeks**
Storage Stability	**D4625**			
Total Insolubles		mg/100 mL	1.75	1.79
Mass of Filterable Insolubles		mg/100 mL	1.16	1.20
Mass of Adherent Insolubles		mg/100 mL	0.59	0.59
RSSOT - Oxidation Stability of Gasoline	**D7525**	min	116	106
Gum Content	**D381**			
Unwashed		mg/100 mL	6.50	4.50
Washed		mg/100 mL	2.50	1.00
Jet Fuel Thermal Oxidation Test	**D3241**			
Test Temperature		°C	240	240
ASTM Code		--	--	--
Maximum Pressure Drop		mm/hg	250	250
Ellipsometer		nm	106.754	79.625
Total Volume		cm³	1.70E-05	1.30E-05

ASTM D3241 Testing Effort (Thermal Stability Testing)

For this testing, the first step was to establish an appropriate test temperature. To do that, we measured the fuel Breakpoint (per Appendix X2 of ASTM D3241), using the standard setup, and at a starting temperature of 280 °C. The test fuel was the Pump Gasoline (CL17-0556). This initial effort failed badly as the standard Pump Gasoline seriously eroded the aluminum heater tube. Since these tubes are known to work with a broad range of hydrocarbons, it is assumed that the corrosion was generated by the presence of ethanol in the fuel (see Appendix A).

Alternative heater tubes, made of 304SS (stainless steel) alloy, are also available and were selected for this program. This change made the JFTOT II a fortuitous choice because ferrous alloy tubes have a different temperature profile for the heated tube and the temperature profile of the heated tube does effect the measured Breakpoint. Of the various ASTM D3241 devices available in the market, only the JFTOT II has a movable thermocouple. This allows correct placement of the thermocouple and correction for temperature profile.

The use of stainless steel tubes has a potential effect on the tube rating methodology. Accurate ellipsometric measurements depend on having the appropriate physical properties, refractive index and absorption coefficient for both the film and the substrate. For this program, it was assumed that film constants would fall within the model developed for the ASTM D3241 Annex A3 ellipsometer (ETR). The substrate change from an aluminum alloy to a stainless steel would obviously make a difference in those parameters. Determining those values and changing the instrument programming to account for them was beyond the scope of this program. The decision was to proceed with the program as written, hoping it would provide reasonable relative results. If it did not, the fallback effort would be to use the ASTM D3241 Annex A1 visual tube rating (VTR) method. Fortunately, we were able to obtain useful results from the ellipsometric method.

Experience suggests that, in many instances, the use ferrous alloy heater tubes results in significantly more deposit for a given temperature than the use of aluminum alloy tubes. So a more conservative starting temperature of 220 °C was chosen for Breakpoint measurement with stainless steel tubes. By convention, the Breakpoint is the last passing temperature in series (with the minimum increment of 5 °C). The Breakpoint of the Pump Gasoline sample was 240 °C. Figures 3 and 4 illustrate the deposition effects.

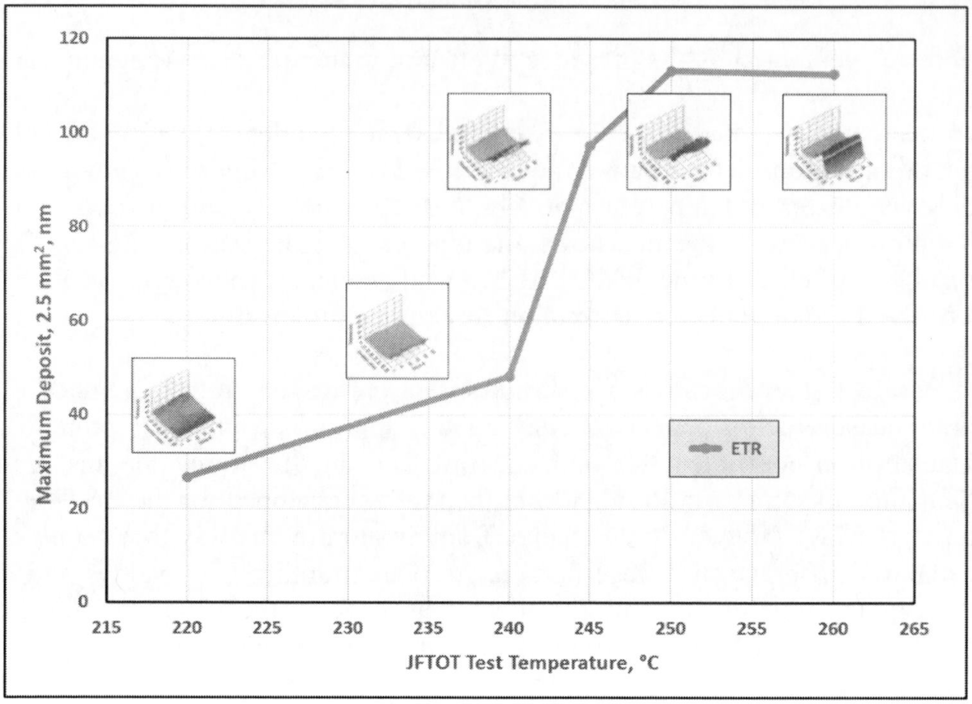

Figure 3. Breakpoint Graph for Pump Gasoline

For this series the standard metric for ASTM D3241 deposition, the maximum deposit depth in nanometers, averaged over 2.5 mm^2, was used to rate each test. The standard aviation turbine fuel pass/fail limit of 85 nm was used to determine the Breakpoint. For reference, the three-dimensional graph of the deposits for the tests in this series are associated with the appropriate data point.

In general, the data were slightly offset upwards from what would be expected from deposits on an aluminum tube. The form, location and increase with temperature of the deposits all appeared to be reasonable. The tubes were examined in the Visual Tube Rater, but the color and surface finish of the tube made such examinations difficult because there was no ability to discern subtle distinctions. For this reason, the remainder of the program was conducted using the ETR and standard ETR settings.

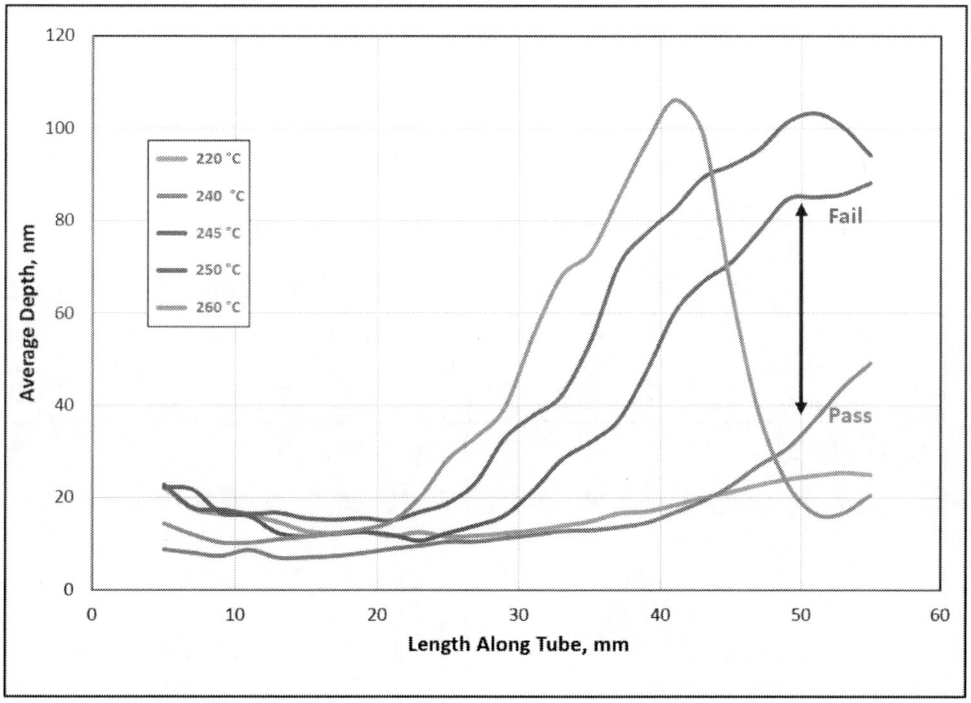

Figure 4. Average Radial Deposition for Pump Gasoline Breakpoint Tests

The deposits in a test such as ASTM D3241 can vary in symmetry, but it is instructive to look at the average radial (around the tube) deposition values for each test in the Pump Gasoline Breakpoint series. In Figure 4, note the large increase in total deposit between the test at 240 °C and the one a 245 °C. This is a clear indication of a stability limit for the Pump Gasoline.

Since this Pump Gasoline is representative of commercially available fuel, its results were used as the bench mark for the other tests. Therefore, all the remaining JFTOT tests in this program were conducted at the Pump Gasoline Breakpoint temperature of 240 °C.

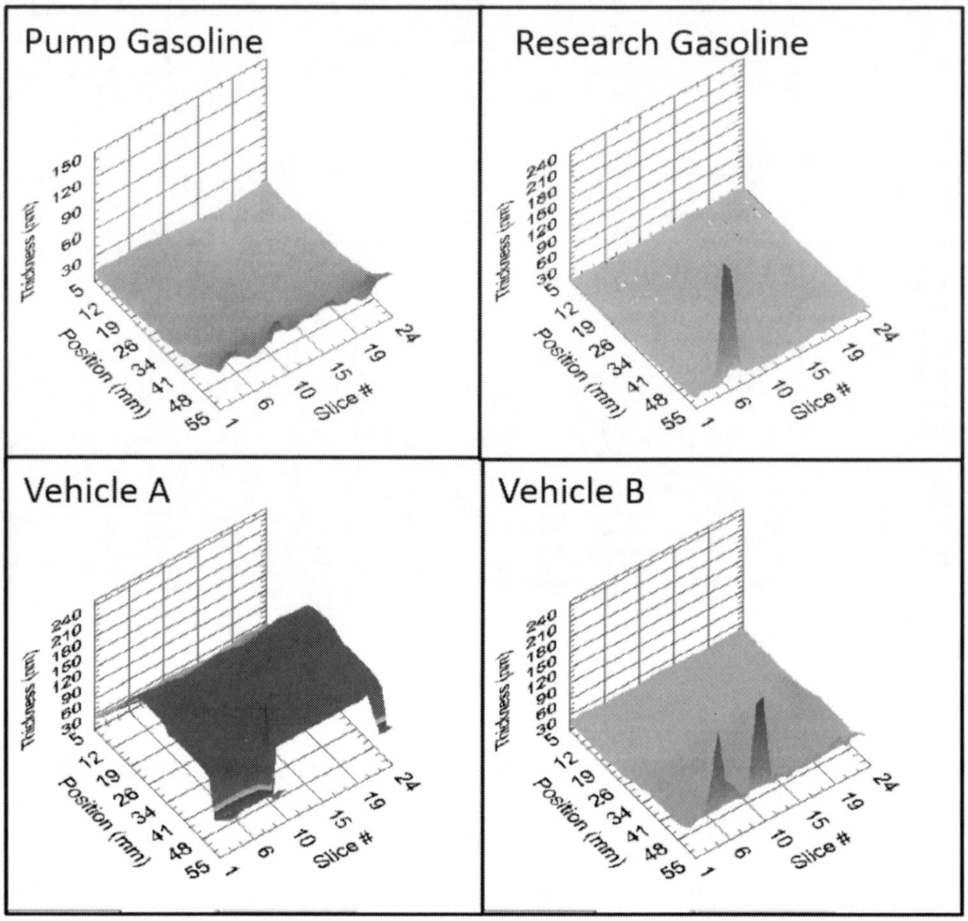

Figure 5. Base Fuel Thermal Stability (ASTM D3241)

This program had two main fuels for analysis, a commercially-available, fully additized Pump Gasoline and a custom blended, marginal stability, Research Gasoline. In the top two frames of Figure 5, the difference in stability, measured at 240 °C, between these two samples appears to be slight. Note, however, the large spike in deposit near the 55 mm position on the Research Gasoline graph. This shows there is potential for heavier deposition. That one spike tops out near 240 nm of deposit, even though the depth reported in Table 2 is much lower. This is an artifact of the way the method is used for aviation turbine fuel. The standard ETR methodology, derived from aviation turbine fuel experience, normalizes the data to provide the highest average of six contiguous data points (essentially 2.5 mm^2). Doing so downplays the effect of the data spikes seen on the Reference Gasoline graph. The question might be, is that difference as much as we might expect in this testing?

As a check on that question, the two vehicle fuel samples were tested at the 240 °C reference temperature using ASTM D3241. The results of the testing are shown in the lower two frames of Figure 5. One fuel, from Vehicle A, showed a significant loss of thermal stability compared to the fuel in Vehicle B. Note, however, the higher deposit thickness scale for Vehicle B similar to that for the Reference Gasoline sample. These results indicate that testing a gasoline type fuel may require a different approach than used for aviation turbine fuel.

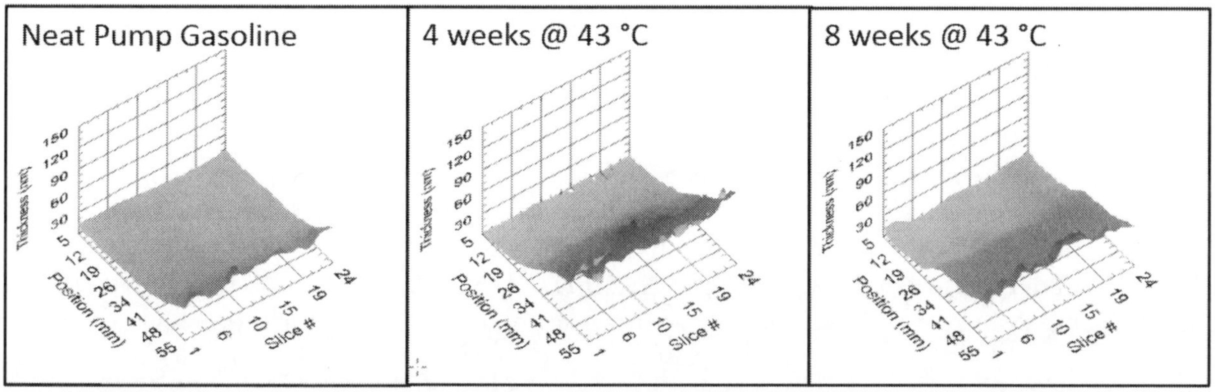

Figure 6. Storage Related Pump Gasoline Thermal Stability (ASTM D3241)

The Pump Gasoline was aged under the conditions of ASTM D4625 and the aged fuels were then tested at 240 °C using ASTM D3241. The results are shown in Figure 6. Interestingly, the ETR data show a reduced deposit at 8-weeks, compared to the 4-week sample. However, this is not unheard of in this type of testing. Some of the reactant materials that formed and caused heavy deposits at 4-weeks may have been spent by the time the 8-week testing was conducted. A complete study of the effects of long term storage will take analysis of many more fuels. But, these results do indicate that long term storage could be significant issue for gasoline fuel deposition.

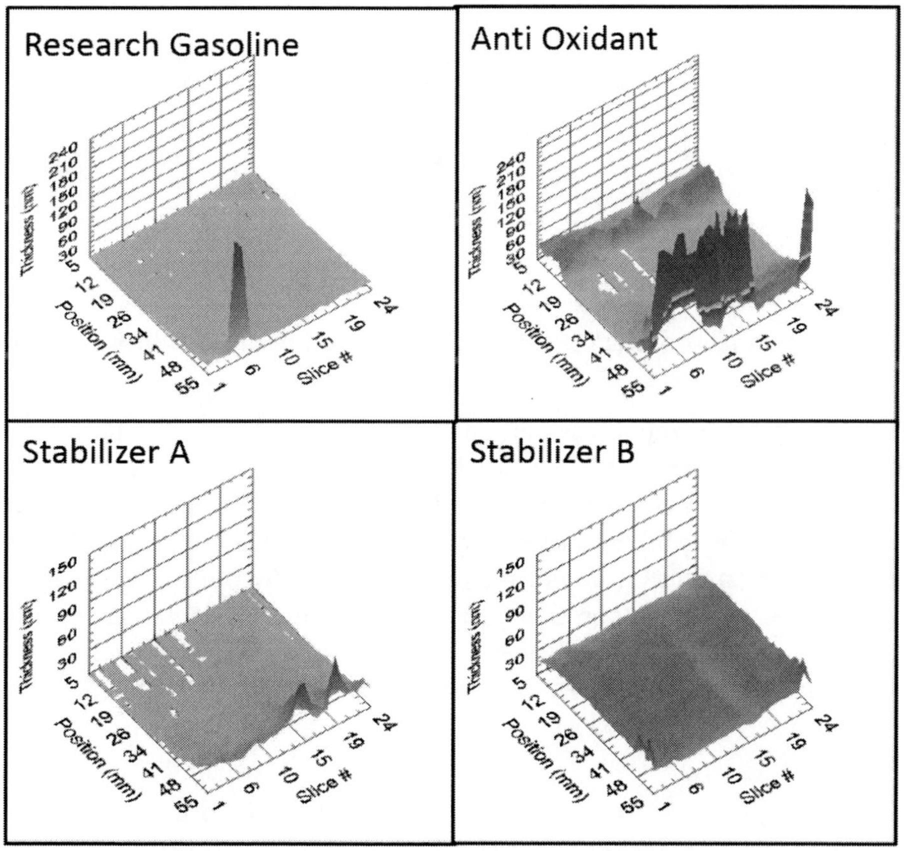

Figure 7. Additive Impact on Research Gasoline Thermal Stability (ASTM D3241)

In previous testing, it was demonstrated that extended ambient storage (Figure 5) and laboratory accelerated storage (Figure 6) could have a negative effect on thermal stability. Figure 7 presents results for unstable fuel (Research Gasoline) with additives meant to improve the stability of the fuel. In this testing, three additives, a typical hindered phenol antioxidant (AO) and two commercial, after-market, fuel stabilizers, were added to the Research Gasoline. The AO product, surprisingly, had a decided negative effect on the thermal stability of the product. This was unexpected as experience has shown that AO has little effect on the thermal stability measured under these conditions. Typically, this AO would be expected to reduce/eliminate future degradation of the fuel but not cause an immediate change for either the better or the worse.

The commercial fuel additives, Stabilizers A and B, both showed significant improvement in the fuel thermal stability based on maximum deposit depth. However, for these stabilized samples, there is a notable difference between the two. Stabilizer B has a higher baseline deposit value than does Stabilizer A; so much so, that its' total deposit volume is similar to that of the AO treated sample. It is known that ASTM D3241 has some imprecision in the measurement of the Breakpoint. For aviation fuel Breakpoint measurements, there is a generally accepted repeatability of 5 °C. It may be that the similarity of all four of the results in Figure 7 is representative of this precision and not necessarily indicative of actual additive effects.

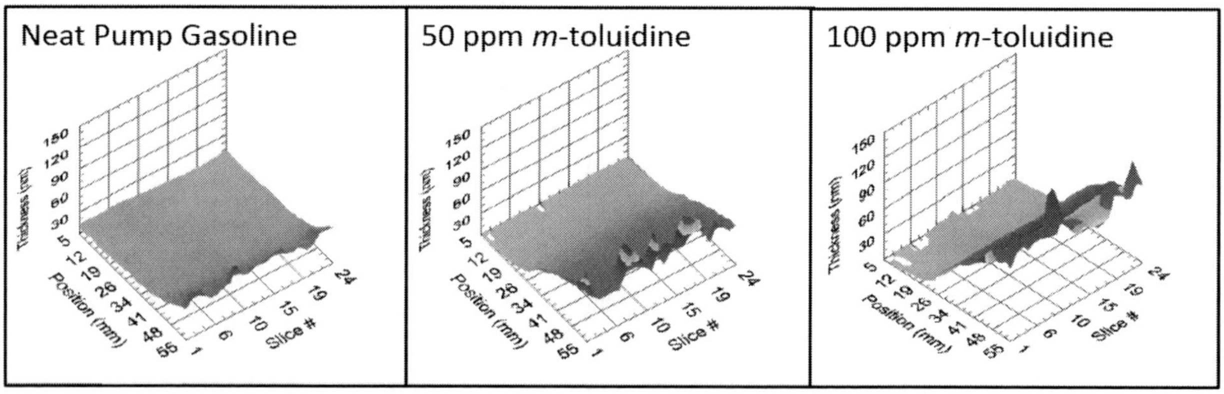

Figure 8. Contamination Effect on Pump Gasoline Thermal Stability (ASTM D3241)

The final thermal stability testing looked at the potential deposition effects of fuel contamination. Based on previous experience with artificially poisoning aviation turbine fuel,[16] the Pump Gasoline was contaminated with two levels, 50 ppm and 100 ppm, of *m*-toluidine. As can be seen in the data and the graphs (see Figure 8) these additions had a nearly linear increase in deposition based on the amount of contaminant added. While purposeful contamination of fuel is not a normal use issue, it can come accidentally through normal fueling practice. In a fuel system with potential extended storage issues, the appropriate additive package may also need to guard against contaminated fuel. This kind of test could be used for such a proactive evaluation.

CONCLUSIONS

Oxidation Stability, Gum Content, and Accelerated Storage Stability

As stated earlier, the main objective of this work was to evaluate a proposed new method for assessing gasoline deposit-forming characteristics. In order to test gasolines with a range of possible deposit-forming characteristics, we obtained base fuels then used additives and accelerated aging to vary their stability properties. The oxidation stability testing (RSSOT) and gum content measurement were used as quantitative indicators of the changes in the fuels. The accelerated aging under the conditions of ASTM D4625 was a means to age the fuel under mildly accelerated, yet more realistic, conditions.

The RSSOT results did document changes in the oxidation stability of the fuel. The ASTM D3241 testing was found to be equally, if not more so, sensitive to these fuel differences. The gum content results did not provide reliable assessments of the deposit forming characteristics or oxidation stability of the test fuels.

Thermal-Oxidative Stability (ASTM D3241) and Deposit Measurement

The primary difference between the ASTM D3241 type testing and the other tests done in this program is that ASTM D3241 is a dynamic test. The RSSOT and Gum tests are static tests, in that the same aliquot of fuel is acted upon for the duration of the test. In ASTM D3241, the test fuel is continuously flowing over the heater tube. The average residence time for the sample in the test section is approximately 13 seconds. The difference between static and dynamic testing should be considered in evaluating the different apparent outcomes in this program. While there are reasons to be concerned about thermal issues affecting stagnant fuel, the experience with aviation turbine fuel has been that controlling thermal stability in a dynamic test provides the best protection for the equipment. The key questions about this type of testing for this application are:

1. Can it differentiate good and bad fuels?
2. Will it respond to the loss of quality for a give fuel?
3. Does it show the efficacy of fuel treatment?
4. Is it sensitive to contamination?

The ASTM D3241 test clearly shows the ability to differentiate the quality of a fuel, as shown in Figure 5 above. The initial assumption was that the Breakpoint (Figure 3) for the fully additized Pump Gasoline sample would suffice as the reference temperature for all of the comparative testing. The potential problem with that assumption was that the temperature might be too low for any fuel to fail. That was considered unlikely but the comparative testing of the four base samples was necessary to validate the process. The change in tube material from that used in the standard test further reinforced the need to follow the Breakpoint protocols.

The use of additives was investigated in this program. Not all gasoline is equal in additive response. Automotive industry research shows that gasolines rated as "Top Tier" are clearly superior in stability. It is not guaranteed, however, that a given vehicle will always be fueled with Top Tier fuel. So, there is a need to understand the potential improvement that can be made by introducing quality additives to poor fuel. Such a scenario is illustrated in Figure 7. In an unanticipated turn, adding a simple hindered phenol anti-oxidant resulted in a significant loss in thermal stability. Introducing two commercial stabilizer additives had an apparent marginal effect on improving the fuel stability (the elevated baseline with one additive suggests additional research). But, it may well be the results of these four tests represent the statistical variation of this fuel at the selected temperature. This topic needs additional study.

Finally, the sensitivity of the method to contamination was established (see Figure 8). This is important as fuel contamination is an ongoing issue with fuel in storage and transportation. Even an additized fuel, like the Pump Gasoline, can be adversely effected by contamination. If a specific contamination type becomes a problem, then this methodology could be used to evaluate treatment options.

Thus, most of the questions posed have been answered positively. The ASTM D3241 type method can differentiate fuels by thermal stability, it can monitor the loss of thermal stability in fuel, it may be able to evaluate the efficacy of stability additives and it can evaluate the effect of contamination. This work suggests that an ASTM D3241 derivative test and the appropriate operating hardware would be very useful for evaluating the thermal stability characteristics of gasoline type fuels and the associated additives.

RECOMMENDATIONS

Recommendations for additional work and future efforts in several areas related to this project are given below.

ASTM D3241 Test Equipment

For use of this method in the future, the correct heater tube substrate properties to use for the ETR measurement need to be determined. The data obtained in this study are good but not as accurate as they could be with the actual metallurgy parameters. Fortunately, ellipsometry is a non-destructive test methodology so all of the test tubes can be evaluated with alternative physical parameters as desired, multiple times if necessary.

Regarding ellipsometric instruments, SwRI should survey the currently available equipment to determine if a more suitable device(s) is available. If so, consult the company(s) involved to determine if they would like to have SwRI run a validation test program with their equipment.
Testing should be conducted to evaluate the effects of the other commonly available heater tube metallurgies. Besides the 304SS stainless steel alloy used in this program, there are (not including failed aluminum) tubes available made from 1018 carbon steel and 316SS stainless steel. The carbon steel tube would be considered a more reactive surface for oxidation. The primary difference between 304 and 316 stainless steels is the latter include a significant percentage of molybdenum to prevent corrosion by chlorides. While these alternatives might not have an application advantage for automotive uses over 304 stainless steel, they could be valuable for extending ellipsometry based research into other areas involving refined products and chemical.

Further testing to evaluate the surface finish of the test coupons would be useful. This testing would focus on the degree of polish on the heater tube surface. The surface finish of the test tube can have an impact on the formation of deposits. Another potential study would be to evaluate the use of platings and coatings that could have deposition resisting characteristics.

Gasoline

Based on the results of this project, an appropriate next step would be to analyze the typical fuel systems in modern vehicles to determine the type of testing, static or dynamic, that would be most likely to yield the best information on fuel related issues.

Using the data from this study, SwRI will contact the Top Tier gasoline certification program personnel to discuss the possibility that there may still be stability issues to address. There may be an opportunity to generate a new Top Tier gasoline classification. A potential program could be conducted using the Southwest Research Institute owned Chevrolet Volt, wherein it would be driven routinely and exclusively on electric power with the quality of the fuel in the tank being monitored.

A follow up to the above would be to equip the Volt tank with an on-board inert gas generating system (OBIGGS). A significant reduction in oxygen in the ullage space could have significant effect on reducing fuel reactions with a secondary benefit of reducing the potential for a static discharge explosion. The long term impact of the OBIGGS on fuel quality would be monitored by the techniques in this report.

KEYWORDS

Deposits, Ellipsometry, Gasoline, Spark-Ignition

ACKNOWLEDGMENTS

This work was conducted for the Southwest Research Institute, Advisory Committee for Research under project 08.R8741. The authors wish to also thank David Torres, Kristi Rutta, Carly Currier, Valerie Rios, and Rita Sanchez for their contributions.

REFERENCES

[1] ASTM D4814, "Standard Specification for Automotive Spark-Ignition Engine Fuel," ASTM International, West Conshohocken, PA, 19428.

[2] ASTM D525-14, "Standard Test Method for Oxidation Stability of Gasoline (Induction Period Method)," ASTM International, West Conshohocken, PA, 19428.

[3] D7525-14, Standard Test Method for Oxidation Stability of Spark Ignition Fuel-Rapid Small Scale Oxidation Test (RSSOT), ASTM International, West Conshohocken, PA, 19428.

[4] ASTM D381, "Standard Test Method for Gum Content in Fuels by Jet Evaporation," ASTM International, West Conshohocken, PA 19428.

[5] *Arrhenius, S.A., "Über die Dissociationswärme und den Einfluß der Temperatur auf den Dissociationsgrad der Elektrolyte". Z. Phys. Chem. 4: 96–116, 1889.*

[6] ASTM D4625, "Standard Test Method for Middle Distillate Fuel Storage Stability at 43°C (110°F)," ASTM International, West Conshohocken, PA 19428.

[7] Schwartz, F.G., et.al., "Storage Stability of Gasoline, Development of a Stability Prediction Method and Studies of Gasoline Composition and Component Reactivity," Bulletin 660, U.S. Department of the Interior, Bureau of Mines, 1972.

[8] Bartleson, J.D. and Shepherd, C.C, "How to Select Gasoline Antioxidants," Hydrocarbon Processing & Petrol. Refiner., Vol. 43, No. 8, August 1964, pp. 153-158.

[9] Bowden, J.N., "Stability Survey of Hydrocarbon Fuels, Final Report," DOE/BETC/1778-4, Southwest Research Institute, San Antonio, TX, October 1979.

[10] Stavinoha, L. L. and Westbrook, S. R., "Accelerated Stability Test Techniques for Middle Distillate Fuels," and Garner, M. Q. and White, E. W., "Correlation of Long-Term Storage and Accelerated Stability Tests," Distillate Fuel Stability and Cleanliness, ASTM STP 751, L. L. Stavinoha and C. P. Henry, Eds., ASTM International, 1981, pp. 3 and 34 respectively.

[11] ASTM D3241, "Standard Test Method for Thermal Oxidation Stability of Aviation Turbine Fuels," ASTM International, West Conshohocken, PA 19428.

[12] JFTOT is a registered trademark of the Petroleum Analyzers Company.

[13] From D3241: "Ellipsometry-a technique used for measuring the optical properties of surfaces (refractive index and absorption coefficient) based on changes in the polarized state of light upon reflection from the surface."

[14] Wilson, G.R. and Jeyashekar, N., "Metal Deactivator Additive (MDA) Impacts on Thermal Stability," Coordinating Research Council, CRC Contract No. AV-6-06, June 2010.

[15] ASTM D4814, "Standard Specification for Automotive Spark-Ignition Engine Fuel," ASTM International, West Conshohocken, PA, 19428.

[16] Wilson, G.R. and Jeyashekar, N., "Metal Deactivator Additive (MDA) Impacts on Thermal Stability," Coordinating Research Council, CRC Contract No. AV-6-06, June 2010.

IASH 2017, the 15TH INTERNATIONAL CONFERENCE ON STABILITY, HANDLING AND USE OF LIQUID FUELS
Rome, Italy
10-14 September 2017

OXIDATION STABILITY OF AVIATION GASOLINES FOR GENERAL AVIATION

Cunping Huang*, Sneha Gollamudi and Keith Roller

Aviation Fuels Research Laboratory, FAA William J. Hughes Technical Center, Atlantic City International Airport, NJ 08405, USA

ABSTRACT

The oxidation stability of an aviation gasoline (AVGAS) for piston aircraft is a more important research area than for ground-based motor gasolines because (1) on average there are only about 50 flight hours per year for a piston aircraft and (2) the precipitation of lead particles from tetraethyl lead (TEL) decomposition can catalyze fuel oxidation. This research reports the kinetics of a 100 low lead (100LL) avgas and an unleaded high aromatic aviation fuel blend. Fuel oxidation kinetics were carried out based on the ASTM D873 and oxidation products were characterized using FT-IR and comprehensive Two-Dimensional Gas Chromatography/Time of Flight Mass Spectrometry (GCxGC/TOFMS GCxGC/FID) for detailed hydrocarbon analyses. Results proved that tetraethyl lead in 100LL avgas can be oxidized and precipitated onto a copper surface during the fuel copper strip corrosion tests. In comparison with unleaded high aromatic fuel blend, 100LL fuel has a higher tendency for oxidation. Detailed hydrocarbon analyses by comprehensive GCxGC/TOFMS analyses further proved this finding via the detection oxygenated hydrocarbons. Effects of avgas oxidation on a fuel's physical and chemical properties, such as fuel net heat of combustion, viscosity and vapor pressures before and after oxidation, were also carried out based on related ASTM standards to investigate the effect of fuel oxidation.

KEYWORDS: aviation gasolines, oxidation stability, oxidation kinetics, GCxGC-FID/TOFMS, ASTM standards.

I. INTRODUCTION

Fuel's oxidation stability or storage stability is one of the most important research areas for aviation gasolines. In comparison with motor gasoline, aviation gasolines suffer more storage stability issues because non-commercial piston engine aircraft are used relatively infrequently while on-the-ground vehicles see much higher usage rates. It is estimated that the average in-flight time per year for a non-commercial piston engine aircraft is only 50 hours. This means that aviation gasoline will be stored in an aircraft for a much longer time than the fuel in an automobile. Additionally, aviation gasolines require higher motor octane numbers to overcome engine detonation. Currently 100 low lead (100LL) aviation gasoline is the only leaded aviation fuel used in general aviation. Also, as a very effective additive to boost a fuel's motor octane number (MON), tetraethyl lead ($(C_2H_5)_4Pb$, TEL), is unstable under heat, light and oxidizing conditions. The decomposition or oxidation of TEL results in lead particle precipitation that can serve as a catalyst center for the promotion of fuel oxidation. Combining long term storage and lead particle precipitation, the oxidation stability of aviation gasolines can be one of the most important research topics for general aviation. It is noted that the effects of metal particle contaminants, including lead (Pb), Fe, Ni, Mn Co and Cu, on the stability of biofuels and other fuels used for on-ground vehicles have been widely reported in literature. However, reports of research into the in-storage stability of aviation gasolines are rarely found in open literature.

This research investigates the oxidation stability of two aviation gasoline fuels: very commonly used 100LL aviation gasoline and an unleaded high aromatic fuel blend prepared in lab. Fuel oxidation kinetics and fuel physical and chemical property changes after oxidation are the main research focuses of this paper. Comprehensive GCxGC/TOFMS has been used to identify oxidation products and explain the fuel oxidation mechanisms.

II. EXPERIMENTAL

Two fuel samples were employed for this research. A 100 low lead aviation gasoline (100LL) was purchased from the Atlantic City International Airport. The fuel met the ASTM D910-17a [1] specification for leaded aviation gasolines, with a motor octane greater than 99.6 and tetraethyl lead (TEL) less than 0.53 mL/L of fuel or 0.56 gram of Pb/L of fuel. The second fuel was a lab blended unleaded high aromatic fuel with high toluene and trimethylbenzene content.

Prior to oxidation stability testing, the vapor pressure, net heat of combustion, density and viscosity of each fuel was verified based on ASTM D5159-04(2014) [2], or ASTM D6378 [3], ASTM D4809-13 [4], ASTM D7042-16e3 [5], respectively. Fuel total hydrogen content (mass%) was determined using a Carbon (C), Hydrogen (H), Nitrogen (N), Sulfur (S) and Oxygen (O) (CHNS/O) elemental analyzer based on ASTM D5291-16 [6] and ASTM D5622-12 [7]. A microbalance (Mettler Toledo) was used for the determination of sample mass. Fuel copper strip corrosion tests were performed based on ASTM D130-12 [6]. A Phenom XL scan electron microscope (SEM) was used to observe surface contamination and lead particle deposition on the surface of copper strips.

A PerkinElmer's Clarus® 580 gas chromatogram (GC) equipped with Clarus® SQ8T mass spectrometer (MS) and a flame ionized detector (FID) was applied for the determination and quantification (ASTM D6377-01 (2016) [8]) of a fuel's chemical compositions. The GC column, 50 meter, 0.20 mm inner diameter and 0.5 μm df, was purchased from Restek with a maximum temperature of 340 °C and minimum bleeding at 300 °C.

The oxidation stability of fuel samples was carried out using ASTM D873-12 [9]. In detail, a high pressure stainless steel bomb with 50 mL of a fuel sample was pressurized to 700 kPa with high pure oxygen. After that, the bomb was heated at 100 °C for 1 to 5 days (24 to 120 hours). A pressure sensor recorded oxygen pressure change as a function of time. After oxidation, fuel samples were subjected to tests of vapor pressure, net heat of combustion, density and viscosity in order to investigate the effects of oxidation on each fuel's physical and chemical properties. Fourier-transform infrared spectroscopy (FT-IR) (PerkinElmer) was applied for the determination of the formation of organic functional groups after fuel oxidation. The liquid cell used potassium bromide (KBr) as window material. Each sample was scanned 32 times with a spectrum resolution of 4 cm^{-1}.

The comprehensive two dimensional GCxGC analysis was conducted using an Agilent 7890B GC system. The outlet of fuel vapor from the second GC column was split equally to an FID detector and a JEOL accurate time of flight mass spectrometer (TOFMS) operated under electron ionization mode (EI) at 70 eV. The modulation system used Zoex's ZX1 thermal modulator operated under liquid nitrogen conditions. The first column of the GCxGC system was a 30 m non-polar column with 0.25 mm ID. The second column was a 3 m by 0.1 mm inner diameter polar column. The GC oven heating rate was set at 2.0 °C/min from 70 oC to 280 °C. The heat jet heating pulsation was controlled at a 6.00 second stop time and a 0.35 second on time.

III. RESULTS AND DISCUSSION

III-1. Testing of Fuel Stability on Copper Strip

A polished copper strip described in ASTM D130 was place in a beaker containing 40 mL of test fuel. The beaker was then placed in a stainless steel bomb that was soaked in an aqueous ethylene glycol solution at 100 °C for 72 hours. After testing, the fuel sample was used in the oxidation stability tests to investigate the effect of copper ions, Cu(II), on the oxidation kinetics. A SEM was used to observe the surface of the copper strip after testing.

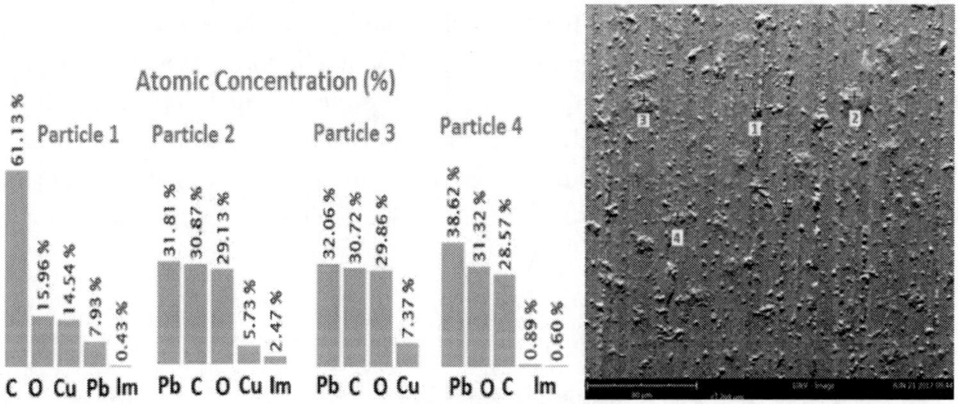

Figure 1. Lead deposition on the surface of a copper strip during 100LL fuel copper strip corrosion testing (Left: Energy-Dispersive X-ray elemental analyses (Im: impurity), Right: SEM image of copper surface and EDX analysis locations).

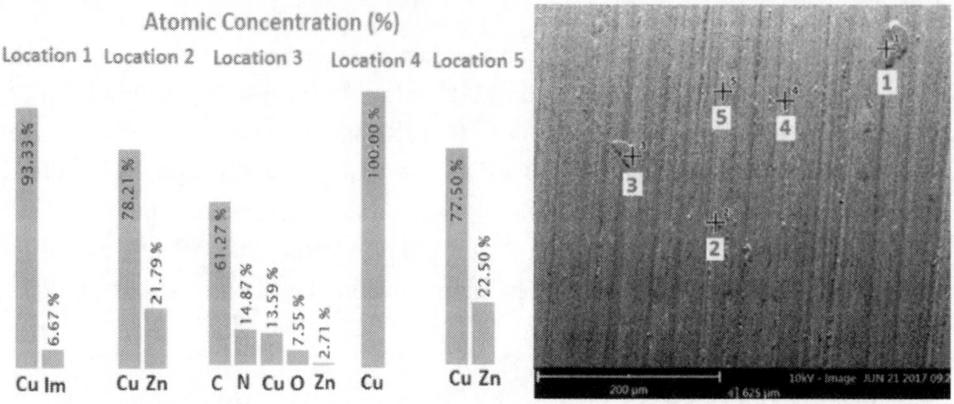

Figure 2. Copper strip surface during high aromatic fuel copper strip corrosion testing (Left: Energy-Dispersive X-ray elemental analyses (Im: impurity), Right: SEM image of copper surface and EDX analysis locations).

Figure 1 shows that lead particles are precipitated and deposited onto the surface of the copper strip during the corrosion test for 100LL fuel. This result indicates that, although a very effective octane boosting additive, TEL is not stable for a longer storage time or at an elevated temperature. A similar result is also observed when a 100LL fuel is exposed to the light. The precipitation of lead particles from leaded fuel leads to a decrease in fuel MON, which would be an unsafe factor for general aviation. Compared to the 100LL fuel, unleaded high aromatic fuel is much more stable in a copper corrosion test. As indicated in Figure 2, the surface of the copper strip after testing is clean and free of particulate deposition. The EDX analyses of four locations on the copper strip surface (Figure 2 (left)) show mainly copper and a small amount of Zn. Comparing these results with SEM analyses, elemental carbon and oxygen in Location 3 of the copper surface are a result of operational contamination.

III-2. Fuel Oxidation Kinetics

Figure 3. Fuel oxidation tester (Left: apparatus and Right: stainless steel high oxygen pressure bomb).

Figure 3 illustrates the fuel oxidation stability tester. The oxygen pressure within a high-pressure bomb can be recorded above 480 kPa. As shown in Figures 4 and 5, when a bomb is heated from room temperature to 100 °C the bomb pressure increases from 700 kPa to about 1000 kPa due to the increase of fuel vapor pressure at the higher temperature. Comparing with 100LL and unleaded high aromatic fuel, this pressure increase between room temperature and 100 °C is very similar. The significant differences can be observed between the original fuel samples and copper spiked fuels. Figure 5 shows that 100LL avgas can be stable for up to 24 hours. That can be termed as fuel stability induction time. However, the pressure of copper spiked 100LL avgas starts dropping after only 2 to 3 hours. As the pressure decreases to 480 kPa, the un-treated 100LL avgas requires 70 hours but the copper treated 100LL fuel requires only 50 hours. This result indicates that the presence of copper ions (Cu(II)) in 100LL avgas can facilitate fuel oxidation kinetics. The lower oxidation stability of leaded avgas may also result from the lead particle precipitation from the bulk fuel that catalyzes the oxidation of fuel components.

Compared to the stability of 100LL avgas, the unleaded high aromatic fuel blend (Figure 5) shows higher oxidation stability. The virgin high aromatic fuel can be stable for more than 42 hours without a significant pressure drop. The pressure of copper spiked fuel, however, can remain constant for over 72 hours. The lower vapor pressure for copper spiked high aromatic avgas is mainly due to the light components lost during copper strip corrosion testing. But, since some less stable components in unleaded high aromatic fuel are precipitated as fuel gum onto the surface of copper strip or testing beaker, the copper spiked aromatic avgas shows higher stability than un-treated virgin fuel.

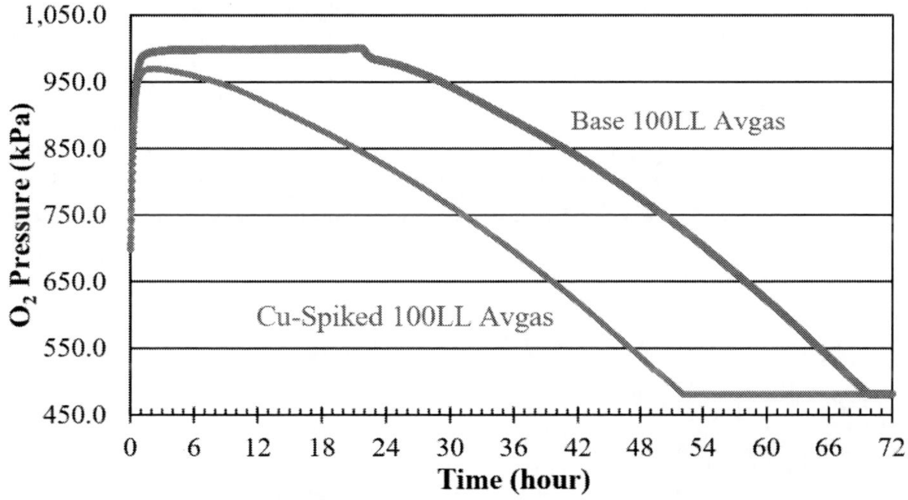

Figure 4. Kinetics of 100LL fuel oxidation

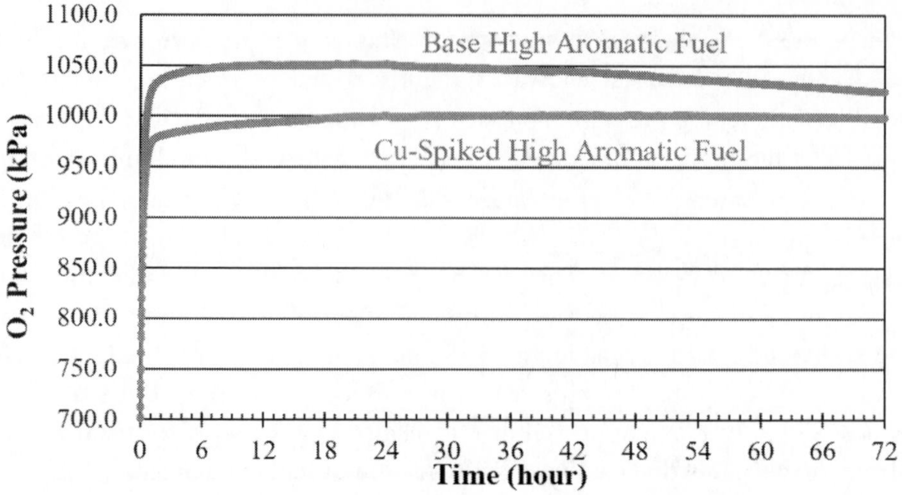

Figure 5. Kinetics of high aromatic fuel oxidation.

III-3. Qualitative and quantitative analyses of oxidized aviation gasolines

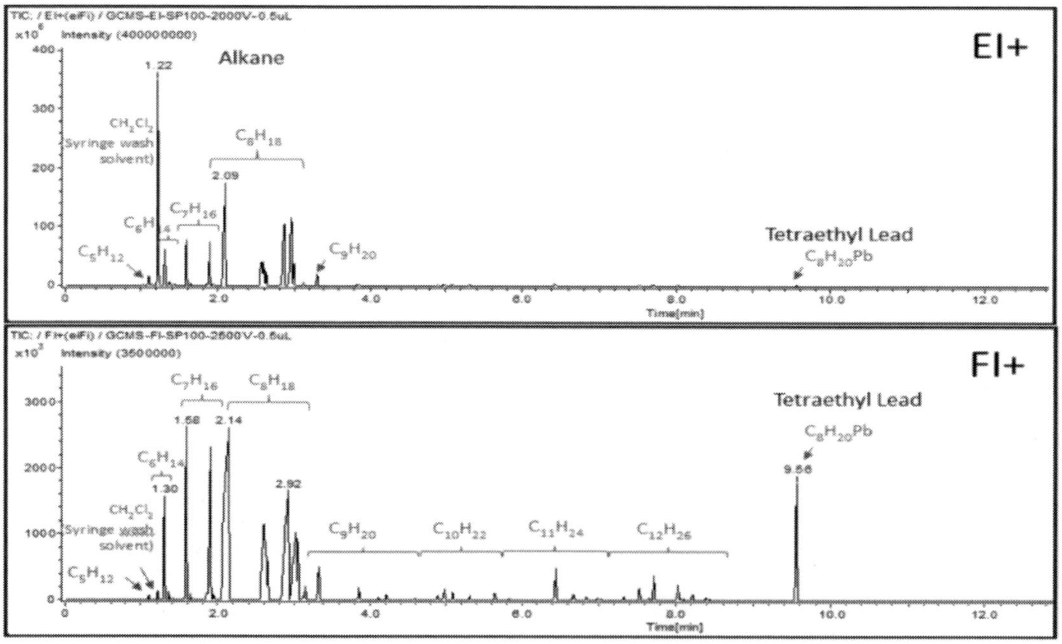

Figure 6. GC/MS analyses for 100 low lead aviation gasoline: Electron ionization detector (up) and field ionization detector (down).

Figure 6 shows GC/MS analyses with electron ionization detector (up) and field ionization detector (down) for the 100LL aviation gasoline sample. 100LL fuel consists mainly of iso-alkanes centered at C_7 to C_8. No normal alkanes are included in the fuel because of the requirement of fuel octane rating. Low concentrations of higher carbon number alkanes (C_9 to C_{12}) can also be found as fuel impurities. Tetraethyl lead (TEL) with a concentration lower than 500 ppm by mass is detected as a fuel additive to

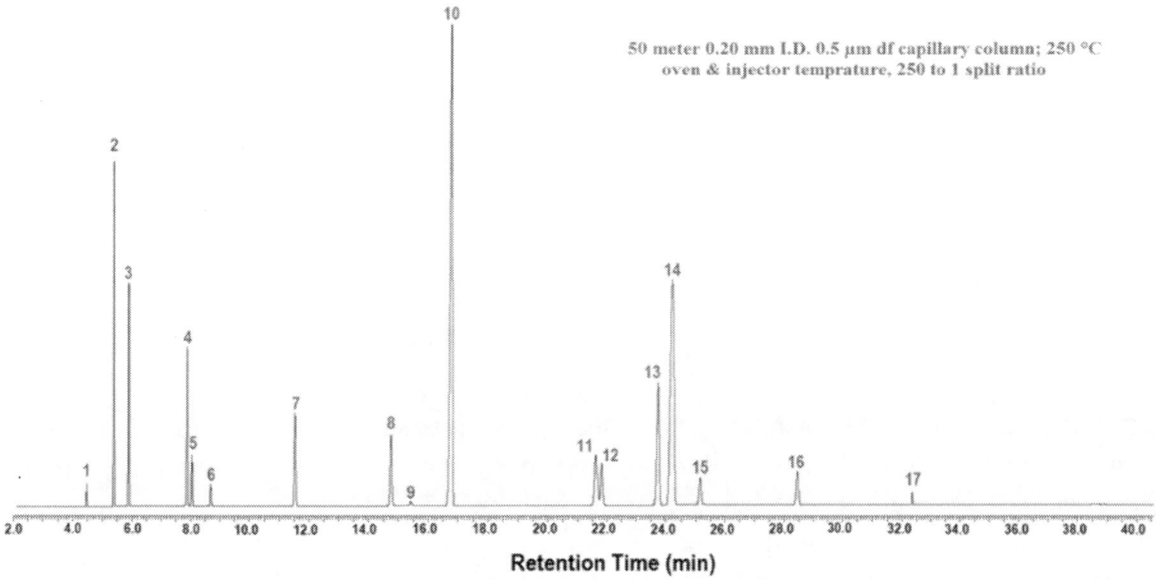

Figure 7. GC/FID analysis for 100 low lead aviation gasoline.

Table 1. Quantitative analyses (GC/FID, ASTM D6733) for 100LL avgas before and after oxidation

#	Component	Concentration (mass%)		
		Base 100LL	72-hour Oxidation	72-hour Oxidation w/Cu
1	***i*-Butane**	0.36	1.06	-
2	2-Methylbutane	11.87	4.09	10.33
3	***n*-Pentane**	4.53	-	-
4	2,3-Dimethylbutane	4.5	4.71	4.14
5	2-Methylpentane	1.41	1.09	1.32
6	n-Hexane	0.65	0.60	0.53
7	2,4-Dimethylpentane	3.65	5.77	3.89
8	**2,2,3-Trimethylbutane**	-	0.22	0.12
9	**2-Methylhexane**	-	0.24	0.22
10	2,3-Dimethylpentane	3.13	6.18	3.03
11	3- Methylhexane	0.19	0.23	0.18
12	**2,2,4-Trimethylpentane**	29.77	33.05	32.82
13	2,5-Dimethylhexane	3.62	5.33	3.95
14	2,4-Dimethylhexane	1.97	2.85	1.91
15	**2,3,4-Trimethylpentane**	7.85	13.94	8.82
16	Toluene	22.55	14.55	24.68
17	2,3-Dimethylhexane	1.58	2.91	1.70
18	**3,4-Dimethylhexane**	-	0.44	-
19	2,2,5-Trimethylhexane	2.14	2.57	2.12
20	2,3-Dimethylheptane	0.24	0.19	0.25
	Total	100.01	100.02	100.01

boost fuel octane number. Comparing the GC/EID spectrum, GC/FID is more suitable and sensitive for the detection of low concentration impurities and additives.

Table 2. Summary of quantitative analyses (GC/FID, ASTM D6733-01) for 100LL avgas (Peak numbers correspond to Figure 7)

Peak no.	Component	Mass%	Peak no.	Component	Mass%
1 (1)	i-C_4H_{10}	0.254	11-17 (7)	i-C_8H_{18} + (C_7H_8)	75.710
2 (1)	n,i-C_5H_{12}	2.443	18-20 (3)	i-C_9H_{20}	1.653
3-5 (3)	i-C_6H_{14}	6.107	21 (1)	i-$C_{10}H_{22}$	0.230
6-10 (4)	i-C_7H_{16}	13.889		Total	100.286

Figure 7 depicts the GC/FID spectrum of the 100LL aviation gasoline. Table 1 summarizes the detailed quantitative fuel composition analyses for base and oxidized fuel. No oxygenated hydrocarbons were detected via GC/MS and GC/FID because of the instrument's sensitivities. To facilitate engine cold ignition small amounts of i-C_4H_{10} and i-C_5H_{12} are needed for the aviation gasoline evaporation requirement. During oxidation processes, C_4 and C_5 components are being lost due to fuel vaporization. Small amounts of 2,2,3-Trimethylbutane and 2-Methylhexane are formed after fuel oxidation both with and without copper spike. Interestingly, about 3 mass% increase in 2,2,4-trimethylpentane was found in two oxidation fuel samples. The concentrations of other heavy fuel components remain almost unchanged. Table 2 summarizes that the major components of a 100LL fuel contain about 14 mass% of i-C_7H_{16} and 76 mass% i-C_8H_{18}. Hydrogen content analysis found that the average molecular formula can be written as C_7H_{16} with overall hydrogen concentration of 16.156 mol.%.

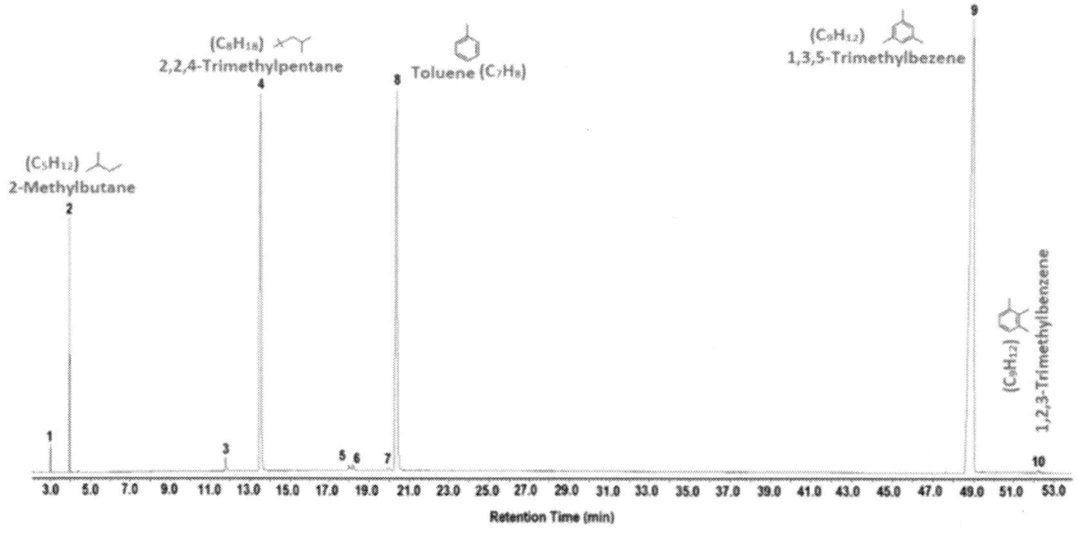

Figure 8. GC/MS analysis of an unleaded high aromatic lab fuel blend.

Figure 8 depicts the GC/MS spectrum of an unleaded high aromatic lab fuel blend with toluene ($C_6H_5CH_3$) and 1,3,5-trimethylbenzene as major aromatic components. Isobutane and 2-methylbutane are used to increase fuel vapor pressure for the engine cold start need. The results of fuel quantitative analyses (GC/FID, ASTM D6733-01) are listed in Table 3. Similar to the 100LL fuel oxidation, except for the small amount of C_4 and C_5 losses there are no significant composition changes for the fuel blend after two oxidation processes. These results match with the kinetics observed for high aromatic fuel (Figure 5).

Table 3. Detailed quantitative analyses (GC/FID, ASTM D6733-01) of an unleaded high aromatic blend

#	Component	Concentration (mass%)		
		Base High Aromatic Fuel	72-hour Oxidation	72-hour Oxidation w/Cu
1	Isobutane	1.05	0.27	0.33
2	2-Methylbutane	6.45	4.77	4.93
3	2,3-Dimethylpentane	0.42	0.36	0.37
4	2,2,4-Trimethylpentane	22.00	21.56	21.75
5	2,5-Dimethylhexane	0.19	0.18	0.14
6	2,4-Dimethylhexane	0.15	0.18	0.17
7	2,3,4-Trimethylpentane	0.07	0.06	0.09
8	Toluene	19.66	21.23	21.03
9	1,3,5-Trimethylbenzene	49.74	51.21	50.96
10	1,2,4-Trimethylbenzene	0.26	0.18	0.24
	Total	99.99	100.00	100.01

III-4. FT-IR observation for the formation of hydroxide and carbonyl groups from oxidized fuels

Figure 9 illustrates the FT-IR spectra for unleaded high aromatic fuel blends before and after oxidation. In comparison with the FT-IR spectrum for the base fuel, there are no new absorption bands found for the fuel oxidized from 24 to 72 hours. This observation indicates that high aromatic fuel can be stable under current oxidized conditions. This result agrees well with the kinetics of unleaded high aromatic fuel blend (Figure 5).

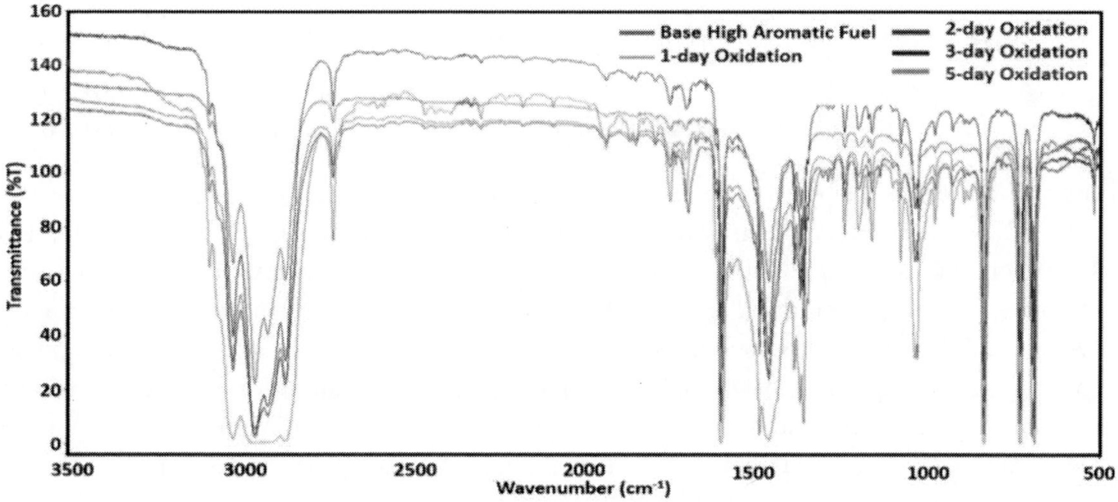

Figure 9. FT-IR spectral analyses for unleaded high aromatic fuel blend.

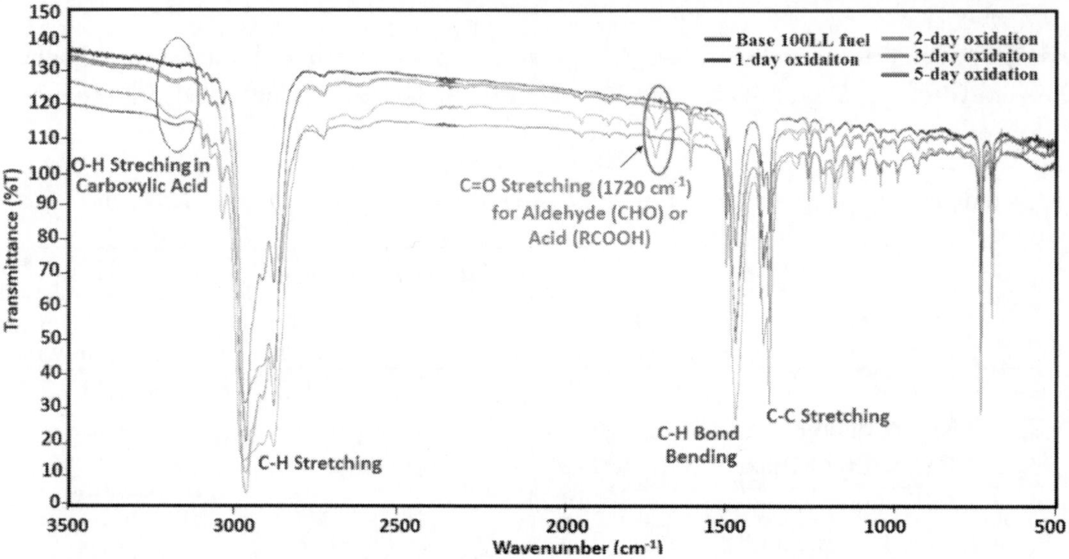

Figure 10. FT-IR spectral analyses for 100 low lead aviation gasoline (500 to 3500 cm^{-1}).

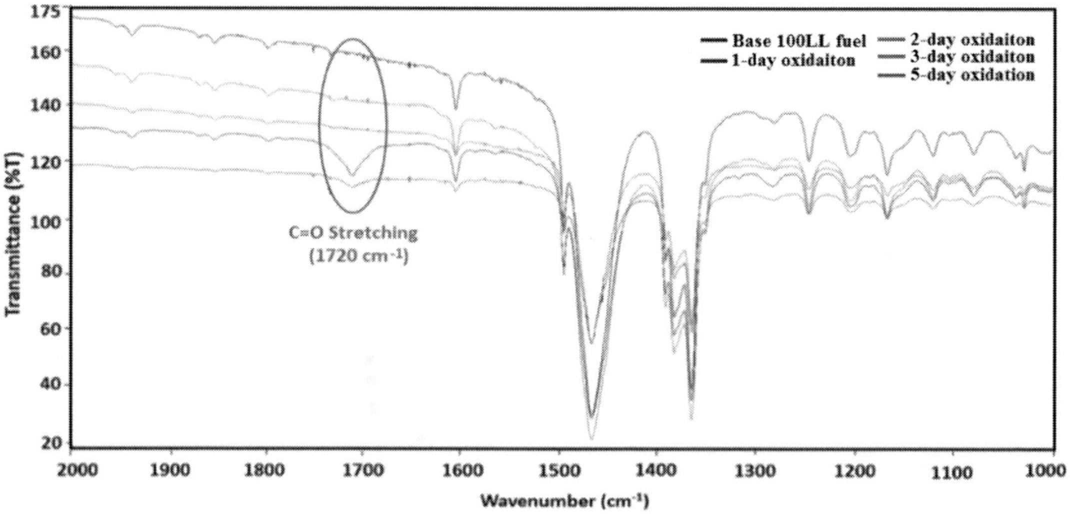

Figure 11. FT-IR spectral analyses for 100 low lead aviation gasoline (1000 to 2000 cm^{-1}).

Compared with the FT-IR spectrum of base 100LL fuel two significant absorption bands are found for oxidized fuel samples (Figures 10). The 1720 cm^{-1} absorption is the C=O stretching for aldehydes (RCHO) or organic acids (RCOOH). When the oxidation time increases, the 1720 cm^{-1} absorption intensity increases, suggesting that the concentrations of aldehydes or acids increase with the oxidation time. O-H stretch banding located at about 3200 cm^{-1} on the other hand responds to the oxidation time differently. When absorption intensity appears for the 24-hour oxidation sample and then it reaches a maximum and at 72 hours. But in the case of the 5-day oxidation sample, this O-H band intensity reduces to the base fuel level. Figure 11 highlights the formation of carbonyl at the 1720 cm^{-1}. The significant absorption band of C=O can only be observed after three days' oxidation. At five days the C=O absorption band can be very significant.

III-5. The changes of fuel vapor pressure, net heat of combustion, density and viscosity

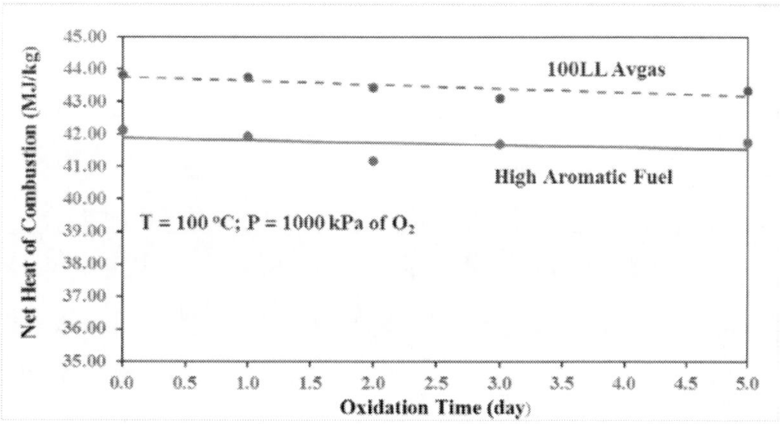

Figure 12. Fuel net heat of combustion for oxidized 100LL and unleaded fuels.

Figure 12 shows the net heat of combustion (ASTM D4809, MJ/kg) after oxidation tests. The hydrogen contents of the two fuels are determined using a CHNS/O elemental analyzer (ASTM D5291). Results of the GC/FID (ASTM D6733) quantitative analyses (Tables 1 and 3) are used to confirm the hydrogen contents of the two test fuels in this research. The net heat of combustion can be calculated as:

Net Heat of Combustion at 25 °C (MJ/kg) = Gross heat at 25 °C (MJ/kg) – 0.2122 x H (mass%)

As shown in Figure 12, the net heat of combustion for both 100LL avgas and unleaded high aromatic fuel blend decreases with the oxidation time. The decreasing tendency for 100LL fuel is slightly higher than that of unleaded high aromatic fuel.

Fuel density as a function of temperature for base and oxidized 100LL fuels is illustrated in Figure 13. Fuel density is linearly related to temperature. The oxidized fuel has shown much higher density at all testing temperatures in the range -50 to 30 °C. Similarly, the density of base unleaded high aromatic fuel is also linearly related to temperature. The density of oxidized aromatic fuel, however, does not show a significant density increase as compared with that of base aromatic fuel blend because of its higher oxidation stability.

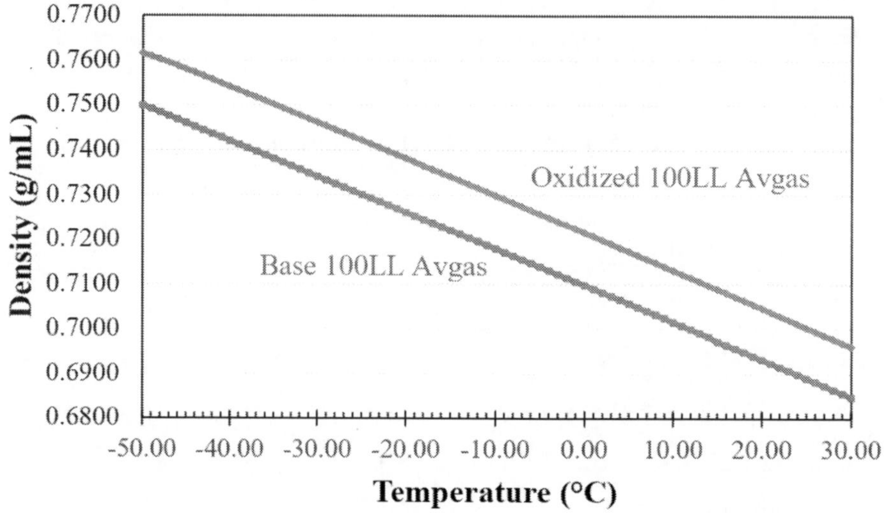

Figure 13. Fuel density as a function of temperature for 100LL fuels.

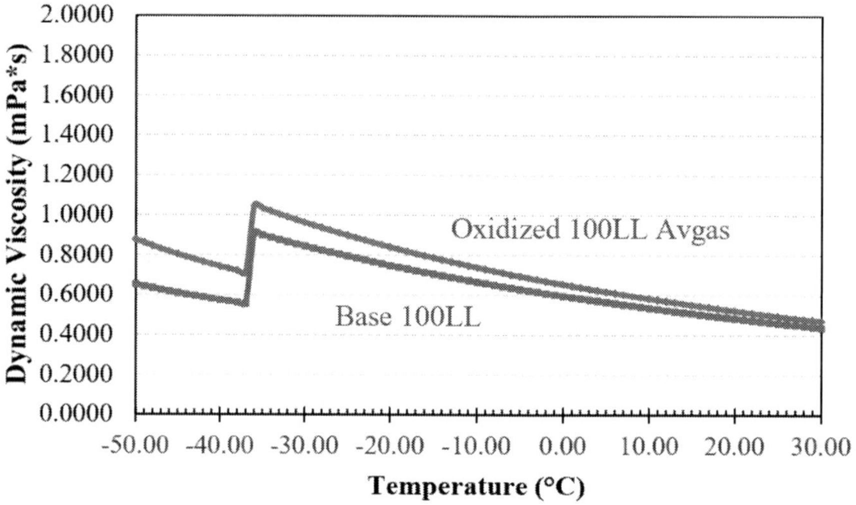

Figure 14. Fuel dynamic viscosity as a function of temperature for 100LL fuels.

Fuel dynamic viscosity, measured based on ASTM D7042), is depicted in Figure 14. A slight dynamic viscosity increase from 30 °C to -50 °C can be observed in Figure 14. Unlike 100LL fuel, the viscosity of unleaded high aromatic fuel does not show significant differences between base and oxidized fuels. The densities of the base and oxidized high aromatic fuels were also nearly identical. The abrupt dynamic viscosity change in Figure 14 can be ascribed to the precipitation of trace amounts of water (ppm level) crystals from the fuel sample, leading to the viscosity drops. It is found that all aviation gasoline samples encounter the same viscosity changes in the range from -30 °C to -40 °C. The decrease is not significant, but because gasoline samples do not have very high viscosity, the small amount of viscosity change appears notable. The amount of viscosity change while dropping from -30 °C to -40 °C can be used as an indicator of water content in a gasoline sample.

Table 3. Fuel vapor pressure change after oxidation (ASTM D6378)

100 low lead aviation gasoline (kPa at 33.8 °C (100 °F)			Unleaded high aromatic fuel blend (kPa at 33.8 °C (100 °F)		
Base	Oxidized	Oxidized w/Cu	Base	Oxidized	Oxidized w/Cu
39.20	32.00	30.86	31.03	21.44	22.88

Table 3 Summarizes tested fuel vapor pressure based on ASTM D6378. The significant fuel vapor pressure decreases are due to the light component losses during fuel oxidation processes. The vapor pressure loss for the unleaded high aromatic fuel blend is more significant than that of 100LL fuel. For example, comparing vapor pressures of base fuels and oxidized fuel (without copper ions participation), the relative losses of vapor pressure of for 100LL avgas are 18.4% ((39.2 – 32.0)/39.2*100) and for unleaded high aromatic fuel they are 30.9% ((31.03-21.44)/31.03*100). Tables 1 and 3 list the fuel component changes after oxidation. The inclusion of C_4 and C_5 components in avgas is to improve fuel vapor pressure so that the fuel can have a good cold start in winter time. For the 100LL avgas the total concentration losses of C_4 and C_5 can be calculated as 13.01%, meaning that each one percent loss of $C_4 + C_5$ will result in a 1.41% decrease in vapor pressure. For the unleaded aromatic fuel, on the other hand, the total $C_4 + C_5$ loss is 2.46%. This causes a 30.9% vapor pressure decrease, or 12.6% vapor pressure decrease for each one percent total $C_4 + C_5$ loss. Table 4 summarizes the detailed calculation results. All these results indicate that the vapor pressure of the unleaded aromatic fuel is very sensitive to its C_4 and C_5 concentrations, which suggests, when comparing with 100LL fuel, the unleaded aromatic fuel is more unstable in terms of maintaining fuel vapor pressure during storage.

Table 4. Fuel vapor pressure losses related to the change of a fuel's C_4 and C_5 components

	100 low lead avgas			Unleaded high aromatic fuel blend		
	Base	Oxidized	Change	Base	Oxidized	Change
Vapor Pressure (kPa)	39.2	32.0	18.4%	31.03	21.44	30.9%
i-Butane (mass%)	1.06	0.36	0.7	1.05	0.27	0.78
2-Methylbutane (mass%)	11.87	4.09	7.78	6.45	4.77	1.68
n-Pentane (mass%)	4.53	0.00	4.53	0.00	0.00	0.00
Total C_4+C_5 loss (mass%)			13.01			2.46
Vapor pressure/C_4+C_5 loss	0.55 kPa/mass%			0.76 kPa/mass%		
Relative pressure change/C_4+C_5 loss	1.41			12.6		

III-6. Comprehensive GCxGC/TOFMS and GCxGC/FID analyses

In this research, comprehensive GCxGC/TOFMS and GCxGC/FID were operated at the same time. The TOFMS spectra are used for the identification of chemical structures of unknown fuel components. GCxGC/FID spectra were recorded for use in quantitative analyses which will be completed in the future. Figures 15 and 16 show GCxGC/FID spectra for the unleaded aromatic base fuel and a 3-day oxidized fuel sample. No significant spectral differences were observed for these two fuel samples except that (1) iso-butane is lost in the oxidized fuel and (2) two new peaks (Peaks 10 and 11 in Figure 16) are detected that can be ascribed to 1,4-Dimethylbenzaldehyde ($C_9H_{10}O$) and 3,4-Dimethylbenzylalcohol ($C_9H_{12}O$), respectively. These two oxygenated aromatic hydrocarbons, which cannot be detected via low resolution GC/MS, indicate that more or less aromatics, although slow, can still be oxidized via fuel oxidation processes in this research. With the presence of metal catalyst particles, the oxidation process can be greatly accelerated.

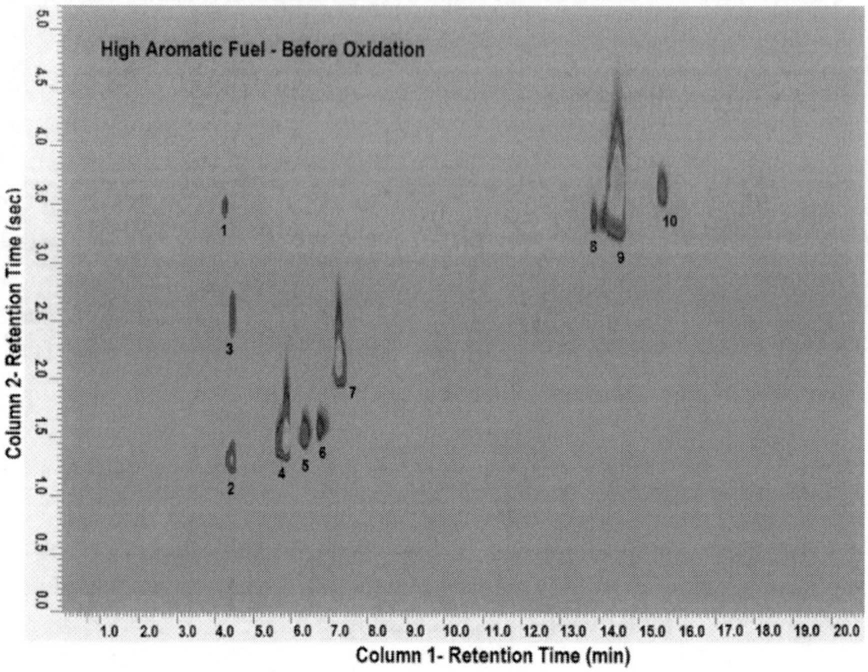

Figure 15. GCxGC/FID Spectrum for unleaded high aromatic blend (base fuel).

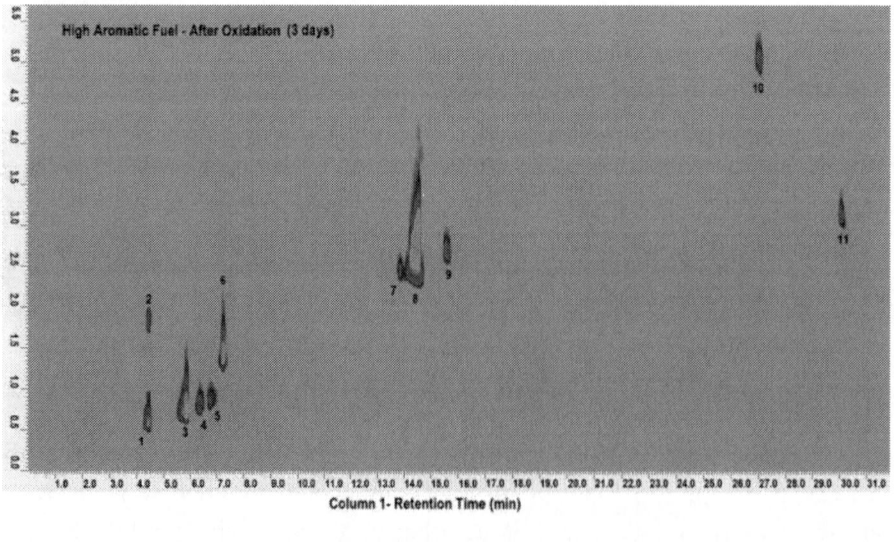

Figure 16. GCxGC/FID Spectrum for unleaded high aromatic blend (3-day oxidized fuel).

The related detailed hydrocarbon analysis results for the unleaded aromatic fuel blend before and after oxidation are listed in Tables 5 and 6.

Table 5. Detailed hydrocarbon analysis of unleaded high aromatic blend (base fuel)

Peak Number	*Retention Time*		*Peak Area*	*Component*
	Column 1 (min)	*Column 2 (sec)*		
1	4.20	3.41	64.82	Isobutane
2	4.40	1.38	1408.09	2-Methylbutane
3	4.40	2.57	384.44	2,3-Dimethylpentane
4	5.80	1.47	15243.56	2,2,4-Trimethylpentane
5	6.30	1.51	276.01	2,4-Dimethylhexane
6	6.70	1.54	72.18	2,3,4-Trimethylpentane
7	7.20	2.26	12483.32	Toluene
8	13.80	3.39	81.65	Propyl benzene
9	14.40	3.43	6067.40	1,3,5-Trimethylbezene
10	15.50	3.62	56.54	1,2,4-Trimethylbenzene

Table 6. Detailed hydrocarbon analysis of unleaded high aromatic blend (3-day oxidized fuel)

Peak Number	*Retention Time*		*Peak Area*	*Component*	*Molecular Formula*
	Column-1 (min)	*Column-2 (sec)*			
1	4.40	1.82	60.89	2-Methylbutane	C_5H_{12}
2	4.40	0.58	1817.78	2,3-Dimethylpentane	C_7H_{16}
3	5.80	0.76	17455.46	2,2,4-Trimethylpentane	C_8H_{18}
4	6.30	0.84	186.79	2,4-Dimethylhexane	C_8H_{18}
5	6.70	0.86	88.20	2,3,4-Trimethylpentane	C_8H_{18}
6	7.20	1.42	8308.74	Toluene	C_7H_8
7	13.80	2.44	69.01	Propyl benzene	C_9H_{12}
8	14.50	2.66	3628.63	1,3,5-Trimethylbezene	C_9H_{12}
9	15.50	2.68	116.38	1,2,4-Trimethylbenzene	C_9H_{12}
10	27.00	5.02	117.78	1,4-Dimethylbenzaldehyde	$C_9H_{10}O$
11	30.00	3.12	163.56	3,4-Dimethylbenzylalcohol	$C_9H_{12}O_3$

Figures 17 and 18 depict GCxGC/FID spectra for the base 100LL avgas and 5-day oxidized fuel. Detailed hydrocarbon analyses are summarized in Tables 7 and 8. Compared to the GCxGC/FID spectrum, oxidized 100LL avgas shows multiple oxygenated hydrocarbon peaks (Figure 18 and Table 6) of alcohols, aldehydes and organic acids that match the observation of FT-IR (Figures 10 and 11). It is noted that the oxidation mechanism of a hydrocarbon fuel consists of (1) free radical initiation; (2) free radical propagation; (3) termination of free radicals and (4) chain branching. The observation of O-H and C=O bonds and the related oxygenated hydrocarbon compounds further proves free radicals related mechanisms for fuel oxidation. Because one MS spectral peak may represent several different hydrocarbons or oxygenates, more accurate species identifications must rely on the soft ionization mass spectral analyses. We will perform GCxGC/TOF Field Ionization MS analyses in our future work.

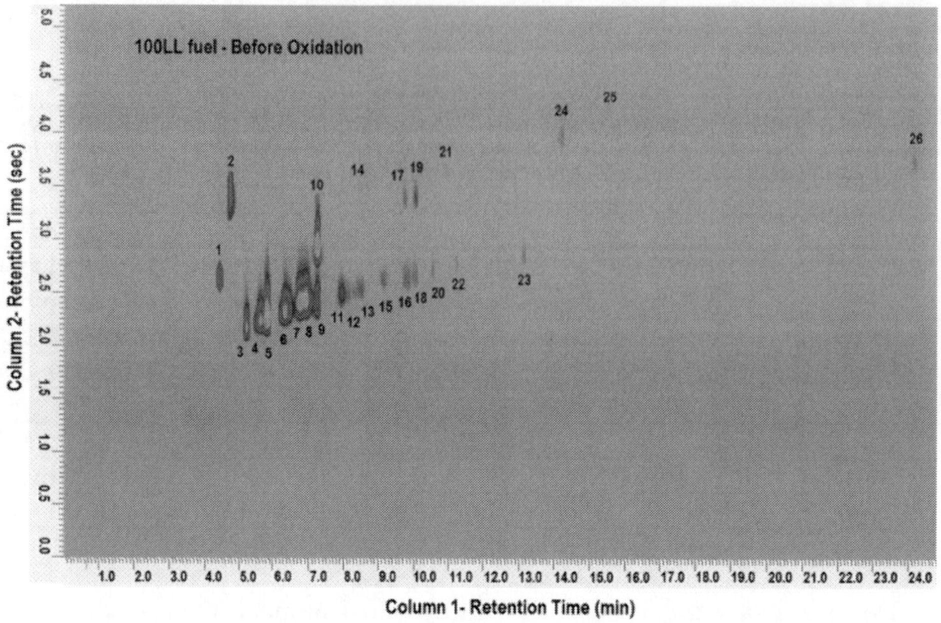

Figure 17. GCxGC/FID Spectrum for 100LL avgas (base fuel).

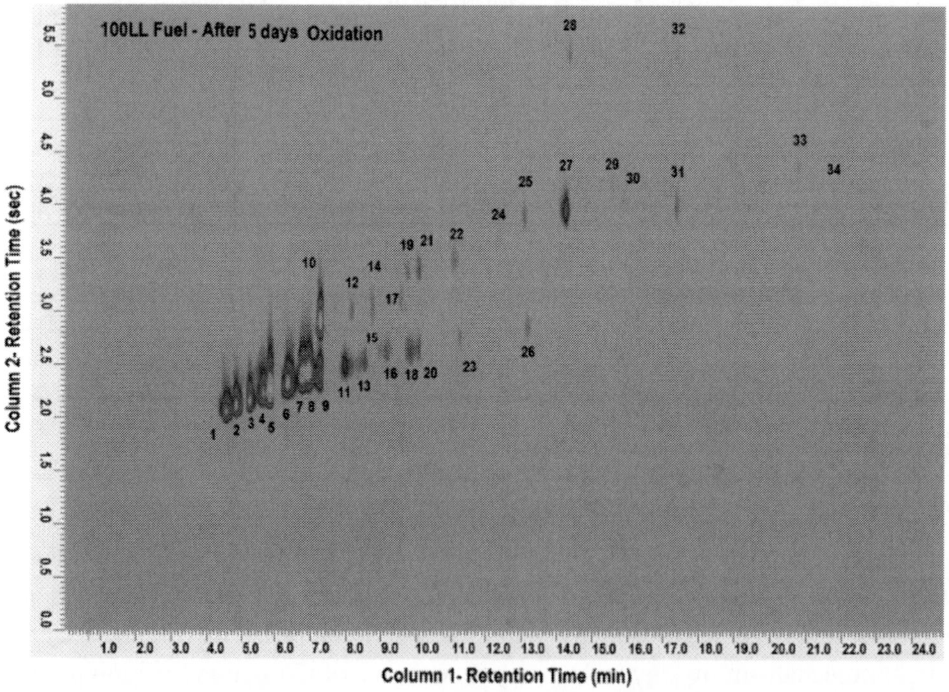

Figure 18. GCxGC/FID Spectrum for 100LL avgas (5-day oxidized fuel).

Table 7. Detailed hydrocarbon analysis of 100LL avgas (base fuel)

Peak Number	*Retention Time*		*Peak Area*	*Component*
	Column-1 (min)	*Column-2 (sec)*		
1	4.40	2.46	655.54	2-Methylbutane
2	4.70	3.42	173.73	2,3-Dimethylbutane
3	5.20	2.18	2058.85	2,4-Dimethylpentane
4	5.50	2.20	217.28	2,3-Dimethylpentane
5	5.80	2.28	17388.01	2,2,4-Trimethylpentane
6	6.30	2.34	2860.31	2,4-Dimethylhexane
7	6.70	2.40	4424.54	2,3,4-Trimethylpentane
8	6.90	2.44	3191.17	3.3-Dimethylhexane
9	7.20	2.40	1119.03	2,2,3-trimethylhexane
10	7.20	2.94	8021.03	Toluene
11	7.90	2.48	113.41	2,4-Dimethylheptane
12	8.10	2.50	17.67	2,2-Dimethylheptane
13	8.40	2.54	20.17	2,5-Dimethylheptane
14	8.40	3.50	5.34	1,2-Dibromoethane
15	9.10	2.62	21.33	2,3-Dimethylheptane
16	9.70	2.64	18.38	2,2,4-Trimethylheptane
17	9.70	3.38	21.83	Ethylbenzene
18	10.00	2.66	21.55	1,3-Dimethylbenzene
19	10.00	3.40	42.25	2,2,5-Trimethylheptane
20	10.50	2.74	11.09	3,3-Dimethyloctane
21	10.90	3.68	5.13	1,4-Dimethylbenzene
22	11.20	2.76	7.74	2,3,5-Trimethylbenzene
23	13.10	2.88	12.12	2,2,6-Trimethyloctane
24	14.20	3.96	12.07	1,3,5-Trimethylbezene
25	15.50	4.18	4.55	1,2,4-Trimethylbenzene
26	24.10	3.74	10.32	Tetraethyllead

Table 8. Detailed hydrocarbon analysis of 100LL avgas (5-day oxidation)

Peak Number	Retention Time Column-1 (min)	Retention Time Column-2 (sec)	Peak Area	Component	Molecular Formula (MS matching factor)
1	4.50	2.04	2582.71	2-Methylbutane	C_5H_{12}
2	4.80	2.10	3369.41	2,3-Dimethylbutane	C_6H_{14}
3	5.20	2.16	2765.15	2,4-Dimethylpentane	C_7H_{16}
4	5.50	2.20	304.49	2,3-Dimethylpentane	C_7H_{16}
5	5.80	2.26	22137.58	2,2,4-Trimethylpentane	C_8H_{18}
6	6.30	2.30	3522.01	2,4-Dimethylhexane	C_8H_{18}
7	6.70	2.40	5214.87	2,3,4-Trimethylpentane	C_8H_{18}
8	6.90	2.42	4306.56	2,3,3-Trimethylpentane	C_8H_{18}
9	7.20	2.42	1284.60	2,2,5-Trimethylhexane	C_9H_{20}
10	7.20	2.90	5305.67	Toluene	C_7H_8
11	7.90	2.46	137.74	2,3,4-Trimethylhexane	C_9H_{20}
12	8.10	2.98	17.71	2-Methyl-3-pentanol	$C_6H_{14}O$ (728)
13	8.40	2.50	21.64	1,2-Dibromoethane	$Br(CH_2)Br$
14	8.70	3.08	10.03	2,2,4-Trimethyl-3-pentanone	$C_8H_{16}O$ (903)
15	8.90	2.62	7.90	2,2,3,4-Tetramethylpentane	C_9H_{20}
16	9.10	2.62	24.64	2,3,5-Trimethylhexane	C_9H_{20}
17	9.50	3.08	9.60	Tert-Butyl formate	$C_5H_{10}O_2$ (769)
18	9.70	2.62	21.76	Ethylbenzene	C_8H_{10}
19	9.70	3.40	13.37	1,3-dimethylbenzene	$C_9H_{12}O$
20	10.00	2.62	29.76	2,2,4-Trimethylheptane	$C_{10}H_{22}$
21	10.00	3.40	30.08	p-Xylene	C_8H_{10}
22	11.00	3.50	11.26	Tetramethyloxirane	$C_5H_{10}O$ (792)
23	11.20	2.74	10.35	3-ethyl-2,5-dimethylhexane	$C_{10}H_{22}$
24	12.30	3.72	4.92	2-Methyl-4-octanol	$C_9H_{20}O$ (783)
25	13.00	3.88	13.70	2-Methylpentanal	$C_6H_{12}O$ (756)
26	13.10	2.84	17.94	2,2,6-Trimethyloctane	$C_{11}H_{24}$
27	14.20	3.94	184.61	Mesitylene	C_9H_{12}
28	14.20	5.46	0.57	Benzaldehyde	C_7H_6O (864)
29	15.40	4.16	0.13	6-ethyl-2-methyldecane	$C_{13}H_{28}$
30	15.80	4.02	5.82	di-n-Propylether	$C_6H_{14}O$ (804)
31	17.30	3.98	18.80	2,4,4-Trimethyl-1-pentanol	$C_8H_{18}O$ (820)
32	17.30	5.40	5.16	Benzylalcohol	C_7H_8O (739)
33	20.70	4.36	4.96	3-Methylbutanone	$C_5H_{10}O$ (657)
34	21.70	4.10	4.13	5-Methyl-3-hexanol	$C_7H_{16}O$ (768)

IV CONCLUSIONS

This research reports the oxidation stability of a 100 low lead aviation gasoline and an unleaded high aromatic fuel blend. The unleaded high aromatic aviation fuel blend is stable, both from copper strip corrosion and oxidation in a pure oxygen environment. During 72 hours of testing no significant oxygen pressure decrease was observed for either base or copper spiked fuel samples. The leaded avgas, however, suffered lead particle depositions on the surface of a copper strip during fuel copper strip corrosion testing. No oxygenated hydrocarbons in oxidized 100LL fuel can be found in a low sensitivity GC/MS or GC/FID. However, significant hydroxide and carbonyl groups were found in FT-IR spectra for oxidized 100LL fuels with or without the involvement of copper ions. Comprehensive two-dimensional GCxGC/TOFMS analyses have confirmed the formation of alcohols, aldehydes and organic acids after fuel oxidation. The GCxGC/TOFMS analyses also found the loss of tetraethyl lead after fuel oxidation. Fuel oxidation product identification via GCxGC/TOFMS analyses confirm that the mechanism of 100LL fuel oxidation is a chain reaction that consists of free radical initiation, propagation, termination and chain branching. Both lead and copper ions can catalyze the fuel oxidation process and decrease fuel stability during long-term storage

ACKNOWLEDGMENT

The authors of this paper acknowledge financial support from the FAA Piston Aviation Fuels Initiative (PAFI) program. The authors also thank Mr. Dave Atwood, Manager of the Fuels and Energy Section at the FAA William J. Hughes Technical Center for his support and invaluable comments on this research.

REFERENCES

1. ASTM D910-17a: Standard Specification for Leaded Aviation Gasolines.
2. ASTM D5159 - 04(2014): Standard Guide for dusting Attrition of Granular Activated Carbon.
3. ASTM D6378 - 10(2016): Standard Test Method for Determination of Vapor Pressure (VPX) of Petroleum Products, Hydrocarbons, and Hydrocarbon-Oxygenate Mixtures (Triple Expansion Method).
4. ASTM D4809 - 13: Standard Test Method for Heat of Combustion of Liquid Hydrocarbon Fuels by Bomb Calorimeter (Precision Method).
5. ASTM D7042 - 16e3: Standard Test Method for Dynamic Viscosity and Density of Liquids by Stabinger Viscometer (and the Calculation of Kinematic Viscosity).
6. ASTM D873 - 12: Standard Test Method for Oxidation Stability of Aviation Fuels (Potential Residue Method)
7. STM D5622 - 17: Standard Test Methods for Determination of Total Oxygen in Gasoline and Methanol Fuels by Reductive Pyrolysis
8. ASTM D130 - 12: Standard Test Method for Corrosiveness to Copper from Petroleum Products by Copper Strip Test.
9. ASTM D6733 - 01(2016): Standard Test Method for Determination of Individual Components in Spark Ignition Engine Fuels by 50-Metre Capillary High Resolution Gas Chromatography.
10. ASTM D873-12: Standard Test Method for Oxidation Stability of Aviation Fuels (Potential Residue Method).

IASH 2017, the 15TH INTERNATIONAL SYMPOSIUM ON STABILITY, HANDLING AND USE OF LIQUID FUELS
Rome, Italy
10-14 September 2017

EXPLIC PROGRAM
IMPACT OF AROMATIC TYPES AND QUANTITIES ON O-RING POLYMERS

Mickaël Sicard, Bruno Raepsaet, Jacques Ancelle, Frédéric Ser

ONERA, The French Aerospace Lab, Centre de Palaiseau, BP 80100, Palaiseau F-91123, France, mickael.sicard@onera.fr

ABSTRACT

In 2015, during the previous IASH symposium, we presented the results obtained on the impact of the jet fuel chemical composition on O-ring polymers. 5 fuels obtained by blending a Jet A-1 with a SPK fuel with aromatic rates from 2.5 up to 20.6 mass % were prepared and 3 O-ring types (nitrile, fluorosilicone and fluorocarbon) from the ASTM material list were selected. The test procedure described in the ASTM D4054 was followed. In addition to the standard 28-day soak period, a longer test (125 days) was also performed. After that, the mechanical properties of the O-rings were evaluated through hardness, tensile strength, elongation and volume swell tests. But, additional tests and characterizations (density, XRD, SEM…) were performed in order to highlight the chemical and structural modifications suffered by the gaskets. Moreover, the liquid phases were collected and analyzed by GC, FTIR, particles counting measurements… In the case of the nitrile O-rings, the aromatic content has an important impact on the mechanical properties. More surprising, we showed evidence of the same matter loss for all the tested fuels, it is mainly obtained during the 14 first days of the soaking tests. We also highlighted particulates in the liquid phase. For the fluorosilicone and fluorocarbone O-rings, the aromatic content seems to have less influence.

Blending fossil jet fuels with SPK fuels allows decreasing the aromatic rates but it also impacts the amount of other products like cycloalkanes, sulphur compounds, phenols… Thus, with these blends not only the rate of aromatic compounds is studied. That is the reason why a second approach was also proposed. SPKs were blended with an aromatic cut representative of the one found in jet fuels but also with a mono-aromatic cut, a di-aromatic cut and a bicyclic aromatic compound. The aim of this work was to determine the impact of the quantity and type of the aromatic compounds on polymers.

The results show that the nitrile O-rings are the most affected by the fuels. The more the quantity of bicyclic compounds is, the more the volume swell and the mass gain are. Moreover, a partial dissolution of the o-rings is observed with the apparition of particles in the fuel liquid phase.

KEYWORDS: alternative jet fuel, material compatibility

INTRODUCTION

From 2007, ONERA has become involved in national and European programs dedicated to alternative fuels for aircrafts such as CALIN, DREAM, ALFA-BIRD and MOCCASSIN and was the coordinator of SWAFEA. It was decided in 2013 to self-finance a program on alternative jet fuels. The main objectives of this ONERA funded program, called EXPLIC, is to better understand the consequences of the chemical modification of a jet fuel on its performance criteria (physicochemical properties, material compatibility, combustion and emissions) and to determine the acceptable limits of these modifications.

Among the different options, the two principal lines of research proposed are:

- To understand the aromatic role, the influence of their chemical composition on the jet fuels properties and the minimum concentration required,
- To study the impact of the modification of the molecules distribution in the jet fuel.

These two strands of work are consistent with two significant issues of alternative jet fuel development.

The aim of this paper is to present the results of material compatibility for O-ring soaked in fuels with various aromatic rates and types of aromatics. The mechanical properties of the O-rings are evaluated.

EXPERIMENTAL SECTION

Fuel test matrix

The low aromatic content limit imposed in standards (ASTM D7566 and DEF STAN 91-91) is 8 %. The objectives of the project are to better understand the consequences involved by the modification of the aromatic rates and of the aromatic types.

Blending fossil jet fuels with SPK allows decreasing the aromatic rates but it also impacts the amount of other products like cycloalkanes, sulphur compounds, phenols... Thus, with these blends not only the rate of aromatic compounds is studied. That is the reason why a second approach is also proposed. SPKs are blended with an aromatic cut representative of the one found in jet fuels. This aromatic surrogate cut was obtained by blending 3 aromatic fluids as it is described in the ASTM 4054 and considered as Jet Reference Fuel (JRF). In this paper this aromatic cut is referenced as Aromatic 450 (CA450). Thus, to study the impact of the aromatic compound concentration, a first matrix is proposed (Figure 1). The essence of the work concerns the values below 8 % as the fuels above this limit have already been certified. However, tests are also performed at 8 and above 16 % aromatic rates for comparison with the results in the literature and to follow the evolution of the properties above the overall aromatic rates.[1-7]

In jet fuels, it can be found mono aromatics (alkyl-benzenes) and di-aromatics (alkyl-naphthenes) compounds. It has already been shown that these types of compounds have effects on exhaust emissions and material compatibility. Based on these findings, a second matrix is proposed (Figure 1). The SPK fuel is blended with mono-and di-aromatic cuts at 2 different rates: 8 % and the percentage of aromatics in the jet fuel which has been purchased. The mono-aromatic cut contains alkylbenzenes with 9 and 10 carbon atoms. It is named Aromatic cut 100 (CA 100). It was not possible to find a "pure di-aromatic cut". Thus we had have to content ourselves with a 71.5 % mass of di-aromatic and 28.5 % mass of mono-aromatic cut whose name is Aromatic cut 250 (CA 250). Moreover, tests were also performed with the methylnaphthalene (MNPH), chosen as representative of the di-aromatic compounds.

The SPK fuel was purchased from SkyNRG. A volume of 1600 L was obtained. The fuel aromatic rate was 2.52 %.

The kerosene Jet A-1 (F-35) was obtained from the French Army. The aromatic rate was significantly higher, but in the specifications, than expected: 20.6 against 16 %.

A fruitful collaboration was concluded with the petroleum company TOTAL Marketing Services which provided us various fuels and hydrocarbon cuts. Among them, there were aromatic streams, which were blended to form an aromatic cut representative to the one of the Jet A-1.

Fuels
Jet A-1 (aro = 20,6 %)
~ 70 % SPK/~ 30 % Jet A-1 (aro = 8 %)
~ 81 % SPK/~ 19 % Jet A-1 (aro = 6 %)
~ 92 % SPK/~ 8 % Jet A-1 (aro = 4 %)
SPK (t_{aro} = 2,52 %)
Aromatic cut
SPK+ aromatic cut (aro = 4 %)
SPK+ aromatic cut (aro = 6 %)
SPK+ aromatic cut (aro = 8 %)
SPK+ aromatic cut (aro = 20,6 %)

Fuels
Mono-aromatic cut
SPK+ mono-aromatic cut (aro = 8 %)
SPK+ mono-aromatic cut (aro = 20,6 %)
Di-aromatic cut
SPK+ di-aromatic cut (aro = 8 %)
SPK+ di-aromatic cut (aro = 20,6 %)
1 aromatic
SPK+ 1-aromatic cut (aro = 8 %)
SPK+ 1-aromatic cut (aro = 20,6 %)

Figure 1: Fuel matrixes on the impacts of the aromatic concentration and type

O-ring gaskets

The 3 O-ring types (nitrile, fluorocarbone, fluorosilicone) from the material list of the ASTM D4054 are tested. They are purchased from Parker.

Experimental procedure

The soaking tests to be comparable with those performed during the fuel certifications are done following the ASTM procedure described in the ASTM D4054 standard (Table 1). Thus, each material is tested with the appropriate temperature. The soak tests last 28 days and a replacement of the old fuel with a fresh one is performed after 14 days. After ageing, the mechanical properties of the O-rings are evaluated.

Table 1: ASTM D4054 experimental procedure

Experimental conditions			
Gasket, (O-ring) 25.12 x 1.78 (mm)	T (°C)	Ageing time (days)	Mechanical properties
Nitrile Fluorosilicone Fluorocarbon	80 110 160	28 (Fuel change out after 14 d)	Hardness, shore M Tensile strength Elongation Compression Set Volume swell

Mechanical tests are carried out in the ASTM procedure. Despite their interests, they do not explain the failure reasons that can be observed. To go further and to understand the evolution highlighted, we propose in the framework of the EXPLIC program other characterizations on the materials like mass evolution, density, X-ray diffraction, microscope observations (Table 2). The study of the fuel can be also interesting as it is in direct interaction with the O-ring. Any evolution of the fuel could be linked with a potential O-ring alteration. Thus, fuels after the tests are characterized by techniques like gas chromatography, particulate measurement...
Moreover, the ageing time in the ASTM procedure lasts 28 days and in order to magnify the phenomena, a longer ageing time, 125 days, is used. However, these results should be interpreted with care as we do not know if this ageing is representative of what could be observed in reality.

Table 2: EXPLIC experimental procedure

Experimental conditions					
Gasket, (O-ring) 25.12 x 1.78 (mm)	T (°C)	Ageing time (days)	O-ring		Kerosene characterizations
			Mechanical properties	Other characterizations	
Nitrile Fluorosilicone Fluorocarbon	80 110 160	28 (Fuel change out after 14 d) 125 (Fuel change out after 14 d)	Hardness, shore M Tensile strength Elongation Compression Set Volume swell	Mass evolution Density (pycnometer) XRD SEM ...	FTIR, GC, GC-MS, HPLC... Particulates

Each soaking test is performed on 12 O-rings, 10 of them are used for the mechanical tests.

Tensile strength and Ultimate elongation

They are measured with a DY 25 device from the MTS Company equipped with a 100 DaN force sensor. The procedure and the calculations described in the ASTM D1414 are followed. The machine is set for a speed of 450 mm/min. The average value is obtained with 10 O-rings.

Volume Swell

The thickness is determined by laser measurement. The device is an opto NCDT model from the Micro-Epsilon company. Internal diameters are evaluated with a calipers square. The volume measurement of the 12 O-rings is done. The final volume is an average of all the measurements. The volume variation is obtained by comparison with unaged O-rings.

RESULTS

In this article, only the results of the mechanical properties are presented. The other results will be given elsewhere.[8]

Tensile strength and elongation

Nitrile O-rings

After 28 days and whatever the blend, the tensile strength of the soaked O-rings compared to the unaged ones show a decrease (Figure 2). The more the aromatic content is, the more the decrease is. For example, the tensile strength of the unaged O-ring is 11.69 MPa and the tensile strengths of the O-rings soaked in SPK and Jet A-1 are respectively 11.8 and 9.4 MPa. The aromatic type does not seem to have an impact as the same trade is observed for all the blends. After 125 days, a supplementary decrease of around 10-15 % of the tensile strength values is observed.

A significant decrease of the maximum elongation values is noticed for the O-rings after 28 days of soaking (Figure 3). They drop from 300 % (unaged sample) to around 200 %. After 125 days, the elongation values decrease by around 30 %. They are around 150 %. The aromatic content and type have no significant influence on the maximum elongation.

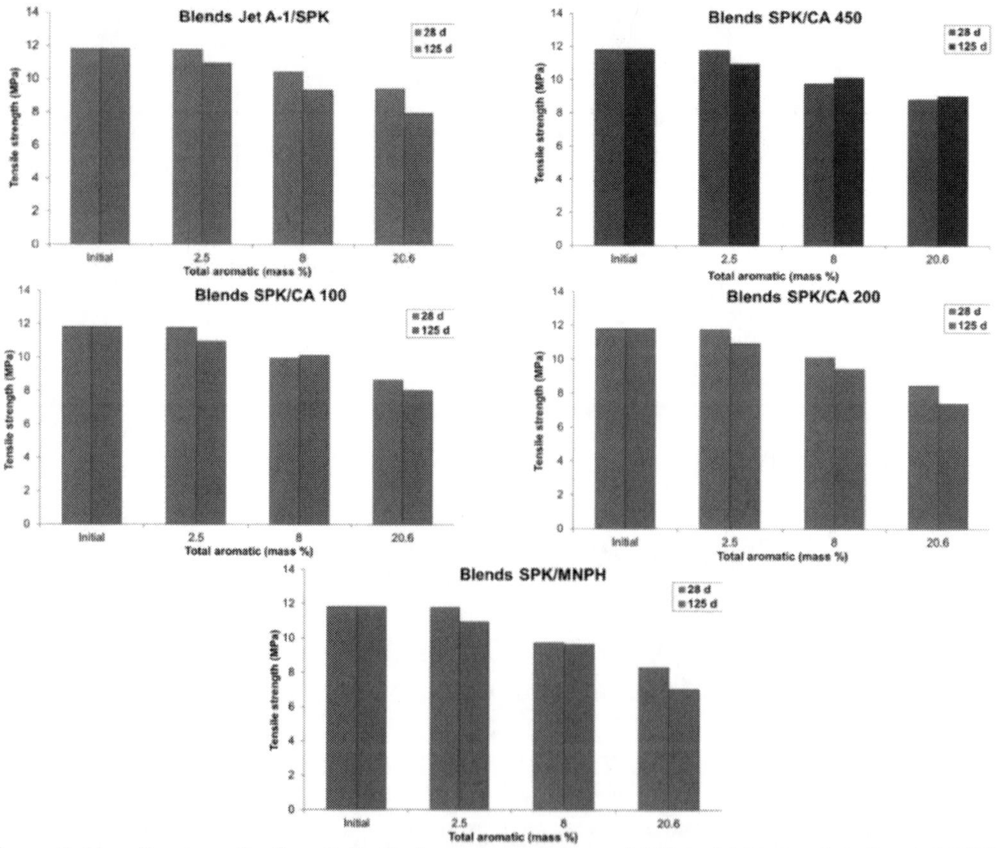

Figure 2: Tensile strengths for nitrile O-rings unaged or aged (28 and 125 days) in Jet A-1/SPK, SPK/CA450, SPK/CA100, SPK/CA200, and SPK/MNPH blends (aromatic % 2.52, 8 and 20.6)

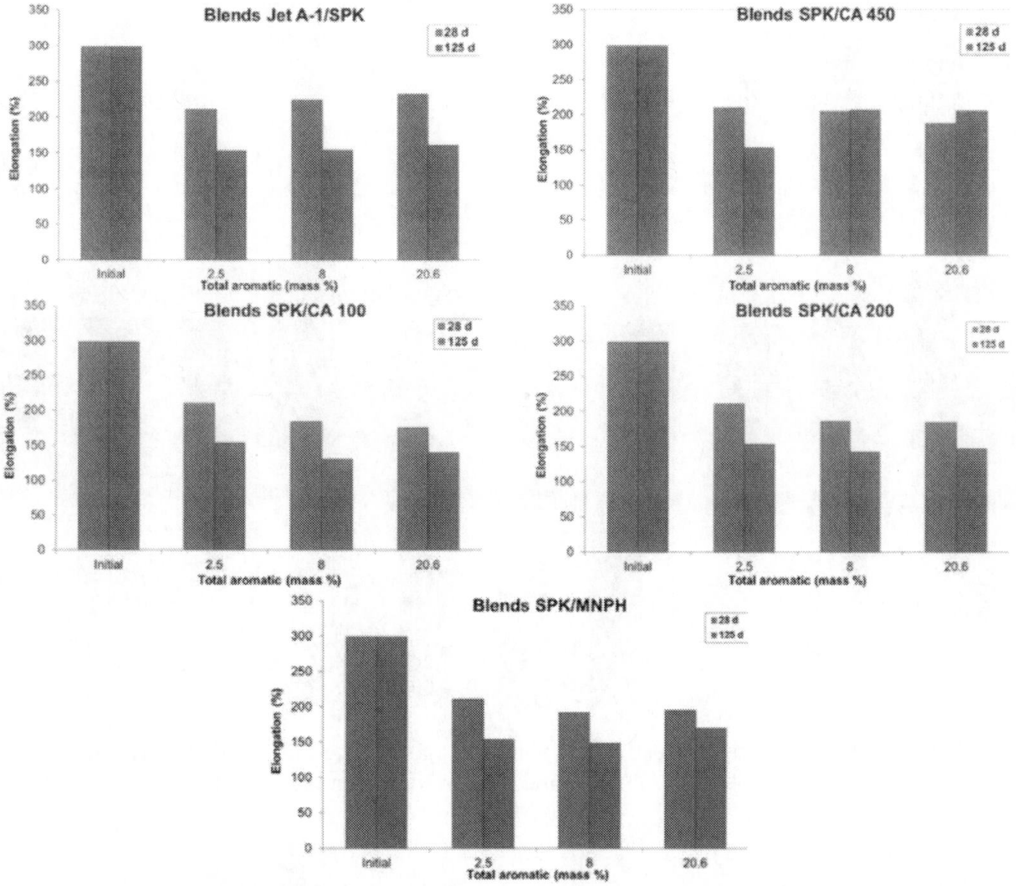

Figure 3: Elongations for nitrile O-rings unaged or aged (28 and 125 days) in Jet A-1/SPK, SPK/CA450, SPK/CA100, SPK/CA200, and SPK/MNPH blends (aromatic % 2.52, 8 and 20.6)

Fluorosilicone O-rings

The tensile strength of the O-rings soaked in the blends Jet A-1/SPK shows a very particular evolution (Figure 4). It falls dramatically with the increase of Jet A-1. After 125 days, the O-rings become so friable that it was not possible to do the measurement. This surprising but repeatable results cannot be explained by the increase of the aromatic content. The blends SPK/CA 450, SPK/CA 100, SPK/CA 200 do not show a same evolution. Indeed, we observed a decrease of the tensile strength after a soaking period, it is linked neither with time nor the aromatic content. In the case of the blends SPK/MNPH, a significant decrease of the tensile strength is highlighted with for the MNPH highest content and duration while this might not compare to the Jet A-1 results.

Obviously, the elongation results obtained for the O-rings soaked in the Jet A-1 are not good. It was not possible to measure it. For the other blends and after 28 days, the maximum elongation values decrease with the aromatic content. After 125 days, the values are roughly stable as if a stable state was reached. The blends SPK/MNPH set itself apart. The maximum elongation values decrease with the aromatic content.

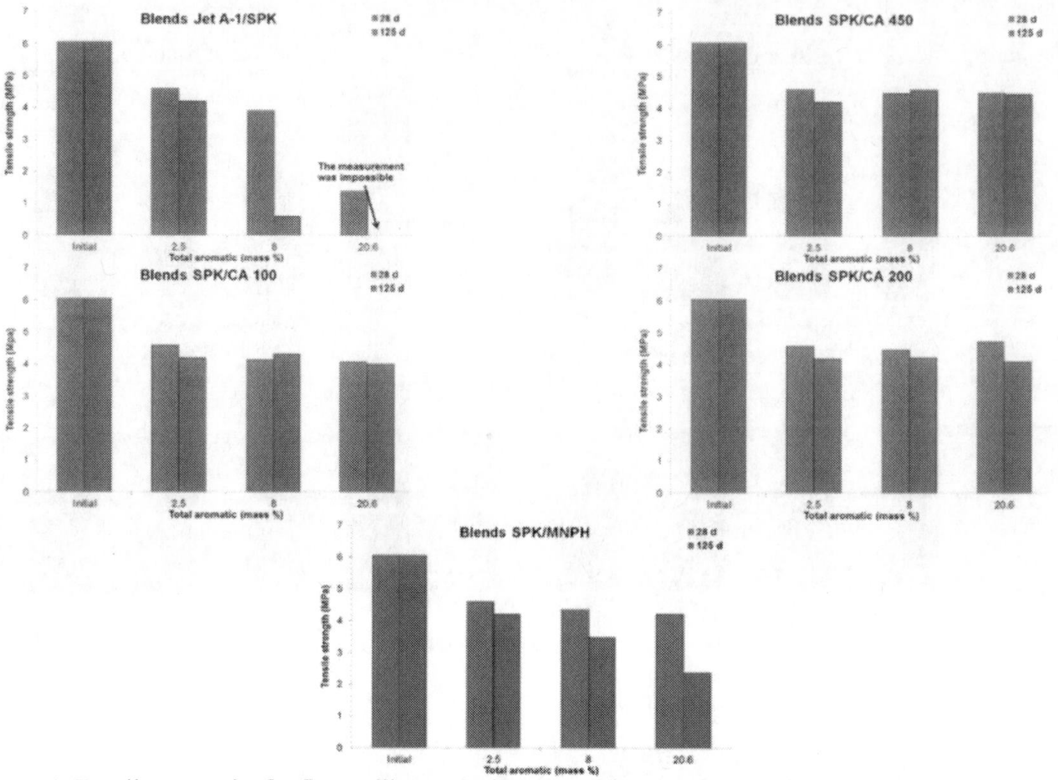

Figure 4: Tensile strengths for fluorosilicone O-rings unaged or aged (28 and 125 days) in Jet A-1/SPK, SPK/CA450, SPK/CA100, SPK/CA200, and SPK/MNPH blends (aromatic % 2.52, 8 and 20.6)

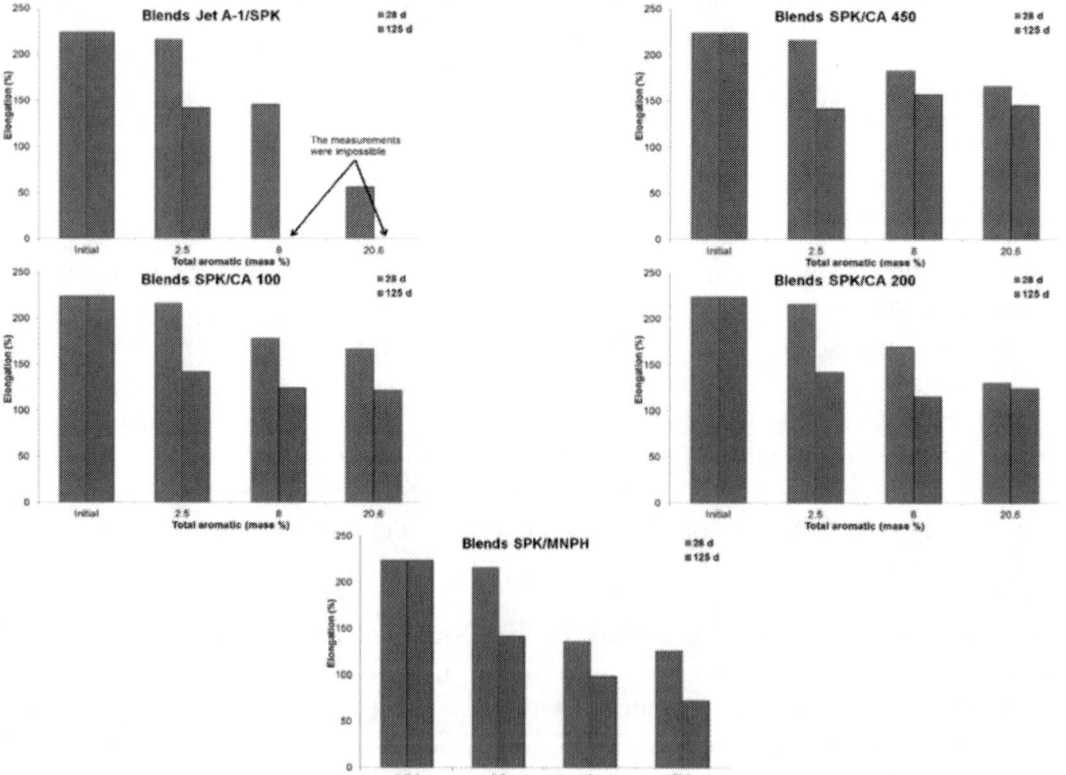

Figure 5: Elongations for fluorosilicone O-rings unaged or aged (28 and 125 days) in Jet A-1/SPK, SPK/CA450, SPK/CA100, SPK/CA200, and SPK/MNPH blends (aromatic % 2.52, 8 and 20.6)

Fluorocarbone O-rings

After the tests and whatever the fuels (expect the blends SPK/MNPH), the tensile strength and elongation maximum for all the O-rings are divided by two compared to unaged O-rings. The results show a slight dependence with time.
The results obtained with the blends SPK/MNPH are different. The increase of the amount of MNPH is followed by an increase of the tensile strength values (125 days) and decrease of the elongation maximum values (both test durations). Even if such levels of di-aromatics are not representative of the amount found in a jet fuel, di-aromatics seem to have an impact on the structure of the fluorocarbone O-rings.

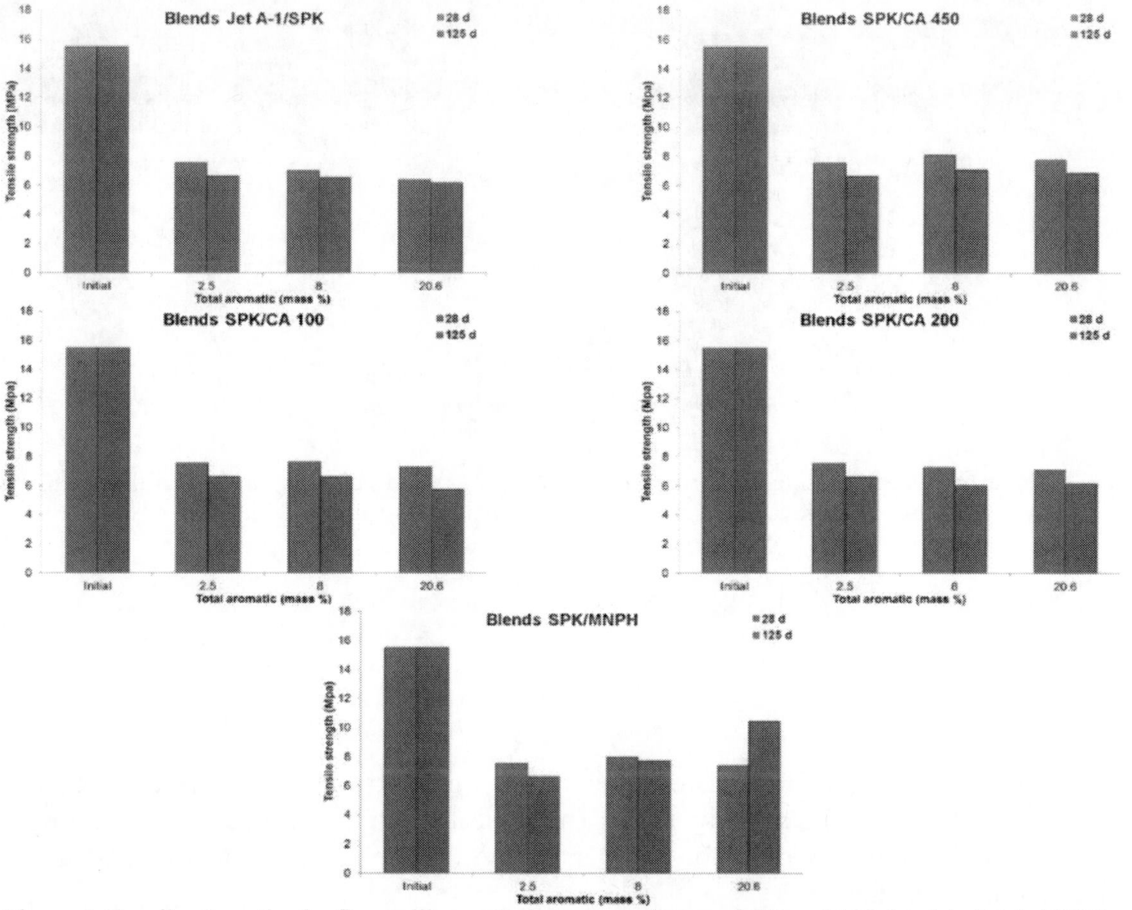

Figure 6: Tensile strengths for fluorosilicone O-rings unaged or aged (28 and 125 days) in Jet A-1/SPK, SPK/CA450, SPK/CA100, SPK/CA200, and SPK/MNPH blends (aromatic % 2.52, 8 and 20.6)

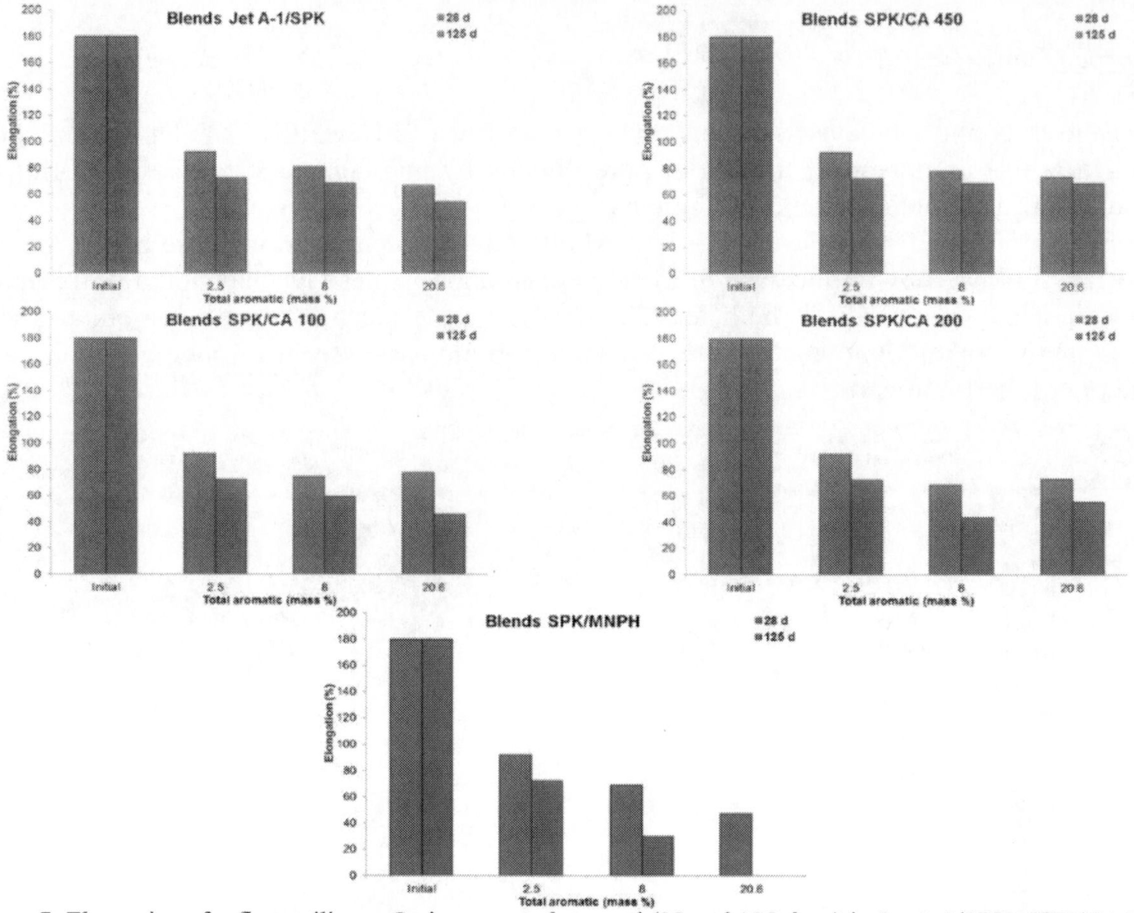

Figure 7: Elongations for fluorosilicone O-rings unaged or aged (28 and 125 days) in Jet A-1/SPK, SPK/CA450, SPK/CA100, SPK/CA200, and SPK/MNPH blends (aromatic % 2.52, 8 and 20.6)

Volume Swell

Nitrile O-rings

The influence of the aromatic rate is obvious on the volume swell: the O-rings swell more and more with the aromatic content (Figure 8). The di-aromatic content has clearly an impact. The highest volume swell are obtained for the blends with the highest di-aromatic contents. The test duration has no real influence.

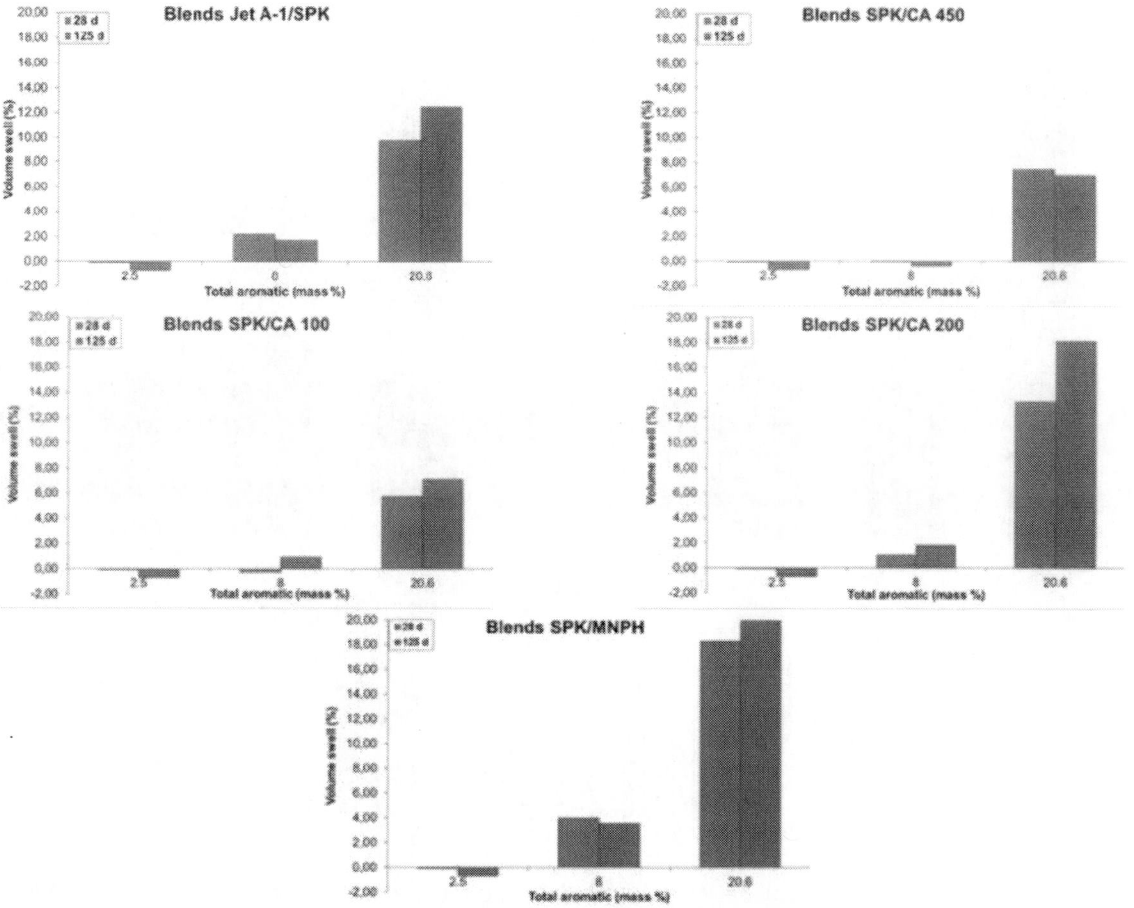

Figure 8: Volume swells for nitrile O-rings unaged or aged (28 and 125 days) in Jet A-1/SPK, SPK/CA450, SPK/CA100, SPK/CA200, and SPK/MNPH blends (aromatic % 2.52, 8 and 20.6)

Fluorosilicone O-rings

After being soaked for 28 days in the SPK, the O-rings swelled around 6.8 % (Figure 9). The volume variation increases progressively with the aromatic rate of the blends. The phenomenon seems to be more apparent with the blends with the highest di-aromatic rates.
The result obtained after 125 days is very surprising for the blends Jet A-1/SPK (Figure 9). The opposite phenomenon is observed: the O-rings contract themselves more and more with the aromatic rate. This result is really specific to the blends Jet A-1/SPK as the others blends react "normally": an increase of the volume swell with the aromatic rate.

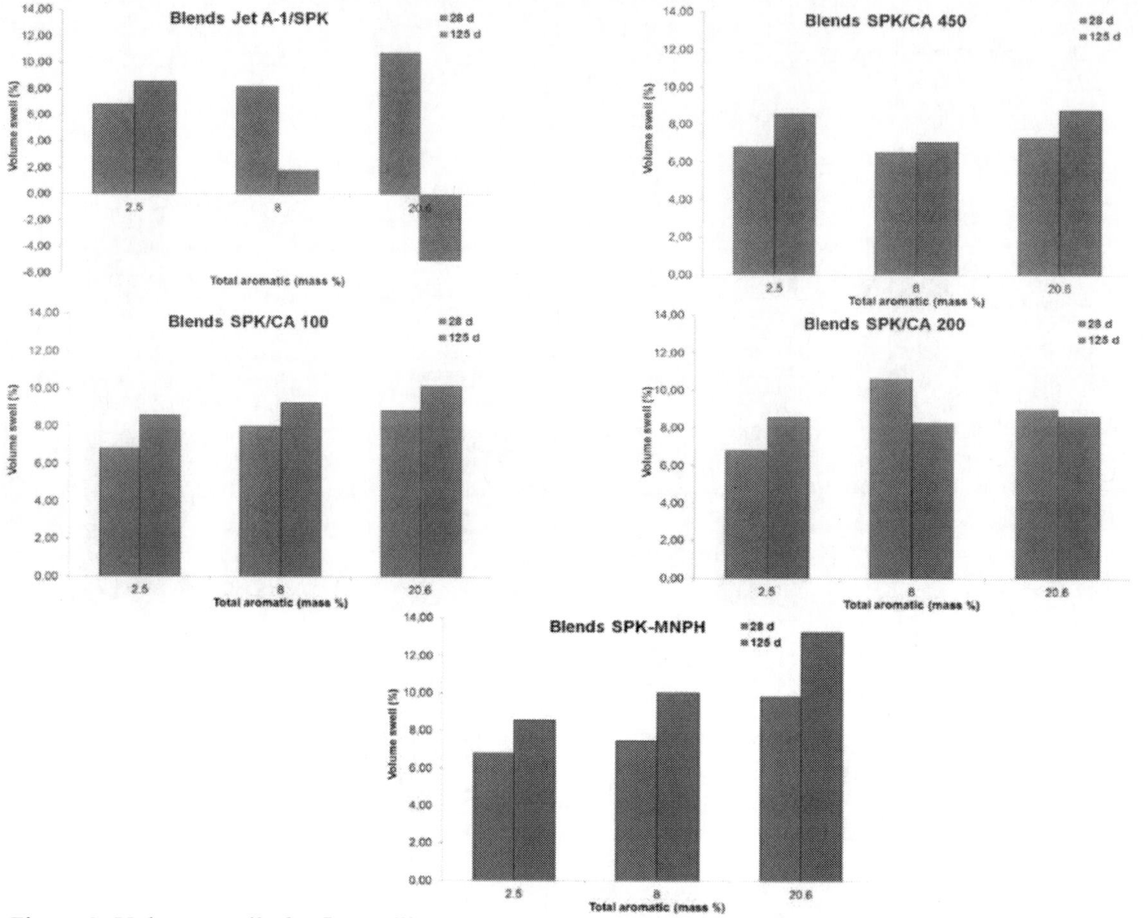

Figure 9: Volume swells for fluorosilicone O-rings unaged or aged (28 and 125 days) in Jet A-1/SPK, SPK/CA450, SPK/CA100, SPK/CA200, and SPK/MNPH blends (aromatic % 2.52, 8 and 20.6)

Fluorocarbone O-rings

The fluorocarbone O-rings swell with the aromatic rate for the blends Jet A-1/SPK, SPK/CA 450, CA 100 and CA 200. The volume variations are moderate: between 4 and 7 % (28 days) and between 4.7 and 8 % (125 days).

The behaviour of the blend SPK/MNPH with 20.6 % aromatic content is remarkably different. The volume swell is significantly higher for this blend and especially after 125 days.

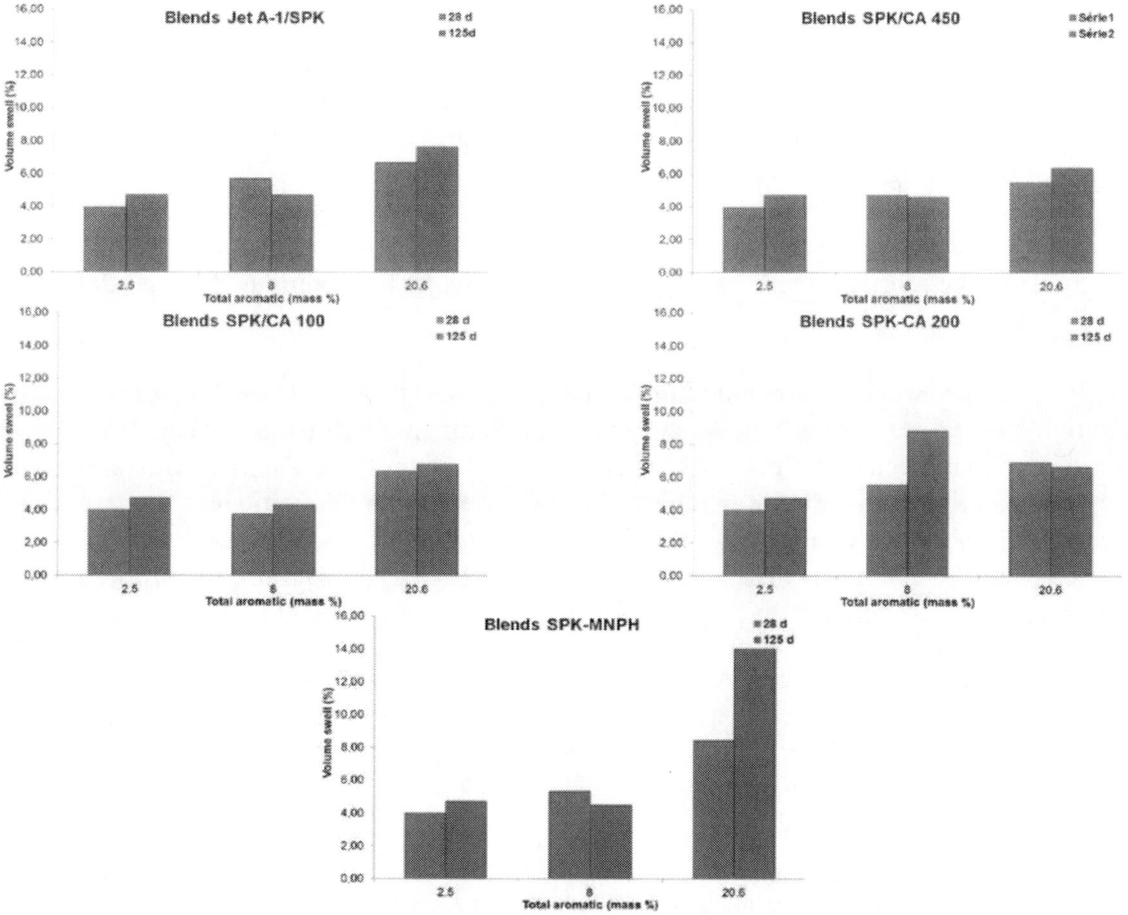

Figure 10: Volume swells for fluorocarbone O-rings unaged or aged (28 and 125 days) in Jet A-1/SPK, SPK/CA450, SPK/CA100, SPK/CA200, and SPK/MNPH blends (aromatic % 2.52, 8 and 20.6)

DISCUSSION AND CONCLUSION

In the case of the nitrile O-rings, the aromatic content has an impact on the mechanical properties (tensile strength and elongation). The fuel diffusion (volume swell) is related to the total aromatic content, but mainly correlated to the amount of di-aromatics (Figure 11).

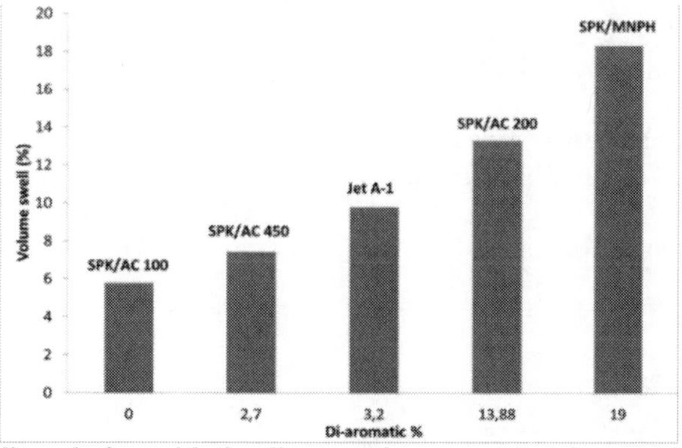

Figure 11: Volume swell evolution with the di-aromatic content (total aromatic content: 20.6 %mass)

The aromatic % has also an impact on the mechanical properties (tensile strength and elongation) on the fluorosilicone O-rings (28 days). These results are coupled with O-rings swell. The impact of the duration test on the O-rings behaviour is very surprising and unexpected for the blends Jet A-1/SPK. After 125 days, the O-rings became breakable when soaked in fuels with high Jet A-1 %. As a result, the tensile strength and elongation values are very low. This is accompanied by a decrease of the O-ring swellings.

The tensile strength and the elongation values of the fluorocarbon O-rings are halved after the tests whatever the fuel and the test duration. The impact of the aromatic % is moderate with this type of O-rings.

Generally, whatever the O-rings, the higher the aromatic rate is the lower the tensile strength and elongation values are. Nevertheless, it is very difficult to establish a logical link between the evolutions of one mechanical property with the values of another. The mechanical properties are a key point to verify the material compatibility with new fuels. Nevertheless, they do not explain why an O-ring adopts a certain type of behaviour. To do that, it will be very helpful to use other characterization techniques. This is carried out in the EXPLIC program, the whole of the results will be presented elsewhere.[8]

REFERENCES

1: Corporan, E., Edwards, T., Shafer, L., DeWitt, M.J., Klingshirn, C., Zabarnick, S., West, Z., Striebich, R., Graham, J., and Klein, J., Chemical, *"Thermal Stability, Seal Swell, and Emissions Studies of Alternative Jet Fuels,"* Energy & Fuels, 25, 955-966, 2011.

2: DeWitt, M.J., Corporan, E., Graham, J.L. and Minus, D. "Effects of Aromatic Type and Concentration in Fischer-Tropsch Fuel on Emission Production and Material Compatibility," Energy & Fuels, 22, 2411-2418, 2008.

3: Moses, C.A., *"Comparative Evaluation Of Semi-Synthetic Jet Fuels,"* CRC Project No. AV-2-04a, September 2008.

4: Moses, C.A., "Development Of The Protocol For Acceptance Of Synthetic Fuels Under Commercial Specification," CRC Contract No. AV-2-04, September 2007.

5: The Boeing Company, and University of Dayton Research Institute, "Impact of Alternative Jet Fuel and Fuel Blends on Non-Metallic Materials Used in Commercial Aircraft Fuel Systems," CLEEN Project Final Report, July 2011.

6: Moses, C.A., Roets, P.N.J., *"Properties, Characteristics, and Combustion Performance of Sasol Fully Synthetic Jet Fuel,"* Journal of Engineering for Gas Turbines and Power, Vol. 131 / 041502, pp. 1-17, July 2009.

7: Muzzell, P., Stavinoha, L., Chapin, R., *"Synthetic Fischer-Tropsch (Ft) Jp-5/Jp-8 Aviation Turbine Fuel Elastomer Compatibility,"* Final report, February 2005.

8: Sicard, M., Ancelle, J., Guérin, S., Raepsaet, B., Ser, F., an article on this work is in preparation.

IASH 2017, the 15TH INTERNATIONAL SYMPOSIUM ON STABILITY, HANDLING AND USE OF LIQUID FUELS
Rome, Italy
10-14 September 2017

EXPLIC PROGRAM: IMPACT OF AROMATIC CONTENT IN DIFFERENT ALTERNATIVE AERONAUTIC FUEL EMISSIONS

D. Delhaye[1], J. Ancelle[1], M. Sicard[1], L. Jing[2], and I.K. Ortega[1]

[1]*ONERA, The French Aerospace Lab, Centre de Palaiseau, BP 80100, Palaiseau F-91123, France, david.delhaye@onera.fr*

[2]Jing Ltd., Im Park 4, CH-3052 Zollikofen, Switzerland

ABSTRACT

In the period up to 2030, global aviation is expected to grow by 5 % annually according to International Air Transport Association IATA[1]. The aviation industry has identified the development of biofuels as one of the major ways it can reduce its emissions from fossil origin. With an increasing number of alternative fuels available, there is a need for a systematic method to evaluate how the chemical composition of different fuels affects their emissions. Complete engine tests are technically and economically challenging.

In this work, we have used a novel Combustion Aerosol STandar (CAST) Generator especially designed to work with few amount of liquid fuel[2] to study the emission generated by up to 15 different fuels: Jet A-1, Synthetic Paraffinic Kerosene (SPK), Aromatic Cut (CA) 450 (a mixture of 3 aromatic representative cuts found in a jet fuel, AF 100, 150 and 200), SPK/JetA-1 blends (4 and 8 % aromatic contents (AC)), SPK/CA 450 blends (4, 6, 8 and 20.6 % AC), SPK/AF 100 blends (8 and 20.6% AC), SPK/CA 150 blends (8 and 20.6 % AC) SPK/CA 200 blends (8 and 20.6 % AC) and SPK/1-Methyl-naphthalene blends (8 and 20.6 % AC). We have measured the total number and size distribution of particles produced by these fuels at different flame richness (between 0.5-1.6), both at the raw exhaust and after treatment by a catalytic stripper coupled with a Dekati Engine Exhaust Diluter (DEED, Dekati oy) to remove the volatile fraction.

KEYWORDS: alternative jet fuel, emission, soot

INTRODUCTION

Aviation industry clearly has set their efforts for reducing aircraft emissions on the development of biofuels as an alternative for standard jet fuel. Currently there are only six production pathways technically certified, but 16 more certifications are in preparation. Design of new generation biofuels is challenging. First of all, they need to fulfil the highly demanding aeronautic technical standards. Secondly, testing of biofuels so far requires costly and technical complete engine tests; the high fuel consumption (~600 lpm) and the low availability of biofuels impose strict time limits for these tests, thus systematic experiments beyond the standard certification purposes are rarely available. This lead to a really small number of essays focused on studying the emissions of different biofuels. In addition, individual testing of different parameters of the fuel, like sulfur content, percentage of aromatics, family of aromatic compounds, etc. are not feasible.

In this work we have used a liquid CAST generator as a source of surrogates for engine emissions. Thanks to the low fuel consumption of CAST generator, we have been able to test large number of fuels. The fuel matrix chosen for the work was focused to study the impact of aromatic compounds in the emissions. The study has been not only limited to evaluate the effect of the aromatic content of different fuels, but also we have studied different families of aromatic compounds, to determine how the precise chemical composition of the aromatic compounds present in the fuel can impact their emissions.

EXPERIMENTAL SECTION

Fuel test matrix

The main objective of this work is to evaluate the consequences of the modification of the aromatic compound rate into the jet fuel. The low aromatic content limit imposed in standards (ASTM D7566 and DEF STAN 91-091) is 8 %. Blending fossil jet fuels with SPK allows decreasing the aromatic rates but it also impacts the amount of other products like cycloalkanes, sulphur compounds, phenols, etc. Thus, with these blends not only the rate of aromatic compounds is studied. That is the reason why a second approach is also proposed. SPKs are blended with an aromatic cut, called CA450, representative of the one found in jet fuels.

The kerosene Jet A-1 (F-35) was obtained from the French Army. The aromatic rate was 20.6 %.

The SPK fuel was purchased from SkyNRG. The fuel aromatic rate was 2.52 %.

A fruitful collaboration was concluded with the petroleum company TOTAL Marketing Services which provided us various fuels and hydrocarbon cuts. Among them, there were aromatic streams, which were blended to form an aromatic cut representative to the one of the Jet A-1.

To investigate the impact of the aromatic compound concentration we use a fuel matrix composed of jet A-1 with 20.6 % of aromatic compounds, Jet A-1 blend with SPK with 8 % of aromatic compounds, SPK blend with either mono aromatics (alkyl-benzenes di-aromatics (alkyl-naphthenes), or methyl-naphtalene compounds, from now on we will refer to these as CA100, CA200 and MNPH respectively with 8 and 20.6 % of aromatic compounds (Figure 1).

Fuels
Jet A-1 (aro = 20,6 %)
~ 70 % SPK/~ 30 % Jet A-1 (aro = 8 %)
~ 81 % SPK/~ 19 % Jet A-1 (aro = 6 %)
~ 92 % SPK/~ 8 % Jet A-1 (aro = 4 %)
SPK (t_{aro} = 2,52 %)
Aromatic cut
SPK+ aromatic cut (aro = 4 %)
SPK+ aromatic cut (aro = 6 %)
SPK+ aromatic cut (aro = 8 %)
SPK+ aromatic cut (aro = 20,6 %)

Fuels
Mono-aromatic cut
SPK+ mono-aromatic cut (aro = 8 %)
SPK+ mono-aromatic cut (aro = 20,6 %)
Di-aromatic cut
SPK+ di-aromatic cut (aro = 8 %)
SPK+ di-aromatic cut (aro = 20,6 %)
1 aromatic
SPK+ 1-aromatic cut (aro = 8 %)
SPK+ 1-aromatic cut (aro = 20,6 %)

Figure 1: Fuel matrixes on the impacts of the aromatic concentration and type

Only the results obtained with the aromatic cuts matrix are presented in this paper (Figure 1).

Set-up

To burn liquid fuel, we used a Combustion Aerosol Standard CAST[2]. This device vaporized the liquid fuel with a propane flame to allow its combustion. Parameters controlled for the combustion are liquid fuel flow, dilution air flow and propane flame air/fuel ratio. The setting of the propane flame has been chosen to not produce soot particles[3]. The main advantage of this burner is the low consumption of fuel in counterpart, the main drawback is that the combustion is only at atmospheric pressure leading to a high production of volatile organic compounds (COV). The particulate number concentration was measured with a Condensation Particle Counter (CPC), a Grimm® CPC 4503. The size distribution was established with a Scanning Mobility Particle Sizer (SMPS) composed of an Annular Dielectric Barrier Discharge (ADBD), a nano Vienna Differential Mobility Analyser (DMA) and a Condensation Particle Counter (CPC) Grimm® CPC 4503. The ADBD is used to neutralize and give a monocharge to the particulate. Then the sample is classified by the DMA. The monocharged particles follow a mass to charge ratio selection through an electric field. Finally, selected particles are counted with the CPC. A CPC is composed of two stages. The first one corresponds to a bath ofsursatured of buthanol vapor that allows particulate to growth in size enough to be measured with the second stage made of laser detectors.

Between the CAST and CPC and SMPS, we used a dilution box, a Palas® VKL10, to dilute the sample with compressed air filtered at 60 °C to avoid condensation of vapor and to reduce temperature. The dilution ratio is 1:10. The sample coming out of this device is composed of soot and VOC. In order to determine properties of only the soot part, we employed a Dekati® Engine Exhaust Diluter coupled to a Catalytic Stripper (CS-DEED). The CS-DEED first destroy VOC with the CS heated at 350 °C, the sample is diluted by using two stages of ejector both with filtered compressed air. The first stage is heated at 450 °C and the second one at room temperature. The total dilution ratio is closed to 1:30. The Figure 2 summarizes the set-up used.

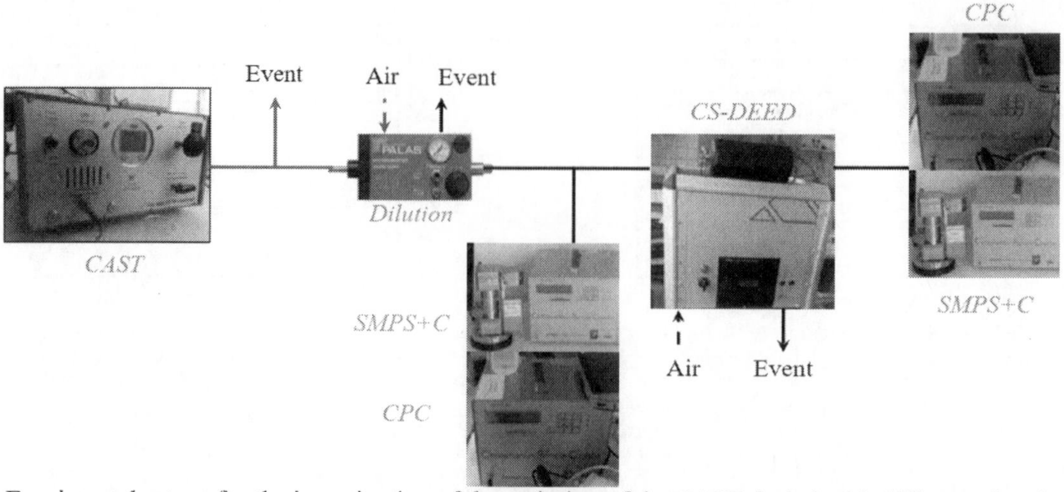

Figure 2: Exprimental set-up for the investigation of the emission of the CAST fueled with different chemical composition of fuel. The raw emission, i.e. soot + volatile organic compounds, and the emission passed through CS-DEED, i.e. only soot, are monitored in terms of particulate size distribution (with SMPS) and number concentration of particles (with CPC).

RESULTS

In this article, only the results of the number concentration of particles emitted are presented. The other results will be given elsewhere[4].

The Figure 2 represents number concentration of particles measured for various richness at the exit of the CAST fueled with the different compositions of fuel but with a fixed rate of aromatic compounds (20.6 %). On the left side of the Figure 2, the raw emission data are exhibited (soot and Volatile Organic Compound) and in the right side we are shown the emission after CS-DEED treatment (only soot). For both cases, under a richness of 1.2, jet-A1 emission is a combination of mono and di-aromatic values. The mono-aromatic compounds seem to be responsible of the level of the number concentration of particles encounter. Over 1.2 of richness, for the raw signal the jet-A1 emission is higher than other fuel cases. At this point, we don't have explanation. For the sample that passed through CS-DEED, with a richness higher than 1.2, the mono-aromatic compounds seem to drive the emission. For both cases, of we compare the Jet A-1 emission and the SPK/CA 450 emission, we observed that under a richness of 1.1, values measured are similar while over 1.1 differences appear. We are still investigating this effect.

We performed the same work but with fuels containing an 8% aromatic compounds rate (Figure 3). For the raw emission case and the emission passed through CS-DEED case; under 0.8 of richness, emission of jet-A1 follow SPK+di-aromatic compounds (CA 200) curve, then between 0.8 and 1.1, Jet A-1 emission evaluates as the SPK+mono-aromatic compounds (CA 100) one. Finally, for richness higher than 1.1, the jet-A1 emission follow the trends of the emission of SPK blend with methyl-naphthalene (MNPH).

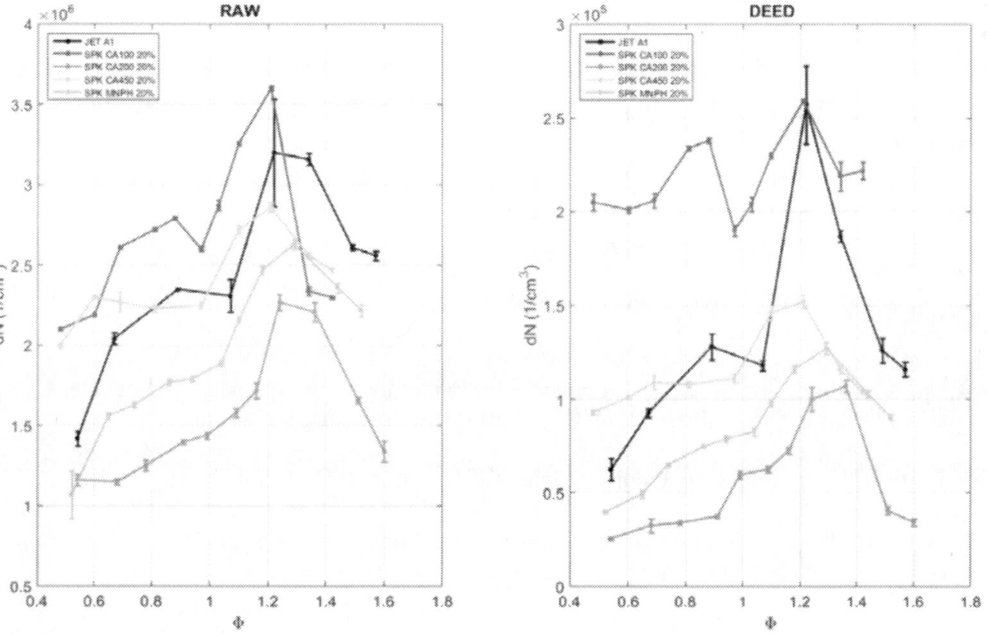

Tensile strength Elongation

Figure 3: Number concentration of particles measured for various richness at the exit of the CAST fueled with the different compositions of fuel but with a fixed rate of aromatic compounds (20.6 %). On the left side the raw emission data are exhibited (soot and Volatile Organic Compound) and in the right side are shown the emission after CS-DEED treatment (only soot)

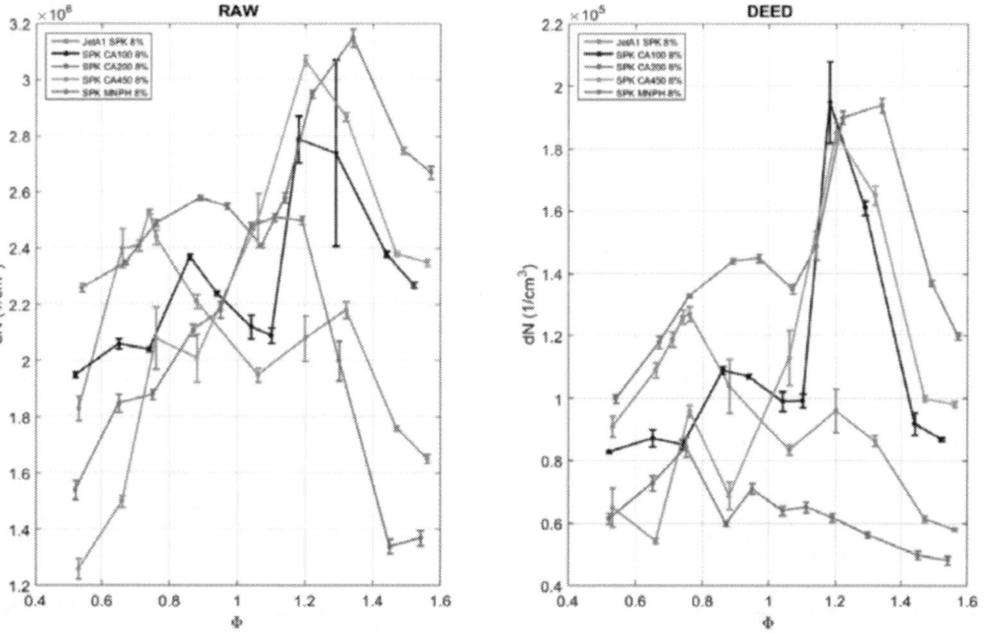

Figure 4: Number concentration of particles measured for various richness at the exit of the CAST fueled with the different compositions of fuel but with a fixed rate of aromatic compounds (8 %). On the left side the raw emission data are exhibited (soot and Volatile Organic Compound) and in the right side are shown the emission after CS-DEED treatment (only soot)

CONCLUSION

In conclusion, understanding contribution of mono and di-aromatic compounds on the emission observed for the combustion of Jet A-1 aircraft fuel is complicated. Moreover, it appears that for rich condition of burning, something more than aromatics play a complementary role. More investigation is needed. To perform that, it will be very helpful to use other characterization techniques and the other different SPK blinding case burned. This is carried out in the EXPLIC program, the whole of the results will be presented elsewhere[4].

REFERENCES

1: Lee, D. S., Pitari, G., Grewe,V.,Gierens, K;, Penner, J.E., Petzold, A., Prather, M.J., Schumann, U., Bais, A;, Berntsen,T., Lachetti,D., Lim, L.L., Sausen R. (2010) Atmos. Environ., 44, 4678-4734.

2: Jing, L. (2003) 7th ETH Conference on Nanoparticle Measurement, ETH Hönggerberg Zürich.

3: Ortega, I., Delhaye D., Jing L., Ledur D., Ferry D., Irimea C., Pirim C., Chazallon B., Focsa C., Ouf F.-X., Salm F., Vancassel X., (2016) European Aerosol Conference EAC, Tours, France

4: Ortega, I., Delhaye D., Ancelle, J., Raepsaet, B., Sicard, M., a journal article on this work is in preparation.

IASH 2017, the 15TH INTERNATIONAL SYMPOSIUM ON STABILITY, HANDLING AND USE OF LIQUID FUELS
Rome, Italy
10-14 September 2017

EVAPORATION AND IGNITION OF ALTERNATIVE JET FUELS

O. Rouzaud[(1)], J. Garraud[(2)], R. Lecourt[(2)], C. Lempereur[(1)], M. Orain[(2)], G. Illac[(1)], M. Sicard[(3)]

(1): ONERA, The French Aerospace Lab, Centre de Toulouse, BP74025, FR-31055 TOULOUSE, France, rouzaud@onera.fr
(2): ONERA, The French Aerospace Lab, Centre du Fauga-Mauzac, FR-31410 Mauzac, France
(3): ONERA, The French Aerospace Lab, Centre de Palaiseau, BP 80100, Palaiseau F-91123, France

ABSTRACT

Within the in-house EXPLIC project, ONERA studied the behavior of several alternative Jet Fuels. A fuel matrix has been built according to two main parameters: the influence of the aromatic compound and the influence of adding single or dual paraffinic cut in a blend. In this paper, we only consider the aromatic influence both in terms of aromatic quantity or aromatic type. In the first case, the aromatic cut composition is kept constant while its mass fraction within a blend may vary. In the second case, the aromatic mass fraction in the blend is kept constant while its composition between mono- and di-aromatic may vary.

Although the project covers the experimental characterization of a large range of fuel properties, we focus on the evaporation and ignition behavior of the fuels. Indeed, ignition is a major concern in aeronautical applications due to safety and operational capabilities reasons. Additionally, amongst the different phenomena participating in ignition process, evaporation is one of the most important.

For evaporation, the experimental characterization rests on a Monodisperse Droplet Generator, using an optical device to determine fuel behavior. Evolution of the droplet surface is a good indicator of the burning rate. The influence of the aromatic quantity is not clear. Conversely, in the aromatic type case, the burning rate decreases as the di-aromatic mass fraction increases. In addition, a surrogate modeling has been proposed based on the evaporation properties of the fuels.

For ignition, the alternative jet fuels have been tested in a specific ignition test rig (MERCATO). The operating conditions correspond to a low pressure and cold gas flow, and the liquid is at ambient temperature. The ignition equivalence ratios of the alternative jet fuels are quite comparable to the Jet A-1 value but no clear trend has been identified relative to the aromatic quantity or the aromatic type.

KEYWORDS : alternative jet fuel, evaporation, ignition, surrogate modeling

INTRODUCTION

The introduction of alternative fuels in the aeronautical market raises many questions about their compatibility with the airplane fuel system or the engine system. One of the most important questions is related to the ignition problem. Ignition is particularly critical in cruise flight or cold start conditions both for safety and operational capabilities reasons. Such a process involves many basic phenomena like liquid atomization, droplet dispersion, turbulent mixing or evaporation.

Within the in-house EXPLIC project [Sicard], we aim at studying the behavior of several alternative fuels for specific topics like thermal oxidation or ignition. The fuel matrix has been built according to two different parameters: aromatic compounds and addition of specific paraffinic cuts to an existing fuel. In this paper, we only focus on the first parameter and address either the influence of the aromatic compound mass fraction, or the influence of the aromatic compound type (mono-aromatic, di-aromatic) on evaporation and ignition problems. A complementary task consists in building surrogate models of the different fuels with respect to evaporation.

EVAPORATION

The present study aims at experimentally characterizing the evaporation of the different fuels, and proposing an associated surrogate model for each tested fuel. Table 1 gathers the different fuels under investigation, detailing the fuel names, associated aromatic mass fraction and the influence under consideration. The acronyms SPK, AC and Mnph stand respectively for Synthetic Paraffinic Kerosene, Aromatic Cut and Methyl-naphthalene. Three different aromatic cuts have been employed. The AC100 blend consists of a mono-aromatic compound, the AC200 blend is made of mono-aromatics (28.5%m) and di-aromatics (71.5%m), and the AC450 is similar to the aromatic composition found in the Jet A-1 used in this work.

The SPK fuel was purchased from SkyNRG. The kerosene Jet A-1 was obtained from the French Army. A fruitful collaboration was concluded with the petroleum company TOTAL Marketing Services which provided us various fuels and hydrocarbon cuts.

Table 1: original fuels

Fuel (%mass)	Aromatic (%mass)	Influence
Jet A-1	20.6	Aromatic content
SPK	2.5	
SPK + AC450	20.6	
AC450	100	
SPK + AC100	20.6	Aromatic type
SPK + AC200	20.6	
SPK + Mnph	20.6	

Experimental apparatus

A Monodisperse Droplet Generator (MDG) is employed to characterize the fuel evaporation at ambient conditions. In the present configuration, the droplet stream is injected vertically upwards, in still air at ambient conditions. The reader will refer to Ashgriz et al. [Ashgriz] for the basic principles of the MDG. To observe a noticeable evaporation, the liquid has to be heated prior to its injection into air and the droplet has to be ignited once formed above the MDG. In most of the cases, the injection temperature is fixed to 343 K. However, for the AC450 blend, such a temperature produced an unstable stream and a sooty flame, thus leading to non-usable results.

Table 2 summarizes the initial conditions for all tested fuels specifying the frequency F of the ceramic plate, the injection temperature Ti of the liquid, the initial diameter Do and velocity Vo of the droplet at the beginning of the measurement region.

The measurement technique is based on an optical system with a CCD camera and a laser diode. The pulse duration, equal to 20 nanoseconds, enables to get still images of the droplets (Figure 1). A telecentric microscopic lens allows to obtain a high spatial resolution and to measure the droplet diameter with an accuracy of 0.5 micrometers. Since the laser is dual-pulse, a droplet appears twice on the same image but at two different locations, providing the droplet velocity. Thus, the spatial evolution of the droplet diameter and velocity of the droplet is obtained by displacing the optical system along the stable part of the droplet stream. Finally, the surface evolution of the droplets versus time can be derived (Figure 2).

Table 2: initial conditions (MDG)

Fuel	Initial conditions			
	F (Hz)	Ti (K)	D_0 (μm)	V_0 (m/s)
Jet A-1	21100	343	127.6	11.7
SPK	21510	343	126.5	11.6
SPK + AC450	20900	343	129.1	12
AC450	19500	313	124.4	10
SPK + AC100	21100	343	127.6	11.7
SPK + AC200	21510	343	126.5	11.6
SPK + Mnph	20900	343	129.1	12

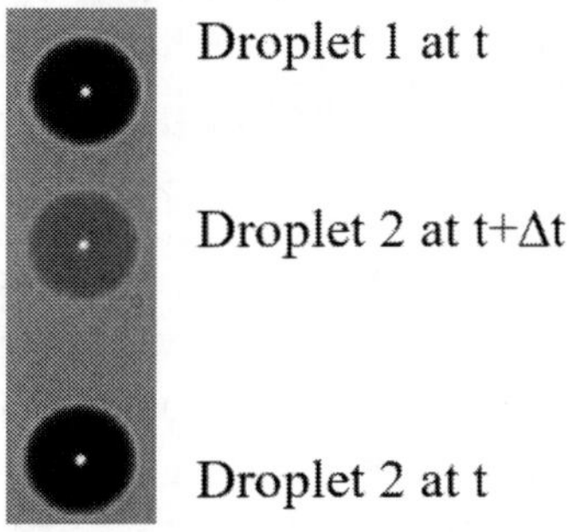

Figure 1 : Image of droplets with dual-pulse illumination

Table 3: representative species

Family	Species
n-Paraffin	n-decane, n-dodecane
iso-Paraffin	dimethyl-heptane, dimethyl-octane, dimethyl-nonane, dimethyl-decane
cyclo-Paraffin	decalin, dimethyl-ethyl-cyclohexane
Mono-Aromatic	dimethyl-ethyl-benzene
Di-Aromatic	naphthalene, methyl-naphthalene

Surrogate modeling

The definition of a surrogate modeling is based on a limited choice of representative species gathered in Table 3. These species have been selected using chemical analysis performed on original fuels. They represent the main families from n-paraffins to di-aromatics.

Furthermore, the chemical analysis also provides the mass fractions for each representative species with respect to each surrogate fuel. Table 4 presents their compositions. Assessment of the surrogate fuels is based on a comparison between the MDG experimental data and numerical results provided by a theoretical approach, described hereafter.

Theoretical model

The original model refers to the isolated droplet theory proposed by Godsave [Godsave] and Spalding [Spalding]. Its use on a droplet stream necessitates accounting for droplet interaction which can modify the vaporization/burning rate or the force exerted on the droplet by the gas [Virepinte]. In the former case, a correction factor η is introduced as:

$$\eta = 1 - 0.42\left[1 + \frac{e^{-0.19(C-5)} - e^{0.6(C-5)}}{e^{-0.19(C-5)} + e^{0.6(C-5)}}\right]$$

It depends on the distance parameter C, which is the ratio between the droplet distance and the droplet diameter. The relationship is valid for a parameter distance C>2 and a Reynolds number between 5 and 55. For a large value of C (C>15), the droplets behave like isolated droplets. The interaction becomes stronger as the parameter distance decreases. Due to the complex composition of the fuels, a multi-component aspect has also been added to the original model.

The next stage consists in generating the thermo-physical properties for each representative species. Numerical simulation may require mass density, specific heat, thermal conductivity and dynamic viscosity for the gaseous and liquid phases. For multiphase flows, surface tension, heat of vaporization and vapor pressure are also necessary. In the case of fuel compounds, several models exist in the literature such as in the reference book of Poling et al. [Poling]. A careful assessment of the properties has been done through comparison with experimental data [NIST] when possible. Once thermo-physical properties of each representative species have been established, mixture rules are employed to define the properties of surrogate fuels [Poling].

Results

At first, we present comparisons between the experimental data, using Jet A-1 as the reference fuel. Figure 2 (left) yields the evolution of the reduced droplet diameter with respect to a reduced time for the Jet A-1, the SPK and the blend SPK + AC450. The stream formed by the AC450 fuel was too unstable to retrieve usable data. The initial diameter Do is measured at the beginning of the measurement region, once spherical droplets are formed. The SPK + AC450 behaves like the Jet A-1 fuel while the evaporation of the pure SPK fuel seems to be less important. Thus, the burning rate of this fuel is reduced compared to the first two. Since the compositions of Jet A-1 and SPK + AC450 blends are similar in terms of aromatic quantity and aromatic type, the different behavior of the pure SPK blend may be due to the aromatic quantity. Next figure (Figure 2 right) presents the same kind of results for the second group of fuels. The comparisons of the three SPK-based blends clearly demonstrate the influence of the aromatic type on the burning rate. It decreases with an increasing proportion of di-aromatic components. On the other hand, the comparison with the Jet A-1 behavior can not exclusively rely on the aromatic compounds and involves the other families.

Figure 3 presents comparisons between experimental curves and their numerical counterpart for Jet A-1, SPK and SPK + AC200 blends. Since the droplet initial temperature is not known in the experiment, a preliminary numerical simulation is performed with an approximate value of temperature. The droplet surface evolution usually shows that its surface increases, a fact attributed to the droplet's heating phenomenon. Since the experimental behaviors demonstrate that the droplets are no longer heated when they enter into the measurement region, we choose, as an initial temperature for our final calculation, the temperature corresponding to the maximum dilatation of the droplet. Comparisons range from quite fair for the SPK and SPK + AC200 blends (two rightmost images) to quite poor (leftmost image) for Jet A-1. Such results provide some insight on the quality of the surrogate model with respect to droplet evaporation. Indeed, the burning rate is associated to the slope of the droplet surface curve. We note that the surrogate model associated, for instance, to the SPK blend will provide a correct value while the model associated to Jet A-1 will underestimate the burning rate as time increases. The discrepancies are due to uncertainties on the initial conditions, approximations of the theoretical model and composition of the surrogate model.

Table 4: surrogate compositions

Fuel	Surrogate composition (%mass)
Jet A-1	dodecane (20) dimethyl-decane (40) decalin (10), dimethyl-ethyl-cycloxexane (10) dimethyl-ethyl-benzene (16.9) naphthalene (3.1)
SPK	decane (15), dodecane (5) dimethyl-decane (68) decalin(5), dimethyl-ethyl-cycloxexane (5) dimethyl-ethyl-benzene (2)
SPK + AC450	decane (12.22), dodecane (4.08) dimethyl-decane (55.42) decalin(4.08), dimethyl-ethyl-cycloxexane (4.08) dimethyl-ethyl-benzene (17.4) naphthalene (2.72)
AC450	dimethyl-ethyl-benzene (85.3) naphthalene (14.7)
SPK + AC100	decane (12.15), dodecane (4.05) dimethyl-decane (55.1) decalin (4.05), dimethyl-ethyl-cycloxexane (4.05) dimethyl-ethyl-benzene (20.6)
SPK + AC200	decane (12.15), dodecane (4.05) dimethyl-decane (55.1) decalin (4.05), dimethyl-ethyl-cycloxexane (4.05) dimethyl-ethyl-benzene (5.87) naphthalene (14.73)
SPK + Mnph	decane (12.15), dodecane (4.05) dimethyl-decane (55.1) decalin (4.05), dimethyl-ethyl-cycloxexane (4.05) methyl-naphtalene (20.6)

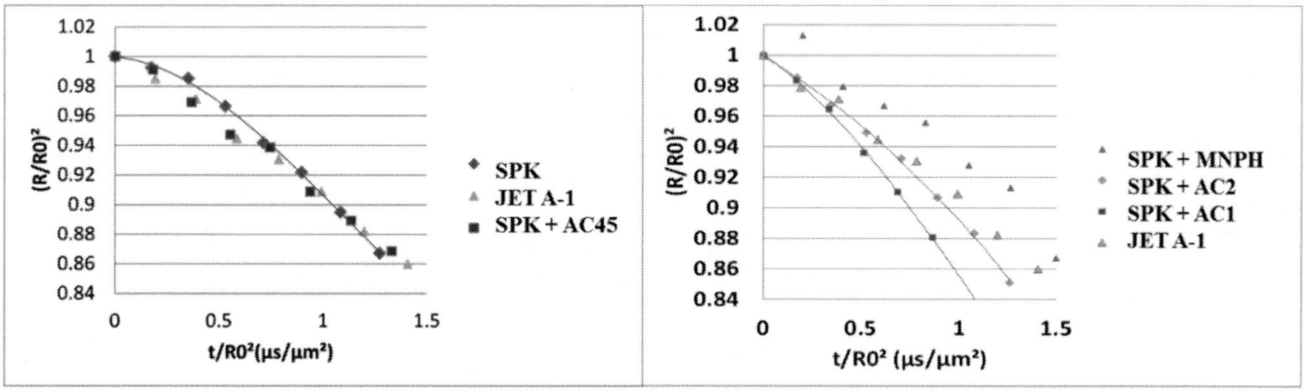

Figure 2 : droplet surface evolution – Aromatic quantity (left) and Aromatic type (right) influences

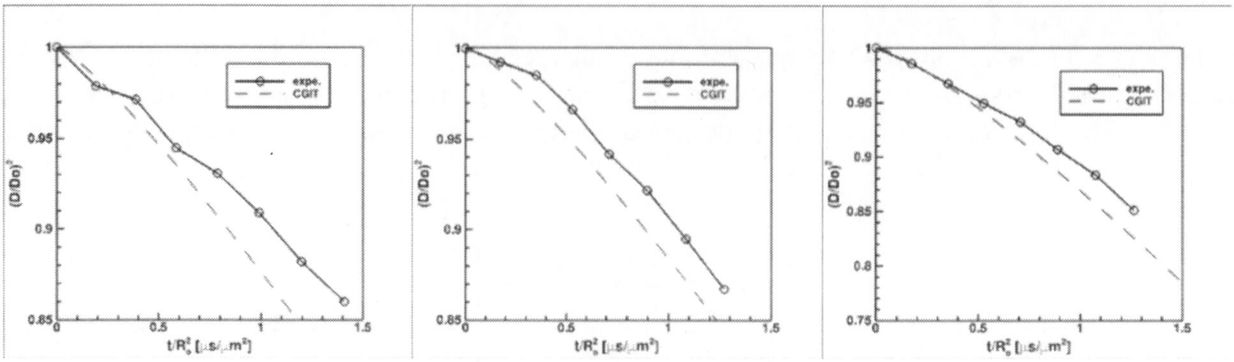

Figure 3 : droplet surface evolution (from left to right: Jet A-1,SPK and SPK + AC200)

IGNITION

The objective of this work is to qualify ignition behavior of several alternative fuels relative to Jet A-1.

Experimental apparatus

The MERCATO test rig is an ONERA experimental bench that enables study of the ignition phenomenon for helicopter (cold start conditions) and airplane (medium altitude relight conditions) applications. Basically, pressure inside the test chamber varies between 0.5 and 1.1 bar, air inlet temperature between 233 and 500 Kelvins and liquid fuel temperature between 233 and 373 Kelvins. The maximal air mass flow rate is equal to 100 g/s.

The test rig (Figure 4 left) consists in a plenum containing the injection system for both air and liquid fuel, the combustor equipped with multiple optical access locations and a supersonic ejector in order to reach sub-atmospheric pressure in the combustor. The injection system (Figure 4 right) is based on an industrial design MAKILA-DLN from Safran-HE with a Delavan atomizer. Air flows into the test chamber through a radial vane swirler while a pressure-swirl atomizer injects the liquid fuel. Ignition is ensured by a spark system equivalent to an onboard helicopter system.

Table 5 describes the thirteen fuels tested during the EXPLIC project. The reference fuel is the Jet A-1 with an aromatic mass fraction of 20.6, another Jet A-1 having also been tested with a lower amount of aromatic and noted "Jet A-1". The meaning of the acronyms is the same as previously.

Table 5: fuel matrix – Ignition part

Aromatic quantity influence	Aromatic type influence
Aromatic %mass = 2.5	
SPK	
Aromatic %mass = 8	
Jet A-1 + SPK	SPK + AC100
SPK + AC450	SPK + AC200
	SPK + Mnph
Aromatics %mass = 16.	
"Jet A-1"	
Aromatic %mass = 20.6	
Jet A-1	SPK + AC100
SPK + AC450	SPK + AC200
	SPK + Mnph
Aromatic %mass = 100	
AC450	

Table 6 presents the operating conditions used within the framework of the EXPLIC project. These conditions are quite critical since we operate at low pressure and cold air temperature. The usual ignition process is divided in four steps. At first, the cold air flow is injected into the test rig. Once the conditions are stabilized, the supersonic ejector is set to reach the desired pressure in the combustor. Liquid fuel is then injected and, after one second, spark ignition proceeds with a frequency of 6 Hz for a duration of 5 seconds. The values of air and liquid flow rates are controlled by flowmeters for each ignition sequence and the values are recorded along with air and liquid fuel temperatures and combustor pressure by means of a custom-made software.

In the present work, the idea was to compare the mean equivalence ratio limit of the fuels. For each fuel, air flow rate is fixed and different liquid flow rates are employed in order to vary the equivalence ratio φ which is calculated based on the common Jet A-1 value (fuel mass flow rate / air mass flow rate = 1/14.725 at stoichiometry). The determination of the minimum equivalence ratio at ignition requires, at least, ten valid trials for each fuel and is defined by the following relationship:

$$\varphi = \frac{1}{2}\left(MAX(\varphi)_{no\ ignition} + MIN(\varphi)_{ignition}\right)$$

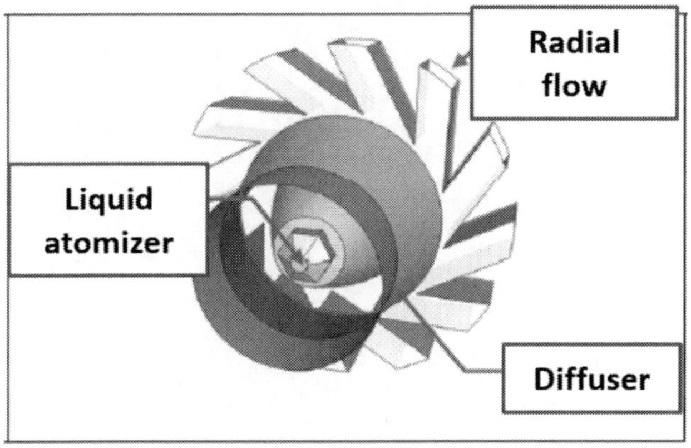

Figure 4 : side view of the test rig (left) sketch of the injection system (right) – MERCATO test rig

Table 6: operating conditions for ignition experiments

Operating Conditions
$\dot{m}_{air} = 0.192\,\frac{kg}{s}$; $P_{chamber} = 0.06$ MPa ; $T_{air} = 233$ K ; $W = \frac{\dot{m}_{air}\sqrt{T_{chamber}}}{P_{chamber}} = 0.488\,\frac{kg.\sqrt{K}}{bar.\ s}$
$T_{liquid} = 290$ K

Results

Figure 5 presents ignition results for the influence of the aromatic content and associated minimum equivalence ratios at ignition are given in Table 7. The different fuels are placed on the horizontal axis as a function of increasing aromatic mass fraction. Equivalence ratio is shown along the vertical axis. Open and solid symbols stand respectively for the unsuccessful and successful ignitions while the red bar represents the minimum equivalence ratio at ignition according to the previous relationship. The reference value is equal to 2.09 (Jet A-1 with 20.6 %m aromatic). At first glance, all the fuels show comparable minimum equivalence ratio at ignition with the best performance for the SPK + AC450 blend and the worst for the AC450 blend. The results also demonstrate that all the alternative fuels, except for AC450, have a slightly better minimum equivalence ratio at ignition than Jet A-1. Concerning the influence of the aromatic content, no clear trend is obvious. For instance, while pure SPK and AC450 have respective minimum equivalence ratios at ignition equal to 2.03 and 2.30, the minimum equivalence ratio at ignition of their mixture is outside of this interval, precluding any linear relation between the minimum equivalence ratio at ignition and the aromatic content in this case. Another example is provided by the comparisons between the blends at 8 %m aromatic and at 20.6 %m. In the former case, the two blends have a similar minimum equivalence ratio at ignition. In the latter case, the discrepancy is much larger.

Table 7 summarizes the results for the influence of the aromatic content and the aromatic type. A similar comparison between the fuels under consideration leads to the same conclusion, no clear trend being associated to the aromatic type.

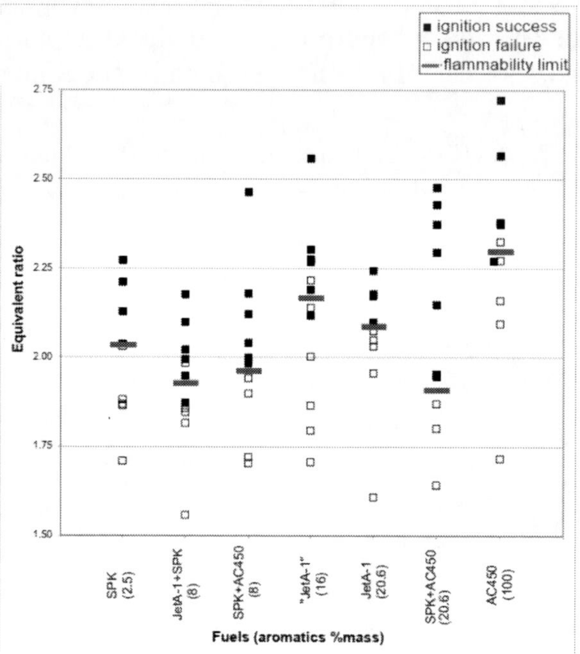

Figure 5 : Minimum equivalence ratio at ignition – Aromatic content influence

Table 7: minimum equivalence ratio at ignition

Fuel	Aromatic (%mass)	Minimum equivalence ratio at ignition	Influence
SPK	2.5	2.03	Aromatic content
Jet A-1 + SPK	8.0	1.93	
SPK + AC450	8.0	1.96	
Jet A-1 v1	16.0	2.17	
Jet A-1 reference	20.6	2.09	
SPK + AC450	20.6	1.91	
AC450	100	2.30	
SPK + AC100	8.0	1.90	Aromatic type
SPK + AC200 (a)	8.0	2.25	
SPK + AC200 (b)	8.0	2.19	
SPK + Mnph	8.0	2.14	
SPK + AC100	20.6	2.08	
SPK + AC200 (a)	20.6	2.19	
SPK + AC200 (b)	20.6	2.10	
SPK + Mnph	20.6	2.21	

CONCLUSION

The paper presents a study on the evaporation and the ignition of some alternative fuels within the framework of the ONERA EXPLIC project. The aim is to experimentally characterize the influence of the aromatic mass fraction and aromatic type of a selection of alternative fuels on these two phenomena. A secondary objective is to assess surrogate models with respect to the evaporation behavior.

On the evaporation side, the experimental characterization stands on a Monodisperse Droplet Generator. Droplets are ignited, and their diameter and velocity measured. At equivalent mass fraction of aromatic, the aromatic type plays a role on the burning rate since it decreases while increasing the di-aromatic mass fraction. On the other hand, the influence of the aromatic mass fraction level shows no clear trend. Assessment of the surrogate modeling provides contrasted comparisons with the experimental results.

On the ignition side, the experimental characterization is performed on a specific test-rig called MERCATO. The minimum equivalence ratios at ignition of the alternative fuels are quite comparable to Jet A-1 value. No clear trend can be observed for the aromatic mass fraction value or the aromatic type.

Future work will consist in investigating the effect of adding specific paraffinic cuts –single or dual- to a specific fuel on the evaporation or the ignition phenomena. Additional work will also be performed on the present study to complement the current results.

Acknowledgments
The authors would like to thank Mrs. J. Methel and V. Duplessis for their help in preparing this paper.

REFERENCES

N. Ashgriz et al., "Handbook of Atomization and Sprays", N. Ashgriz Ed., Springer-Verlag, 2011.

G.E. Godsave, "Studies of the combustion of drops in a fuel spray – the burning of single drops of fuel", Symposium (international) on combustion, 1953.

B.E. Poling, J.M. Prausnitz, J.P. O'Connell, "The properties of gases and liquids", McGraw-Hill,5th Ed., 2004.

M. Sicard, J. Ancelle, S. Guérin, B. Rapsaet, F. Ser, "Impacts of the Jet fuel chemical compositions on O-rings", Proceedings IASH, Roma, 2017.

D. B. Spalding, "*The combustion of liquid fuels*", Symposium (international) on combustion, 1953.

J.-F. Virepinte, Ph.D Thesis, ENSAE, 1999.

USE OF ILLUMINA 16S rRNA NEXT GENERATION SEQUENCING TO INVESTIGATE ANAEROBIC BACTERIAL COMMUNITY COMPOSITION IN ENVIRONMENTAL AND FUEL ASSOCIATED WATERS

Giovanni Cafà[1], Lisa C. Offord[1], Alan G. Buddie[1] and Joan Kelley[2]

[1] CABI, Bakeham Lane, Egham, Surrey, TW20 9TY, UK. g.cafa@cabi.org
[2] Conidia Bioscience Ltd, Bakeham Lane, Egham, Surrey, TW20 9TY, UK. Joan@conidia.com

ABSTRACT

Sulphate-reducing Bacteria (SRB) are anaerobic microorganisms that use sulphate as an electron acceptor, generating H_2S as the product of respiration, leading to acidification and, ultimately, corrosion. The oil industry can be affected by unexpected generation of H_2S by SRB in locations where fuel becomes contaminated with water, e.g. fuel storage tanks, fuel tanks of marine crafts, or legs of oilrigs. Therefore, the detection of SRB-contaminated fuel sources becomes critical, requiring analytical and quantitative assays to detect the microbial content. The aim of the present study was to describe the Illumina 16S rRNA Next Generation Sequencing (NGS) performed on seven samples that were chosen in two different combinations, focusing on two types of observations. The first objective was to investigate the differences in the bacterial community composition (BCC) of the three hypoxic samples derived from: estuarine mud (EST), the silt layer of a shallow pond (POND) and from an old diesel tank from a farm (PHIL4). The second part of the study was concerned with the fuel sample PHIL4, and investigated the effect on the BCC of two SRB-enhancing liquid media: Postgate's B (PGB) and Postgate's C (PGC) at two different incubation times (2 and 9 days). Samples PHIL4 and EST shared a similar BCC, and this differed from that of POND. Overall, POND was the least diverse of the three samples, and mostly consisted of Clostridia (74.5%). Clostridia and Gammaproteobacteria were the most variable taxa observed in PHIL4, EST, and POND. Similarly, Clostridia and Gammaproteobacteria showed large variability in PHIL4 with PGB and PGC. Longer incubation time (9 days vs 2 days) resulted in a decrease of Gammaproteobacteria and an increase of Clostridia. An increase in relative richness of Clostridia was due to the increased development of *Blautia* spp. This trend was seen in POND, and similarly in PHIL4 with PGB after 9 days of incubation.

Keywords: SRB, NGS, Postgate's B (PGB), and Postgate's C (PGB)

1. INTRODUCTION

Sulphate-reducing Bacteria (SRB) are anaerobic microorganisms that use sulphate as an electron acceptor. They can grow heterotrophically and rely on small organic molecules, but can also grow autotrophically. In the latter case, they use H_2 as the electron donor, and CO_2 as the carbon source (Muyzer and Stams, 2008; Ben-Dov et al., 2009). Historically, SRB have been considered to be strict anaerobes. However, several studies have shown that certain species of SRB can tolerate high concentrations of oxygen, and also utilize oxygen as a terminal electron acceptor (Jonkers et al., 2005), thereby indicating considerable adaptability of their metabolism. Such adaptability is not unexpected given the extensive phylogenetic variability of taxa that are considered to comprise this group of organisms. SRB are known to be present in complex consortia of microorganisms involved in anaerobic processes used in municipal and industrial wastewater treatment (Jørgensen, 1982). A major indicator of sulphate reduction is the production of the toxic odorant H_2S, which is also an agent that significantly enhances microbially-mediated corrosion (Muyzer and Stams, 2008; Ben-Dov et al., 2009).

The oil industry can be affected by the long-term generation of H_2S by SRB. The growth of SRB is a particular problem in tanks used to store middle distillate fuels, including kerosene, gas oil, marine and automotive diesels, and heating oils. Their growth may also occur in unleaded gasoline tanks, although it is exceedingly rare in leaded gasoline tanks (Energy Institute, 2015). Therefore, it is of considerable commercial interest to establish whether, or not, microbial contamination by SRB is present in a fuel source. This requires analytical and quantitative tests of potentially contaminated samples, with determination of microbial content (Energy Institute, 2008). Growth of other organisms such as yeasts, moulds and non-SRB prokaryotes has been detected in refinery blend and product storage tanks, terminal storage tanks, distribution facilities such as pipelines, product tankers and barges, loading racks, road and rail tankers, and in tanks at privately owned and retail sites. This is not a new phenomenon, being first recorded in international literature in 1895 (Energy Institute, 2008).

The water phase of any fuel, or fuel component, provides a suitable environment for microbial degradation. The susceptibility of any fuel or fuel component to microbial degradation and spoilage is dependent on its chemical composition, the chemistry of its additives and their concentration. Some components of gasoline may support microbial growth at low concentrations but inhibit or even kill micro-organisms at higher concentrations (Energy Institute, 2008). In all these cases, however, water is an essential component for microbial growth, with some environmental water that can also provide a source of additional soluble nutrients. (Energy Institute, 2008). The H_2S generated by SRB is present in the water phase, and in partitions between oil, water and gas. Its control, therefore, can be predicted, based on the fluid phase ratios, temperature, pressure, pH of the aqueous phase and some other factors such as the type of oil and the average molecular mass of the gas. (Energy Institute, 2015).

The objective of this investigation was to describe the Illumina 16S rRNA Next Generation Sequencing (NGS) analysis performed on selected samples with two different combinations, to focus on two observations. The first was to investigate the differences of the bacterial community composition of the three samples PHIL4, Estuary (EST), and Pond (POND). The second focused on the fuel sample PHIL4, to investigate the effect of the two liquid media: Postgate's B (PGB) and Postgate's C (PGC) at two different incubation times, after 2 (P4B2 and P4C2) and 9 days (P4B9 and P4C9).

2. METHODS

Cells were harvested from Postgate's C liquid medium (PGC) and Postgate's B liquid medium (PGB) (Postgate, 1979). Inocula of the samples were originally subcultured into 100 ml of PGC, and overlaid with 90 ml of sterile mineral oil in 250 ml glass screw-top Erlenmeyer-style flasks. Aliquots of 10 ml of PGB overlaid with 5 ml of sterile mineral oil in a glass universal bottle were used to test SRB activity, which was indicated by the production of a black FeS precipitate.

2.1. Cells harvest from Postgate's liquid media

Cultures of bacterial consortia were harvested via a sterile separating funnel. The funnel was set up in a retort stand, with a sterile funnel placed at the top. Liquid cultures were poured into the separating funnel, and left to stand for 15 minutes. This stage allowed the aqueous and oil layers to separate. The aqueous layer was dispensed into 50 ml sterile centrifuge tubes by opening the tap of the separating funnel and allowing the aqueous layer to fall into the tube, leaving the layer of oil in the separating funnel. The centrifuge tubes containing the aqueous fraction were centrifuged at 1,200 x *g* for 15 minutes at 15°C. Supernatant was discarded without disturbing the pellet. Cells were re-suspended in the remaining supernatant, and spun at 4,500 x *g* for 7 minutes. A wash step was included, where cells were re-suspended in 1 ml of sterile distilled water and spun at 4,500 x *g* for 7 minutes; the supernatant was then removed with a sterile pipette. Aliquots of the obtained pellets of cells were frozen at -20°C for molecular use, or freeze-dried for long term storage. Aliquots preserved at -20°C were used for the DNA extraction.

Additional cells of PHIL4 were harvested from PGB and PGC after 2 (P4B2 and P4C2) and 9 days (P4B9 and P4C9). Cultures were grown in glass universals containing 10 ml of PGB and PGC, respectively, overlaid with mineral oil. Eight aliquots of 1ml from each universal were dispensed into sterile 1.5 ml microcentrifuge tubes, avoiding the sediment in the media. Microcentrifuge tubes were centrifuged at 4,500 x *g* for 7 minutes, and the supernatant removed. The pellet was re-suspended in 1 ml of sterile distilled water, and spun at 4,500 x *g* for 7 minutes; most of the supernatant was then removed with a sterile pipette – leaving just enough to re-suspend the cells. The contents of the eight microcentrifuge tubes from each medium type were then combined in one microcentrifuge tube. These were spun at 4,500 x *g* for 7 minutes and the supernatant was removed. Microcentrifuge tubes were preserved at -20°C until required.

2.2. DNA template preparation

Aliquots of 5 µl of the harvested cells were added to 20 µl of MicroLYSIS®-PLUS solution (MLP; Microzone Ltd., Haywards Heath, UK). DNA was released by subjecting the samples to the following heat profile, as recommended by the manufacturer, in a Hybaid PCR Express thermocycler (Thermo Hybaid, Ashford, UK): 45 min at 65°C, 2 min at 96°C, 4 min at 65°C, 1 min at 96°C, 2 min at 65°C, 30 sec at 96°C. Samples were finally cooled to 20°C and either used immediately for PCR amplification, or else stored at -20°C until required. Cells were obtained from PHIL4, Estuary (EST), Pond (POND), P4B and P4C. The latter two were isolated after 2

(P4B2 and P4C2) and 9 days (P4B9 and P4C9) from inoculation.

2.3. Illumina library generation for MiSeq

The regions V3 and V4 of the 16S rRNA gene were amplified with two modified primers as follows (Klindworth et al., 2013): Forward Primer = 5'-TCGTCGGCAGCGTCAGATGTGTATAAGAGACAGCCTACGGGNGGCWGCAG-3'; Reverse Primer = 5'-GTCTCGTGGGCTCGGAGATGTGTATAAGAGACAGGACTACHVGGGTATCTAATCC -3'. The primer pair contains the Illumina adapter overhang nucleotide sequences to the 5'-end, and was selected according to the Illumina 16S Metagenomics Sequencing Library preparation protocol (http://www.illumina.com/content/dam/illumina-support/documents/documentation/chemistry_documentation/16s/16s-metagenomic-library-prep-guide-15044223-b.pdf). PCR was undertaken in an Eppendorf Mastercycler pro S vapo.protect (Eppendorf, Hamburg, Germany) with a reaction mix containing 5 µl of each primer at the concentration of 1 µM, 2.5 µl of template DNA at the concetration of 5 ng/µl, and 12.5 µl of KAPA HiFi HotStart ReadyMix (KAPA Biosystems Ltd., London, UK) to a final volume of 25 µl with PCR grade water. PCR reactions were preincubated for 3 min at 95°C followed by 25 cycles of 30 sec at 95°C, 30 sec at 55°C, and 30 sec at 72°C. Samples were finally incubated for 5 min at 72°C, followed by chilling to 10°C.
Aliquots of 1 µl of amplification products were assessed for quality with an Agilent Tapestation 4200 (Agilent Technologies Ltd., Cheadle, UK).
A clean-up was carried out on PCR products with AMPure XP beads (Beckman -Coulter, High Wycombe, UK), following the manufacturer's instructions. Purified products were resuspended in 25 µl of 10 mM Tris pH 8.5.

Index PCR was carried out on purified PCR products for the attachment of Illumina sequencing dual indices with the Nextera XT Index Kit (Illumina Ltd., Cambridge, UK). The PCR reaction was undertaken in an Eppendorf Mastercycler pro S vapo.protect with a reaction mix containing 5 µl of template purified PCR product, 5 µl of Nextera XT Index-1, 5 µl of Nextera XT Index-2, 25 µl of KAPA HiFi HotStart ReadyMix, and to a final volume of 50 µl with PCR grade water. The reactions were preincubated for 3 min at 95°C followed by 8 cycles of 30 sec at 95°C, 30 sec at 55°C, and 30 sec at 72°C. Samples were finally incubated for 5 min at 72°C, followed by chilling to 10°C.

A second clean-up was carried out on the above products with AMPure XP beads (Beckman-Coulter), following the manufacturer's instructions. Purified products were resuspended in 50 µl of 10 mM Tris pH 8.5.

Aliquots of 1 µl of index PCR products were assessed for quality with an Agilent Tapestation 4200 (Agilent Technologies), and quantified on a Qubit™ 3.0 Fluorometer (Life technologies, Warrington, UK). Concentration in nM was calculated based on the size of DNA amplicons as determined by Agilent Tapestation 4200. Concentrated libraries were diluted to 4 nM with 10 mM Tris pH 8.5, and aliquots of 5 µl of each library were pooled.

Five µl of pooled DNA libraries were combined with 5 µl of 0.2 N NaOH, and incubated for 5 minutes at room temperature to denature the DNA into single strands. Denatured DNA was then diluted to a final concentration of 15 pM with Illumina Hybridization buffer (Illumina). In addition, this denaturation step was carried out on 15 pM PhiX control (Illumina). Finally, 30 µl of denatured PhiX control library, and 570 µl of denatured amplicon library were combined in a microcentrifuge tube. This ratio determined a 5% PhiX control in the mix. The combined library and PhiX control tube was incubated at 96°C for 2 minutes, mixed, and placed in an icewater bath for 5 minutes. The mix was then loaded on an Illumina V3 cartridge and transferred to the Illumina MiSeq for sequencing.

2.4 Data analysis

After Samples were loaded on the MiSeq, the system provided secondary analysis with the Metagenomics workflow of the MiSeq Reporter software (MSR). The Metagenomics workflow classified organisms based on the V3 and V4 amplicon of the 16S rRNA, based on the Greengenes database (http://greengenes.lbl.gov/). The output of the workflow was a classsification of reads at several taxonomic levels: kingdom, phylum, class, order, family, genus, and species.

3. RESULTS

Illumina NGS of 16S rRNA

Samples from which bacterial communities were investigated are shown in Table 1, and were obtained in two separate runs of the Illumina MiSeq (Table 1). The total number of DNA reads of each of the two runs are shown in Table 2.

3.1 PHIL4 vs EST vs POND

Classes of bacteria that showed differences in relative richness of DNA reads of > 1% across the three samples were: Bacilli, Betaproteobacteria, Clostridia, Deltaproteobacteria, Flavobacteria, Gammaproteobacteria, Sphingobacteriia, Spirochaetes, and Thermotogae (Table 3).
Overall, POND was less diverse than each of PHIL4 and EST, and mostly consisted of Firmicutes (87.8%). Specifically, 74.5% were Clostridia, and 13.3% Bacilli (Figure 1).

Further analyses were carried out on the above classes of bacteria, which showed a difference in relative richness of DNA reads of > 1% among the three samples (data not shown).

Bacilli were more abundant in POND than in PHIL4 or EST. Bacilli ranged from 1.2% in PHIL4, to 9.6% in EST, and 13.2% in POND. In PHIL4, the most abundant of the Bacilli was *Planococcus maritimus* with 2,749 DNA reads (0.3%), in EST *Enterococcus gallinarum* with 24,672 DNA reads (5.2%), and in POND *Bacillus cereus* with 39,609 (5.4%). Betaproteobacteria were the 13.8% of the bacteria in PHIL4, and <1% in EST and POND. In PHIL4, the most abundant genus was *Rhodocyclus* with 86,487 DNA reads (10.5%), of which 6,132 were identified to the species *Rhodocyclus purpureus* (0.7%). Betaproteobacteria detected in EST and POND were all unclassified. Clostridia showed the largest differentiation in the 3 samples: from 6.6% in EST, to 15.7% in PHIL4, and 74.5% in POND (Table 3). In the latter sample, the most abundant species was *Blautia coccoides* with 233,762 DNA reads (32%). Additionally, 168,454 DNA reads were unclassified *Blautia* (23%), making up 55% of *Blautia* out of the 74.5% of Clostridia. In PHIL4,

Blautia were detected with 7,708 DNA reads (<1%), while in EST *Blautia* were not detected. Deltaproteobacteria made up 14.8% of the bacteria in EST, and <1% in PHIL4 and POND. In EST, the most abundant genus was *Desulfomicrobium* with 64,481 DNA reads (13.6%), of which 63,563 were identified to the species *Desulfomicrobium norvegicum* (13.4%). The most abundant Deltaproteobacteria in PHIL4 were *Desulfovibrio oryzae* with 1,922 DNA reads (0.2%), while all of those detected in POND were unclassified. Flavobacteria were higher in EST (5.8%) than PHIL4 (2.8%), with the lowest percentage in POND (<0.1%). The most abundant Flavobacteria in PHIL4 were unclassified *Flavobacterium* with 21,706 DNA reads (2.6%), while in EST were unclassified Flavobacteriaceae with 12,402 DNA reads (2.6%). Gammaproteobacteria showed the second largest differentiation among the three samples: from 54.2% and 52% of bacteria in PHIL4 and EST, respectively, to <1% in POND (Table 3). The three most numerous DNA reads for a species in PHIL4 were *Stenotrophomonas retroflexus* with 84,251 (10.3%), *Pseudomonas azotoformans* with 58,900 (7.2%), and *Pseudomonas marginalis* with 37,750 (4.6). In EST, *Pseudomonas* spp. were the most abundant taxon with 174,847 DNA reads (36.8%). The two species *Thioalkalimicrobium sibiricum* and *Microbulbifer okinawensis* were the first and second most abundant species with 17,308 (3.6%) and 16,833 (3.5%) DNA reads, respectively. The 4,600 DNA reads detected in POND were unclassified. Sphingobacteriia were more abundant in PHIL4 than EST and POND. In PHIL4, The most abundant Sphingobacteriia was *Pedobacter* with 29,902 DNA reads (3.6%), while in EST was *Lewinella antarctica* with 1,547 DNA reads (0.3%). All the 507 DNA reads detected in POND were unclassified. Where noted, Spirochaetes were mostly detected in EST, with only a minor number of DNA reads in PHIL4 and POND. The most abundant taxon in EST was the genus *Sphaerochaeta* with 18,712 DNA reads (3.9%), of which 9,299 were *Sphaerochaeta pleomorpha* (2%). In PHIL4 and POND, 35 DNA reads (<0.01%) and 28 (<0.01%), respectively, were unclassified. Thermotogae were more abundant in POND than PHIL4 and EST. All the detected DNA reads were of the genus *Marinitoga*, with 62,103 (8.5%) in POND, 1,363 (0.2%) in PHIL4, and 522 (0.1%) in EST (data not shown).

3.2. P4B2 vs P4C2 vs P4B9 vs P4C9

The major changes in relative richness of bacteria were seen in PGB, as compared to PGC (Table 4, Figure 2).

PGB vs PGC

Gammaproteobacteria and Clostridia showed major changes in relative richness among the 4 samples (Figure 2). Overall, Gammaproteobacteria were the most abundant class of bacteria, with a maximum of 97.9% of DNA reads in P4C2, and a minimum of 45.4 % P4B9 (Table 4). Also Clostridia showed a large variation in relative richness, with 0.7% of DNA reads in P4B2, and 44.3% in P4B9. Other bacteria changed their relative richness of DNA reads by more than 1%, such as Bacilli, Bacteroidia, and Deltaproteobacteria. Additional minor differences (changes in relative richness of DNA reads between 0.5% and 1%) were shown by Alphaproteobacteria, Sphingobacteriia, and Thermotogae (Table 4).

PGB after 2 and 9 days

Major differences caused by a longer incubation time in PGB were reflected in a decrease in Gammaproteobacteria from 88.3% to 45.4%, which were replaced by Clostridia (0.7% to 44.3%). The major changes in relative richness for Gammaproteobacteria were seen for *Gluconacetobacter* and *Yersinia kristensenii*. The former decreased from 232,731 DNA reads (25.8%) to 6,127 (0.8%), and the latter decreased from 113,101 DNA reads (12.5%) to 4,881 (0.6%). For Clostridia, *Blautia coccoides* increased from 1,563 DNA reads (0.2%), to 48,559 of DNA reads (6.1%). Other Clostridia accumulated after 9 days of incubation, with the most dramatic changes observed for *Clostridium chromoreductans* with 203,851 DNA reads (25.8%), and *Clostridium frigidicarnis* with 7,393 DNA reads (0.9%). Longer incubation in PGB decreased the relative richness of Bacilli (8.7% to 0.8%), and it was mostly caused by *Paenibacillus* spp. An increase in bacteroidia (<1% to 4.1%), was mostly caused by *Parabacteroides goldsteinii* with 27,021 DNA reads (3.4%). Finally, a minor increase was seen of Sphingobacteriia (0.09% to 1%) with *Pedobacter* spp., and of Thermotogae (0.04% to 0.9%) with *Marinitoga* spp. (data not shown).

PGC after 2 and 9 days

Major differences caused by a longer incubation in PGC were shown by a decrease in abundance of Gammaproteobacteria (97.8% to 85.3%), and an increase in Clostridia (0.5% to 4.2%). The decrease in Gammaproteobacteria was mostly caused by *Gluconacetobacter* spp. which decreased from 458,814 DNA reads (37.1%) to 283,847 (23.5%). The increase in Clostridia was caused by *Blautia* spp. with 42,089 (3.5%), of which *Blautia coccoides* had 23,646 DNA reads (2%). Finally, Deltaproteobacteria increased from <1% to 9%, and such change was caused by *Desulfovibrio* spp. Specifically, *Desulfovibrio oryzae* had 38,542 DNA reads (3.2%), *Desulfovibrio vulgaris* 32,295 (2.7%), *Desulfovibrio oxamicus* 5,079 (0.4%), and unclassified *Desulfovibrio* 30,586 (2.5%) (data not shown).

4. DISCUSSION

The two samples PHIL4 and EST showed a similar bacterial community composition (BCC), which was different from the BCC of POND. Overall, POND was the least diverse of the three samples screened, and mostly consisted of Clostridia (74.5%) (Figure 1). A large relative richness of Clostridia was also observed in PHIL4, but only when incubated for 9 days in Postgate's B medium (P4B9) (Figure 2).

Clostridia and Gammaproteobacteria were the most variable taxa among PHIL4, EST, and POND. Clostridia showed the largest differentiation in the three samples, changing their relative richness from 6.6% in EST, to 15.7% in PHIL4, and 74.5% in POND. Gammaproteobacteria showed the second largest differentiation, changing from 54.2% and the 52% in PHIL4 and EST, respectively, to <1% in POND. Similarly, Clostridia and Gammaproteobacteria showed large variability in PHIL4 with PGB and PGC. Longer incubation in the two media (2 days vs 9 days) determined a decrease of Gammaproteobacteria and an increase of Clostridia. However, changes were more distinctive in PGB as compared to PGC. For example, Gammaproteobacteria in P4B9 decreased from 88.3% to 45.4%, and were replaced by Clostridia (0.7% to 44.3%). In PGC, Gammaproteobacteria decreased from 97.8% to 85.3%, and Clostridia increased from 0.5% to 4.2%.

A consistent trend was observed in Clostridia. An increase in their relative richness was caused by overdevelopment of *Blautia* spp. This trend was seen in POND, and similarly in P4B9. In the former, the increase was caused by *Blautia coccoides,* while in the latter, such changes were mostly caused by the combination of *Blautia coccoides* and *Clostridium chromoreductans*. Such results support experimental observations, in which a strong blackening of PGB was observed after 9 days of incubation as opposed to 2 days.

Cloning results from previous studies showed low recovery of "Typical" SRB in the fuel sample PHIL4, for which only a small percentage of Clostridia were detected. Specifically, two sequences were identified as *Clostridium* sp., closely related to *C. xylanolyticum*. Firmicutes such as *Pelosinus fermentans* and *Psychrosinus fermentans*, lactate reducers, were detected with cloning, but not with NGS. "Typical" SRB were considered from the five lineages of Eubacteria: Deltaproteobacteria, Nitrospirae, Clostridia, Thermodesulfobiaceae, and Thermodesulfobacteria, and the two lineages of Archea: Euryarchaeota and Crenarchaeota (Muyzer and Stams, 2008). NGS of the 16S rRNA showed large quantities of "Typical" SRB taxa, particularly in sample POND, where Clostridia made up the 74.5% of the total. Other "Typical" SRB such as Deltaproteobacteria comprised 14.8% of DNA reads in EST, of which the 13.4% were *Desulfomicrobium norvegicum*. In PHIL4, data indicate that Typical" SRB such as Clostridia accumulate on PGB after 9 days of incubation. Previous experiments have shown that PHIL4 and EST were H_2S producers.

Data show that the chosen growth media can alter the bacterial community composition, and that incubation time can drastically change the population components with differences in the bacterial population before and after the black FeS precipitate becomes visible in PGB. This was shown by a large increase of Clostridia in PGB from 0.7% after 2 days to 44.3% after 9 days of incubation. It also supports the idea that SRB can be preferentially selected in Postgate's media.

REFERENCES

Ben-Dov E, Kushmaro A, and Brenner A (2009). Long-term surveillance of sulfate-reducing bacteria in highly saline industrial wastewater evaporation ponds. Saline Systems 5:2

Energy Institute (EI) (2015). Health and safety issues in the oil industry related to sudden or unexpected generation of H_2S by micro-organisms, EI, London URL http://www.energyinst.org.

Energy Institute (EI) (2008). Guidelines for the investigation of the microbial content of petroleum fuels and for the implementation of avoidance and remedial strategies, EI, London URL http://www.energyinst.org.

Jonkers HM, Koh I-O, Behrend P, Muyzer G, de Beer D (2005). Aerobic organic carbon mineralization by sulfate-reducing bacteria in the oxygen-saturated photic zone of a hypersaline microbial mat. Microbial Ecology 49: 291-300.

Jørgensen BB (1982). Mineralization of organic matter in the seabed – the role of sulphate reduction. Nature 296: 643-645.

Klindworth A, Pruesse E, Schweer T, Peplles J, Quast C, Horn M, and Glöckner FO (2013). Evaluation of general 16S ribosomal RNA gene PCR primers for classical and next-ge neration sequencing-based diversity studies. Nucleic Acids Research 41.

Muyzer G, De Wall EC, Uitterlinden AG (1993). Profiling of Complex Microbial Populations by Denaturing Gradient Gel Electrophoresis Analysis of Polymerase Chain Reaction-Amplified Genes Coding for 16S rRNA. Applied and Environmental Microbiology 59: 695-700

Muyzer G, and Stams AJM (2008). The ecology and biotechnology of sulphate-reducing bacteria. Nature Reviews Microbiology 6: 441-454

Postgate JR (1979). The Sulphate reducing bacteria. Cambridge: Cambridge University Press. 151

TABLES

Table 1. Selected samples analysed with illumina 16S rRNA NGS

Sample	16S run 1	16S run 2
PHIL4	*v*	x
Estuary	*v*	x
Pond	*v*	x
P4B2	x	*v*
P4C2	x	*v*
P4B2	x	*v*
P4C9	x	*v*
Positive CTRL	*v*	*v*
Negative CTRL	*v*	*v*

v = performed
x = not performed

Table 2. Number of DNA reads and ratio of DNA reads passing quality filter of samples

Sample	Total Reads	Reads Passing Quality Filtering	% Reads Passing Quality Filtering
PHIL4	1,281,316	821,893	64.1 %
Estuary	732,980	474,775	64.8 %
Pond	1,100,588	730,653	66.4 %
P4B2	958,547	903,250	94.2 %
P4C2	1,311,691	1,236,410	94.3 %
P4B2	839,148	789,802	94.1 %
P4C9	1,267,656	1,207,879	95.3 %

Table 3. Number of sequences of each class of bacteria and relative richness of DNA reads of 16S rRNA gene sequencing of the 3 samples PHIL4, EST, and POND.

Class of bacteria	PHIL4		EST		POND	
	Reads	%	Reads	%	Reads	%
Acidimicrobiia	20	0.002	35	0.007	16	0.002
Acidobacteria	30	0.004	39	0.008	8	0.001
Actinobacteria	2,735	0.333	1,823	0.384	2,695	0.369
Alphaproteobacteria	6,256	0.761	3,128	0.659	2,012	0.275
Anaerolineae	341	0.041	63	0.013	47	0.006
Armatimonadia	1	0.000	ND		1	0.000
Bacilli	9,964	1.212	45,702	9.626	96,819	13.251
Bacteroidia	1,679	0.204	2,317	0.488	5	0.001
Betaproteobacteria	113,250	13.779	579	0.122	710	0.097
Brachyspirae	9	0.001	5	0.001	3	0.000
Brocadiae	43	0.005	16	0.003	31	0.004
Caldisericia	1	0.000	ND		1	0.000
Caldithrixae	56	0.007	17	0.004	22	0.003
Chlamydiia	16	0.002	28	0.006	12	0.002
Chlorobia	43	0.005	8	0.002	2	0.000
Chloroflexi	14	0.002	10	0.002	8	0.001
Chrysiogenetes	1	0.000	ND		ND	
Chthonomonadetes	4	0.000	2	0.000	ND	
Clostridia	128,994	15.695	31,277	6.588	544,033	74.458
Deferribacteres	17	0.002	6	0.001	3	0.000
Deinococci	82	0.010	47	0.010	42	0.006
Deltaproteobacteria	5,773	0.702	70,073	14.759	511	0.070
Epsilonproteobacteria	161	0.020	129	0.027	51	0.007
Erysipelotrichi	5	0.001	139	0.029	2	0.000
Fimbriimonadetes	1	0.000	ND		ND	
Flavobacteriia	22,724	2.765	27,768	5.849	128	0.018
Fusobacteria	67	0.008	21	0.004	7	0.001
Gammaproteobacteria	445,999	54.265	246,644	51.950	4,600	0.630
Gemmatimonadetes	6	0.001	6	0.001	4	0.001
GroupII	20	0.002	9	0.002	4	0.001
Holophagae	16	0.002	1	0.000	17	0.002
Ignavibacteria	5	0.001	1	0.000	ND	
Ktedonobacteria	123	0.015	20	0.004	37	0.005
Leptospirae	14	0.002	52	0.011	6	0.001
Methanomicrobia	1	0.000	ND		ND	
Methylacidiphilae	77	0.009	37	0.008	35	0.005
Mollicutes	150	0.018	313	0.066	57	0.008
Nitriliruptoria	49	0.006	19	0.004	100	0.014
Nitrospira	290	0.035	160	0.034	118	0.016
Nostocophycideae	449	0.055	48	0.010	71	0.010
Opitutae	226	0.027	142	0.030	76	0.010
Oscillatoriophycideae	557	0.068	56	0.012	45	0.006
Pedosphaerae	8	0.001	22	0.005	5	0.001
Planctomycetia	63	0.008	14	0.003	38	0.005
Rubrobacteria	2	0.000	ND		2	0.000
Solibacteres	5	0.001	5	0.001	8	0.001
Spartobacteria	63	0.008	29	0.006	71	0.010
Sphingobacteriia	31,142	3.789	4,121	0.868	507	0.069

Spirochaetes	35	0.004	20,789	4.379	28	0.004
Synechococcophycideae	22	0.003	107	0.023	3	0.000
Synergistia	227	0.028	1,366	0.288	124	0.017
Thermobacula	8	0.001	47	0.010	4	0.001
Thermodesulfobacteria	7	0.001	18	0.004	5	0.001
Thermoleophilia	33	0.004	15	0.003	42	0.006
Thermotogae	1,577	0.192	581	0.122	62,860	8.603
Verrucomicrobiae	49	0.006	19	0.004	39	0.005
Unclassified	48,383	5.887	16,902	3.560	14,578	1.995
∑	**821,893**	**100%**	**474,775**	**100%**	**730,653**	**100%**

* ND = Not Detected

Table 4. Number of sequences of each class of bacteria and relative richness of DNA reads of 16S rRNA gene sequencing of the fuel samples PHIL4 in Postgate's B and Postgate's C after 2 and 9 days incubation.

	P4B2		P4C2		P4B9		P4C9	
Class of bacteria	**Reads**	**%**	**Reads**	**%**	**Reads**	**%**	**Reads**	**%**
Acidimicrobiia	3	0.000	5	0.000	3	0.000	4	0.000
Acidobacteria	87	0.010	90	0.007	45	0.006	75	0.006
Actinobacteria	1,151	0.127	2,016	0.163	703	0.089	1,793	0.148
Alphaproteobacteria	6,467	0.716	3,585	0.290	3,915	0.496	2,670	0.221
Anaerolineae	46	0.005	104	0.008	157	0.020	50	0.004
Bacilli	78,963	8.742	2,123	0.172	6,216	0.787	1,714	0.142
Bacteroidia	596	0.066	7	0.001	32,584	4.126	9	0.001
Betaproteobacteria	985	0.109	865	0.070	866	0.110	628	0.052
Brachyspirae	3	0.000	5	0.000	1	0.000	1	0.000
Brocadiae	28	0.003	52	0.004	11	0.001	21	0.002
Caldithrixae	12	0.001	19	0.002	14	0.002	4	0.000
Chlamydiia	25	0.003	32	0.003	17	0.002	18	0.001
Chlorobia	3	0.000	5	0.000	7	0.001	3	0.000
Chloroflexi	2	0.000	ND		ND		ND	
Chrysiogenetes	ND		1	0.000	ND		1	0.000
Chthonomonadetes	4	0.000	2	0.000	2	0.000	1	0.000
Clostridia	6,768	0.749	6,822	0.552	350,156	44.335	51,736	4.283
Deferribacteres	5	0.001	4	0.000	10	0.001	7	0.001
Deinococci	76	0.008	79	0.006	70	0.009	54	0.004
Deltaproteobacteria	1,398	0.155	778	0.063	2,371	0.300	107,969	8.939
Epsilonproteobacteria	38	0.004	41	0.003	64	0.008	35	0.003
Erysipelotrichi	1	0.000	ND		ND		ND	
Fimbriimonadetes	9	0.001	4	0.000	1	0.000	2	0.000
Flavobacteriia	404	0.045	71	0.006	1,245	0.158	42	0.003
Fusobacteria	11	0.001	9	0.001	3	0.000	5	0.000
Gammaproteobacteria	798,008	88.349	1,209,979	97.862	358,920	45.444	1,030,085	85.280
Gemmatimonadetes	4	0.000	2	0.000	ND		3	0.000
GroupII	17	0.002	14	0.001	8	0.001	5	0.000
Holophagae	4	0.000	6	0.000	3	0.000	5	0.000

Ignavibacteria	4	0.000	9	0.001	19	0.002	2	0.000
Ktedonobacteria	29	0.003	40	0.003	9	0.001	26	0.002
Leptospirae	26	0.003	3	0.000	1	0.000	4	0.000
Methylacidiphilae	17	0.002	20	0.002	5	0.001	14	0.001
Mollicutes	66	0.007	49	0.004	520	0.066	28	0.002
Nitriliruptoria	26	0.003	34	0.003	16	0.002	40	0.003
Nitrospira	101	0.011	76	0.006	168	0.021	88	0.007
Nostocophycideae	46	0.005	42	0.003	92	0.012	19	0.002
Opitutae	65	0.007	94	0.008	44	0.006	50	0.004
Oscillatoriophycideae	39	0.004	26	0.002	279	0.035	9	0.001
Pedosphaerae	36	0.004	75	0.006	20	0.003	54	0.004
Planctomycetia	38	0.004	49	0.004	15	0.002	29	0.002
Solibacteres	5	0.001	8	0.001	3	0.000	7	0.001
Spartobacteria	15	0.002	13	0.001	6	0.001	10	0.001
Sphingobacteriia	779	0.086	407	0.033	7,724	0.978	243	0.020
Spirochaetes	115	0.013	19	0.002	12	0.002	17	0.001
Synechococcophycideae	5	0.001	6	0.000	2	0.000	7	0.001
Synergistia	367	0.041	30	0.002	76	0.010	21	0.002
Thermobacula	4	0.000	1	0.000	1	0.000	2	0.000
Thermodesulfobacteria	4	0.000	4	0.000	9	0.001	1	0.000
Thermoleophilia	26	0.003	26	0.002	9	0.001	25	0.002
Thermoprotei	ND		2	0.000	1	0.000	ND	
Thermotogae	315	0.035	722	0.058	7,142	0.904	2,453	0.203
Verrucomicrobiae	20	0.002	41	0.003	3	0.000	11	0.001
Unclassified	5,984	0.662	7,894	0.638	16,234	2.055	7,779	0.644
∑	**903,250**	**100%**	**1,236,410**	**100%**	**789,802**	**100%**	**1,207,879**	**100%**

* ND = Not Detected

FIGURES

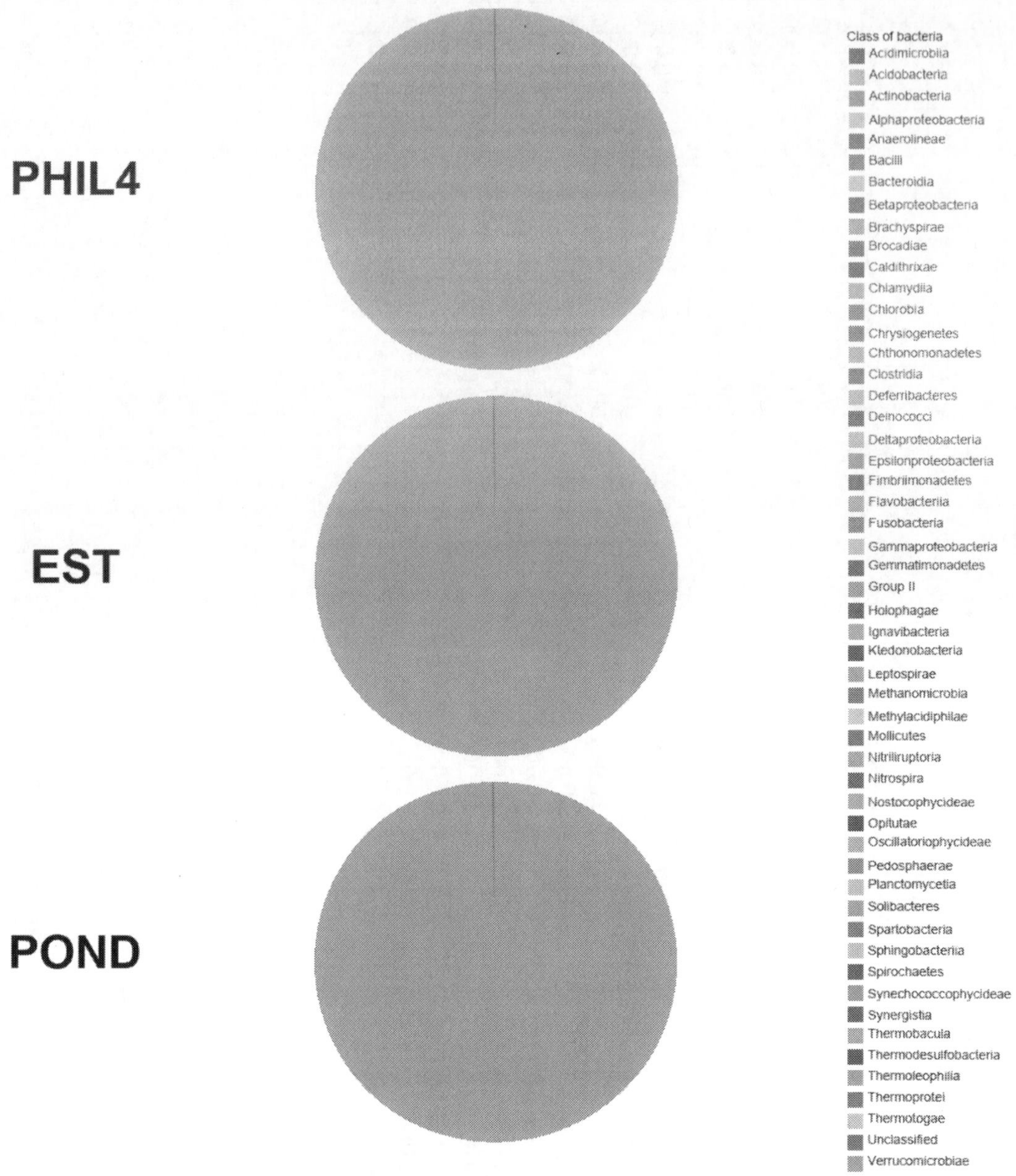

Figure 1. Taxonomic affiliations of 16S rRNA gene sequencing of the 3 samples PHIL4, EST, and POND.

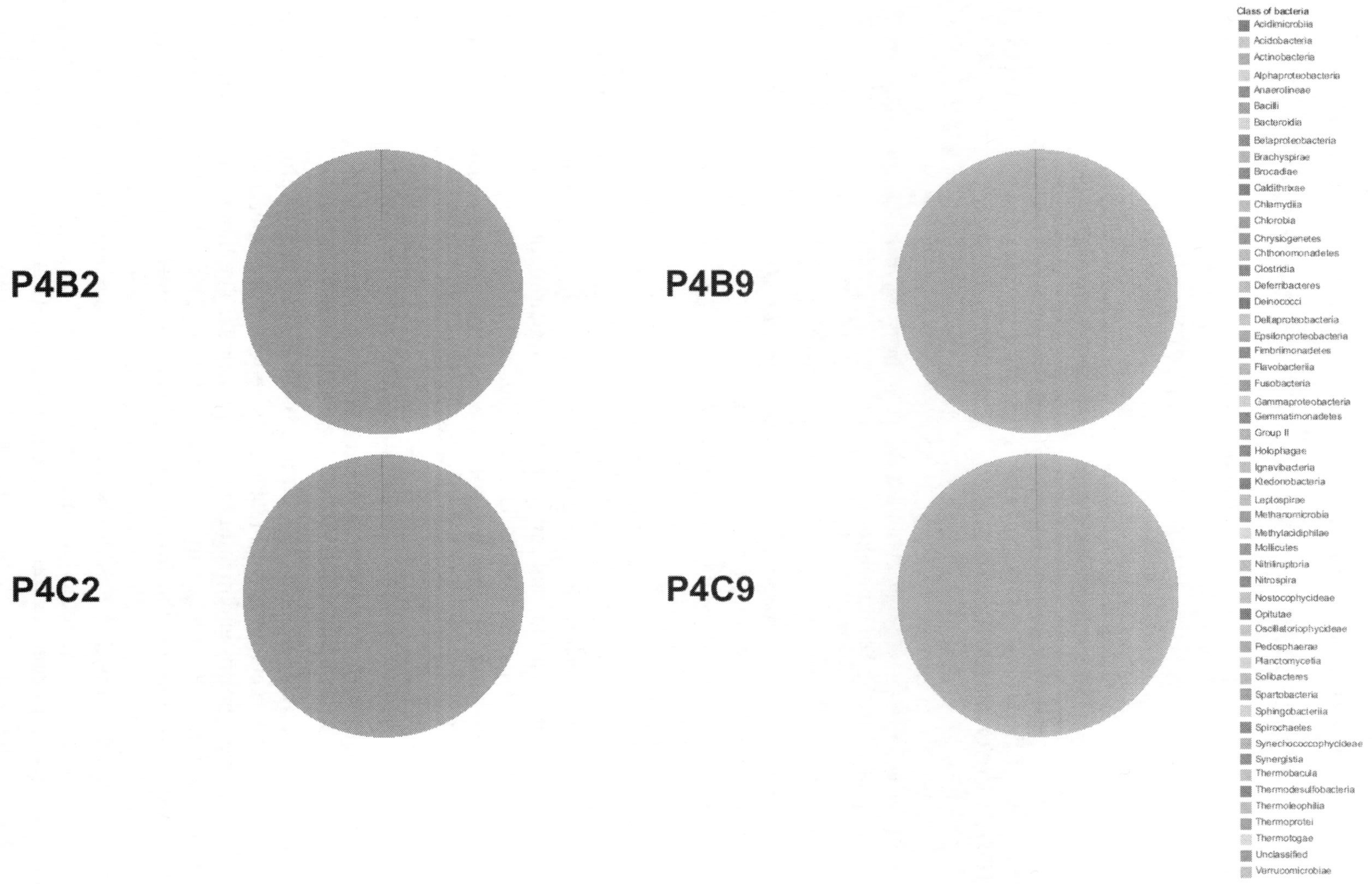

Figure 2. Taxonomic affiliations of 16S rRNA gene sequencing of the fuel samples PHIL4 in Postgate's B and Postgate's C after 2 and 9 days incubation.

IASH 2017, the 15TH INTERNATIONAL CONFERENCE ON

STABILITY, HANDLING AND USE OF LIQUID FUELS

Rome, Italy

10-14 September 2017

ADVANCED MOLECULAR TOOLS FOR THE DETECTION AND MITIGATION OF FUEL BIODETERIORATION

Oscar N. Ruiz[1] and Thusitha S. Gunasekera[2]

[1]*Fuels and Energy Branch, Aerospace Systems Directorate, Air Force Research Laboratory, Wright-Patterson AFB, Ohio 45433, United States. oscar.ruiz@us.af.mil*
[2]*University of Dayton Research Institute, Dayton, Ohio 45469, United States.*

Microbial growth in fuel is a complex issue that negatively affects the fuel's quality and properties, and degrades fuel system performance and lifetime. Thus, it is important to develop effective methods for detection, prevention, and mitigation of biocontamination in fuel. Here, we discuss a three prong approach that focuses on early detection, nanofiltration and advanced biocides to prevent fuel biodeterioration. In the area of detection and risk assessment we present the development and use of quantitative lateral flow assay based on broad-range fluorescent peptide biorecognition elements (BREs) for detection of microbial contamination in fuel. BREs were designed to target conserved cell surface determinants of hydrocarbon-degrading microorganisms. The assay was simple to perform and demonstrated good linearity and high sensitivity. The BREs may also be used to biofunctionalize optical and electrical transducers to develop effective biosensors for fuel applications. The second approach entails the development and use of an advanced biocide that targets adaptive mechanisms required for microbial proliferation in fuel including efflux pumps, porins and membrane phospholipids. The biocide is effective at very low concentrations (≤1ppm in the fuel), can be applied directly to either the fuel or the tank water bottom, and is safe to people and the environment. Finally, to prevent microbes from entering the fuel system, strategies based on filtration present a potentially attractive approach. However, to trap microorganisms filters with sub-micrometer size pores are required and they can dramatically decrease fuel flow rates and increase backpressure. We have developed a graphene oxide (GO)-based filtration media that relies on microbe-nanomaterial surface interaction, not on size exclusion, to filter out microbes in fuel with a 99.97% efficiency while maintaining rapid fuel flow. By combining better detection methods with advanced filtration and biocides we can ensure the future integrity of our fuel supply.

IASH 2017, the 15th INTERNATIONAL SYMPOSIUM ON STABILITY, HANDLING AND USE OF LIQUID FUELS
Rome, Italy
10-14 September 2017

FACTORS AFFECTING THE PRECISION OF FUEL MICROBIOLOGY TEST METHODS

Frederick J. Passman, PhD[1], Joan Kelley, PhD[2], and Pat Whalen, PE[3]

[1]*Biodeterioration Control Associates, Inc., PO Box 3659, Princeton NJ 08543-3659, USA, fredp@biodeterioration-control.com*
[2]*Conidia Bioscience Ltd, Bakeham Lane, Egham, Surrey, TW20 9TY, UK, j.kelley@conidia.com*
[3]*LuminUltra Technologies, Ltd., 520 King Street, Fredericton, NB E3B 6G3, Canada*

ABSTRACT

Fuels and fuel-associated waters present unique challenges to the microbiologist. The heterogeneity of microbial contamination distribution in fuel systems makes it difficult to obtain samples with heavy bioburdens, even when a system is heavily contaminated. Moreover, the historical fuel microbiology literature is based largely on results obtained using test methods that were borrowed from other applications – particularly drinking water testing. This presentation reports the results of a study to compare two non-culture test methods: a lateral flow device and an adenosine triphosphate (ATP) method. The results illustrate the degree of agreement between the two methods. Eight laboratories tested a total of 144 fuel and fuel-associated water samples by each of the two methods. Raw data were converted into *negligible*, *moderate*, and *heavy*, attribute scores (1, 3 and 5, respectively). The two methods' attribute scores agreed for 122 of the 144 samples (83%). This degree of agreement is a very strong indicator that the two methods provided an accurate indication of microbial contamination.

1. INTRODUCTION

The test methods used to determine whether fuels are fit for use are specified by consensus standards organizations such as ASTM International (ASTM), the Energy Institute (EI) and the International Standards Organization (ISO). Within each organization, final approval of a given test method requires the completion of an interlaboratory study (ILS). A full ILS includes the participation of at least six different laboratories. One analyst from each participating laboratory minimally runs duplicate analyses of each sample provided. ASTM Practice D6300 (ASTM, 2017a) stipulates that each specimen be identified by a random number. The intention is to minimize the risk of bias introduced by an analyst who knows that they are testing replicate specimens. This works satisfactorily when the parameter to be tested is both stable and homogeneous. Fuel and fuel-associated water bioburdens are neither stable nor homogeneous.

This paper discusses the impact that strict adherence to ASTM Practice D6300 can have on microbiological test method ILS results.

2. METHODS

2.1 ATP testing

ASTM Method D7687 (ASTM, 2017b) is a protocol for quantifying the concentration of cellular adenosine triphosphate ([cATP]) in fuels and fuel-associated waters. Briefly, either 20 mL of fuel or 5 mL of fuel-associated water is filtered. The filter is then washed with a proprietary solution to remove interferences (Passman et al. 2009) and then air dried. Cells trapped on the filter are then lysed to release cATP. The cATP is diluted in a buffer. The diluted cATP is mixed with a solution of Luciferase enzyme and Luciferin substrate in a 10 mm x 50 mm cuvette. The cuvette is placed into a luminometer which detects the intensity of light emitted by the ATP mediated, Luciferin-Luciferase reaction. The luminometer output is relative light units (RLU). Raw RLU values are compared against an ATP reference standard to obtain [cATP] in pg mL^{-1}. All reagents, consumable supplies and luminometers were provided by LuminUltra Technologies, Ltd, Fredericton, New Brunswick, Canada.

2.2 Lateral flow device immunoassay

ASTM D8070 (ASTM 2016a) has two protocols; differing in the specific volumes used, depending on whether the sample contains water. In both protocols, a volume of sample is transferred to a sample bottle that contains a proprietary extraction solution. The extraction bottle is shaken vigorously and the phases are given time to separate. Four drops of extraction fluid are then placed into each of six, lateral flow device (LFD) wells. The fluid then wicks across the LFD; coming into contact with a proprietary antibody mixture. The six LFD are pairs designed to detect negligible (in fuel: $\leq 150\ \mu g\ l^{-1}$; in water: $\leq 33\ \mu g\ l^{-1}$), moderate (in fuel: $>150\ \mu g\ l^{-1}$ to $<750\ \mu g\ l^{-1}$; in water: $>33\ \mu g\ l^{-1}$ to $<166\ \mu g\ l^{-1}$), and high concentrations (in fuel: $\geq 750\ \mu g\ l^{-1}$; in water: $\geq 166\ \mu g\ l^{-1}$) of bacterial, general fungal and *Hormoconis resinae* antigens (fig 1). Nominal results are converted to attribute scores: negligible = 1; moderate = 3, and heavy = 5. All reagents and supplies for the LFD test were supplied by Conidia Bioscience Ltd., Egham, UK.

2.3 Preliminary interlaboratory studies

For D7687, three analysts each performed triplicate analyses of X fuel and fuel associated water samples per the test matrix shown in Table 1. Preliminary test method repeatability and reproducibility were determined per Practice D6300.

Similarly, for D8070, two analysists tested 12 fuel and fuel-associated water samples (negligible, moderate, and heavy bioburdens in Jet A, ultra-low-sulfur diesel – ULSD, marine diesel and marine diesel bottoms-water). Preliminary reproducibility was determined by comparing the percent agreement between each analyst's duplicate test results, and between observed versus expected results.

2.4 Interlaboratory studies

Two months before the ILS, preliminary microcosms were prepared in duplicate. For each fuel grade, 100 mL of fuel was added to 100 mL of microbiologically bottoms-water samples known to be contaminated (uncharacterized bacterial population, augmented with *Yarrowia lipolytica*, and *Hormoconis resinae* in order to ensure the spectrum of microbial contaminants needed to test each of the three LFD) in 100 mL Bushnell Haas Salts Medium (BHSM: in g L^{-1}: magnesium sulfate, 0.2g; calcium chloride, 0.02; monopotassium phosphate, 1.0; dipotassium phosphate, 1.0; ammonium nitrate, 1.0; ferric chloride, 0.01; pH 7.2). Preliminary microcosms were stored in the dark for two weeks at 25±2° C. After two weeks, microcosm bottoms-water [cATP] was determined. If [cATP] ≥ 1,000 pg L^{-1}, 200 mL was transferred to a secondary microcosm. If [cATP] < 1,000 pg l^{-1}, the microcosm was reinoculated with fuel-grown contaminants and incubated for another two weeks. Secondary microcosms were 10L containers to which the pooled contents of one, primary microcosm; uncontaminated BHSM (to bring aqueous-phase volume to 1,700 mL); and nominally clean fuel (to bring fuel-phase volume to 3,300 mL) were added. Secondary microcosms were incubated in the dark for at least four weeks at 25±2° C. For each fuel grade, secondary microcosms were prepared. Fuel grades used were: diesel (EN590; Shell Oil, Virginia Water, UK), jet A (Rizon Jet, Kent, UK), marine fuel oil (MFO; ExxonMobil, Fawley, UK), and petrol (spark-ignition fuel/gasoline; Shell Oil, Virginia Water, UK). ILS samples were prepared by pooling the duplicate secondary microcosms from each fuel grade. The pooled aqueous and fuel phases were each tested for [cATP]. Nominally clean fuels, to be used to dilute the test fuels to create high, medium, and low bioburdens, were also tested. Fuel phase fluid was siphoned into two fresh containers. Each grade of undiluted microcosm fuel was then dispensed into 24, 0.25 L metal containers (UN Tinplate, Oblong Can, Non-Lacquer, with lined 26mm cap; fig 1). Jet A, EN590 and MFO were each diluted 1:10 to prepare medium bioburden fuels. These, and nominally clean fuels were each dispensed in the same manner as the undiluted microcosm fuels. Similarly, heavy, moderate, and negligible bioburden EN590 bottoms-water samples were prepared from undiluted, microcosm bottoms-water, diluted microcosm bottoms-water and fresh BHSM. Each container was labelled with a randomly generated code. The ILS sample matrix is shown in Table 2.

Table 1. ASTM D7680 Preliminary ILS Test Matrix

Sample Type	Bioburden Levels [a]	Replicates
E-0 gasoline	N, M, H	3
E-10 gasoline	N, M, H	3
ULSD	N, M, H	3
Jet A	N, M, H	3
B-5	N, M, H	3
B-20	N, M, H	3
E-10 Bottom water	N, H	3
ULSD Bottom water	N, H	3
Jet A Bottom-water	N, H	3

Note a) N – negligible: [cATP] ≤ 50 pg mL^{-1}; M – moderate: 50 pg mL^{-1} < [cATP] ≤ 5,000 pg mL^{-1}; H – high: [cATP] > 5,000 pg mL^{-1}.

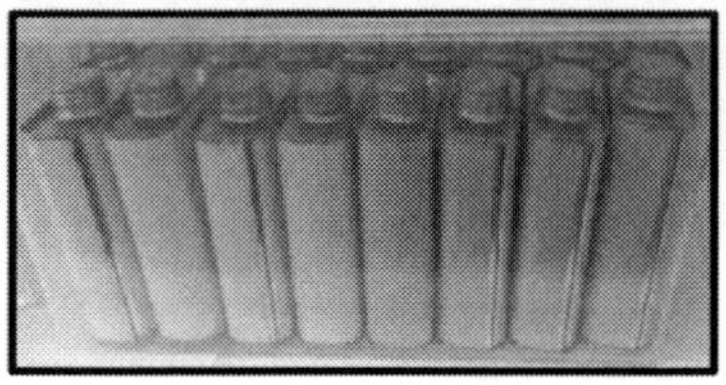

Fig 1. Fuel and bottoms-water sample set for one ILS participant.

The two ILS were performed on successive days at Conidia Bioscience's facilities. Eight independent work stations were set up in Conidia's laboratories. Each participant received a set of 15 samples (fig 1) to test in order of sample identification number. The ASTM D7687 ILS was performed on one day and the ASTM D8070 ILS was performed on the following day. Each participant entered their test data in to an Excel spread sheet which they sent directly to ASTM upon completion.

Table 2. ILS Sample matrix.

	Bottoms-water	Petrol	EN590	Jet	MFO		Sums
LOW	1	1	1	1	1		5
MEDIUM	1	1	1	1	1		5
HIGH	1	1	1	1	1		5
# SAMPLES	3	3	3	3	3	# TESTS	15
REPLICATES	3	3	3	3	3	# TESTS	15
SAMPLES/LAB	9	9	9	9	9	# TESTS	45
# LABS	8	8	8	8	8	# LABS	8
TOTAL # SUBSAMPLES	72	72	72	72	72	TOTAL TESTS	**360**

The sample preparation process was completed 15h to 18h before the first ILS began.

2.5 Data analysis

ASTM D7687 precision was analyzed using ASTM's software for repeatability and reproducibility determination. ASTM D8070 precision was analyzed for within-operator consistency, between-operator consistency, and accuracy; using proprietary software that was developed by ASTM consultant, Alex Lau. The D7687 and D8070 data sets were compared using Excel add-on software. D7687 data were converted to attribute scores that had been developed during the original D7687 validation effort:

- [cATP] <10 pg mL^{-1} in fuel; <100 pg mL^{-1} in fuel-associated water: attribute score = 1.
- [cATP] ≥ 10 pg mL^{-1} to <100 pg mL^{-1} in fuel; ≥ 100 pg mL^{-1} $<1,000$ pg mL^{-1} in fuel-associated water: attribute score = 3.
- [cATP] $>1,000$ pg mL^{-1} in fuel; $>10,000$ pg mL^{-1} in fuel-associated water: attribute score = 5.

Agreement was computed based on the number of samples from which both methods had the same attribute score.

3. Results and Discussion

3.1 Preliminary ILS; ASTM D7687

The preliminary D7687 ILS average [cATP] and standard deviations (*s*) are shown in Table 3. The method's reproducibility (r) was determined to be:

$$r = 0.313\,X \tag{1}$$

Where X = [cATP] in pg mL^{-1}. The overall coefficient of variation (CV%) for the 72-sample data set was 11±2%.

Preliminary ILS; ASTM D8070

Two analysts participated in the ASTM D8070 preliminary ILS. As shown in Table 4, all test results accurately reflected the sample bioburdens. There was 100% within-operator consistency and between-operator consistency. ASTM D8070 is three complementary tests; two LFD each for bacteria, fungal (general) and *H. resinae* antigen detection. All three tests were run. The results were identical for all three data sets. Consequently, only the bacterial LDF results are presented here.

3.2 ASTM D7687 ILS

The complete ASTM D7687 data set is provided in ASTM Research Report D02-1854 (ASTM, 2017c). The ILS repeatability and reproducibility statistics were:

$$Repeatability\ (r) = 6.1387 * X^{0.8559}\ pg/mL \quad (2)$$

$$Intermediate\ Reproducibility\ (R) = 6.7528 * X^{0.8559}\ pg/mL \quad (3)$$

In ASTM jargon, the term *intermediate* is used when participants are not performing the ILS at separate locations. Both r and R are quite large. Comparison of equation (2) with equation (1) suggested that some factor other than inherent test method variability contributed substantially to the variability of the ILS data. It was hypothesized that the most likely source of variability was differences in actual biomass among replicate sample containers. ASTM Practice D6300 dictates that, to minimize the risk of operator bias, each replicate specimen must be drawn from a unique sample container. Although a preliminary test indicated that the ATP-bioburden would remain stable for at least seven days, it is possible that, during the 15h to 18h between sample preparation and testing, the bioburdens in each container changed to varying degrees. Perhaps even more likely was the impact of biomass distribution heterogeneity in the bulk fluids from which ILS samples were prepared. Although the bulk fluids were agitated continuously during the dispensing process, that agitation might not have been sufficient to ensure biomass homogeneity.

Table 3. ASTM D7687 Preliminary ILS Precision Data

Bioburden [a]	Fuel Grade	$[cATP]_{AVG}$	$[cATP]_s$
Low	E0 gasoline	30	3
	E10 gasoline	31	7
	ULSD	37	5
	Jet A	28	4
	B5 ULSD	39	4
	B20 ULSD	40	5
	E10 Bottoms-water	34	5
	ULSD Bottoms-water	40	4
	Jet A Bottoms-water	41	7
Medium	E0 gasoline	1,200	115
	E10 gasoline	1,000	90
	ULSD	1,020	121
	Jet A	1,200	173
	B5 ULSD	1,100	132
	B20 ULSD	1,200	115
High	E0 gasoline	18,000	3,200
	E10 gasoline	18,000	1,500
	ULSD	24,000	1,500
	Jet A	20,000	1,000
	B5 ULSD	22,000	1,150

B20 ULSD	21,000	3,200
E10 Bottoms-water	21,000	1,000
ULSD Bottoms-water	17,000	1,500
Jet A Bottoms-water	18,000	2,100

Note a) L – low: [cATP] ≤ 50 pg mL^{-1}; M – moderate: 50 pg mL^{-1} < [cATP] ≤ 5,000 pg mL^{-1}; H – high: [cATP] > 5,000 pg mL^{-1}.

3.3 ASTM D8070 ILS

The ASTM D8070 ILS outcome was quite similar to that of the D7687 ILS. In stark contrast to the preliminary ILS, the full ILS data set indicated substantial probabilities of within-operator and between-operator disagreement. The full data set was provided in Research Report D02-1844 (ASTM 2016b). Moreover, 14%, 12% and 11% of the participant bacterial, fungal, and *H. resinae* LFD, respectively, disagreed with results obtained by a test method expert (Table 5; % misclassification). It should be noted that the expert observed only one sample. Photos of each LFD panel were taken as one participant ran the test. The photos were subsequently shown to a D8070 expert. All participant LDF observations were compared against the expert's observations to test % misclassification.

Table 4. ASTM D8070 Preliminary ILS Precision Data, Bacteria.

		Analyst and Replicate			
Fuel Grade	Bioburden [a]	1		2	
		1	2	1	2
Jet A1	N	1	1	1	1
	M	3	3	3	3
	H	5	5	5	5
EN590	N	1	1	1	1
	M	3	3	3	3
	H	5	5	5	5
MGO	N	1	1	1	1
	M	3	3	3	3
	H	5	5	5	5
EN590 Bottoms-water	N	1	1	1	1
	M	3	3	3	3
	H	5	5	5	5

Note a) N – negligible: [antigen] ≤150 µg l^{-1} in fuel; ≤33 µg mL^{-1} in the aqueous phase; M – moderate: 150 µg l^{-1} < [antigen] ≤750 µg l^{-1} in fuel; 33 µg mL^{-1} [antigen] ≤166 µg mL^{-1} in aqueous phase; H – heavy: [antigen] >750 µg l^{-1} in fuel; [antigen] >166 µg mL^{-1} in aqueous phase.

Table 5. ASTM D8070 ILS Summary Statistics.

Parameter	Bacteria	Fungi	*H. resinae*
Within-operator disagreement probability	16%	13%	20%
Between-operator disagreement probability	28%	23%	26%
% misclassification between operators	14%	12%	11%

In Table 5, within-operator disagreement reflects the percentage of samples for which the replicate results (each specimen taken from a different container) yielded different LFD results. Again, the considerable differences between the preliminary and full ILS results suggested that variation between duplicate samples was indicative of actual bioburden differences between the replicates.

3.4 ILS Result Comparison

To test the hypothesis that the observed variabilities reflected actual bioburden differences, we compared the D7687 ILS dataset with the D8070 bacterial LFD data set. Table 6 shows the comparison results. Cells with a 1 indicate that both methods had the same attribute scores. Cells with a 0 indicate that the attribute scores differed between methods. A total of 128 samples were compared. The two methods' attribute scores agreed for 84% of the samples. This strong agreement between two methods that detect different indicator molecules supports the hypothesis that each of the methods accurately detected actual variation in bioburdens among replicate samples. Had the observed repeatability and reproducibility variability been due to either method's inherent variability, the number of samples for which the attribute scores did not agree would have been closer to 50% (likelihood of random agreement).

Although the observed agreement between methods was indicative of bioburden variability among replicate samples, it did not differentiate between variability due to population changes that occurred in individual sample containers after the samples had been dispensed and heterogeneity of the bioburden distribution in the bulk fluids from which subsamples had been dispensed. There were data to suggest that the bioburden variability was due to the interaction of these two factors. Comparing the [cATP] in the bulk fluids before dispensing them into individual sample containers with the $[cATP]_{AVG}$ from the ILS samples, we observed that the direction and magnitude of $[cATP]_{AVG}$ varied with sample type. The percent change ranged from a 91% decrease (EN590-M) to a 940% increase (MFO-M). The $[cATP]_{AVG}$ increased in all but two of the sample types. Given the data variability, and the absence of replicate tests on the predispensed samples, it is impossible to assess how many of these $[cATP]_{AVG}$ changes were statistically significant. However, the strong agreement between the D7687 and D8070 data sets (D8070 was run the day after D7687) indicated that the bioburdens did not change substantially during the 24h period between the start of the two successive ILS. The variability of the predispensed Jet A [cATP] results (500 ± 320 pg mL-1 in Jet A-M and 600 ± 330 in Jet A-H) suggest that sample heterogeneity was substantial in the bulk fluids before they were dispensed into individual samples containers.

Table 6. ASTM D7687 and D8070, Attribute Data Results Comparison; [cATP] v. Bacterial LDF.

Lab	REP	EN590-N[a]	EN590-H	Jet A-N	Jet A-H	MFO-N	MFO-H	BHSM-N	Water-H	Totals
1	1	1	1	1	1	0	0	1	1	6
	2	1	1	1	0	1	1	1	1	7
2	1	1	1	1	1	1	1	1	1	8
	2	1	1	1	1	1	1	1	1	8
3	1	1	1	1	0	1	0	1	1	6
	2	1	1	1	1	0	1	1	1	7
4	1	1	1	1	1	1	1	0	1	7
	2	1	1	0	1	0	1	0	1	5
5	1	1	1	0	1	1	1	1	1	7
	2	1	1	0	1	0	1	1	1	6
6	1	1	1	1	0	1	1	1	1	7
	2	1	1	1	0	1	1	1	1	7
7	1	1	1	1	1	1	1	1	1	8
	2	1	1	1	1	0	1	1	1	7
8	1	1	0	1	0	1	1	1	1	6
	2	1	1	0	0	1	1	1	1	6
n = 128	Totals	16	15	12	10	11	14	14	16	108

Note: a) Suffixes: N – negligible contamination (unchallenged), M – moderately contaminated; H – heavily contaminated.

4. CONCLUSIONS

Both ASTM D7687 and D8070 had years of field use before the two ILS were performed. When used to monitor [cATP] in fuel systems, ASTM D7687 precision is consistently CV% ≤ 20. Similarly, replicate D8070 tests typically yield identical results. The historically observed precision of both methods was consistent with their respective preliminary ILS precision statements. The magnitude of variability observed from the full ILS data sets was unexpected. There were insufficient pre-dispensed, bulk fluid data to support analysis of variance to determine whether heterogeneous bioburden distribution, post-dispensing proliferation or the interaction of these two phenomena contributed to the observed variability. Preliminary assessment of ATP-bioburden changes was inconclusive. Moreover, bioburdens did not appear to change between the time the D7687 ILS was run and the following day when the D8070 ILS was run. There might be an interaction effect, but bioburden heterogeneity seems to have had an overall greater impact than either sample aging or any interaction effect.

Although other fuel properties for which ASTM standard test methods have been developed are perishable, microbial bioburdens are unique. First, microbes tend to occur in masses (flocs) that can vary substantially in the number of cells. Second, the microbial population in a given microcosm can proliferate, remain in dynamic equilibrium or die off. Recognizing the unpredictable population dynamics in fuel system samples, ASTM Practice D7847 (ASTM 2017d) prescribes that microbiological tests be initiated within 4h (unrefrigerated) to 24h after sample collection.

Table 7. $[cATP]_{AVG}$ Changes between Dispensing and ASTM D7687 ILS.

Sample	[cATP] pg mL^{-1}				Change	% Change
	Before Dispensing		ILS Results			
	AVG	*s*	AVG	*s*		
EN590-N[a]	1	N.R.[b]	2	2	1	100
EN590-M	3,300	N.R.	300	420	-3000	-91
EN590-H	81,000	N.R.	80,000	53,000	-1000	-1
EN590-AQ	100,000	N.R.	180,000	81,000	80000	80
EN590-AQ (10%)	13,000	N.R.	60,000	26,000	47000	362
MFO-N	1	N.R.	2	4	1	100
MFO-M	385	N.R.	4,000	3,700	3615	939
MFO-H	17,000	N.R.	20,000	11,000	3000	18
JET A - N	1	N.R.	4	7	3	300
JET-A - M	500	320	3,000	2,900	2500	500
JET-A - H	600	330	4,000	4,600	4000	667

Notes:

a) Suffixes: N – negligible contamination (unchallenged), M – moderately contaminated; H – heavily contaminated.

b) N.R.: No replicate tests were performed.

It is most likely that a variance from ASTM D6300 would substantially minimize the contribution of bioburden heterogeneity. As noted above, any ILS that is not performed at different facilities yields an intermediate precision statement. If all participants in a fuel or fuel-associated water microbiology test method ILS drew specimens from a single, well-mixed vessel, the ILS results would be more likely to isolate test method variability from other sources of variation.

ACKNOWLEDGMENTS

The two ILS reported in this study were supported by generous funding from ASTM International, Conidia Biosciences Ltd, and LuminUltra Technologies, Ltd. The ILS would not have been possible without the tremendous preparatory and logistics support provided by CABI staff members: Anthony Kermode, Sharon Lawrence, Emma Thompson, and Gina Watling. The authors also thank the ILS participants: David Armitage, Dr. Christine Gaylarde, Stephen Hornsby, David Humphreys, Andriea Jorge, Alexander McFarlane, Ryan Morris, Neil Reynolds, Soren Sabroe, Jean-Yves Soulard, and Patrick Taylor.

REFERENCES

ASTM D8070-16, 2016a. Standard Test Method for Screening of Fuels and Fuel Associated Aqueous Specimens for Microbial Contamination by Lateral Flow Immunoassay, ASTM International, West Conshohocken, PA, 2016, www.astm.org, DOI: 10.1520/D8070-16.

ASTM, 2016b. Research Report: D02-1844. Interlaboratory Study to Establish Precision Statements for ASTM D8070-16, Test Method For Screening of Fuels and Fuel Associated Aqueous Specimens for Microbial Contamination by Lateral Flow Immunoassay, ASTM International, West Conshohocken, PA, 2011, www.astm.org, 23 pp.

ASTM D6300-17, 2017a. Standard Practice for Determination of Precision and Bias Data for Use in Test Methods for Petroleum Products and Lubricants, ASTM International, West Conshohocken, PA, 2016, www.astm.org, DOI: 10.1520/D6300-17.

ASTM D7687-17, 2017b. Standard Test Method for Measurement of Cellular Adenosine Triphosphate in Fuel, Fuel/Water Mixtures, and Fuel-Associated Water with Sample Concentration by Filtration, ASTM International, West Conshohocken, PA, 2011, www.astm.org, DOI: 10.1520/D7687-17.

ASTM, 2017c. Research Report D02-1854. Interlaboratory Study to Establish Precision Statements for ASTM D7687-17, Test Method for Measurement of Cellular Adenosine Triphosphate in Fuel, Fuel/Water Mixtures, and Fuel-Associated Water with Sample Concentration by Filtration, ASTM International, West Conshohocken, PA, 2011, www.astm.org, 64 pp.

ASTM D7847-17, 2017d. Standard Guide for Interlaboratory Studies for Microbiological Test Methods, ASTM International, West Conshohocken, PA, 2012, www.astm.org, DOI: 10.1520/D7847-17.

Passman, F. J., Egger, G. II, Hallahan, S., Skinner, B., and DeSchepper, M., 2009. Real-time Testing of Bioburdens in Metalworking Fluids Using Adenosine Triphosphate. Tribol. Trans. 52(6): 288-292.

IASH 2017, the 15TH INTERNATIONAL CONFERENCE ON STABILITY, HANDLING AND USE OF LIQUID FUELS
Rome, Italy
10-14 September 2017

THE RELATIONSHIP BETWEEN MICROBIOLIGICAL CONTAMINATION IN WATER PHASE AND FUEL PHASE IN JET FUEL SYSTEMS

Gareth J. Williams & Graham C. Hill

ECHA Microbiology Ltd., Unit 22&23, Willowbrook Technology Park, Llandogo Road, Cardiff, CF3 0EF, WALES, UK ghill@echamicrobiology.co.uk

ABSTRACT

Microbial growth may occur wherever any water accumulates in aviation fuel tanks and systems and it can lead to operational problems. Because it is difficult to keep aircraft fuel tanks completely free of water, IATA recommends microbiological monitoring to ensure systems remain free of contamination. Similarly, guidance from organisations such as JIG and API relating to aviation fuel supply and distribution recommend some extent of microbiological monitoring. These guidance documents generally recommend use of IATA-recommended microbial test kits for detection of microbial contamination in fuel samples. The three most widely used IATA recommended kits are also IP and or ASTM standard methods with designations IP 613/14 / ASTM D7978-14, ASTM D7463-16 and ASTM D8070-16. Each of the kits detects and reports microbial growth in a different way and varies in its approach to testing water phase if present in a sample, for example either testing water and fuel phase combined or testing each phase independently. IATA guidelines provide contamination limit values for each method, for fuel and water phase tests, which are indicative of negligible, moderate or heavy contamination. IATA recommend that every effort is made to recover and test water in tank sump drain samples as water phase samples will provide the best indication of any microbial growth present. JIG, however, recommends that fuel phase samples are tested and that water is first flushed away before taking samples for microbial analysis. This paper reports on work using laboratory simulants and field samples to investigate the correlation between detectable microbial contamination in fuel and water phases. It aims to provide insight into the basis for establishing contamination limit guidelines for fuel phase and water phase samples drawn from fuel tanks.

KEYWORDS

Microbial growth, microbial contamination, ASTM D6974, IP 385, IP 613, ASTM D7978, ASTM D7463, ASTM D8070, water content, settling.

INTRODUCTION

The routine microbiological monitoring of aviation fuel is an important quality assurance and problem avoidance measure for aircraft fuel tanks[1] and aviation fuel supply facilities[2-4]. Microbiological tests are conducted on samples taken from fuel tank or filter vessel drains at a frequency appropriate to the perceived or established risk and, depending on circumstances, these samples may contain fuel and/or water.

Active microbial growth only occurs in pockets of water or in condensate films adhered to fuel tank surfaces. Whilst there will generally be far more microbial cells and microbial biomass in water phase within fuel tanks and systems, microbial cells and spores from the active growth will become dispersed in fuel phase with agitation. For routine monitoring of aircraft fuel tanks, IATA guidelines recommend that 1 litre samples from each drain valve are drawn and that water should be tested if it is present[1]. In this context, water will primarily be derived from condensation and therefore have very low microbial content unless active microbial proliferation is occurring. Microbial tests of water phase in drain sump samples will provide the best indication that active growth is occurring. Ideally for most reliable indication, these samples will be of free water or fuel containing suspended free water.

Sampling recommendations made by the Joint Inspection Group (JIG) for aviation fuel supply facilities[4] differ from the IATA guidelines in that they recommend that samples should be taken from tank and filter vessel drain points after flushing away any visible free water or hazy product and that the fuel phase should be tested. This is because water in aviation fuel supply will primarily be derived from ingress of environmental water during fuel distribution and therefore may have higher microbial content without necessarily indicating significant contamination of fuel. Additionally, because aviation fuel tanks are sumped frequently to remove water, it is often not possible to recover free water in drain samples from storage tanks. Testing fuel phase samples therefore enables more consistency in the sampling procedure and the ability to directly compare results obtained at different time points for trend analysis. Therefore, for reliable indication of tank contamination, the microbiological test used should be able to detect contamination in both water phase samples and fuel phase samples which contain no free water. However, it must be recognised that any sample, fuel or water, will have limitations in its ability to truly reflect all microbial activity within a tank and notably that occurring in biofilms on internal tank surfaces.

Most industry guidelines advise the use of IATA recommended microbiological methods for testing fuel. ASTM D6974 / IP 385 has historically been the laboratory reference method and uses the conventional microbiological technique of culture of Colony Forming Units (CFU). Whilst modern molecular techniques have demonstrated that, in many environments, culture techniques only detect a small percentage of all viable microbial species present, they are still widely recognised as reasonably reliable indicators of fuel tank contamination status. White et al.[5] used the laboratory reference culture method, ASTM D6974 / IP 385, and two culture independent molecular techniques, DGGE and 16S rRNA V6 gene pyrosequencing, to evaluate bacterial contamination in a range of fuel samples. The study showed that at the phylum and class levels there was concordance between the cultivation-dependent and -independent approaches. In most fuel system scenarios, there will always be some culturable microorganisms present in contaminated tanks and these will generally reflect the species of most operational significance so the numbers recovered by culture will generally provide good indication of systems status. The reliability of culture methods is particularly dependent on the ability of the culture medium utilised to provide a nutrient profile which meets the growth requirements of any significant microbial contaminants present. The ASTM D7978 / IP 613 thixotropic gel culture method is a field or laboratory culture method which employs the fuel sample itself to provide a significant part of the nutrient requirement of the microbial contaminants it is intended to detect.

The principal disadvantage of culture methods is that they take several days to produce results. Where operational decisions are made on the consideration of results of two or three samples, or from trend analysis data, the speed of the test result is not time critical. However, in some circumstances, for example where decisions need to be made about the microbiological quality of fuel on an aircraft prior to defueling it, a rapid test result is desirable. Consequently, a number of rapid methods have been developed, two of which are IATA approved. The ASTM D8070 immunoassay utilises Lateral Flow Devices (LFD) for detection of microbial antigens, providing semi-quantitative assessment of the presence of three categories of microbial antigens (defined as bacteria, *Hormoconis resinae* and other fungi); if detected, each antigen category is assessed as being present at high or low level, pre-defined by the sensitivity of the LFD's. To test fuel phase samples, microbial antigens have first to be recovered from the fuel by shaking it with an aqueous extractant solution. If a relatively large amount of water is present in the sample this can be tested by direct addition to the aqueous extractant without the need to extract antigen from fuel. The ASTM D7463 method detects Adenosine Tri-Phosphate (ATP) using a luminometer to quantify the amount of light (Relative Light Units, RLU)) emitted when any ATP in the sample reacts with a luciferase enzyme. This method also uses a process of shaking with an aqueous extractant to recover ATP from fuel phase samples. Unlike the other methods, the same technique and reporting of RLU is used regardless of whether water is present and regardless of the proportion of water to fuel. This is because, in theory, utilising the extraction technique prescribed, results will not be dependent on the proportion of water to fuel, if the assumption is made that in typical samples there is 1000 times more ATP in water phase compared to fuel phase.

It must be emphasised that the methods differ fundamentally in the way they detect microbial contamination and direct comparison of results of the same sample by different methods is not always possible. For example, whereas culture methods will detect both culturable viable microbial cells and their dormant spores (which tend to be more readily dispersed in fuel due their hydrophobicity), tests for ATP will be dependent on the activity state of the microbes present and will generally not detect spores (which contain little or no ATP). Although due to their small size, spores have little direct spoilage significance, because spores are hydrophobic and disperse readily in fuel, their presence may be indicative that active microbial growth is occurring within the system at a point distant from the point sampled.

The principles of the tests methods and the way results are expressed are described in Table 1.

Table 1: IATA recommended methods

Test Method	Principle of Test Method	Expression of Result
IP 385 / ASTM D6974	Recovery of bacteria and fungal CFU from fuel (1 to 100 mL) on 0.45 μm membrane filter. Filter is washed and placed on solid growth media and incubated.	Colony Forming Units (CFU) per Litre of Fuel
IP 613 / ASTM D7978	Direct addition of fuel (0.5 mL) OR water (0.01 or 0.1 mL) to thixotropic growth media which is then shaken and incubated.	Colony Forming Units (CFU) per Litre of Fuel OR per mL of Water
ASTM D7463	Adenosine Tri-Phosphate (ATP) recovered by shaking 1 mL aqueous extractant (Capture solution) with 1 L sample. Capture solution is reacted with luciferase in "pen" and light emitted is measured in a luminometer.	Relative Light Units (RLU) in 1 Litre Sample (fuel, fuel-water mix or water only)
ASTM D8070	Microbial antigens are recovered by shaking 3 mL aqueous extractant with 150 mL fuel or fuel/water mix sample or 15 mL water sample. Extractant is added to Lateral Flow Device (LFD) and visible lines develop when antigens are below specified levels.	Level of antigens (High / Low) of bacteria, *Hormoconis resinae* and Other Fungi in fuel, fuel-water mix or water only

Both JIG[4] and IATA[1] recommend that preventative and remedial actions are instigated according to the interpretation of test results against warning (moderate) and action (heavy) contamination limits. The IATA guidance on contamination limits[1] are shown in Table 2. These limits are specifically applicable to drain samples from aircraft fuel tanks and represent a very high degree of microbiological cleanliness. It should be noted that heavy contamination indicates that significant microbial growth is occurring which ultimately, if unchecked, could have operational implications. However, heavy contamination does not necessarily mean that contamination is presenting an imminent risk to aircraft operations nor that fuel is unfit for use.

The contamination limits provided in the JIG Technical Information Document[4] are shown in Table 3. The JIG Technical Information Document uses the terms "Action Level" and "Warning Level" in defining limits which are applicable to bottom or drain samples from product tanks, import tanks and filter water separators. For the conventional culture methods, ASTM D6974 / IP 385 and ASTM D7978 / IP 613, the JIG limits are less stringent than the IATA limits for aircraft tanks. This reflects the fact that fuel in airport or supply and distribution facilities will pass through several stages of filtration before it is uplifted onto aircraft. Also, any water in supply and distribution facilities could be derived from environmental sources and thus have an inherently higher microbial bioburden. Adherence to the JIG limits is intended to ensure microbe levels at the point of delivery to aircraft are significantly lower than the IATA recommended maximum levels for aircraft fuel tanks. Limits for microbial antigen detected using the ASTM D8070 are pre-set by the test configuration, and therefore contamination limits quoted for microbial antigen are the same for all applications (i.e. both IATA limits for aircraft fuel tanks and JIG limits for samples from airport tanks and supply and distribution). The limits recommended by the manufacturers of ASTM D7463 for detection of ATP are also the same for both aviation fuel supply and aircraft tanks.

Table 2: IATA microbiological contamination limits[1]

Test Method	Sample type	Normal (Negligible) Level Contamination	Moderate Level Contamination	Heavy Level Contamination
IP 385 / ASTM D6974	Fuel	<4000 CFU/Litre	4000 – 20,000 CFU/Litre	>20,000 CFU/Litre
IP 613 / ASTM D7978	Water	<1000 CFU/mL	1000 – 10,000 CFU/mL	>10,000 CFU/mL
ASTM D7463	Fuel and/or Water	<1000 RLU	1000 – 5000 RLU	>5000 RLU
ASTM D8070	Fuel	Negligible (<150 µg/L)	Low (150 - 750 µg/L)	High (>750 µg/L)
	Water	Negligible (<33 µg/mL)	Low (33-166 µg/mL)	High (>166 µg/mL)

Table 3: JIG microbiological contamination limits[4]

Test Method	Sample type	Normal Level Contamination	Warning Level Contamination	Action Level Contamination
IP 613 / ASTM D7978	Fuel	<10,000 CFU/Litre	10,000 – 100,000 CFU/Litre	>100,000 CFU/Litre
	Water	<100,000 CFU/mL	100,000 – 1,000,000 CFU/mL	>1,000,000 CFU/mL
ASTM D7463	Fuel and/or Water	<1000 RLU	1000 – 5000 RLU	>5000 RLU
ASTM D8070	Fuel	Negligible (<150 µg/L)	Low (150 - 750 µg/L)	High (>750 µg/L)
	Water	Negligible (<33 µg/mL)	Low (33-166 µg/mL)	High (>166 µg/mL)

In setting these contamination limits, assumptions are made about the correlation between fuel and water phase contamination levels and the manner in which microbial contamination disperses in fuel;

- There is a reasonable assumption that there will be far more microorganisms in water phase than in fuel phase.
- It is assumed there is consistent correlation between the detectable level of contamination in water phase and in associated fuel phase (see Table 4 below). In fact this is likely to be highly dependent on the degree of agitation of both the sample and the contents of the system sampled immediately prior to sampling and also upon the proximity of fuel to active microbial growth in water. It will also be dependent on the source of water and whether it will have an inherent microbial bioburden (e.g. condensate water would be expected to be virtually sterile whereas even clean environmental water may have in excess of 1000 microbial CFU/mL).
- For the methods which recover microbes from fuel phase by aqueous extraction (ASTM D7463 and ASTM D8070), it is assumed the contaminants suspended in fuel will be readily attracted and retained by the droplets of the aqueous extractant when it is shaken in the fuel. In practice, this is likely to be influenced by the extent to which microbial contamination in the fuel phase is associated with suspended free water droplets or hydrophilic material; these will readily coalesce with the aqueous extractant droplets whereas freely suspended cells, fungal fragments or spores may not.
- For the methods which recover microbes from fuel phase by either filtration (IP 385 / ASTM D6974) or direct addition to the detection (culture) medium (IP 613 / ASTM D7978), it is assumed that a sufficient sample aliquot is analysed to be representative of the total volume of fuel phase sampled. In fact, contamination will be discontinuously and non-uniformly dispersed through the fuel.

Table 4. The ratio of fuel phase to water phase contamination assumed by IATA and JIG contamination limits

Test method used to determine contamination	Assumed Ratio of contamination			
	IATA Limits		JIG Limits	
	Fuel	Water	Fuel	Water
ASTM D7978 / IP 613	1	250 to 500	1	10,000
ASTM D7463	1	1000[a]	1	1000[a]
ASTM D8070	1	220	1	220

[a] Although separate limit values for fuel phase and water phase are not given for ASTM D7463, a ratio of 1:1000 is assumed as an approximation to enable similar interpretation of analysis of samples containing fuel only, fuel and water or water only.

This paper presents the findings from a laboratory study investigating the correlation between levels of fuel phase and water phase contamination in simulants of aviation fuel storage tanks. It employed the most widely used IATA recommended industry standard test methods for detection of microbial contamination. The study aimed to provide insights into whether current assumptions on the correlation between fuel and water phase contamination are correct. It is recognised that small scale laboratory studies will have limitations in their ability to fully reflect the complex and highly heterogenous dispersion of microbial contamination in large storage tanks. Consequently, the simulants aimed to reflect conditions in the bottom portion of fuel tanks where fuel is in close association with microbial growth in a free water phase. It investigated levels of water phase and fuel phase contamination after agitation followed by application of a settling time corresponding to standard industry guidance. It is hoped that the results obtained from the investigation will provide a more informed basis for establishing guideline contamination limits.

PROCEDURE

Set up of Jet fuel storage tank simulants

Two microcosms (labelled tank simulant A & C, shown in Figure 1) were set up to contain 9.75 litres of Jet A-1 and 250 mL aqueous phase (water with 1/4 strength Bushnell-Haas mineral salts medium). The water added included a mixed suspension of bacteria (approximately 10^5 Colony Forming Units (CFU)/mL), yeasts (approx. 10^3 CFU/mL) and moulds (approx. 10^3 CFU/mL) derived from stock collections and a variety of field samples (see Table 5). The microcosms were held in the dark at 25°C for 2 months and agitated vigorously 3 times per week throughout this period to simulate regular disturbance of tank contents (e.g. due to product receipts).

Table 5. List of microorganisms added to simulants A and C (one isolate per species added unless stated otherwise).

Bacteria	Yeasts	Moulds
Acinetobacter sp.	*Candida boidinii*	*Acremonium strictum* (ATCC 34063)
Burkholderia cepacia (2 isolates)	*Candida famata*	*Hormoconis resinae* (4 isolates, incl. ATCC 22711, ATCC 20495)
Burkholderia vietnamiensis	*Candida* spp.	*Paecilomyces* spp.
Kocuria sp.	*Candida zeylanoides*	*Penicillium* spp. (3 isolates)
Pseudomonas aeruginosa (5 isolates, incl. ATCC 15442, ATCC 33988).	*Galactomyces* spp.	
Pseudomonas spp.	*Rhodotorula glutinis*	
Pseudomonas oleovorans	*Yarrowia tropicalis* (ATCC 48138)	
Pseudomonas putida (DSM 6899)	*Yarrowia lipolytica* (ATCC 20177)	
	Cryptococcus spp.	

Set up of jet fuel storage tank bottom field sample composite

A composite of bottom and drain samples from six Jet A-1 storage tanks and a Jet A-1 product recovery tank (all tanks located on the same terminal) was produced comprising of 9.9 litres of jet fuel and 90 mL of water phase (labelled simulant B, shown in Figure 2). Prior testing of fuel using the IP 613/14 / ASTM D7978-14 test method had established that some degree of viable (live) microbial contamination was present in some of the tanks; notably heavy microbial contamination was detected in the sample from the product recovery tank. Other than compositing, no processing or incubation was conducted and no additional nutrients or microbes were added (i.e. this was effectively an evaluation of field samples as received).

Figure 1. Jet fuel simulants A and C

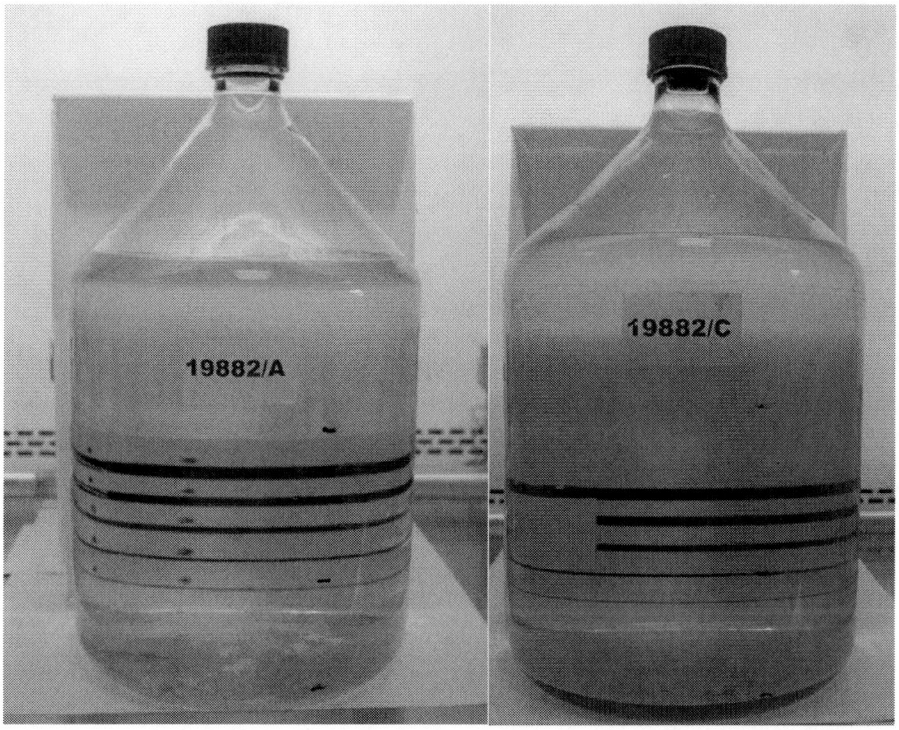

Figure 2. Simulant B: composite of bottom and drain samples from Jet A-1 storage tanks.

Sampling

Samples from jet fuel storage tank simulant A and C were collected for analysis after 2 months incubation. Samples from the jet fuel tank bottom field sample composite (simulant B) were collected approximately 20 hours after the composite was prepared.

Prior to sampling, the simulants were shaken vigorously for 10 seconds and then allowed to settle for 54 minutes which equates to a settling time of 1 hour per 30 cm of height. This settling time reflects common industry guidance and, in theory, would allow settling of most suspended free water from the fuel phase. Thus, the conditions prior to sampling aimed to reflect those which would typically occur if a contaminated jet fuel storage tank was filled creating turbulence, and then given a settling time prior to sampling. The sampling regime aimed to investigate the microbiology of the water phase and the fuel phase just above the water after settling had been applied.

The sampling procedure was standardised to minimize the disturbance of fuel layers as far as possible. Samples were collected by means of a sterile polytetrafluoroethylene (PTFE) siphon tube and transferred into sterile high-density polyethylene (HDPE) containers. Fuel samples (2.5 Litres) were drawn off at the three heights (18 cm from the bottom, 9 cm from the bottom and approx. 1-2 cm from the bottom just above settled free aqueous phase), starting with the highest sampling height and ending with the lowest sampling height. The siphon tube had a right angle bend at the end with a 6 mm orifice which was positioned horizontally at the required height and a 2.5 litre sample drawn. This equated to the removal of a 6.3 cm deep band of fuel from the microcosm at each designated sampling height; due to the low flow rate used when taking samples it is believed that most of the fuel was drawn from above the suction point.

All water phase from the bottom of simulants was then collected.

Test methods

Visual assessment

A visual assessment was made of each microcosm prior to sampling to note aqueous phase clarity, fuel phase clarity, fuel / water interface sharpness and cleanliness and the presence of any particulates (specifically particulate typical of microbial biomass).

Analysis of fuel phase samples collected from simulants A, B and C

The fuel samples taken from 18 cm and from 9 cm above the bottom of the simulants A, B and C were tested in duplicate by;

- ASTM D6974-16 / IP 385/99 standard laboratory method for total viable count (TVC) of aerobic heterotrophic bacteria and fungi.
- ASTM D7978-14 / IP 613/14 total viable count (TVC) (combined total aerobic heterotrophic bacteria and fungi) thixotropic gel culture method.
- ASTM D8070-16 immunoassay for microbial antigens.
- ASTM D7463-16 for ATP content.
- IP 438/01 water content by Karl Fischer Coulometry; one replicate was tested after centrifuging at 800 rcf for 10 minutes as per modified ASTM D2709-96; this enabled determination of total water (before centrifugation) and approximate determination of "dissolved" water (after centrifugation) and, by calculation, free suspended water.

Note; a total of 2.5 litres was taken at each layer, vigorously shaken and decanted into 2 x 1 Litre HDPE containers (for each replicate D7463-16 test) and 1 x 500 mL HDPE container (for other tests; aliquots for tests were collected from the mid-point of containers after agitation).

Analysis of water phase samples collected from simulants A, B and C

Water phase subsamples were tested in duplicate using the same microbiological test methods as for fuel phase. However, instead of filtering sample as specified for ASTM D6974-16 / IP 385/99, ten-fold serial dilutions of water were plated directly into or on the same nutrient media. The pH of water phase was also measured using indicator strips.

Analysis of fuel-water mix samples collected from simulants A, B and C

The ASTM D7463-16 and ASTM 8070-16 methods enable testing of fuel-water mixes. Fuel taken 1 - 2 cm from above the bottom of simulants was mixed with the water phase at various ratios and tested using these methods. Fuel and water phase samples were vigorously shaken prior to the production of the fuel-water mixes. Two different fuel:water ratios were investigated for each of these test methods; due to limited sample availability only singletons could be tested for each ratio.

RESULTS

Visual assessment

After 2 months, the fuel in jet fuel tank simulants A and C was clear and bright. The water phase in these simulants was turbid and there was some floccose material (typical in appearance to fungal biomass) and a translucent polymeric material (typical in appearance to bacterial Extracellular Polymeric Substance (EPS)) at the fuel:water interface. The water phase in simulants A was a different colour to that of C despite the same microorganisms being added and them being held under the same conditions; water in simulant A was cream-coloured and water in simulant C developed a pink-red colour (Figure 1) . We suspect that a red pigment producing *Pseudomonas aeruginosa* proliferated in simulant C, but it did not proliferate as well in simulant A.

For the jet fuel tank bottom field sample composite (simulant B), the fuel as received was hazy and the water phase comprised of turbid brown water. The water was visibly contaminated with brown particulate which readily suspended in the fuel.

When simulants A and C were shaken prior to sampling, it was noted that the water phase became suspended in the fuel phase but all visible water settled out within 10 minutes (Figure 3) and fuel above the water was observed to be clear and bright. In the case of the tank field sample composite (simulant B), non-aqueous particulate material was readily suspended in the fuel and remained suspended for longer than the prescribed 54 minute settling time (Figure 4).

Figure 3: Simulant A (a) immediately after shaking and (b) after 54 minutes settling.

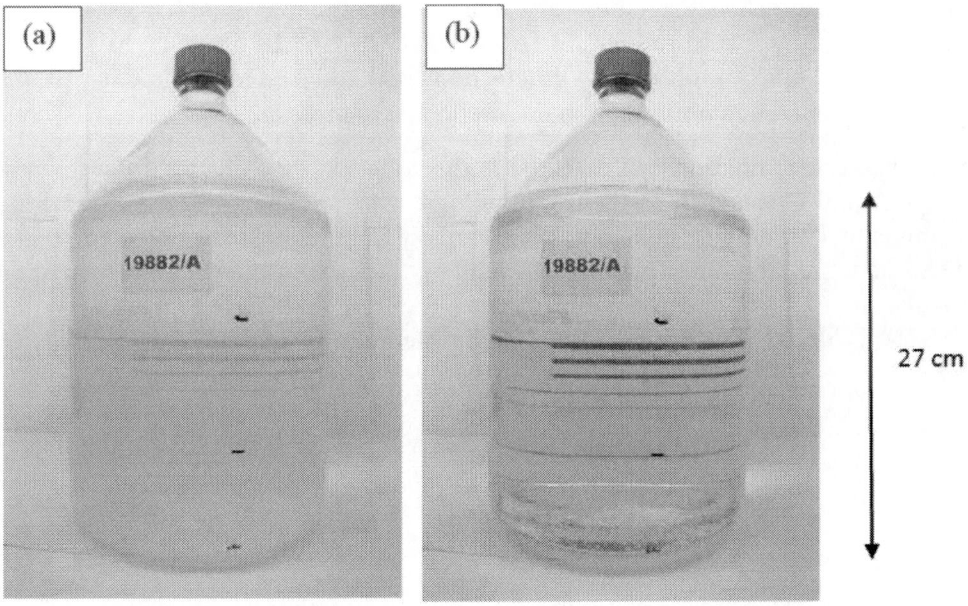

Figure 4: Simulant B (a) immediately after shaking and (b) after 54 minutes settling.

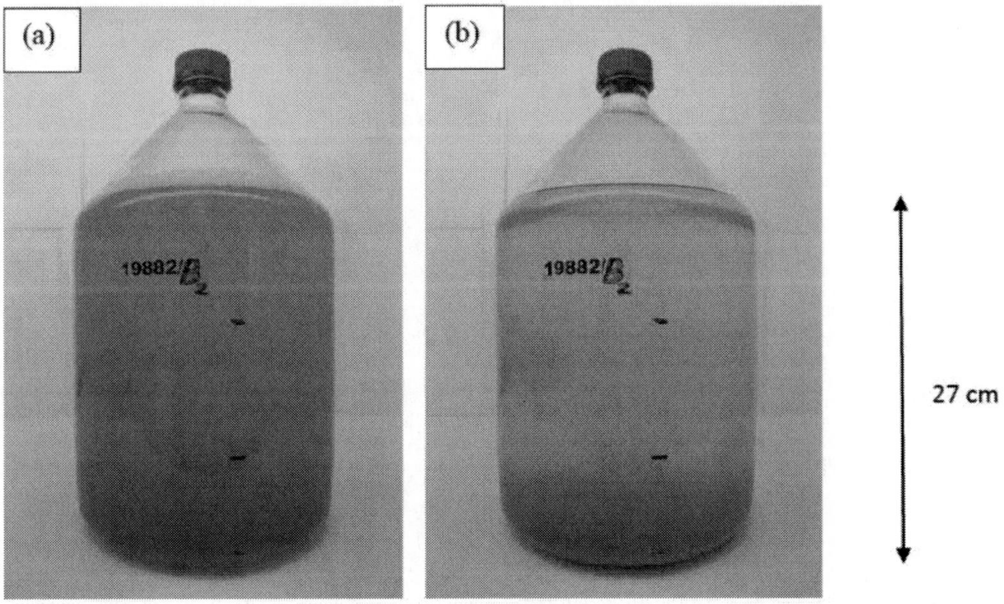

Microbiological Analysis Results

Table 6 shows the results of the microbiological tests conducted on the fuel phase and water phase samples collected from the simulants. According to the JIG guidance limits[4] (see Table 3), Action or Warning level microbial contamination was detected in all fuel phase samples from simulants A, B and C using the laboratory reference method (ASTM D6974-16 / IP 385/99) and the thixotropic gel culture method (ASTM D7978-14 / IP 613/14). In contrast, the immunoassay (ASTM D8070-16) and ATP detection method (ASTM D7463-16) results indicated negligible microbial contamination in the fuel phase samples.

Heavy microbiological contamination was detected in the water phase collected from jet tank fuel tank simulants A and C using all of the microbiological test methods. In the case of the field sample composite (simulant B), heavy microbial contamination was detected in the water phase by the standard laboratory plate count method, the thixotropic gel culture method (ASTM D7978-14 / IP 613/14) and the immunoassay (ASTM D8070-16). However, the ATP method (ASTM D7463-16) results indicated negligible microbial contamination.

Table 6. Results of microbiological tests performed on the fuel phase and water phase from simulants A, B and C.

Notes to Table 6 a) to c)

Mean of duplicate tests presented for the standard laboratory culture methods (ASTM D6974 / IP 385), thixotropic gel culture method (ASTM D7978 / IP 613) and ATP content method (ASTM D7463). For ASTM D6974 / IP 385, the mean counts for bacteria, yeasts and moulds are presented, and the total combined mean count.

Results of both immunoassay tests (ASTM D8070) are presented (N = negligible, L = low positive, H = high positive).

Red, yellow and green boxes indicate Action Level, Warning Level and negligible microbial contamination, respectively, according to JIG guidance limits[4].

* Fuel phase analysis by ASTM D6974 / IP 385. Water phase analysis by serial dilution and standard plate count.

** According to IATA guidance limits for aircraft fuel tanks[1] these results would be considered indicative of heavy microbial contamination.

*** 0.1 ml sample added directly to extractant in test pen as described in the ASTM D7463 procedure and RLU expressed as equivalent for neat sample.

CFU = Colony Forming Units, RLU = Relative Light Units, *H. resinae* = *Hormoconis resinae*

(a) Jet Fuel Tank Simulant A

Sample	Standard laboratory culture method (ASTM D6974 / IP 385)*	Thixotropic gel culture method (ASTM D7978 / IP 613)	Immunoassay (ASTM D8070)	ATP content (ASTM D7463)
Fuel phase 18 cm	Bacteria: 3.40×10^4 CFU/L Yeasts: 1.19×10^4 CFU/L Moulds: 5.80×10^3 CFU/L Total: 5.17×10^4 CFU/L**	6.64×10^5 CFU/L	Bacteria: N / N Fungi: N / N *H. resinae*: N / N	66 RLU
Fuel phase 9 cm	Bacteria: 1.6×10^4 CFU/L Yeasts: 7.2×10^3 CFU/L Moulds: 6.15×10^3 CFU/L Total: 2.94×10^4 CFU/L**	2.51×10^5 CFU/L	Bacteria: N / N Fungi: N / N *H. resinae*: N / N	87 RLU
Water phase	Bacteria: 1.63×10^8 CFU/mL Yeasts: 7.15×10^3 CFU/mL Moulds: 2.40×10^3 CFU/mL Total: 1.63×10^8 CFU/mL	$\geq 10^6$ CFU/mL	Bacteria: H / H Fungi: N / N *H. resinae*: L / N	169 750 RLU***

(b) Jet Fuel Tank Bottom Field Sample Composite Simulant B

Sample	Standard laboratory culture method*	Thixotropic gel culture method (ASTM D7978 / IP 613)	Immunoassay (ASTM D8070)	ATP content (ASTM D7463)
Fuel phase 18 cm	Bacteria: 2.75×10^3 CFU/L Yeasts: <100 CFU/L Moulds: 6.5×10^4 CFU/L Total: 6.78×10^4 CFU/L**	1.52×10^5 CFU/L	Bacteria: N / N Fungi: N / N *H. resinae*: N / N	59.5 RLU
Fuel phase 9 cm	Bacteria: 1.73×10^4 CFU/L Yeasts: <100 CFU/L Moulds: 4.45×10^4 CFU/L Total: 6.18×10^4 CFU/L**	1.56×10^5 CFU/L	Bacteria: N / N Fungi: N / N *H. resinae*: N / N	67 RLU
Water phase	Bacteria: 5.75×10^6 CFU/mL Yeasts: 4.40×10^2 CFU/mL Moulds: 1.09×10^4 CFU/mL Total: 5.76×10^6 CFU/mL	$\geq 10^6$ CFU/mL	Bacteria: N / N Fungi: H / H *H. resinae*: H / H	878 RLU***

(c) Jet Fuel Tank Simulant C

Sample	Standard laboratory culture method*	Thixotropic gel culture method (ASTM D7978 / IP 613)	Immunoassay (ASTM D8070)	ATP content (ASTM D7463)
Fuel phase 18 cm	Bacteria: 5.0×10^3 CFU/L Yeasts: 9.0×10^2 CFU/L Moulds: 3.25×10^4 CFU/L Total: 3.84×10^4 CFU/L**	6.70×10^4 CFU/L**	Bacteria: N / N Fungi: N / N *H. resinae*: N / N	192 RLU
Fuel phase 9 cm	Bacteria: 3.45×10^3 CFU/L Yeasts: 1.70×10^3 CFU/L Moulds: 2.65×10^4 CFU/L Total: 3.17×10^4 CFU/L**	4.40×10^4 CFU/L**	Bacteria: N / N Fungi: N / N *H. resinae*: N / N	50 RLU
Water phase	Bacteria: 1.30×10^8 CFU/mL Yeasts: 1.43×10^3 CFU/mL Moulds: 1.15×10^4 CFU/mL Total: 1.30×10^8 CFU/mL	$\geq 10^6$ CFU/mL	Bacteria: H / H Fungi: N / N *H. resinae*: N / N	247 300 RLU***

Table 7 shows the results of the tests conducted on fuel-water mixes. The result of the immunoassay test of 14 mL of water phase and fuel from jet fuel tank simulant A indicated the presence of Warning Level / moderate microbial contamination. When 10- fold less water from simulant A was tested the immunoassay indicated Action Level / heavy microbiological contamination. The ATP method indicated heavy microbial contamination was present in a fuel:water mix from simulant A comprising of 990 mL fuel and 10 mL of water. The ATP method indicated Warning Level / moderate microbial contamination when a fuel water mix comprising of 999 mL fuel and 1 mL water from simulant A was tested.

The results of the immunoassay tests on the fuel-water mixes from simulant B indicated heavy microbial contamination. In contrast, the results of the ATP tests on fuel-water mixes from simulant B indicated negligible microbial contamination.

The result of the immunoassay test of 1.4 mL of water phase and fuel from simulant C indicated the presence of heavy microbial contamination. When 10-fold less water from simulant C was tested the immunoassay test indicated negligible microbial contamination. The results of the ATP tests on fuel-water mixes from simulant C indicated heavy microbial contamination.

It must be stressed the results of the fuel:water mix samples represent single assays for each fuel:water ratio, for each simulant and each test method, and thus any conclusions which can be drawn are limited.

Table 7. Results of microbiological tests performed on the fuel-water mixes from simulants A, B and C.

<u>Notes to Table 7</u>

Single tests were conducted on each fuel-water mix. Results of immunoassay tests (ASTM D8070) presented as N = negligible, L = low positive, H = high positive. Red, yellow and green boxes indicate Action Level, Warning Level and negligible microbial contamination, respectively, according to JIG guidance limits. For these methods, these also equate to heavy, moderate and negligible contamination in accordance with IATA guidance limits.

* Fuel volume is actually approximate; water phase was measured to the sample preparation bottle provided with the kit and then filled with fuel to the top line (total liquid in sample preparation bottle for testing is approx. 165 mL).

** Neat sample gave overread and thus sample was diluted as described in the ASTM D7463 procedure and RLU expressed as equivalent for neat sample.

Simulant	Immunoassay (ASTM D8070)*	ATP content (ASTM D7463)
A	**14 mL water: 151 mL fuel**	**10 mL water: 990 mL fuel**
	Bacteria: L Fungi: N *H. resinae*: N	230,800 RLU **
	1.4 mL water: 163.6 mL fuel	**1 mL water: 999 mL fuel**
	Bacteria: H Fungi: N *H. resinae*: N	2100 RLU
B	**14 mL water: 151 mL fuel**	**10 mL water: 990 mL fuel**
	Bacteria: N Fungi: L *H. resinae*: H	370 RLU
	1.4 mL water: 163.6 mL fuel	**1 mL water: 999 mL fuel**
	Bacteria: N Fungi: L *H. resinae*: H	170 RLU
C	**1.4 mL water: 163.6 mL fuel**	**10 mL water: 990 mL fuel**
	Bacteria: H Fungi: N *H. resinae*: N	87,000 RLU
	0.14 mL water: 164.86 mL fuel	**1 mL water: 999 mL fuel**
	Bacteria: N Fungi: N *H. resinae*: N	175,800 RLU **

Table 8 shows the combined (both free and dissolved) and free water content for the fuel samples drawn 18 cm and 9 cm from the bottom of each simulant. Free water was assessed as the difference between the measured water content before and after centrifugation of replicate sub samples at 800 rcf for 10 minutes. The tests on the fuel sample taken 18 cm above the bottom of simulant B produced an anomalous negative result (i.e. the baseline water content measured before centrifugation was lower than the measurement after centrifugation). Overall, for all other tests, the majority of water appeared to be contributed by dissolved water with less than 14 ppm m/m free water measured in the fuel samples.

Table 8. Total (both free and dissolved) and free water content for the fuel samples drawn 18 cm and 9 cm from the bottom of each simulant.

Simulant	18 cm from bottom		9 cm from bottom	
	Total water (ppm m/m)	Free water (ppm m/m)	Total water (ppm m/m)	Free water (ppm m/m)
A	57.6	12.3	57.3	13.4
B	65	-6.2	84.8	5.5
C	64.5	13.2	63.6	9.8

pH

The water phase in jet fuel tank simulants A and C had a neutral pH. The water in field tank composite simulant B was acidic (pH 4.5 to 5).

DISCUSSION

The established ratios of water phase to fuel phase microbial contamination detected by the laboratory reference culture method (ASTM D6974 / IP 385), as shown Table 9 below, were significantly higher than assumed by the IATA and JIG contamination limits (see Table 4 above).

Table 9. The ratio of fuel phase vs water phase contamination detected in the simulants by standard laboratory culture (ASTM D6974 / IP 385)

Simulant	Ratio of fuel phase* to water phase contamination	
	Fuel	Water
A (Jet fuel tank simulant)	1	4,020,000
B (Field sample composite)	1	88,900
C (Jet fuel tank simulant)	1	3,710,000

* The mean of total CFU detected at 9 cm and 18 cm was used to compare with total CFU detected in water phase CFU.

The laboratory culture tests indicated bacteria and fungi were present in both fuel phase and water phase of the microcosm simulants and the storage tank field sample composite. However, particularly high levels of bacteria proliferated in the water phase of the jet fuel storage tank simulants A and C. In the field sample composite B, moulds formed a higher proportion of the total water phase contamination; this was probably a reflection of the lower pH which tends to favour fungal growth and is not optimum for growth of many heterotrophic aerobic bacteria. Because of the very high numbers of bacteria in water phase of A and C, the ratios of water phase to fuel phase microbial contamination are also particularly high, at least two orders of magnitude higher than the ratio derived from consideration of the respective guidance limits for fuel and water phase microbial contamination in JIG guidance. For the field sample composite (B), the established ratio of water phase to fuel phase microbial contamination is slightly higher than the ratio derived from consideration of the JIG limits but is within one order of magnitude. The finding suggests that when bacterial contamination is dominant, the JIG microbial limits for water phase might be considered too stringent. However, given that in our experience fungal contamination in fuel tanks is still generally of most operational significance, we would not recommend alteration. The finding does highlight the need to investigate the extent of penetration of contamination into fuel phase in the event of detecting Action Level water phase contamination. The finding also suggests that the current JIG recommended practice of draining tanks to remove water prior to sampling for routine microbiological testing is well advised, as detection of Action Level contamination in water phase may not necessarily indicate significant impact on fuel. Analysis of water phase samples may be a prudent measure when investigating contamination incidents but may not be a reliable indicator for routine trend monitoring.

The ratios of fuel to water phase microbial contamination established in this study are very much higher, by nearly four orders of magnitude, than the ratios derived from consideration of IATA guidance limits. However, it must be appreciated that the microcosms A and C provided a simulation of conditions in ground storage tanks rather than of aircraft tanks, given the water phase included inorganic nutrients (Bushnell-Haas mineral salts). Because, water in aircraft tanks is likely to be primarily derived from condensation, it will normally contain few inorganic nutrients and will be less likely to support microbial growth. Therefore, more stringent limits on water phase from aircraft tanks are probably appropriate. However, the study does suggest that for conventional culture methods, the current IATA limits for water phase may be slightly too stringent; there is a risk of unnecessarily instigating remedial actions for aircraft when the water phase limits are exceeded. We do however, endorse the IATA recommendation that water phase should be tested whenever it is recovered from fuel tanks, given that monitoring is relatively infrequent and given in the interests of operational safety, operators need to have the best chance of detecting any contamination present. However, if possible it might be advisable to consider the extent of contamination in fuel phase before implementing remedial actions such as biocide treatment and tank cleaning, as typically stipulated in aircraft maintenance manuals. The study does support the contention that the IATA limits are only appropriate for aircraft drain tank samples and the same limits are likely to be too stringent for bottom or drain samples from land storage tanks or filter vessels.

The results of fuel and water phase tests using the thixotropic gel culture method (ASTM D7978 / IP 613) were broadly consistent with the results obtained using the laboratory reference culture method. Overall, slightly higher CFU counts were detected in fuel phase analyses using the thixotropic gel culture method, ASTM D7978 / IP 613, compared with the laboratory reference method, ASTM D6974 / IP 385. Consequently, if applying JIG limits, results fell into the "Action Level" band for the former method whereas they equated to "Warning Level" contamination for the latter. This is probably because the thixotropic gel culture method involves vigorously shaking the sample into the culture medium thus resulting in the dispersion of microbial cell aggregates and consequential a higher number of colony forming units. Additionally, because in the ASTM D7978 / IP 613 procedure, fuel sample is incorporated into the growth medium, the test may provide a more complete nutrient source for the contaminants present and consequently higher microbial recoveries. Because contamination detected in water phase exceeded the quantifiable level for the ASTM D7978 / IP 613 test (more than 10^6 CFU/mL detected), only semi-quantitative results could be expressed, meaning exact ratio values could not be calculated. As far as is possible from consideration of the quantitative data for fuel phase tests and semi-quantitative data for water phase, the ratios of microbial contamination detected appear to be comparable to those established by the laboratory standard tests. Thus, the same comments in respect of ratios of fuel phase to water phase microbial contamination and the implications for the validity of current JIG and IATA limits apply.

No microbial contamination was detected in the fuel phase of any of the storage tank simulants or the storage tank field sample composite using the immunoassay (ASTM D8070). This was despite the presence of high levels of viable bacteria and fungi, including *Hormoconis resinae*, as indicated by the laboratory standard culture method (ASTM D6974 / IP 385) and also the thixotropic gel culture method (ASTM D7978 / IP 613). For the fuel-water mixes analysed, contamination was detected at either Warning or Action Levels in accordance with JIG limits. For the water phase samples, the results were generally consistent with the laboratory plate culture method and indicated Action Level (heavy, in accordance with IATA limits) contamination.

In the case of the ATP method (ASTM D7463), all fuel phase samples gave very low RLU readings, indicative of negligible contamination in accordance with IATA limits, again, contrary to the indications of the culture methods. A known cause of variance between ATP bioluminescence and culture methods is the detection of dormant fungal spores by the latter. Due to their hydrophobicity, spores are readily suspended in fuel but they are generally not detected by ATP bioluminescence as they do not contain significant amounts of ATP. However, this cannot solely explain the discrepancy in this study as it was established that bacterial and fungal contamination was suspended in fuel phase as cells and mycelial fragments at levels which would generally be considered heavy. Further testing of simulants A and C confirmed that fuel suspended contamination was predominantly *Pseudomonas aeruginosa*, a bacteria which does not form spores. It should be pointed out that the ASTM D7463 method does clearly state that the method is only designed to detect fuel phase contamination when it is associated with suspended water or hydrophilic material. In considering ratios of fuel phase to water phase microbial contamination, it should be remembered that the ASTM D7463 method utilises the same test protocol and interpretation regardless of the volume of water, if any, present in the sample. This assumption holds true if it is assumed that the ratio of fuel to water phase microbial contamination is approximately 1:1000 (see manufacturers technical information for further details). If the assumption is correct, one would expect the same RLU readings regardless of whether the sample was fuel or water or a mix of the two phases; this was not found in our study as significant contamination was only detected when samples contained some amount of visible water phase. For simulant C, broadly similar RLU readings, within one order of magnitude, were obtained regardless of the proportion of water. For water phase samples from jet fuel tank simulants A and C, the RLU readings were broadly consistent with the laboratory reference culture method and indicated Action Level / heavy contamination. However, the field sample composite simulant B produced unexpectedly low RLU readings in water phase, inconsistent with the other methods and with visual and microscope examination which indicated active contamination by viable bacteria and fungi. Further analysis suggested that an unidentified component in the water phase from simulant B was interfering with ATP detection; the interference could not be diluted out using the procedure stipulated in Appendix X4 of the standard.

Despite the confirmed presence of significant levels of microbial contamination in all test set ups, under our study conditions where a settling time was applied prior to sampling, analysis of fuel phase samples by the immunoassay (ASTM D8070) and ATP (ASTM D7463) methods did not indicate presence of contamination at levels which would be considered Warning or Action Level in accordance with JIG Limits. The fuel samples were drawn from only a few centimetres above water phase supporting extensive active growth and they reflect a practical reality that it is often not possible to recover water phase when sampling fuel systems. Both these methods utilise an aqueous extraction step for recovery of microbes from fuel and thus recovery is likely to be dependent on the extent to which microbial contamination is associated with suspended water or hydrophilic material and the extent to which it can be coalesced with the extractant. Whilst the levels of suspended free water measured in the fuel samples from simulants would be considered to be low (<14 ppm m/m), in theory this would still equate to a very high number of water droplets; approximately 10^7 water droplets if they were all 10 μm in diameter. The results of this investigation and other recent studies of settling times[6] indicate that microbial contamination in fuel is not always associated with larger (e.g. 20 μm and above) suspended free water droplets and this might explain why the immunoassay and ATP method did not detect significant microbial contamination in the fuel phase samples. On considering the settling time allowed prior to sampling, the free water remaining in the fuel phase in this study would in probability have been dispersed in droplets of 20 μm or less and some would likely remain suspended for example due to Brownian motion[7]. The fact that similar free water concentrations were observed at both 9 and 18 cm levels suggests this is the case. In the case of the field sample composite (B), the contamination was probably also associated with non-aqueous particulate material which remained visibly suspended in the fuel for longer than the prescribed settling time. It is possible that the ASTM D8070 and ASTM D7463 extraction procedures allow insufficient dispersion of aqueous extractant to coalesce with smaller suspended free water droplets or hydrophilic particles and hence do not recover associated microbial ATP and microbial antigens when the number of freely suspended water droplets is relatively small. Additionally, the amount of fuel suspended ATP or antigen recoverable, even in heavy contamination situations, may be below detection levels of the methods.

SUMMARY

A laboratory study was undertaken using simulants to investigate how the extent of microbial contamination in fuel and water phase influences the results obtained using industry standard test methods. The study investigates whether assumptions on the correlation between fuel and water phase microbial contamination in setting contamination limits values, for example IATA and JIG, are correct.

The established ratios of water phase to fuel phase microbial contamination for the laboratory reference culture method (ASTM D6974 / IP 385) and the thixotropic gel culture method (ASTM D7978 / IP 613) were slightly higher than the ratio derived by consideration of JIG contamination limits and considerably higher than the ratio derived by consideration of IATA contamination limits. Nevertheless, all things considered, the JIG contamination limits for both fuel and water phase would appear to be reasonable. Although this study did not attempt to replicate conditions in aircraft fuel tanks, it could be argued that the water phase IATA contamination limits for these conventional culture methods are too stringent and could instigate unnecessary remedial action.

No microbial contamination was detected in the fuel phase of any of the simulants using the immunoassay method (ASTM D8070). Tests of water phase and fuel water mix samples by the immunoassay method gave results broadly consistent with the culture methods.

Using the ATP method (ASTM D7463), all fuel phase samples gave very low readings which would be considered indicative of negligible contamination in accordance with JIG and IATA Limits. Tests of water phase of laboratory simulants gave results broadly consistent with the culture methods and the immunoassay but suspected interference meant the test failed to indicate heavy contamination in the storage tank field sample water phase.

The study findings indicate that, in consideration of testing fuel phase samples, a number of factors may influence the ability of various methods to detect microbial contamination. The extent of free water suspended in fuel samples is likely to influence the level of detectable contamination regardless of the method used . Consequently, it should be appreciated that all test methods have limitations and selection of the most appropriate should take into account the specific operational circumstances and particularly whether free water is likely to be present in samples taken.

REFERENCES

[1] Guidance Material on Microbiological Contamination in Aircraft Fuel Tanks, International Air Transport Association (IATA), Montreal & Geneva, 5th Edition. (2015).

[2] EI/JIG STANDARD 1530 Quality assurance requirements for the manufacture, storage and distribution of aviation fuels to airports. 1st Ed. (October 2013).

[3] API RP 1595, Design, Construction, Operation, Maintenance, and Inspection of Aviation Pre-Airfield Storage Terminals, 2nd Ed. (October 2012).

[4] Joint Inspection Group (JIG) Bulletin 83 & Technical Information Document, Microbial Monitoring Strategies (October 2015).

[5] J. White, J. Gilbert, G. Hill, E. Hill, S.M. Huse, A.J. Weightman & E. Mahenthiralingam. 2011. Culture-independent analysis of bacterial fuel contamination provides insight into the level of concordance with the standard industry practice of aerobic cultivation. Applied and Environmental Microbiology. 77: 4527-4538.

[6] G. Williams, L. O'Malley & Graham C. Hill. Investigation of relationship between water content in biodiesels and microbial growth and contamination in *Proceedings of the 14th International Conference on Stability and Handling of Liquid Fuels*, IASH, Charleston, 2015.

[7] S. D. Anderson. A. Q. Clarke, A. Glendinning & G. Norris. A theoretical and full scale investigation into the use of tank settling to remove water and solid contaminants from aviation fuel in Proceedings of the *15th International Conference on Stability and Handling of Liquid Fuels*, IASH, Rome, 2017.

IASH 2017, the 15TH INTERNATIONAL CONFERENCE ON STABILITY, HANDLING AND USE OF LIQUID FUELS
Rome, Italy
10-14 September 2017

COMPARISON OF PARTICULATE CONTAMINATION MEASUREMENT TECHNIQUES IN DISTILLATE FUELS

Thomas G. Smagala, Camden Cook, Andy Ye Yuan Chen, Krege M. Christison

Chevron Corporation, 100 Chevron Way, Richmond, CA 94801 USA, tsmagala@chevron.com

As technology for vehicles, aircraft, and engines has advanced, original equipment manufacturers have pushed for tighter specifications on the cleanliness of the fuels that go into such equipment. The size distribution of particulate matter in fuel is important because many sophisticated modern engines have smaller and smaller clearances for the fuel to pass through at higher speeds and pressures, in particular the diesel injectors. There are existing test methods for measuring the total amount of particulate contamination in distillate fuels, such as ASTM D6217 Standard Test Method for Particulate Contamination in Middle Distillate Fuels by Laboratory Filtration" and ASTM D5452 "Standard Test Method for Particulate Contamination in Aviation Fuels by Laboratory Filtration." However, these methods fail to indicate the particle size distribution. Various laboratory instruments and test methods exist for determining particle counts in distillate fuels, but they fail to indicate the total gravimetric particulate contamination. With some assumptions about particle geometry, one can estimate the total particulate concentration in distillate fuels from the integration of the particle size distribution as determined by particle counting techniques. These estimates come quite close to the measured total gravimetric techniques which require the filtration of one liter to one gallon of fuel. Thus, a particle counter test, which consumes much less fuel, can provide an estimate of the total particulate contamination in addition to the particle size distribution.

IASH 2017, the 15th INTERNATIONAL SYMPOSIUM ON STABILITY, HANDLING AND USE OF LIQUID FUELS
Rome, Italy
10-14 September 2017

EVALUATION OF PARTICLE COUNTING AS A VALID TOOL TO DETERMINE THE FILTRATION EFFICIENCY

Robert Pawlik[1], Tom Muzik[2], and Lewis Wolfe[3]

[1] *Parker Hannifin Corp., 1210 Garden of the Gods Rd, CO, USA,* robert.pawlik@parker.com
[2] *Parker Hannifin Corp., 1210 Garden of the Gods Rd, CO, USA,* tom.muzik@parker.com
[3] *Parker Hannifin Corp., 1210 Garden of the Gods Rd, CO, USA,* lewis.wolfe@parker.com

This study evaluates the potential of using particle counters as a useful tool to determine the efficiency of filters in aviation fuels. Currently, there are two ASTM standard methods used to evaluate the cleanliness of aviation fuel. *ASTM D2276*(1) is a gravimetric method and *ASTM D7619*(2) method uses particle counting to determine the fuel contamination level. The study looked at the correlation between the theoretical contamination level of jet-A fuel, gravimetric and particle counting methods while flowing fuel at 50 gpm through a filter separator. Parker Hannifin's automatic particle counter (ACM20) was used for the study. Water was added upstream of the pump to increase the free water content to 5 ppm, 10 ppm, 15 ppm and 20 ppm. Also, the contamination levels were increased to 0.25 mg/L, 0.50 mg/L, 1.00 mg/L, 1.50 mg/L. The sampling ports were upstream and downstream of the vessel allowing for evaluation of the filter/separator system. In the second part of the study two particulate filters with nominal rating of 1 micron and 10 micron respectively were used. The correlation between the standard methods and filter efficiency was determined.
During the study it was observed that the results from automatic particle counter correlated well with the increase in the contamination level of dry fuel. It was observed that when dry fuel was used the particle counter was a good instrument to use to observe and quantify the difference in the efficiency of the two filters used.

References:

1. ASTM International, https://www.astm.org/Standards/D2276.htm, retrieved on 1 Feb 2017
2. ASTM International, https://www.astm.org/Standards/D7619.htm, retrieved on 1 Feb 2017

IASH 2017, the 15TH INTERNATIONAL SYMPOSIUM ON STABILITY, HANDLING AND USE OF LIQUID FUELS
Rome, Italy
10-14 September 2017

What new methods for evaluating D3241 heater tubes deposit thickness may mean to the development of new D3241 methods.

Michael C Croudace[1],

[1.] *Compass Instruments, 1020 Airpark Dr., Sugar Grove, IL, USA, MCroudace@Compass-Instruments.com*

Since 2014, D3241 has introduced new standards for evaluating heater tube deposits in Annex 2 and 3 of the method. Rather than a simple visual comparison of the deposit's color to ASTM color codes in a visible tube rater box, the tubes are analyzed systematically for deposit thickness via interferometer or ellipsometer analyses. These results are graphically represented and digitally stored for general comparison by experts. The graphs are precise enough to see the variability from test to test and instrument to instrument. This paper will describe the types and size of these variations with variety of fuels and instruments and what these observations mean about instrument controls that will improve consistency.

IASH 2017, THE 15th INTERNATIONAL CONFERENCE ON STABILITY, HANDLING AND USE OF LIQUID FUELS
Rome – Italy, 10 – 14 September 2017

Novel Automated Systems for On-site Fuel Stability and Compatibility Testing according to ASTM D4740:

Didier Pigeon

CEO AD Systems, P.A. Portes de la Suisse Normande, Allée de Cindais, 14320 Saint André sur Orne France, dpigeon@adsystems-sa.com

Bulk fuel stored for long periods can become unstable the asphaltene content can precipitate out of solution causing the formation of sludge. This has the potential to block filters and pipes, leaving tanks with an unpumpable residue. Once a fuel has chemically broken down there is no way to satisfactorily reverse the process. Precipitated asphaltene cannot be re-dissolved.

The marine fuel market is undergoing complex and far-reaching change, largely as a result of a growing body of environmental legislation. Issues surrounding fuel stability and compatibility have never been more relevant, especially with the new 0.10 per cent sulphur level for fuels used in Emission Control Areas (ECAs).

Although incompatibility is not a common phenomenon its likelihood increases in tandem with fuel switching, such as when entering and leaving an Emission Control Areas (ECA). Compatibility problems must be treated as a critical concern as the fuel systems can become paralyzed.

The ability to test on-site, at the point of use, enables engineers and facilities managers to conduct fuel analysis quickly and easily. Detecting out-of-spec fuels can identify potential problems before equipment damage occurs.

Dedicated compatibility / stability kits exist but those kits are manual. Those kits can be used on-site. They are operator time consuming and the result relies on the operator skill.

The presentation will describe a new automated system that perfectly mimics the difference phases described in D4740. The automated instrument is contained in a rugged case. The test can be carried on-site by unskilled personal in strict accordance with the test method. The built-in computer and software analyses the spot image and reports quick and solid results.

IASH 2017, the 15th INTERNATIONAL SYMPOSIUM ON STABILITY, HANDLING AND USE OF LIQUID FUELS
Rome, Italy
10-14 September 2017

FAME IN JET FUEL: A NOVEL ANALYTICAL TEST METHOD BASED ON LASER MID-INFRARED SPECTROSCOPY

Bernhard Siebenhofer, Michael Martl, and Wolfgang Ritter

QuantaRed Technologies GmbH, Columbusgasse 1-3/54, A-1100 Vienna, Austria, b.siebenhofer@quantared.com

ABSTRACT

The recent development of Quantum Cascade Laser (QCL) technology gives rise to new analytical methods in the field of mid-infrared laser spectroscopy. The high spectral power density of QCLs enables the analysis of liquid fuels with enhanced sensitivity and selectivity, compared to conventional methods such Fourier Transform Infrared (FTIR) spectroscopy. This paper introduces an innovative approach to the detection of low levels of Fatty Acid Methyl Ester (FAME) in jet fuels. FAME or biodiesel has been in focus of the aviation industry and pipeline operators in the last years. The surface-active FAME has become a major concern due to increased usage in the supply chain and the potential cross-contamination of subsequently transported jet fuel. In this paper, a novel jet fuel test method for the concentration range from 10 mg/kg up to 400 mg/kg FAME is presented. This broad range will facilitate measurements in respect to changing FAME limits for Jet-A and Jet-A1 standards. Furthermore, FAME from different feed stocks was reliably detected, while jet additives at their specified limits did not lead to false positive FAME detection. For industrial application, the test method was implemented in a rugged and portable stand-alone FAME analyzer with high sample throughput and rapid measurement capabilities. This fully automated device allows on-site jet fuel screening, as well as equipping laboratory staff with a reliable tool to quantify FAME levels in jet fuels. Data from preliminary industry studies and method development is presented in this paper.

KEYWORDS

Fatty Acid Methyl Ester (FAME); Jet Fuel; mid-infrared spectroscopy; Quantum Cascade Laser (QCL); test method

INTRODUCTION

The present and growing international governmental requirements to add fatty acid methyl esters (FAME) to diesel fuel had the unintended side-effect of potential FAME cross-contamination of jet fuel. This problem arises when using multiproduct transport facilities such as tanker ships, trucks or pipelines. FAME has led to industry wide concerns, because it negatively influences the thermal stability of jet fuel. Furthermore, it can cause performance deterioration of aircraft engines by coke deposits and filter blockages. Therefore, the FAME content in jet fuel is limited to < 50 mg/kg in international jet fuel specifications (Def.-Stan. 91-91 and ASTM D1655). The Joint Inspection Group is providing guidelines on required and recommended FAME measurement points in the jet fuel procurement chain in their bulletin number 75. Certain stakeholders push for an increased FAME limit of < 100 mg/kg, which is currently extensively discussed by the aviation industry.

FAME, or biodiesel, is a class of substances derived from different natural feedstocks by esterification. Possible feedstocks are soy- (SME), rapeseed- (RME), palm- (PME), coconut- (CME) or used cooking oil (UCOME). Additional options are lard or tallow (TME). These different types of feedstock result in biodiesel comprising of FAME with different carbon chain lengths and molecular weight. This makes the detection and quantification of FAME in jet fuel a challenging task, since both the matrix and the analyte are substance classes consisting of a variety of different types of molecules. Furthermore, several classes of additives are approved for use in jet fuel, that can have a negative impact on the methods used to quantify FAME.

Analytical methods, based on gas or liquid chromatography (IP 585, IP 590 and IP 599), have been developed with the capability of measuring down to < 5 mg/kg levels of FAME. However, these chromatographic methods are laboratory based and it is not possible to detect short chain methyl esters, like coconut methyl ester (CME). A FTIR based rapid screening method (IP 583) has been developed for use in the supply chain by non-specialized personnel to cover the range of 10 mg/kg to 150 mg/kg. This method requires single-use consumables like solid phase extraction cartridges and filters.

These shortcomings lead to an industry demand for an easy to use, portable, fully automated analyzer with high sample throughput and wide measurement range, while using less consumables. The proposed FAME in jet fuel test method discussed in this paper addresses these needs. It is based on Quantum Cascade Laser mid-infrared absorption spectroscopy, which features high sensitivity and precision compared to conventional infrared techniques. FAME is selectively removed via a multi-use column to avoid single-use consumables. The selectivity of the FAME removal results in a high specificity of the method for a wide range of jet fuel, including hydrotreated (HT), non-HT and Merox-type jet fuels. The method detects all types of FAME and offers fast measurement times between 10 and 24 minutes. The preliminary target measurement range is 10 to 400 mg/kg, which already spans possible higher FAME specification limits in the future. The use of QCL technology allows compact design and portable devices. Furthermore, the method can be fully automated with rapid sample measurement capabilities as there is no manual sample preparation necessary.

MATERIALS AND METHODS

The first Quantum Cascade Laser (QCL) was presented in 1994,[1] with room temperature operability of QCLs demonstrated in 1996.[2] The first QCLs were commercially available in 1999 from "Alpes Laser SA". In 2000 the first successful application of QCLs for liquid phase transmission measurements was published.[3] In QCLs, photon emission occurs within the conduction band via intersubband transitions as shown in Figure 1. In contrast, Figure 2 shows the principle of conventional semi-conducting diode lasers. Here photon emission is due to the recombination of electrons and holes from the conduction to the valence band. The QCL intersubband transitions can be controlled by careful band-structure engineering at monolayer resolution, thereby forming an artificial semi-conducting crystal with a heterostructure of alternating "well" and "barrier" layers. These layers can be grown by molecular beam epitaxy techniques and recently by metal oxide chemical vapor deposition. The intersubband nature of these optical transitions has several key advantages: Firstly, the laser emission wavelength is primarily a function of the thickness of the quantum well layers. Hence, the wavelength can be tailored over a wide spectral region. In conventional diode lasers, the intrinsic band gap of the active material determines the emission wavelength, resulting in the so-called "band gap slavery".[4] Secondly, the multistage cascade nature permits electrons to be recycled as they remain inside the conducting band, contributing in every cycle to the gain and photon emissions. Furthermore, the wavelength of the QCL emission can be precisely fine-tuned by changing the laser current or operation temperature. In recent years, important advances concerning wavelength tuneability of QCLs have been made. External cavity QCLs are commercially available and can be tuned ~ 15 % with respect to the central laser wavelength.[5] Dimensions of the active material of QCLs are in the range of a few tens of micrometers in width and height and a few millimeters in length. Therefore, QCLs are ideal components for the development of portable analyzers.

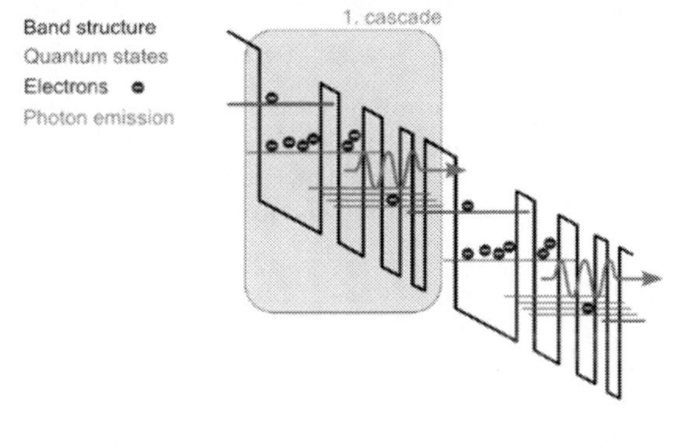

Figure 1: QCL band structure and intersubband transitions. The cascade structure is repeated, leading to multiple photon emissions per electron.

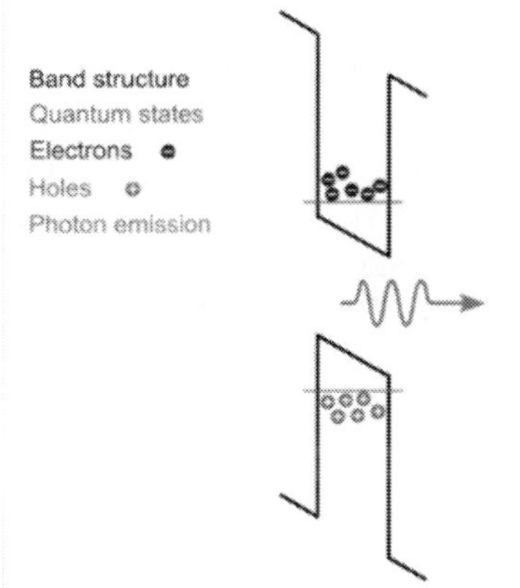

Figure 2: Conventional diode laser with recombination of electrons and holes over the band gap.

QCLs have been employed as light source for determining sub-ppm concentrations of hydrocarbons in water in ASTM D7678, an infrared absorption method developed by "QuantaRed Technologies GmbH" in cooperation with the "Vienna University of Technology". This method employs a liquid-liquid extraction step which is followed by absorbance measurement in the region between 1350-1500 cm^{-1} using mid-IR QCLs.

In absorption spectroscopy, Beer-Lambert's law

$$A = \log\frac{I_0}{I} = \varepsilon * c * d,$$

(where A is the absorbance, I_0 and I the intensity before and after attenuation, ε is the molar extinction coefficient, c the concentration of the absorbing molecules and d the optical path length of the sample medium) is used to calculate analyte concentration from measured absorbance. From this equation, it is obvious that the superior output power of QCLs can be used to increase the sensitivity by allowing the measurement of smaller concentrations by using longer optical path lengths.

In the novel method to measure FAME in jet fuel, the analyte is quantified by specifically measuring the carbonyl-stretching vibration in methyl-ester configuration (see Figure 3) excited by mid-infrared lasers. In general, it is very challenging to measure analytes in complex hydrocarbon matrices like jet fuel. Jet fuel is a mixture of paraffines, naphthenes and aromatics in different ratios. Additives are used to guarantee that the mixture complies with jet fuel specifications Def.-Stan. 91-91 or ASTM D1655. To measure the FAME absorbance in these conditions, a differential approach is necessary. Figure 4 shows the schematics of the measurement setup. A pump is used to draw the test specimen and circulate it through a multi-use column, where the FAME is specifically removed in less than 20 minutes. Absorbance is measured employing a QCL, an infrared detector and a flow cell. The FAME content in the sample is measured in mol/L by comparison of the detector signal before and after the removal of FAME from the sample. A calibration curve is then used to convert this value to a FAME concentration in mg/kg.

Figure 3: FAME (Methyl palmitate, C16) molecule consisting of carbon chain, carbonyl group and methyl-ester group. Carbon chain length composition varies with FAME feedstock, but is generally between C12 and C20.

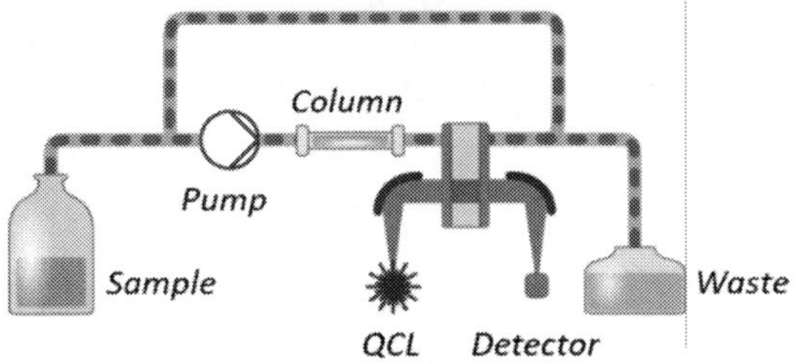

Figure 4: Schematics of the proposed method. FAME is removed by pumping the jet fuel specimen through a multi-use column. Absorbance is measured using a QCL light source, flow cell and infrared detector.

RESULTS AND DISCUSSION

Figure 5 shows preliminary results of samples spiked with different concentrations of a FAME mixture (SME, RME, TME, PME and UCOME, 20% each in B7 diesel) compared to a ten-point calibration with pure substances. The calibration shows a good correlation between measured and expected concentrations. The calibration curve is linear from 0 to 500 mg/kg, yielding a preliminary quantification range of 10 to 400 mg/kg for FAME in jet fuel. The results of jet fuels from different point of origins and different treatments, spiked with the same FAME mixture to yield 100 mg/kg FAME, are shown in Figure 6.

As discussed above, the carbonyl chain length of FAME depends on the feedstock used for esterification. Therefore, the method is evaluated using jet fuel spiked with B7 produced from specific FAME types. The results of these measurements are shown in Figure 7. Figure 8 shows the respective chain length composition measured with EN 14103. As expected from Beer-Lambert's law, the calculated FAME concentration decreases with increasing molecular weight of the individual FAME molecules. This is due to the fact that longer carbon chains entail increased molecular weight, without affecting the amount of carbonyl groups that contribute to the measured absorbance. Therefore, a sample spiked with 100 mg/kg of CME contains significantly more FAME molecules than a sample spiked with 100 mg/kg of the other, longer-chained types. As a result, CME is overestimated when evaluated with a calibration carried out with the FAME mixture mentioned above. This bias is imminent to all infrared absorption methods (like IP 583), but can be overcome by using a calibration mixture tailored to the expected FAME types used in specific geographical regions. There are distinct advantages in reporting the amount of FAME molecules in mol/L, compared to a concentration in mg/kg. The carbonyl and methyl-ester groups are what differentiate FAME from the hydrocarbon components of the jet fuel. If the problems that FAME causes in jet fuel are attributed to those functional groups, an overestimation of short chain length methyl esters can be advantageous. An overestimation of CME does not pose a safety concern for aircraft operation.

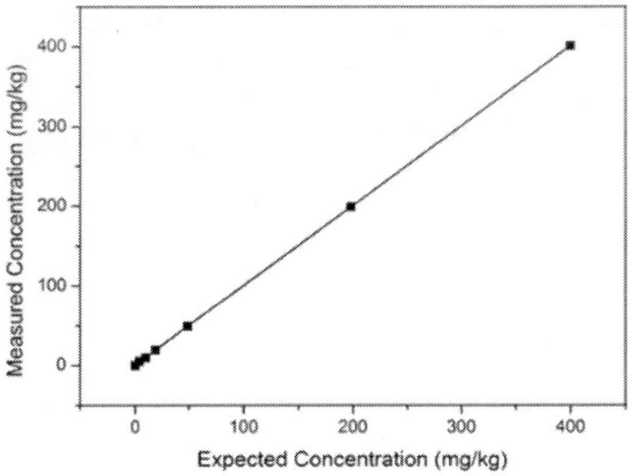

Figure 5: Jet fuel spiked with different concentrations of a FAME mixture. The calibration shows good correlation between expected and measured concentrations.

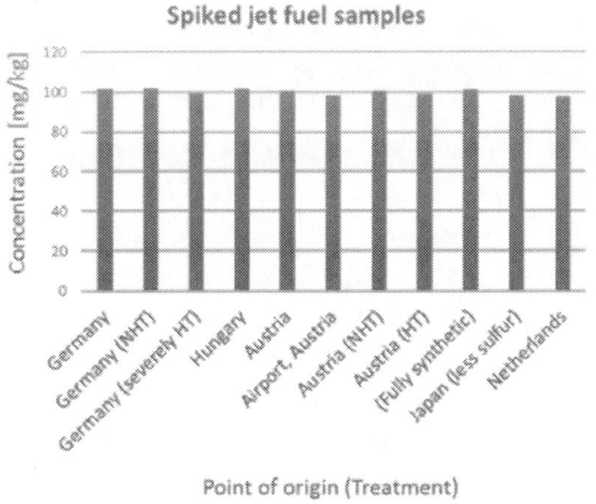

Figure 6: Jet fuel samples from different point of origins and different treatments shows negligible deviations from the spiked concentrations.

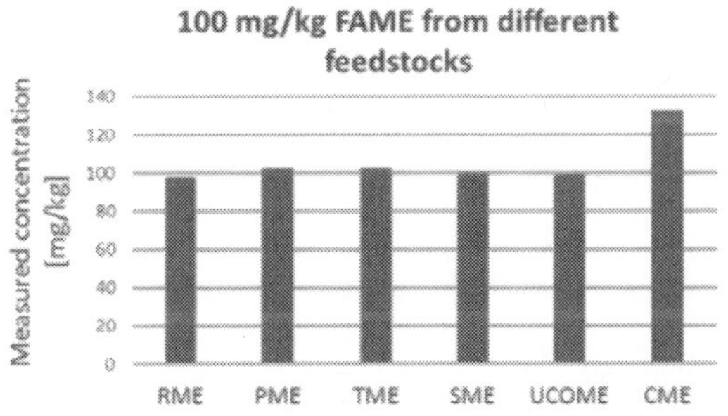

Figure 7: Jet fuel spiked with different types of FAME. The calculated concentration varies with the carbonyl chain length and therefore with the molecular weight of each FAME molecule. Note that the short chain methyl ester CME was not part of the FAME mixture (20 % RME, PME, TME, SME and UCOME) used for calibration.

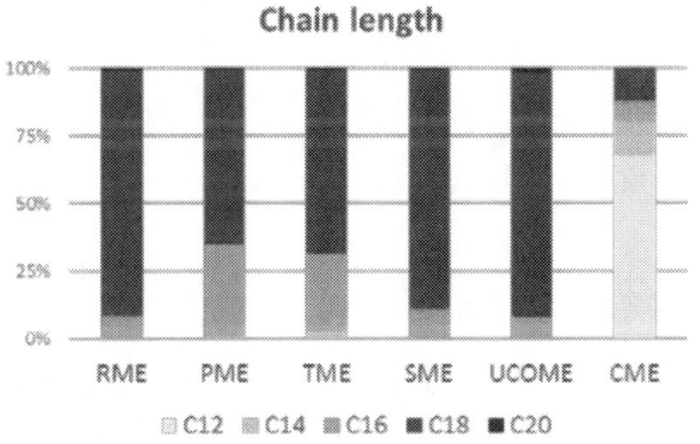

Figure 8: Carbonyl chain length, as measured by EN 14103, composition of different types of FAME. Chain length compositions are in good agreement with the measured concentrations, with lower molecular weight resulting in higher FAME levels in mg/kg.

To further assess the robustness of the proposed method, jet fuel additives were tested at their maximum allowed concentrations. The influences of metal deactivators, static dissipators, anti-oxidants, lubricity improvers, fuel system icing inhibitors (FSII) and biocides were found to be negligible, as shown in Figure 9. Additionally, 200 ppm of dispersed water have been shown to not affect the FAME measurement.

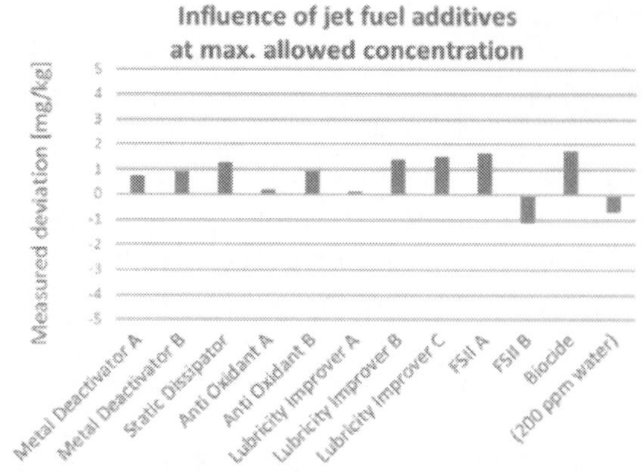

Figure 9: Influence of jet fuel additives at their maximum allowed concentration on the FAME in jet fuel measurement. 200 ppm dispersed water has negligible influence on the measured FAME concentration.

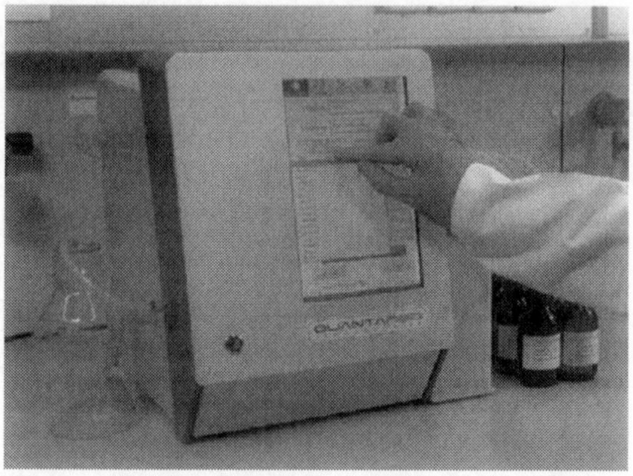

Figure 10: The ***FameSpec*** FAME in jet analyzer prototype. The sample is drawn from the right side into the device where it is automatically analyzed. There is no need for manual sample preparation or single-use consumables.

The test method has been incorporated in a fully automated, portable prototype device, the ***FameSpec*** FAME in jet analyzer shown in Figure 10. It detects all types of FAME and has a wide measurement range, suitable for possible future adaptations of the FAME in jet fuel specification limit. The use of QCL technology leads to high sensitivity and precision. The device features a ten-port autosampler for high sample throughput and removes the necessity for manual sample preparation or single-use consumables. This prototype was used in a refinery for six months to test jet fuel for FAME.

CONCLUSIONS

This paper shows the feasibility of a novel analytical test method based on laser mid-infrared spectroscopy for FAME in jet fuel. QCL technology for liquid sensing with improved sensitivity enables the development of portable high precision instruments. Key features of the method are its high sensitivity and precision, with a wide measurement range and the capability to detect all types of FAME. The method can be fully automated as there is no manual sample preparation necessary.

REFERENCES

[1] J. FAIST et al. “Quantum Cascade Laser”, Science, 264 (1994) 553

[2] J. FAIST et al. “High power mid-infrared (λ ~5 μm) quantum cascade lasers operating above room temperature”, Applied Physics Letters 68 (1996), 3680

[3] B. LENDL et al. “Mid Infrared Quantum Cascade Lasers for Flow Injection Analysis”, Anal. Chem 72 (2000) 1645.

[4] G. TAUBES “A New Laser Promises to Put an End to Band Gap Slavery”, Science, 264 (1994) 508

[5] R. MAULINI et al. "Broadband tuning of external cavity bound-to-continuum quantum-cascade lasers", Applied Physics Letters. 84 (2004) 1659

IASH 2017, the 15TH INTERNATIONAL SYMPOSIUM ON STABILITY, HANDLING AND USE OF LIQUID FUELS
Rome, Italy
10-14 September 2017

Understanding Precision Issues with Single Temperature Testing by ASTM D3241

George R. Wilson, III[1],

[1] *Southwest Research Institute, San Antonio, Texas, USA, george.wilson@swri.org*

When the industry last visited the precision of single temperature testing by ASTM D3241, the results were very discouraging. Despite efforts by the precision study participants to quantify the results as much as possible, the aviation industry decided to maintain the method in a form without a stated precision. The ensuing quarter century has seen significant improvements in the test equipment and, more importantly, a revolution in deposit rating that is moving the evaluation process from subjective to objective. This paper discusses the roots of the precision issue, observes how the methodology has steadily improved and offers fresh insight in how to make the single test temperature evaluations more meaningful. The key factor in understanding single temperature precision for the test will be shown to be limiting the analysis to the temperature range up to, and including, the Breakpoint temperature.

IASH 2017, the 15TH INTERNATIONAL SYMPOSIUM ON STABILITY, HANDLING AND USE OF LIQUID FUELS
Rome, Italy
10-14 September 2017

APPLICATION OF CHEMOMETRIC METHODS TO DEVOLVE CO-ELUTING PEAKS IN GC-MS DATA

Jeffrey A. Cramer, Mark H. Hammond, Thomas N. Loegel

Naval Research Laboratory, Code 6181, 4555 Overlook Ave. SW, Washington, DC, USA 20375

ABSTRACT

Fuel chromatography is inherently limited by the high complexity of petroleum fuel compositions. In practice, almost none of the fuel constituents are fully resolved from other components. This is due to a combination of 1) insufficient peak capacity for the large number of individual components within time and chromatographic efficiency constraints and 2) insufficient resolving power of the stationary phase in the gas chromatography column relative to the many structurally similar isomers or homologs present in fuel. Multidimensional approaches, longer columns and slower heating rates can offer some benefits but will not necessarily fully resolve previously co-eluting fuel compounds. Furthermore, the manual selection and fine-tuning of existing data deconvolution algorithms to optimize results for any given data collection would be overly burdensome to end-users in most realistic applications. This presentation will cover the development of novel peak deconvolution strategies suitable for automated use with a variety of data sets.

INTRODUCTION

Gas chromatography (GC) remains a primary tool for the analysis and compositional characterization of mobility fuels. As the requirement for accurate analytical data increases, the performance boundaries of fuel chromatography become a significant limiting factor. This limitation becomes increasingly important in fuel failure investigations, where minor or trace compositional artifacts, characterized by means of mass spectroscopy (MS), are often responsible for poor fuel stability or performance. Fuel chromatography is inherently limited by the high complexity of petroleum fuel compositions. In practice, almost none of the fuel constituents are fully resolved from other components due to a combination of insufficient peak capacity for the large number of individual components within time and chromatographic efficiency constraints, and insufficient resolving power of the stationary phase in the gas chromatography column relative to the many structurally similar isomers or homologs present in fuel. Multidimensional approaches, longer columns and slower heating rates can offer some benefits but will not necessarily fully resolve co-eluting fuel compounds.

The Fuel Composition and Screening Tool (FCAST), a software package developed in-house that characterizes fuel compositions via GC-MS data, sends MS mass fragment data from individual GC time slices to a National Institute of Standards and Technology / Environmental Protection Agency / National Institute of Health (NIST/EPA/NIH) Mass Spectral Library for the purposes of peak identification. When multiple compounds co-elute, the GC peaks overlap and the time-slices sent to the mass analyzer will contain contributions from all co-eluting compounds. Depending on the degree of overlap, the library search algorithm either yields incorrect peak assignments due to comingled mass spectral data, or results in information loss via the discarding of poorly resolved mass spectral data within overlapped peaks. The end-user will not see the discarded compounds in the library search report, but will see occurrences of misidentified compounds. Often, these mangled mass spectra will return compound identifications that are nonsensical and generally discarded, but there are also instances where high Match Factor (MF) values are obtained from erroneous library matches. Co-elution is also responsible for the reporting of multiple instances of the same compound throughout a single analysis.

The impact of co-elution is illustrated in Figure 1. The left of Figure 1 shows a portion of a fuel total ion chromatogram (TIC), where two peaks seem to be coinciding. However, as shown on the right side of Figure 1, this composite doublet is actually three different compounds. The chromatography was not adequate in resolving the two naphthalene isomers.

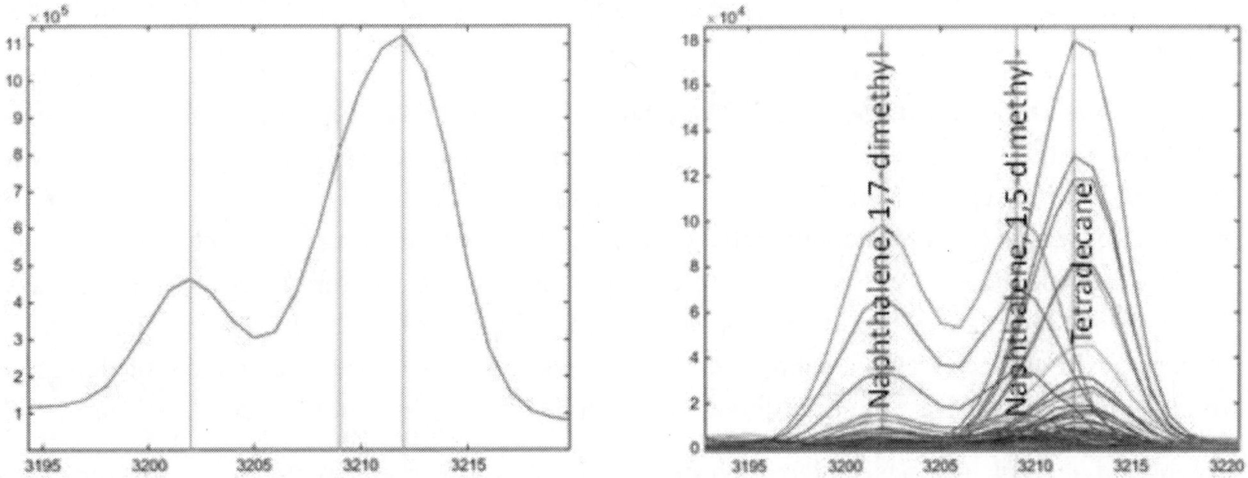

Figure 1. Portion of a fuel chromatogram showing two adjacent peaks (left) and individual ions showing two unresolved naphthalene isomers (right).

More importantly, when the unresolved peaks shown in Figure 1 were sent to the aforementioned NIST/EPA/NIH Mass Spectral Library for identification, an aromatic compound (2,6-dimethylnaphthalene) was not detected or reported. In other words, co-elution masked the presence of a potentially important analyte in the fuel. Fuel stability or performance problems that arise are often due to the presence of trace levels of contaminants or other minor changes in composition, so detailed compositional analyses of suspect fuel samples are often critical to determinations of the cause(s) of the problem(s) at hand. Sensitive methods to compare fuel compositions by GC-MS are available, both within the FCAST software and otherwise, but the sensitivity and accuracy of these comparisons depend upon the quality of the library searches.

The present work focuses on the algorithmic development of evolving factor analysis (EFA)[1] and multivariate curve resolution (MCR)[2] into an integrated technique suitable for subsequent implementation in automated software packages, such as the FCAST, in a manner that results in optimal performance while remaining user-friendly. The approach will be to use a modified version of EFA, evolving window factor analysis (EWFA), to initially estimate the number and location of individual chemical components within local regions containing convolved chromatographic peaks. This will be followed by MCR to refine estimates of the shapes of pure-component mass spectra, which will be used to provide more effective library searches and thus more accurate compositional profiles.

EXPERIMENTAL

Software. Synthetic data production and data analysis algorithms were developed using functionalities provided by MATLAB (MathWorks, Inc., Natick, MA), the PLS_Toolbox for MATLAB (Eigenvector Research, Inc.), and the MATLAB-based Calibration and Standard Toolboxes (FABI / ChemoAC Consortium). When necessary and appropriate, both mass spectral data and derived mass spectral loadings were submitted to the NIST Mass Spectral Search Program for the NIST/EPA/NIH Mass Spectral Library (NIST 11, version 2.0g) for identification purposes.

Mass Factor Determinations. The FCAST software includes an improved version of the fuel profiler algorithm,[3] which operates on mixture separations by GC-MS, that is a part of the Navy Standard Work Packages for alternative jet[4] and diesel[5] fuel certification. This algorithm has been demonstrated to be effective for the rapid compositional profiling of complex mixtures that would otherwise take unreasonable times to manually analyze. Nevertheless, one limitation of this profiler algorithm is that it reports relative contribution of each class of compounds as a percentage of the total area counts measured in each analysis, thus neglecting the effect of differing response factors among different compounds within the context of GC-MS data.

Compound-specific response factors observed with GC-MS are dependent on the molecular ionization efficiency of the compound, and to some extent, the fragmentation pattern induced by the mass spectrometer. This makes them highly dependent on molecular structure in ways that are difficult to generalize across a wide range of possible mixture constituents.[6] Additionally, mass spectrometric detection provides less dynamic range than other detection methods commonly used with GC separations, such as flame ionization detection (FID). Detection by FID, on the other hand, demonstrates exceptional dynamic range and, importantly, response factors that vary more predictably as they are largely dependent on the number of carbons in the molecule[7] and can be derived from molecular combustion enthalpies.[8]

Compound class response factors were thus empirically measured on in-house instrumentation to more accurately inform not only post-deconvolution results, but also the synthetic GC-MS data production to be described below. This was done by collecting parallel MS and FID data on a GC that was modified to send the column effluent to both detectors simultaneously. Quantitative MS measurements were obtained from blends of the following compound classes: n-alkanes (C_5-C_{16}), iso-alkanes (C_5-C_{10}), cycloalkanes/naphthenes (C_5-C_{12}), aromatics (benzene to benzene + C_6), and olefins (C_5-C_{10}). From these measurements, two mass correction factors were derived (Eq. 1):

$$\text{FID Peak Area} = \text{MS Peak Area x} \left[\frac{1}{\text{k x C}}\right] \qquad \text{Eq. 1}$$

Where, k = compound class factor; and C = carbon number correction factor. The compound class factor (k) represents the ionization efficiency of that particular class. The carbon number correction is necessary to correct for the impact of carbon number on the FID response.

Assembling Synthetic GC-MS Data. During the course of the present work, useful (i.e. both realistic and challenging) synthetic data sets were produced for the purposes of algorithm development. As will be shown later, this allowed for rigorous determinations of the conditions under which automated EFA-based peak component identifications and MCR decompositions become unreliable for fuel profiling, as well as the selection of built-in threshold values suitable for the automated quality control of algorithmic results.

The initial, unblended mass spectra for the compounds upon which to base convoluted data sets were first located in the NIST/EPA/NIH Mass Spectral Library, normalized to account for inter-instrument variances in this reference data, and adjusted by the most appropriate mass spectral response factors determined in the manner indicated previously. This last step was undertaken simply to ensure that arbitrated blend ratios would accurately translate to what one would see if an actual blend possessing these ratios were to be analyzed using in-house GC-MS instrumentation.

After individual mass spectral data vectors are thus preprocessed, they can then be individually cross-multiplied by vectors containing Gaussian curves to produce data matrices consisting of mass spectra and retention time axes. The use of Gaussian curves accurately represents the normally distributed manner in which mass spectral data can appear across multiple retention times.[9] Because chromatographic columns can vary a great deal in separation characteristics, the locations of the Gaussian curves in the chromatographic vector can be arbitrated by the end-user. This, in turn, very easily allows end-users to arbitrate that the two-dimensional data sets for individual compounds will possess identical or nearly-identical synthetic column retention times to result in co-elution. Once the appropriate two-dimensional data sets are constructed, they are then multiplied by the desired blend ratios, then added to a separate noise matrix, itself being a convolution of random noise, which can be represented by a normalized distribution of random numbers,[9] and a Gaussian distribution that simulates the hump-like background seen in realistic fuel samples. Initial development work only required the production of data sets possessing a single convoluted peak of interest, such as are represented by the example total ion chromatographs (TICs) seen in Figures 2 and 3, because convoluted peaks centralized at the apex of the hump-like background noise provided the greatest analysis challenge given any given set of data production parameters, but multi-peak data sets could easily have been produced as well.

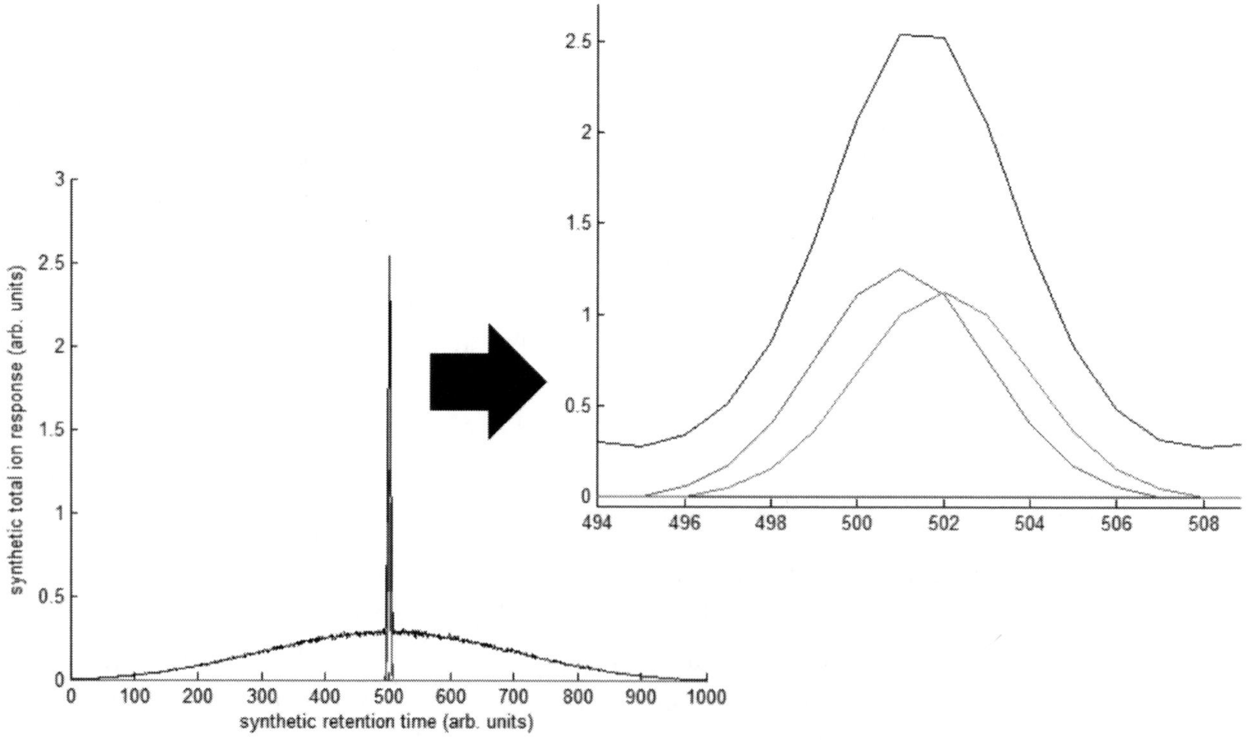

Figure 2. Example synthetic data set TIC (black), consisting of a 50/50 blend of two constituents and noise. The TIC of the p-tert-butylphenol without noise is in red, the corresponding 2-propylphenol TIC is in magenta. The p-tert-butylphenol component is higher in magnitude due to instrument response factors. Both components have Gaussian widths of 11 variables, and individual component peaks are only separated by a single retention time variable.

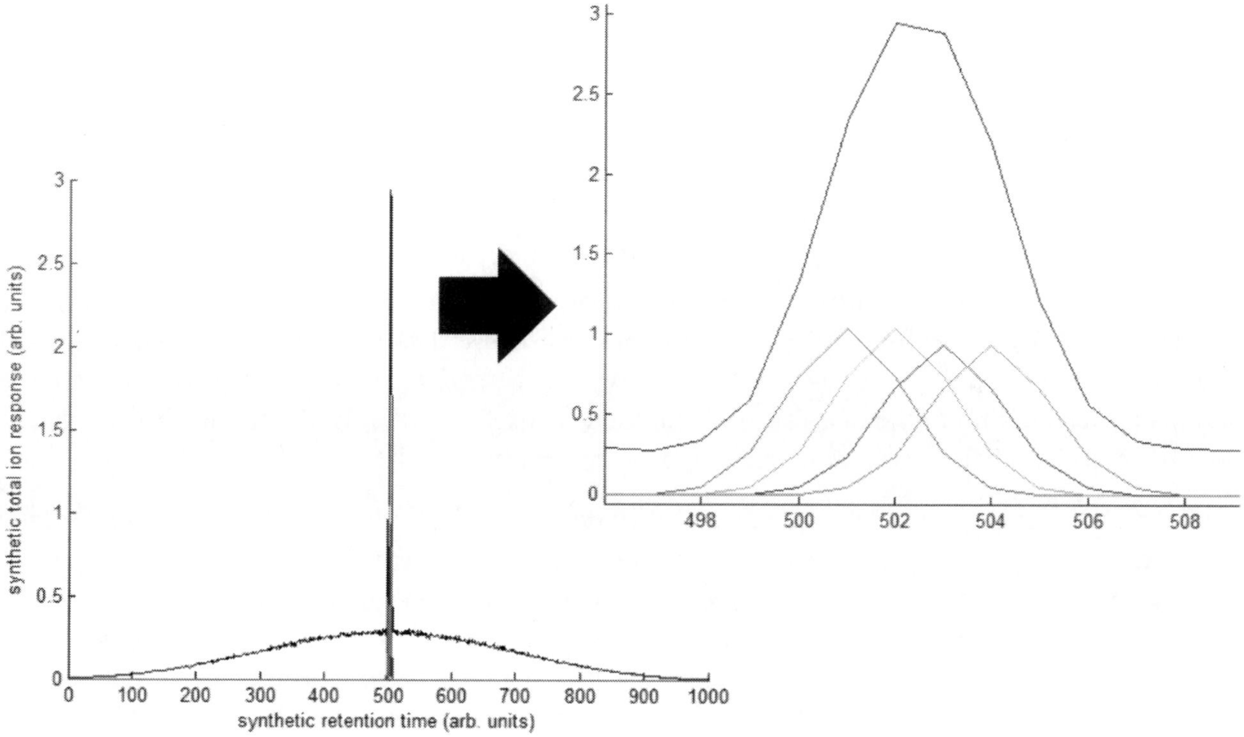

Figure 3. Example synthetic data set TIC (black), consisting of a 25/25/25/25 blend of four constituents and noise. The noiseless TIC of the p-tert-butylphenol is plotted in red, the noiseless TIC of the 2-(1,1-dimethylethyl)phenol is plotted in green, the noiseless TIC of the 2-(1-methylethyl)phenol is plotted in blue, and the noiseless TIC of the 2-propylphenol is plotted in magenta. Component magnitudes differ due to the instrument response factors. All components have Gaussian widths of 7 variables, and the individual components are staggered by a single retention time variable from each other.

GC-MS Data Acquisition. For non-synthetic data evaluations, GC-MS data sets were retrieved, from our internal fuel archives, for four real-world petrochemical fuel samples: a JP-5 jet fuel, an F-76 diesel fuel, a marine gas oil (MGO) fuel obtained from a location associated with Miami, and an MGO fuel obtained from a location associated with Singapore. These GC-MS data sets were originally acquired on an Agilent 5890 GC with an Agilent 5971 mass selective detector. Injections (1.0 μL) of the neat fuels were made with an autoinjector into a split/splitless inductor at a split ratio of 200:1. An AT-1 cross-linked polysiloxane capillary column (50 m x 0.25 mm ID, 0.20 μm film thickness) was used with an oven temperature program that initiated data collection at a temperature of 40 °C and ramped the temperature at 10 °C/min to 290 °C, holding this temperature for the remainder of the data collection. Data were acquired at column retention times from 5.1 to 70.0 minutes, at a rate of about two mass spectra per second, from 35 to 357 m/z. The chromatographic data, of course, possessed peak widths exceeding one variable along the retention time axis, which primarily accounts for the multiple singe-fuel identifications of the same compound that can be found within the results herein. These results were not collected into overall peak identity assignments to more clearly indicate the utility of the developed peak deconvolution algorithms.

RESULTS AND DISCUSSION

Initial Phenol-Based EWFA Development Work. EWFA is a modified version of EFA that basically functions by performing EFA, in the form of repeated singular value decomposition (SVD) operations, on subsets of the overall data set. These data subsets, in the case of GC-MS data, are subdivided along the retention time axis in the present work because the decomposed results corresponding to the mass spectra can then be used in NIST/EPA/NIH library searches to identify the chemical sources of underlying variance within the data.

SVD mathematically breaks a given data matrix down into its underlying linear factors, or latent variables (LVs), which are typically represented as scores, loadings, and singular values. The singular values in particular can be used to indicate how many of the LVs in the decomposed results can be considered prominent, which is important to know in most applications of EFA. Specifically, in EFA, SVD is performed on increasingly large portions of a data matrix, proceeding in both the forward and reverse directions. In the forward direction, SVD is performed on a data subset, initially defined starting from the first row/column, which increases in size by one row/column per SVD operation until the last row/column is included in the SVD. In the reverse direction, the same stepwise increase in data subset size proceeds in the opposite direction from the last row/column instead.

The theory behind EFA is that the number of prominent singular values found through SVD will increase when the ever-increasing subsets of data being analyzed possesses more prominent underlying factors than previous subsets. By going in both a forward and reverse direction, one can thus pinpoint where LVs within a data set appear, disappear, and overlap, all of which being important information when assessing convoluted GC-MS data. In addition, the loadings collected along with these singular values are also valuable information, as they will closely correspond to the actual shapes of the underlying factors themselves, i.e. the mass spectra required to identify individual components by means of a NIST/EPA/NIH library search. The use of EWFA further refines EFA results by moving a window within which to perform EFA across a larger data set, allowing for a more nuanced understanding of these appearances, disappearances, and overlaps.

MCR can also be included in the use of EWFA, functioning as a means by which to refine the shapes of the loadings produced using EWFA, typically by means of an iterative alternating least squares (ALS) algorithm. This allows the shapes of loadings of lower prominence to more accurately reflect underlying chemical phenomena, thus allowing more fully obfuscated chemical components to be effectively identified via the NIST/EPA/NIH library searches performed by the FCAST software.

The EWFA-MCR algorithm produced herein has also been designed to utilize the information produced via NIST/EPA/NIH library searches in a novel manner compared to how EFA is typically performed. This was done because NIST/EPA/NIH library searches provide an internal metric that can be used to interrogate loadings in a manner more directly related to composition than that which would be provided by their corresponding singular values. Specifically, because NIST/EPA/NIH library searches provide Match Factor (MF) values to quantify identification quality, the loadings can effectively be ranked by how accurately they represent the mass spectra contained within convoluted data, under the assumption that high-quality matches are unlikely to be primarily superfluous data artifacts. Although the manner in which underlying factors evolve within the data is thus not explicitly being modeled thanks to this disregard of singular values, the fact of their evolution remains critical to the analysis, hence this novel automated algorithm's continued designation as an EFA variation.

It had previously been known internally that phenols tend to convolute with one another during the GC-MS analysis of fuel samples, so work initially focused on convoluting, in various proportions, the spectra of four phenols possessing similar chemical structures: p-tert-butylphenol, 2-(1,1-dimethylethyl)phenol, 2- (1-methylethyl)phenol, and 2-propylphenol. Phenol-based data such as that previously shown in Figure 3 (peak width 7 variables, peak separation 1 variable) can be considered very convoluted data because the maximum values of the four individual component peaks all fall within at least some portion of each of the other three component peaks, all within the space of ten total retention time variables. Preliminary investigations was thus undertaken to deconvolve not only phenol-based data with these parameters, but also similar data produced using peak widths of 3 and 5 (expected to be representative of more routine convoluted data).

These investigations, at least at first, focused upon what EWFA window size would be most appropriate for an automated methodology (window sizes thus evaluated are all odd numbers to ensure that a central variable is available for each window to allow the results collected within said window to be easily correlated to a distinct central retention time). However, as can be seen in Figure 4, assigning a peak width parameter is not as straightforward as it might initially seem because larger peaks require larger window sizes to effectively deconvolve. While this might mean that the safest and most thorough course of action would be to simply apply the maximum possible window size to the entire EWFA analysis area, some consideration should be given to automated calculation times, which, as can be seen in Figure 5, increase significantly when larger window sizes are employed. Fortunately, the FCAST software already quantifies peak lengths along the retention time axis. This information can be leveraged to dynamically adjust window sizes to each peak, which will, in turn, provide for an adaptive strategy suitable for automation.

When applying MCR, a minimum window size of 11 is necessary to acceptably analyze the four-component peak with a total width of 8 variables (Gaussian Width 5 data, 11-8=3), and a window size of 13 is necessary to acceptably analyze the four-component peak with a total width of 10 variables (Gaussian Width 7 data, 13-10=3). The Gaussian Width 3 data analysis seems to be a bit more forgiving and is thus not being considered at present. Although adding 3 to the width of dynamically-selected window sizes would thus seem to be in order, peak lengths would necessarily be defined as odd numbers within the algorithm (rounding even numbers up to odd numbers if necessary) due to the aforementioned use of central variables for indexing purposes. Because adding an odd number to such a length would eliminate these central variables, dynamically adjusting window sizes of both 2 and 4 plus the underlying TIC peak length should be evaluated here to compare their relative merits in the context of this automated analysis.

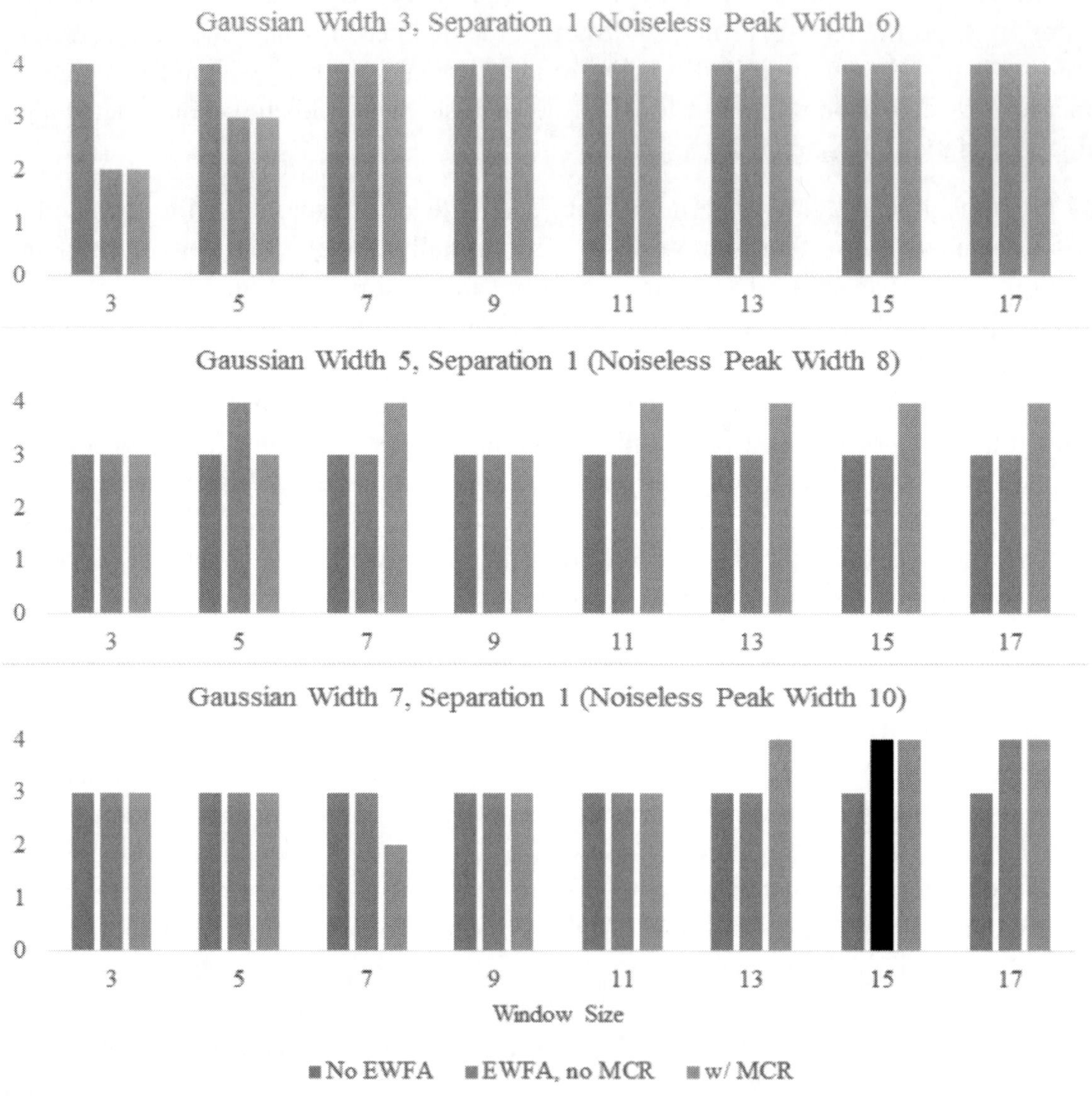

Figure 4. Compound identification results after applying EWFA and MCR, using multiple analysis window sizes, to convoluted synthetic data sets. These results represent the number of phenols detected out of four possible. Black column additional information: The compound 4-(1-methylethyl)phenol was detected in lieu of 2-(1-methylethyl)phenol, which is considered acceptable in the present circumstances given how similar the mass spectra of these two compounds would be to one another even in non-convoluted data.

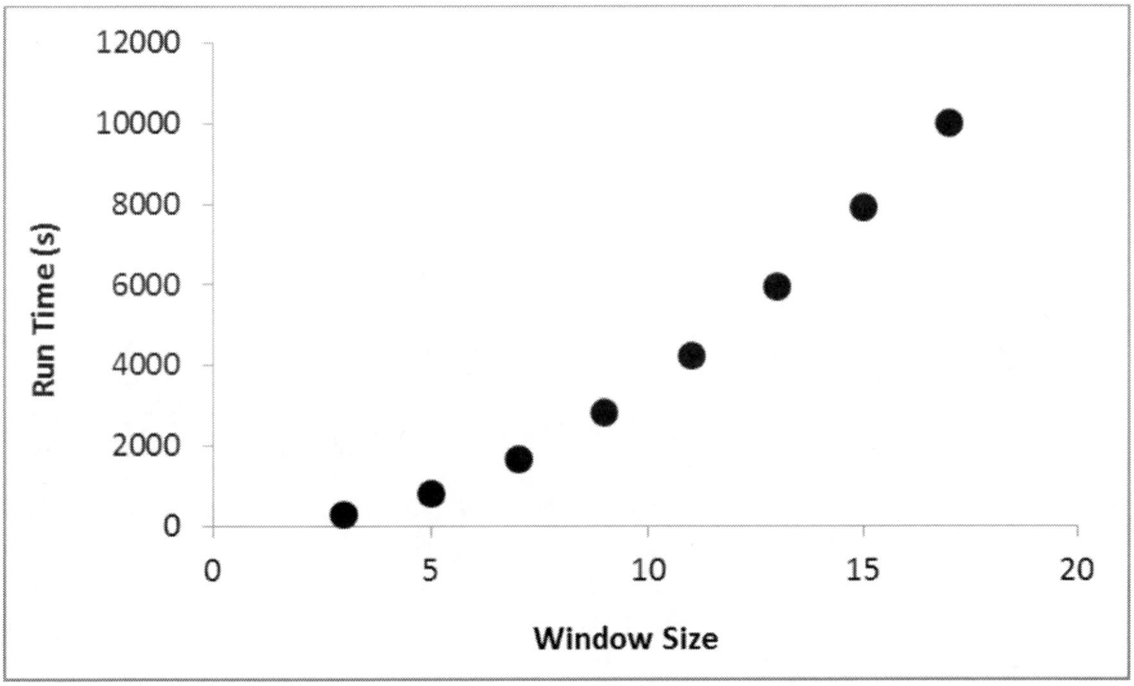

Figure 5. Example run times of representative iteration of deconvolution algorithm, as applied to the Gaussian Width 5 data shown in Table 1, with the MCR option employed, versus arbitrated window size. The same computer was used for all run time measurements.

The +2 and +4 versions of the dynamic-window EWFA-MCR algorithm were applied to synthetic data sets similar to those produced previously, except that the individual chemical component peaks began at a distance of 5 variables from each other and were brought closer together in increments of 1 variable to determine if and when the algorithm would become ineffective at uncovering underlying chemical information. It should be noted that these updated compound identification results, found in Figure 6, do not evaluate EWFA without MCR because the results in Figure 5 indicate that applying MCR reliably improves upon EWFA results, at least in the context of the present work. As one might expect, unacceptable amounts of data convolution do not truly appear until the individual component peaks are quite close to one another, though the results seen when using peak separations of 2 through 5 are still of some use insofar as they show that the deconvolution algorithm is not inadvertently causing compound identifications to decrease in quality through data overprocessing. Interestingly, the use of either +2 or +4 window size additions produces almost identical results, though the use of the +2 version of the algorithm does uncover one additional compound identity and is thus considered superior in the context of a dynamically-adjusting EWFA-MCR algorithm for which calculation times are a consideration. Although Figure 6 does show that the algorithm cannot reliably deconvolve peaks that are completely overlapping with one another (i.e. Peak Separation 0), the +2 dynamic-window algorithm is capable of extracting three chemical identities that would have remained unavailable otherwise.

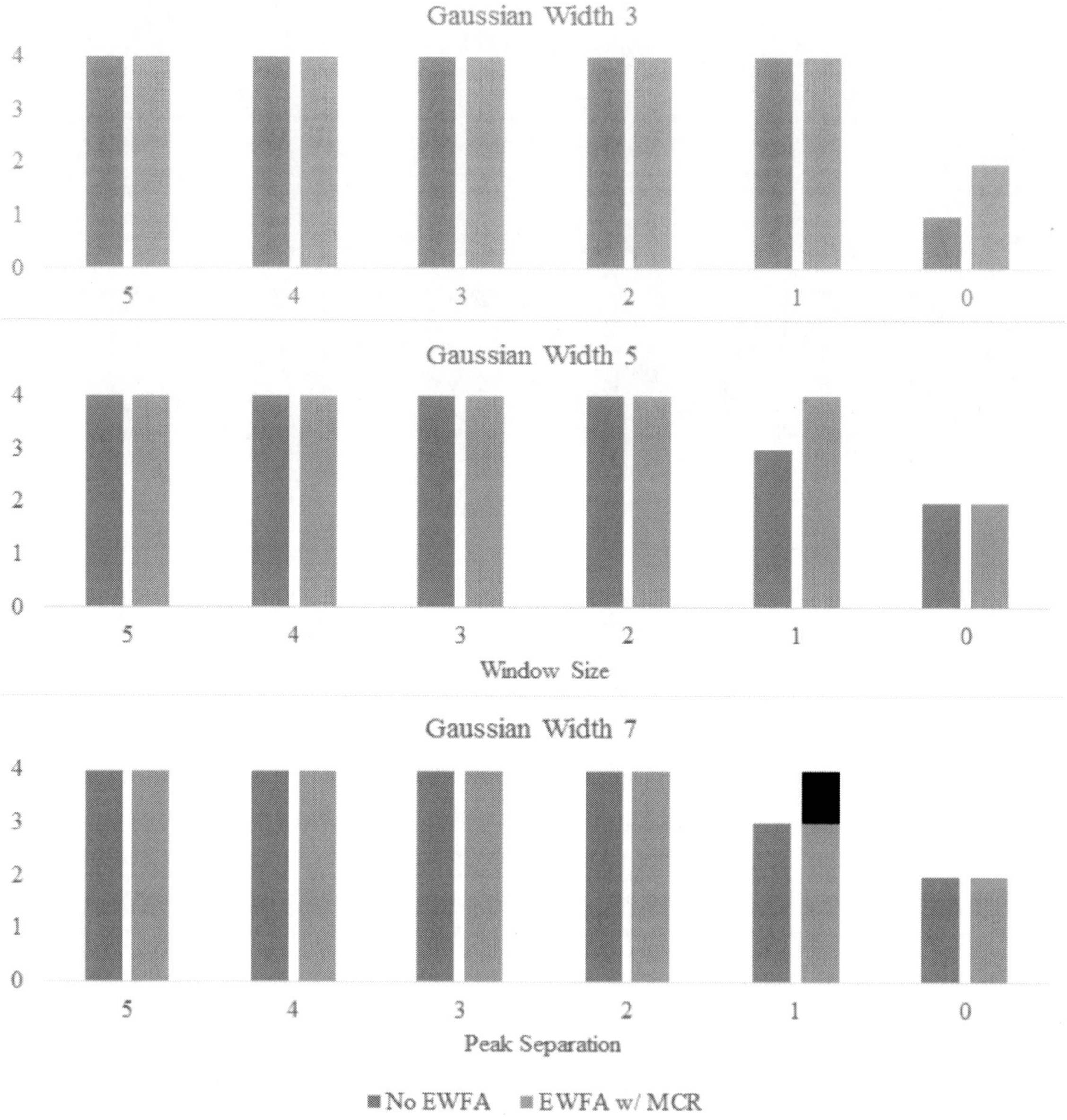

Figure 6. Compound identification results after applying EWFA and MCR, using a dynamically-adjusting window analysis size, to data sets similar to those shown in Figure 1. These results represent the number of phenols detected out of four possible. The only difference between using a size addition of +2 and +4 is indicated by the partially blackened column, which indicates where the +2 size addition identifies 4 phenols and the +4 size addition identifies 3 phenols.

Synthetic Data Trace Component Detection Trials. The results produced thus far would seem to indicate that the developed EWFA-MCR algorithm is a reliable and effective GC-MS data deconvolution method suitable for automated applications, such as would be necessary in the context of user-friendly software such as the FCAST. However, the combined algorithm has, up to this point, only been trained upon a single, albeit difficult, family of analysis challenges involving multiple convoluted phenols. In order to thoroughly evaluate the algorithm, a much wider range of convoluted data sets must be evaluated. Also, in addition to simply varying the compounds to be convoluted with one another, this wide-ranging synthetic data-based evaluation work should focus upon the algorithm's abilities to ascertain the existence of trace components from larger parent peaks, obviously an important capability for the deconvolution algorithm to possess.

Synthetic data sets were thus produced each consisting of a four-component blend of various organic chemical compounds at a combined blend ratio of 30/10/30/30, with the 10% component being defined as the trace component to explicitly be detected. To further convolute the trace component, the two 30% component peaks immediately surrounding the trace component peak were given a width of 5 and the fourth component peak was given a width of 7, thereby allowing all three non-trace components to interfere with the width-3 trace peak. Figure 7 shows an example of what such a four component TIC peak would look like. Although considerations were given to producing data sets with multiple peaks along the full lengths of their TICs for simultaneous analyses, it was quickly determined, as implied previously, that peaks appearing at locations further from the middle of the major noise hump, such as that which is apparent in Figure 7, simply possessed lower amounts of noise, with no additional analysis challenges that required overcoming. Therefore, each synthetic data set shown in the present work was made to possess only a single, centrally-located TIC peak requiring deconvolution in order to ensure that interferences from noise were not mitigated.

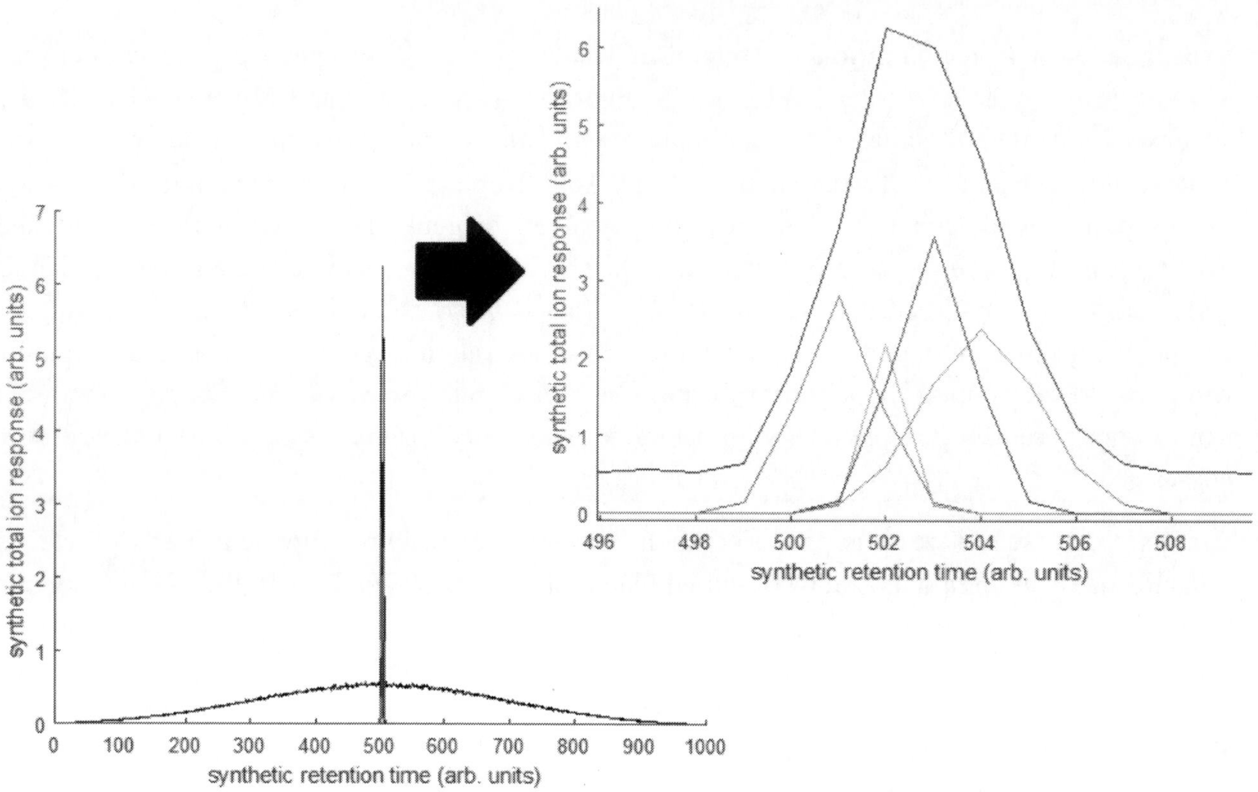

Figure 7. Example noisy synthetic data set TIC (black), consisting of a 30/10/30/30 blend of four constituents, plotted with no noise. The TIC of the trace component is plotted in green.

The specific chemical compounds upon which to base these convoluted data sets were determined, for each data set, by first randomly selecting an initial compound from the same NIST/EPA/NIH Mass Spectral Library referenced previously. These randomly selected compounds were restricted to those composed only of any number of carbon atoms, any number of hydrogen atoms, one oxygen atom, one nitrogen atom, and/or one sulfur atom, as chemical compounds possessing other potential components, such as azo compounds and halogenated compounds, were deemed to be outside of the primary scope of the present fuel analysis-based work. The chemical formula for the initial compound thus selected was then used to randomly select three additional isomers of the initial compound from the same library. These four compounds were then used in subsequent convoluted data constructions, with the actual trace component to be detected being randomly selected from the four compounds. Once 100 data sets were created under these conditions, the combined EWFA-MCR algorithm was applied to them to determine its effectiveness at determining the presence of the trace component. Furthermore, multiple MF acceptance/rejection thresholds were applied to determine their effects on the overall analysis strategy, with the best (i.e. highest) MF value being selected if multiple identifications are made. In addition, as a further validation of the work detailed previously, a +4 version of the program was again evaluated alongside to the primary +2 version. A summary of the results of this operation can be found in Table 1.

It should, of course, first be noted that the EWFA-MCR algorithm is not completely effective at uncovering the presence of all 100 trace components, as one might expect given the challenging nature of the parent data sets. Nonetheless, the algorithm allows for significant improvements in trace component detection. This is particularly striking in the case of non-heteroatomic compounds (i.e. those containing only carbon and hydrogen atoms), where an almost trivial ability to detect trace components (7% detection rate) is improved to the point of becoming decidedly non-trivial (61% detection rate). Also, consistent with the previous results, employing the +4 version of the algorithm results in no clear improvements in trace detection in the present evaluation. As the +4 version of the algorithm additionally increases calculation times by roughly 50% per data set over the +2 version of the algorithm, the decision arrived at previously to proceed with the +2 version as the primary EWFA-MCR algorithm appears to be well-founded.

Interestingly, as can again be seen in Table 1, the use of either 600 or 700 as an MF threshold value produces identical trace detection results when applying the +2 algorithm. It is only when the value of 800 is employed that a higher level of discrimination becomes apparent, though a threshold of 800 would already be a rather stringent threshold to employ when using software such as the FCAST. The fact that the results obtained using the deconvolution algorithm all possess MF values high enough to pass a reasonable detection threshold of 700 indicates that even lower threshold values could be employed with this algorithm while retaining confidence that the results obtained from the data deconvolution accurately reflect underlying chemical compositions.

	Best MF > 600			Best MF > 700			Best MF > 800		
<u>Trace Component</u>	<u>w/o EWFA-MCR</u>	w/ EWFA-MCR (+2 ver.)	w/ EWFA-MCR (+4 ver.)	<u>w/o EWFA-MCR</u>	w/ EWFA-MCR (+2 ver.)	w/ EWFA-MCR (+4 ver.)	<u>w/o EWFA-MCR</u>	w/ EWFA-MCR (+2 ver.)	w/ EWFA-MCR (+4 ver.)
only carbon/hydrogen									
Total Identified (out of 28):	2 (7%)	17 (61%)	***18 (64%)***	2 (7%)	17 (61%)	17 (61%)	1 (4%)	15 (54%)	***16 (57%)***
with nitrogen									
Total Identified (out of 17):	4 (24%)	11 (65%)	11 (65%)	4 (24%)	***11 (65%)***	10 (59%)	2 (12%)	***9 (53%)***	8 (47%)
with oxygen									
Total Identified (out of 38):	11 (29%)	29 (76%)	***30 (79%)***	10 (26%)	29 (76%)	***30 (79%)***	4 (11%)	24 (63%)	***28 (74%)***
with sulfur									
Total Identified (out of 2):	1 (50%)	2 (100%)	2 (100%)	1 (50%)	2 (100%)	2 (100%)	1 (50%)	2 (100%)	2 (100%)
with nitrogen and oxygen									
Total Identified (out of 8):	3 (38%)	***8 (100%)***	7 (88%)	2 (25%)	***8 (100%)***	7 (88%)	0 (0%)	***6 (75%)***	4 (50%)
with nitrogen and sulfur									
Total Identified (out of 2):	2 (100%)	2 (100%)	2 (100%)	2 (100%)	2 (100%)	2 (100%)	2 (100%)	2 (100%)	2 (100%)
with oxygen and sulfur									
Total Identified (out of 1):	1 (100%)	1 (100%)	1 (100%)	0 (0%)	**1 (100%)**	0 (0%)	0 (0%)	***1 (100%)***	0 (0%)
with nitrogen, oxygen, and sulfur									
Total Identified (out of 4):	2 (50%)	3 (75%)	3 (75%)	2 (50%)	3 (75%)	3 (75%)	0 (0%)	***3 (75%)***	2 (50%)

Table 1. Summary of trace compound identification results before and after applying EWFA-MCR. In those cases in which the +2 and +4 versions of the combined EWFA-MCR algorithm produce different results using the same match factor (MF) threshold, the result indicating a greater trace detection capability is highlighted in bold italics.

EWFA-MCR with Dynamic Window Sizing Applied to Petrochemical Fuel Data Sets. While promising, the combined algorithm, thus far, has only been trained and evaluated with synthetic data sets. Though these data sets have clearly presented the algorithm with non-trivial challenges during development, it remains to be seen how well the combined deconvolution algorithm performs with experimental GC-MS data collected from real-world fuel samples.

Determining whether or not the previous synthetic data sets were productively deconvolved was relatively straightforward, as the synthetic data were constructed using known quantities of known compounds. However, real-world fuels are typically never interrogated for similarly precise compositional information, simply because this information is rarely required during the course of routine fuel handling and use. Two indirect methodologies were thus developed to assess algorithm effectiveness with real-world fuel samples:

1. It can be reasonably assumed that the petrochemical jet fuel contains dodecane and petrochemical diesel and MGO fuels contain tetradecane. This assumption is consistent with the fact that, at least in certain circumstances, dodecane can function as a surrogate jet fuel and tetradecane can function as a surrogate diesel fuel. It can thus be determined whether or not the deconvolution algorithm allows these compound identities to be assigned to a larger number of mass spectra in their respective data sets and/or causes these assignments to be more accurate. In the latter case, accuracy can be quantified by applying thresholds to the Match Factor (MF) values existing between the experimental mass spectra and the NIST/EPA/NIH reference mass spectra.

2. The NIST/EPA/NIH Mass Spectral Library contains several mass spectra correlated to bioactive molecules that are most likely not present in petrochemical fuels to a significant extent. However, during the course of previous work, these bioactive molecules were identified as being present in fuel samples a non-trivial number of times. It is speculated that these dubious identifications were made because GC-MS data convolution results in mangled mass spectra that happen to look something like the Mass Spectral Library's reference mass spectra for these bioactive molecules. Data deconvolution should thus reduce the number of such identifications.

With respect to the first evaluation methodology, it will be clarified here that the EWFA-MCR algorithm, based on repeated iterations of singular value decomposition (SVD), is capable of producing multiple mass spectrum-like loadings per iteration, each corresponding to a different latent variable (LV). This, of course, allows for the deconvolution of multiple compounds from even closely overlapping data, which is a necessary part of the algorithm's core functionality. However, for the purposes of a single-compound evaluation that compares the number of times said compound is identified with or without EWFA-MCR, it is more appropriate to evaluate the results obtained from only the first, most prominent LV in each iteration, utilizing multiple MF threshold values, as summarized in Figure 1 for both the jet and the diesel fuel data.

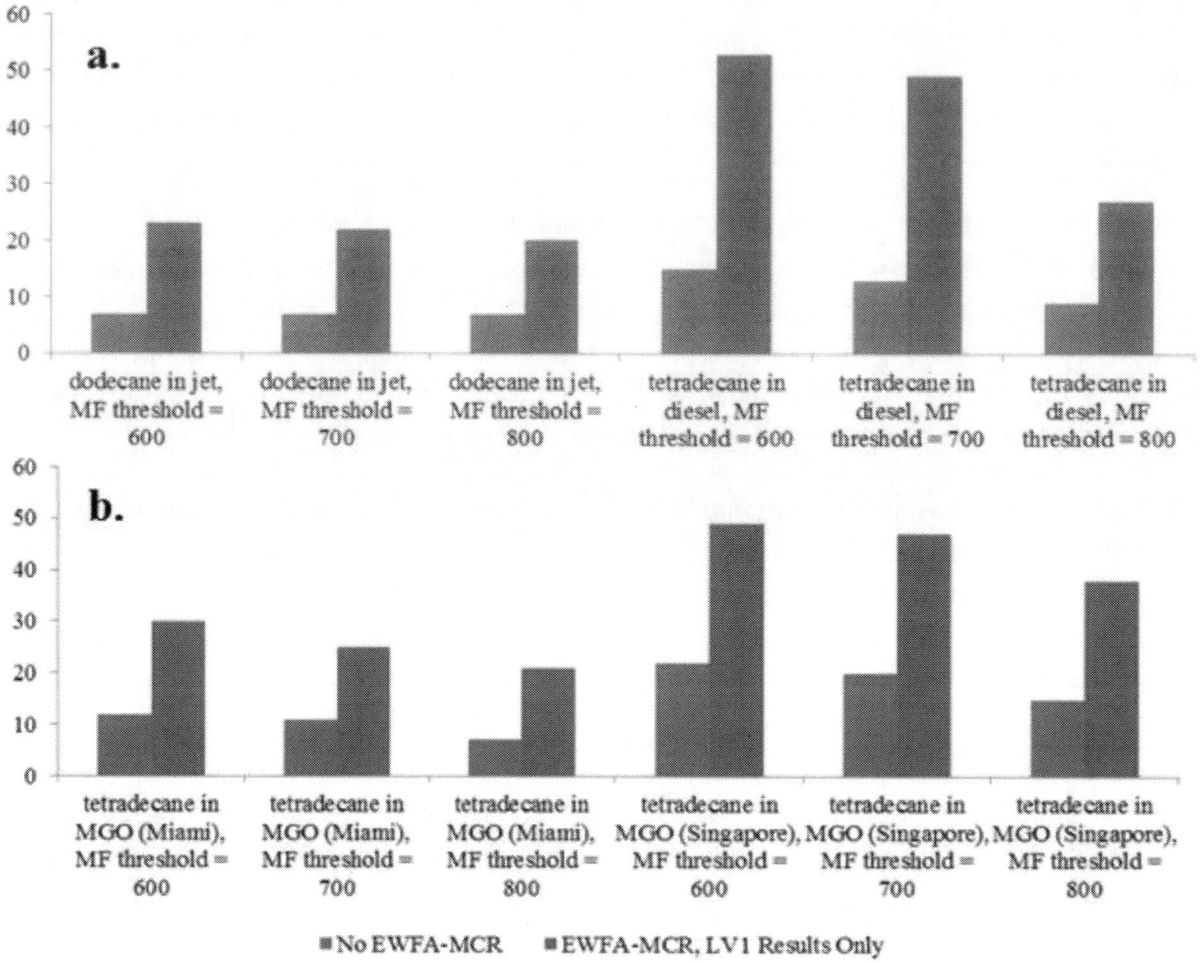

Figure 8. Summary of **a.)** the number of dodecane identifications obtained from the GC-MS data collected from a petrochemical jet fuel and the number of tetradecane identifications obtained from the GC-MS data collected from petrochemical diesel fuel, and **b.)** the number of tetradecane identifications obtained from the GC-MS data collected from two petrochemical MGO fuels, all with and without the application of the EWFA-MCR deconvolution algorithm. EWFA-MCR results are collected only from the compounds identified in the first, most prominent LV in each SVD iteration.

Figure 8 shows that the EWFA-MCR algorithm allows more mass spectra in the parent GC-MS data to be identified as a compound that is likely to be present in the underlying fuel. This is promising, as is the fact that this improvement is based on deconvolved results yielding sufficiently high MF values that the overall trend remains consistent when utilizing the various MF threshold values. Although the use of EWFA-MCR is not required to find these compounds in the first place, the increased number of identifications across all retention time axes would seem to indicate that convoluted data, perhaps existing at the edges of convoluted dodecane-heavy and/or tetradecane-heavy chromatographic peaks, is effectively deconvolved using the algorithm, allowing for more accurate and thorough compositional assessments of fuel samples.

Figure 9, meanwhile, shows similar results for the numbers of identifications obtained for twelve bioactive molecules. In this figure, however, the results from the first five LVs are shown, as opposed to only the results from the first LV as shown in Figure 8. This was done to more clearly indicate that each of these groupings of bioactive molecule identifications weren't simply made less prominent, but were eliminated almost entirely, through the use of the EWFA-MCR algorithm, regardless of the MF threshold used. The primary exception to this trend is in the case of pulegone identifications, but this exception does not overly detract from the demonstrated effectiveness of the EWFA-MCR algorithm with respect to the deconvolution of real-world fuel data. It should further be noted that the repeated identifications of pulegone made during the present work, across multiple petrochemical fuel samples, might indicate that this molecule is actually present in petrochemical fuels to a non-trivial degree.

CONCLUSIONS

The combined EWFA-MCR algorithm with dynamically adjusting window sizes developed during the present work has been shown to be effective at deconvolving the various synthetic and real-world GC-MS data sets evaluated. The synthetic data has been used to confirm that the combined algorithm improves upon trace component detection capabilities such as can be provided by software such as the FCAST, and the real-world data shows that applying the algorithm increases the number of identifications of compounds likely to be present in a petrochemical fuel while also decreasing the number of identifications of compounds not likely to be present in a petrochemical fuel.

Any investigation that depends upon the gas chromatographic analysis of fuel constituency would benefit from the improved compound-level peak resolution that this combined EWFA-MCR algorithm provides. This includes case-by-case fuel failure analysis, the elucidation of contaminants, the detection of adulterants, and compound class profiling. In a more general sense, this automated algorithm has been shown to provide for a more sensitive and accurate fuel analysis capability to solve downstream fuel quality issues without requiring specialized end-user knowledge.

ACKNOWLEDGEMENTS

This work was funded by the Defense Logistics Agency Energy (DLA-E, program manager Philip Chang).

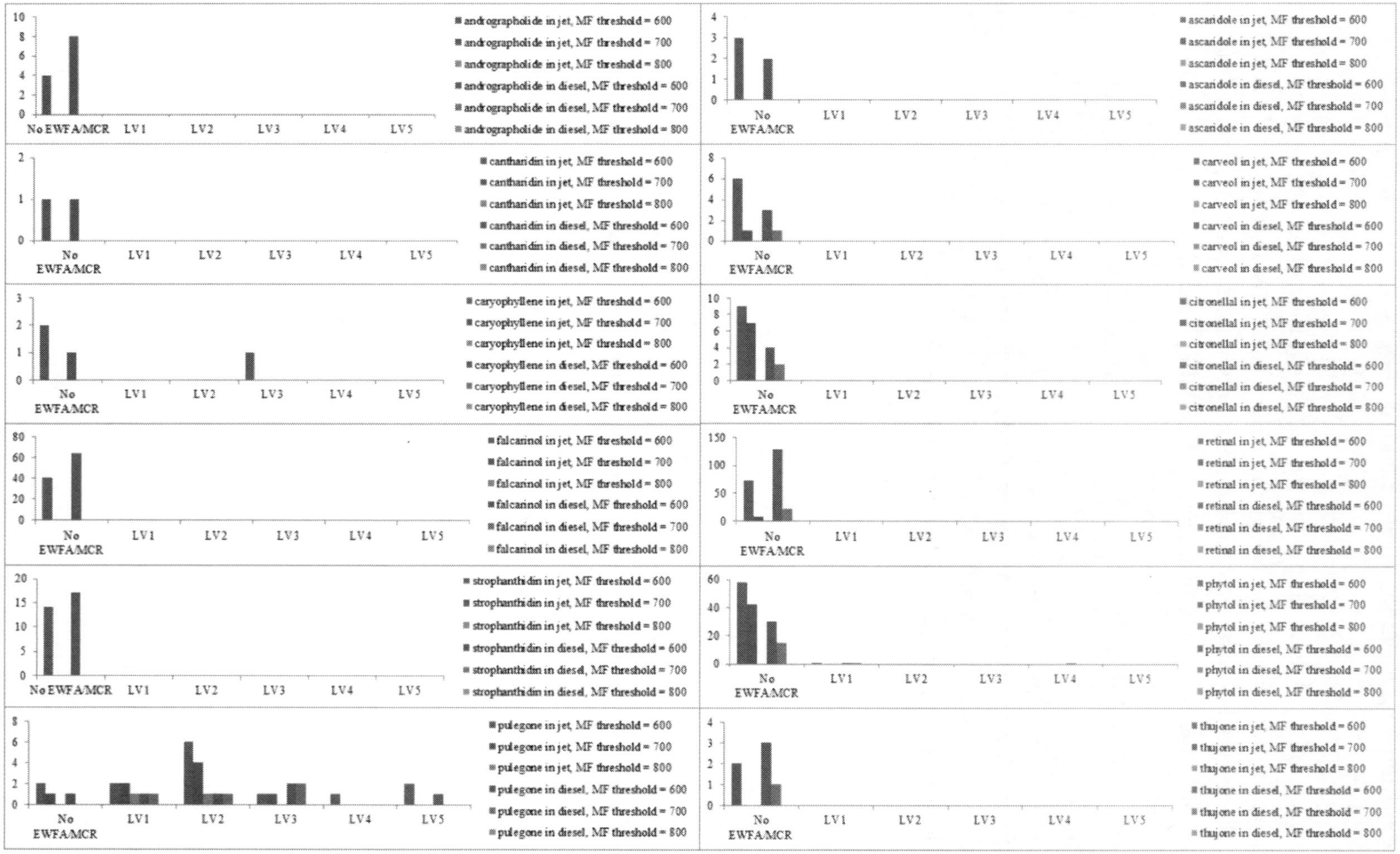

Figure 9a. Summary of the number of various bioactive molecule identifications obtained from the GC-MS data collected from a petrochemical jet fuel and a petrochemical diesel fuel, both with and without the application of the EWFA-MCR deconvolution algorithm. EWFA-MCR results are collected from the compounds identified in the first five LVs in each SVD iteration and grouped separately.

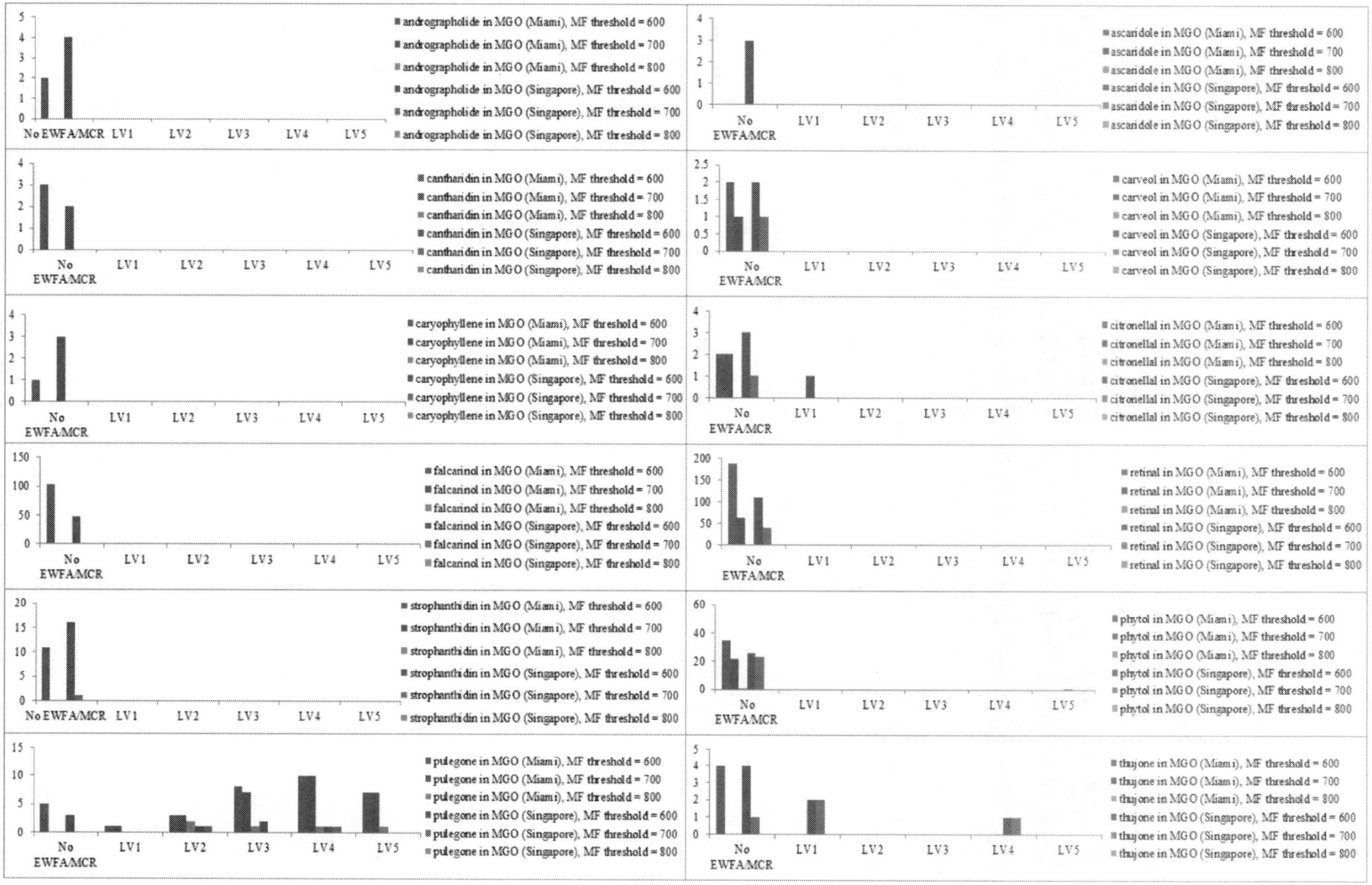

Figure 9b. Summary of the number of various bioactive molecule identifications obtained from the GC-MS data collected from MGO fuels, both with and without the application of the EWFA-MCR deconvolution algorithm. EWFA-MCR results are collected from the compounds identified in the first five LVs in each SVD iteration and grouped separately.

LITERATURE CITED

[1] Maeder, M. *Analytical Chemistry*, **1987**, *59*, 527-530.

[2] Parastar, H; Tauler, R. *Analytical Chemistry*, **2014**, *86*, 286−297.

[3] Begue, N.J.; Cramer, J.A; Von Bargen, C.; Myers, K.M.; Johnsn, K.J.; Morris, R.E. *Energy and Fuels*, **2011**, *25*, 1617-1623.

[4] Standard Work Package, Naval Fuels & Lubricants CFT Shipboard Aviation Fuel, JP-5 Qualification Protocol for Alternative Fuel / Fuel Sources. SWP44FL-006 04 March 2011.

[5] Standard Work Package, Naval Fuels & Lubricants CFT Shipboard Qualification Protocol for Alternative Fuel / Fuel Sources. SWP44FL-005 16 February 2011.

[6] Fitch, W.; Sauter, A. *Analytical Chemistry*, **1983**, *55*, 832-835.

[7] Jorgensen, A., Picel, K., Stamoudis, V., *Analytical Chemistry*, **1990**, *62*, 683-689.

[8] Saint Laumer, J., Cicchetti, E., Merle, P., Egger, J., Chaintreau, A., *Analytical Chemistry*, **2010**, *82*, 6457-6462.

[9] Dixon, S.J.; Brereton, R.G.; Soini, H.A.; Novotny, M.V.; Penn, D.J. *Journal of Chemometrics*, **2006**, *20*, 325-340.

IASH 2017, the 15th INTERNATIONAL SYMPOSIUM ON STABILITY, HANDLING AND USE OF LIQUID FUELS
Rome, Italy
10-14 September 2015

Selective Isolation of Cyclic Sulfides from Jet Fuel and Their Contribution to Fuel Oxidation Rates.

Paul Rawson[1,2], and Sylvester Abanteriba[2],

[1] *Defence Science and Technology Group, 506 Lorimer St Fishermens Bend, Victoria, Australia, Paul.Rawson@dsto.defence.gov.au*

[2] *School of Aerospace, Mechanical and Manufacturing Engineering, RMIT University, 124a Latrobe Street, Melbourne, Victoria, Australia*

ABSTRACT

A novel method for selective isolation of cyclic sulfides from aviation fuels has been developed. The method is based on ligand exchange chromatography using silica gel loaded with copper sulfate. The distribution and type of cyclic sulfide was further elucidated using gas chromatographic techniques with the dominant species observed being thiophanes, thianes and dihydro benzothiophenes. Two representative saturated cyclic sulfides were selected for thermal stability assessment using a quartz crystal microbalance technique, where rapid oxidation rates in a model fuel was observed.

IASH 2013, the 15TH INTERNATIONAL CONFERENCE ON STABILITY, HANDLING AND USE OF LIQUID FUELS
Rome, Italy
10-14 September 2017

DEVELOPMENT OF A NEW AUTOMOTIVE FUEL FILTER TEST METHOD INCORPORATING VIBRATION AND CYCLIC FLOW TEST PARAMETERS

Larry Hollingsworth[1] and Gary Bessee[2]

[1] Southwest Research Institute, 6220 Culebra Road, San Antonio, TX 78238
USA lhollingsworth@swri.org
[2]Southwest Research Institute, 6220 Culebra Road, San Antonio, TX 78238
USA. gbessee@swri.org

Key words: Filtration, vibration, cyclic flow, particle counting, standardized test methods

ABSTRACT

SAE and ISO committees are pursuing new test methods to simulate "real-world" applications for fuel filters to include vibration and cyclic flow operating conditions. Currently, most fuel filters are evaluated using *ISO 19438, Diesel fuel and petrol filters for internal combustion engines* or *SAE J1985, Fuel Filter - Initial Single-Pass Efficiency Test Method*. Similar methods exist for hydraulic filters (ISO 16889) and engine oil filters (ISO 4548-12, SAE HS-806). These methods evaluate the test filter under steady conditions, e.g., steady flow rate, no controlled vibration input, and constant temperature. The test filter is challenged with a standardized test dust with a known particle size distribution and contamination level. In most filtration standards, on-line particle counters are used to count the dust particles upstream and downstream of the test filter to determine the filter's particle removal efficiency, down to 4-µm. OEMs in particular are looking for improved filter test methods to include dynamic inputs because they are experiencing wear issues and do not feel the current methods provide adequate assessment of the filters for selecting the proper ones for their vehicles and equipment.

Attempts have been made to create filtration efficiency test standards that include cyclic flow, such as *ISO 23369 Hydraulic fluid power -- Multi-pass method of evaluating filtration performance of a filter element under cyclic flow conditions.* This test method was initially drafted in 1999, and the most recent draft was recently withdrawn before it reached publication. Presently, an industry based round robin evaluation of the most recent draft using a specific filter is being planned, but the standard has never been approved and formally published. Another cyclic flow test standard, *ISO 14085-3:2015 Aerospace series -- Hydraulic filter elements -- Test methods -- Part 3: Filtration efficiency and retention capacity*, is currently being evaluated in a round robin to determine repeatability and reproducibility. A third test method, *ISO 19612 Diesel engines – fuel filters —Multi pass method of evaluating filtration performance of a fuel filter under cyclic flow conditions in combination with mechanical vibration*, has been in the draft stage for over two years with little progress made during that period. Finally, the hydraulic filter efficiency tests for U.S. Army helicopters, MIL-DTL-8815/31, 32, and 33, incorporates both cyclic flow and controlled mechanical vibration. SwRI has been a major contributor to the development of these test specifications with the U.S. Army and their suppliers. SwRI is one of the two U.S. companies that have achieved approved status for performing these tests for the filter element manufacturers.

The objective of this research is to complete a test standard development for heavy duty diesel fuel filters that includes cyclic flow and vibration by leveraging our knowledge of both filtration efficiency testing and standards creation. From our experience with MIL-DTL-8815/31-33 and the technical advisory committee meetings for ISO and SAE, test standard development for filters that includes cyclic flow and vibration has been inefficient for two main reasons. First, filtration manufacturers and associated testing companies do not have the expertise in structural dynamics or the capital equipment necessary to attempt the development. Second, standards organizations such as ISO or SAE can take multiple years to complete the process of drafting a standard, obtaining member agreement, validating the standard through laboratory round robin evaluations, and publishing a completed document. Both of these inefficiencies can be reduced at SwRI.

OEMs have requested filter evaluations that are more representative of the application because they are experiencing wear issues and do not feel the traditional methods provide adequate assessment for selecting filters for their vehicles and equipment. A new fuel filter test method is under development by SwRI for the heavy-duty vehicle class. This method includes multipass filtration efficiency with particle counting, an analysis of the structural dynamics of a spin-on fuel filter, vibration input to the filter, and flow rate changes. An envelope of operating parameters for this class of filters was first defined. Second, a dynamic model of the filter was created and verified with modal testing to identify structural properties. Third, effects of identified structural resonances and flow cycling on filtration efficiency were evaluated using multipass filtration efficiency tests. The culmination of the project will be to complete a test procedure and evaluate candidate filters.

INTRODUCTION

Southwest Research Institute (SwRI) has been performing research on the effects of vibration on filtration in diesel engines since 2000. The original research was performed as part of a consortium evaluating filtration performance on a CAT 3406E and a Cummins N14 engine. The fuel systems for these engines used unit injectors with injection pressures of approximately 1200 bar (18,000 psi). Figure 1 provides an example of the accelerometer data achieved during this program comparing on-road vibration to a non-firing engine in a test cell. In this study, the vibration from the two applications were found to be very similar. Further research was performed to determine if the engine vibration affected the filtration performance. Two tests were performed at the same operating conditions with a difference in the filter mounting. One filter was remote mounted and the second was engine mounted. As shown in Table 1, the engine mounted configuration had approximately twice as many particles pass through the filter than remote mounted.[1,2,3]

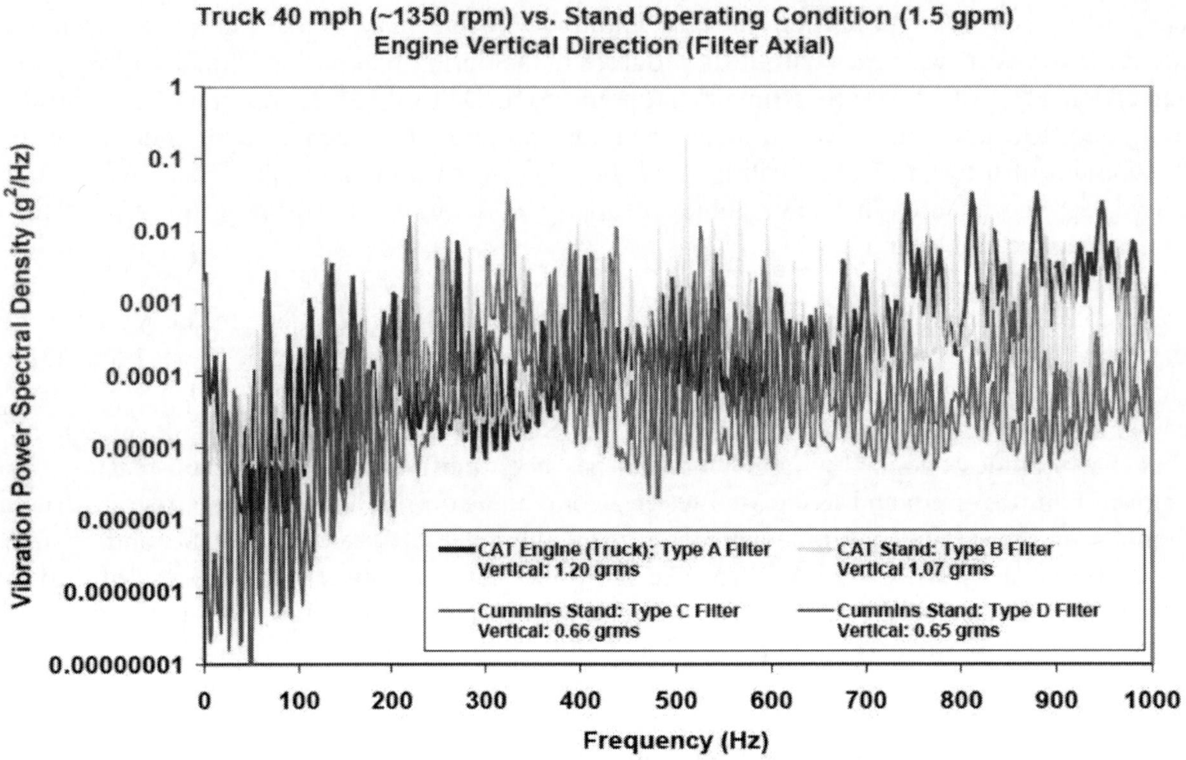

Figure 1. Vibration Comparison between On-road vehicle and Test stand

Table 1. Comparison of Filtration Efficiency between Remote Mounted and Engine Mounted Filters Under Dynamic Load, Counts per 100 mL

Time, Hrs.	Engine Mounted			Remote Mounted		
	6-µm (c)	10-µm (c)	14-µm (c)	6-µm (c)	10-µm (c)	14-µm (c)
5	254,496	10,996	648	162,052	5,704	200
10	272,980	17,120	1,212	191,076	8,164	376
15	233,468	16,464	1,188	170,536	8,392	648

Subsequent research was performed once high pressure common rail (HPCR) fuel systems were developed to meet to meet new emission requirements.[4,5] This research also demonstrated the differences in filtration performance under dynamic loads and the reduction of critical abrasive particle size being approximately 2-3µm (c) versus 6-7 µm (c) for unit injectors.

As HPCR fuel systems became standard, Bosch published International Standards Organization (ISO) 4406[6] fuel cleanliness code requirements for fuels used in the HPCR fuel systems to be 13/10/5 with no limits on operation and nothing greater than 18/15/10 for any application. This research demonstrated real-world results, but industry had difficulty accepting data from an engine that wasn't theirs and tested only at a constant speed. Therefore, this concept was abandoned.

Seventeen years later, similar activities have been ongoing at the Society of Automotive (SAE) and ISO technical committees to develop a new test method for evaluating filtration performance under vibration and cyclic flow that should improve the original concept but will likely take years to accomplish.

OBJECTIVE

Realizing the industry will not have a standardized method for years and there is an immediate need for a test protocol, SwRI is working with some of the heavy-duty original equipment manufacturers (OEM) to determine their specific needs. Using their input, SwRI has initiated a program to 1) translate these parameters to an actual filter and test stand, 2) create a model of a filter to approximate structural resonance frequencies, 3) evaluate instrumented filters to verify the model predictions are appropriate, 4) draft a proposed test method, and 5) verify the test method will differentiate filtration performance under the proposed test conditions. Instead of waiting years for the industry to agree on a method, SwRI will have a method the industry can use within one year. Because this is an on-going project, the final test method and verification results are not yet available. The goal for the project is to be completed by the end of the 2017 calendar year.

This paper will present results from the 2nd quarter of this project where the scope of work was identification of filter system resonance modes/frequencies using FEA for a select filter type including mounting structure resonances, inner filter core resonances, and outer shell resonances. Verification of the model's predictions using an instrumented filter on a shaker system is also being conducted. Extending the knowledge of the structural resonances to a filter efficiency test, various controlled vibration inputs that include and exclude system resonance were used to assess their effect on filtration efficiency. Preliminary test results from this evaluation are included.

RESULTS - Finite Element (FE)

The first step of the Finite Element (FE) and experimental testing phase of this program was to design a mounting fixture to support the filter on the shaker table. The fixture will be used during the model verification effort of the filter assembly's dynamic response as well as the filter efficiency testing with vibration. The fixture material is 6061-T6 aluminum, and the current fixture design alone weighs approximately 16 lbs. The FE model of the fixture included a lump mass of 12.2 lbs representing a commercially available filter head with the filter full of fluid. A schematic of the mounting fixture, filter head, and filter is shown in Figure 1.

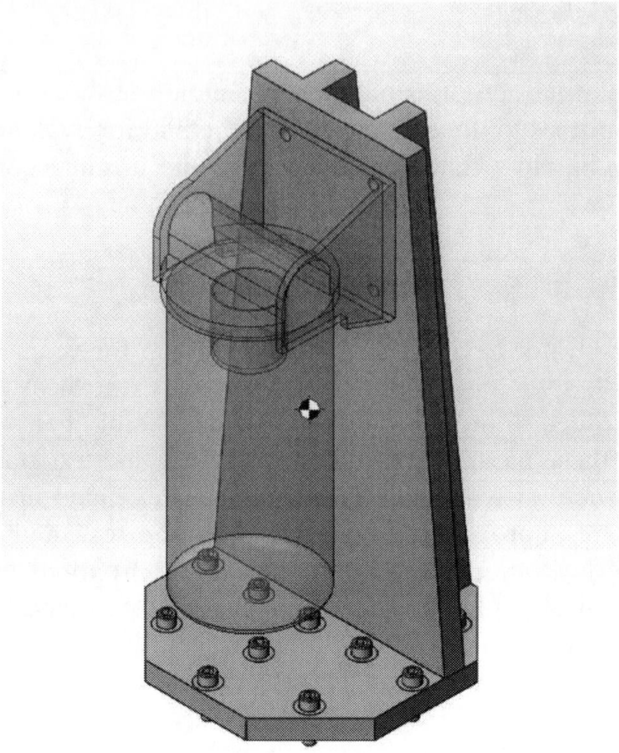

Figure 1. Fixture for Mounting Filter during Vibration Testing

The fixture design was modified iteratively based on FE results with a target of the first critical resonance mode being above 500 Hz. This threshold frequency is above the vibration input frequencies for the class of engines associated with the filter application. The fixture was also designed so that the system center of gravity, including the installed and filled filter, would be over the center of the vibration table when installed.

The FE model of the initial design concept of the fixture predicted the first and second order bending response modes close the 500 Hz target. However, when the mass of the filter attached to the fixture was included, both of these modes dropped to the 2-300 Hz range. Had this fixture been built and used blindly without knowledge of the filter to be installed, the motion of the filter would have been extreme if it was driven by an input with the frequency content that aligned with this natural resonance. Given the expected vibration inputs for the class of engine in this application, this would very likely be the case. Hence, the need for determining the structural properties of the system before building and testing the filter performance were reinforced early in this project.

The results of the FE modal analysis for the current fixture design are included in Table 1. The only mode below 500 Hz was found at 417 Hz, but this is a torsion mode and not expected to be excited directly by the vertical vibration input from the shaker. The first mode expected to be excited by a vertical input is at 604 Hz. This mode is a first order bending response as depicted by the deflection map in Figure 2. A second fore-aft bending mode is predicted at 661 Hz, and its deflection shape is illustrated in Figure 3. The maximum response associated with this mode occurs at the top of the fixture and is due to the requirement that this region be open to accommodate filter mounting hardware. It should be noted the FE analysis does not account for potential stiffening due to the installation of the filter head, thus these frequencies could be higher.

Table 1. Modal Analysis Results for Filter Vibration Fixture

Mode ID	Modal Frequency (Hz)	Description
1	417	Torsion about Y-Axis (vertical)
2	604	Bending about X-Axis
3	661	Combined Bending about X- and Y-Axis
4	1203	Torsion about Y-Axis with Bending about Z-Axis
5	2127	Bending about Y-Axis
6	2366	2nd Order Torsion about Y-Axis

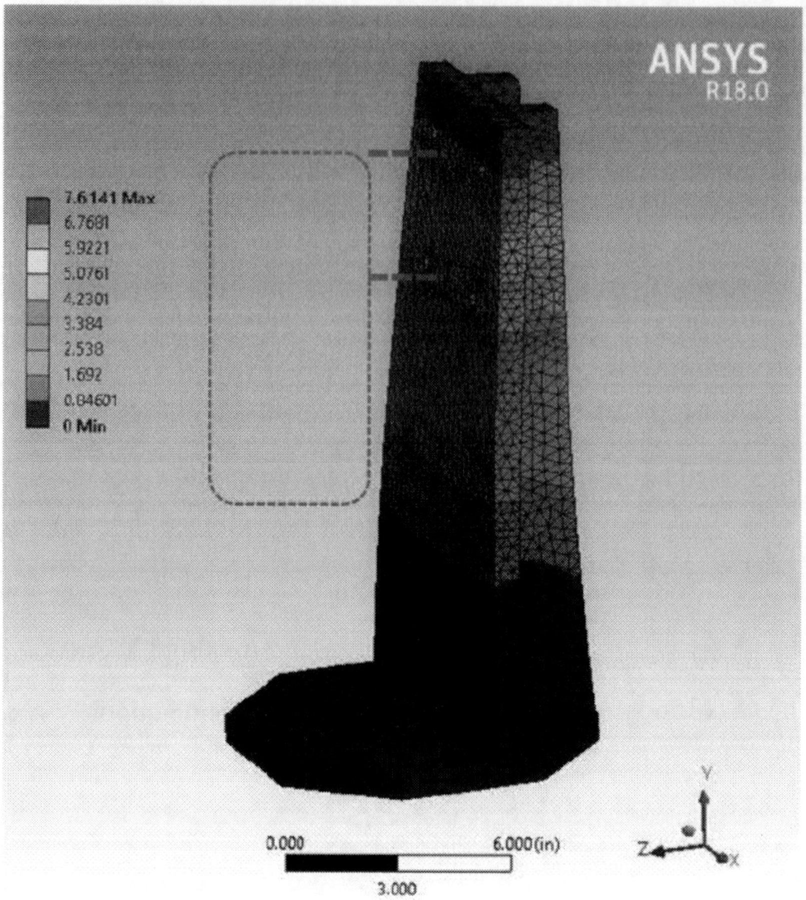

Figure 1. Vibration Fixture FE Modal Analysis Results – Mode #2

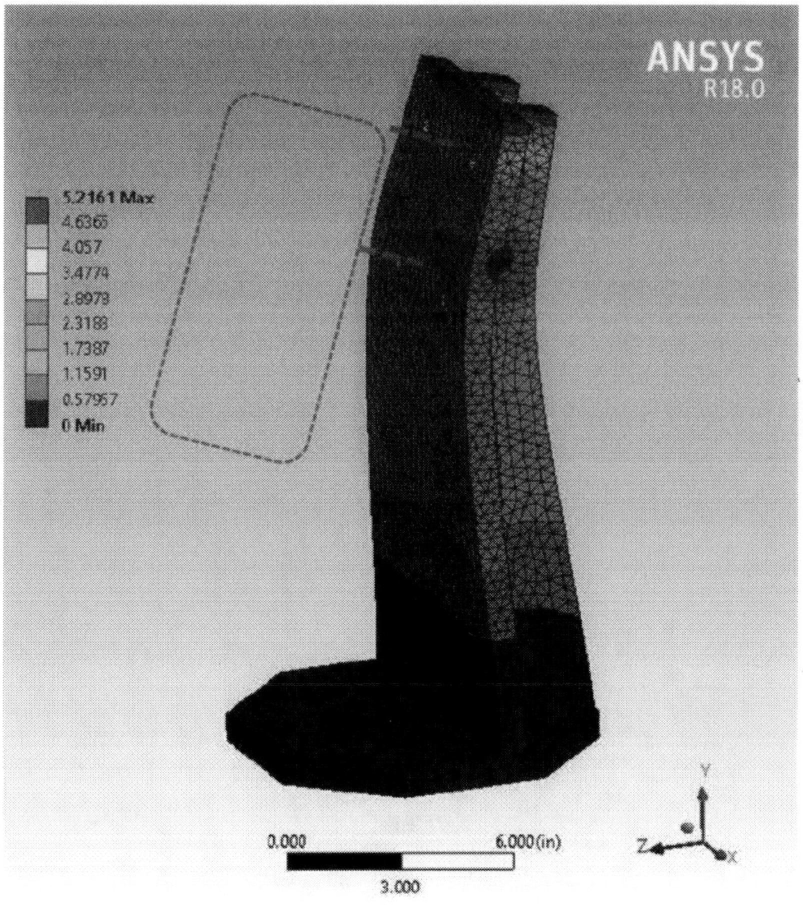

Figure 2. Vibration Fixture FE Modal Analysis Results – Mode #3

The next objective of the FE analysis was to build a model of a representative fuel filter and investigate the overall dynamic response of this filter assembly and its individual internal components. This information will prove insight into the vibratory motion of the filter media and support tube and outer housing in the frequency range of the expected input. The model will also help guide the experimental vibration mode verification phase by predicting mode shapes and frequencies and giving direction in selecting accelerometer locations for mode shape sensing.

A screenshot of the generic fuel filter model is shown in Figure 4. The model was constructed in ANSYS R18.0 using geometry provided by a SolidWorks CAD model. All filter components were modeled as solid bodies with bonded contact between the components. The component materials used in this model are listed in Table 2. The properties selected for these materials were based on various manufacturer input, published information, and are not unique to the filter being modeled. The FE mesh used to perform the modal analysis is shown in Figure 5. The mesh sizing parameters were chosen to minimize model solution times, yet obtain accurate results for a modal analysis of fundamental modes. The resulting mesh is not suitable for stress calculations. The boundary conditions for the model were fixed displacements and rotations at the threaded attachment which simulate in-service conditions.

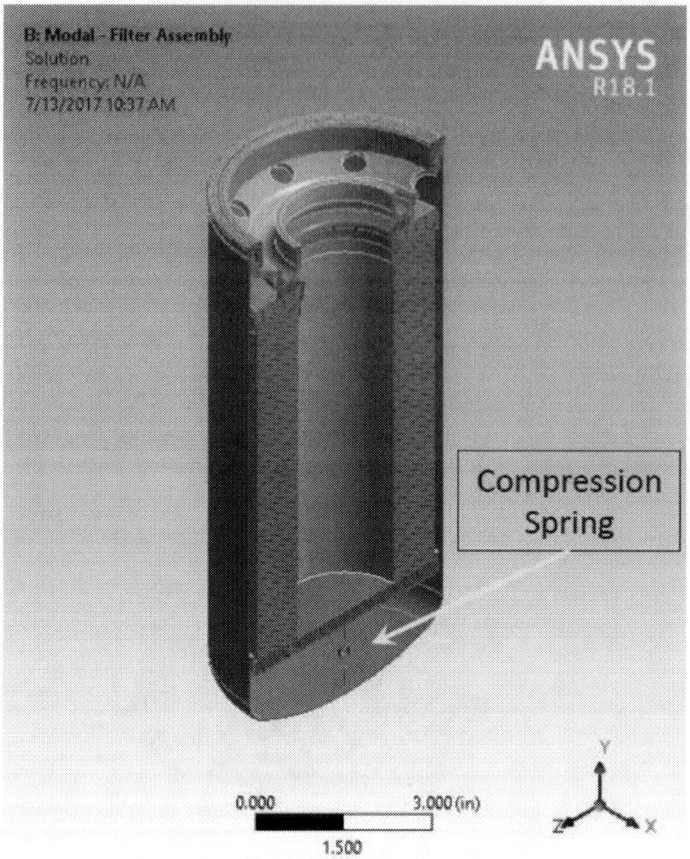

Figure 4. Generic Diesel Fuel Filter – FEA Model – Dry Filter

Table 2. Filter Component Material Properties

Components	Material
Threaded attachment, outer filter screen, outer can	Steel 1010
Filter elements	Cardboard
Inner filter screen	Nylon
Internal assembly endcaps	Polyurethane
Gaskets	Nitrile Rubber

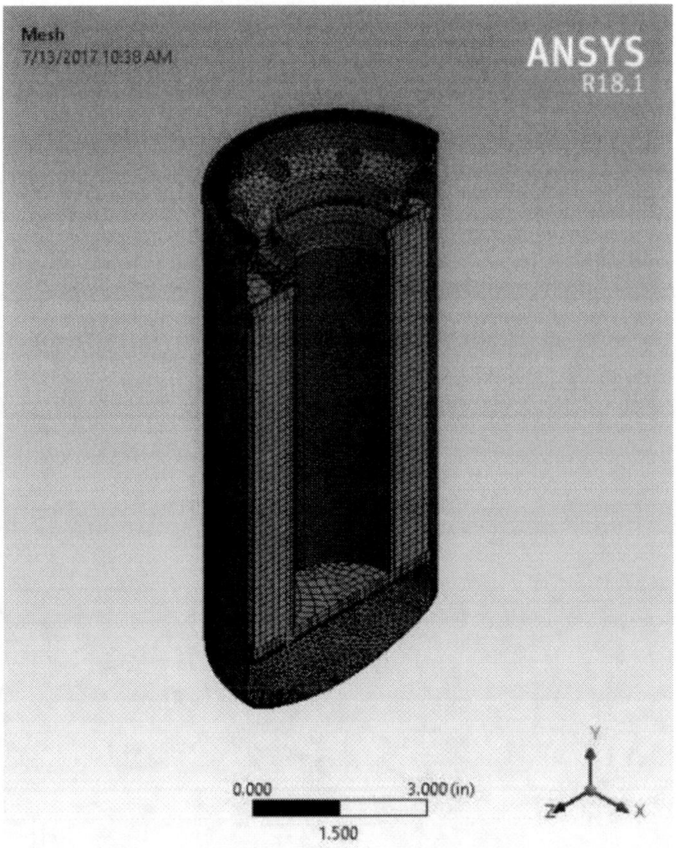

Figure 5. FE Mesh Used for Modal Analysis

Modal analysis results for five of the significant modes for the dry filter case are given in Table 3 with the corresponding mode shapes depicted in Figures 6 through 15. In these figures, a color scale is used to represent the deflection amplitudes of the mode shape. Gray represents no deflection, royal blue represents minimal deflection, and red represents maximum deflection. Mode 1 at 406 Hz is described as a side-to-side bending of the center tube as illustrated in Figures 6 and 7. For this mode, the filter media and the inner steel tube are exhibiting a first-order, cantilever bending response. The first mode involving the outer housing is predicted at 1047 Hz, a shell mode with a circumferential wavenumber of four. This mode is illustrated in Figures 8 and 9. The next mode involving the outer housing is predicted at 1161 Hz and illustrated in Figures 10 and 11. This mode is a first-order, cantilever bending shape about the fixed end of the filter assembly. A higher order mode of the internal filter elements is illustrated in Figures 12 and 13. This mode a torsional mode about the longitudinal axis of the center tube. The final mode included is the first longitudinal mode of the internal tube at 2017 Hz and is illustrated in Figures 14 and 15. In this dry filter case, this natural frequency is well above the expected 0-500 Hz frequency range of interest.

Table 3. Modal Analysis Results for Dry Filter Assembly

Mode ID	Modal Frequency (Hz)	Description
1	406	Cantilever Bending Mode of Inner Tube + Media
2	1047	4-Node Outer Shell Mode
3	1101	Cantilever Bending Mode of Outer Shell
4	1470	Torsional Mode of Inner Tube + Media
5	2189	Longitudinal Mode of Inner Tube + Media

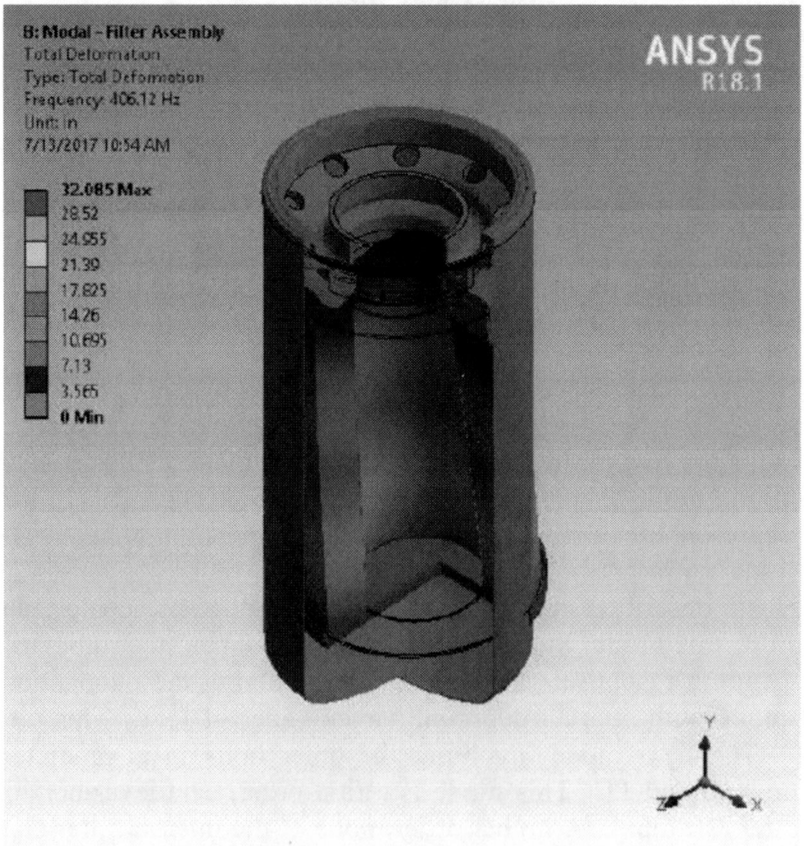

Figure 6. Dry Filter Assembly FE Modal Analysis Results – Mode 1 Bending of Filter Elements

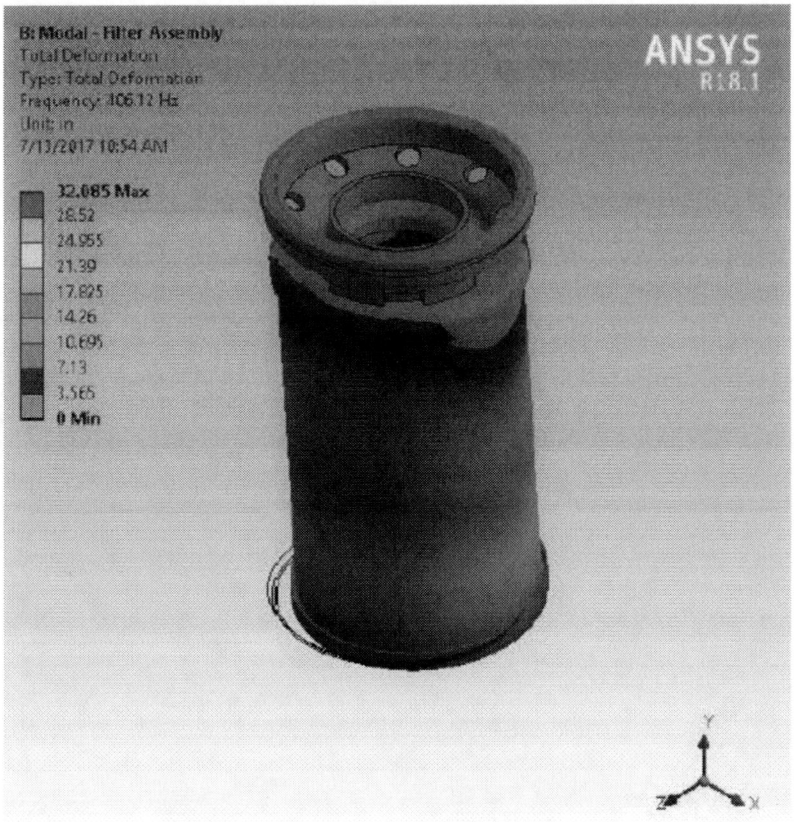

Figure 7. Dry Filter Assembly FE Modal Analysis Results – Mode 1 Bending of Filter Elements

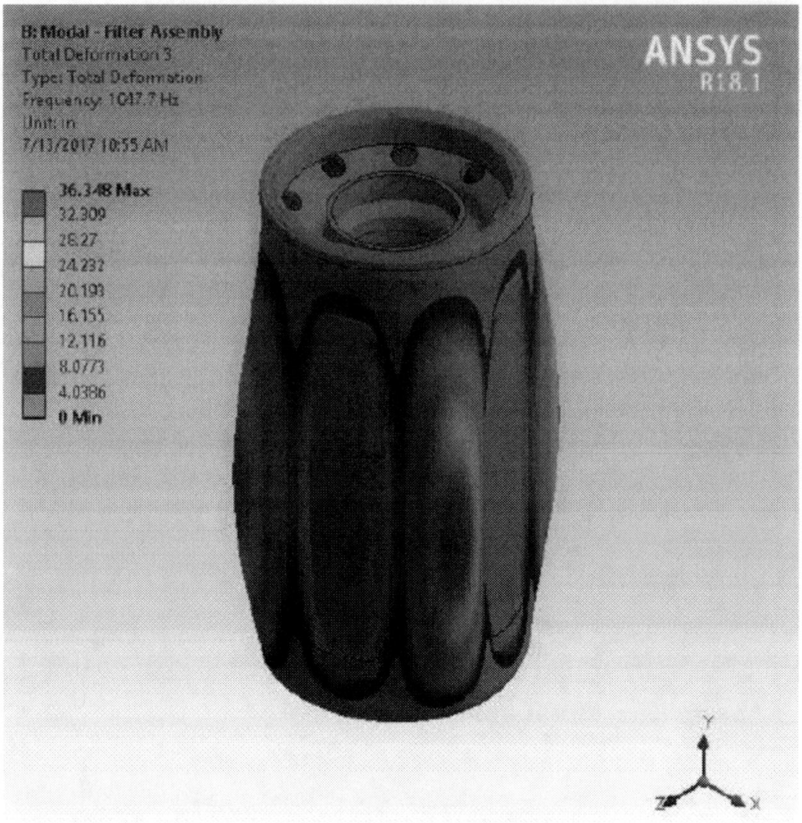

Figure 8. Dry Filter Assembly FE Modal Analysis Results – Mode 2
4-Node Shell Mode of Canister

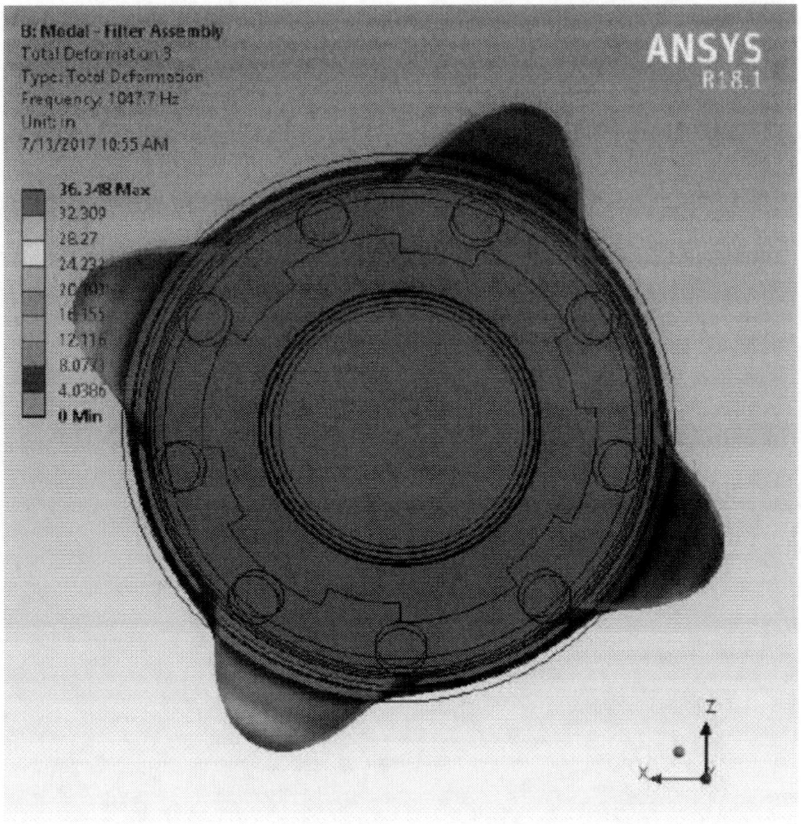

Figure 9. Dry Filter Assembly FE Modal Analysis Results – Mode 2
4-Node Shell Mode of Canister

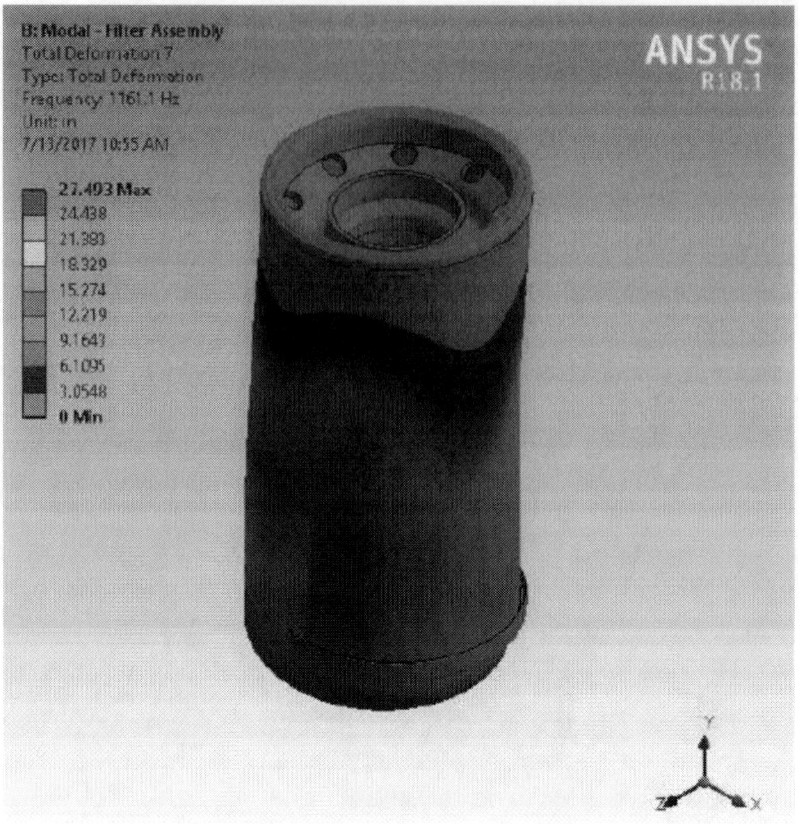

Figure 10. Dry Filter Assembly FE Modal Analysis Results – Mode 3 Bending of Canister

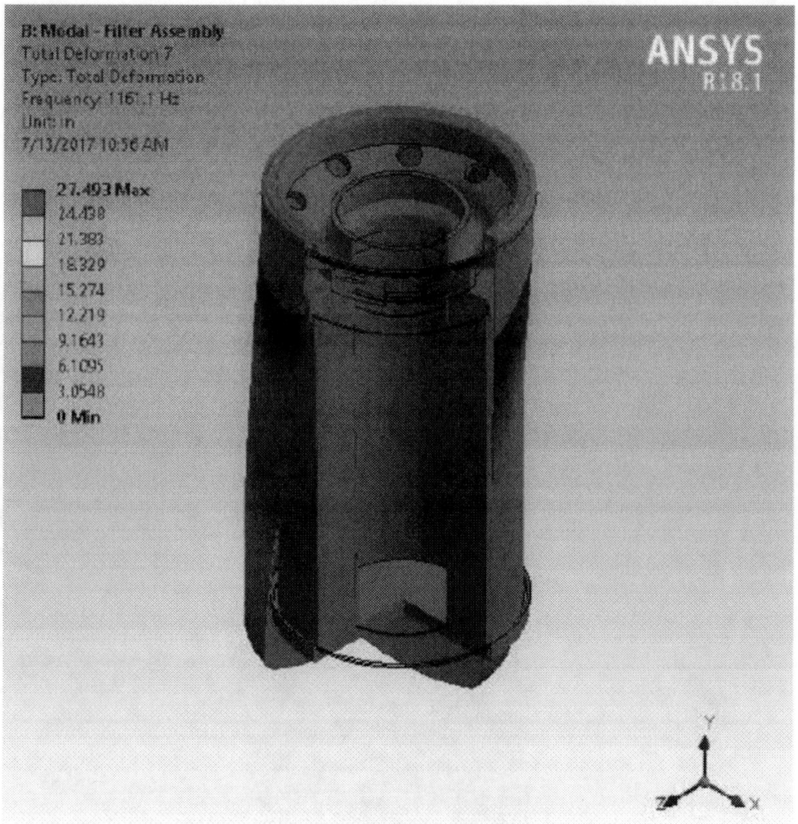

Figure 11. Dry Filter Assembly FE Modal Analysis Results – Mode 3 Bending of Canister

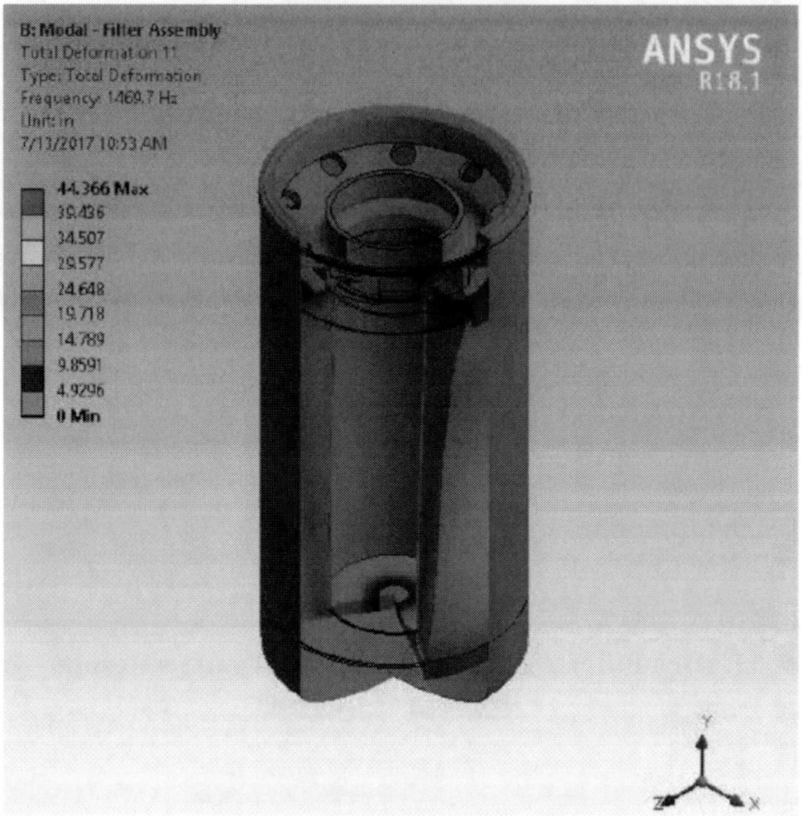

Figure 12. Dry Filter Assembly FE Modal Analysis Results – Mode 4 Torsion of Filter Element

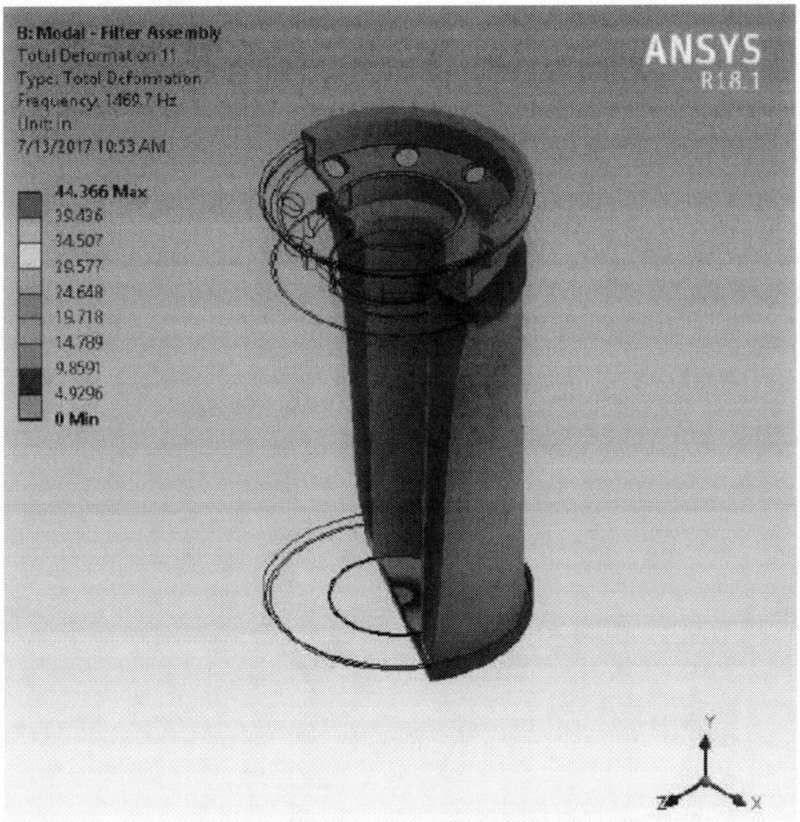

Figure 13. Dry Filter Assembly FE Modal Analysis Results – Mode 4 Torsion of Filter Element

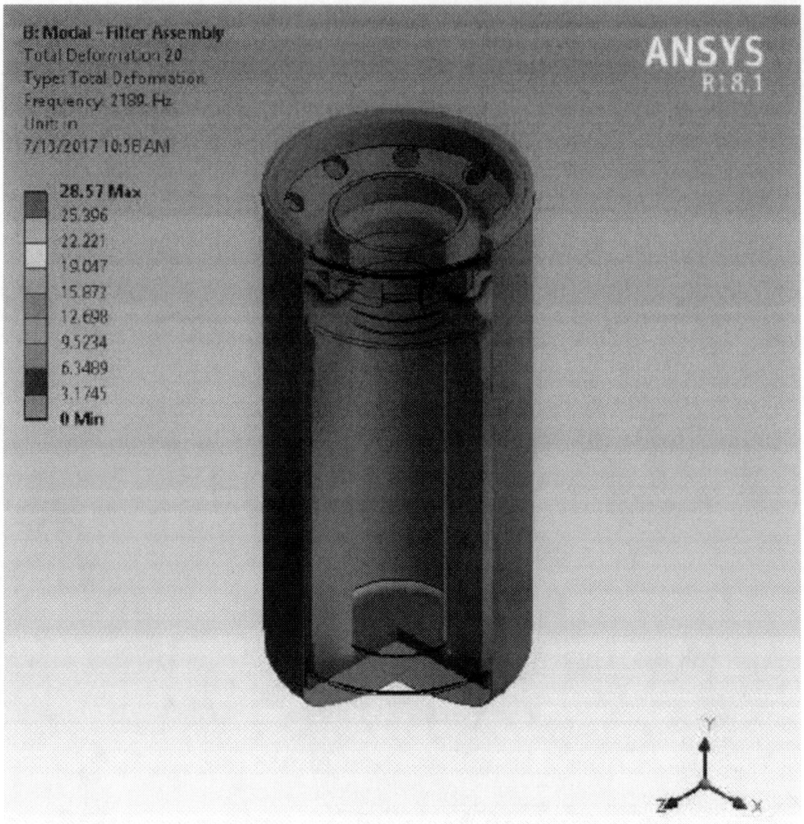

Figure 14. Dry Filter Assembly FE Modal Analysis Results – Mode 5 Axial Bending (Pogo) of Element

Figure 15. Dry Filter Assembly FE Modal Analysis Results – Mode 5 Axial Bending (Pogo) of Element

The frequencies of these modes are expected to lower when the inertial effects of the liquid are added to the filter assembly, but the mode shapes should be similar. This dry filter FE results will be compared in an upcoming experimental investigation. Starting with a comparison of the dry filter model and modal test results, an initial assessment of the filter dynamics can be made without the complicating effects of the fluid. This will allow for more confidence in the accuracy of the resulting filter FE model and predicted modal responses before the more complex case of the liquid filled filter is analyzed.

To extend the modal understanding gained from the finite element (FE) model, filters instrumented with accelerometers and filled with fluid were evaluated on a shaker system using a sinusoidal input swept across a frequency range from 10Hz to 1500Hz. The accelerometer response plots identify system resonance frequencies of the actual hardware. Because these measurements were done with a fluid-filled filter, the measured resonant frequencies are expected to be different than those predicted by the dry filter FE models presented in Table 3. The effort to identify the mode shapes of the measured resonances is planned for a later date.

RESULTS – Preliminary Efficiency Testing

The effect of the measured system resonances on filtration performance was studied using a multipass efficiency stand with the filter mounted to a shaker table as shown in Figure 16. Vibration input during the efficiency test was controlled by a vibration controller and power spectral density (PSD) curves that purposefully included or excluded the measured system resonances. Five different random vibration input PSD curves were defined that have different amplitudes and/or frequency content. The five PSD curves are shown in Figure 17. Three of the five PSD curves have 2 G-RMS amplitudes, and the other two had 4 G-RMS amplitudes. The other difference between these PSD curves is the frequency content. Of the five, only the 2 G-RMS and 4 G-RMS PSD's with frequency content between 425 to 625 Hz contains an identified major system resonance at 500Hz.

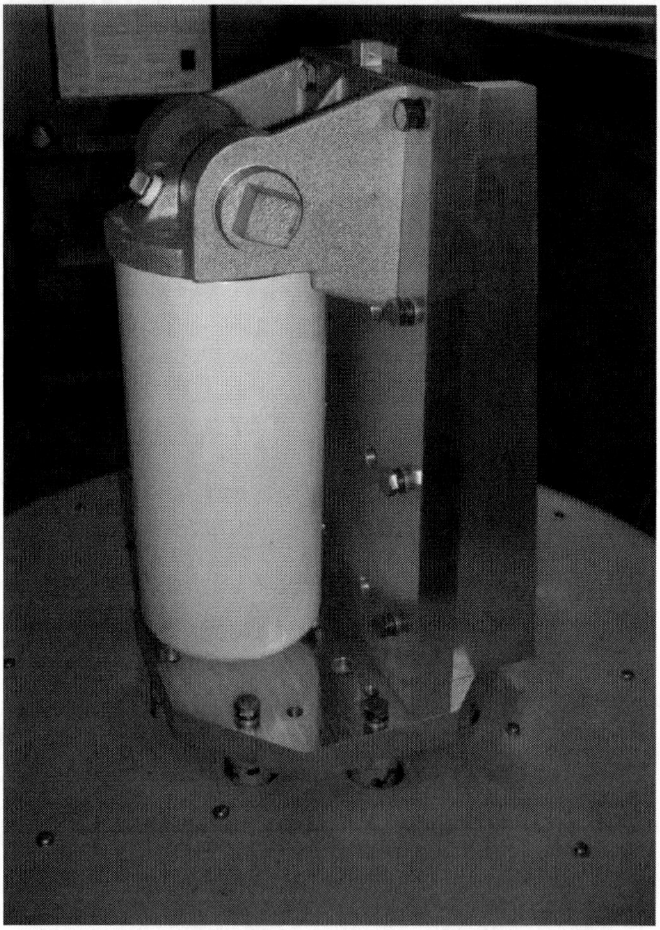

Figure 16. Multipass Efficiency Test Filter on Shaker Table.

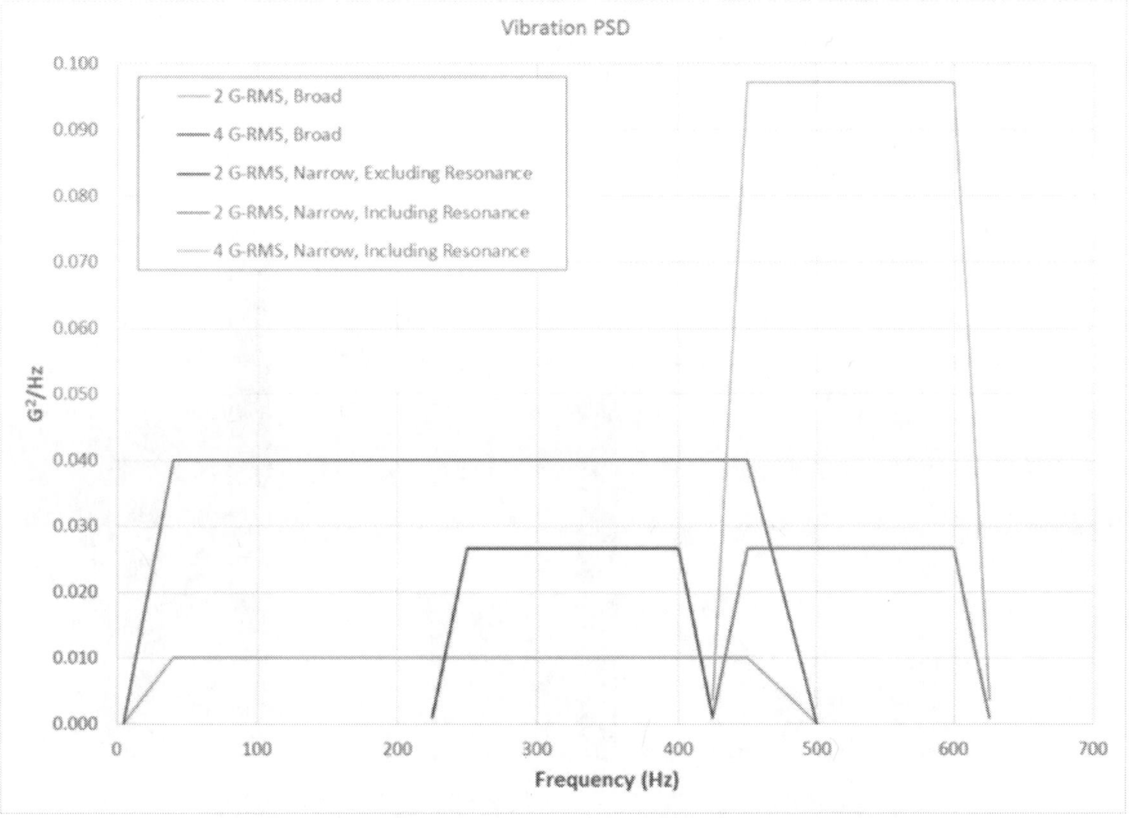

Figure 17. Selected Random Vibration Input PSD's.

Figure 18 contains filter performance in terms of "filtration ratio" at a constant flow rate for a single filter model subjected to the five separate vibration inputs in Figure 17. Also shown is the filtration ratio for the no input vibration case for this filter. The amplitudes in this bar chart are the filtration ratio (or "beta") values during the constant flow condition (Note: The color of the lines in Figure 17 and bars in Figure 18 correspond to the same test). Higher filtration ratios indicate the filter is capturing more particles, and in equation form:

Filtration Ratio = Beta = (Particles Counted Upstream) / (Particles Counted Downstream)

The importance of the system resonance on filter efficiency is seen in at least three ways from Figure 18. First, the vibration input of 2 G-RMS from 225 Hz to 425 Hz, which is in the range containing no system resonance, had the same effect as zero vibration input on the filter's efficiency. Second, the 4 G-RMS spectrum that excluded resonance resulted in the same filtration ratio as the 2 G-RMS spectrum that included resonance. In other terms, an equivalent degradation in filter performance was obtained with half the vibration energy input by selecting a frequency range that included resonance. Third, doubling the amplitude from 2 G-RMS to 4 G-RMS in the range including resonance decreased the filtration ratios by more than half.

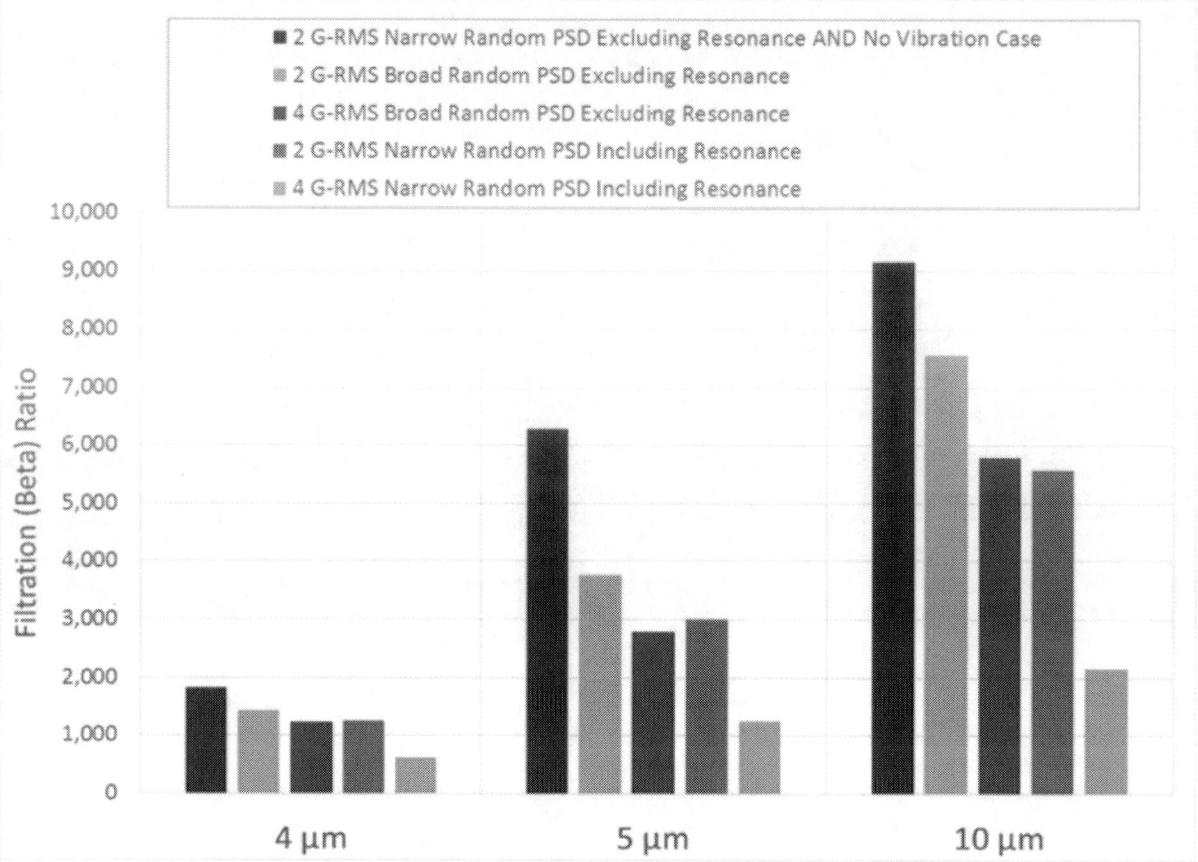

Figure 18. Filtration Ratios for Various Input Vibration.

CONCLUSIONS

Filter efficiency was found to be very sensitive to the inclusion or exclusion of a system resonance in the driving input frequency range. By holding vibration energy levels constant and varying the frequency range of that energy, filtration ratios could be as high as the no vibration case or cut in half for various particle sizes. This highlights the importance of having a detailed understanding of the structural dynamic properties of the filter mounting structure on both the test stand and in application. Modal finite element models and structural tests are important and useful tools in evaluating filter structures. Filtration standards that are considering adding controlled vibration input would benefit from further study of filter test structures using these methods.

For more information, please contact Gary Bessee, Southwest Research Institute, gbessee@swri.org, 001 210-522-6941 or Larry Hollingsworth, Southwest Research Institute, lhollingsworth@swri.org, 001 210-522-5711.

REFERENCES

1. "Wear Index Fuel Filter Rating System", Gary Bessee, Southwest Research Institute, 4th International Filtration Conference, San Antonio, TX, January 16-18, 2001.
2. "Fuel Filter Vibration Measurement Under Dynamic Simulated Loads", Steve Gomez-Leon, Southwest Research Institute, 4th International Filtration Conference, San Antonio, TX, January 16-18, 2001.
3. "Fuel Filter Wear Index – Pressure-side Applications", Gary Bessee, Southwest Research Institute, 5th International Filtration Conference, Stuttgart, Germany, October 15-17, 2002.
4. "Effects of Filtration Performance High Pressure Common Rail Fuel Systems", SwRI Consortium, May 2011.
5. "Diesel Fuel Filtration Needs for High Pressure Common Rail Fuel Systems," 8th International Filtration Conference, San Antonio, TX, January 16-18, 2007.
6. ISO 4406 – Hydraulic fluid power – Fluids – Method for Coding the level of contamination by solid particles

IASH 2013, the 15TH INTERNATIONAL CONFERENCE ON STABILITY, HANDLING AND USE OF LIQUID FUELS
Rome, Italy
10-14 September 2017

COMPARISON BETWEEN SAE J1488 AND ISO/DIS16332 DIESEL FUEL WATER SEPARATION TEST METHODS

Gary Bessee[1] and Kristi Rutta[2]

[1]Southwest Research Institute, 6220 Culebra Road, San Antonio, TX USA. gbessee@swri.org
[2] Southwest Research Institute, 6220 Culebra Road, San Antonio, TX USA. krutta@swri.org

ABSTRACT

Currently, ISO/DIS 16332.2[1] is in the final approval stages, while SAE J1488[2] is a mature test method. These methods are the two standards for automotive diesel fuel water separation utilized by most OEMs and diesel fuel water separation manufacturers. This paper will discuss the differences in philosophy for the two methods and share comparison data, as well as highlight the major differences in the performance and water removal results. Differences in test fuel and creating "standardized" biodiesel test fuel will be discussed as well as the development background for the creation of the standardize biodiesel fuel. The paper will also discuss differences in interfacial tension (IFT), microseparometer, water challenge, test time, and water droplet size distributions.

INTRODUCTION

Currently, two fuel water separation test methods are available for evaluating diesel fuel water separators for emulsified fuel water removal efficiency. SAE J1488 – Emulsified Water/Fuel Separation Test Procedure was developed in 1985 and has had four revisions for its 5-year reviews. In 2010, the document had a major change in the system to better control the water droplet distribution using a slip stream flow loop. The document is currently up for its next 5-year review. SAE J1488 is the recognized test method for evaluating emulsified fuel water separation in the US and Japan.

ISO/DIS 16332.2 – Diesel engines – fuel filters – method for evaluating fuel/water removal efficiency has been in development since 1999 and currently is being balloted as an ISO document. ISO/DIS 16332.2 is being developed by the German delegation and will be highly utilized in Europe.

Water removal in diesel fuel systems is becoming more of an issue. High pressure common rail fuel systems (HPCR) have little tolerance for water contamination. Bosch currently limits the total water content in diesel fuel to be less than 200-ppm which includes diesel fuels containing up to 5% biodiesel (B5). It has been shown that some B5 fuels have water saturation limits greater than 200-ppm total water. Therefore, the limits for any free water challenging their pumps are virtually zero.

Between biodiesel and other fuel additives used in diesel fuel and the potential of biodiesel contamination, the need for improved water separation test method is required to improve the filtration systems used in both diesel applications.

Although there are many differences in the test methods, Table 1 lists the most significant parameters which will be discussed individually below. Although test fuel is probably the biggest variable in fuel water separation testing, it will not be covered in this discussion due to the complexity of the issue and that the industry currently does not have a good handle how to resolve this issue.

Table 1. Comparison of Emulsified Diesel Fuel Water Separation Test Methods

Parameters	SAE J1488	ISO/DIS 16332
Water Challenge	2,500 ppm	1,500 ppm/2%
Droplet Size	Included in next review	$D_{3,50}$: 10 ± 1.5 μm $D_{3,90}$: ≤ 30 μm
Interfacial Tension (IFT) Measurement	ASTM D971	ISO 9101
Test fuel IFT	17 mN/m	13±2 mN/m or 22 ± 2 mN/m
Emulsion generation	Centrifugal pump	Orifice plate
Separability measurement	ASTM D7261 DB cell	ASTM D1401
Test time, minutes	150	90

RESULTS AND DISCUSSION

Water Challenge and Test Time

For SAE J1488, the water challenge is 2,500 ppm for 150 minutes. This water challenge and test time has been shown to demonstrate if there is a drop in water removal performance towards the end of the life of a fuel water separator, Figure 1.

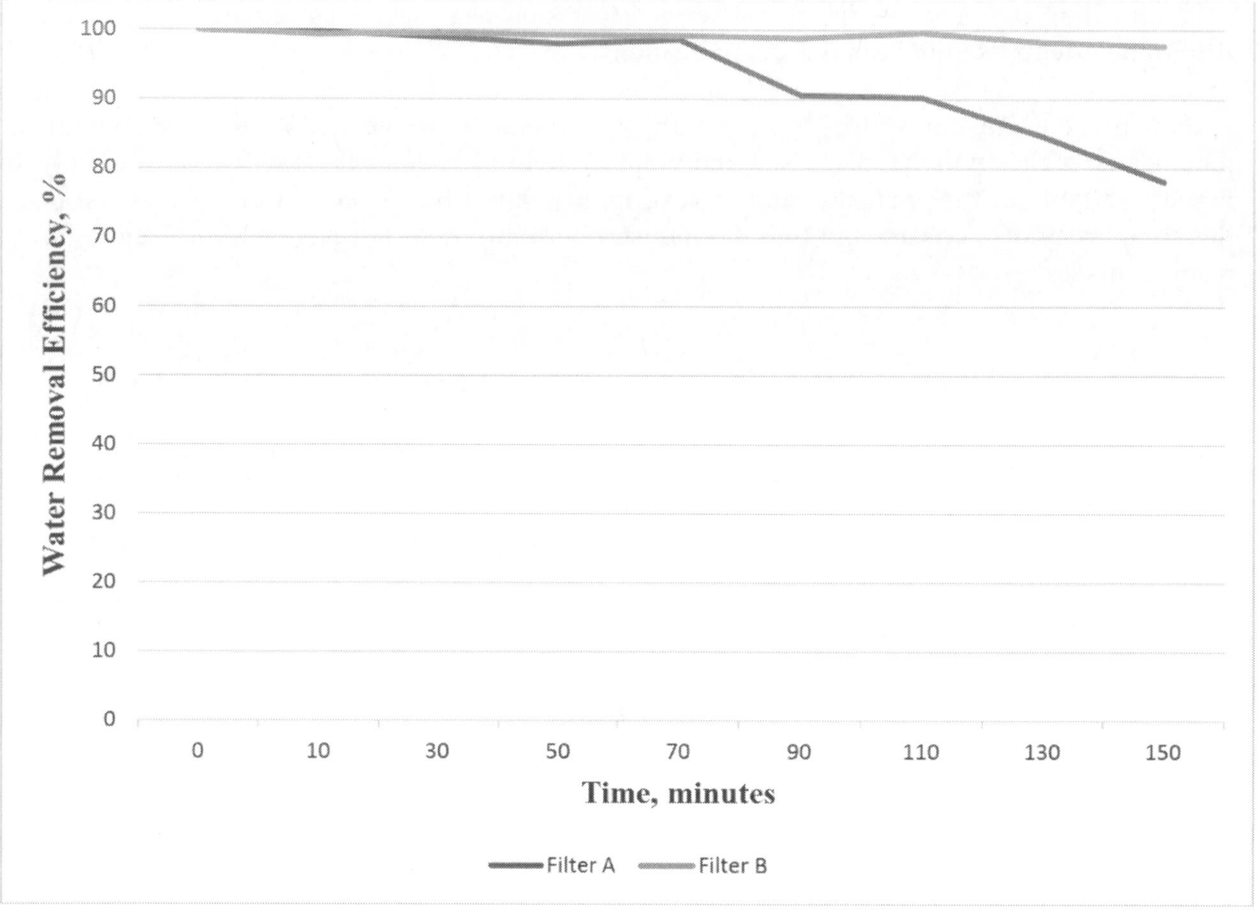

Figure 1. Comparison of fuel water separators using SAE J1488

ISO/DIS 16332.2 has two different water changes: 1,500 ppm and 2% and is defined by what the client requires. The test time for ISO/DIS 16332.2 is 90 minutes which raises concerns that the test will not differentiate between excellent and good fuel water separators. As shown in Figure 1, Filter A starts deteriorating at between 70 and 90 minutes. With most of the efficiency results being approximately 100%, the time weighted water removal efficiency will be higher than including the continued decline in performance if using SAE J1488. Simply stated, if the client has a 95% emulsified water removal efficiency requirement, Filter A would pass and Filter B would fail.

Water Droplet Size Distribution

In solid removal testing, standardized test dust is utilized to insure the upstream solid challenge is similar for all tests and laboratories. Until the last few years, SAE J1488 assumed the emulsions were created the same since we thought the centrifugal pump was defined well enough. After further research, although all the test stands evaluated used 3,500 rpm centrifugal pumps, it was determined that the angular velocity varied due to differences in the diameter of the centrifugal pump. Also, due to the internal bypass in the pump, the emulsions will vary as different quantities of fuel remain in the bypass loop, thereby more energy is absorbed creating a tighter emulsion at lower flow rates.

ISO/DIS 16332.2 has established droplet size distributions (DSD) for the test fuel and SAE will be incorporating similar guidelines in SAE J1488 in the next method review. For standard test conditions, ISO/DIS 16332.2 defines the DSD for emulsified fuel/water as $D_{3,50}$ 10 ± 1.5 µm $D_{3,90} \leq 30$ µm. An example of a passing fuel is shown in Figure 2. Once the ISO document is finally approved and the industry gathers more DSD data, these DSD limits most likely to need to be revised.

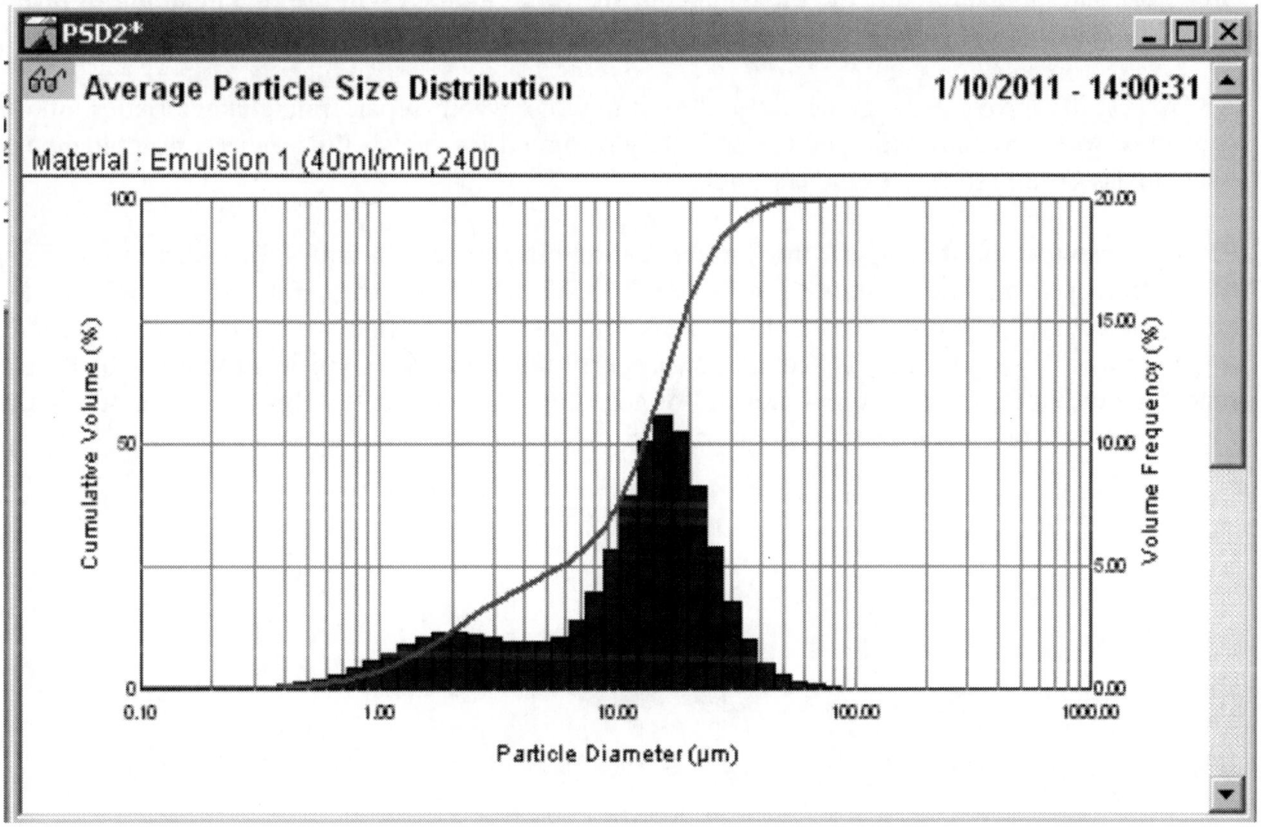

Figure 2. Fuel/water droplet size distribution

Interfacial Tension (IFT) And Separability

In order to have repeatable and reproducible test results, two key parameters for the test fuel are the interfacial tension (IFT) and some measure of fuel water separability. Testing has been shown that definition of both parameters are required to have a well-defined test fuel. As shown in Table 2, using the same test filters, fuels with the same IFT values can generate different time weighted water removal efficiencies (TWA) and vice versa.

Fuel	IFT, mN/m	TWA Efficiency
Jet A	36.0	97%
Low sulfur diesel fuel A	36.89	Terminated after 20 minutes
Low sulfur diesel fuel B	25.27	Terminated after 20 minutes
Low sulfur diesel fuel C	23.80	65%
Low sulfur reference diesel fuel	35.44	98.6%
Low sulfur diesel fuel D	25.64	99.3%

SAE J1488 uses the ASTM D7261 (DB cell) to determine the fuel water separation versus ISO/DIS 16332.2 using ASTM D1401. ASTM D7261 is still be evaluated to determine the limits for our application and the method scope of work claims it will provide meaningful results for B1 to B20.

ASTM D1401 provides a guide for determining the water separation characteristics of oils subject to water contamination and is not really designed for fuels. Plus there is no guidance on what a pass or fail criteria for the test fuel.

Plenty of research has been performed investigating difference techniques for measuring IFT and the differences in the results, Figure 3.[3,4, 5] Most of the research measures the IFT "instantaneously" or within 60 seconds. However, as shown in Figure 4, the IFT as measured by a goniometer, illustrates IFT is a dynamic property that can take significant amount of time to reach an equilibrium point; sometimes up to more than 60 minutes.[6,7] This creates another issue for the testing world insuring all laboratories are measuring the fuel properties the same way.

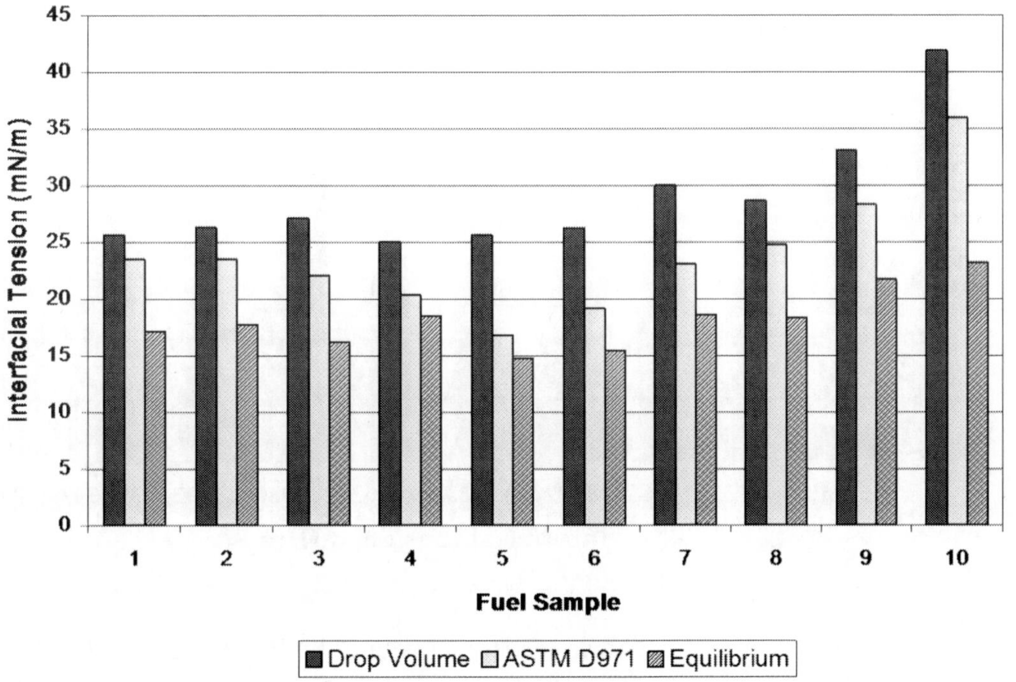

Figure 3. IFT Measurement of Various Fuels Samples Using Different Measurement Techniques

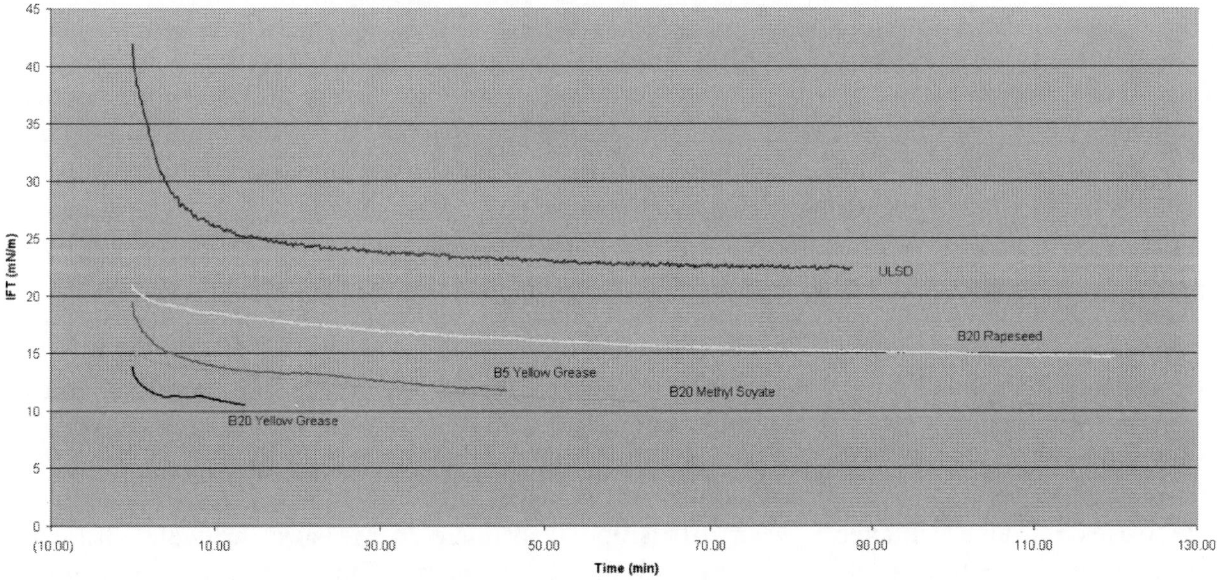

Figure 4. IFT Measurements of Various Types of Diesel Fuels Using a Goniometer

Another issue to resolve for standardized testing is what range of IFT is required for the test method. The OEM wants something relatively severe to cover worse case scenarios in the field but the filtration manufacturers want a test that will be more representative of the mean. Since fuel chemistry and additives often makes this a moving target and creates a large range of IFT values in the field, Figure 5.[8]

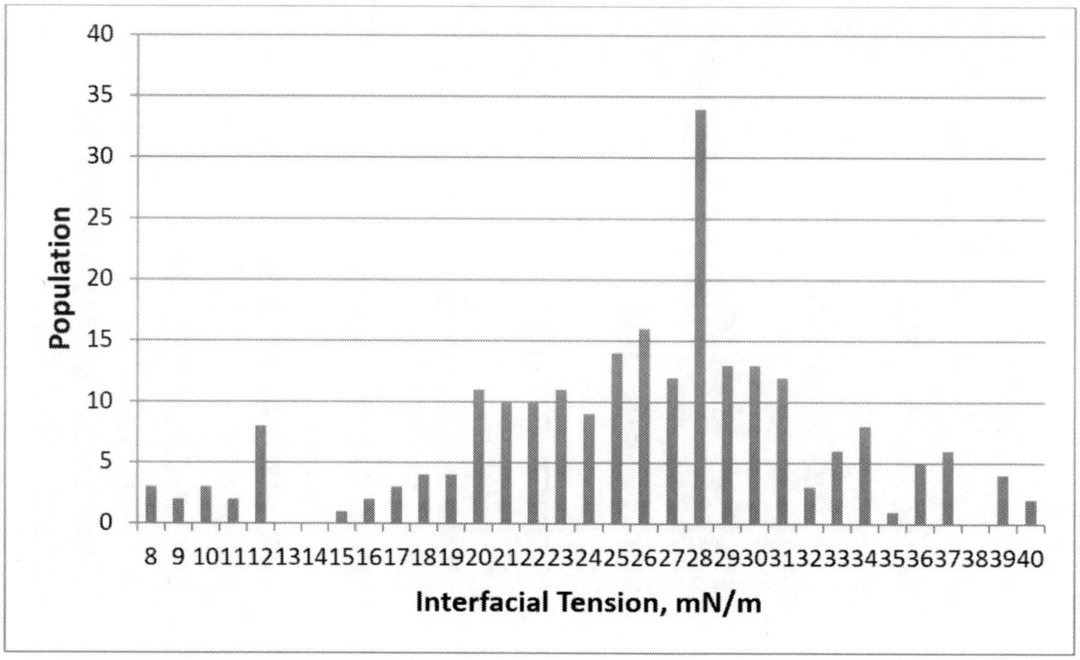

Figure 5. IFT of Fuel Samples from Around the World, Circa 2014

Emulsion Generator

SAE J1488 and ISO/DIS 16332.2 utilize different methods for generating fuel water emulsions. SAE J1488 generates a 2,500-ppm water emulsion challenge using a 3,500-rpm centrifugal pump. ISO/DIS 16332.2 uses a series of orifice plate to generate the water emulsion depending upon the system flow rate and differential pressure. The current emulsion requirements are to have a $D_{3,50}$ of 10 ± 1.5-μm and a $D_{3,90} \leq 30\ \mu m$.

In all of these mature test methods, the water droplet size distribution has never been characterized to determine what it is and how it changes with the addition of additives, biodiesel, or contamination from trail-back or other sources. Recent changes in SAE J1488 and ISO 16332 has initiated determination of the water droplet size distribution to improve the methodology to improve the repeatability of the methods and also to gain a better understanding of the fuel water removal process.

SwRI uses a Malvern Insitec L laser diffraction instrument to measure the water droplet size distributions for both test methods. The first interesting point of the data is typically the water size distribution is not a normal distribution but is bi-modal, Figure 2.

As shown in Figure 6 and 7, as the IFT is lowered, the DSD shifts towards the smaller droplet sizes creating a more severe challenge for the test filter. This demonstrates why having consistent measurement of the IFT for the test fuel is so important.

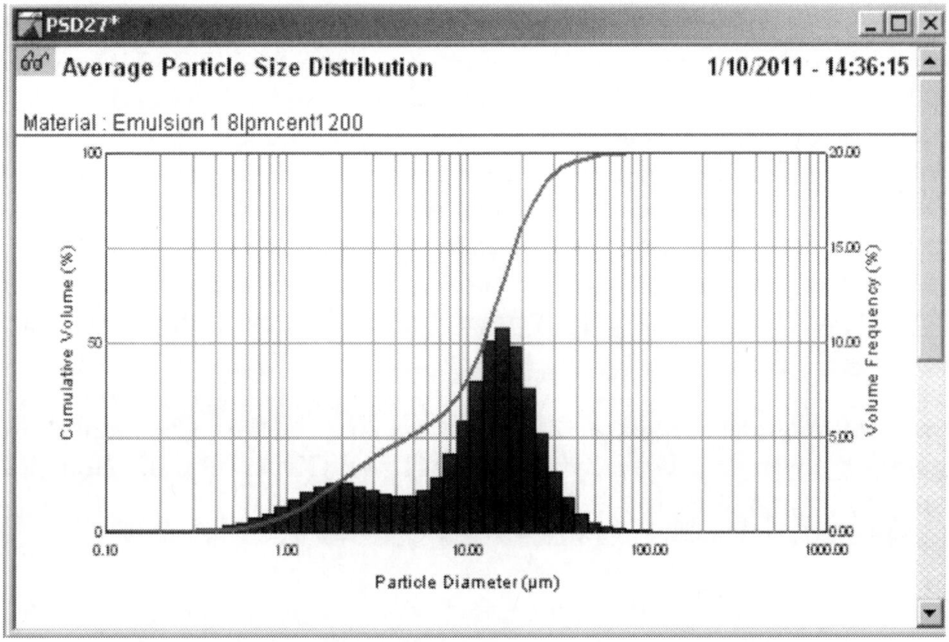

Figure 6. SAE J1488 Droplet Size Distribute for a Test Fuel With an IFT Value of 31 mN/m

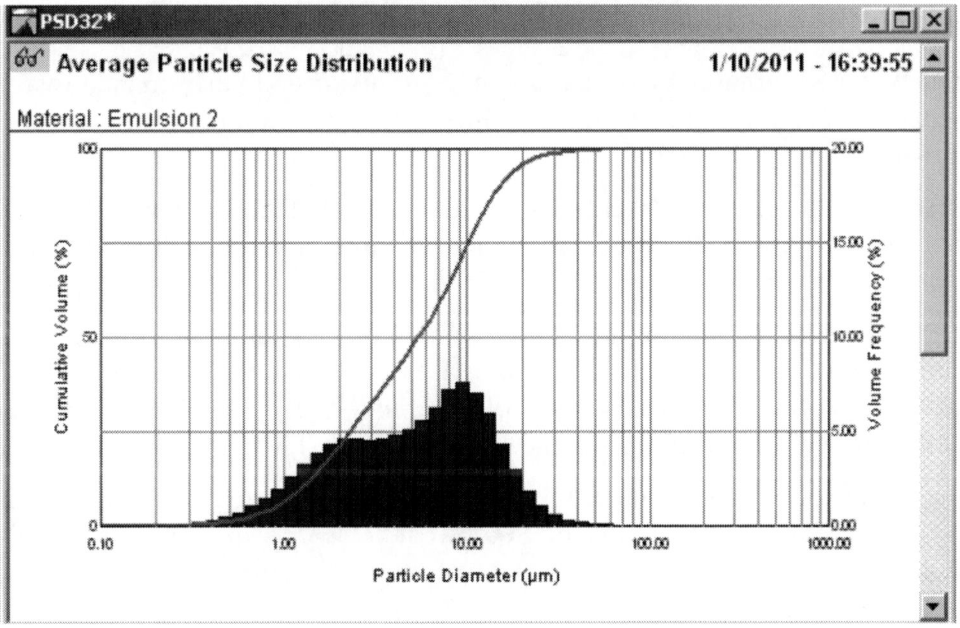

Figure 6. SAE J1488 Droplet Size Distribute for a Test Fuel With an IFT Value of 14 mN/m

CONCLUSIONS

Diesel fuel water separation technologies are very complicated due to not knowing what the challenge in the real world is. Fuels vary greatly from location to location and with different providers and additives (including biodiesel). This creates a significant challenge to the automotive industry to create a standardized test method for evaluating products to go into the field and properly protect the pumps and injectors. Currently SAE J1488 is used in the US and a variation in Japan and ISO/DIS 16332.2 should be approved by the end of 2017. There are significant differences in the methods so the users need to understand the two methods may generate different results. Preliminary data (not shown in this paper) on high efficiency fuel water separators shows good similarity. But that was testing performed on one filter in the same laboratory by the same technician. Expect changes in both documents as both SAE and ISO learn more of the pros and cons for both methods and improves the respective documents during their 5-year reviews. The biggest improvement to be expected is the droplet size distribution required for future testing.

For more information, please contact Gary Bessee, Southwest Research Institute, gbessee@swri.org, 001 210 522-6941 or Kristi Rutta, Southwest Research Institute, krutta@swri.org, 001 210 522 2102.

REFERENCES

1. ISO/DIS 16332.2 – Diesel engines – fuel filters – method for evaluating fuel/water removal efficiency, 2017.
2. SAE J1488 – Emulsified Water/Fuel Separation Test Procedure, 2012.
3. "Fuel Water Separation – Droplet Diameter Analysis and Performance Evaluation", Mark Wieczorek, Cummins Filtration, 10th International Filtration Conference, San Antonio, TX, September 28-30, 2010.
4. "Effects of Biodiesel on Interfacial Tension", Scott Hutzler, Gary Bessee, and Kavitha Moorthy, Southwest Research Institute, 9th International Filtration Conference, San Antonio, TX, November 11-13, 2008.
5. "A Proposed 'Slip Stream' Emulsion Generation System for More Consistent Droplet Size Distribution in SAE J1488", Gary Bessee, Southwest Research Institute and Dr. Christine Stanfel, Ahlstrom Filtration, LLC, and Wally Stone, Parker Hannifin – Racor, 9th International Filtration Conference, San Antonio, TX, November 11-13, 2008.
6. "Generation of Water Emulsions for Standardized Testing – What We Don't Know", Gary Bessee and James Johnson, Southwest Research Institute, and Vic Hughes, Vic Hughes and Associates, 7th International Filtration Conference, San Antonio, TX, May 10-12, 2005.
7. "Factors of Influence of Diesel Fuel Water Separation Performance", Dr. Christine Stanfel, Ahlstrom, LLC, 8th International Filtration Conference, San Antonio, TX, January 16-18, 2007.
8. ISO/TC22/SC34/WG1, San Diego, CA, October 13, 2015.

IASH 2015, the 15th INTERNATIONAL SYMPOSIUM ON STABILITY, HANDLING AND USE OF LIQUID FUELS
Rome, Italy
10-14 September 2017

Fuel Contamination Specifications by ASTM D5452 Gravimetric, MIL-DTL-2261D Light Obsurance (JF-WA1-NP), and Laser Particle Counter IASTM D7619, Comparisons and Performance Review, Future Potentials.

Alan J. Fougere

D-2 Incorporated, 23 Edgerton Drive, North Falmouth, MA, USA, afougere@d-2inc.com

In 2016 D-2 Inc. developed a new version of its JF-WA1 Free Water in fuel by chemical pad detection instrument to also allow a provision for measuring solid contaminates in fuel. The modified instrument incorporates a second detector below the original filter pad plane allowing measures of the amount of light that can be passed through a filter membrane. A differential dual filter technique is then used to determine the "total particulate" in mg/L. This type of light obscurance measurement of particulate in light distillates has an established history in the US DOD, reference MIL-DTL-2261D. The original technique also used two filters which are placed together through which a known volume fuel sample is passed. The particulate contamination is collected on the top filter while both filters are color stained by the fuel. Subsequent to fuel filtering a light is shown through each pad separately and the total transmission radiance is measured, the radiance difference through the two filters is then computed. The difference in radiance has been shown to be directly proportional to total particulate (mass) loading of the fuel sample.

The new D-2 unit has been optimized around a 47 mm 0.65 um pore size cellulose filter, which yields higher resolution at lower contamination levels. This presentation includes trial data and precision estimates from this new instrument. The trial data includes multiple fuel types, along with data from other particulate contamination methods including laser based particle counters and the ASTM D5452 gravimetric method.

A recent study performed under the auspices of the US Army TARDEC group at the SWRI laboratory in St. Antonio, TX using the three method illustrates the performance of each method against know water/dirt fuel concentrations and how the three measure perform in relation to each other in terms of setting fuel contamination levels.

IASH 2017, the 15th INTERNATIONAL SYMPOSIUM ON STABILITY, HANDLING AND USE OF LIQUID FUELS
Rome, Italy
10-14 September 2017

NEW SULFUR REDUCTION TECHNOLOGY AS SOLUTION TO INCREASINGLY STRINGENT SULFUR REGULATIONS

Zhenning Gu[1], Ksenija Babic, Paul Biggerstaff, Don Wolfe, Jerry Weers, Waynn Morgan

Baker Hughes, a GE Company. 12645 West Airport Blvd. Sugar Land, Texas, USA,
[1]Zhenning.Gu@bhge.com

ABSTRACT

Sulfur concentrations in liquid fuels are more stringently regulated all over the world. An example of this is the recent implementation of the US Environmental Protection Agency's (EPA) Tier 3 gasoline regulations requiring that the average sulfur content of the refiners' gasoline pool be reduced from 30 ppm to 10 ppm, or less. Regulatory requirements like Tier 3 force refiners to explore more flexible sulfur reduction technologies beyond hydrodesulphurization.

This paper introduces newly developed sulfur reduction technology that combines chemical solutions with mechanical processes to achieve effective sulfur reduction in various light hydrocarbon streams. The innovative approach enables complex sulfur compounds, not removed by conventional refinery caustic washing, to be effectively extracted in a customized combination of specialty chemicals and mechanical equipment. The technology provides enhanced flexibility in blending light hydrocarbons into gasoline while remaining compliant with the Tier 3 gasoline regulation. Results from case studies will be shared to present efficacy in total sulfur reduction, based on individual sulfur species and provide an overview of how the technology is applied in the field. The unique chemical-mechanical modular technology provides a flexible and effective sulfur reduction solution for liquid fuel producers.

Keywords

Sulfur reduction, Tier 3 gasoline, complex sulfur compounds, sulfur extraction, desulfurization

INTRODUCTION

To eliminate SO_x pollution from burning fuels as energy source, environmental organizations of governments and global fuel regulation organizations have been continuously restricting the sulfur contents in major fuel categories including gasoline, diesel, and marine fuels. In 2009, as specified in EN590[1], EU established the 10 ppm sulfur specifications for diesel fuels, both on-road and off-road. In 2011, as specified in EN228[1] (under EN ISO 20846), EU established the 10 ppm sulfur specifications for gasoline, both on-road and off-road.

The US Clean Air Act Amendments of 1990 (CAAA90)[2] regulated the sulfur limit at 15 ppm in ULSD ultra-low sulfur diesel (ULSD, or highway diesel). The final NRLM ("non-road, locomotive, and marine") rule established a new ULSD limit of 15 ppm for non-road diesel by mid-2010. For locomotive and marine diesel, the rule established an ULSD limit of 15 ppm in mid-2012.

US EPA established US Tier 3 Gasoline specification to further regulate the sulfur content in gasoline. According to EPA Tier 3 Sulfur Standard[3] for gasoline refiners and importers, starting Jan. 1, 2017, gasoline must contain no more than 10 ppm sulfur, on an annual average basis. This is a significant tightening from the previous specification of 30 ppm sulfur for gasoline.

The recent IMO (International Maritime Organization) regulation on "Cleaner Fuels for Ships" that will come into force in 2020 requires that sulfur emission from marine ship fuels go down from 3.5% to 0.5%, over 85% sulfur reduction[4].

On the other hand, average sulfur content from crude oil sources for US refiners has increased from <1% back in 1984 to 1.4 – 1.5%, on average, since 2001 (Figure 1). Sour crudes with low to medium API could contains up to 3%+ sulfur. Sour crudes such as Mexico Maya usually contain over 3% sulfur. Mars crude from US typically has a sulfur content about 2%. Heavy crudes from Mideast such as Arab Heavy have an average sulfur level of 2.8 – 2.9% (Figure 2). While discounts on some of these opportunity crudes are economically attractive, they do present higher challenge for refineries regarding desulfurization. To stay in compliance with more stringent sulfur regulations on finished fuel including gasoline, diesel, and fuel oil, refiners globally must continuously advance their processing technologies on sulfur removal during crude oil refining, and explore alternative technologies for sulfur removal from refinery process streams.

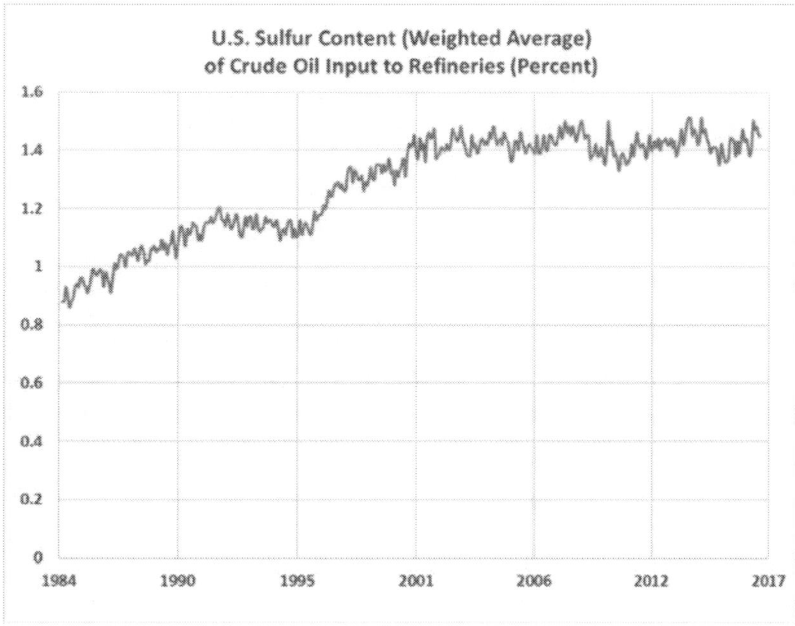

Figure 1 US Sulfur Content of Crude Oil input to Refineries[5] (Percent)

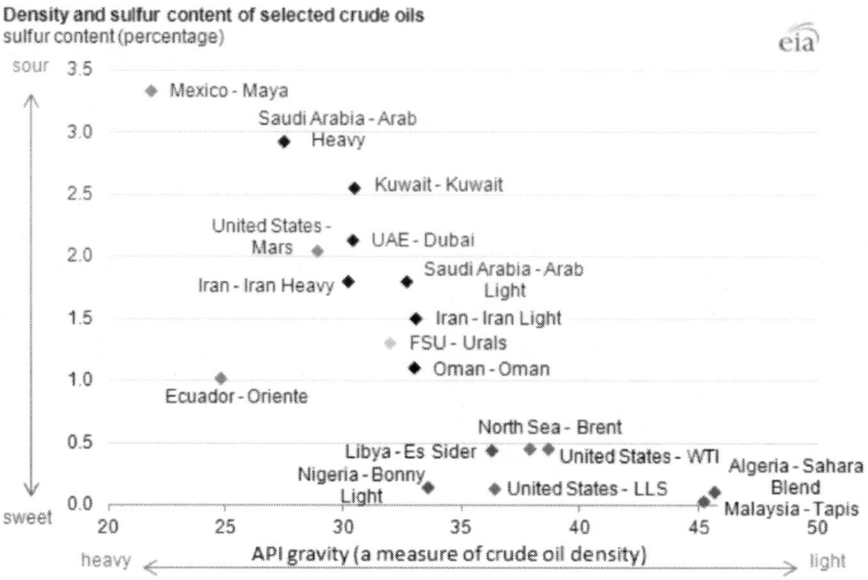

Figure 2 Sulfur Content of Selected Crude Oils[6]

Traditional processing technologies for sulfur removal from hydrocarbon streams in petroleum refining industries mainly include catalytic hydrodesulfurization (HDS or Hydrotreater), and caustic (or amine) scrubber unit. Designed in alignment with refinery capacity, both technologies are usually capable of processing large volumes, and they provide a certain level of operation flexibility for constantly changing feedstocks regarding sulfur species and sulfur levels. But with higher desulfurization demands on both throughput and sulfur removal efficiency, both processes possess fundamental gaps in responding to the sulfur removal needs. For example, due to aromatics and olefin saturation during catalytic hydrotreating, it's been observed that the octane rating declines with high severity hydrotreating[7]. Lower octane rating would devalue a hydrocarbon stream for gasoline blending and result in loss of profit for the refineries. Increasing throughput and severity in the hydrotreating process also lead to high hydrogen consumption and frequent fouling in hydrotreating equipment[8]. The caustic scrubber in a refinery is usually dedicated to an alkylation unit, with limited flexibility for desulfurization of other process streams. Also, from a sulfur removal perspective, the caustic washing process is efficient at removing hydrogen sulfide and mercaptans with lower molecular weight, but it has very limited impact on other sulfur compounds.

Realizing there is a gap in sulfur removal technologies within the industry motivated the development of EXALT™, a new sulfur extraction technology, which combines specialty chemicals with mechanical equipment to achieve effective extraction and removal of challenging sulfur species from light hydrocarbons.

EXPERIMENTAL

The technology development was approached from two directions: identifying the effective sulfur extraction chemistries, and designing a modular mechanical process with optimal extraction efficiency and design flexibility.

Sulfur Containing Reagents

To determine the representative target sulfur compounds for evaluation, sulfur speciation (per ASTM D5623) results from multiple light hydrocarbon streams were studied and a list of model sulfur compounds was generated for lab extraction testing, including carbon disulfide, diethyl sulfide, dimethyl disulfide and n-butyl mercaptan.

Lab Benchtop Extraction Procedure

A liquid-liquid extraction procedure was developed to simulate the sulfur extraction on a lab bench. 50 ml of the light hydrocarbon sample to be treated was added to a 500 ml glass separatory funnel, to which 50 ml of the extraction formula solution was also added. The mixture was agitated for 1 – 2 minutes to allow adequate contacting (with frequent pressure release to avoid vapor pressure buildup). The mixture was then allowed to settle undisturbed to enable complete phase separation.

Sulfur Removal Efficiency Determination

Sulfur content in the hydrocarbon phase before and after the extraction procedure was determined by ASTM D7039, using a Sulfur Analyzer (XOS SINDIE 2622-10) with LOD of 0.4 ppm at 300 s. The difference in sulfur contents post extraction process are used to calculate the sulfur removal efficiency.

RESULTS AND DISCUSSION

Function of the Specialty Sulfur Extraction Chemistries

Different from the function mechanism of a H_2S or mercaptan scavenger, sulfur extraction chemistry functions by forming adducts with the sulfur species in the hydrocarbon upon liquid phase contact. The higher aqueous affinity of the sulfur-containing adducts retain sulfur in the water phase while phase separation occurs, thus achieving sulfur extraction from the hydrocarbon (Figure 3).

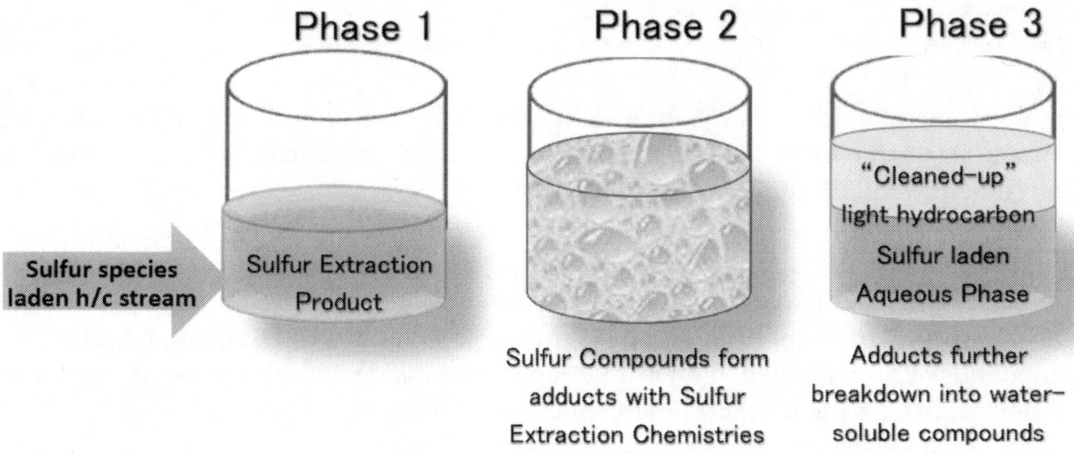

Figure 3 Function Mechanism: Sulfur Extraction Chemistries

Lab testing results from hydrocarbon samples pre-dosed with the model sulfur compounds (Table 1 and Figure 4) indicated that both EXAL 1000 and PRODUCT B have high sulfur extraction efficiency on the main sulfur compounds seen in light hydrocarbons from refinery. Additionally, PRODUCT B showed extraction efficiency on challenging sulfur compounds such as diethyl sulfide and n-butyl mercaptan.

Table 1 Lab Results: Efficiency of the Sulfur Extraction Chemistries

Model Sulfur Compound Type	Model Sulfur Compound conc. (ppm)	Sulfur Extraction Chemistry	Sulfur content (ppm)	S Removal Efficacy, %
Carbon Disulfide	1000	PRODUCT A + Filter	135.4	86%
Carbon Disulfide	1000	PRODUCT B + Filter	112.5	89%
Dimethyl Disulfide	1000	PRODUCT A + Filter	37.9	96%
Dimethyl Disulfide	1000	PRODUCT B + Filter	34.5	97%
Mixed Sulfur Compounds**	4000	PRODUCT A + Filter	2020	50%
Mixed Sulfur Compounds**	4000	PRODUCT B + Filter	1090	73%

** Mixed Sulfur Compounds: include 1000 ppm each carbon disulfide, diethyl sulfide, dimethyl disulfide and n-butyl mercaptan.

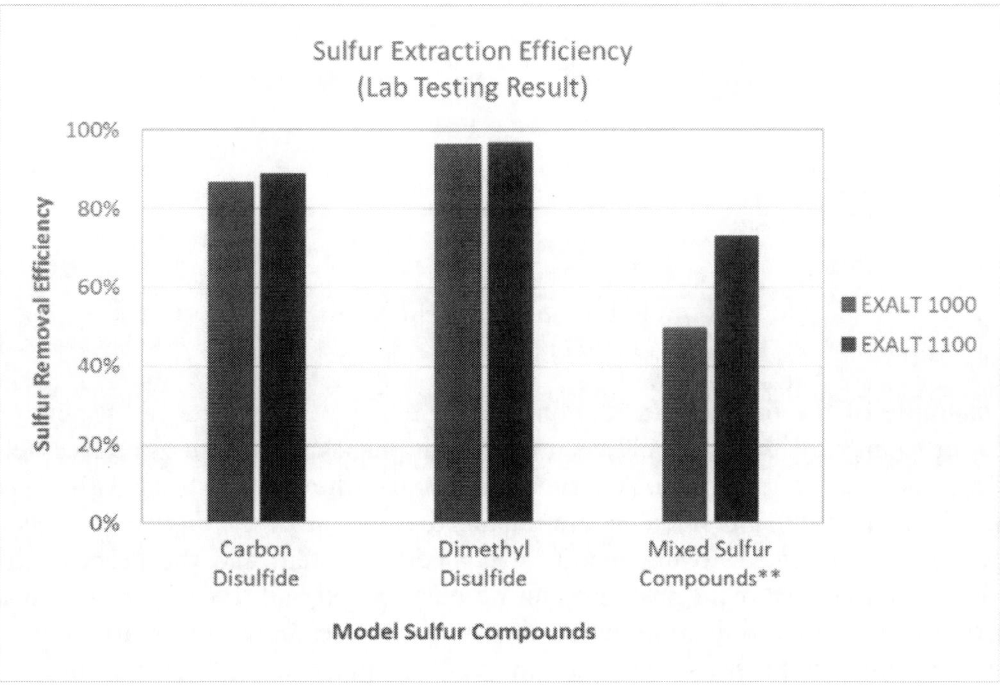

Figure 4 Sulfur Extraction Chemistries Lab Evaluation Results

The key for an effective extraction process lays not only on the extraction chemistries, but also the mechanical design. To fill the gap in sulfur removal efficiency, a flexible, modular system with multiple equipment elements was designed in response to specific application requirement. An example pilot extraction system is illustrated in Figure 5. The core components are the two-stage extraction vessels that contains different chemistry formulations to target H_2S and C1-C4 mercaptans in the first stage, and C4+ mercaptans and other heavier sulfur species in the second stage. Additional components including particulate filters, recirculation filters, recirculation pumps, and coalescers can be arranged in various sequence as well as in multiple locations to maximize the flexibility of the overall system to meet quality and spec requirements of the hydrocarbon stream to be processed.

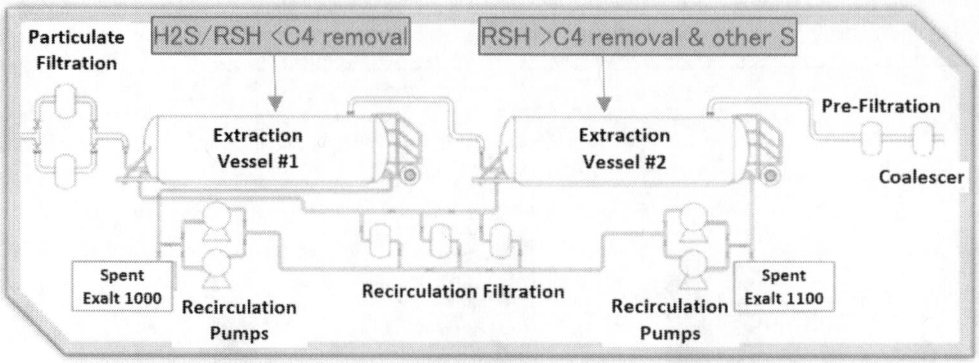

Figure 5 Example Layout of the Sulfur Extraction System

Another example of the modular extraction system that is used in the field includes a vertical contact tower to maximize contact between the liquid phases, and a horizontal vessel that allows optimal phase separation to happen. A series of specialty filters downstream of the second vessel are designed for removing residual contaminants before tankage (Figure 6). By performing upfront testing and analytics to understand the specific system and the process streams being processed, an optimal treatment program can be customized and developed. It can also include specialty chemical additive injection before the extraction vessels as pretreatment option (Figure 6). A full-size unit of the two-vessel design (vertical plus horizontal) was built as a mobile system for field evaluation (Figure 7).

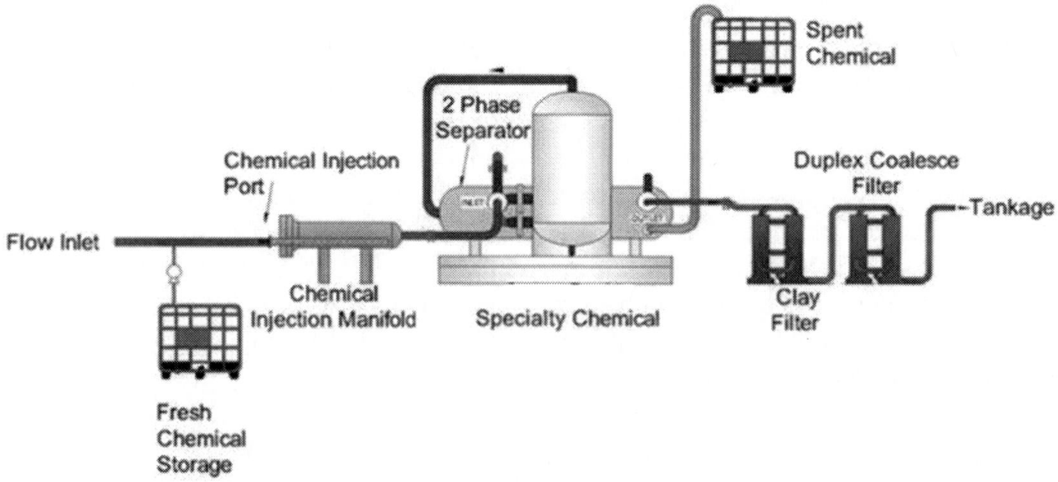

Figure 1 Modular Extraction System Example

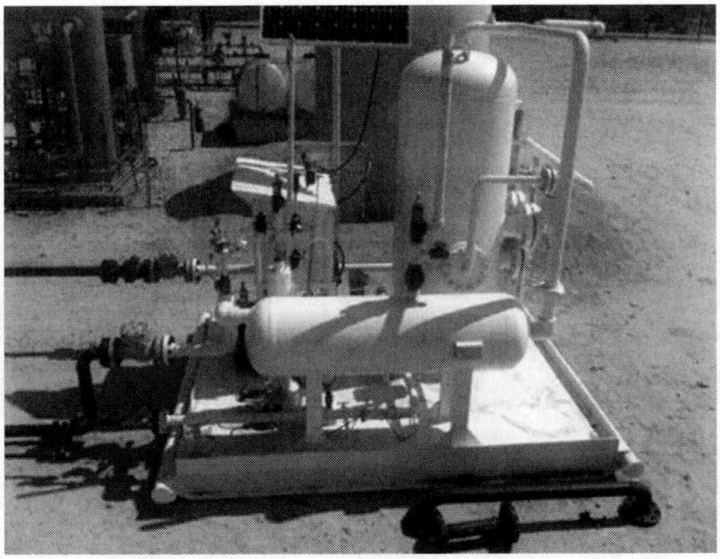

Figure 2 Pilot Sulfur Removal Modular System

Case Study I

A US Gulf Coast refinery faced several challenges in producing on-spec naphtha and propane as a result of reactive sulfur contamination. The propane had to be sold at a discount, because it failed the ASTM D-1838 copper corrosion specification due to the presence of reactive sulfur. In addition, valuable naphtha was being lost into the propane stream, causing the light naphtha to be downgraded. Looking at long-term solutions, the refiner considered installing a hydrotreater, but a hydrotreater for this type of operation would require large capital investment, as well as additional operating expenses.

After surveying the system, the sulfur extraction services were recommended to the refinery. It delivered a turnkey solution that removes the sulfur compounds and brings the streams back on spec. The treatment included a combination of sulfur removal chemistries paired with proprietary equipment for optimal extraction efficiency. The extraction system was installed on the Y grade propane stream to separate the light naphtha and heavier sulfur compounds, generating a new liquid light naphtha stream. The treatment results showed more than 98% removal of the total sulfur compounds (Figure 8), including reduced levels of methyl mercaptan, carbonyl sulfide, and carbon disulfide, which could not be removed by caustic treatment alone. It is important to note that this treatment did not affect the characteristics of the product (color, volatility, etc.) and the propane successfully passed the copper corrosion tests. This solution also helped reduce other extraneous contamination, limiting future catalyst and corrosion issues. This novel technical approach helped the refiner to bring the propane stream back on-spec, and recovered significant revenues that were lost due to off-spec propane. This on-spec product combined with the new naphtha stream allowed the refinery to debottleneck operations, realizing a 20% increase in overall process efficiencies and valuable products with high profit.

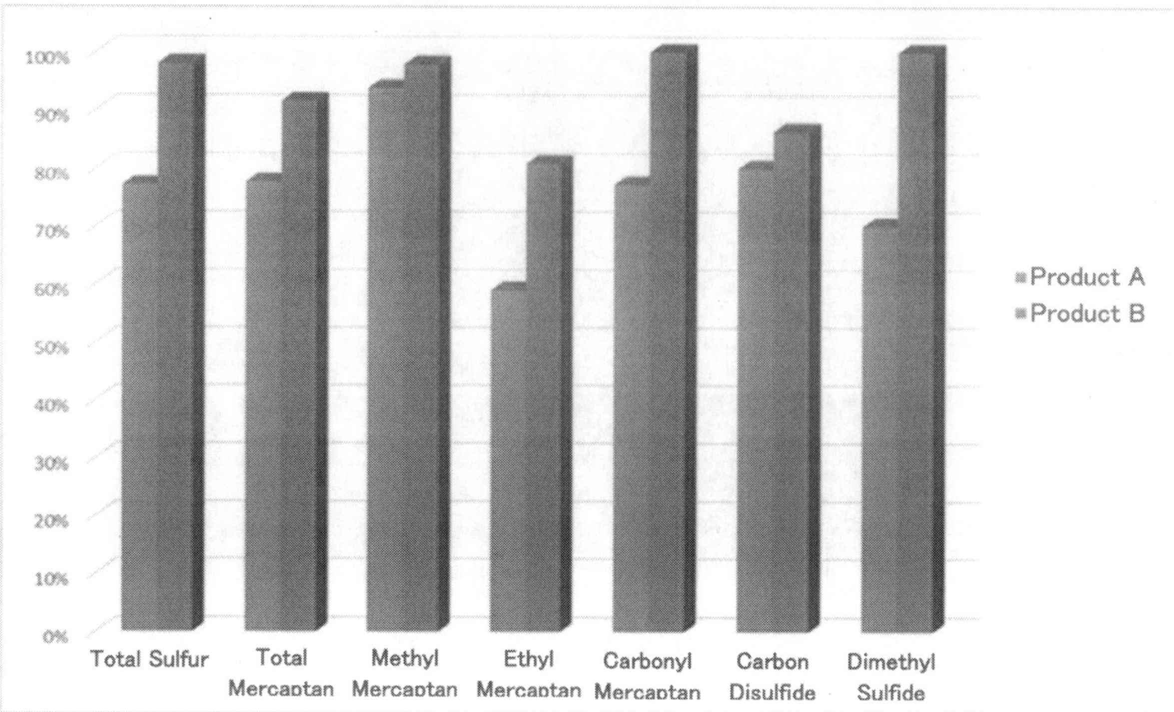

Figure 3 Sulfur Compounds Removal Efficacy

Case Study II

A petroleum refiner experienced atypically high level of mercaptans (over 1000 ppm) in its shale oil feedstock. The traditional mercaptan scavenger approach was able to reduce the mercaptan content to 190 ppm. However, operation requirement for total mercaptan is less than 20 ppm. Lab evaluation using the new sulfur reduction technology indicated that PRODUCT A was able to bring down the final mercaptan level to less than 10 ppm (Table 2). A full scale system is being developed to apply technology at field scale on-site.

Table 2 Mercaptan Reduction from Shale Oil

Sample	Treatment Chemistries	Mercaptan (ppm)	% Mercaptan Removal
Shale Oil	-	1020	-
Shale Oil	Traditional Mercaptan Scavenger	190	81.4
Shale Oil	PRODUCT A	9	99.1
Shale Oil	PRODUCT B	46	95.5
Shale Oil	1:1 PRODUCT A: PRODUCT B	7.7	99.2

CONCLUSIONS

A new approach for sulfur removal from petroleum refining process streams, Sulfur Extraction technology, was developed by combining patent pending sulfur extraction chemistries with modular mechanical designs that brings flexibility and customizable solution to meet sulfur spec for many fuel blending components. Successful field applications in Tier 3 gasoline blending components and high efficiency mercaptans removal from shale oil were discussed. Continuous development from both chemistry and mechanical aspect will expand the capability of the technology to wider range of sulfur removal challenges.

REFERENCES

[1] European Committee for Standardization. https://standards.cen.eu

[2] US EPA. https://www.epa.gov/clean-air-act-overview/evolution-clean-air-act#caa90

[3] US EPA. https://www.epa.gov/regulations-emissions-vehicles-and-engines/final-rule-control-air-pollution-motor-vehicles-tier-3

[4] International Maritime Organization. http://www.imo.org/en/MediaCentre/PressBriefings/Pages/MEPC-70-2020sulphur.aspx

[5] U.S. Energy Information Administration (EIA) Open Database.

[6] U.S. Energy Information Administration, based on Energy Intelligence Group—International Crude Oil Market Handbook.

[7] Andy Huang and co-authors. Driving Octane in an Ultra-Low Sulfur Gasoline Market. AFPM 17-75.

[8] G. Medine and B.E. Wright. Distillate Hydrotreater Fouling. AIChE 2008 Spring National Meeting. April 9, 2008.

IASH 2017, the 15TH INTERNATIONAL CONFERENCE ON STABILITY, HANDLING AND USE OF LIQUID FUELS
Rome, Italy
10-14 September 2017

Evaluation of Aviation Fuel Property Influences on Diesel Engine Performance

Andy McDaniel

U.S. Navy, NAVAIR 4451 - Fuels & Fuel Systems, AndrewMcDaniel@navy.mil

The United States Navy has completed qualification testing on several non-petroleum based fuels and fuel blends which exhibit a wide range of physical and chemical properties. Since Navy operations require the use of JP-5 in diesel engines during sea operations, testing experience has accumulated on a diverse set of diesel engines and engine components. Data from this wide range of hardware technology has facilitated a deeper understanding of performance risk caused by fuel properties. Identifying these performance sensitivities is a crucial step to evolving the efficiency and technical breadth of future programs that qualify changes in fuel properties.

This paper focuses on the relationship between aviation fuel properties and diesel engine performance, with a primary emphasis on engine start-ability. A compilation of four non-petroleum fuel data sets derived from testing on five diesel engine models highlights the most influential combination of fuel properties affecting diesel engine performance. These data sets also highlight the sensitivity of engine designs to "worst case" fuel properties. To understand which environmental and diesel engine design factors drive susceptibility to fuel properties, an empirical model has been developed through the combination of compression ignition combustion theory and empirical test results. This empirical application allows comparison of diesel engine sensitivity to fuels based on a truncated set of engine design factors.

IASH 2017, the 15th INTERNATIONAL SYMPOSIUM ON STABILITY, HANDLING AND USE OF LIQUID FUELS
Rome, Italy
10-14 September 2017

PENTAMETHYL HEPTANE AS A PRIMARY REFERENCE STANDARD FOR CETANE NUMBER.

Indresh Mathur[1] and Mical Renz[2]

[1]*Haltermann Solutions™, 15635 Jacintoport Blvd. Houston, Texas 77015, USA,* *imathur@jhaltermann.com*

[2]*Dixie Services Incorporated, 1706 First Street, Galena Park, Texas 77547,* *mrenz@dixieservices.com*

Today ASTM D613 method, the engine test method, is the gold standard for determining the ignition quality of diesel fuels. When the arbitrary cetane number scale was adopted, alpha-methylnaphthalene (AMN) was assigned a cetane number value of zero and n-cetane was assigned a cetane number of 100. In 1962 a change was made to adopt 2, 2, 4, 4, 6, 8, 8 - heptamethylnonane (HMN) as the low cetane reference standard because of the issues related to the availability of pure AMN. HMN was determined to have a cetane number accepted reference value (ARV) of 15 and today the cetane number scale is defined as:

Cetane Number = vol. % n-cetane + 0.15 (vol. % HMN)

The low Cetane Number Primary Reference Standard, HMN (98% minimum purity) that is required is not currently available in large quantities. Also, the very small laboratory quantities that are available are cost prohibitive. The diesel fuels industry has therefore resorted to using Secondary Reference T and U Fuels which can be unstable and require a 16 lab round-robin evaluation every one to two years. Ideally, the industry requires a substitute for HMN that can be produced in large quantities at reasonable cost. Our study has shown that 2, 2, 4, 6, 6-pentamethylheptane (PMH) can be the HMN substitute. PMH can be used not only to standardize the T and U Secondary Fuel blends, but users will also have the option the to utilize blends of Primary Reference Fuel, n-cetane and PMH for cetane number determination.

The authors will present the results of the study undertaken to compare the combustion characteristics of blends of HMN and n-cetane with blends of PMH and n-cetane in CFR engine using the ASTM D613 method. It has been found that PMH behaves very similar to HMN. Results of an inter-laboratory study will be presented with data to arrive at the ARV of PMH relative to HMN to justify that PMH and n-cetane can be used as Primary Reference Fuels to determine Cetane Number (CN).

IASH 2017, the 15TH INTERNATIONAL SYMPOSIUM ON STABILITY, HANDLING AND USE OF LIQUID FUELS
Rome, Italy
10-14 September 2017

Design of an on board Aviation Fuel Deoxygenator Unit for Improvement of Fuel Thermal Stability

Ehsan. Alborzi [1] , Matthew Dwyer[2], Simon G Blakey[1] and A H J M Meijer[2]

[1]*The University of Sheffield, Department of Mechanical Engineering, Sheffield, S1 3JD, UK,* *e.alborzi@sheffield.ac.uk*

[2]*The University of Sheffield, Department of Chemistry, Sheffield, S3 7HF, UK*

ABSTRACT

Gas turbine fuels are thermally stressed on route from the fuel tank, through the engine, to the combustion chamber. The temperature rise in the fuel as it passes through the engine initiates a range of chemical reactions in bulk fuel which are manifested as the formation of soluble and insoluble particulates. These ultimately participate in surface carbonaceous deposition in fuel system. These changes in chemical composition can block fuel nozzles, cause disruption to the flow of fuel, and result in the breakdown in operation of specific engine components. Surface deposition initiates from bulk fuel autoxidation reactions at t< 300°C.

This work presents the experimental investigation of the use of an optimal on-board oxygen adsorbent unit to reduce the propensity of surface carbonaceous deposit formation. Improved thermal oxidation stability by oxygen removal offers the fuel to serve as a better heat sink, absorbing more waste heat from "Very High By-Pass Ratio (VHBR) engines and using the heat to benefit in the engine performance cycle. The application of oxygen adsorbents requires a careful assessment of fuel chemical composition following on fuel deoxygenation. This is primarily due to the fact that there is a trade-off between thermal stability enhancement via oxygen adsorption separation and fuel lubricity decrease as a result of polar species separation. The experimental work is supported by using quantum chemistry calculations to investigate the selective binding of oxygen by adsorbents.

IASH 2017, the 15th INTERNATIONAL SYMPOSIUM ON STABILITIY, HANDLING AND USE OF LIQUID FUELS
Rome, Italy
10-14 September 2017

ENDOTEHRMIC REACTIVTY OF HYDROCARBONS UNDER CONDITIONS RELEVATN TO HIGH-SPEED FLIGHT SYSTEMS

Matthew J. DeWitt[1], Donald K. Phelps[2], Theodore Williams[1], Tyler Hendershott[1], Quinn Casselberry[1], Rich Striebich[1], Linda Shafer[1], Steven Zabarnick[1], Zachary West[1], Tim Edwards[2]

[1]University of Dayton Research Institute, 300 College Park, Dayton, OH 45469-0043, USA, matthew.dewitt@wpafb.af.mil
[2]AFRL/RQTF, Wright Patterson AFB, OH 45433

The development and implementation of advanced combustion and high-Mach propulsion systems requires advanced thermal management strategies due to high heat loads generated during flight. A potential strategy is to use hydrocarbon fuels as the primary on-board heat sink. A portion of the heat sink can be provided via fuel sensible heating (e.g., $Cp\Delta T$), which can be supplemented with deliberate endothermic (e.g., heat-absorbing) reactions via thermal and/or catalytic cracking decomposition chemistry. Although the endothermic decomposition of the fuel can provide a significant amount of cooling capacity, an undesirable consequence is the formation of carbonaceous deposits (also known as coking) which can reduce fuel flow, increase resistance to heat transfer and foul injector nozzles. Therefore, a primary need is the development of viable fuel chemistry and engineering strategies to provide the required heat sink while mitigating undesirable deposit formation. Improved understanding of the relationships between fuel chemical and physical properties and engineering implementations on the corresponding reactivity and deposition characteristics is needed to provide the requisite basis for the development of advance high-speed aerospace platforms. Experimental and computational efforts are being performed using single components, model solvents and fully-formulated jet and rocket fuels to improve the understanding of the correlations between fuel chemistry and reaction conditions, including the use of catalysts, on fuel performance. The experimental, analytical and computational methodologies employed and relevant findings will be presented.

THE ROLE OF HYDROCARBON COMPOSITION ON THE THERMAL STABILITY OF AVIATION FUELS

Matthew R. Dwyer[1,2], Simon G. Blakey[1], Ehsan Alborzi[1], Anthony J.H.M. Meijer[2]

[1]*Department of Mechanical Engineering, The University of Sheffield, Sheffield, UK, S1 4ET.*
[2]*Department of Chemistry, The University of Sheffield, Sheffield, UK, SJ 7HF.*

ABSTRACT

Advances in jet engine efficiencies and environmental concerns have lead to increased demands being placed on aviation fuel. This has driven interest in the use of alternative fuels, such as those derived from synthetic processes, in the aviation sector. This development however requires a greater understanding of the dependence of a fuels chemical composition on its thermal stability. Previous work on the oxidation mechanisms of hydrocarbons has been restricted to treating them as a general class of species. With the increased control of fuel composition that alternative fuels can achieve, it is necessary to research the role that chemical composition has on the oxidation process, allowing for the development of fuels with increased thermal stability, for use in more efficient engines. This work investigates the oxidation mechanisms for three classes of hydrocarbon commonly found in fuel with quantum chemistry techniques, using methods established for use in mechanistic organic chemistry studies. A straight chain alkane, dodecane, a cyclic alkane, decalin, and an aromatic hydrocarbon, toluene, were used to model the three classes of hydrocarbons. Our calculations indicate that aromatic and aliphatic hydrocarbons oxidise through different routes. Aromatic hydrocarbons act as antioxidants in the fuel, donating hydrogen to other hydrocarbons, thus stabilising radicals formed during the oxidation process. However this increases their susceptibility to oxidation, and as a consequence they undergo aromatic substitution reactions to form deposits, while aliphatic alkanes oxidise through different mechanisms. The theoretical modelling work has been supported by experiments carried out on small-scale thermal stability test devices.

KEYWORDS: Jet fuel, Alternative fuel, Thermal stability, Hydrocarbon oxidation, Autoxidation, Computational modelling

Nomenclature

Hartree-Fock theory - The first method developed for quantum chemistry
MP - Moller-Plesset perturbation theory, A post Hartree-Fock method to describe correlation
CCSD - Coupled Cluster theory, a post Hartree-Fock method to describe correlation
CASSCF - A quantum chemistry method which describes a system with many slater determinants
DFT - A quantum chemistry method which used the electron density to find the systems energy
Aromatics - Unsaturated cyclic hydrocarbons
Alkanes - Saturated hydrocarbon which can be either chains, branched or straight, or cyclic
Radical - A chemical species with an unpaired electron, which is very reactive
Electron spin - Electrons can be up or down spin
Singlet - Where there is no unpaired electrons, or the unpaired electrons have opposite spin
Triplet - Two unpaired electrons with the same spin, molecular oxygen is a triplet
PES - Potential energy surface, the energy landscape a reaction proceeds through

INTRODUCTION

The jet engines developed for use in aviation have increased greatly in both efficiency and the power output they produce over the last few decades. This has increased the operating temperature of the engine dramatically, which if uncontrolled would lead to system failures. As such aviation fuel has a secondary use, supplementary to being the primary energy source, as a heat sink to remove this increased temperature from the engine.[1,2] However the excess heat being moved into the fuel has the negative effect of causing the fuel to breakdown leading to thermal stability issues. This occurs through autoxidation reactions which rapidly increase in rate at these higher temperatures.[3-6] These mechanisms proceed through radical species formed from the fuel reacting with the dissolved oxygen.[7-12] The oxidised species then react further through radical pathways to form partially soluble polars species, which will then agglomerate and react together, causing insoluble gums and deposits to form on the the fuel systems. The deposits formed lead to reduced performance as a heat exchange, and inservice issues, reducing the profits of the airlines by causing the aircraft to spend less time in the air. This means that the thermal stability of aviation fuel is a major issue for airline operators, engine manufacturers and fuel producers, and is one which needs to be understood better and controlled.

During the early developments of the jet turbine engine it was thought, and demonstrated in use, that aviation was insensitive to the fuel being used for combustion. As such a wide cut of kerosine was used. However due to the increased development and specialisation of jet engines it has become important to regulate the properties of the fuels in service, and thus provide controls on its chemical composition. This is carried out by the fuel specification documents released by aviation regulation bodies around the world. The most followed fuel specifications are controlled by the American Society for Testing and Materials international (ASTM) in the USA and the Defence Standard (DEFSTAN) in the UK as well as European union regulations.[13-15] The fuel specifications set out the property requirements and methods of production for aviation fuel. The physical properties controlled by the fuel specification include freezing point, appearance, distillation curves and density to name a few. These requirements have an impact on the chemical compositions of the fuel that can be used, as straight chain alkanes will have have freezing points to high for use in cold environments, while highly branched alkanes will lead to densities which are too low. As well as these implied controls on fuel composition, the specifications also provide explicit controls on the concentrations of specific species in the fuel. The volume of aromatics in aviation fuels is limited to 25% and the volume of naphthalenes to 3%. The fuel specifications also limit the amount of impurities which can be found in the fuel, such as copper and sulfur species.

Despite these controls over the properties and composition of aviation fuel, a number of petroleum and alternatively derived products are suitable for use.[16-18] Fuel produced from petroleum sources are referred to as conventional fuels and can be described as falling in to three categories. Straight run fuels from factional distillation, fuels which have undergone post distillation hydrotreating at the refinery and hydrocracked heavy oil residues. These three types of fuels have different compositions and also demonstrate markedly different thermal stabilities during use. Straight run kerosine, produced from fractional distillation, is the easiest fuel to manufacture as no post distillation treatment occurs. Straight run fuels are formed from the middle distillate fraction of the oil, and have an average carbon chain length of 10-16. However the hydrocarbon and chemical composition is controlled by the oil source used, with this effecting the properties of the fuel. Due to the lack of control that the distillation has over the fuel they can have high iso-alkane, alkene and aromatic concentrations compared to fuels produced using other methods. Due to this broad ranges of fuel compositions, there is no single way a straight run fuel behaves at high temperatures, with those from Brent crude or west Texas sources demonstrating better thermal stability that those from Venezuelan or Indonesian sources.[19] The breakdown of the oil products obtained from the major oil fields of the world can be seen in figure 1. The dark blue segment shows the amount of heavy oil waxes and sediments found in the oil, and is usually a good indication of the concentration of sulfur species in the fuel. Understanding the composition of the fuel that can be got out of each oil field is important, as Venezuela now has the largest reserves of oil in the world. This as well as the increased utilisation of shale derived oils, and all of the issues being reported in association of there use in service, means that these inservice issues are likely to increase in the future even using the current architecture. So while the middle distillate of light oils from the middle east and the north sea are very good, in general straight run kerosenes have lower thermal stability than fuels produced from other methods due to the high levels of polar species and unsaturated hydrocarbons which are not removed from these fuels.[20-25]

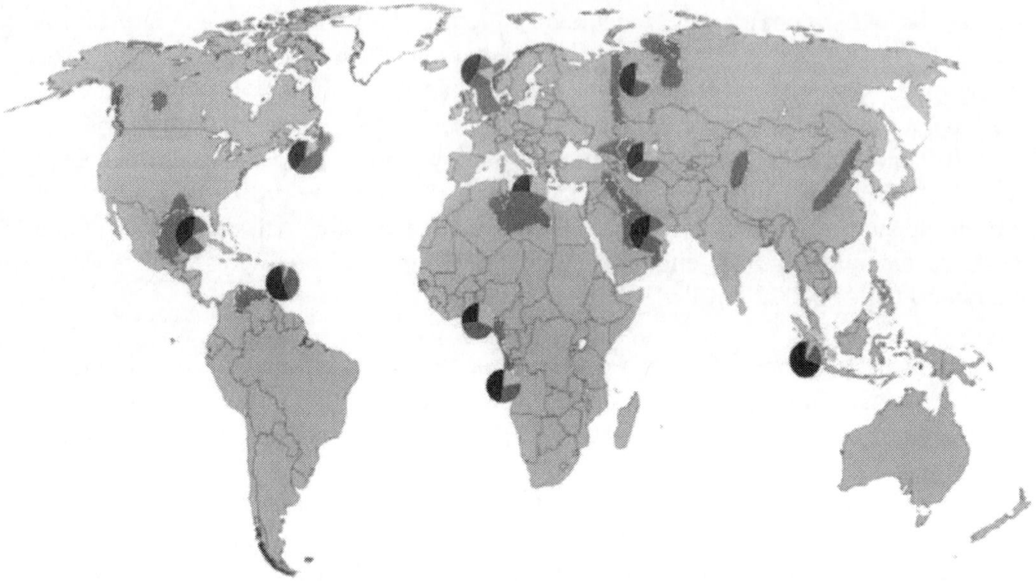

Figure 1: Breakdown of the oil quality of the major oil field, the dark blue segment indicates the heavy waxes which require hydrocracking for use in aviation fuel.[19]

To improve the control over the composition of the fuel, and thus provide improvements of its thermal stability, the straight run kerosine can undergo post distillation processing. This is done through hydrotreating the middle distillate using catalysts. Hydrogenation catalysts are used to chemical reduce the heteroatomic species found in the fuel, such as sulfur and phenolic species, into hydrocarbons. Previous studies have shown that changing the extent of hydro treatment of a straight run fuel impacts its thermal stability.[1] Mildly hydrotreating a fuel improved its breakpoint by 68°C to 354°C, a significant increase, so clearly these polar species have a negative effect on the thermal stability of a fuel. Further hydrotreatment can occur if more specialised catalysts are used to further alter the fuel. They hydrogenate the unsaturated hydrocarbons, such as alkenes and aromatics, to alkanes. These processes have the effect of producing a fuel which is more chemically pure, and they demonstrate improved thermal stability. This more severe hydrotreatment increased the fuels stability of the same fuel to >371°C. This demonstrates that a fuels chemical composition plays a major role in its thermal stability. The polar species found in a fuel clearly reduce the breakpoint of the fuel, but the hydrocarbon composition of the fuel also plays an important part in the formation of oxidation products and insoluble gums even in the single fuel that was used in these tests.

The third class of conventional fuel is manufactured from the heavy oil residues left over after the distillation process. The heavy waxes can be found in greater concentration in poorer quality oil reserves which are harder to exploit, such as in Venezuela or Indonesia as can be seen in figure 1. However due to the economic drive for oil they are increasingly being exploited. To be used these heavy oils require greater processing in order for them to have the correct physical properties. They are hydrocracked to lower molecular weight produces, using metal catalysts. If untreated these fuel have poor thermal stability, as they have higher metal contamination and polar species concentrations. This means that fuels produced using this method usually undergo post production treatment to remove the high concentration of impurities. These production methods to improve the fuels thermal stability however increase the cost of production and the fuel. These changes also have the result of negatively effecting other physical properties such as reducing the lubricity of the fuel. A summary of these processes can be seen in the figure 2 below.

Aviation fuel can also be produced from hydrocarbon sources other than petroleum, these fuels are referred to generally as alternative, or synthetics fuels. This can include fuels derived from hydrotreated bio-oils, esters of fatty acid, as well as those produced from Fisher-Tropsch processes or other carbon sources. Due to this wide variety of hydrocarbon sources these fuels can have varied compositions.[26-30] However synthetic fuels generally undergo post production processing. This improves the chemical purity of these fuels by removing the polar species found in the feedstock. This means that alternative fuels can potentially have very high thermal stability compared to straight run kerosine. This is highly dependent on the feedstock used and the manufacturing processes.

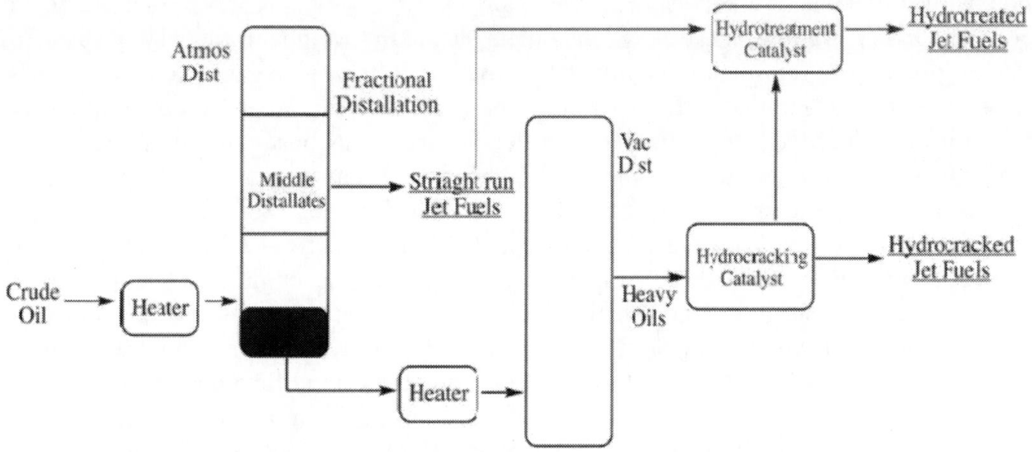

Figure 2: Routes towards conventional fuels

Fisher-Tropsch fuels are produced from hydrocarbon sources, such as coal and natural gas, which are converted into syngas using the steam reforming reactions. For this process any hydrocarbon source can be used, but these fuels have developed in regions where oil is not readily available, as such coal is usually used as the hydrocarbon source. The syngas are used in the Fisher-Tropsch reaction to form long chain hydrocarbons, known as Fisher-Tropsch waxes, and water as the byproduct. The Fisher-Tropsch waxes are hydrocracked to form lighter kerosenes suitable for use in aviation. The composition of the fuel produced can be controlled by the selection of catalyst. The gas-water shift reaction is used to produce more hydrogen for use in the Fisher-Tropsch reaction. Fisher-Tropsch fuels are very chemically pure and demonstrate high thermal stability. However the reduced polar and aromatic concentrations in the fuel can have a poor effect on its other physical properties. This can be solved by the use of fuel additives to improve lubricity. The major draw back of Fisher-Tropsch fuels is the high costs involved in their production due to the number of processes, high operating temperatures and expensive catalysts needed. This has limited their use outside of areas where political situations have made oil alternatives necessary.

$H_2O + CH_4$ --- $CO + 3H_2$	Steam reforming reaction
$H_2 + nCO$ --- $(CH_3)_n + H_2O$	Fisher-Tropsch reaction
$H_2O + CO$ --- $H_2 + CO_2$	Gas-Water shift reaction

The other common route to alternative fuels is to use bio-oils and carbohydrates from other sources, such as sugars and farming waste, to convert into hydrocarbons suitable for use in aviation. The feedstock undergoes trans-esterification with alcohols which converts these highly branched carbohydrates into smaller straight chain molecules. These esters are usually hydroprocessed into hydrocarbons with the correct molecular weight for use as an aviation fuel. These products are known as Hydro-processed Esters and Fatty Acids (HEFA), or as bio-fuels. Their thermal stability and impurity levels is heavily linked to the feedstock, and the level of post production processing such as hydrotreatments. If they are from a clean source, with low metal and polar contamination, and undergo hydrotreatment these fuels can show very high thermal stability.

Literature Review

From this overview of the types of fuels used in aviation it is clear that a fuel can meet specification and have a varied chemical composition and properties. As such it is important to discuss what species are in fuel, how this effects its properties, and what consequences this has on its chemistry. The major component of all aviation fuel is the hydrocarbon component, but the nature of this component can be vary while still meeting the specification. These hydrocarbons all have different effects on the fuel properties. The hydrocarbon species found in jet fuel include straight chain n-alkanes, branched iso-alkanes, cyclic alkanes of varied sizes, aromatics as well as potentially unsaturated alkenes. The n-alkanes form the majority of the hydrocarbon fuel, they have a high hydrogen to carbon ratios, increasing the fuels heat to weight ratio. They also offer good combustion properties for the fuel, reducing the polluting emissions and the fuels negative environmental impacts. However due to n-alkanes uniform structure they have high freeze points, as they can pack tighter, forming stronger Van der Waals forces between the chains. They also have high viscosity at low temperatures, reducing the ability to pump the fuel to the engines. To combat these issues iso-alkanes can be used to reduce the freezing point of the fuel and improve their viscosity. However this reduces the energy density and the maximum range of the fuel. With the drive for efficiency in aviation these properties are vital for the engine to get the most out of the fuel. To improve the energy density of the fuel aromatic hydrocarbons can be added to the fuel. They are hydrogen deficient compared to saturated hydrocarbons, meaning they have very poor combustion properties, and form a lot of smoke and carbon particulates when burned. For this reason aromatic concentrations by volume is limited to 25% in aviation fuel. It is clear that a number of combinations of these hydrocarbons would meet the fuel specifications and the ratio of them is determined by the feedstock.

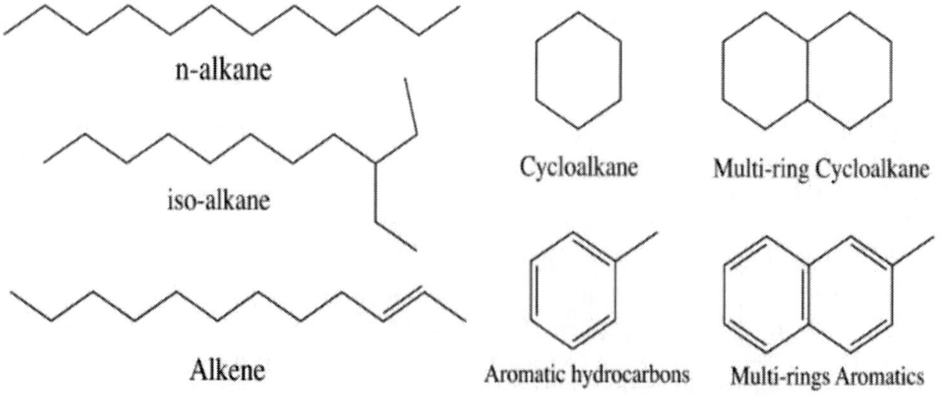

Figure 3: Hydrocarbon species found in aviation fuels

The hydrocarbon component of the fuel controls its bulk properties, but a number of fuel specification properties are controlled by the trace species found in the fuel. These include the fuels lubricity and cleanliness. These minor components of the fuel include polar oxygenated species, sulfur and nitrogen containing compounds and metal impurities.[31,32] The exact composition and concentrations of these species is determined by a fuels feedstock and the post production processing that it undergoes. Metal contaminants, especially copper, have a negative effect on the fuels thermal stability and cleanliness. This is thought be because they facilitate the fission of peroxides, either through catalysis or by reacting with the peroxides in a one electron reaction.

Nitrogen and oxygen containing species, particularly heteroatomic aromatic species, have been suggested to have negative effects on the fuels thermal stability and cleanliness They are thought to react with radical species to form the sediments and deposits in the fuel. Not all of these species have a negative effect on the fuel. Sulfur species improve the lubricity of the fuel by binding to the metal surfaces, improving the slip between the fuel and the surface. Phenolic species found in the fuel can improve the storage stability of the fuel, by acting as antioxidants donating a hydrogen atom to the radical species, slowing the oxidation process. However both have been linked in to reduce the thermal stability of a fuel as sulfur and oxygen have been found in the deposits in high concentrations.[33-35]

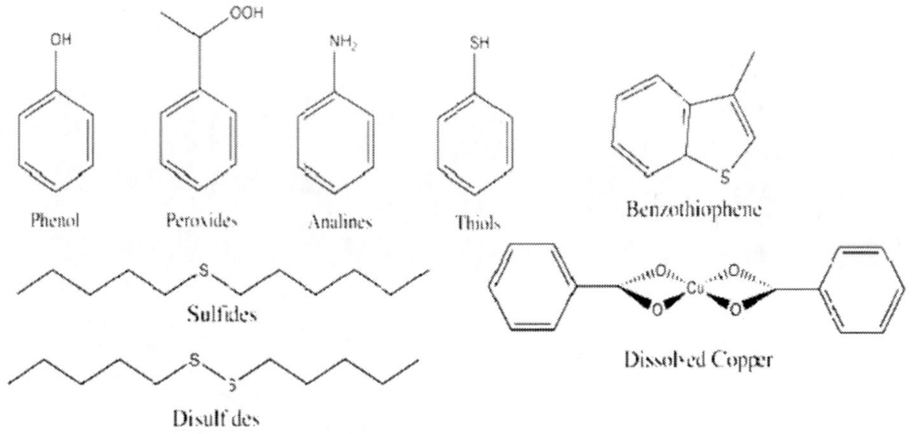

Figure 4: Heteroatomic species found in fuel

Phenolic compounds are found naturally in the fuel but are also added to some fuels, to improve their storage stability, in the form of additive packages. The antioxidant used is butylated hydroxytoluene (BHT), a highly sterically hindered phenol. The hindrance means that the radical formed from its donation of a hydrogen to a hydrocarbon radical is stable to further reactions by the steric hinderance of the radical site and the aromaticity of the phenol, slowing the rate of formation deposit. Other additive packages can be added to the fuel to improves it properties. Static dissipators are used to improve the conductivity of the fuel, stopping the accidental sparking and ignition of the fuel. Fuel system icing inhibitors are added to stop the formation of water ice in cold climates. For fuel with poor lubricity, lubricity improvers can be added, and metal deactivators can be used to chelate to copper and stop it from reacting with the fuel.

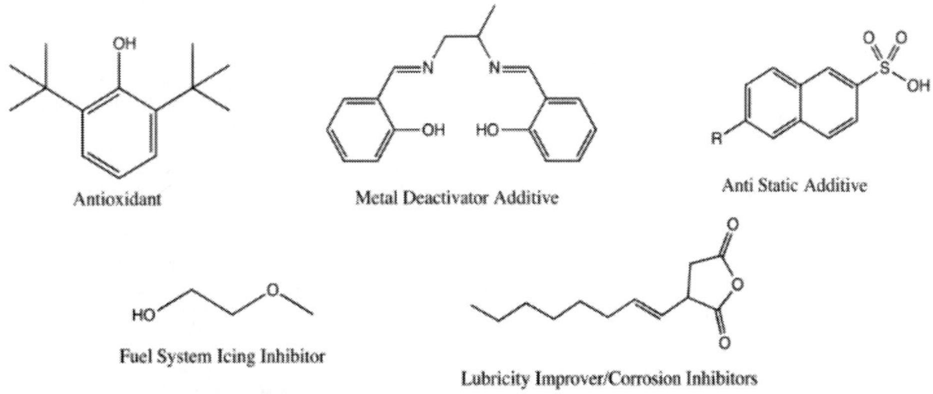

Figure 5: Additives used in aviation fuels

It is clear the aviation fuel is a complex mix of chemical species, which can react to form insoluble sediments and deposits, leading to inservice problems. This occurs through the hydrocarbon species reacting with the dissolved oxygen in the fuel.[36] These reaction proceed through radical chemistry mechanisms to form peroxides and other hydrocarbon radicals.[37,38] The oxidation reaction is initiated by the abstraction of a hydrogen from a fuel hydrocarbon. The hydrogen abstraction can be caused by the metal surface, dissolved metals or peroxides found in the fuel. The hydrocarbon radical can now react with the dissolved oxygen to form the peroxide radical which can propagate the oxidation reactions. The reason that the hydrocarbon can not react directly with oxygen present in the fuel is that they are closed shell singlet species, while the oxygen is in a triplet spin state with two unpaired electron.[39,40] The reaction between two species in different spin states is slow, as reactions must obey the law of the conservation of momentum, this includes the angular momentum of electrons. This also effects the rate of fission of hydrogen peroxide as the reaction involves the unpairing of two electrons.[41]

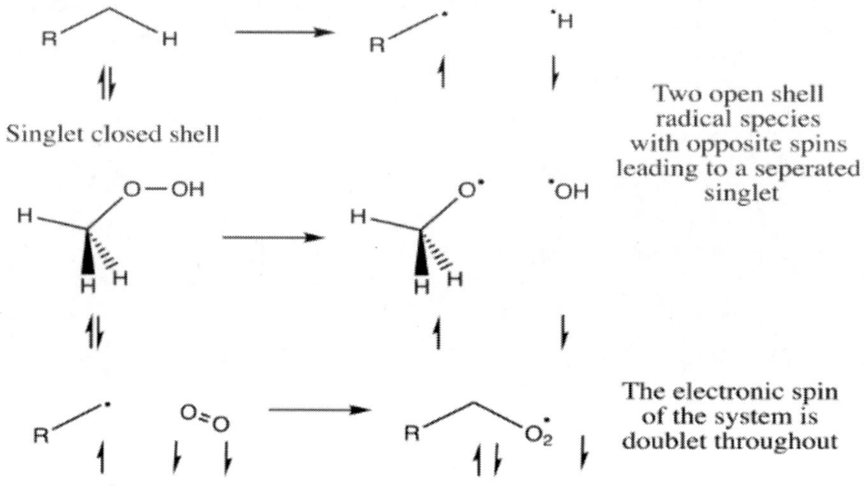

Figure 6: fission of the fuel species, while still conserving the momentum of the electrons. The open shell hydrocarbon radical can then react with the triplet dissolved oxygen.

The hydrocarbon radical and peroxide can react further to form the primary oxidation products, through hydrogen abstraction and radical termination reactions. A number of mechanisms have been developed to describe the processes that form the primary oxidation products. One of these mechanism was developed by Zabarnick *et al*, and includes the roles that metals and sulfur species play in the oxidation of fuel.[42-46] The products of these reaction are alcohols, aldehydes, ketones and ethers. These species are more polar than the majority hydrocarbon fuel, and as such will agglomerate together to minimise the negative enthalpic interactions with the apolar hydrocarbons.

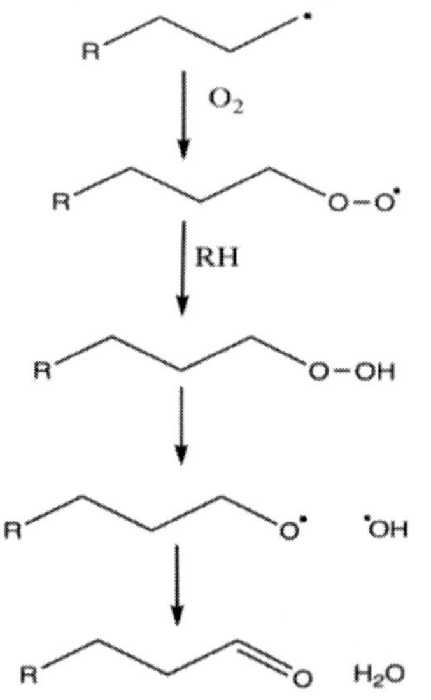

Figure 7: Formation of the initial oxidation products, such as alcohols and aldehydes, these products can then go on to react further.

$$I^\bullet \rightarrow R^\bullet$$
$$R^\bullet + O_2 \rightarrow RO_2^\bullet$$
$$RO_2^\bullet + RH \rightarrow RO_2H + R^\bullet$$
$$RO_2^\bullet + RO_2^\bullet \rightarrow \text{termination}$$
$$RO_2^\bullet + AH \rightarrow RO_2H + A^\bullet$$
$$A^\bullet + RH \rightarrow AH + R^\bullet$$
$$A^\bullet + RO_2^\bullet \rightarrow \text{Products}_{AH}$$
$$R^\bullet + R^\bullet \rightarrow R_2$$
$$RO_2H \rightarrow RO^\bullet + {}^\bullet OH$$
$$RO^\bullet + RH \rightarrow ROH + R^\bullet$$
$$RO^\bullet \rightarrow R^\bullet_{prime} + \text{carbonyl}$$
$${}^\bullet OH + RH \rightarrow H_2O + R^\bullet$$
$$RO^\bullet + RO^\bullet \rightarrow RO^\bullet_{term}$$
$$R^\bullet_{prime} + RH \rightarrow \text{alkane} + R^\bullet$$
$$RO_2H + SH \rightarrow \text{Products}_{SH}$$
$$RO_2^\bullet \rightarrow R^\bullet + O_2$$
$$RO_2^\bullet + R^\bullet \rightarrow \text{termination}$$
$$RO_2H + M \rightarrow RO^\bullet + {}^\bullet OH + M$$

Figure 8: The Zabarnick psuedo-detailed autoxidation mechanism, used to understand and model the oxidation of jet fuel.

This agglomeration brings the primary oxidation products close together. The oxygenated species then react with the heteroatomic species found in the fuel to form larger partially soluble species. These reactions proceed to form higher molecular weight, more polar species until they are no longer soluble in the jet fuel. These reactions are thought to progress through a mix of radical reaction mechanisms and electrophilic aromatic substitution reactions with the aromatic oxidation products and heteroatomic species this is known as the SMORS mechanism and was proposed by Hardy *et al* and developed by Beaver *et al*.[47-51]

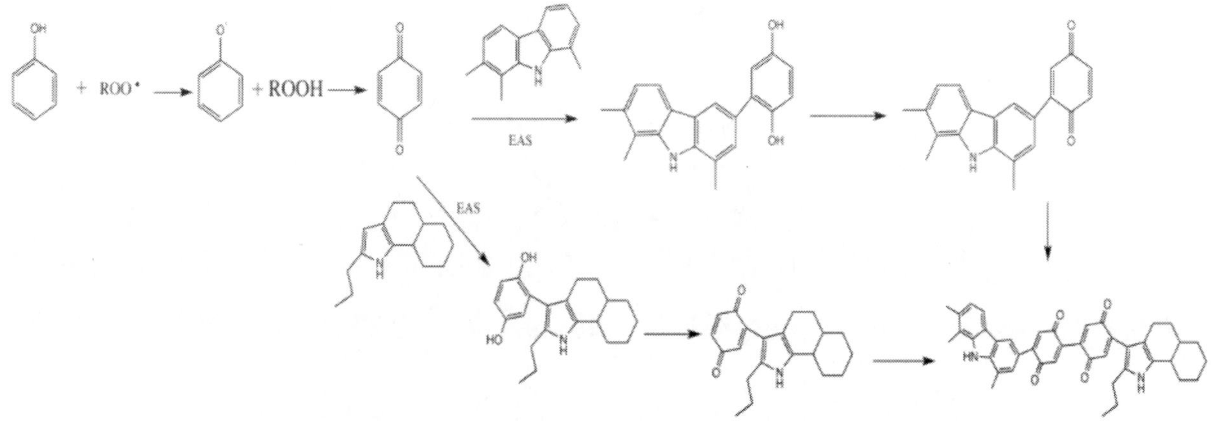

Figure 9: The SMORS mechanism proposed by Beaver *et al*, the products are high molecular weight polar species which can form deposits and sediments.[47-51]

These high molecular weight polar species, have high heteroatomic and aromatic content. They agglomerate together, forming insoluble sediments at storage temperatures. At increased operating temperatures these reaction not only proceed at a much higher rate, but also causes the products to bind to the surface forming insoluble deposits on the fuel system working parts. This suggest that binding to the surface is not barrierless, and requires energy to be put into the system to form the chemical bond to the surface, in this case by an increase in temperature and their by the kinetic energy.[52-54] As the literature indicates the initial deposit on the surface has a high concentration of sulfur. Thus at increased temperature the sulfur in the insoluble sediments must be able to chemisorbed to the surface and form strong covalent bonds. The initial deposit it then able to react through the same mechanisms as in the liquid phase increasing the weight of deposit over time. All of these processes leads to a full picture of a global oxidation mechanism, which describes the formation of sediments at storage temperatures, and the deposits found at higher temperature synonymous with thermal stability issues. An outline of this global oxidation mechanism can be seen below.

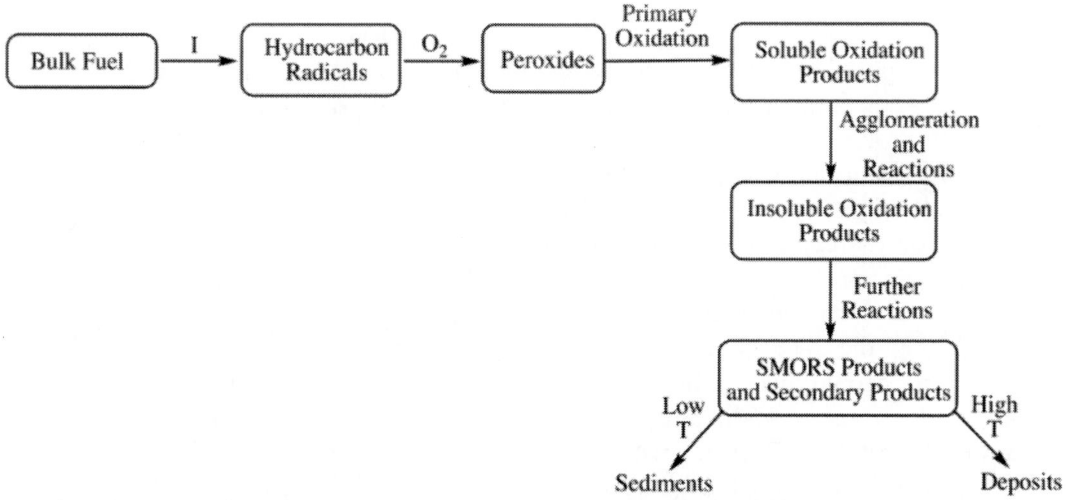

Figure 10: The overall global oxidation mechanism linking the initial oxidation mechanisms and sediment and deposit forming processes

Early work on the effect that a hydrocarbons chemical structure has on its oxidation rate was carried out in the polymer industry, investigating their use as monomers as well as in relation to their thermal stability as fuel. The work by Mayo *et al* and Taylor *et al* investigated the rate of oxidation of a number of saturated and unsaturated hydrocarbons including aromatic species.[55,56] They found that olefins, particularly styrene, was very reactive to radical species and formed large molecular weight products. Under oxidation condition these products would also have high amounts of oxygen present in them. It is unclear if the styrene monomers are oxidised prior to polymerisation or if the extended oligomer system, with high aromatic stabilisation of oxidisation products, are oxidised post polymerisation. Due to the fast rate of polymerisation of the styrene monomers once initiated it would be suggested that the oligomer is oxidised after formation.

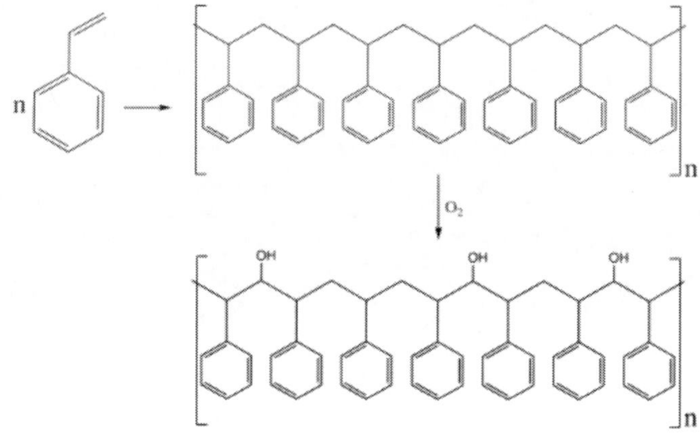

Figure 11: Styrene is a reactive species, which rapidly reacts to form oligomers, which then oxidise rapidly.

The literature indicates that all alkene have a very poor effect on the fuels thermal stability. The alkenes can form oligomers in the fuel, through reacting with the initiated radical and propagating the reaction. The alkenes are also easily oxidised, as the hydrogen next to an unsaturated double bond can be easily abstracted and the radical formed is stabilised by conjugation with the double bond. Alkenes are also labile towards addition reactions with the oxygenated products of the fuel, yielding extended oxygenated compounds. For this reason alkenes are seen as undesirable in the fuel, as they do not provide any benefits to the fuel other properties, and are excluded from the fuel.

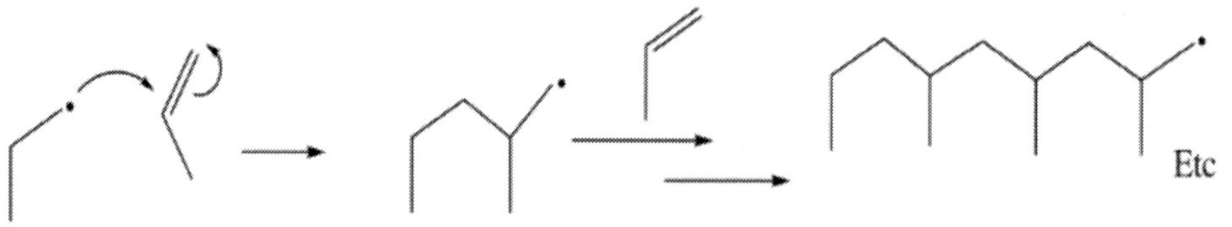

Figure 12: reaction of the alkene monomers to a long chain polymer, similar to the mechanisms observed for polymerisation.

Aromatic hydrocarbons allowed in aviation fuel as they improve the energy density both in terms of by volume and weight which are controlled by the fuel specifications. They however have the negative effect on the combustion properties of the fuel, hence limits on the concentration allowed are imposed. Their effect on the thermal stability of aviation fuel is complex and not completely understood. The literature indicates that the aromatic hydrocarbons can improve the stability of the fuel, by reducing the rate of the deposition, while increasing the amount of deposit formed overall.[57-64] Work carried out at IFPEN indicated that the addition of aromatic increased the induction period of the oxidation of the fuel. This increased induction period was found to be particularly effective for larger aromatic hydrocarbons with benzylic carbons. This supports the earlier work carried out by Igarashi *et al* and Taylor *et al*.[10,59,60] That work showed that 1-methyl-naphthalene (1-MN) had a dramatic effect in increasing the induction period of the fuel. It is suggested that this is caused by the stabilisation of the radical formed on the benzylic carbon by conjugation through the two aromatic rings

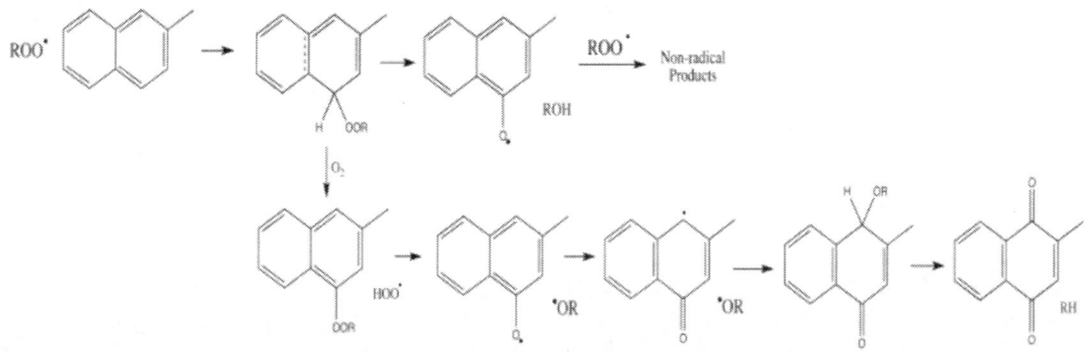

Figure 13: The mechanism proposed by Igarashi *et al* for the 1-MN aromatic hydrocarbon.[59,60]

The effect was seen by Taylor in a number of aromatic hydrocarbons, with the increased aromaticity of the hydrocarbon used leading to decreasing rates of deposition. This reinforces the theory that the aromaticity of the hydrocarbon increases the number of resonance structures of the radical, stabilising the radical through conjugation. This stability of the radicals of aromatic hydrocarbons means they can act as antioxidants in aviation. They behave as a sink for the radicals formed during oxidation by donating their hydrogen to the radicals products of other hydrocarbons. Similar to the mechanism that phenols exhibit.

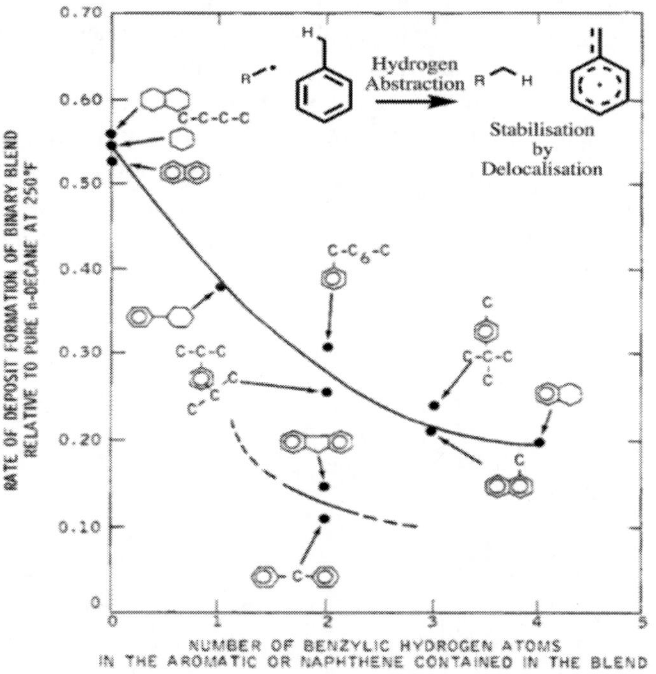

Figure 14: Aromatic hydrocarbon acting as a antioxidants in the fuel donating hydrogens to radicals formed, with the data originating from work carried out by Taylor *et al.*[10]

Despite increasing the induction period of oxidation in aviation fuels, aromatic hydrocarbons increase the total amount of deposit formed during the oxidation process. When the size of the aromatic species added to a test fuel is increased the amount of deposit also increases, with 1-MN leading to a dramatic increase in deposit over single ringed aromatics. This leads to a picture where aromatics are easier to oxidise and form more deposits, but slow the rate of deposition. This is highlighted by Heneghan *et al* where the the addition of antioxidant to a fuel leads to slow deposition but more deposit over all.[8] This would suggest that the product forming reactions for the AO_2 recombination are much slower than that for RO_2 recombination, but the products must have at least a higher tendency to form deposit. Similar to the homolytic fission of peroxides this is an interesting problem, as it is a radical reaction which involve a number of complex spin states and potential pathways to the products. The rate of recombination and product formation could be controlled by steric and electronic factors which need further investigation.

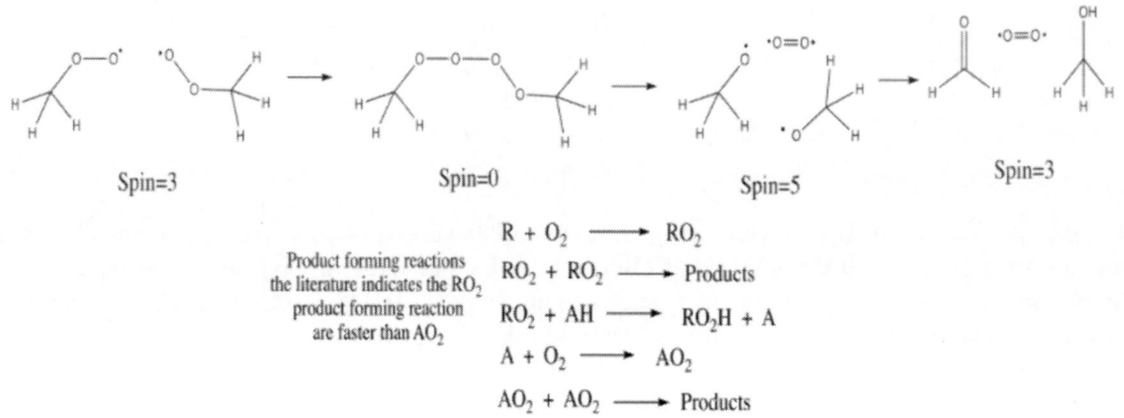

Figure 15: Peroxide product forming reactions. The literature has indicated that the antioxidant peroxides form products at a slower rate.[8]

Alkanes fall in to two branches, cyclic and noncyclic, which can be branched or unbranched. Previous work investigating the effect that branching has on the abstraction has shown that tertiary carbons react quicker, forming the radical products.[65] This is caused by the adjacent carbons donating electron density to the electron deficient radical centre, stabilising the radical through inductive effects. Experimental work carried out at IFPEN found that increasing chain length caused a reduction in the induction period of the fuel.[63] This continued up to a plateau reached at 12 carbons chains and longer. Mixtures of hydrocarbon by chain lengths showed a linear response behaviour, indicating that it is to do with the ratio of terminal and chain carbons in the hydrocarbon. This supports the theory that the rate of hydrogen abstraction is linked to stabilisation of the radical through inductive effects.

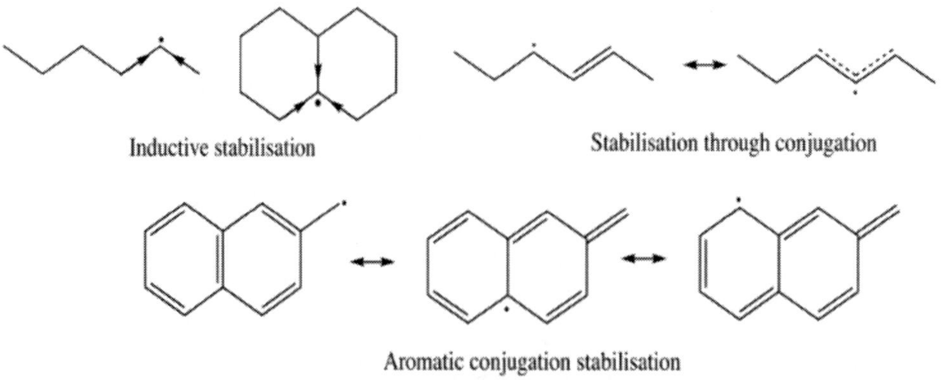

Figure 16: Mechanisms that hydrocarbons stabilise the radicals formed during oxidation

The roles that cyclic alkanes have on the thermal stability is open to debate. They have increased density comparatively to iso-alkanes. This means that if they show improved thermal stability compared to aromatics they could be used to improve the range of a fuel while not reducing its thermal stability. Previous work indicates that they do not increase the inductions period, so are not acting as an antioxidant in the fuel. Nor do they increase the deposit noticeably when compared to aromatic species. However the literature shows that not all aromatics and cyclics behave the same, so the exact nature of these mechanisms need to be clarified before development of the fuel can proceed.

A review of the literature clearly indicates that there is still a number of unanswered questions about the autoxidation mechanisms of aviation fuel, and the effect that chemical composition plays in these processes. All mechanisms and models developed up to this point have treated fuel as a generic hydrocarbon species, represented as RH, which undergo the same oxidation reactions at the same rate. Experimental evidence however clearly indicates that not all hydrocarbons react in the same way, nor at the same rate. As such it would make sense to develop a more tailored mechanism to appreciate these nuances. In this work the oxidation mechanism for dodecane, decalin and toluene have been investigated, as well as other reactions of interest. These mechanisms will focus on the production of the initial oxidation products, such as alcohol, ethers and ketones. The purpose of this work is to demonstrate the effect that changing hydrocarbon composition has on the mechanism and kinetics of oxidation.

Description of Methods

Investigating the individual oxidation mechanisms in fuel is challenging because it is difficult to experimentally follow a single species throughout a reaction. As such a different approach is required to research the role that a specific hydrocarbon species has in the oxidation mechanism. On these small scales its common to use modelling techniques, as they allow for the experimental and analytical research to be directed and for experimental handles to be highlighted. The work carried out in this paper used quantum chemistry techniques to construct detailed mechanism for dodecane, a n-alkane, decalin, a cyclic alkane and toluene, an aromatic hydrocarbon. Calculations were carried out using the Gaussian 09 software on the University of Sheffield High Performance Computer (HPC) Iceberg and the chemistry departments HPC Sol. The calculations were carried using Density functional theory (DFT) at the B3LYP/cc-pVTZ/IEF-PCM (heptane) level of theory unless stated otherwise. The B3LYP refers to the functional used for the calculations, and is a hybrid functional that includes a proportion of the Hartree-Fock energy, which corrects the errors known in pure functionals. The cc-pVTZ is a correlation consistent triple zeta basis set developed by Dunning and describes the wavefunction of the system. The IEF-PCM is the solvent model used to simulate the bulk fuel treating it as polarised continuous medium applied to the calculations. The thermochemical and kinetic data for the reactions were calculated. The difference in energy between the optimised structures of the reactants and the products gives the thermochemical data for the reactions and the difference in energy between the reactant and the transition states gives the kinetics of the reactions. The transition states are considered optimised when they have a single imaginary frequency, which has a vibration along the reaction coordinate of interest. This theoretical work has been supported by experiments previously carried out in the group using the PetroOxy small scale static thermal stability test device. This test devices investigates the rate of oxidation by measuring the pressure drop in the reactor as the sample oxidises. As the fuel reacts, the dissolved oxygen concentration in the fuel drops, leading to the oxygen in the air above the sample moving across the phase barrier into the liquid. From this the rate of oxidation can be obtained. These methods help to build up a picture of the effect that hydrocarbon composition has on it thermal stability

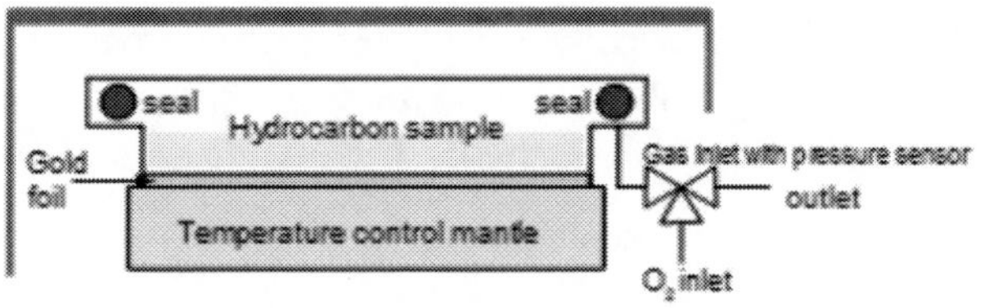

Figure 17: PetroOxy thermal stability test device.

RESULTS AND DISCUSSION

The literature indicates that aromatic hydrocarbons are linked to increasing the overall amount of deposit formed during autoxidation process. However they have also been shown to slow the rate of deposit formation, confusing the picture of their role in the autoxidation mechanism. The role that different hydrocarbon building blocks play in the oxidation of aviation fuel is important to understand. To aid in the development of future fuels with improved thermal stability without negatively effecting their physical properties. The work presented in this paper will look at the simple oxidation mechanisms of dodecane, decalin and toluene, to investigate the oxidation of these hydrocarbons. If the cyclic hydrocarbons do not behave noticeably different from the straight chain alkanes under oxidation then it would suggest that they can be added to the aviation fuel instead of aromatics, to improve its density, while not effecting it thermal stability. The stability of the radical formed and the ease at which the hydrocarbons can be oxidised was investigated by calculating the bond dissociation energies of abstracting a hydrogen from the three hydrocarbons. These calculated bond dissociation energies were compared to experimental results found in the literature.[66] It is clear that the energy to abstract toluene's benzylic hydrogen is significantly lower than that for either the cyclic or normal alkane. This would lead to it being the easiest to oxidise, and thus would then demonstrate lower thermal stability when added to the fuel. However the literature clearly shows that the addition of aromatic hydrocarbons slows the rate of deposition, particularly at storage temperatures. This disparity between the ease at which a molecular species is oxidised and the inverse rate of deposition and sedimentation it shows when added to fuels, can also seen by the use of antioxidant. This indicates that these species are stabilising the radical formed during the oxidation and thus are slowing the rate of product formation. Analysis of the electronic spin density of the radical shows that the toluene radicals spin density is spread over the aromatic carbon atoms, where as the alkanes can only move a small amount of spin density onto the adjacent carbons through electron inductive effects.

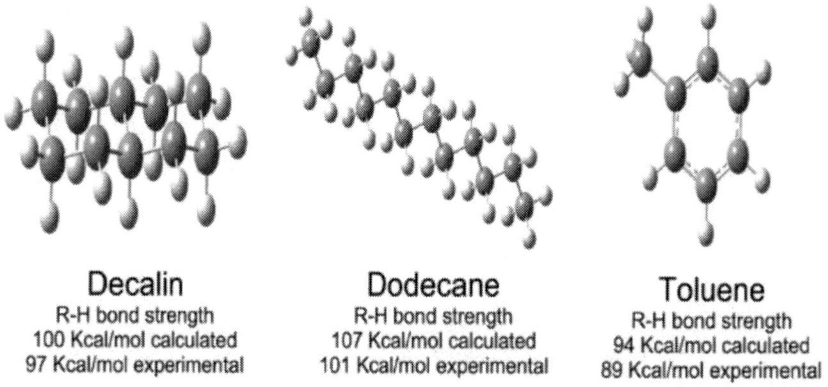

Figure 18: Bond dissociation energy for the three hydrocarbons investigated
the bond dissociation methods were calculated using DFT in the Gaussian 09 software.[66]

The difference between the experimental and calculated dissociation energy can be attributed to a number of reason. One of the overall aims of computational chemistry is to achieve accuracies of 1 kcal mol^{-1} in predicting barrier heights and dissociation energies. These calculations demonstrate far larger inaccuracies with them being between 5-10 kcal mol^{-1}. This difference is often attributed to the correlation energy not being accounted for properly or described correctly. The general term correlation energy is attached to any difference between the calculated energy of a system and the true energy of that system. However correlation energies can have real physical meanings. There are two forms of correlation energy, static and dynamic correlation, and refer to the failures to describe the true nature of physical system in the correct mathematical formalisation.

Dynamic correlation refers to the movement of the individual electrons in the system not being properly described. This is because the electrons in most computational chemistry methods are described using mean field approximations. In this approach the individual electrons feel an average of the repulsions of all of the other electrons rather than the point charges of the individual electrons in the real system. Treating the electrons as moving in a mean field causes them to approach to closely to one another. This increases the energy of the system as the electrons then repulse each other too strongly. This is a major flaw in Hartree-Fock theory causing the barrier heights to be overestimated. This has lead to the development of more accurate post Hartree-Fock theories such as CCSD(T) and MP at the expense of greater computation times. Density functional theory underestimates the dynamic correlation, so to overcome this error hybrid functionals include some Hartree-Fock energy in them, this is what B3LYP does. As such dynamic correlation is unlikely to have a large impact on the accuracies of the calculations, and more computationally expensive post Hartree-Fock method do not need to be used.

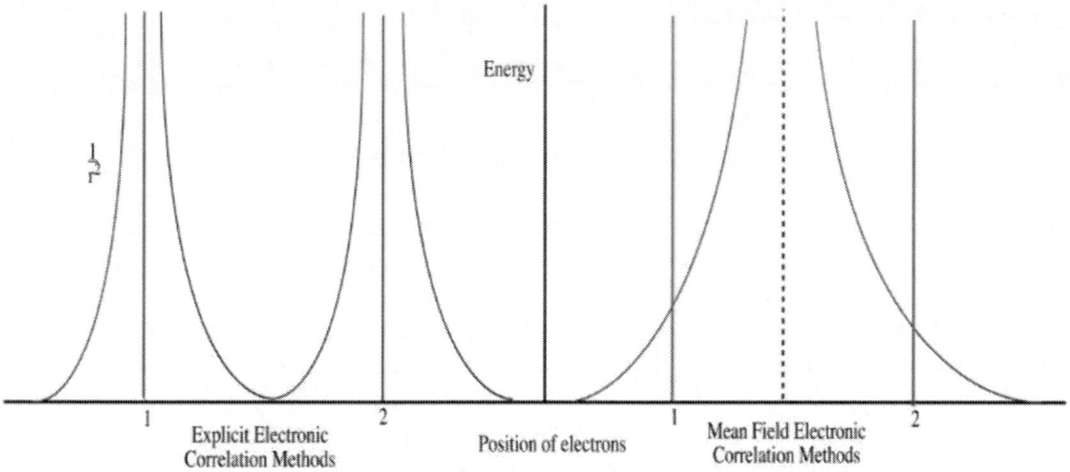

Figure 19: Dynamic correlation, showing the flaws of mean field theories

Static correlation relates to the electronic structure of the system and the orbitals being occupied not being described properly. It is related to the spin states of the systems being investigated and the excitation of the electrons into excited states. This becomes a major problem during calculations investigating the homolytic fission of a bond. This is due to DFT, Hartree-Fock and single reference post Hartree-Fock methods not describing the spin state and electronic structure of the system correctly as the bond is lengthened. In the bound structure the electrons in the bond form a pair and are in the singlet configuration in the same electronic orbital. As the bond breaks physically these electrons need to become unpaired. However there is no flip in the electron spin states, as such it is still a singlet this time range separated. Single reference methods struggle to describe this properly so the result is that the unbound system need to be incorrectly described as a triplet or as an ionic singlet. This lead to errors in the energy of dissociation when compared to the real system. To overcome this inaccuracy, multi-reference methods can be used which allow dissociated product to be described as a range separated singlet with two unpaired electrons in opposite spin states through the use of multiple Slater determinants.

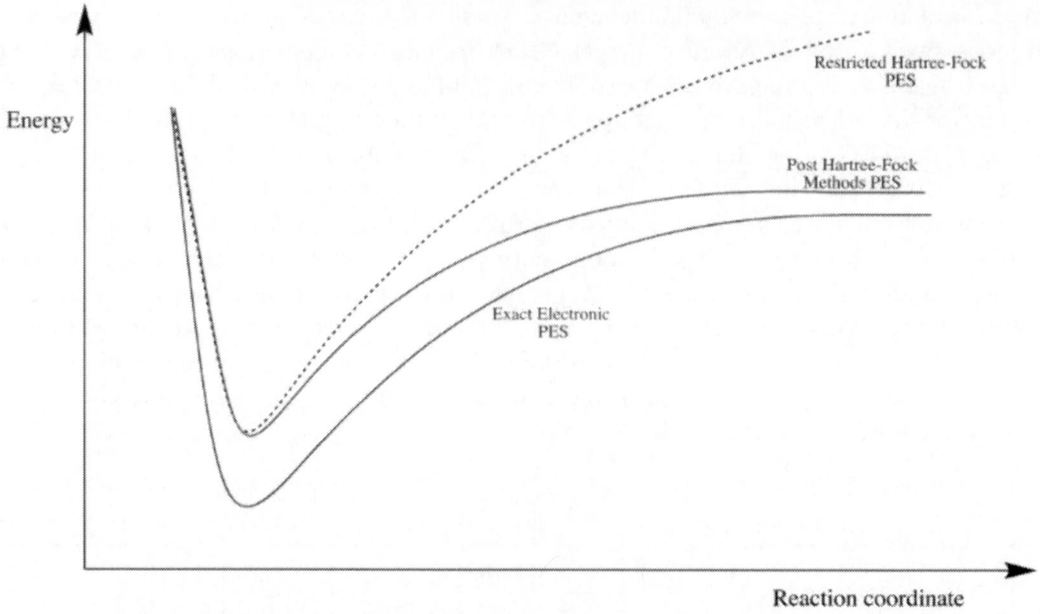

Figure 20: Static correlation, single reference methods fail to describe the electronic structure of dissociated bonds.

These inaccuracies can be found in the homolytic fission of the peroxide bond, and calculations investigating the homolytic fission of peroxide give poor chemical agreement with experiments. Calculation of the homolytic fission of the dodecane peroxide gives a barrier of 36 kcal mol^{-1}, compare this to the experimental value of 45 kcal mol^{-1} shows that the calculations do not show chemical accuracy. The cause of this can be seen in the scans of the spin state potential energy surfaces. In these calculations the O-O bond was extended from bonded to unbonded, using a relaxed scan along the reaction coordinate, in the triplet and singlet spin states. These scan show that the triplet product is a stable state forming an asymptote, but does not form a stable bound structure with the electrons of the same spin repulsing each other leading to very high energies. The singlet scan does not form a stable dissociated state, as the lack of an asymptote indicates the wavefunction is not stable when calculated using a single reference method.

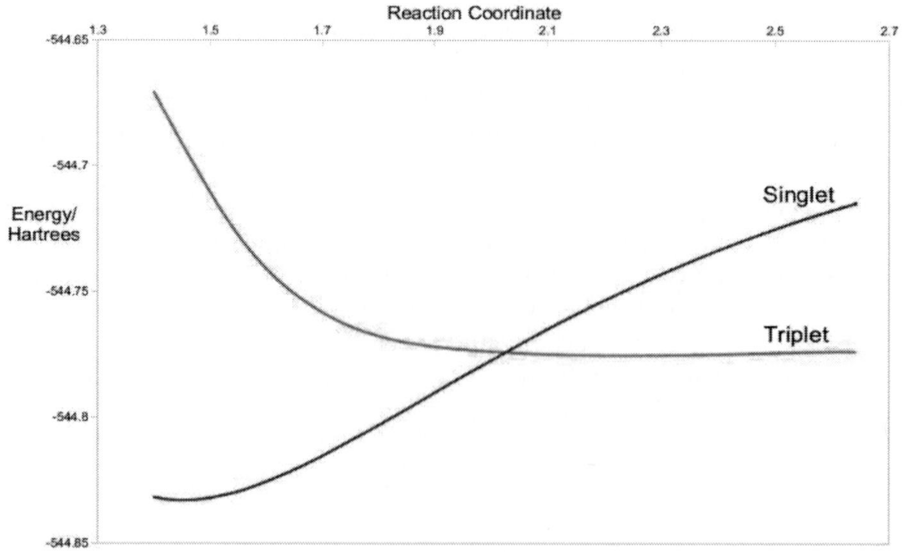

Figure 2l: Potential energy scan of the triplet and singlet surfaces for the homolytic fission of a peroxide.

To investigate the effect that electronic correlation has on the homolytic fission of peroxide, further calculations were carried out. These calculations used the method Complete Active Space Self Consistent Field theory (CASSCF), which is a multi-reference methods which use more than a single Slater determinant to describe the wave function. These were carried out using the MOLPRO electronic structure code. To test the methods for use in further investigations a small system was selected, with the methyl-peroxide molecule used to carry out the investigations. The geometries used for the these calculations were optimised in Gaussian 09 using DFT. The results can be seen in the figure below.

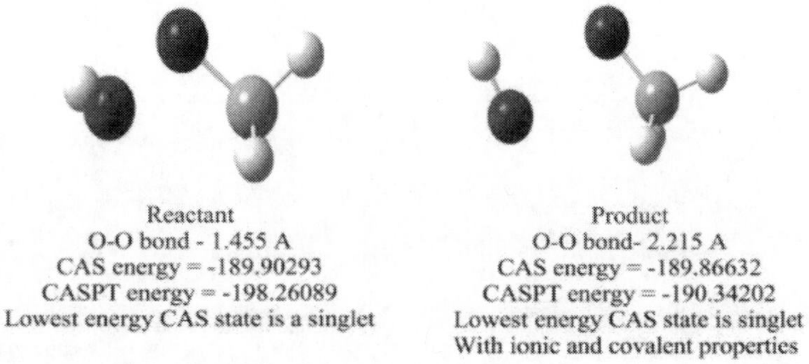

Figure 22: Homolytic fission of a peroxide, the CASSCF calculations indicating that the product is a separated singlet. Geometries calculated in Gaussian 09 and the energies calculated in MOLPRO

The CASSCF calculations indicate the the dissociated product is indeed a range separated singlet spin state product. The character of this state indicates that it has both covalent and ionic contributions. The product has two degenerate states on the oxygens, each partially occupied by an electron. Despite the CASSCF calculation qualitatively describing the system correctly is also underestimates the barrier to homolytic fission predicting it to be 23 kcal mol^{-1}. The reason for this is likely to be due to the size of active space selected during the calculations as these calculations are known to strongly depend on not only the basis set selected but also the size of active space. These methods are important to use further, as the product forming reactions between two peroxides are known to have a number of electron spin issue and can only be described properly using multi-reference methods.

The initial oxidation mechanisms for the three hydrocarbons were calculated. All of the hydrocarbons radicals react through a barrierless exothermic reaction with the oxygen, forming a peroxy radical. This reaction will be limited by the concentration of the oxygen found in the fuel. As the oxygen in the fuel is consumed the hydrocarbon radical will react with the other species found in the fuel. The alkane hydrocarbons react in similar ways, with the hydrocarbon radical being unstable compared to all species it reacts with. This means that the radicals of the alkane hydrocarbons react rapidly in the fuel, propagating the oxidation mechanism and forming insoluble products quicker than the more stable antioxidant radicals formed. This can be seen in the reaction energies profiles shown in figure 23 below. The alkanes preferentially react with the antioxidants found in the fuel which stabilise the radical, slowing the product formation reactions. When the other fuel species, such as alcohols or peroxides formed during the oxidation process, react with the alkanes they donate their hydrogen to the alkane radical. This reaction is favourable and has low barriers compared to other reactions. This would lead to the radical chain mechanism reaching a bottle neck and the induction period increasing when antioxidants are added to the fuel. However when the alkane radical reacts with alcohols formed during oxidation, the process is slow and there is no thermodynamic push, as the products have similar stabilities to the reactants. Thus the oxidation mechanism will continue to proceed.

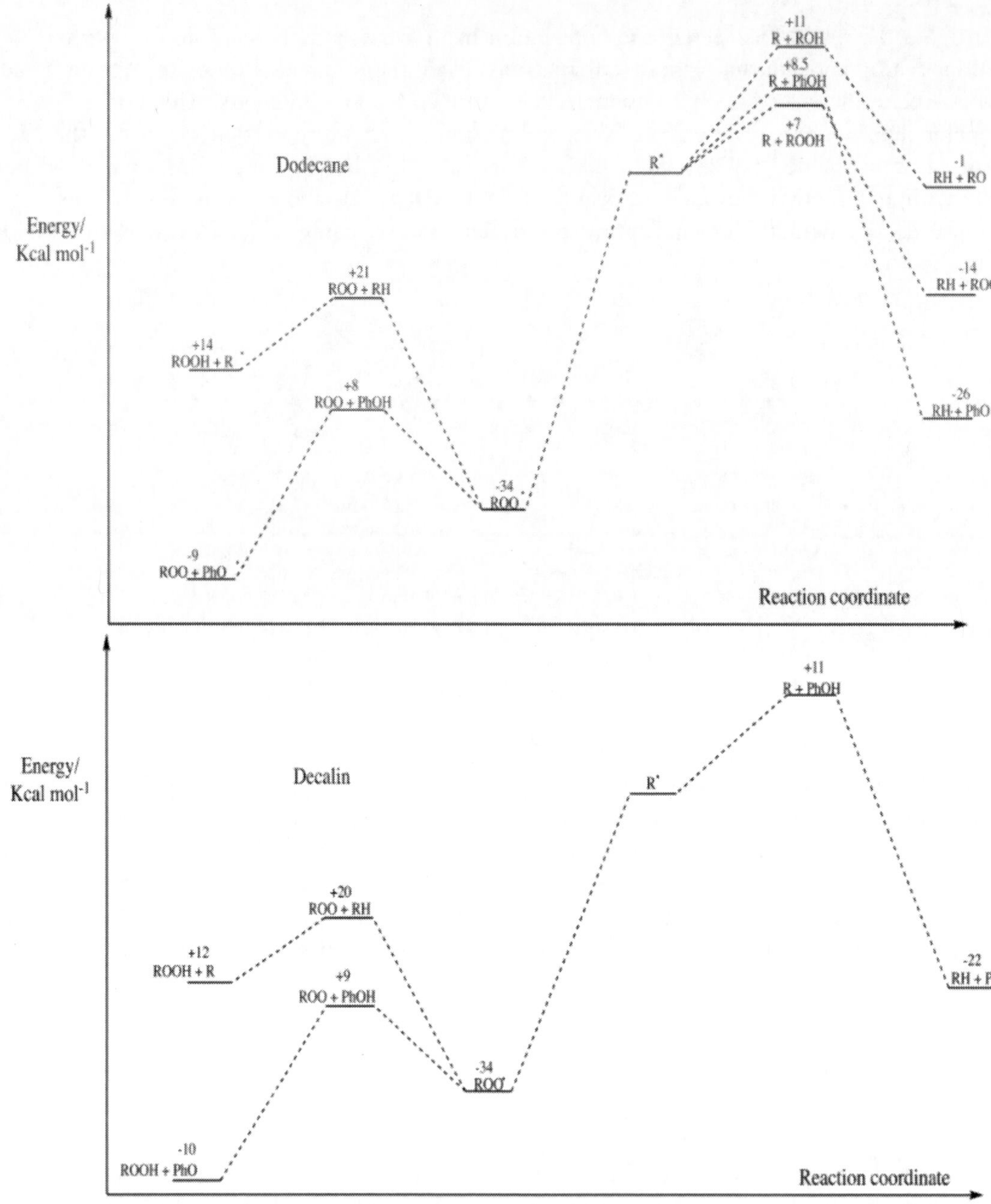

Figure 23: Oxidation of alkanes investigated using quantum chemistry methods. Both classes of alkanes behaved in a similar way

When the alkanes reaction surface is compared to that of the aromatic hydrocarbon toluene it is clear that the aromaticity of toluene stabilises the radical produced during oxidation. When the hydrocarbon radical reacts with fuel species the products produced are higher in energy than for the alkane species. This indicates that the aromatic is a more stable radical species and will form preferentially in a mixture of hydrocarbon compounds. It will then react slower and there is a lower thermodynamic push and a higher barrier to reaction. If this pattern is followed throughout the rest of the oxidation mechanism this would account of the observations in the literature which shows that aromatics are more easily oxidised, indicated by the greater amount of deposits formed, but form these insoluble gums more slowly. When the energies of the alkane radicals reacting with the alcohol are compared for the two classes of hydrocarbons, the aromatic hydrocarbon is clearly more stable than the alcohol and as such in a mixture of the fuel would form preferentially.

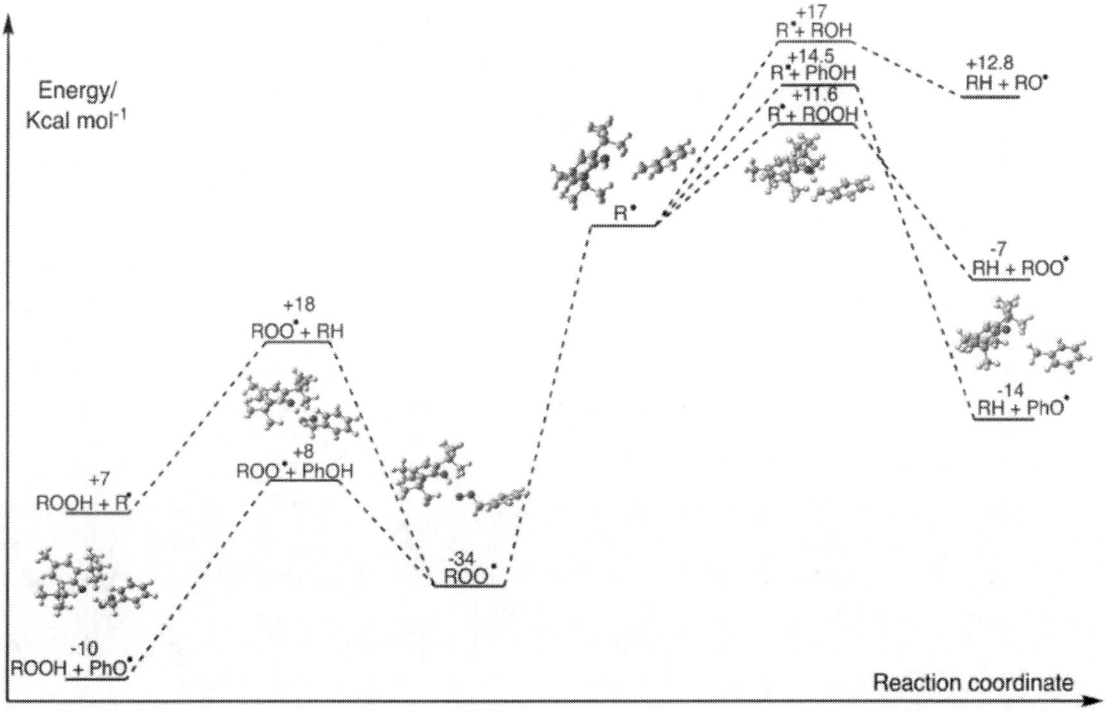

Figure 24: Potential energy surface of the initial oxidation of toluene, demonstrating different reactivity to alkanes

The effect that changing the hydrocarbon structure has on the oxidation mechanism is more clear when you compare the potential energy surface of the reaction directly. The dodecane hydrocarbon reacts quicker than the toluene, propagating the radical oxidation mechanism. The toluene however will only react with the antioxidant, which stabilises the radical formed during oxidation, or the radical is stabilised on the aromatic hydrocarbon itself through delocalisation and thus can react to form peroxides and deposits.

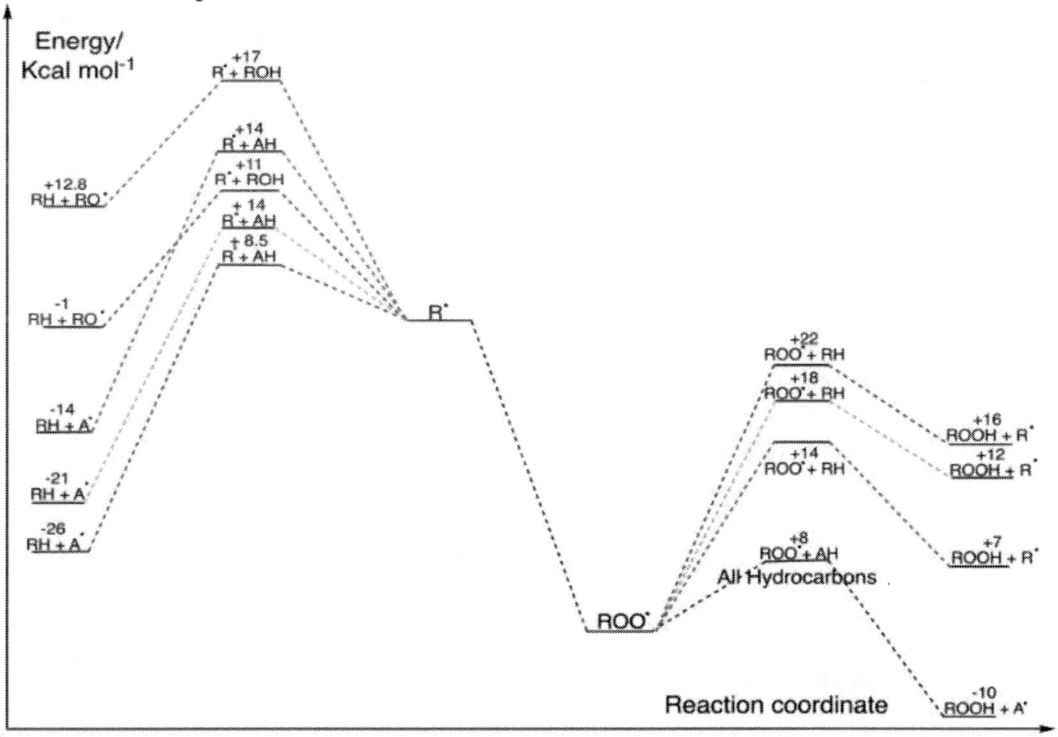

Figure 25: Potential energy surface comparing dodecane and toluene oxidation. The blue line indicates the reaction surface for dodecane, yellow is the decalin reaction surface and the red line is for toluene.

This effect is lost when the peroxide radical is formed, as the radicals electron density is centralised on the oxygen atom. This means the hydrocarbons react with the antioxidant at the same rate. However during the formation of products it would be expected that this would not always be true as the products would benefit from delocalisation. These early results suggest that cyclic alkanes have similar reactivity as straight chain alkanes, as such increasing their use in future designer aviation fuels over aromatic hydrocarbons could improve their thermal stability, while maintaining the high energy density. However the effect that the increased steric effects of cyclic hydrocarbons compared to n-alkanes in the product forming reactions still needs investigating, as does the storage stability and antioxidant behaviour or aromatics. Steric effects refers to the role that a molecules structure has on the way that it reacts and the rate at which it reacts. This can be seen in the use of butylated hydroxytoluene (BHT) as an antioxidant, as the large t-butyl groups around the phenol helps to stabilise the radical formed by blocking the reaction site, thus slowing the rate of further reactions.

This theoretical work has been supported by experiments carried out in the group, this work was carried out using the PetroOxy small scale test device.[67] The finding that cyclic hydrocarbons are more reactive to oxidation is consistent with previous experimental findings, comparing decalin with decane, which found that increasing the percentage of decalin increased the rate of oxidation. These experiments were carried out previously in the group and demonstrate that as the amount of cyclic hydrocarbon increases the rate of oxidation also increases. However for the sample with 25% decalin and 75% dodecane the rate is slowed compared to the pure dodecane sample. This could be caused by slight antioxidant behaviour by the decalin in the fuel, as the hydrogen on the tertiary carbon in the bridged position could form a more stable radical product. This needs to be investigated further, as do the observations in the literature that different aromatics contribute to the fuel thermal stability to different extents. This is the next step of the project to explain the observations in the literature using theoretical means, and to understand the effect this has on the product forming reactions of radicals and peroxides formed.

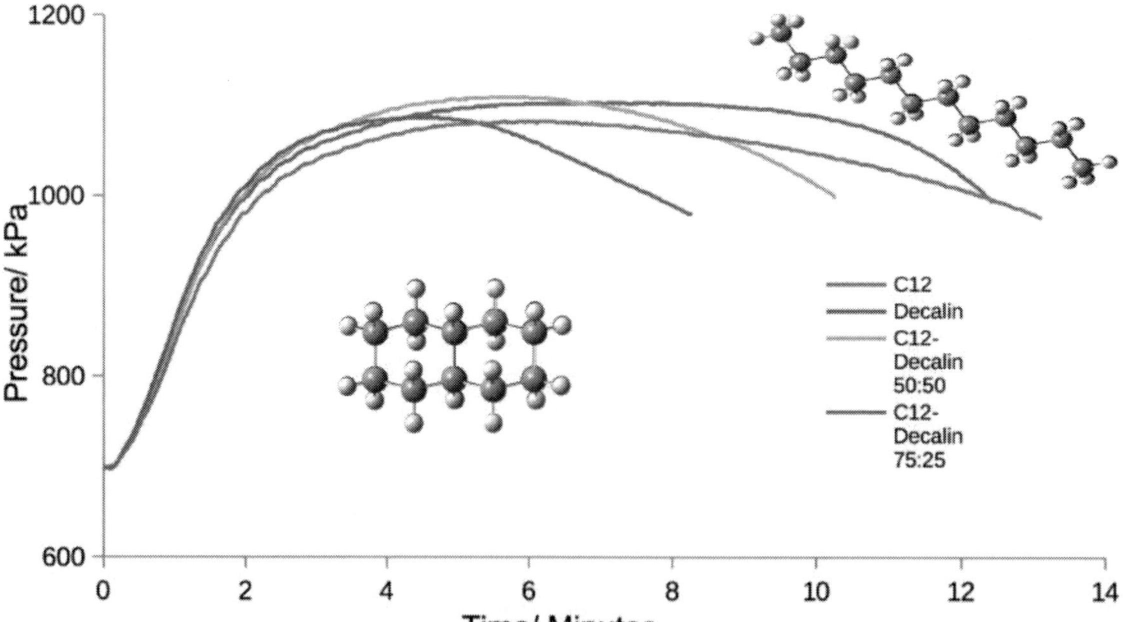

Figure 26: Thermal stability tests carried out of the PetroOxy test device investigating the effect the decalin concentration has on the oxidation rate.[67]

This work is part of an ongoing project with an entire oxidation mechanism being built for the three hydrocarbons being investigated and for the alkanes reacting with the toluene, where the two R groups are not the same. This is a mechanism which leads to the formation of the primary oxidation products such as alcohol and ketones, and does not focus on the formation of insoluble gums as yet. These calculations are underway, and the kinetic and thermochemical data obtained from the calculations will be used in kinetic solver and CFO codes to predict the rate of oxidation of simple fuel surrogates with specific hydrocarbon composition. The results from these mechanisms will be compared to the results found in the literature and through experimentation. Further to the development and optimisation of a mechanism to predict and understand the primary oxidation of hydrocarbons, the effect that polar species have on the thermal stability of the fuel needs to be considers and incorporated into the final mechanism. The phenols, sulfur and additives found in the fuel are the most obvious areas to expand the mechanism. The final aim of this would be to form a unified mechanism for oxidation and product formation to predict how a fuel with a specific chemical composition will oxidise and behave under thermally stressed conditions. The literature and the investigations into the homolytic fission also raises an interesting question. The reactions of peroxides, particularly the product forming reaction can only be correctly described using multi-reference methods such as CASSCF. As these reactions are very important to understand the mechanism and rate of oxidation, they will be investigated using these methodologies to understand why a species which is easier to oxidise is slower to form oxidised products.

CONCLUSIONS

Research into the effect that changing chemical composition has on a fuels thermal stability is an active topic of interest. This paper has reviewed the composition of aviation fuel and discussed its effects on the thermal stability and properties of the fuel. The paper has also reported some early results from work investigating the effect that changing the hydrocarbon composition of a fuel has on its thermal stability. The conclusion drawn from this work are as follows

- From the energy of bond dissociation of C-H fission for each hydrocarbon, aromatic hydrocarbons are the most easily oxidised hydrocarbon. Both cyclic and straight chain alkanes behaved very similarly, with the slight differences in the dissociation energy attributed to stabilisation of the radical on the bridged carbon by three carbons adjacent.
- Inaccuracies found in these calculations and of those for the Homolytic fission of peroxides can be attributed to OFT failing to describe the electronic structure correctly, as such the solution demonstrated significant errors in correlation energies. This has been investigated using multi-reference methods, such as CASSCF, with the early results being able to describe the system properly. However more work is required to optimise this method.
- The higher stability that cyclic alkanes demonstrate compared to aromatic hydrocarbons could lead them to be used to improve a fuels energy density while not leading to a reduction in its thermal stability and increased deposition.
- Aromatic hydrocarbons appear to act as antioxidants in the fuel, stabilising the radicals produced during oxidation through delocalisation. This would leads them to slow the rate of deposition, but deposit more overall as they are more easily oxidised.

REFERENCES

1. R. N. Hazlett, *Thermal Oxidation Stability of Aviation Turbine Fuels*, **1991**, ASTM, Phildelphia PA
2. G. Heminghaus, T. Boval, J. Bacha, F. Barnes, M. Franklin, L. Gibbs, N. Hogue, J. Jones, O. Lesnini, J. Lind and J. Morris, *Aviation Fuels Technical Review*, **2007**, Chevron Products Company, San Ramon CA
3. Zuhdan Fathoni and B. O. Batts, *Energy & Fuels*, **1992**, 6, 681-693
4. Zuhdan Fathoni and B. O. Batts, *Energy & Fuels*, **1991**, 5, 2-21
5. P. Watkinson and O. I. Wilson, *Experimental Thermal and Fluid Science*, **1997**, 14, 361-374
6. F. Pradella, S. L. Braga, A. F. A. Martins, F. Turkovics and R. N. C. Pradella, *Energy and Fuels*, **2015**, 29, 7753-7770.
7. J. S. Ervin and T. F. Williams, *Ind. Eng. Chem. Res,* **1996**, 35, 899-904
8. S. P. Heneghan and S. Zabarnick, *Fuel*, **1994**, 73, 35-43
9. F. R. Mayo, *Accounts. Chem. Res,* **1968**, 1, 193-201
10. W. F Taylor, *I & EC Prod. Res. & Dev,* **1969**, 8, 375-380
11. W. F. Taylor and J. W. Frankenfeld, *IASH Proceeding,* **1986**, San Antonio, Texas
12. P. Oagaut and M. Cathonnet, *Prog. Energy & Comb. Sci,* **2006**, 32, 48-92
13. M. Rumizen, *ASTM D4054 Users Guide*, **2013**, ASTM International, West Conshohocken, Pennsylvania, United States
14. Standard Specification for Aviation Turbine Fuels, *ASTM D1655-09a*, **2010**, ASTM International, West Conshohocken, Pennsylvania, United States
15. G. Cook, *Defence Standard 91-91*, **2015**, MOO, Bristol UK.
16. G. Liu, B. Yan and G. Chen, *Renew & Sustain. Energy. Rev,* **2013**, 25, 59-70
17. S. Blakey, L. Rye, C. W. Wilson, *Proceeding of Comb. Inst,* **2011**, 33, 2863-2885
18. L. Rye, S. Blakey, C. W. Wilson, *Energy & Environ. Sci,* **2009**, 3, 17-27
19. O. J. Hadaller and J. M. Johnson, *World Fuel Sampling Program*, **2006**, CRC, Alpharetta GA
20. O. Altin and S. Eser, *Prepr. Pap. Am. Chem. Soc. Div. Fuel Chem,* **2004**, 49, 764-766
21. E. G Jones, L. M. Balster and W. J. Balster, *Energy & Fuel*, **1996**, 10, 509-515
22. E. G. Jones and L. M. Balster, *Energy & Fuel,* **1999**, 13, 796-802
23. W. F. Taylor, *Ind. Eng, Chem. Prod. Res, Develop,* **1974**, 13, 133-138
24. W. F. Taylor, *Ind. Eng, Chem. Prod. Res, Develop*, **1976**, 15, 64-68
25. W. F. Taylor and J. W. Frankenfeld, *Ind. Eng, Chem. Prod. Res, Develop*, **1978**, 17, 86-90
26. J. W. Frankenfeld, W. F. Taylor and O. W. Brinkman, *Ind. Eng, Chem. Prod. Res, Develop,* **1983**, 22, 608-614
27. J. W. Frankenfeld, W. F. Taylor and O. W. Brinkman, *Ind. Eng, Chem. Prod. Res, Develop,* **1983**, 22, 622-627

28. E. E. Elmalik, B. Raza, S. Marrag, H. Ramadhan, E. Alborzi and N. O. Elbashir, *Ind. Eng. Chem. Res.* **2014**, 53, 1856-1865

29. P. M. Rawson, CA. Stansfield, R. L. Webster, O. Evans and U. Yildirim, *Fuel*, **2015**, 161, 97-104

30. E. Corporan, T. Edwards, L. Shafer, M. J. OeWitt, C. Klingshirn, S. Zabarnick, Z. West, R. Striebich, J. Graham and Klein, *Energy & Fuels*, **2011**, 25, 955-966

31. L. M. Balster, S. Zabarnick, R. C. Striebich, L. M. Shafer and Z. J. West, *Energy & Fuel*, **2006**, 20, 2564-2571

32. L. M. Shafer, R. C. Striebich, S. Zabarnick and Z. J. West, *IASH proceedings*, **2015**, Charleston, South Carolina

33. M. A. Wechter and O. R. Hardy, *Energy & Fuel*, **1989**, 3, 461-464

34. D. R. Hardy and M. A. Wechter, *Energy & Fuel,* **1990**, 4, 270-274

35. D. R. Hardy and M. A. Wechter, *Energy & Fuel*, **1994**, 8, 782-787

36. J. W. Frankenfeld and W. F. Taylor, *Ind. Rng. Chem. Prod. Res. Rev,* ***1980**, 19, 65-70*

37. F. R. Mayo and A. A. Miller, *J. Am. Chem. Soc,* **1956**, 78, 1023-1034

38. J. A. Howard, K. U. Ingold and M. Symonds, *Can. J. Chem,* **1968**, 46, 2661-2666

39. Szabo and N. S. Ostlund, *Modern Quantum Chemistry,* **1996**, Oover pubilications incorporated, Mineola NY

40. N. Levine, *Quantum Chemistry, 4th edition,* **1991**, Prentice Hall, Englewood cliffs NJ

41. T. S. Oibble, *Atmospheric Environment,* **2008**, 42, 5837-5848

42. S. Zabarnick, *Ind. Eng. Chem. Res,* **1993**, 32, 1012-1017

43. S. Zabarnick, *Energy & Fuels*, **1998**, 12, 547-553

44. J. S. Ervin and S. Zabarnick, *Energy & Fuels,* **1998**, 12, 344-352

45. N. J. Kuprowicz, S. Zabarnick, Z. J. West and J. S. Ervin, *Energy & Fuels*, **2007**, 21, 530-544

46. Z. H. Sander, Z. J. West, J. S. Ervin and S. Zabarnick, *Energy & Fuels*, **2015**, 29, 7036-7047

47. Beaver, L. Gao, C. Burgess-Clifford and M. Sobkowiak, *Energy & Fuels*, **2005**, 19, 1574-1579

48. M. Sobkowiak, J. M. Griffith, B. Wang and B. Beaver, *Energy & Fuels*, **2009**, 23, 2041-2046

49. P. Aksoy, O. Gul, R. Cetiner, O. A. Fonseca, M. Sobkowiak, S. Falcone-Miller, B. G. Miller and B Beaver, *Energy & Fuels*, **2009**, 23, 2047-2051

50. O. Gul. R. Cetiner, J. M. Griffith, B. Wang, M. Sobkowiak, O. A. Fonseca, P. Aksoy, B. G. Miller and B. Beaver, *Energy & Fuels,* **2009**, 23, 2052-2055

51. G. Kabana, S. Botha, C. Schmucker, C. Woolard and B. Beaver, *Energy & Fuels*, **2011**, 25, 5145-5157

52. R. E. Kauffman, *Journal of Engineering for Gas Turbines and Power*, **1997**, 119, 322-327

53. W. F. Taylor and T. J. Wallace, *I & EC Prod. Res. & Dev,* **1968**, 7, 198-202

54. R. Epping, S. Kerkering and J. T. Andersson, *Energy & Fuels,* **2014**, 28, 5649-5656

55. J. A. Howard and K. U. Ingold, *Can. J. Chem,* **1965**, 43, 2729-2736

56. Miller and F. R. Mayo, *J. Am. Chem. Soc*, **1956**, 78, 1017-1023

57. M. J. OeWitt, Z. West, S. Zabarnick, L. Shafer, R. Striebich, A. Higgins and T. Edwards, *Energy & Fuels*, **2014**, 28, 3696-3703

58. J. A. Howard, K. U. Ingold, *Can. J. Chem,* **1966**, 44, 1119-1130

59. J. Igarashi, J. Lusztyk and K. U. Ingold, *Journal of the American Chemistry Society*, **1992**, 114, 7719-7726.

60. J. Igarashi, R. K. Jensen, J. Lusztyk, S. Korcek and K. U. Ingold, *Journal of the American Chemistry Society*, **1992**, 114, 7727-7736.

61. J. A. Howard and K. U. Ingold, *Can. J. Chem,* **1967**, 45, 793-802

62. B. Amara, S. Kaoubi and L. Starck, *Fuel*, **2016**, 173, 98-105.

63. 63.K. Chatelain, A. Nicolle, A. B. Amara, L. Catoire and L. Starck, *Energy adn Fuels*, **2016**, 30, 1294- 1303.

64. S. Blakey, A. Yusuf, C. W. Wilson and S. Oufferweil, *IASH Proceedings,* **2013**, Rhodes, Greece

65. S. Zabarnick and O. K Phelps, *Energy and Fuels*, **2006**, 20, 488-497.

66. S. J. Blanksby and G. B. Ellison, *Accounts of Chemical Research*, **2003**, 36, 255-263.

67. Alborzi, S. Blakey, T. Keane and A. Meijer, *IASH Proceedings*, **2015**, Charleston, South Carolina

IASH 2017, the 15TH INTERNATIONAL SYMPOSIUM ON STABILITY, HANDLING AND USE OF LIQUID FUELS
Rome, Italy
10-14 September 2017

THE IMPACT OF ORGANONITROGEN COMPOUNDS ON THE STORAGE STABILITY OF MIDDLE DISTILLATE FUELS

Robert E. Morris[1], Thomas N. Loegel[2], Kristina M. Myers[1], Iwona A. Leska[1], Christopher J. Katilie[1], Alison E. Metz[3]

[1]*NOVA Research, Inc., 1900 Elkin Street 230, Alexandria, VA, USA 22308*
[2]*U.S. Naval Research Laboratory, 4555 Overlook Ave. S.W., Washington, D.C., USA 20375*
[3]*Naval Air Systems Command, Patuxent River, MD 20670*

ABSTRACT

Storage stability of middle distillate diesel fuels is of utmost importance, as poor storage stability can result in formation of gums and sediment that seriously impair the usability of the fuel. Chemically reactive fuels, either by virtue of composition, incompatibility with comingled fuels, or contamination, are generally identified by standard laboratory test methods, including ASTM D2274 and D5304. However, there have been instances where diesel fuels that pass all specification test requirements have nonetheless failed in service. One such event was reported at the 14th IASH Symposium, where diesel fuels in one region were forming high levels of particulate materials (PM) in storage. The results of that investigation were reported to indicate that while those fuels passed all specification requirements, there may have been a causal link between the total nitrogen content and the incidence of high PM formation.

Subsequent to that work, a study has been undertaken to ascertain the role of the organonitrogen compounds in impairing the storage stability of these diesel fuels. A liquid-liquid phase extraction was developed to isolate the basic and non-basic nitrogen compounds from those fuels and the extracts analyzed by GC-NPD (nitrogen-phosphorous detection) and GC-MS. The findings of this study reveal a strong correlation between the occurrence of certain basic nitrogen species and the formation of high particulates in those fuels. As a consequence of these findings, it has been recommended that a focused analysis for basic nitrogen compound content be conducted whenever storage stability is a concern.

Approved for Public Release

INTRODUCTION

It is recognized that high nitrogen content, in itself, is not necessarily a prerequisite for poor stability, but the deleterious impact of any particular organonitrogen compound will be determined by the structure of that compound, with some being very reactive, and others fairly innocuous. In a series of pure compound studies[1,2,3] conducted by Frankenfeld and Taylor, they found that pyrroles were very reactive, followed by indoles, whereas quinolones were relatively unreactive, with respect to their ability to induce the formation of insoluble gums and sediment in middle distillate fuels. The relative reactivities within each class are shown in Figure 1, where it is clear that resonance stabilization at the nitrogen atom determines the reactivity of the molecule. Pyrroles, particularly when containing electron releasing groups adjacent to the nitrogen, are generally very reactive and have the potential to degrade fuel stability. This difference is a consequence of the availability of the lone electron pair on the nitrogen atom to undergo electron transfer reactions, whereas in a six-membered ring (e.g. quinolones, pyridines) the nitrogen electron pair is not available and thus those compounds are relatively innocuous.

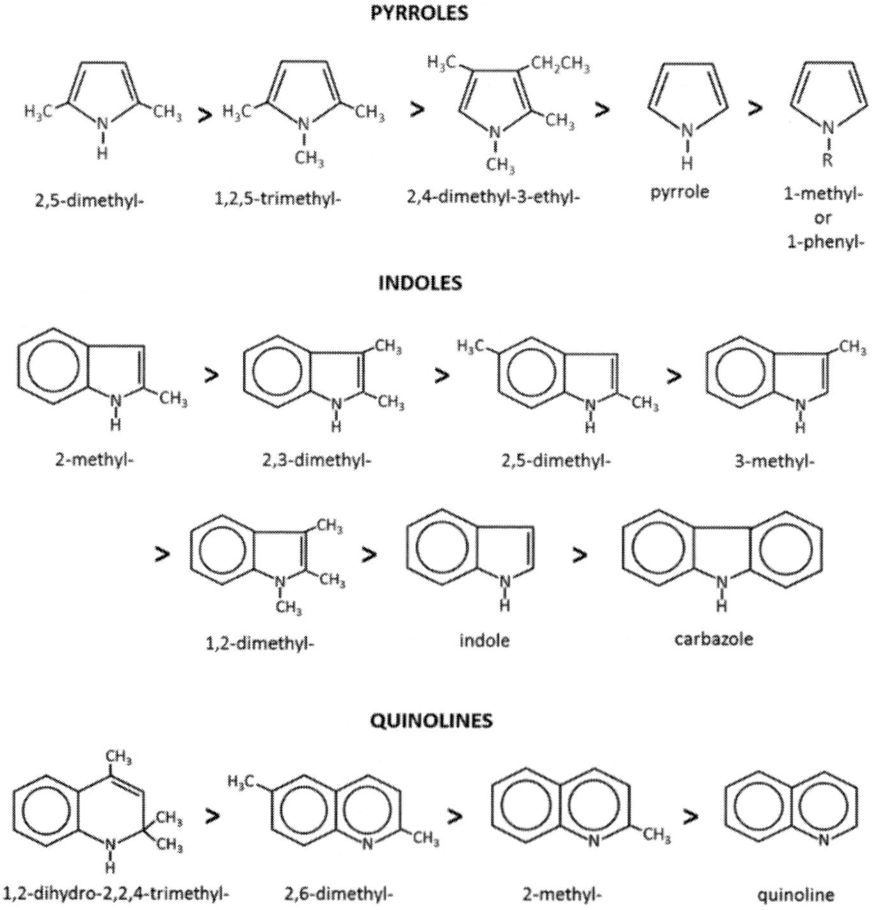

Figure 1. Relative reactivity of organonitrogen compounds with pyrroles > indoles >> quinolines to form insolubles in fuel (from Frankenfeld and Taylor[2]).

Acid or base catalyzed condensation has been identified as a major pathway by which these heteroatomic compounds can react to form large insoluble reaction products. Thus, the impact of these various organonitrogen compounds on storage stability can depend on other constituents that are present in the fuel. For example, strong synergism has been found[2,3] to exist between carbazoles and some organosulfur compounds. Schrepfer et al. reported[4] that storage instability of diesel fuels occurs through three processes: 1) acid-base condensation reactions[5] between organic acids and basic nitrogen compounds; 2) free-radical autoxidation; and 3) esterification reactions between aromatics, heterocyclic nitrogen compounds and aromatic thiols. Condensation reactions between organic acids and basic nitrogen compounds were described[6] as requiring hours to weeks to occur. This was cited as most often occurring when product streams are blended from separate processes or sources that provide the necessary reactants to cause sediment formation up to several weeks later.

This may appear to be in conflict with the established efficacy of hindered basic amine additives to improve storage stability of middle distillate fuels. However key to this difference is the fact that the most common additive, N,N'-dimethylcyclohexylamine, is a hindered tertiary amine. Tertiary amines are generally more resistant to oxidation at the nitrogen atom due to the absence of any hydrogen atoms attached to nitrogen which are prone to oxidation due to the electron releasing character of the nitrogen atom. Thus, unlike primary amines, tertiary amines are not easily deaminated or oxidized, but they have been described[7] as being capable of acting as an antioxidant, a neutralizing agent and dispersant. N,N'-dimethylcyclohexylamine is a strong base in organic media and will neutralize strong sulfonic acids and weak carboxylic acids to form soluble salts, thus mitigating acid-catalyzed condensation reactions and that lead to insoluble product formation. They have also been thought to possess some dispersant capability by binding with acidic by-products, preventing agglomerization allowing those particles to pass through filters. The mechanism of antioxidant action by tertiary amines has not been completely resolved, although they appear to quench free-radical chain reactions by donation of a hydrogen atom.

Historical Precedents. Storage stability of middle distillate diesel fuels is of utmost importance, as poor storage stability can result in formation of gums and sediment that seriously impair the usability of the fuel. Chemically reactive fuels, either by virtue of composition, incompatibility with comingled fuels, or contamination, are generally identified by standard laboratory test methods, including ASTM D2274[8] and D5304[9]. However, there have been instances where diesel fuels that pass all specification test requirements have nonetheless failed in service.

An example of this was found in 1982, where a marine F-76 diesel fuel was found to be unstable in storage, despite passing all specification tests on receipt. The unstable fuel, identified as diesel fuel, marine (DFM) obtained from a particular refinery, was isolated and placed in a holding tank. Two unstable fuel samples, reported to have been sampled from storage tanks at two different naval stations, were subsequently found to contain this unstable fuel. A stable shale-derived F-76 marine diesel fuel sampled from naval storage was also examined for comparison. Laboratory stability studies[10,11] were conducted in an effort to ascertain the root cause(s) of the developing poor storage stability of these fuels. The poor storage stability was evident from stressing in vented borosilicate glass bottles at 43 °C in accordance with procedures in ASTM D4625[12], for various times. The results of this testing are summarized in Tables 1 and 2. None of the fuels examined contained high concentrations of dissolved metals (Pb, Cu, Fe, Si, Mg).

Table 1. ASTM D4625 Testing at 43 °C of the unstable diesel fuel.

Weeks of Storage	Total Insolubles, mg/100 mL	Filterable Particulates, mg/100 mL	Soluble Gum, mg/100 mL
0	0.9	0.4	0.5
6	23.1	23.1	6.7
12	26.5	26.5	5.2

Table 2. ASTM D4625 testing at 43 °C of a stable diesel fuel.

Weeks of Storage	Total Insolubles, mg/100 mL	Filterable Particulates, mg/100 mL	Soluble Gum, mg/100 mL
0	3.2	2.1	1.1
6	3.3	2.6	0.7
12	4.4	3.1	1.3

In comparison with stable fuel, the unstable fuels were found to contain higher concentrations of aromatics, higher bromine numbers, and higher asphaltene and bitumen contents by the procedure in Appendix A. An empirical Bureau of Mines test (Appendix B) indicated that the unstable fuels had been produced primarily by catalytic cracking, whereas the stable fuels contained much lower levels of cracked stocks.

Total nitrogen content determined before and after stressing at 150 °C for 8 weeks, indicated that not only did the unstable fuels have higher nitrogen content, but the nitrogen content decreased to a much greater extent than the stable diesel fuels as shown in Table 3. Subsequent analyses showed that most of the nitrogen was incorporated in the filterable sediments formed. The basic nitrogen compound (BNC) and non-basic nitrogen compound (NBNC) fractions were isolated by a liquid-liquid extraction procedure and analyzed by GC-MS, where it was found that the unstable fuels contained higher levels of alkyl pyrroles.

Table 3. Total nitrogen content of stable and unstable before and after thermal stress.

Fuel	Total Nitrogen, mg/kg	
	Unaged	Aged
Unstable Fuel #1	263	6
Unstable Fuel #2	313	6
Stable Fuel #1	116	34
Stable Fuel #2	81	69

On the basis of the analytical data, a major factor in the instability of these diesel fuels was ascribed to the higher levels of alkyl pyrroles and the presence of catalytically cracked stocks.

The approach to isolate a particular compound class of interest by liquid-liquid extraction or column chromatography was also used in 1993 during an investigation[13] of unstable JP-5. The basic nitrogen compounds (BNC) were isolated and found to be present at 0.02 – 0.05 % (w/w). Analysis of these BNC extracts revealed them to consist primarily of alkyl pyridines, indoles, tetrahydroquinolines and carbazoles, which were deemed typical of JP-5 fuels.

More recently, there were multiple reports of high particulates in F-76 diesel in the supply chain and onboard naval vessels in the western pacific region. As in the previous two examples, all of these fuels passed specification requirements for stability upon receipt. A substantial amount of work was undertaken to identify the root cause of this degradation and assist in developing a remediation strategy. The results of that investigation indicated that while those fuels passed all specification requirements, including ASTM D5304, there may have been a causal link between the total nitrogen content and the incidence of high PM formation.

Two exceptions to this were two highly reactive samples that were found to contain high levels of dissolved copper, at 3700 and 2440 ppb. Subsequent studies were conducted to probe the sensitivity of diesel fuels to copper. The findings of that study indicated that highly hydrorefined diesel fuels are generally more susceptible to copper-initiated autoxidation, which can lead to the formation of insoluble reaction products. Thus, high sulfur (<1 % S) F-76 is relatively insensitive to copper, while the storage stability of low sulfur (0.05 % S) and ultralow (15 ppm S) diesel fuels can be degraded in the presence of dissolved copper. Since these fuels were low sulfur diesel fuels, it is likely that in those two fuel samples, copper was a contributing factor. However, dissolved copper was not found to be a factor in the poor storage stability exhibited by the other unstable fuels. The particulate formation rates in some of the fuel samples were observed to decrease after filtration, so those fuels were used as quickly as possible after filtration. Other samples continued to produce insoluble products and were downgraded.

Subsequent to that investigation, a number of incidents were reported in the same Western Pacific region where significant amounts of gum were observed in the bottoms of glass bottles used to store retain samples. There were also instances where some fuels continued to produce insoluble products. For example, one line retain sample, which was clean and bright upon receipt, was found to have 34.7 mg/L particulates after only 6 weeks of storage. A cursory survey indicated that out of a random sampling of 69 retain fuels, 46 had visible gum in the glass storage bottles. Analysis of fuel samples that did and did not aggressively form gums in storage were thus rigorously examined.[14] The findings of that study (Table 4) did not reveal any detectable contaminants or compositional features that could be attributed to the cause of poor storage stability.

Table 4. Summary of measured compositional properties in the diesel fuel gum formation investigation (from ref. 12).

NRL I.D.	Sample	Cu, µg/kg	Total N, mg/kg	Cetane Improver[a], mg/kg	Total N mg/kg, corr.[b]	FAME, vol%	Existent Gum, mg/100mL
BASELINE SAMPLES							
15-09	West Coast Refinery Retain	1.9	510	623	455	<0.025	4.6
15-10	Gulf Coast Refinery Retain	< 0.5	15	<10	15	<0.025	5.8
DEFENSE FUEL SUPPLY POINT #1							
15-13	Shipboard Sample*	7.5	300	708	238	<0.025	9.8
15-14	Tank Sample*	8.3	390	694	329	<0.025	9.0
15-15	Truck Sample	15.3	240	614	186	<0.025	2.2
15-16	Tank Sample*	14.5	320	743	255	<0.025	11.0
DEFENSE FUEL SUPPLY POINT #2							
15-20	Tank Bottom Sample	25.2	91	<10	91	<0.025	6.8
15-21	Shipboard Sample*	15.7	440	120	429	<0.025	19.0
15-22	Tank Sample*	12.7	220	72	214	<0.025	11.2
15-23	Tank Sample	6.8	130	<10	130	<0.025	5.8
DEFENSE FUEL SUPPLY POINT #3							
15-26	Tank #1	92.3	230	--	--	<0.025	--
15-27	Tank #2*	34.1	670	--	--	<0.025	--
15-28	Tank #3*	57.4	440	--	--	<0.025	--
15-29	Tank #4*	36.3	580	--	--	<0.025	--
15-30	Refinery sample prior to receipt*	21.6	580	--	--	<0.025	--
SHIPBOARD SAMPLES							
15-24	Sample with Sediment	44.7	390	167	375	<0.025	10.0
15-25	Sample with high Copper*	261.9	230	63	224	<0.025	7.0

*Fuel samples that produced visible ring of gum in glass bottle. [a]2-ethylhexyl nitrate (EHN)
[b]Corrected for 2-EHN content. -- not measured

Given the historical precedent of basic nitrogen compounds being linked to poor diesel fuel stability and the characteristically high nitrogen content of the west coast fuels, this study was undertaken to define the role of organonitrogen compounds in impairing the storage stability of these diesel fuels

EXPERIMENTAL

Extraction of organonitrogen compounds from fuel

The liquid extraction procedure used in previous studies during the 1980's was optimized to reduce sample and reagent volumes and eliminate the need for solvent removal by rotary evaporation. The procedures to isolate the basic and non-basic nitrogen compounds are shown schematically in the flow diagram in Figure 2.

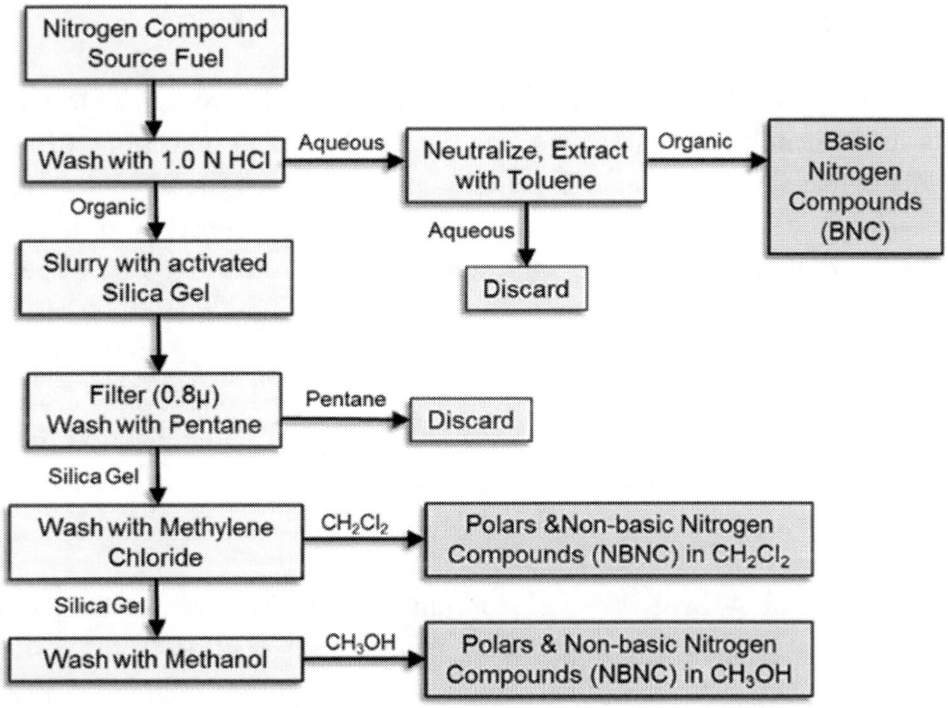

Figure 2. Scheme for isolation of the basic nitrogen compound (BNC) and non-basic nitrogen compound (NBNC) fractions from a fuel sample.

Basic Nitrogen Compounds (BNC). Fuel (25.0 mL) is extracted with aqueous 5.0 mL of 1.0 M HCl and 5.0 mL deionized water. The aqueous layers are combined and extracted with three 6.0 mL portions of toluene. The organic layer is discarded. The aqueous layer is neutralized with solid $NaHCO_3$ to a pH of 8 and extracted with two 15.0 mL portions of toluene. These organic layers are then combined, dried with $MgSO_4$, and concentrated to give a yellow oil, which is dissolved in 2.0 mL of CH_2Cl_2.

Non-Basic Nitrogen Compounds (NBNC). The remaining fuel is extracted with two 5.0 mL portions of deionized water and stirred over silica gel (5.0 g, high purity grade, 60 Å, 70-230 mesh) for 1 h. The silica gel slurry is filtered through a 0.8 µm nylon microporous filter and washed with 150 mL pentane. The pentane wash is discarded and the filter is washed with eight 7 mL portions of methylene chloride. The methylene chloride washes are retained as the NBNC-CH_2Cl_2 extract. The filter is then washed with 25 mL of methanol, and these washes are retained as the NBNC-methanol extract.

Evaporation of the solvents from the NBNC extracts produces dark brown residues. The residue from the methylene chloride extract is dissolved in 2.0 mL CH_2Cl_2. The residue from the methanol extract is dissolved in 2.0 mL of 80:20 CH_2Cl_2:MeOH. Each of these resulting extracts were filtered through an inorganic membrane filter (0.2 µm, 10 mm) before analysis by Gas Chromatography with mass selective detection (GC-MS) and with a nitrogen-phosphorus specific detector (GC-NPD).

GC-MS Analysis. GC-MS data were acquired with an Agilent 7890 gas chromatograph (GC) connected to an Agilent 5975C mass selective detector. Sample injections 0.5 µL were made with an Agilent model G4513A auto injector with a 10 µL Agilent gold series syringe. A split/split-less inlet was used with an Agilent Ultra inert lo-psi drop tapered liner with deactivated glass wool operated with a split ratio of 60:1. A Restek 60 m x 250 µm x 0.5 µm RXI-1ms column was used with a constant helium carrier gas flow rate of 2.0 mL/min. The oven temperature program began at 40 °C held for 2 min, then increased at a rate of 5 °C/min to 165 °C, decreased to 2.5 °C/min to 265 °C, then increased to 10 °C/min to 295 °C, for a total run time of 70 min. The transfer line to the MS was held at 250 °C, the electron impact ionization (EI) source at 230 °C, and the quadrupole analyzer at 150 °C. A mass range of 35 to 400 m/z was scanned with a gain factor of 1.5, and a solvent delay of 5 min.

GC-NPD Analysis. An Agilent 7890 gas chromatograph (GC) was fitted with an Agilent Nitrogen-Phosphorous specific Detector (NPD) using a Agilent Blos Bead. Effluent was split from the GC-MS column using an Agilent three-way capillary flow splitter. A restrictor column of 0.10 mm I.D x 1.03 m was used to connect the MS and a restrictor column of 0.20 mm I.D x 0.25 m was connected to the NPD. Both restrictor columns were bare deactivated GC capillary tubing. Back pressure on the split plate was held at a constant 5.0 psi. The NPD was held at 325 °C with a hydrogen flow rate of 3 mL/min, an air flow rate of 120 mL/min, and a helium makeup of 0.516 mL/min. Bead voltage was automatically adjusted every day. Data were collected at rate of 50 Hz.

FCAST Analysis. Chemical compositions of the fuels and extracts were derived from GC-MS data using the Navy Fuel Composition and Screening Tool[15,16] (FCAST, version 2.9). The FCAST contains an implementation of the NRL compositional profiler, which is an automated chemical component classification developed to provide a classification of all compound classes in a fuel, as an alternative to ASTM D2425[5], which does not function adequately with alternative, non-petroleum derived fuels. The profiler functions by reading the GC-MS data file, identifying each unique compound peak, performing a noise analysis, then sending the peak table to a NIST electron impact mass spectral library. The chemical compounds thus identified are classified with respect to a set of 25 defined compound classes using a set of selection rules that operate on either the molecular formula or IUPAC name. In addition, the profiler calculates and reports carbon number distributions, average carbon number and degrees of unsaturation for each carbon number. The accuracy of the NRL compositional profiler has been verified with known fuels and surrogate fuel blends.

RESULTS AND DISCUSSION

An F-76 fuel, previously reported as being unstable, was used for the initial development of the extraction methodology. While the original method required 2 L of fuel, refinement of the method provided adequate results with 25 mL samples. The GCMS total ion chromatogram (TIC) and the chromatogram from the nitrogen-phosphorous specific detector are shown in Figure 3. Comparison of these two chromatograms indicates that the majority of nitrogen species are in the heavier boiling point region between 40 and 60 minutes.

From comparison of the GC-NPD chromatograms before and after extraction in Figures 3 and 4 it is evident that nearly all of the organic nitrogen compounds in the fuel were isolated by the extraction procedure. The BNC fraction is obtained in the toluene extract and shown in Figure 5, where comparison of the TIC and NPD chromatograms indicate that the BNC fraction constitutes the lighter nitrogen compounds in that fuel. Analysis of the GC-MS data from the BNC fraction (Figure 6) indicates that it is mainly composed of benzeneamines, quinolones and possibly pyridines, although it is known that pyridines are relatively innocuous with respect to fuel stability.

The NBNC fraction is obtained by methylene chloride extraction from the silica gel. As shown in Figure 7, the NBNC fraction constitutes the heavier nitrogen compound fraction. Analysis indicates that the major components of this NBNC fraction are carbazoles, indoles and possibly a benzenamine dimer, as shown in Figure 8.

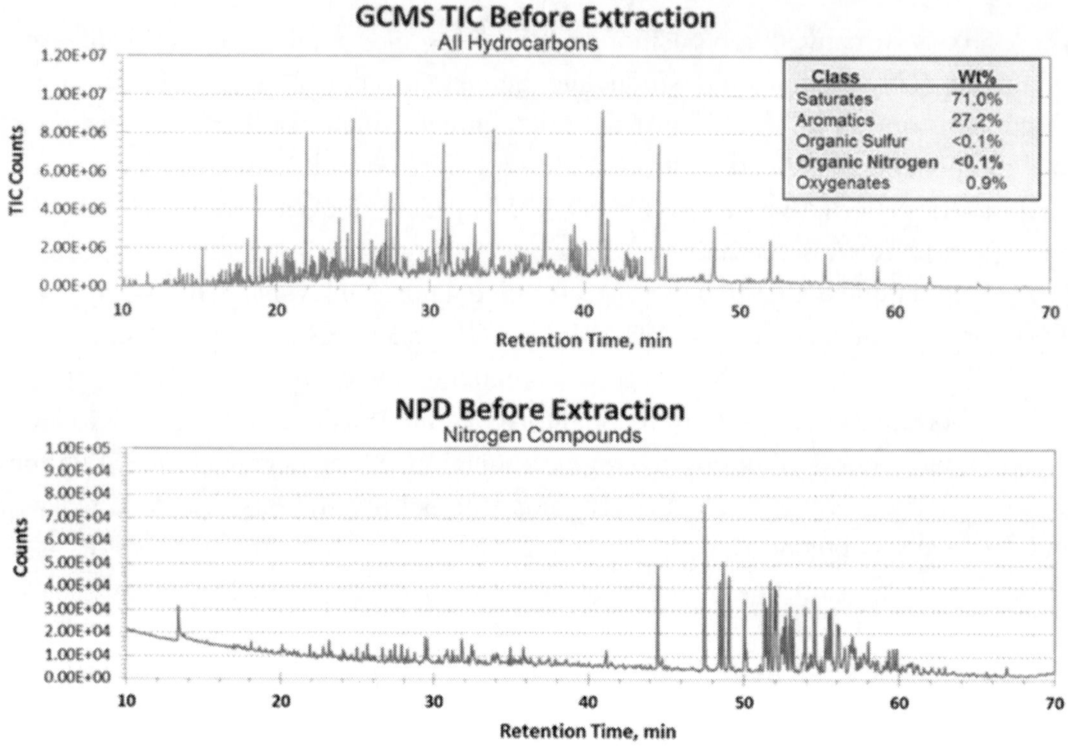

Figure 3. GCMS TIC and GC-NPD chromatogram of an unstable diesel fuel.

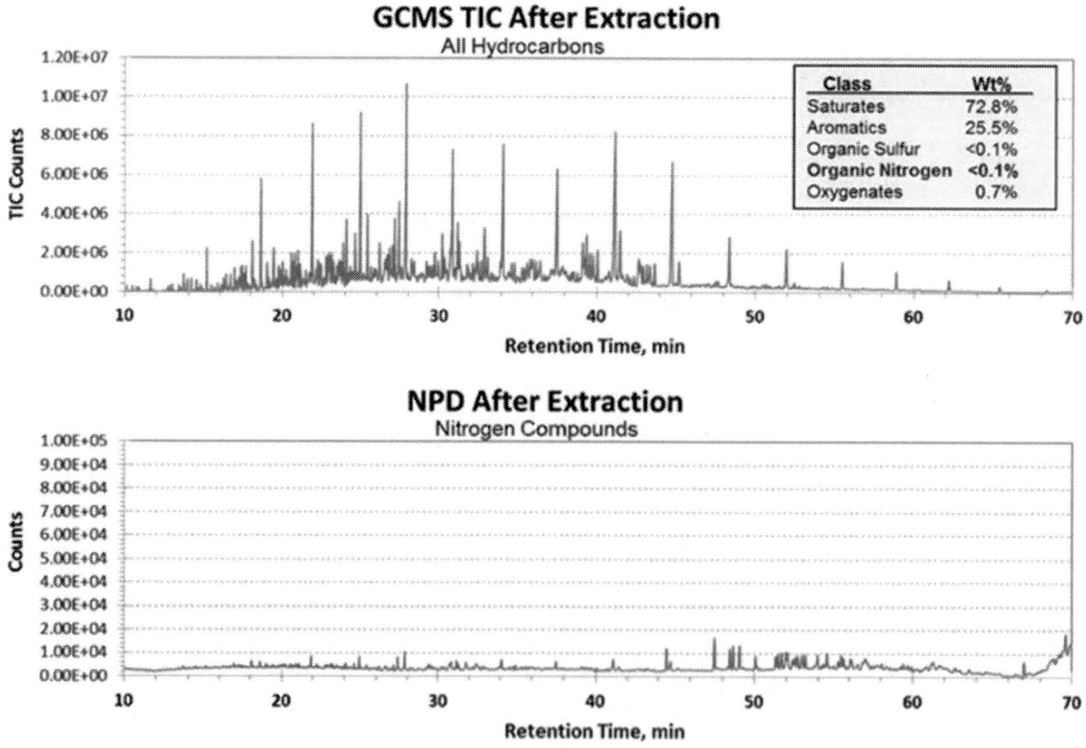

Figure 4. GCMS TIC and GC-NPD chromatogram of base fuel after extraction.

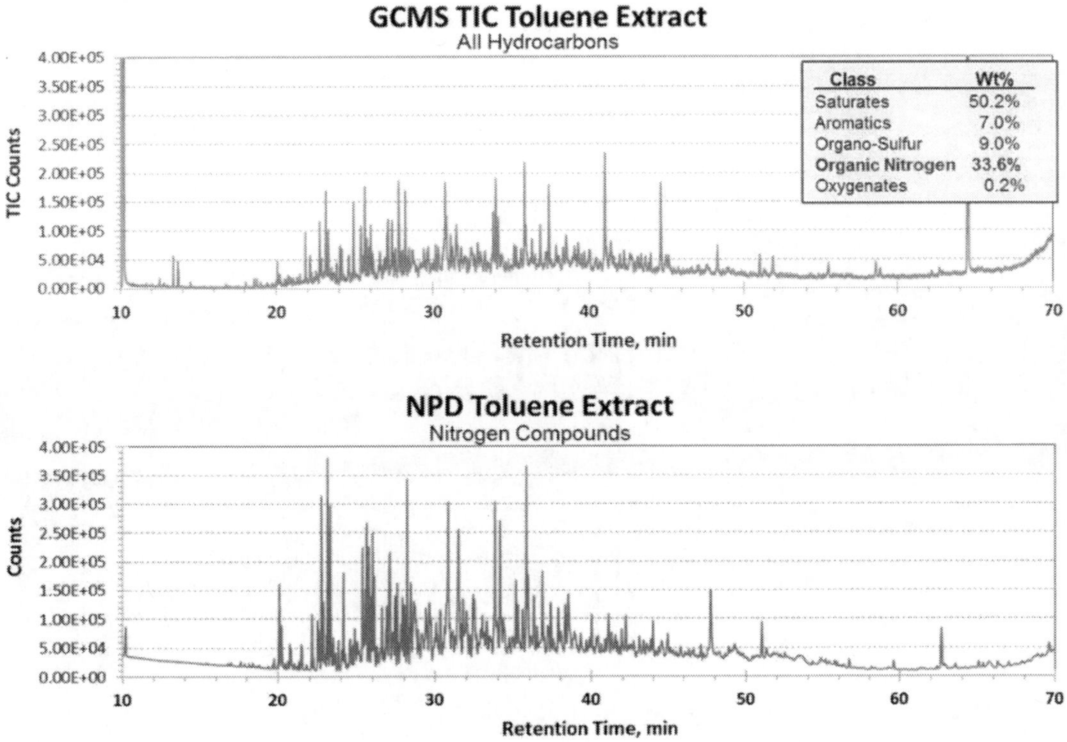

Figure 5. BNC fraction isolated from the diesel fuel.

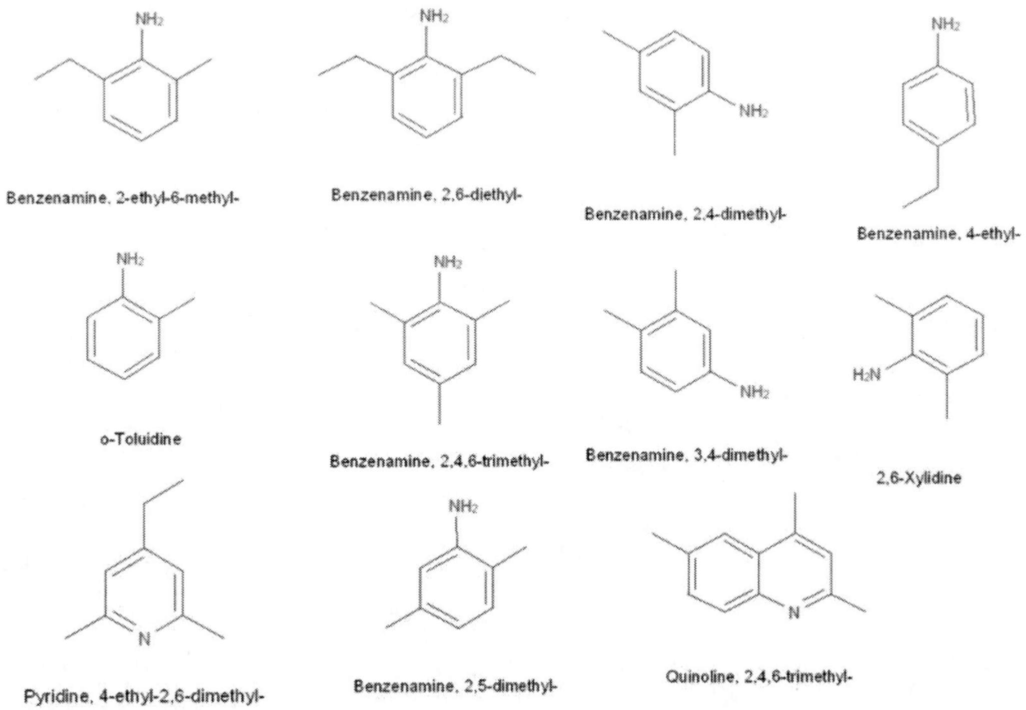

Figure 6. Major components identified in the BNC fraction.

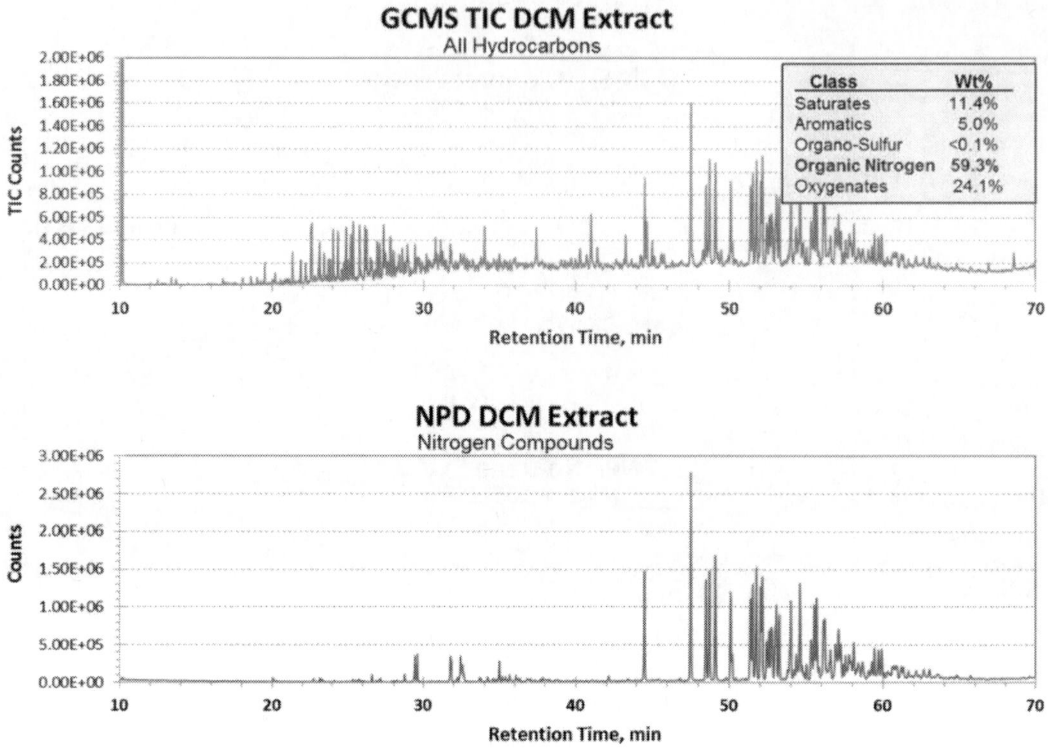

Figure 7. NBNC fraction isolated from the diesel fuel.

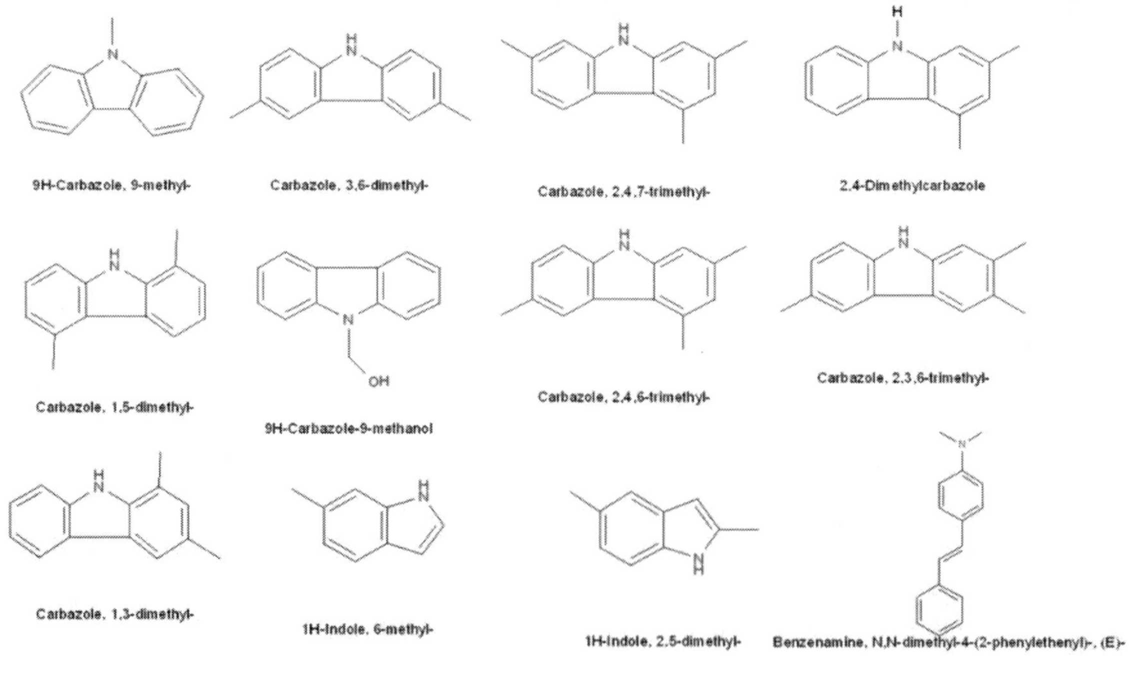

Figure 8. Major components identified in the NBNC fraction in diesel fuel.

Inspection of the nitrogen compounds in the GC-NPD and all compounds in the GC-MS total ion chromatograms from the refinery retains, shown in Figure 9, illustrates the differences in organonitrogen compound profiles between the fuel sample retains from west coast and gulf coast refineries. West coast fuels typically have higher nitrogen content than the gulf coast fuels, and these fuels are no exception, with 510 and 15 mg/kg total N, respectively. Examination of the extracts indicates that both fuels contained non-basic nitrogen compounds, but the west coast fuel contained a significantly greater quantity of the basic nitrogen compounds that have traditionally been associated with poor storage stability.

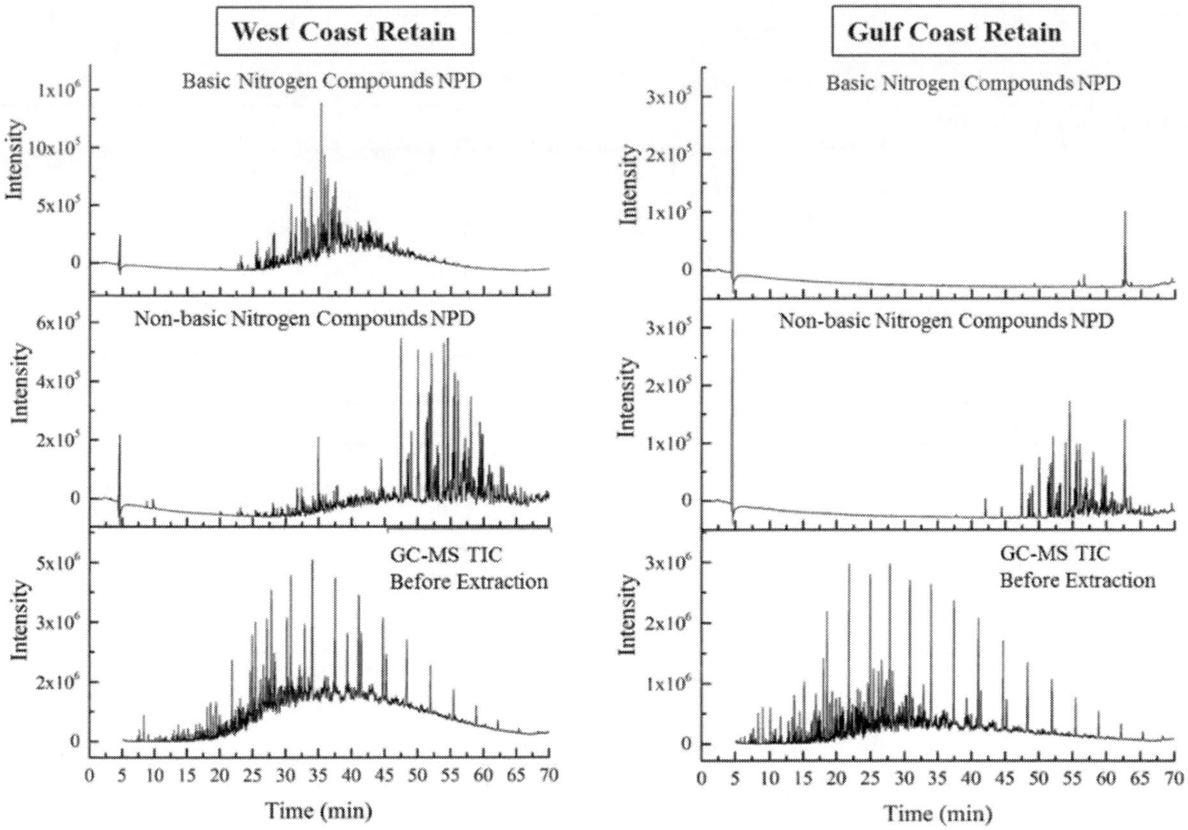

Figure 9. Comparison of the BNC and NBNC fractions between the Gulf Coast and West Coast fuels (note scales are not the same).

Analysis of the unstable western pacific region fuels. The nitrogen compound extraction approach was then extended to the diesel fuels in the western pacific region that were forming high levels of particulates in storage. The qualitative results obtained with FCAST analysis of the extracts, summarized in Table 5, indicate that all the unstable samples contained a BNC fraction composed mainly of benzeneamines. None of the stable fuels contained this BNC fraction. It is noteworthy that subsequent processing changes made at the refinery to produce stable fuel also resulted in the removal of the BNC fraction. The finding that some of the stable fuel samples contained the NBNC fraction suggests that the non-basic organonitrogen compounds, at least in these fuels, were innocuous with respect to storage stability.

Table 5. The presence of basic (BNC) and non-basic (NBNC) nitrogen compounds in the diesel fuel samples. Stable fuels are shown in bold. (X = not found; ✓ = found.)

	NRL #	Total N, ppm	16h PM (mg/mL)	BNC	NBNC
REFERENCE FUELS AND SURROGATE FUEL					
Dodecane	-	–	–	x	x
Ultralow sulfur diesel fuel from local gas station	15-38	–	–	x	✓
Low sulfur diesel fuel from PAX River NAS	15-39	–	–	x	✓
REFINERY RETAINS					
Stable ULSD	14-85	-	-	x	x
Stable Gulf Coast Refinery Retain	15-10	15	1.5	x	✓
Stable West Coast Refinery Retain after Change	16-30	–	–	x	✓
Stable West Coast Refinery Retain #2 after Change	16-31	–	–	x	x
Unstable West Coast Refinery Retain before Change	15-09	510	2.4	✓	✓
UNSTABLE FIELD AND SHIPBOARD SAMPLES					
Unstable West Coast Refinery Retain #1	14-90	–	1.2	✓	✓
Unstable West Coast Refinery Retain #2	14-91	–	1.6	✓	✓
DEFENSE FUEL SUPPLY POINT #1					
Shipboard Sample	15-13	300	2.0	✓	✓
Tank Sample	15-14	390	2.4	✓	✓
Truck Sample	15-15	240	2.4	✓	✓
Tank Sample	15-16	320	1.7	✓	✓
DEFENSE FUEL SUPPLY POINT #2					
Tank Farm Sample	14-87	–	0.2	✓	✓
Shipboard Sample	15-21	440	2.4	✓	✓
Tank Sample	15-22	220	2.0	✓	✓
DEFENSE FUEL SUPPLY POINT #3					
Tank Sample #3	15-28	440	3.2	✓	✓
Tank Sample #4	15-29	580	3.0	✓	✓
West Coast Refinery Sample Prior to Receipt	15-30	580	2.7	✓	✓
SHIPBOARD SAMPLES					
Ship #1 Sample with Sediment	14-122	–	> 20.0	✓	✓
Ship #1 Sample with High Copper	14-123	–	> 20.0	✓	✓
Ship #2 Tank Composite Sample	14-127	–	3.1	✓	✓

CONCLUSIONS

The findings of this study suggest a causal effect between the presence of basic nitrogen compounds and the propensity of a diesel fuel to form unacceptable levels of particulate material in storage. This BNC fraction was composed of various substituted benzeneamines, pyridines and quinolones. Based on previous studies, it is reasonable to conclude that it was the benzeneamines that were promoting sediment formation in storage.

These findings, taken together with the findings of the previous investigations, strongly suggest that the poor storage stability of these fuels were a consequence of the presence of basic benzeneamines. The investigation also revealed that while the presence of these basic nitrogen compounds can cause a fuel to become unstable in storage, they will not necessarily react at high temperatures to produce a failure in the ASTM D5304 and D2274 specification test methods for storage stability.

RECOMMENDATIONS

Until a laboratory test can be developed to detect marginally unstable diesel fuels that contain basic nitrogen compounds that are likely to cause the formation of particulates in storage, it may be prudent to conduct targeted analyses for BNC's in those instances where storage stability is of concern.

Further work is recommended to develop a knowledge base of organonitrogen compound speciation with a broader range of diesel fuels from both gulf coast and west coast refineries, with the ultimate goal of developing a specification test that reliably detects potentially unstable diesel fuels that cannot be detected with current specification test procedures. It became clear from this investigation that the specification test, i.e., ASTM D5304, does not necessarily respond to the compositional features, such as the basic nitrogen compounds, that are responsible for the poor stability during storage.

Since no laboratory method currently exists to replicate long-term formation of particulates in storage, a set of experiments could be conducted with several stable low and ultralow diesel fuels containing various concentrations of BNC, NBNC and a combined BNC+NBNC fraction. These organonitrogen fractions would be derived from a fuel that passes specification, but has a history of producing high levels of particulates in storage. These samples could be stored in the dark in borosilicate glass bottles for a period of time in an attempt to verify the relationship between these organonitrogen fractions and high particulate formation during ambient storage. The samples could also be stressed at 43 °C in accordance with ASTM D4625[17], to determine if that test protocol would detect such marginally stable fuels.

Once an appropriate laboratory test method is obtained, it is also recommended that consideration be given to adding this compositional characteristic to the forensics library currently being developed. This would inform the development of specifications that appropriately limit the allowable level of basic nitrogen compounds that may be tolerated before a fuel is likely to become unstable.

Current efforts are underway at NRL to develop GCxGC methods with nitrogen-specific detection. If successful, it is anticipated that nitrogen speciation maps from this fuel database will further refine our understanding of the basic nitrogen compound influences to target individual compounds or classes of compounds for routine storage stability screening.

ACKNOWLEDGEMENTS

This work was funded by Defense Logistics Agency Energy and the Navy Fuels and Lubricants Crossfunctional Team, through NAVAIR.

APPENDIX A: PROCEDURE FOR ASPHALTENES IN FUEL

1. Add 100 mL of n-pentane or n-hexane per 5 mL of fuel. Allow to stand undisturbed for one week to assure complete precipitation.

2. Filter through a preweighed filter paper (of fine porosity, Whatman No. 50 or equivalent). Air dry filter paper.

3. Weigh paper (Wt. 1)

4. Replace 'filter paper in funnel, same side up. Wash with 25 mL of benzene or carbon disulfide per 5 mL of initial fuel. Air dry filter paper.

5. Reweigh paper (Wt.2)

6. Weight of asphaltenes = Wt. 1 - Wt. 2

7. Weight of bitunenes = Wt. 2 - Initial weight of filter paper

APPENDIX B: PROCEDURE[18] FOR FUEL CHARACTERIZATION

REAGENTS

- Phosphoric Acid, 85\ C.P.
- Glacial Acetic Acid, C.P.
- n-Heptane

APPARATUS

- 125 ml separatory funnel
- S ml and 10 ml pipettes
- 25 ml and SO ml graduate cylinders

PROCEDURE

1. Pipette 10 mL of the fuel sample into the separatory funnel
2. Add 50 mL of n-heptane and swirl to mix well
3. Add 5 mL 85% phosphoric acid
4. Stopper the separatory funnel and shake vigorously for three minutes
5. Add 25 mL glacial acetic acid
6. Shake the contents again for one minute
7. Allow the two layers to separate in the separatory funnel

INTERPRETATION

The color of the acid layer is observed immediately. The immediate color is significant, since the color of the acid layer may change on standing.

- Straight Run: Colorless to pale pink
- Thermally Cracked: Red to reddish brown
- Catalytically Cracked: Yellowish green to deep green.

Experimental blends of suitable samples of the three fuel categories were prepared as comparison standards and the color of the extracts noted are given in Table B-1.

Table B-1. Colors noted of acid extracts in different blends of straight run, catalytically cracked and thermally cracked stocks.

Blend Percentage			Color of Acid Extract
Straight-Run	Catalytically Cracked	Thermally Cracked	
--	10%	90%	red-brown
--	50%	50%	brown
--	90%	10%	deep green-yellow
10%	--	90%	brown-red
50%	--	50%	red-orange
90%	--	10%	pink
10%	90%	--	Green
50%	50%	--	yellow-green
90%	10%	--	pale yellow

LITERATURE CITED

[1] Frankenfeld, J. W.; Taylor, W. F. "Fundamental Synthetic Fuel Stability Study"; First Annual Report for U.S. Dept. of Energy Contract DE-AC19-79BC10045, DOE/BC/ 10045-12, Feb 1981.

[2] Frankenfeld, J. W.; Taylor, W. F. *Ind. Eng. Chen. Prod. Res. Dev.* **1983**, *22,* (4) 608 -614.

[3] Frankenfeld, J. W.; Taylor, W. F. *Ind. Eng. Chen. Prod. Res. Dev.* **1983**, *22,* (4) 615 -621.

[4] Schrepfer, M. W.; Arnold, R. J.; Stansky, C. A. *Oil Gas J.* **1984**, Jan 16, 79-84.

[5] Hazlett, R. N.; *Fuel Science and Technology.* **1988**, 6(*2*), 185.

[6] Batts, B. D.; Fathoni, A. Z. "A literature Review on Fuel Stability Studies with Particular Emphasis on Diesel Oil". *Energy & Fuels*, **1991**, *5*, 2-21.

[7] Banavali, R.; Chheda, B. "Chemical Basis of Diesel Fuel Stabilization by Tertiary Alkyl Primary Amines" *Preprints- ACS Div. Fuel Chem.* **1998**, *43,* 784-787.

[8] ASTM D2274-14, Standard Test Method for Oxidation Stability of Distillate Fuel Oil (Accelerated Method), ASTM International, West Conshohocken, PA, 2014, www.astm.org

[9] ASTM D5304-15, Standard Test Method for Assessing Middle Distillate Fuel Storage Stability by Oxygen Overpressure, ASTM International, West Conshohocken, PA, 2015, www.astm.org

[10] Garner, M. Q.; Morris, R. E. "Laboratory Studies of Good Hope and Other Diesel Fuel Samples". *ARTECH Report No.* J8050.93-FR, March 23, 1982.

[11] Morris, R. E. "Fleet Fuel Stability Analyses and Evaluations". *ARTECH Corp. Report No.* DTNSRDC-SME-CR-01083, September 1983.

[12] ASTM D4625-14, Standard Test Method for Middle Distillate Fuel Storage Stability at 43 °C (110 °F), ASTM International, West Conshohocken, PA, 2014, www.astm.org

[13] Morris, R. E. "Qualitative Analysis of JP-5 Storage Tank Samples", NRL ltr. Rpt. 6180/372.2, 24 June 1993.

[14] Morris, R. E.; Westbrook, S. R.; Baniszewski, D.; Serino, P. "NRL-DLA Fuel Forensics and F-76 Investigation" In *Proceedings of IASH 2015, the 14th International Symposium on Stability, Handling and Use of Liquid Fuels*, October 4 – 8, 2015.

[15] Hammond, M.H.; Morris, R.E.; Cramer, J.A.; Loegel, T.N.; Johnson, K.J.; Myers, K.M. "Navy Fuel Composition and Screening Tool (FCAST) v.2.5" *Naval Research Laboratory Memorandum Report* No. NRL/MR/6180—14-9551. 18 July, 2014.

[16] Morris, R.E.; Cramer, J.A.; Hammond, M.H.; Myers, K.M.; Loegel, T.N. "A Fit-For-Purpose Screening Tool Based on Statistical Modeling of Fuel Composition" *In Proceedings of IASH 2013, the 13th International Symposium on Stability, Handling and Use of Liquid Fuels*. Rhodes, Greece 6-10 October 2013.

[17] ASTM D4625-16, Standard Test Method for Middle Distillate Fuel Storage Stability at 43 °C (110 °F), ASTM International, West Conshohocken, PA, 2016, www.astm.org

[18] Lane, E. C.; Garton, E. L.. "Base of a Crude Oil", U.S. Bureau of Mines—Report of Investigation 3279, Washington, DC, 1935.

IASH 2017, the 15TH INTERNATIONAL SYMPOSIUM ON STABILITY, HANDLING AND USE OF LIQUID FUELS
Rome, Italy
10-14 September 2017

INVESTIGATION OF THERMALLY UNSTABLE AVIATION TURBINE FUELS

Zachary West[1], Linda Shafer[1], Richard Striebich[1], Steven Zabarnick[2], Tim Edwards[3], Milissa Griesenbrock[3]

[1] *University of Dayton Research Institute, 300 College Park, Dayton, OH, 45469 USA*
[2] *University of Dayton Research Institute, 300 College Park, Dayton, OH, 45469 USA, steven.zabarnick@udri.udayton.edu*
[3] *Air Force Research Laboratory, 1790 Loop Road North, Wright-Patterson AFB, OH 45433 USA*

ABSTRACT

The thermal oxidative stability of kerosene based aviation turbine fuels is of critical importance in assuring the longevity and optimal performance of current and future aircraft. The chemical mechanisms producing bulk and surface deposits are known to be governed by trace chemical species, especially heteroatomic compounds, i.e., oxygen, nitrogen, and sulfur containing species. Specific "bad actors," or compounds and/or species classes that negatively impact thermal stability, can vary depending on crude sources and fuel production techniques. Here we apply a number of chromatographic characterization techniques, e.g., SPE-GCxGC, GC-NCD/SCD, and LC-ESI-MS, to examine the total and polar heteroatomic content of a few recent thermally unstable jet fuels from different locations across the United States. The thermal stability of these fuels was assessed using both the JFTOT (ASTM D3241) and QCM (ASTM D7739) methods. Thermal stability evaluations confirm the poor-to-marginal stability quality of the fuels examined. No single chemical class could be identified as the unique cause for the low thermal stability of these fuels, but chemical analyses showed moderate to elevated levels of O/N/S heteroatomic species in all of the fuel samples studied. We also report on the effects of possible additive treatment options, i.e., metal deactivator additive (MDA), antioxidant, and the military +100 additive package to the thermal stability of these fuel samples.

Keywords: thermal stability, polars, heteroatomics, nitrogen, sulfur, oxygen

INTRODUCTION

Reports have surfaced within the past two years of aviation turbine fuels, within the continental United States (CONUS), having poor/failing thermal stability.[1-5] US pipeline operators were the first to notice an unusually high rate of JFTOT failures, which resulted in potential supply disruption concerns.[1-4] Individual accounts have varied as to the cause, severity, and scope of the problem, as well as potential solutions. A known issue with determining a root cause of these thermal stability failures (and a more broad common factors analysis) is the lack of a formalized, systematic approach to sequester, analyze, and report problem fuels.

While the jet fuel specification (ASTM D1655) requires a minimum fuel stability per ASTM D3241, fuel is exposed to a range of temperatures and residence time conditions within aircraft fuel systems. As such, many other methods have been employed to examine the fouling tendencies of jet fuel and middle distillate fuels.[6-8] Other methods of assessing fuel thermal stability are critical to fully realizing the impact of moderate-to-poor stability fuels on real aircraft operations.

It is known that trace, heteroatomic compounds can have a negative impact on jet fuel thermal stability.[6] However, the chemical reactions governing these deposition processes are complex, not well understood, and are further confounded by the variety and scope of species in various fuel samples. The inclusion of S, O, and/or N distinguishes these compounds from the bulk fuel hydrocarbons and provides an opportunity to detect these species using atom specific techniques. However, due to the very low concentrations of these heteroatomic compounds, non-specification, analytical capabilities are typically required for detection.

The purpose of the current work is to report on recent findings of thermal stability and chemical composition of marginal and low thermal stability jet fuel samples received from the field. A focus is placed on results of trace heteroatomic characterizations and overall thermal stability tendencies with both the JFTOT (D3241) and QCM (D7739).

EXPERIMENTAL

Since the heteroatomic compounds of interest were present at trace, e.g., parts per million, levels in fuel samples two general methods of analysis were conducted: 1) pre-separation and concentration of trace compounds using silica gel solid phase extraction (SPE), and 2) sensitive, but very element/molecule specific forms of detection, e.g., GC-NCD/SCD and LC-ESI-MS, on neat fuel.

Compound Detection with Pre-separation

Polar, heteroatomic compounds were isolated from fuel samples using a normal phase, silica gel extraction procedure reported by Shafer et al. (2015).[9] The procedure involves passing fuel sample through a bed of silica gel, washing of the silica with solvent to remove the fuel matrix, and subsequent elution of the polar compounds using methanol. The methanol extract was then subjected to various GC analyses to both qualify and quantify the different types of compounds present in the extract. Sulfur and/or nitrogen specific chemiluminescence detection (SCD and NCD, respectively), coupled with SPE-GC, was used to quantify the polar sulfur and polar nitrogen compound classes. Two dimensional gas chromatography (GC×GC) with dual mass spectrometry (MS) and flame ionization detection (FID) was used to qualify and quantify various oxygenated compound classes and nitrogen containing compound classes, e.g., phenols, alcohols, ketones, indoles, quinolines, and carbazoles.

Sulfur compounds tend to be non-polar and thus go uncaptured in the preceding analysis unless they contain both oxygen and sulfur, e.g., sulfones. Therefore, an alternative technique to classify sulfur species as reactive, e.g., sulfides, disulfides, and thiols was used. The technique to determine reactive sulfur was described in detail by Stumpf et al. (1998), section 2.7.[10] Briefly, the method involves an acid catalyzed oxidation of a jet fuel sample with subsequent reactive sulfur determination via difference between oxidized and neat fuel sample. A GC-SCD was used for quantitation.

Compound Detection without Pre-separation

Total sulfur and total nitrogen levels of fuel samples were quantified using neat injection of the fuel sample on a GC-SCD/NCD system.

Detection of alkyl amine compounds (with molecular formulas of: $C_nH_{2n+z}N$ where n ranged from about 8 to 16 and Z = +1 or +3) was performed using a liquid chromatograph (Agilent 1200 series) with positive electrospray ionization (ESI) and time-of-flight mass spectrometer (Agilent 6210 TOF-MS). All fuel samples were diluted 1:20 by volume in LC-MS grade methanol prior to analysis. An isocratic separation was performed using a C8 column and a methanol/water (95/5%v) mobile phase – with 0.1%wt acetic acid modifier – at a rate of 0.25 mL/min. The ESI source gas temperature was 300°C at 10 L/min and the capillary voltage was 3500 V. A mass range of 100−1000 m/z was scanned.

Quartz Crystal Microbalance (QCM)

An apparatus conforming to ASTM D7739, was used to assess jet fuel thermal oxidative stability. Detailed accounts of the experimental system can be found elsewhere,[11,12] but briefly the QCM was operated as an isothermal, batch reactor. Fuel samples were air saturated at ambient conditions prior to reaction. The samples were stressed at 140°C for 15 hours while surface deposition and headspace oxygen were monitored.

RESULTS & DISCUSSION

Military jet fuel samples (Truck, Tank #1, Tank #2, Tank #3, Tank #4, and AFRL Receipt) and commercial jet fuel samples (Fuel A, Fuel B, Fuel C, and Tank) from the continental US (CONUS) were received by AFRL/UDRI from about June to September of 2016. A set of five, one gallon jet fuel samples from a single military field site were tested for thermal stability using ASTM D3241 at a 260°C set point and the heater tubes were analyzed via ellipsometery for surface deposit depth with the results shown in Figure 1. The five jet fuel samples were all of the same batch/lot of fuel, however, they were sampled at different points in the distribution system at the field site. As the figure shows tank samples #1, #2, & #3 all failed D3241, i.e., gave a deposit standard spot thickness of greater than 85 nm, however, tank sample #2 gave a passing result upon re-testing. Samples of the same fuel from a fourth tank and a dispensing vehicle/truck gave passing results. The high variability of D3241 results shown here is similar to the highly variable JFTOT results reported previously for problematic pipeline fuels.[1-4]

AFRL acquired 5,000 US gallons of the marginal fuel from the same military field site. This later fuel acquisition, along with four of the five original fuel samples, were assessed using the QCM at 140°C, the results are shown in Figure 2. As the figure shows, all five fuel samples display rapid oxidation with complete oxygen consumption occurring within about 2 to 7 hours of thermal stress duration. These fuels also give a rapid rise in deposition, which coincides with the rapid oxygen consumption. Final deposition values (at 15 hours of thermal stress) range from about 4.5 to 9 μg/cm^2. Within our current knowledgebase, deposition values in excess of about 6 μg/cm^2 are considered elevated; the combination of high deposition and fast oxidation results in these fuels exhibiting marginal thermal stability qualities.

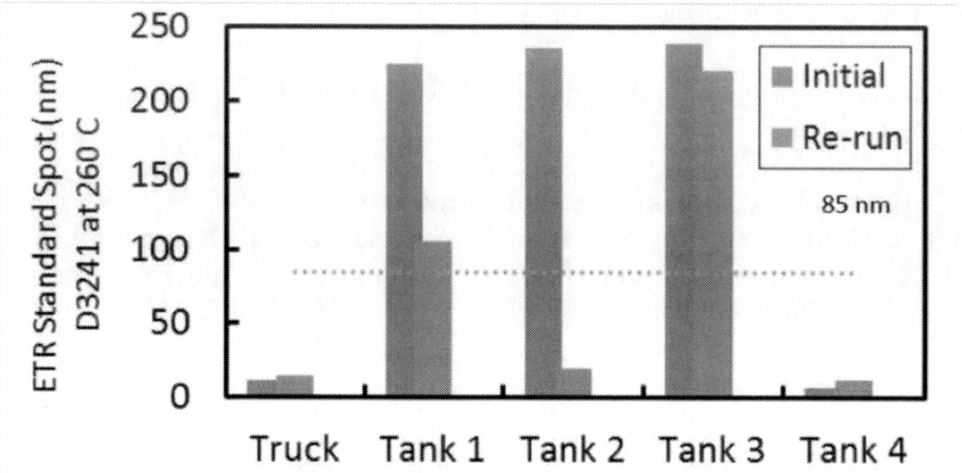

Figure 1. Ellipsometer results for military field site jet fuels.

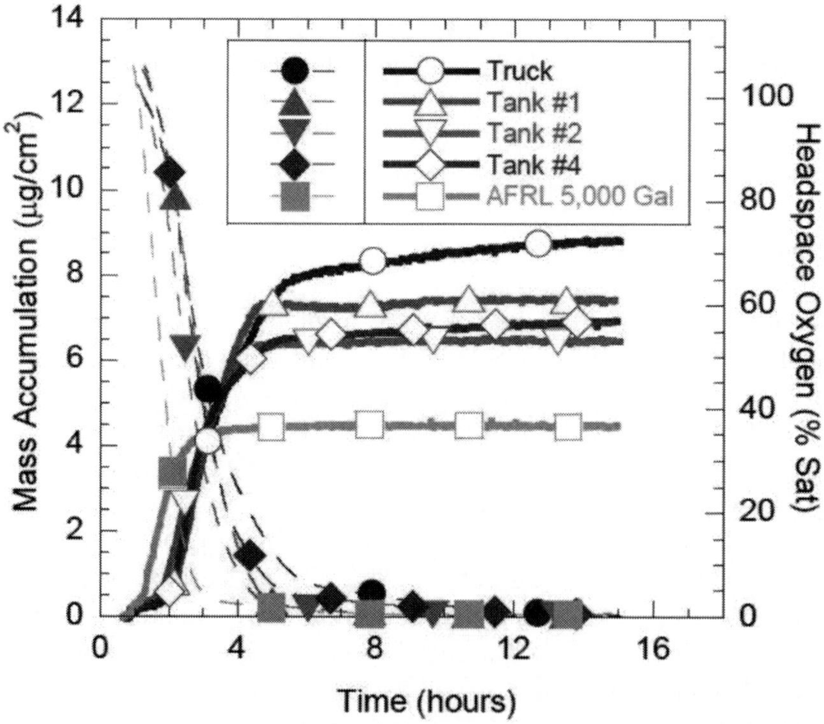

Figure 2. QCM profiles, at 140°C, of military field site jet fuels; solid curves (open markers) show mass accumulation while dashed curves (closed markers) show headspace oxygen.

The trace chemical composition of these military fuels was determined using solid-phase extraction (SPE) with subsequent GC analysis (Table 1). The speciated polars, sulfur classes, and nitrogen classes are consistent for all six samples. The total polars concentrations range from about 660 to 840 mg/L, with most of this attributed to phenols. These polar and phenol concentrations are considered mid- to high-level, with higher levels of these compounds being implicated in causing thermal instability.[13,14] All six fuels exhibit a high level of total sulfur, albeit within specification (~2200 ppm wt); with a reactive sulfur contribution of about 64% of the total, which is typical for jet fuel.[15] The total and polar nitrogen levels are both higher than typical for each of these fuels, with typical levels in the range 0 to 10 mg/L total nitrogen.

Thus while the phenols, sulfur, and nitrogen levels for these fuels are all above typical values, no single factor exhibits unusual levels. Rather, it appears that the combined high levels for each of these species classes may work in concert during thermal oxidative exposure to produce the observed fast oxidation and high deposition.

Table 1. Trace Chemical Compostion of Jet Fuel from Military Field Site

Polars (mg/L)	Truck	Tank 1	Tank 2	Tank 3	Tank 4	AFRL 5,000 Gal Receipt
Phenols	508	560	543	542	482	464
Anilines	17	22	20	20	17	20
Indoles	19	23	20	18	23	16
Quinolines	13	17	15	18	14	15
Pyridines	4	5	4	5	4	4
Carbazoles	5	8	4	9	3	4
Ketones	11	9	9	9	10	8
Cycloketones	10	10	10	9	10	8
Alcohols	15	14	13	11	13	10
Aldehydes	4	4	3	3	3	3
Ethers	<2	<2	<2	<2	2	<2
Esters	<2	11	2	<2	<2	<2
Other	98	141	123	135	107	108
Phthalates	<2	19	3	2	2	<2
Total	704	844	770	780	690	661
Sulfur (mg S/kg)						
Total Sulfur	2330	2360	2280	2350	2260	2090
Reactive Sulfur	1520	1510	1470	1510	1450	1340
Polar - Sulfur	2	3	3	3	2	3
Nitrogen (mg N/kg)						
Total Nitrogen	12	15	15	15	11	15
Polar - Nitrogen	9	12	11	11	9	9

The current pipeline thermal stability issues have resulted in an increased use of metal deactivator additive (MDA) to help failing fuels pass the JFTOT thermal stability requirement.[16] MDA is typically added to jet fuels to improve thermal stability when the fuel is known to have metal contamination, but even in the absence of dissolved metals MDA has been shown to provide antifouling effects under specific circumstances.[17] Here the QCM system was used to evaluate the effects of MDA, the antioxidant BHT (2,6-di-tert-butyl-4-methylphenol), and the JP-8+100 thermal stability improving additive package on these marginal thermal stability fuels. Figure 3 shows the QCM results of stressing this marginal fuel with these additives at 140°C. As the figure shows, addition of MDA (either at 2.0 or 5.7 mg/L) caused a delay in the onset of oxidation and deposition (i.e., increased the lag-time); however, the final absolute amount of deposition was greater than the neat fuel level. This result highlights the complex nature of the MDA effect on thermal stability, as the treated fuel could display a variety of behaviors in a real aircraft depending on actual operating conditions: lower or increased fouling depending on the time/temperature exposure of the fuel. Addition of the +100 additive appears to significantly improve the thermal stability of this fuel for the full 15 hours of thermal stress duration; while the BHT appears to have little effect.

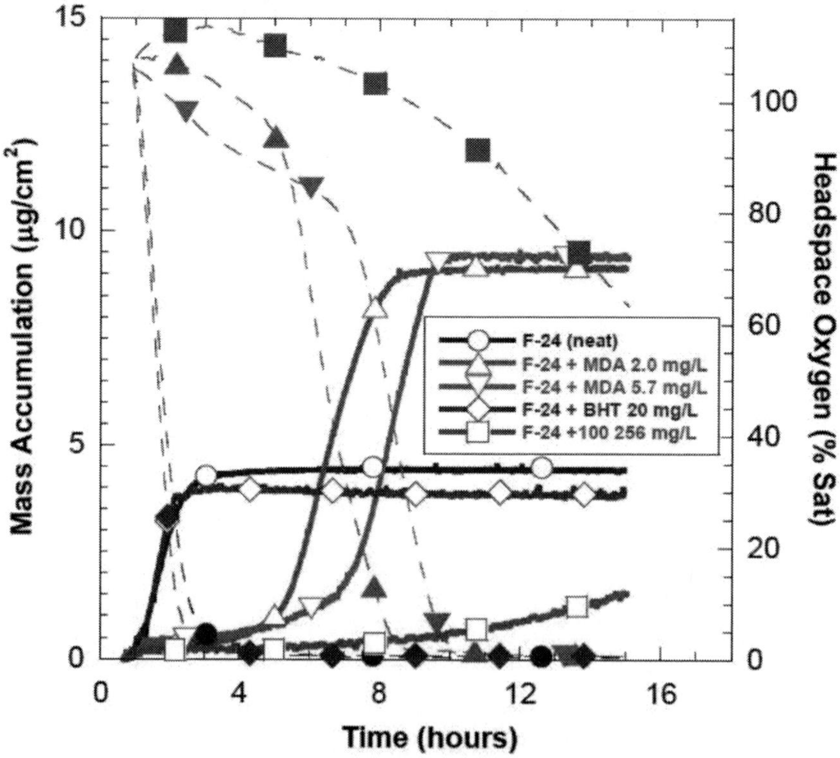

Figure 3. QCM profiles, at 140°C, of Field Site (AFRL receipt) jet fuel with and without potential thermal stability improvement additives; solid curves (open markers) show mass accumulation while dashed curves (closed markers) show headspace oxygen.

The commercial jet fuel samples received (i.e., Fuel A, B, C, and Tank) were analyzed using the same QCM procedure and the results are shown in Figure 4. As the figure shows, these samples display poor thermal stability characteristics, i.e., high levels of deposition (≥6 μg/cm²). It should be noted that fuel samples A, B, and C all passed D3241 at 260°C even though the QCM results indicate these samples might be of concern with respect to thermal stability. While D3241 is by definition the means of assessing thermal stability according to D1655, it is important to consider fully the thermal stability behavior of fuels under conditions to which they are subjected in aircraft fuel systems. The JFTOT is an accelerated test, with higher than expected fuel and surface temperatures, to permit fuel evaluation in a reasonable time period. One should keep in mind that the actual behavior of marginal thermal stability fuels is highly dependent on the time/temperature history of the fuel system.

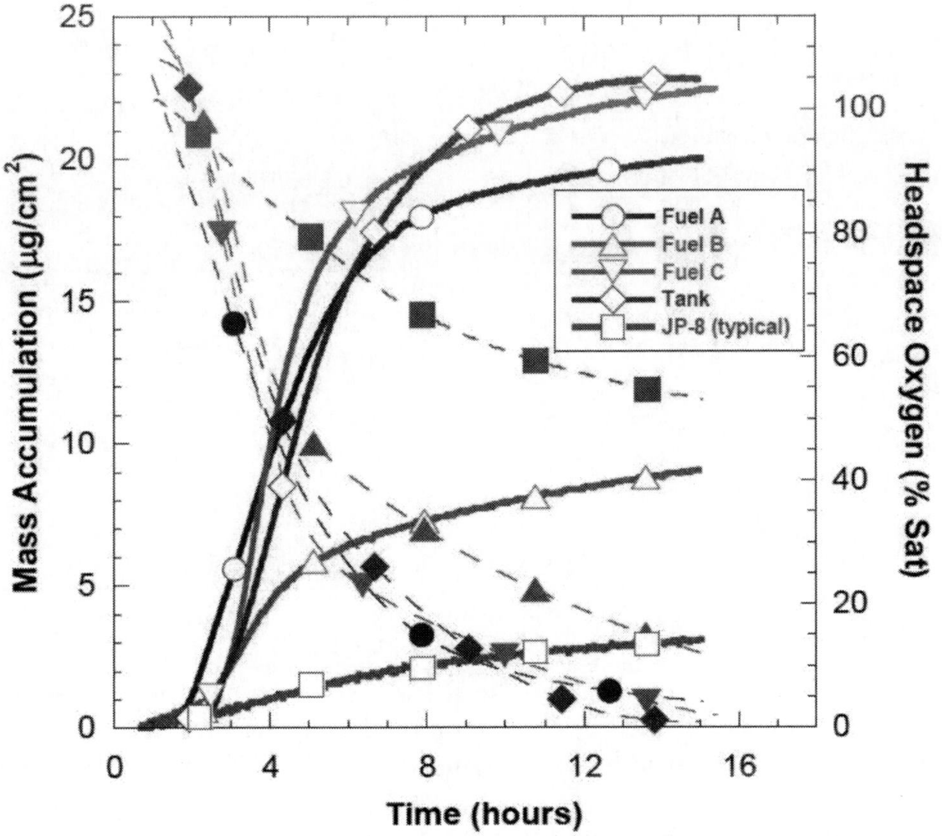

Figure 4. QCM profiles, at 140°C, of various commercial jet fuel samples with a "typical" specification JP-8 shown for reference; solid curves (open markers) show mass accumulation while dashed curves (closed markers) show headspace oxygen.

Additional thermal stability testing was conduct on the military field site fuel sample (AFRL receipt) using the AFRL fuel system simulator (FSS).[18] This rig was designed to simulate heat loads and flows of real aircraft fuel systems and expose the heated fuel to real hardware components, e.g., valves, fittings, and filters. One test metric is that of valve hysteresis; where the flow rate versus pressure drop across a servo valve is evaluated by cycling through flow rates from both low-to-high and high-to-low. Figure 5 shows the pre-test (blue) and post-test (red) servo valve hysteresis results for both the neat fuel and the fuel with either +100 additive or MDA (operating conditions given in figure legends). As the figure shows, significant deviation occurs between pre- and post- test curves for the neat fuel (Figure 5a & b); this is a sign of fouling within the valve and results in poor valve control (which could be detrimental to flight operations). Whereas the pre- and post-test curves for the fuel with JP-8+100 additive package (Figure 5c) show little to no change, therefore, the additive package successfully mitigated the negative effects on this metric under the test conditions. The addition of MDA (Figure 5d) also greatly improved the stability of the fuel, however, not to the same extent as +100 additive. While these results should be more indicative of real aircraft operation, they are acquired at significant expense (e.g., time and fuel volume), thus the need for smaller scale, accelerated test protocol such as D3241 and D7739.

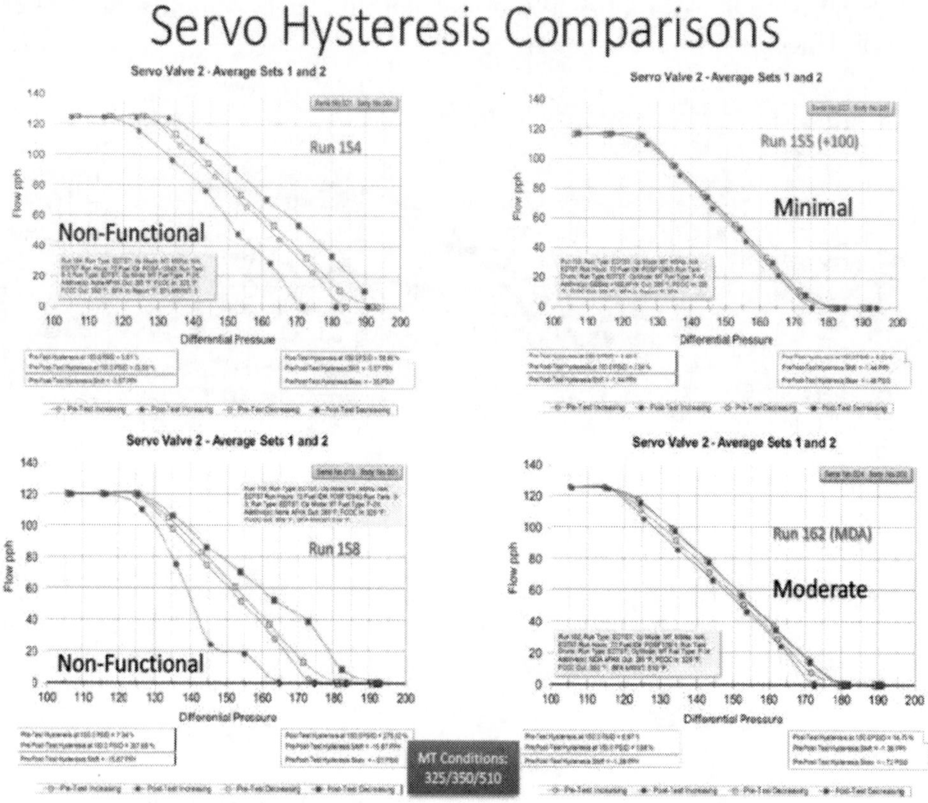

Figure 5. Hysteresis analysis of the servo valve on the FSS using fuel from Field Site #1 (a) neat, (b) replicate neat, (c) with +100 additive, (d) with MDA 5.7 mg/L; blue data are pre-test and red data are post-test.

Finally, we report on the detection of nitrogen containing compounds using an LC-ESI-MS instrument. Recent reports have indicated the possibility that alkyl amine compounds may be occurring in greater levels in recent fuel samples and might contribute to the current thermal stability issues.[20] Two homologous series of amine compounds, i.e., $C_nH_{2n+Z}N$ where Z = +3 or +1, have been identified by Abdallah[20] in current jet fuel samples; however, there has not been independent verification of these species in typical jet fuels nor is there information in the literature on how these compounds affect fuel properties such as thermal stability. The LC-ESI-MS technique given in the experimental section above outlines an alternative means of determining these "Z = +3" and "Z = +1" species compared to the PESI-FTICR-MS technique reported previously.[20] We were able to detect trace amounts of these Z = +3 and +1 amine compounds in both current fuels and historic fuels dating back to ca. 1997. While LC-ESI-MS was able to confirm the presence of these compounds, the technique is unable to resolve molecular structure, e.g., primary, secondary, or tertiary amine. Therefore, a separate method was developed using an alumina solid-phase pre-separation of the fuel sample with subsequent

GC-MS identification. Based on the mass spectral fragments and GC retention time, using the alumina SPE technique, it is determined that the Z = +3 amines have an N,N-dimethylamine structure.

Some initial studies (not shown) were conducted to look at the impact of three Z = +3 amines, i.e., n-octadecylamine, n-tetradecylamine, and N,N-dimethyldodecylamine, on the thermal stability characteristics of two fast oxidizing (low heteroatom-containing) and two slow oxidizing (moderate heteroatom-containing) fuels. Results from the QCM show a small increase in deposition for the fast oxidizing fuels and a large increase for the slow oxidizing fuels for the amine compounds added at 100 mg/L. This observation supports the view that the added nitrogen compounds increase deposition when other heteroatomic compounds are present. However, results using D3241 (at 280°C) gave mixed results, i.e., the n-alkyl amines increased deposition in some, but not all, of the fuels, while the N,N-dimethyldodecylamine caused only minor increases in deposition. These results highlight the complex nature of the current thermal stability issues and the danger in trying to attach a single cause to these issues.

SUMMARY

Current fuels of poor-to-marginal thermal stability were investigated with respect to both composition and thermal stability behavior. Trace chemical analyses indicate an above average level of fuel polars, phenols, sulfur, and nitrogen species in these suspect fuels. While none of these elevated levels is unusual individually, it is thought that the marginal thermal stability of these fuels can be attributed to the combination of elevated levels of all these species types. In QCM studies, MDA improved fuel deposition at short reaction times, but degraded thermal stability at long times. Thermal stability behavior was also assessed using the AFRL fuel system simulator with and without the addition of MDA and JP-8+100 additive package. Both additives appear to improve the thermal stability behavior of the poor fuel, however, JP-8+100 additive package gave superior performance.

Two classes of alkyl amine compounds, i.e., $C_nH_{2n+3}N$ and $C_nH_{2n+1}N$, were verified to exist in various jet fuel samples from ca. 1997 to present. An initial study was conducted to determine the effect of these compounds with mixed results. It is recommended that future thermal studies consider the complex reactions between heteroatomic species classes and their impact on thermal stability.

REFERENCES

1. Woodford, R. Thermal Stability of Jet. Presented at the CRC Aviation Technical Committee Meeting, Alexandria, VA, May 3rd, 2016.
2. Thurston, J. WFS JFTOT Failure Experience in 2015-16. Presented at the CRC Aviation Technical Committee Meeting, Alexandria, VA, May 3rd, 2016.
3. Mengel, J. Recent JFTOT Failures and Corrections. Presented at the CRC Aviation Technical Committee Meeting, Alexandria, VA, May 3rd, 2016.
4. Juan, R. JFTOT Trouble-Shooting Guide. Presented at the CRC Aviation Technical Committee Meeting, Alexandria, VA, May 3rd, 2016.

5. West, Z.; Shafer, L.; Striebich, R.; Zabarnick, S.; Edwards, J.T. Trace Chemical Profiles of Recent Thermally Unstable Jet Fuels. Presented at the ASTM Committee Meeting D02.J0.03, Orlando, FL, December 7th, 2016.

6. Hazlett, R. N. *Thermal Oxidation Stability of Aviation Turbine Fuels;* ASTM monograph 1, PCN 31-001092-12; ASTM: Philadelphia, 1991.

7. Heneghan, S. P.; Zabarnick, S.; Ballal, D. R.; Harrison, W. E. JP-8+100 – The development of high thermal stability jet fuel, Proceedings of the AIAA 34th Aerospace Sciences Meeting and Exhibit, Reno, NV, USA, January 1996.

8. Morris, R. W.; Binns, K. E. Evaluation of Improved Thermal Stability Jet Fuels in an Advanced Reduced Scale Fuel System Simulator, Proceedings of the 5th International Conference on Stability and Handling of Liquid Fuels, Rotterdam, The Netherlands, October 1994.

9. Shafer, L.; Striebich, R.; Zabarnick, S.; West, Z. Identification and Quantitation of Polar Species in Liquid Transportation Fuels by GCxGC, Proceedings of the 14th International Symposium on Stability, Handling and Use of Liquid Fuels, Charleston, SC, USA, October 2015.

10. Stumpf, A.; Tolvaj, K.; Juhasz, M. Detailed analysis of sulfur compounds in gasoline range petroleum products with high-resolution gas chromatography-atomic emission detection using group-selective chemical treatment. *J. Chrom. A* **1998,** *819,* 67−74.

11. Zabarnick, S. Studies of Jet Fuel Thermal Stability and Oxidation Using a Quartz Crystal Microbalance and Pressure Measurements. *Ind. Eng. Chem. Res.* **1994,** *33,* 1348−1354.

12. Grinstead, B.; Zabarnick, S. Studies of Jet Fuel Thermal Stability, Oxidation, and Additives Using an Isothermal Oxidation Apparatus Equipped with an Oxygen Sensor. *Energy & Fuels* **1999,** *13,* 756−760.

13. Balster, L. M.; Zabarnick, S.; Striebich, R. C.; Shafer, L. M.; West, Z. J. Analysis of Polar Species in Jet Fuel and Determination of Their Role in Autoxidative Deposit Formation. *Energy & Fuels* **2006,** *20* (6), 2564–2571.

14. Kuprowicz, N. J.; Zabarnick, S.; West, Z. J.; Ervin, J. S. The Use of Measured Species Class Concentrations with Chemical Kinetic Modeling for the Prediction of Autoxidation and Deposition of Jet Fuels. *Energy & Fuels* **2007,** *21* (2), 530–544.

15. Link, D. D.; Baltrus, J. P.; Rothenberger, K. S.; Zandhuis, P.; Minus, D. K.; Striebich, R. C. Class- and Structure-Specific Separation, Analysis, and Identification Techniques for the Characterization of the Sulfur Components of JP-8 Aviation Fuel. *Energy & Fuels* **2003,** *17,* 1292−1302.

16. *Thermal Stability Failures and Use of Metal Deactivator Additive;* A4A Bulletin 2016.2; Airlines for America: Washington, D.C., July 2016.

17. Wilson, G. R.; Jeyashekar, N. "Metal Deactivator Additive (MDA) Impacts on Thermal Stability," Coordinating Research Council, CRC Report No. AV−6−06, June 2010.

18. Morris, R. W.; Minus, D.; Zabarnick, S.; Balster, L.; Binns, K. E.; Dieterle, G.; Biddle, T. *Protocol f Test Methods for Evaluating High Heat Sink Fuel Thermal Stability Additives for Aviation Turbine Fuel;* AFRL-PR-WP-TR-2002-2037; Report for the

Propulsion Directorate, Air Force Research Laboratory, WPAFB, OH, April 2002.

19. Abdallah, D. East-Central US jet fuel thermal stability issue. Presented at the CRC Aviation Technical Committee Meeting, Portland, OR, May 2nd, 2017.

IASH 2017, the 15TH INTERNATIONAL CONFERENCE ON STABILITY, HANDLING AND USE OF LIQUID FUELS
Rome, Italy
10-14 September 2017

THE DETAILED MEASUREMENT OF FUEL HETEROATOMS BY ELEMENT SPECIFIC DETECTORS AND GCXGC-MS BEFORE AND AFTER STRESSING

Richard C. Striebich[1], Linda M. Shafer[1], Zachary J. West[1], Susan S. Mueller[1], Milissa Griesenbrock[2] and Steven Zabarnick[1]

[1]*University of Dayton Research Institute (UDRI), 300 College Park, Dayton, Ohio 45469-0043, richard.striebich.ctr@us.af.mil*

[2] Air Force Research Laboratory, Fuels and Energy Branch, AFRL/RQTF, Bldg 490, 1790 Loop Road N, Wright-Patterson AFB, OH 45433

ABSTRACT

The polar fraction of aviation fuels contains a variety of sulfur-, nitrogen- and oxygen-containing species. Previous work has shown that removal of this complex collection of low concentration (<1500 mg/L) polar components greatly improves the thermal stability of poor fuels, while addition of the same fraction into stable fuels degrades thermal stability significantly. The small amounts of the heteroatomic components and the complex reactions between them make it difficult to understand the nature of the deposition process. In these experiments, solid-phase extraction (SPE) was used to isolate the polar fraction, followed by analysis using GCxGC and GC with sulfur- and nitrogen-specific chemiluminescent detectors. These investigations were performed in a fuel with a relatively high heteroatom level and as well as in a fuel with a moderate heteroatom level. Results indicated that: 1) the polar fraction contained native nitrogen compounds, some of which were completely reacted after thermal oxidative exposure; and 2) non-polar sulfur compounds were oxidized and were intermediates to deposition reaction during thermal exposures. These polar nitrogen components and oxidized sulfur components as well as other oxygenates were isolated in two separate fractions, added to a solvent and tested in the QCM; this resulted in a similar deposition level compared to the original fuel. This experiment emphasized the importance of these heteroatomic compounds in the deposition process.

INTRODUCTION

Aviation fuels can form solid deposits in hot components of the aircraft, causing both acute and chronic fuel system problems. The chemistry of the trace components in the fuel is suspected to be the primary cause for this thermal instability. Specifically, dissolved oxygen, trace nitrogen-, sulfur- and oxygen-containing species are likely involved. The highest total concentration of sulfur in aviation fuels (by specification limit) is 3000 ppm S for F-24 or Jet A fuels; total nitrogen levels are typically less than 10 ppm of nitrogen and contain a wide variety of compound types [1]. Finally, there is approximately 70 ppm of dissolved oxygen. Deposits formed from typical fuels, however, have much higher levels of nitrogen, sulfur and oxygen components as compared to levels of heteroatoms in fuel. Deposits containing 1-8% sulfur, 1-3% nitrogen and 10-20% oxygen mean that the enhancement in ***deposit*** heteroatom concentration relative to ***fuel*** heteroatom concentration can be > $1x10^5$ for nitrogen, > $2x10^3$ for sulfur and >$1x10^3$ for oxygen [2]. This concentration effect clearly shows the importance of the sulfur, nitrogen and oxygen compounds in thermal oxidation processes and deposition reactions.

Recent field problems involving failures of specification test ASTM D3241, Jet Fuel Thermal Oxidation Tester (JFTOT) for pipeline samples of fuel have increased the desire to understand the influence of heteroatomic components on thermal stability. Initial investigations of pipeline fuel instability problems have suggested that the root cause may be related to the polar fraction (silica gel extractable components) of fuel, as well as incidental contact (in a pipeline or holding tank) between high nitrogen fuels and high sulfur fuels. Previously we have observed that it is possible to add the polar fraction to a thermally stable fuel and render it thermally unstable, and that removing the polar fraction from a poor thermal stability fuel can cause the fuel to be more stable [3]. Thus, we know that 1) the polar fraction of fuel is important in the deposition process because it contains polar sulfur-, nitrogen- and oxygen-containing components; 2) deposits are highly enhanced in heteroatom content as compared to the fuel.

Therefore, the objective of this work is to increase the understanding of deposition by detailed chemical examination of the polar nitrogen- and sulfur-containing components in fuels before and after thermal stressing.

EXPERIMENTAL

Because polar fractions were suspected to be an important consideration in deposition, we isolated polars using silica gel-Solid Phase Extraction (SPE) followed by chromatographic separation and detection [3, 4]. The SPE technique both isolated and concentrated polar components onto a 1000 mg silica gel bed, allowing non-polar fuel to pass through the bed un-adsorbed. After appropriate washing and drying steps, the polar heteroatomic compounds could then be eluted using methanol (a 20:1 concentration effect) and GC analysis conducted. Sulfur-, nitrogen- and oxygen-species were isolated and identified using an Agilent 7890A gas chromatograph with a sulfur and/or nitrogen chemiluminescent detector (GC-SCD or GC-NCD [5], a mass spectrometer (GC-MS) and 2-dimensional gas chromatography (GCxGC) with FID, quadrupole MS, or Time-of-Flight MS (TOFMS) detection [5,6].

Thermal oxidative reactions were examined by subjecting fuels and mixtures of simulated fuels to both the JFTOT and QCM. The JFTOT was modified to bypass the large dead-volume components (gauges, tubing, etc.) at the outlet of the JFTOT tube and was plumbed directly to a sample collector for analysis. The QCM was operated by taking samples before and after batch testing.

A high sulfur and nitrogen fuel (designation-12843), representing the types of fuels which showed failures in the field, were used in QCM and JFTOT testing. A moderately high sulfur and nitrogen fuel was also investigated (designation-10325). Polar extractions taken from these fuels were added back to a mixture of n-hexadecane and butyl benzene (a two-component fuel simulant) to provide insights into fuel stability without the complexities of the hundreds of additional fuel components contained in actual fuels.

RESULTS AND DISCUSSION

Figure 1a shows a GC-MS comparison of fuel (12843) with a similar fuel subjected to a 15 hour, 140°C exposure in the QCM. Even a careful examination of the minor hydrocarbon components does not show any discernable differences between the two traces. Similarly, the total sulfur signal (**Figure 1b**) is devoid of any significant, measurable differences, even though deposits are known to accumulate heteroatom concentration by three orders of magnitude or more. Thus, the gas chromatographic hydrocarbon profile of a fuel before and after stressing is difficult to observe significant changes analytically, without some kind of pre-separation.

Polar components, in these experiments, were compounds extracted using silica gel SPE; polars components were both native in the fuel but also produced by oxidation during thermal stressing [7, 8]. The native polars include nitrogen compounds such as anilines, carbazoles, pyridines, indoles, and amines, as well as oxygenates such as phenols, ketones, aldehydes and esters. **Figures 2a and 2b** show the chromatographic increase in polar sulfur and decrease in polar nitrogen as the fuel is subjected to thermal oxidative exposure in the QCM. In addition, the number of polar hydrocarbons (mainly oxygenates) increases greatly before and after stressing in the QCM. **Figures 2c and 2d** show the "hump" of compounds growing especially in the 25-35 minute portion of the chromatogram. These components were identified using GCxGC as ketones, aldehydes and acids; oxidation products of many hydrocarbons, both aromatic and aliphatic as determined through GCxGC.

The sulfur components changed greatly; the native polar sulfur components comprised only 8 ppm of the almost 2200 ppm S in this particular high sulfur fuel. QCM exposure of the fuel increased the level of polar (oxidized) sulfur by a factor of 17, to 136 ppm S. Compounds tentatively identified include oxides of tetrahydrothiophene and similar cyclic sulfur compounds. Nitrogen, on the other hand, decreased by a factor of 2, from 9.1 to 4.6 ppm N. No chromatographic features could be attributed to oxidized nitrogen; it is possible that nitrogen intermediates were formed which were not chromatographable. **Figure 3** shows similar results for the intermediate-level sulfur fuel, 10325. Both fuels clearly show that polar components are changing greatly as fuel is subjected to thermal oxidative conditions, whether the compounds are sulfur, nitrogen or simply oxygenated hydrocarbons removed in SPE fuel treatment.

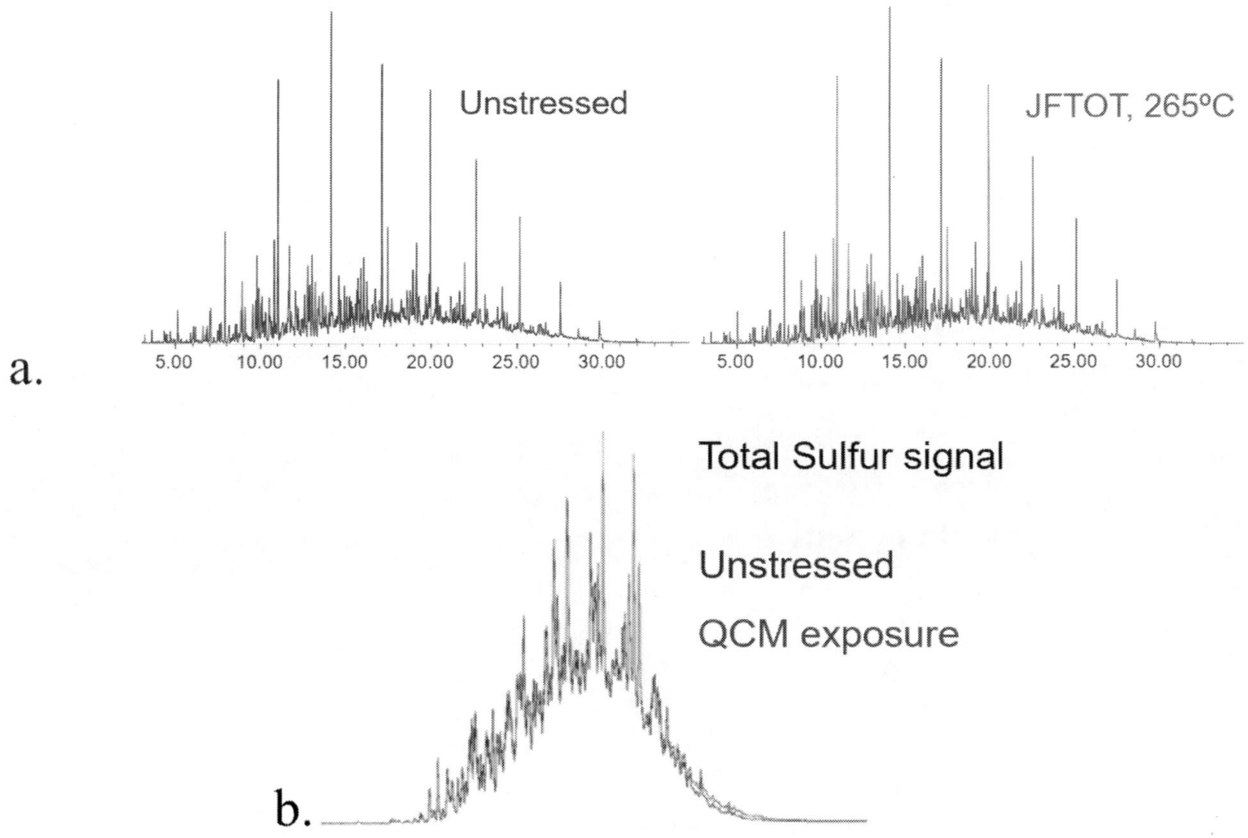

Figure 1. No major or minor differences can be observed before and after stressing in the QCM, either in the **a)** hydrocarbon signal (as measured using GC-MS) or **b)** in the total sulfur signal.

Polar sulfur compounds that formed during stressing were likely oxidized products of sulfur compounds; some of these oxidation products have been tentatively identified using GCxGC-TOFMS as oxides of thianes (tetrahydrothiophenes), but others were certainly being formed. The complexity of the distribution for the oxidized polar sulfur trace from QCM stressing (**Figure 2a**) suggests that many sulfur compounds were being oxidized rather than a few specific ones. The bimodal nature of the distribution may indicate that sulfur- and oxygen-containing primary intermediates continued to react resulting in higher molecular weight secondary intermediates containing sulfur and oxygen.

The purpose of tracking the formation of sulfur and nitrogen intermediates is to attempt to understand whether these heteroatoms intermediates are continuing reactions to actual deposits, and not merely oxidation products. There is, however, a point in the deposition process where gas chromatography may not be able to elute a highly oxygenated, high molecular weight reaction intermediate. However, the formation of reaction intermediates or the disappearance of heteroatomic compounds is certainly an indication of the possible involvement of these important compounds.

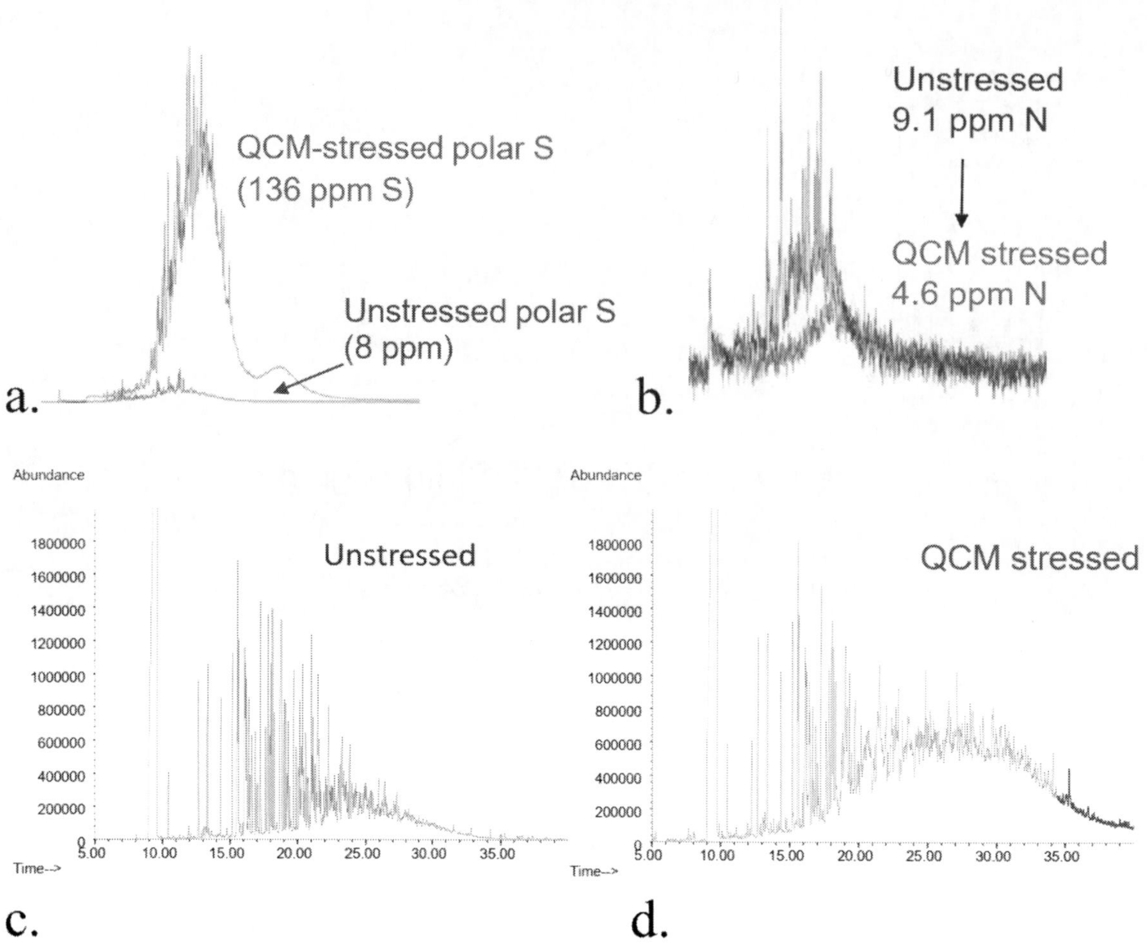

Figure 2. Polars by sg-SPE, the high sulfur fuel, 12843. **a)** sulfur; **b)** nitrogen; **c)** all hydrocarbon polars before stressing; **d)** all hydrocarbon polars after QCM stressing.

In the case of the high heteroatom fuel 12843, it appeared that several nitrogen compounds were eliminated during stressing. In spite of these eliminations of nitrogen containing reactants, the total nitrogen content only changed from 3.3 ppm N to 2.6 ppm N. There was no evidence of higher molecular weight oxidation products of nitrogen in this case. The mid-level sulfur fuel 10325 was slightly different in that at least one sulfur compound decreased in concentration as stressing occurred, although the great majority of polar sulfur compounds increased upon thermal oxidative stressing in the QCM. The sulfur compound that decreased in concentration upon stressing was a dithiazine compound, which may be the reaction product of a hydrogen sulfide scavenger added to fuel at the refinery. Apparently, the available dithiazine was reactive and may have participated in deposition processes. A close examination of the major sulfur compounds in the polar-stressed and polar unstressed fractions indicated that sulfur oxygenates may be common to both samples (**Figure 3a**).

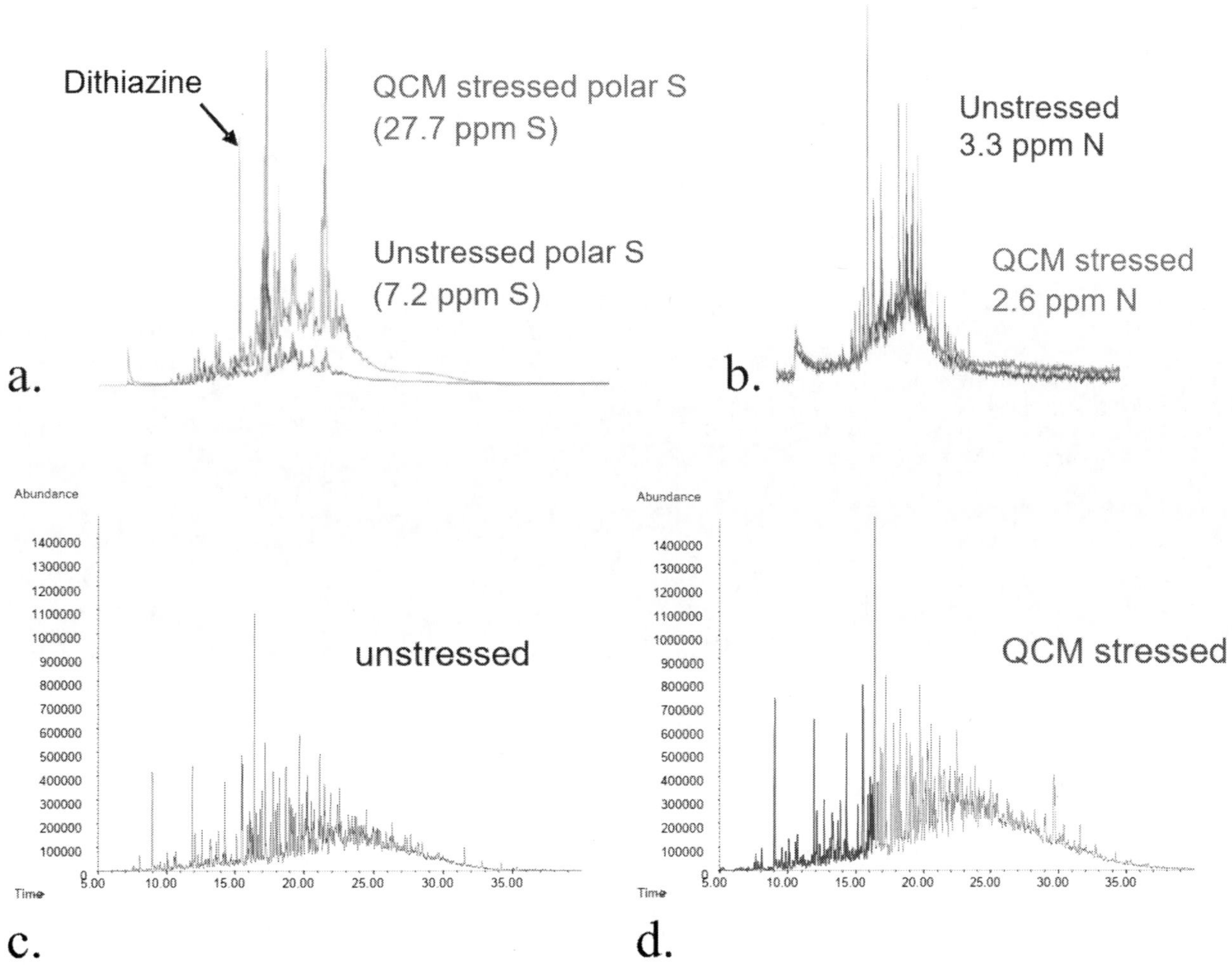

Figure 3. Polars by sg-SPE for 10325, the intermediate sulfur fuel. **a)** sulfur; **b)** nitrogen; **c)** all hydrocarbon polars before stressing; **d)** all hydrocarbon polars after QCM stressing.

GCxGC-TOFMS was used to determine which nitrogen compounds which were completely reacted during thermal oxidative exposure. Several nitrogen-containing species (i.e., pyridines, indoles, anilines, dithiazine) were present in the pre-stressed QCM sample and were not observed in the stressed QCM sample; an example is shown in **Figure 4** for substituted pyridines. This figure shows that the pyridines were completely removed from the QCM-stressed sample, although they were not removed at all from the fuel collected after JFTOT stressing. The JFTOT samples may show less nitrogen reaction because of shorter residence time at temperature or lower oxygen availability compared to the QCM. Carbazole content did not appear to be reduced in JFTOT or QCM stressing in a significant way. The role of carbazole in deposition reactions is still being investigated.

Therefore, in examining the nitrogen and sulfur signals before and after stressing, the hypothesis is: 1) **several polar nitrogen compounds react to form important intermediates that may eventually form deposits, and** 2) **non-polar sulfur compounds oxidize to form polar sulfur compounds** which are then available to form deposits; 3) **nitrogen compounds and oxidized**

sulfur compounds may combine to form deposits[1,5]. Reactions of nitrogen and sulfur containing compounds, as well as oxidation reactions, would certainly increase the composition of heteroatoms in the deposit as compared to the fuel. In order to investigate these hypotheses further, we chose an experiment to remove the unstressed (native) polar fraction from fuel and combine it with an oxidized sulfur fraction (native polars removed before oxidation) to see if, in another matrix, the deposition of the original system could be replicated. This line of experimentation may reveal more about the role of heteroatoms in the deposition process.

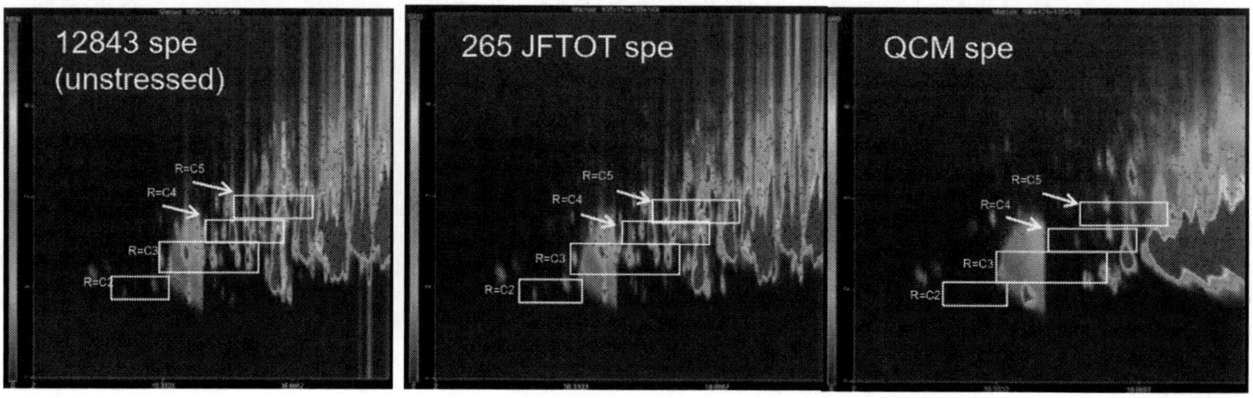

Ions: 107, 121, 135 and 149 amu

Figure 4. Substituted pyridine response is eliminated after QCM stressing, but JFTOT testing does not show any apparent loss of C_2-C_5 pyridines.

The purpose of this experiment was to attempt to account for the QCM deposition from a fuel by extracting polar fractions from the fuel and adding them to a solvent. The important fractions in this experiment were 1) the polar fraction from the neat fuel (60 mL extracted and concentrated to 0.5 mL in methanol; and 2) an extract of the polars from a QCM stressed sample from which the native polars had been removed before stressing (60 mL extracted and concentrated to 0.5 mL in methanol). By taking these important fractions and adding them to a fuel-simulation solvent (80% by volume n-C_{16}/20% butyl benzene), we hypothesized that a QCM stressing of this sample would capture the same deposition as the original fuel. The polar extraction of the fuel (step 1) would add to the solvent the same concentration of reactive nitrogen components and oxygenates as the native fuel. The polar extraction of the stressed fuel (with native polars removed) would capture the sulfur and carbon oxidation products and add these to the solvent. The fuel simulant (n-C_{16}/butyl benzene) model mixture, as well as the model mixture with each fraction and the model mixture with both fractions, would then be exposed to QCM testing.

Figure 5 shows that the neat fuel with its native fractions produced approximately 7 μg/cm^2 of deposition in the 15 hour QCM testing. The solvent C_{16}/BB was also examined under QCM conditions and produced about 1 μg/cm^2. When each of the two fractions (the polar fraction from the neat fuel and the polar fraction from the stressed fuel) were added to the solvent separately and QCM tested, there was a small increase in deposition, but even the sum of each fractions deposition would not account for the same level of deposition as the fuel itself. It was only when the fractions were combined and QCM stressed, that the deposition closely matched the deposition from the original fuel. These results showed that the trace components of the two types of polars (containing nitrogen, sulfur and oxygenates) produced similar deposition characteristics to the original fuel, even though a solvent was used in place of the original fuel.

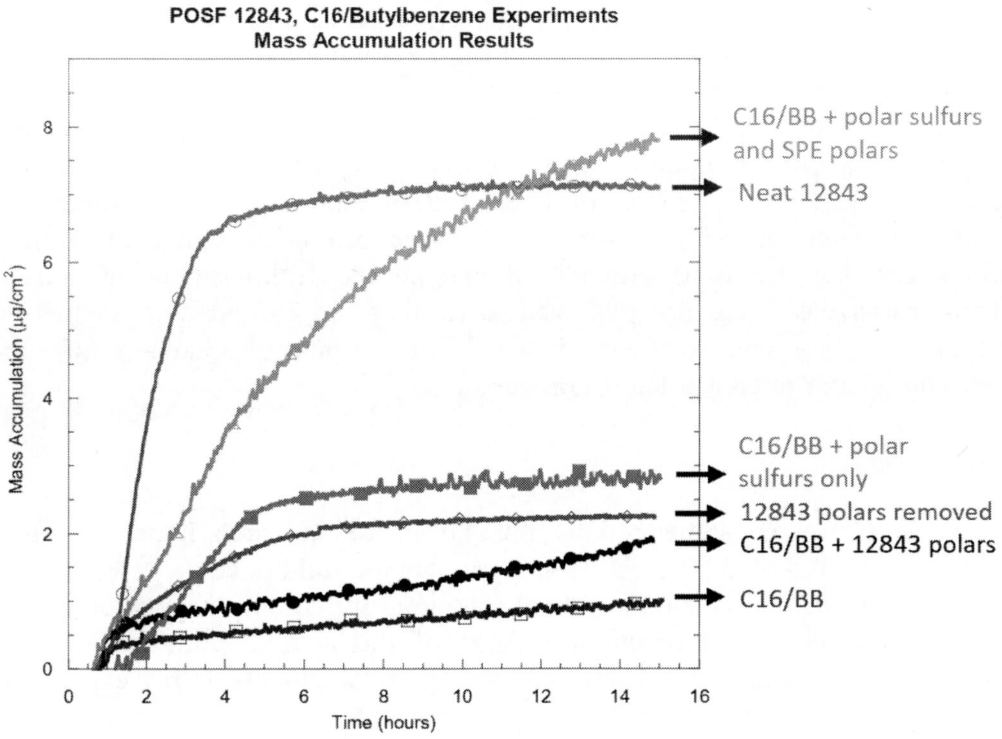

Figure 5. Deposition results of removing polar fractions from 12843 and adding the fractions back into a solvent.

It is clear, and agrees with previous work[3], that the removal of the polar fraction alone significantly reduced deposition. This result, labeled as "12843 with polars removed" shows a deposition level of about 2 µg/cm^2. Putting this fraction (the polar fraction) back into a solvent will increase the amount of deposition, as shown in "C_{16}/BB + 12843 polars", but by itself will not return deposition to the level of the original 12843 fuel. Only by the addition of the oxidized sulfur compounds, obtained by stressing the non-polar fuel components and then extracting these, will the level of deposition be equivalent to the original fuel.

The "polar fraction" contains the polar sulfur and nitrogen compounds, along with polar oxygenates such as phenols. Together, this combination of *native* polar sulfur, *native* polar nitrogen and oxygenates is not sufficient to account for all of the 12843 deposition. It is not until the fuel (and especially the non-polar sulfur compounds) is oxidized that deposition can be more fully accounted for from the original fuel. The native polar sulfur compounds are insufficient by themselves to account for the entire deposit. It is the combination of the *two fractions*, oxidized sulfur fraction and the native nitrogen compounds along with oxygenates, which made the deposition similar to the original fuel.

CONCLUSIONS

There are two important kinds of polar components that influence deposit formation in fuel. The first is the polar fraction obtained from silica gel solid phase extraction of a jet fuel. In this fraction, we have identified certain trace nitrogen compounds which react such that they are no longer detectable in the stressed fuel. The second important fraction is the collection of non-polar sulfur compounds, some of which become oxidized during thermal stressing. These once-non-polar sulfur compounds, now oxidized and therefore polar, can be isolated from other components by a silica gel-SPE procedure after fuel stressing. By combining the fraction containing these oxygenated sulfur compounds with the fraction containing the native nitrogen compounds and certain oxygenates in a solvent, one can obtain deposition levels similar to the fuel from which the nitrogen and sulfur components were originally extracted. Nitrogen compounds which decrease in concentration include dithiazine, indoles, pyridines and anilines. In these experiments, carbazoles were not observed to decrease in composition as fuels are stressed. Future studies will continue to focus on the details of possible reactions of sulfur, oxygen and nitrogen compounds to create deposit precursors.

ACKNOWLEDGEMENTS

This material is based on research sponsored by the Air Force Research Laboratory under agreement number FA8650-16-2-2605. The U.S. Government is authorized to reproduce and distribute reprints for Governmental purposes notwithstanding any copyright notation thereon. The views and conclusions contained herein are those of the authors and should not be interpreted as necessarily representing the official policies or endorsements, either expressed or implied, of Air Force Research Laboratory or the U.S. Government.

REFERENCES

1. G.W. Mushrush, E. J. Beal, D.R. Hardy, J. M. Hughes "Nitrogen compound distribution in middle distillate fuels derived from petroleum, oil shale, and tar sand sources", *Fuel Processing Technology* (**1999**), 61, 197–210.

2. R.N. Hazlett, Thermal Oxidation Stability of Aviation Turbine Fuels, ASTM monograph, PCN 31-001092, ASTM, Philadelphia, PA, **1991**.

3. S. Zabarnick, *Ind. Eng. Chem. Res.*, 33(5), **1994**, p. 1352.

4. R. C. Striebich, L.M. Shafer, R.K. Adams, Z.J. West, M.J. DeWitt, and S. Zabarnick *Energy & Fuels* **(2014)** Vol 28, pp 5696-5706.

5. S. Zabarnick, Z.J. West, L. Shafer, R.C. Striebich, S.S. Mueller, and M.M. Griesenbrock, "Role of Nitrogen and Sulfur Species in Jet Fuel Thermal Stability," Presented at the CRC Aviation Meeting, Alexandria, VA, 2–5 May **2016**.

6. R. Striebich, J. Contreras, L. Balster, Z. West, L. Shafer, and S. Zabarnick, *Energy & Fuels* **(2009)**, 23 (11), pp 5474-5482.

7. S. Zabarnick, R. Striebich, K. Straley, L. Balster, *Petroleum Chemistry Division Preprints* (**2002**), 47(3) 223-226.

8. L. Shafer, R. Striebich, S. Zabarnick, and Z. West, "Identification and Quantitation of Polar Species in Liquid Transportation Fuels by GCxGC," *Proceedings of the 14th International Symposium on Stability, Handling, and Use of Liquid Fuels*, October **2015**.

IASH 2017, the 15TH INTERNATIONAL CONFERENCE ON STABILITY, HANDLING AND USE OF LIQUID FUELS
Rome, Italy
10-14 September 2017

STUDIES OF THE ROLE OF HETEROATOMIC SPECIES IN JET FUEL THERMAL STABILITY: MODEL FUEL MIXTURES AND REAL FUELS

Steven Zabarnick, Zachary J. West, Linda M. Shafer, Susan S. Mueller, and Richard C. Striebich

University of Dayton Research Institute, 300 College Park, Dayton, OH 45469-0043, USA.
Steven.Zabarnick@udri.udayton.edu

ABSTRACT

In this study we utilize oxygen consumption and deposition measurements of model fuel mixtures and real fuels to explore the role that heteroatomic fuel species, and their interactions, play during fuel autoxidation. A range of temperatures, oxygen consumption regimes, and flow environments are employed to provide results applicable over a wide range of fuel autoxidative conditions. The quartz crystal microbalance (QCM) provides a low temperature (140°C) static reactor environment for long reaction times (minutes to hours) with oxygen consumption and sensitive, *in-situ* deposition measurements. The JFTOT system provides a flowing environment at higher temperatures (260 to 300°C) and short residence times (seconds) which is modified with an outlet oxygen sensor measurement and quantitative deposition measurements via ellipsometry. These techniques are used to study model systems (Exxsol D80 with added heteroatom species) and real jet fuels to determine the role of heteroatomic species in jet fuel autoxidation and deposition. The QCM results demonstrate that nitrogen and sulfur species (e.g., indoles/anilines and sulfides) interact during jet fuel autoxidation to encourage deposit formation. The further addition of phenol species, which occur naturally in most petroleum-derived jet fuels, enables even greater deposit production. This behavior is confirmed in the JFTOT via addition of nitrogen and sulfur-containing species to medium and low sulfur jet fuels. Sample collection during autoxidation in the QCM with subsequent gas chromatographic (GC) analysis shows rapid sulfur autoxidation followed by a slow reaction of the nitrogen species to form deposit precursors, implying a step-wise reaction of sulfur oxidation products with nitrogen species to form deposit precursors. These results have important implications for production and mitigation of thermal stability degradation during fuel pipeline transport.

INTRODUCTION

Conventional petroleum-derived jet fuel primarily consists (~99%) of a mixture of hydrocarbons, including n-alkanes, iso-paraffins, cycloparaffins, and aromatics.[1] These species and their relative concentrations determine the bulk properties of the individual fuel sample, such as: density, volatility (e.g., distillation curve), and low temperature properties (e.g., freeze point and viscosity). But there are a number of important properties which are controlled primarily by the identity and concentration of trace fuel species. These properties include lubricity, electrical conductivity, storage stability, and thermal stability. Trace species consist of heteroatomic molecules with very small amounts of dissolved metals (< 1 ppm). These heteroatomic species, containing oxygen, nitrogen, and sulfur atoms, are naturally-occurring species present in the original crude petroleum or products obtained during the refining process from petroleum species.

Previous work has shown the importance and role of oxygen-containing heteroatoms, such as phenols and hydroperoxides, in fuel autoxidative thermal stability.[2] Phenols have been shown to inhibit oxidation but increase surface and bulk deposit formation.[2,3] Hydroperoxides are autoxidation products at relatively low temperatures (<120°C) and rate-controlling reaction intermediates at higher temperatures (≥140°C).[4,5] Sulfur-containing species have been shown to be detrimental to fuel thermal stability but, conversely, provide important fuel lubricity properties.[6-9] Nitrogen species have been studied less frequently with conflicting results – some nitrogen species greatly increase deposition while others are relatively innocuous.[10-13] Most previous work studied heteroatomic species in isolation rather than as mixtures of heteroatomic species classes, as occurs in actual fuels. A relatively small number of studies have explored the interaction between heteroatomic species during fuel autoxidation.[14-18]

In the current paper, we build upon these previous studies and utilize oxygen consumption and deposition measurements to explore the role that heteroatomic fuel species, and their interactions, play during fuel autoxidation. A range of temperatures, oxygen consumption regimes, and flow environments are employed to provide results applicable over a wide range of fuel autoxidative conditions. The quartz crystal microbalance (QCM) provides a low temperature (140°C) static reactor environment for long reaction times (minutes to hours) with oxygen consumption and sensitive, *in-situ* deposition measurements. The JFTOT system provides a flowing environment at higher temperatures (260 to 300°C) and short residence times (seconds) which we have modified with an outlet oxygen sensor measurement and quantitative deposition measurements via ellipsometry. These techniques are used to study a model fuel (Exxsol D80) and real fuels with added heteroatomic species to determine the effect of these species and their interactions on jet fuel oxidation and deposit formation.

EXPERIMENTAL

The QCM batch reactor system with oxygen sensor monitoring has been described in detail previously[19] and is also described in ASTM Standard Practice D7739-11.[20] Standard QCM runs are performed with air saturated fuel at 140 °C. Flow experiments were conducted using an Alcor JFTOT III Jet Fuel Thermal Oxidation Tester Model 230.[21] Details of the system modifications for dissolved oxygen measurements have been detailed previously.[22]

RESULTS AND DISCUSSION

Many previous studies of the effect of heteroatomic species were conducted via addition of representative species to jet fuel samples. As petroleum-derived jet fuel normally contains naturally-occurring heteroatomic species, these studies were complicated by the interactions of the added species with the inherent fuel species. In the current study, we initially employ Exxsol D80 solvent as a model fuel-like hydrocarbon mixture (exhibiting a jet fuel-like distillation range) with insignificant heteroatomic content and low aromatic content (<0.3%). We then move to using real fuels with two different oxidation/heteroatom content characteristics: fast oxidizing fuels with low heteroatom content and slower oxidizing fuels with average heteroatom content.

The 140°C QCM deposition and oxygen profiles for Exxsol D80 are shown in Figure 1. Exxol D80 was run under three different conditions: (1) with normal air saturation, (2) after clay filtration, and (3) with N_2 sparging to remove dissolved oxygen. The plots show that both the air saturated and clay filtered systems show very rapid oxygen consumption and low levels of depositon (~1 $\mu g/cm^2$). Clay filtration has a negligible effect on the oxidation/deposition characteristics confirming that surfactant/polar species are not present in this mixture. The N_2 sparged run shows significantly reduced deposition, demonstrating that oxidation reactions are the cause of the observed low deposit production for the air saturated runs. The Exxsol D80 solvent was selected for this study for its low deposition rate, so that the effects of added heteroatomic species could be readily observed. The air saturated behavior will be used for comparison after addition of nitrogen, sulfur, and oxygen heteroatomic species.

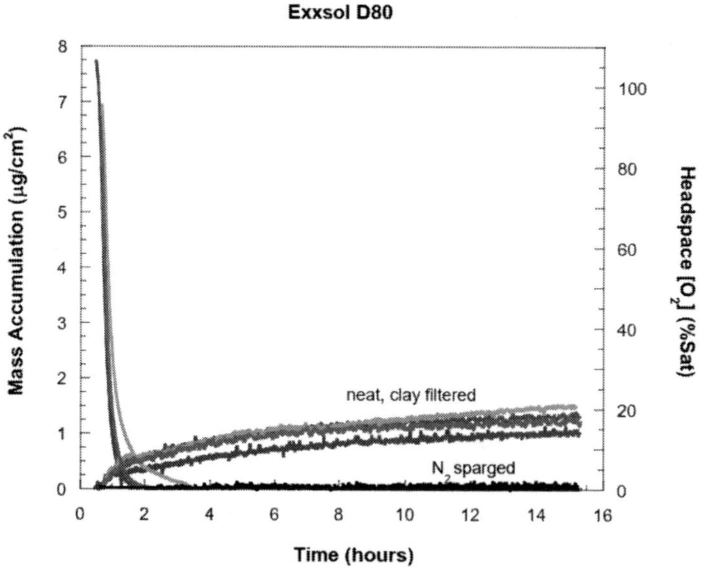

Figure 1. QCM deposition and oxidation plots at 140 °C for Exxsol D80.

The QCM is able to measure very low levels of deposition (down to ~100 ng/cm^2) and thus relatively low levels of added heteroatomic species can employed in the current study, relative to many previous studies which employed much higher levels. A concentration of ~100 mg/L was selected as this level is in the range of the polars content (15 to 2400 mg/L total polars)[3] found in jet fuels and was a level that produced measureable effects on oxidation and deposition. Thus all heteroatomic species additions in this paper were performed at approximately 100 mg/L.

Figure 2 shows the results for the nitrogen species indole added to Exxsol D80. The figure shows that the presence of indole alone had no impact on the rapid oxygen consumption and did not significantly impact the formation of surface deposits. This was a surprising result as many studies have shown greatly increased deposition when nitrogen species, such as indole, are added to jet fuel. But, most fuels have a significant level of naturally-occurring sulfur species present as sulfides, benzothiophenes, and dibenzothiophenes, with the sulfides generally more reactive than the thiophenes. Thus we chose to add hexylsulfide to indole to determine if interactions between these species affects deposit formation. The results show that the presence of both indole and hexylsulfide significantly increases deposition to near 3.2 $\mu g/cm^2$, but the oxidation rate is still unaffected. Figure 3 shows that hexylsulfide alone does not promote deposition. Thus, under these experimental conditions indole and hexylsulfide act synergistically to produce surface deposits. As phenols are also common fuel heteroatomic components which are known to contribute to deposition,[3] we also included the representative phenol species, 2-ethylphenol, as shown in Figure 2. The combination of added indole/hexylsulfide/phenol increases deposition to 7.3 $\mu g/cm^2$. In addition, this combination of three added species also results in a substantial slowing of the fuel oxidation rate. This "antioxidant effect" of phenols has been previously observed and is taken advantage of in the commercial use of hindered phenol antioxidants. Also shown in Figure 3 is the combination of hexylsulfide/2-ethylphenol which shows an oxidation slowing but no deposition increase. Figure 3 shows that 2-ethylphenol alone delays oxidation but only slightly increases deposition, while the combination of hexylsulfide/2-ethylphenol exhibits slow oxidation but no increase in deposits. Thus the increased deposition results from the combination of the indole/hexylsulfide with even greater deposition when 2-ethylphenol is added to the binary combination. The oxidation delay appears to be exclusively due to the presence of 2-ethylphenol.

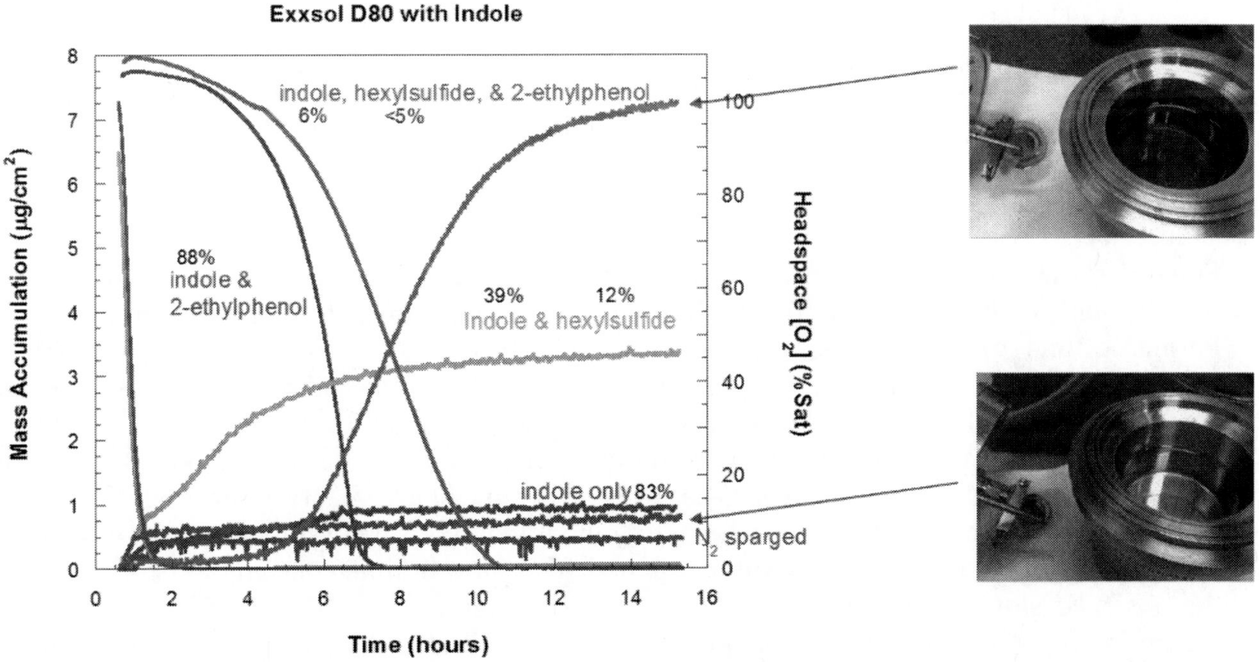

Figure 2. QCM deposition and oxidation plots for Exxsol D80 with addition of 100 mg/L indole, hexylsulfide, 2-ethylphenol, and their combinations. The percentages indicated in the figure are

the percentage of remaining reactant at the end of the run (obtained via gas chromatographic analysis). The photographs on the right show the significant fuel darkening and surface deposit formation in the presence of nitrogen, sulfur and phenol species.

The observation of increased deposition upon the addition of both a nitrogen species and hexylsulfide, and insignificant increases in deposition with only the nitrogen species, was observed for a variety of nitrogen species types including: 3,5-dimethylaniline, 3-methylquinolone, m-toluidine (an aniline species), and 4-ethylpyridine. Thus the identity of the nitrogen component appears to less important than the presence of both nitrogen and sulfur species towards producing deposits. The presence of the phenol species appears to augment the deposition produced by the sulfur and nitrogen species.

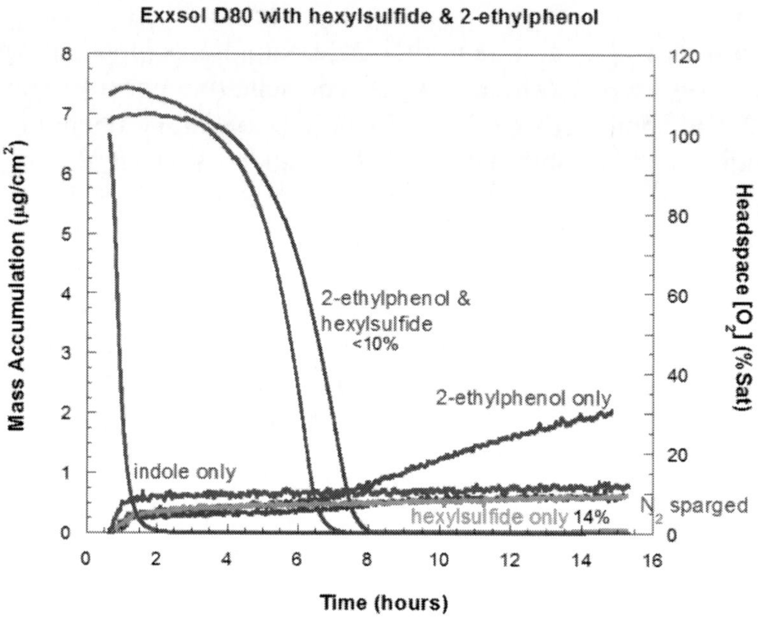

Figure 3. QCM deposition and oxidation plots for Exxsol D80 with addition of 2-ethylphenol, indole, hexylsulfide and their combinations. The percentages indicated in the figure are the percentage of remaining reactant at the end of the run (obtained via gas chromatographic analysis).

Figure 2 shows an interesting difference in reaction time scale for the indole/hexylsulfide system, with a rapid oxidation rate (as for the hexylsulfide alone), but a slow production of deposits. Further efforts to explore the mechanism of oxidation and deposition of nitrogen/sulfur-containing systems were performed with GC analysis of the sulfur and nitrogen reactant species, as well as total sulfur/nitrogen levels, via use of nitrogen and sulfur specific chemiluminescent detectors. One mL samples were extracted during thermal testing for Exxsol D80 with 100 mg/L of 3,5-dimethylaniline and hexylsulfide as shown in Figure 4. The plots show that the hexylsulfide reactant is rapidly removed during the initial oxygen consumption period, while the nitrogen reactant is much more slowly consumed. This slow aniline consumption correlates with the observed slow deposit production for this system (similar to the indole system shown in Figure 2). This appears to imply a step-like mechanism with rapid sulfur oxidation followed by a slow production of deposits upon reaction of the oxidized sulfur with the nitrogen species. This interpretation is supported by the work of Mushrush et al.[14-16] who proposed the reaction of sulfonic acids with nitrogen compounds as an important deposit precursor reaction. Future work needs to determine the identity of the oxidized sulfur compounds produced during oxidation to enable further studies of precursor formation.

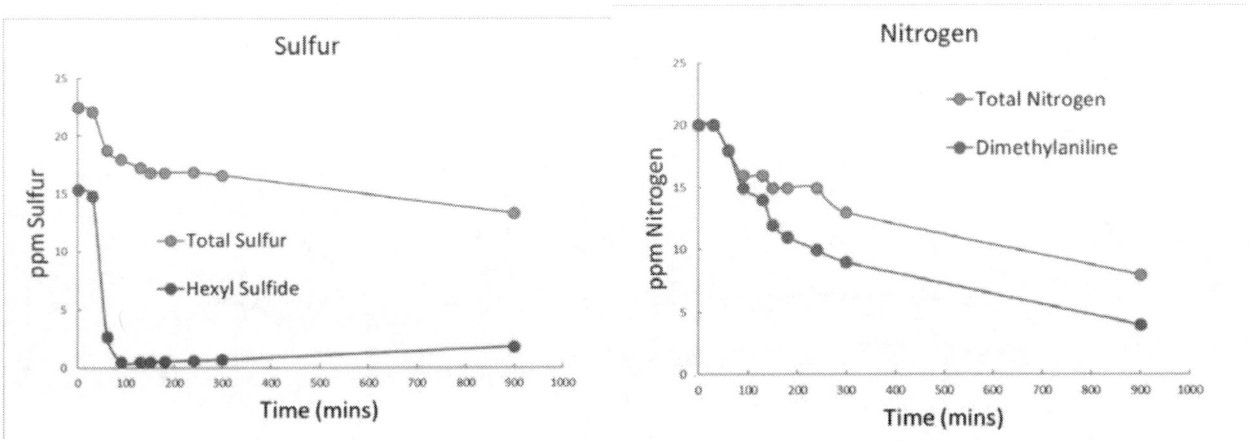

Figure 4. Hexysulfide and 3,5-dimethylaniline reactant and total nitrogen/sulfur levels vs. reaction time during reaction in the QCM at 140°C.

Interestingly, while the sulfur reactant is rapidly reacted, the total sulfur level reduced only by 25% during the run. This shows that while nearly all of the sulfide is rapidly reacted, only 25% of the sulfur is converted to species that are not eluted by the GC system. This indicates that most of the oxidized sulfur products are soluble and chromatographable. Conversely, the aniline reactant and total nitrogen time behavior are similar with relatively small differences in the reaction time scale and amounts, implying that the nitrogen reaction products are insoluble and/or not chromatographable. Future work needs to be performed to help identify the nitrogen oxidation products.

The role of nitrogen species in jet fuel oxidation and deposition was further studied by addition of these species to real fuel samples with oxidation in both the QCM and JFTOT. Two fuels were selected for this study: a low heteroatom (89 ppm sulfur), fast oxidizing fuel (POSF-4877) and a medium heteroatom (420 ppm sulfur), slow oxidizing fuel (POSF-4658). With its low heteroatom content, the fast oxidizing fuel should behave similarly to the Exxsol solvent, while the slow oxidizing fuel may show a greater response to the addition of nitrogen species due to its larger sulfur content. The results for the addition of four different nitrogen species in the QCM are shown in Figure 5. The plots show that the addition of 100 mg/L of these nitrogen species result in only modest increases in deposition for the fast oxidizing fuel. In contrast, the slow oxidizing fuel shows very large increases in deposit production upon nitrogen species addition. The moderate level of sulfur species present in the slow oxidizing fuel appears to be sufficient to encourage deposit increases upon nitrogen addition, while the fast oxidizing fuel contains insufficient sulfur content. An interesting unanswered question is what level of sulfur, and what types of sulfur species, are required to enable nitrogen addition to negatively impact fuel thermal stability? In additional experiments not shown here, we have added 100 mg/L hexyl sulfide to the fast oxidizing fuel with the nitrogen species and have observed a significant increase in deposition, showing that sulfur and nitrogen species act together in real fuel mixtures to encourage deposit formation.

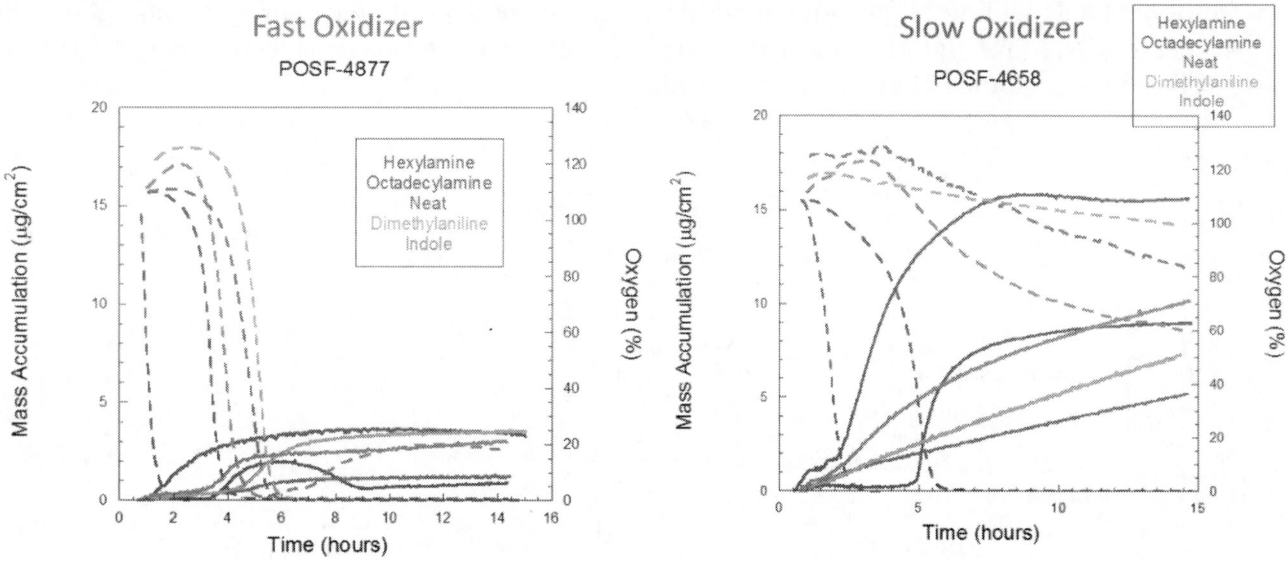

Figure 5. Plots of QCM mass accumulation and oxidation for two different fuels upon addition of four nitrogen species.

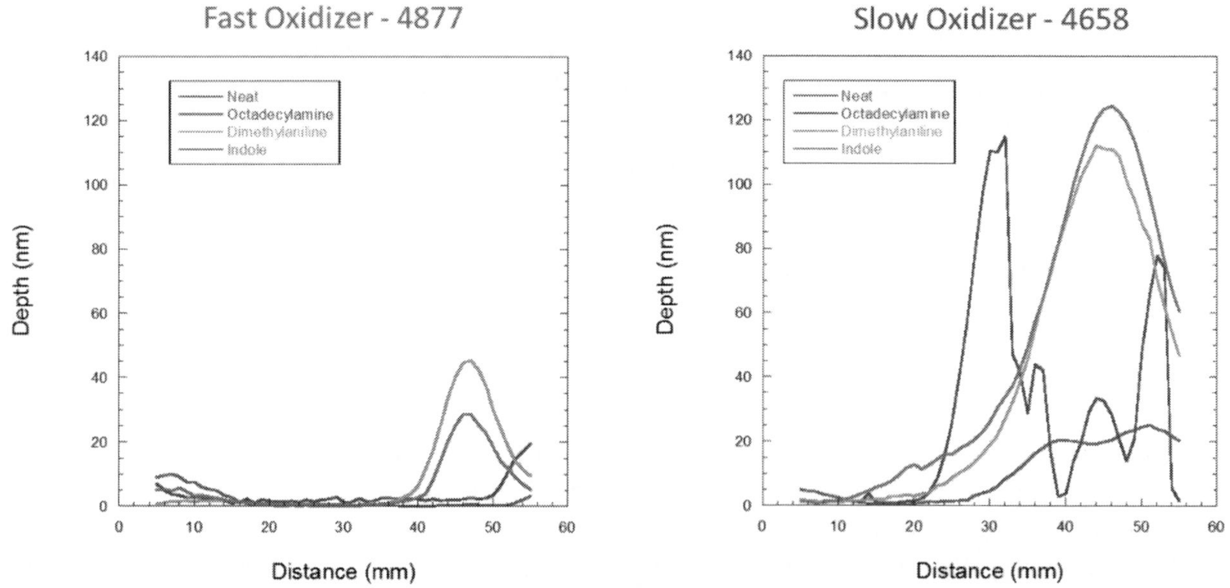

Figure 6. Ellipsometer profiles for JFTOT runs at 280ºC for two fuels upon addition of nitrogen compounds at 100 mg/L. The rise and fall of the octadecylamine deposit profile in the slow oxidizing fuel is an artifact from limitation of measuring thick deposits with the ellipsometer. The decrease is an indication that the deposit thickness is larger than the system can measure.

Three of the same nitrogen species were added to the same two fuels and thermal stability testing was performed on the JFTOT at 280ºC. The circumferentially-averaged ellipsometer deposit depth profiles are shown in Figure 6. In agreement with the QCM observations, the plots show modest increases in deposition upon nitrogen species addition in the fast oxidizing fuel, contrasted by significant increases in the slow oxidizing fuel. The similar behavior in the QCM and JFTOT is striking when considering the significant differences in temperature and reaction time between the two reacting system, i.e. 140°C and 15 hours for the QCM and 280°C and a few seconds residence time for the JFTOT. Thus these observations of the importance of both sulfur and nitrogen species for thermal stability degradation is relevant over a wide range of time and temperature conditions for both simple solvents and complex fuel mixtures.

These observations may have important implications for recently observed thermal stability degradation in the US multiproduct pipeline system. The observations imply that thermal stability degradation might occur in these systems upon fuel blending during transport. The measurements reported here indicate that a high nitrogen fuel with a low sulfur content could degrade significantly (with respect to its thermal stability) when blended with a higher sulfur fuel. Unfortunately, the fungibility of pipeline transported fuel makes it difficult to trace the resulting low thermal stability fuel blend back to the original individual fuel samples that make up the blend. Thus, it is extremely challenging to determine if these nitrogen/sulfur interactions are the cause of these pipeline issues. These experimental observations indicate that it may be useful to monitor fuel nitrogen and reactive sulfur content in individual fuel samples and their blends to help mitigate the observed thermal stability degradation issue.

CONCLUSIONS

This study has shown that the interaction of fuel sulfur and nitrogen species can be an important contributor to the tendency of a fuel sample to produce surface deposits upon autoxidative thermal exposure. The presence of naturally-occurring fuel phenols can significantly enhance this deposit forming tendency. The time scale of oxygen, sulfur, and nitrogen consumption, along with the deposition time scale indicates a possible step-wise process occurs at 140°C, with fast oxidation of the sulfide fuel component following by a slower reaction of these oxidized sulfur products with the nitrogen species to form deposit precursors. These precursors form bulk and surface deposits on the same time scale as the nitrogen removal. These observations may help explain the cause of observed fuel degradation upon pipeline transport in the United States in recent years.

ACKNOWLEDGEMENTS

This material is based on research sponsored by the Air Force Research Laboratory under agreement number FA8650-16-2-2605. The U.S. Government is authorized to reproduce and distribute reprints for Governmental purposes notwithstanding any copyright notation thereon. The views and conclusions contained herein are those of the authors and should not be interpreted as necessarily representing the official policies or endorsements, either expressed or implied, of Air Force Research Laboratory or the U.S. Government.

REFERENCES

1. Hazlett, R. N. *Thermal Oxidation Stability of Aviation Turbine Fuels*; ASTM: Philadelphia, 1991.

2. Kuprowicz, N. J.; Zabarnick, S.; West, Z. J.; Ervin, J. S. *Energy & Fuels* **2007**, *21*, 530-544.

3. Balster, L.M.; Zabarnick, S.; Striebich, R.C.; Shafer, L.M.; West, Z.J. *Energy & Fuels* **2006**, *20*, 2564-2571.

4. Reddy, K.T.; Cernansky, N.P.; Cohen, R.S. *J. Propulsion* **1989**, *5*, 6-13.

5. West, Z.; Adams, R.K.; S. Zabarnick *Energy & Fuels* **2011**, *25*, 897-904.

6. Kauffman, R.E. *J. Eng. Gas Turbines Power* **1997**, *119*, 322-327.

7. Taylor, W.F.; Wallace, T.J. *Ind. Eng. Chem. Prod. Res. Dev.* **1968**, *7*, 198-202.

8. Jones, E.G.; Balster, W.J *ASME Paper* 92-GT-122, **1992**.

9. Taylor, W.F. *Ind. Eng. Chem. Prod. Res. Dev.* **1976**, *15*, 64-68.

10. Taylor, W.F.; Frankenfeld, J.W.; *Ind. Eng. Chem. Prod. Res. Dev.* **1978**, *17*, 86-90.

11. Frankenfeld, J.W.; Taylor, W.F; Brinkman, D.W. *Ind. Eng. Chem. Prod. Res. Dev.* **1983**, *22*, 608-614.

12. Taylor, S.E. *ACS Petroleum Chemistry Division Preprints*, **2002**, *47*, 165-169.

13. Frankenfeld, J.W.; Taylor, W.F; Brinkman, D.W. *Ind. Eng. Chem. Prod. Res. Dev.* **1982**, *22*, 615-621.

14. Mushrush, G.W.; Beal, E.J.; Hazlett, R.N.; Hardy, D.R. *Energy & Fuels* **1991**, *5*, 749-753.

15. Mushrush, G.W. *Fuel Sci. Tech. Int.* **1992**, *10*, 1523-1560.

16. Mushrush, G.W.; Honeychuck, R.V.; Stalick, W.M.; Slone, J.E.; Beal, E.J.; Hardy, D.R. *Fuel Sci. Tech. Int.* **1995**, *13*, 793-805.

17. Reddy, K.T.; Cernansky, N.P.; Cohen, R.S. *Energy & Fuels* **1988**, *2*, 205-213.

18. Stalick, W.M.; Faour, S.; Honeychuck, R.V.; Beal, E.; Hardy, D.R.; Mushrush, G.W. *Petroleum Sci. Tech.* **2000**, *18*, 287-304.

19. Zabarnick, S.; Mick, M.S. *Energy & Fuels* **1999**, *38*, 3557-3563.

20. ASTM D7739-11, "Thermal Oxidative Stability Measurement via Quartz Crystal Microbalance," ASTM International, West Conshohocken, PA, 2014.

21. ASTM D3241-14b, "Standard Test Method for Thermal Oxidation Stability of Aviation Turbine Fuels," ASTM International, West Conshohocken, PA, 2014.

22. Sander, Z.H.; West, Z.J.; Ervin, J.S.; Zabarnick, S. *Energy & Fuels* **2015**, *29*, 7036-7047.

IASH 2017, the 15th INTERNATIONAL SYMPOSIUM ON STABILITY, HANDLING AND USE OF LIQUID FUELS
Rome, Italy
10-14 September 2017

UNRAVELLING THE DETAILED COMPOSITION OF OXIDIZED JET FUELS: AN ORIGINAL DETAILED MECHANISM OF TOLUENE AUTOXIDATION

Matrat, M.*[1] ; Ben Amara, A.*[1]; Mielczarek, D.[1]; Cologon, P.[1]; Bouyou, Y.[1]; Starck, L.[1]; Gandubert, A. D[2].

[1]IFP Energies Nouvelles, 1 et 4 Avenue du Bois-Préau, 92852 Rueil-Malmaison Cedex, France.

[2]TOTAL, 24 cours Michelet, La Défense 10 , 92069 Paris La Défense Cedex, France

ABSTRACT

Aromatics are key components of aviation fuels as they contribute to increase the fuel density and are required for materials compatibility. However, they are being pointed out as precursors of combustion particles and fuel system deposits. The first part of this study was dedicated to the study of the impact of aromatics on the variation of thermal-oxidation stability of commercial jet fuels using Jet fuel Thermal stability Oxidation Tester (JFTOT) associated with quantitative deposit thickness measurement. In the second part of this study, a detailed kinetic mechanism was developed for toluene liquid phase oxidation, selected as a surrogate compound for jet fuels aromatics. The mechanism generation was based on an open source automated mechanism generator (RMG) associated with quantum chemistry calculations for thermochemical parameters estimation. Original experiments were carried out to validate the mechanism, involving induction period measurements using PetroOxy from 120 °C to 160 °C and major liquid-phase oxidation products monitoring using Gas Chromatography-Mass Spectrometry. The model allows to predict the induction period within one order of magnitude and reproduces key oxidation pathways and major toluene autoxidation products, such as benzyl alcohol, benzaldehyde, benzoic acid and cresols, in agreement with literature. This study provides a methodology associating automated mechanism generation and quantum chemical calculations to address aromatics autoxidation. This opens the way for a better assessment of the detailed composition of oxidation products and deposit precursors. Further work will be dedicated to different aromatic surrogates representative of aviation fuels and aromatics / alkanes co-oxidation.

[1] Corresponding author: arij.ben-amara@ifpen.fr; mickael.matrat@ifpen.fr

1.0 INTRODUCTION

Alternative jet fuels are receiving considerable attention for their potential to diversify energy sources and to reduce the impact on environment. National and international bodies have launched initiatives to increase their use in the coming years.

Aside from the conventional pathways, several alternative aviation fuels were recently proposed and approved for use in mixture with conventional jet fuel from up to 10 % to 50 % v/v max according to their pathway. Alternative jet fuels so far approved can be classified into two categories, either paraffins only or mixtures of paraffins and aromatics depending on the process employed. Synthesized Paraffinic Kerosene (SPK) are obtained through different ways such as Fischer Tropsch FT-SPK, Hydroprocessed Esters and Fatty Acids HEFA-SPK, Alcohols to Jet conversion ATJ-SPK, Synthesized Iso-Paraffins from Hydroprocessed Fermented Sugars (SIP). While Synthetic Kerosene with Aromatics pathway (FT-SKA) add alkylated benzenes produced from the processing of coal tar to the FT process coal-to-liquid [1]. Besides, several new pathways are undergoing the certification procedure based on new feedstocks and conversion pathways [1, 2]. The present legislation requires a minimum aromatic content of 8 % wt for any blend with alternative jet fuel (ASTM D7566) to ensure material compatibility.

Although, aromatics are essential components to ensure seal swell in jet fuels, they may be implicated in deposit formation and have a non-linear impact on autoxidation [3–5]. Several mechanisms were proposed in literature to describe the chemical and physical steps leading to oxidation and deposit formation from middle distillate fuels [6–8]. One simple approach was that proposed by Taylor [6] which proceeds through fuel reacting with oxygen leading to soluble oxidation products. Those react further with oxygen producing insoluble products having a molecular weight in the range of 200 g/mol-600 g/mol. These chemical reaction steps are followed by a physical process in which insoluble products aggregate into spherical particles (50 nm-300 nm) and precipitate on the surface. This process may be promoted by fuel volatility and low solvency. Surface particles may then undergo coalescence and surface reactions leading to deposit formation. This mechanism was extended by Watkinson *et al.* [7] which proposed the chemical interaction of non-oxidized fuel with the surface leading to the formation of precursors in the liquid fuel, in the thermal boundary layer or directly in the surface.

Due to an important need for better simulation tools regarding fuel autoxidation as well as experimental data that can be simulated, this study evaluates in a first part, the impact of aromatics on the variation of thermal-oxidation stability of commercial jet fuels using Jet fuel Thermal stability Oxidation Tester (JFTOT) associated with quantitative deposit thickness measurement. In a second part, a modeling effort is made using a pure aromatic compound (toluene) using detailed kinetic modelling in order to better understand the autoxidation kinetics at high temperature in liquid phase and the possible routes leading to deposit formation. Toluene was selected being the simplest alkyl aromatic. The model is experimentally validated using PetroOxy induction period measurement and detailed speciation by gas-chromatography mass spectrometry.

2.0 EXPERIMENTAL AND MODELLING METHODOLOGY

2.1 Experimental methods

In this study, the impact of aromatics on the thermal-oxidation stability of commercial jet fuels is studied using Jet fuel Thermal stability Oxidation Tester (JFTOT) associated with quantitative deposit thickness measurement, then, a combined approach is employed to assess the autoxidation kinetics of toluene based on detailed kinetic modelling and experimental characterization. In order to investigate its reactivity under a normalized test device, its induction period is first characterized at various temperatures. In a second step, the main oxidation products are analyzed thanks to a batch reactor equipped for repeated sampling into the liquid. The obtained samples are then analyzed qualitatively through gas chromatography coupled with mass spectrometry and quantitatively with a flame ionization detector.

2.1.1 Jet Fuel Thermal Oxidation Test (JFTOT)

The studied fuels were tested using JFTOT IV and Aluminium heater tubes purchased from Alcor®. Tests were carried-out for 2.5 h using a constant fuel flow rate of 3 mL/min at different temperatures from 250 °C-305 °C to determine the breakpoint temperature. During the test, the fuel flows over the heater tube. Contact of the fuel with the heated metal surface may lead to fuel degradation and the formation of surface-adherent deposits and/or bulk insolubles. The formation of insolubles is monitored by measuring the pressure difference through a 17 μm filter placed downstream of the heated section, and tube deposits are characterized following the end of the test by a visual rating of the deposit color or by measuring the deposit thickness.

Deposit thickness measurement by Interferometry Tube rating (ITR). Characterization of JFTOT interferometry (ITR) plating is also an accepted method in ASTM D3241. The standard thickness as defined by the specification ASTM D1655 should not exceed 85 nm. The measurement is based on the spectral reflectance measurement principle developed by Morris and Wagner [9]. The apparatus DR10, shown diagrammatically in Figure 1, uses a UV-VIS light source which emits, through an optical probe, an incident ray perpendicular to the surface of the tube. The reflected ray is collected by a spectrometer. The interference fringes created by the deposit are analyzed by a software integrated in the apparatus to calculate the thickness of the deposit from the refractive index (n) and the extinction coefficient (k). The thickness is calculated at 1200 points (50 measurements along the tube and 24 radial positions) [10].

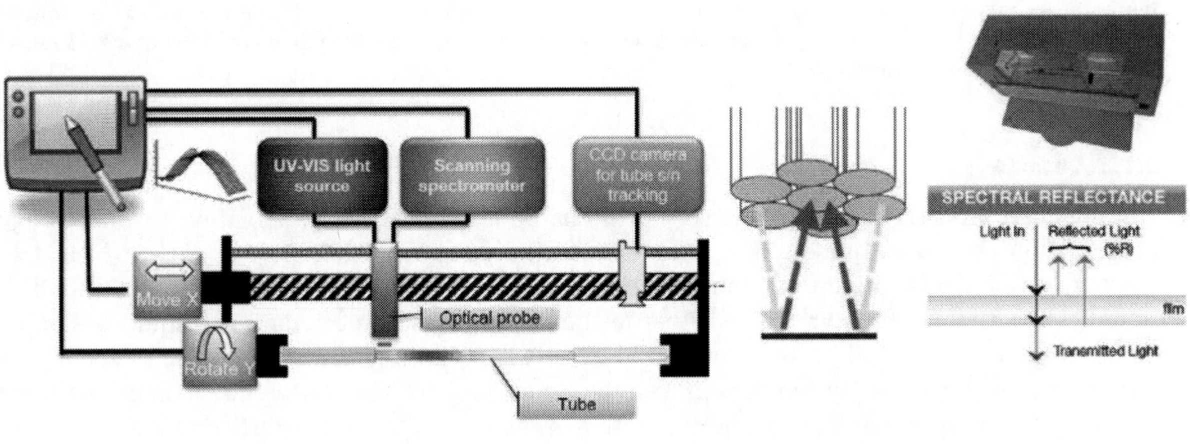

Figure 1: Interferometric apparatus DR10 [10]- General principle.

2.1.2 PetroOxy test

The PetroOxy is a standard test widely used for the oxidation stability of middle distillate fuels (diesel and biodiesel fuels) according to ASTM D7545 or gasoline fuels according to ASTM D7525 [11]. Recently, it has been successfully employed for jet fuel oxidation stability research studies [12, 13]. During this test, a small fuel sample (5 ml) is introduced in a closed and heated test cell. The headspace pressure, measured continuously in the test cell. The test cell is then pressurized with oxygen at ambient temperature to 7 bar. When the target pressure is reached, the induction period measurement starts. The test cell is heated to the set-point temperature. The rise of temperature induces an increase of the headspace pressure function of the volatility of the tested fuel. The fuel oxidation within the test is reflected by oxygen consumption and headspace pressure drop. The induction period (IP) is defined, in the standard method as the time required to reach a 10 % pressure drop as illustrated in Figure 2. The repeatability error, was calculated from 25 different samples tested at IFPEN at different temperatures and repeated 2–3 times each. It can be written as a function of the IP according to the following equation: $\Delta IP = 0.014.IP + 1.229$. In practice, the repeatability error is of about 2 % in agreement with the level indicated in PetroOxy standard for middle distillate fuels (ASTM D 7545) as well as previously published results [14]. In this study, single induction period measurements at three individual temperatures of 120 °C, 140 °C and 160 °C are performed and 7 bar of initial oxygen pressure to characterize the temperature dependence of the system.

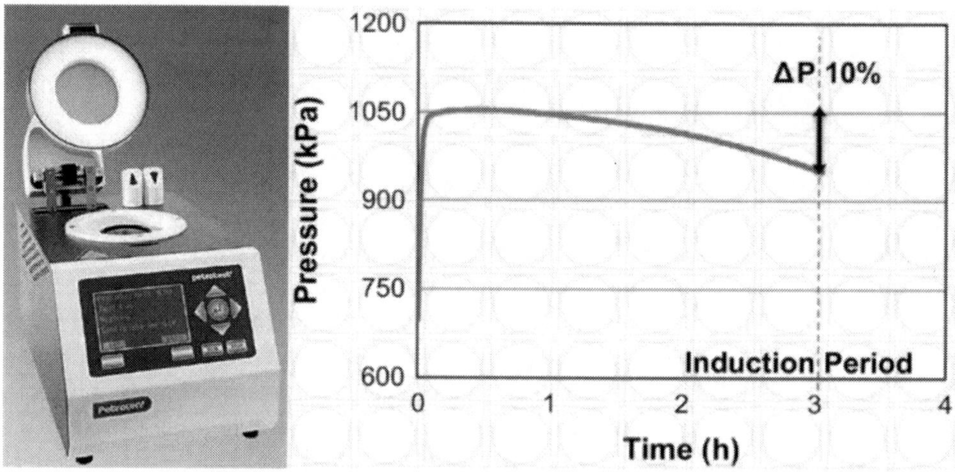

Figure 2: Photo of the PetroOxy test instrument and illustration of headspace pressure variation during the oxidation of n-decane at 140 °C, under 7 bar of initial oxygen pressure. Pressure rises following the temperature increase, then, it decreases due to oxygen consumption by the oxidation process. The induction period (IP) occurs at 10 % of pressure drop and equals 3.05 h. The time is counted starting from the beginning of the temperature rise.

2.1.3 Autoclave

An autoclave illustrated Figure 3 is set up to enable a systematic investigation of intermediate autoxidation products at the highest temperature studied in the PetroOxy. A 250 cm3 Hasteloy reactor is filled with 50 mL of sample and agitated with pure oxygen bubbling through the liquid. The system is pressurized at 10 bar to the targeted temperature and the liquid is sampled at different times to follow products build-up. System homogeneity is characterized by the temperature difference between the gas and the liquid probes. This difference is mainly affected by the rotation speed of the agitator. For this system a 1200 rpm of stirring rate ensures a temperature difference of ± 1.5 °C between the gas and liquid phase. Sampling of the liquid is performed with a manually operated valve. A volume of 0.5 mL is sampled several times during the course of an experiment. The sample is immediately immersed into dry ice to limit further oxidation reactions. The sample is analyzed immediately afterwards to assess qualitatively and quantitatively the intermediate stable products distribution.

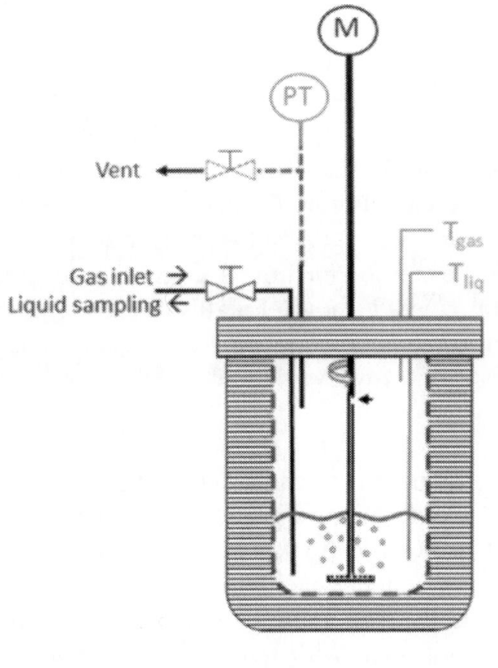

Figure 3: Illustration of the autoclave reactor used to perform liquid phase oxidation under high temperature and high oxygen pressure [15]. "PT" refers to the pressure gauge, "M" to the agitator and T_{gas} and T_{liq} to the measured temperature in each phase.

2.1.4 Gas chromatography analysis

The oxidation products analysis is carried out using gas chromatography coupled with mass spectrometry for identification. A gas chromatograph CP3800 Varian with a FFAP column 30 m x 250 µm x 0.25 µm film thickness is used. Quantification is performed with a flame ionization detector and a different gas chromatograph. An Agilent 6890N with the same FFAP column reference is used. This setup enables retention time comparison and species assignment. In order to perform the FID quantitative analysis of the main oxidation species: benzaldehyde, benzyl alcohol, benzoic acid as well as toluene, both external and internal calibration are performed. For toluene quantification, 0.4 g of sample are diluted into ethanol to reach a total sample mass of 10 g. 1 %m/m of acetonitrile is used as an internal standard as it does not interfere or react with any of the product analyzed here; For the oxidation products, an external calibration is performed and the undiluted sample is used for their quantification with the FID.

2.2 Modelling approach

The Reaction Mechanism Generator RMG is an automated code, designed to propose chemical kinetics mechanisms for both the gas and liquid phase [16]. The generation requires a set of initial conditions including reactant species, their structure and concentration, the temperature as well as, for liquid phases, the solvent used. The mechanism generation is based on a library of reaction types such as hydrogen abstraction or radical recombination, with relevant kinetic parameters, for defined molecular graphs. For every molecule in the mechanism, possible reactions from the library, based on structural information of the molecule are evaluated. The highest reaction flux enables the code to select the next species that moves into the core mechanism. This process is repeated until the defined end criteria is reached. This can be a certain time or reactant conversion for example. Where no kinetic data is available in the database, an average, based on related parameters, is calculated in order to obtain an estimate for the reaction. Besides the kinetic estimation, thermochemical parameters are also estimated based on Benson group additivity for the gas phase. A liquid phase system implies solvation effects, RMG uses corrective methods in order update the thermochemical parameters. The first one is based on a linear solvation energy relationship (LSER) which presents a parametrized description of solvent effects to estimate the Gibbs free energy of solvation. This method takes aspects such as viscosity and polarity into account as well as solute descriptions. This approach has been shown to provide a good accuracy for most solvents at low computational cost [17]. The second correction applied in RMG uses Mintz model which aims to obtain the enthalpy of formation in the liquid phase [18]. This model is quite similar in principle to the LSER parametrization method and has been evaluated for a large set of closed shell species. The remaining liquid phase entropy is then estimated by RMG using the two corrections cited above. Finally, the code also ensures that reactions do not exceed their diffusion limit. This is achieved by estimating an effective rate constant that takes into account the radii of the involved molecules as well as the diffusion rates in the solvent.

In the context of this work, quantum chemical calculations have been performed to asses and improve the quality of the RMG-predicted thermochemical parameters for individual species in the liquid phase. The functional used in this study is M06-2X which has proven its capabilities in terms of accuracy since, including in the application of organic chemistry. A modern Ahlrichs basis set and the RI-J auxiliary basis set are used for lower cost calculations to evaluate a large number of species with reasonable accuracy. Indeed, in the GMTKN30 benchmark, M06-2X allowed better tradeoff of accuracy/calculation time for basic properties with a weighted total mean absolute deviation of 3.2 kcal/mol and is one of the best performing hybrid functional [19]. Finally, solvation is considered using the SMD model [20] which was shown to perform well when compared to other implicit solvation methods. Explicit solvation may provide better predictions however such a calculation is beyond the scope of this study.

A zero-dimensional reactor described elsewhere [21] is used to evaluate the generated mechanism. The higher density within the liquid compare to gas phase is simulated with an elevated pressure to provide correct molar concentrations, and the reactor model also includes a constant-concentration feature to represent the oxygen supply to the liquid from the gas phase during the course of the reaction. In order to discuss the development process for the detailed kinetic mechanism, two intermediate mechanisms are considered. The first one is the initial mechanism provided by RMG and is named “core mechanism''. It is developed with a limit on the maximum carbon number that can be considered along the generation process. This implies that termination reactions are first ignored. The second mechanism, namely the “extended mechanism” includes termination reactions estimated with RMG. Finally, the thermochemistry is updated thanks to the methodology described above and this leads to the final mechanism: “the improved mechanism” consisting into 2309 reactions and 173 species.

2.3 Fuels and reagents

2.3.1 Fossil Jet A-1

The conventional jet A-1 was produced from Merox or sweetening process and supplied by Total

Petroleum Company. GC2D analysis showed a molecules distribution between C8 and C16 centered on C10. It contained about 18 %m/m normal alkanes, 28 %m/m branched alkanes, 21 %wt cyclic alkanes and 22.5 %m/m aromatics (which include 19 %m/m of single- ring aromatics and 3.4 % of poly-aromatics). The distribution of the jet fuel function of carbon number and chemical family is presented in Figure 4.

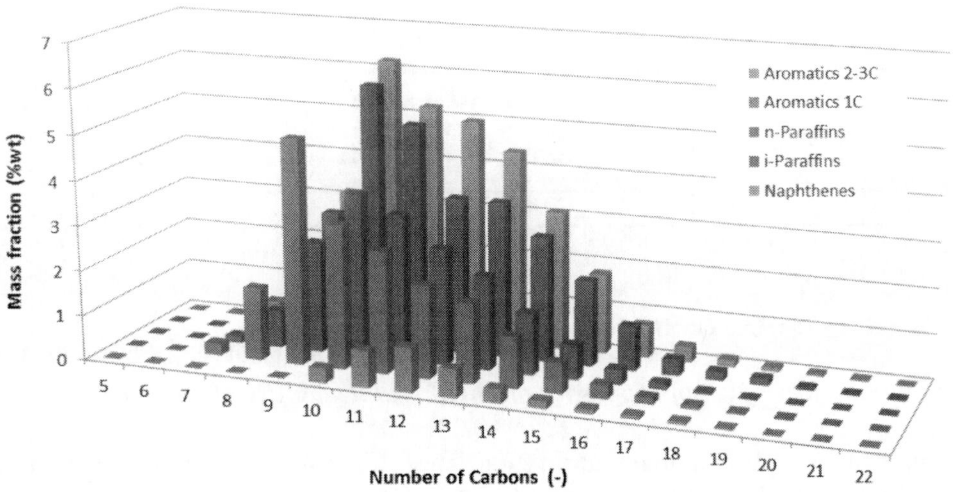

Figure 4: Chemical composition of the conventional fuel given by 2-dimensional gas chromatography

2.3.2 Mono-aromatic solvent

The monoaromatic solvent (MA) is composed mainly of C9 aromatics, namely, of cumene, n-propylbenzene, ethyltoluene and trimethylbenzene. It contains as well several C10 alkylbenzenes and alkanes. In minor fraction, the GC-MS analysis reveals the presence of alkenyl benzenes (methyl styrene, benzene propenyl) and partially hydrogenated bicyclic hydrocarbons (indene) which are likely to accelerate the oxidation. These aspects are discussed further in section 3.

2.3.3 Di-aromatic solvent

1-methylnaphthalene (MN) supplied by Alfa Aesar having ≥96 % of purity was used as a di-aromatic solvent and mixed with the Jet A-1.

2.3.4 Pure reagents

Several pure reagents were used to characterize Toluene autoxidation: Toluene provided by VWR and its purity is ≥ 99.8 %, benzaldehyde (≥ 99.5 %), benzyl alcohol (≥ 99.8 %) and benzoic acid (≥ 99.5 %) were provided by Sigma Aldrich. Figure 5 illustrates the chemical structures of these species.

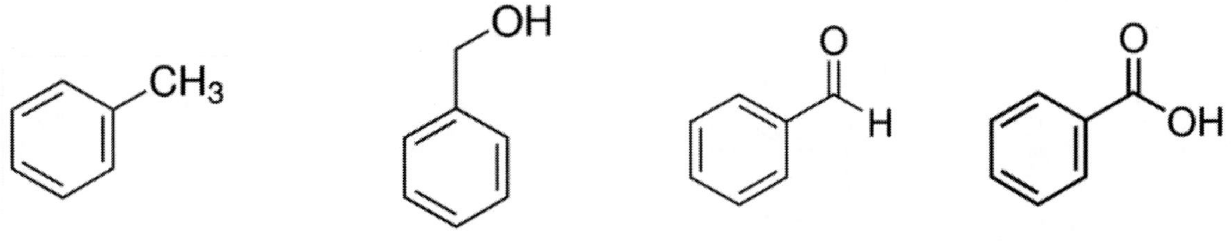

Toluene Benzyl alcohol Benzaldehyde Benzoic acid

Figure 5: Chemical structure of the reactant and the main oxidation products quantified in this study.

2.3.5 Blends

Blends of jet fuel with the mono-aromatic solvent (MA) and the di-aromatic solvent (MN) were formulated in order to reach levels of aromatics and naphthalenes equal to the maximum limit allowed in ASTM D1655: 25%v/v and 3%v/v respectively. First Jet A-1 was blended with 8%v/v MA to reach 25%v/v of aromatics. The resulting blend was further mixed with MN to reach 3%v/v of naphthalenes. The composition of fuel blends is highlighted in Table 1.

Table 1. Composition of the fuel mixtures evaluated in Part 1

	Jet A-1	Jet A-1 + 8%v/v MA	Jet A-1+ 8%v/v MA + 1.6%v/v MN	Limits ASTM D1655
Aromatics (%v/v)	18	26.2	27.8	< 25
Naphthalenes (%v/v)	1.5	1.5	3.1	< 3

3.0 RESULTS AND DISCUSSION

3.1 Evaluation of the impact of aromatics on the thermal stability of commercial jet fuels

Results obtained from the thermal stability oxidation test carried out at different temperature on the fuel matrix are presented in Figure 6. The addition of the aromatic solvent (MA) at a level of 8 %v/v reduces the thermal stability of the conventional jet A-1. The breakpoint temperature is reduced from 300 °C for the Jet A-1 to 280 °C for the Jet A-1+MA. Above 260 °C and for a given temperature, a considerable increase of deposits thickness is observed. For instance, the deposit thickness increases from 10 nm to 71 nm at 280 °C.

The addition of a small fraction (1.6 %v/v) of 1-methylnaphthalene resulted in higher deposit formation especially after 270 °C. For instance, the maximum deposit thickness increased by over 80 % at 280 °C compared to Jet A-1/MA blend. The effect of 1-MN addition on the increase of deposit thickness may be explained by the higher tendency of polynuclear aromatics to form deposits as they require less numerous growth steps to form insoluble compounds in comparison with monoaromatic compounds.

This agrees with recent experimental works by DeWitt et al.[22] that studied the oxidation process of different mono and diaromatic solvents, showing the formation of high molecular weight species, through molecular growth and oligomerization (of the parent fuel or the oxidized compounds), with fewer steps required by the diaromatic solvent compared to the monoaromatic solvent in order to form insolubles.

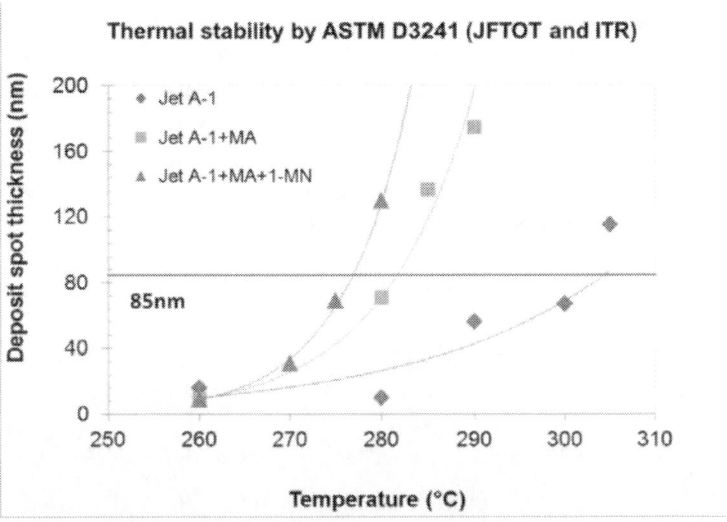

Figure 6: Thermal stability of the jet fuel and its blends with mono-aromatic and di-aromatic solvents (MA, MN) measured by the JFTOT and interferometry tube rating

In order to better understand the reactivity of aromatics in the liquid phase, the next part of this work focuses on the reactivity of toluene, a simple aromatic compound representative of aromatics in fuel. This was conducted through a detailed kinetic modelling of toluene autoxidation associated to induction period measurements and detailed speciation.

3.2 Oxidation of Toluene: Experimental and detailed modelling results

3.2.1 Induction period data

The IP of toluene has been measured at three different temperatures using the PetroOxy apparatus. The results are illustrated Figure 7. The figure also highlights the modeling results for two intermediate mechanisms as well as the improved mechanism.

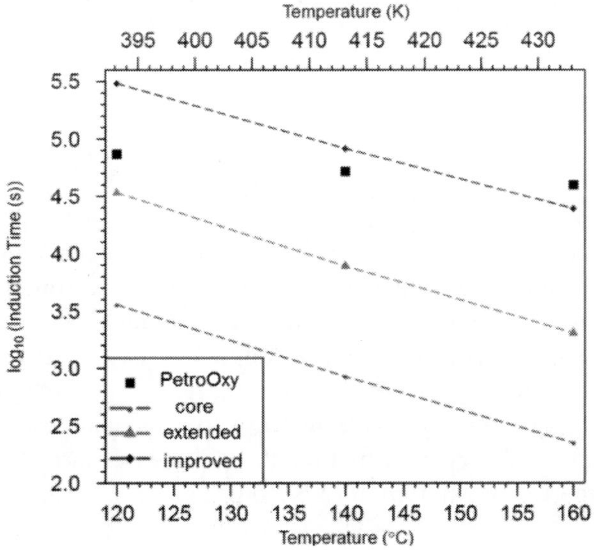

Figure 7: Comparison between the experimental measurement of induction period in the PetroOxy apparatus at 120 °C, 140 °C and 160 °C and the modeling work considering the core, the extended and the improved mechanisms. These three mechanisms consider the raw mechanism from RMG, the one with termination reactions and finally the final mechanism with updated thermochemistry respectively [23].

RMG provides mechanisms for which the thermochemistry can be updated with quantum chemical calculations for thermochemistry parameters. Major changes applied in this work for the extended mechanism to the thermochemistry impacted the equilibrium of key reactions such as the well-known ROO° + RH = ROOH + R° propagation step, that controls the system reactivity at these temperatures. To evaluate the modifications applied to the mechanism, the resulting rate constant for that key step is compared to Denisov's data [24] Figure 8a. The update of the thermochemistry parameters decreases the rate constant by one order of magnitude relative to the extended RMG mechanism, leading to a better agreement with experimental data. This contributes to the fact that the updated mechanism reproduces within less than an order of magnitude the IP obtained with the PetroOxy reactor.

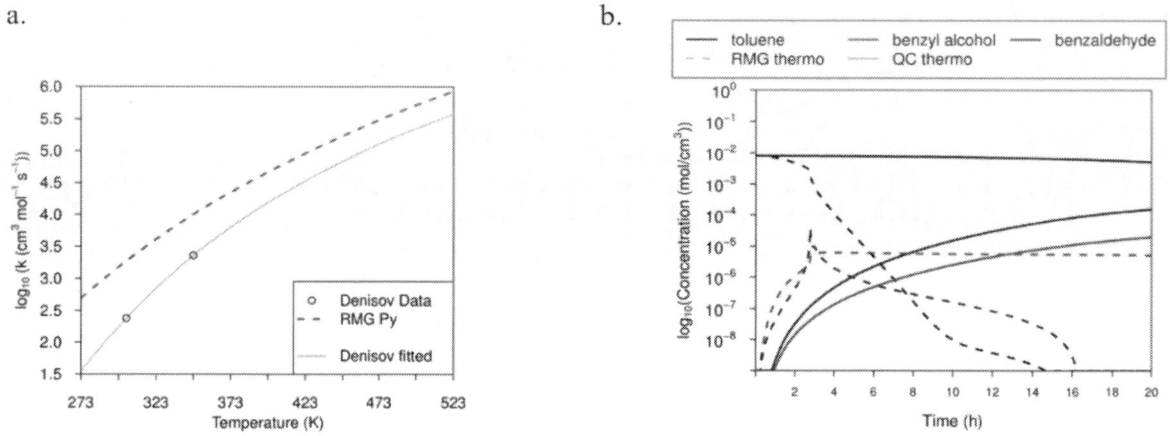

Figure 8: Comparison between the « core mechanism » and the « improved mechanism » for (a) a key propagation reaction for autoxidation systems: ROO°+RH=ROOH+R° where R refers to benzyl radical; (b) main species profiles simulated [23].

Figure 8b. illustrates the effect of the mechanism update on two major species profiles: benzaldehyde and benzyl alcohol. A higher concentration of benzaldehyde is observed compared to benzyl alcohol providing a better agreement with the experimentally observed trend in previous studies [25, 26]. In order to go further into the mechanism evaluation, a dedicated experimental device for liquid sampling is used and the results are described below.

3.2.2 Reactant and Intermediate products analysis

A systematic and robust sampling and analysis method has been applied with the autoclave experiments. Figure 9a. shows the pressure signals obtained during the toluene autoxidation for a repeated experiment. Figure 9b. illustrates the color change of the various samples obtained up to 46 h. A slightly yellow solution after a few hours then evolves to a dark brown fluid. The pressure signal highlights three different stages starting with a dramatic pressure decrease occurring shortly after the introduction of oxygen into the reactor. Then an approximately 3h plateau is observed prior to a slow decrease until the end of the experiment. Even for the longest experiment run which (120 h) oxygen remained into the gas phase. This was identified thanks to an oxygen pressure gauge connected to the reactor at the end of the experiment. The absolute value could not be determined but more than 20 %v/v were present. While the mass transfer between the two phases cannot be elucidated in this work, it is expected that the pressure signal does not only account for oxygen depletion into the liquid. Indeed, toluene reactivity does produce a large range of species including lighter molecules contributing to a pressure increase. A systematic 20 %wt loss is observed when the mass balance is considered indicating that toluene and oxidation products vaporization is significant. Consequently, it is rather difficult to conclude on this pressure profile but the good repeatability of the process offers the possibility to investigate products build up into the liquid. This will help understanding the liquid phase reactivity during this process and provide some hypothesis regarding the pressure trace. A multiphase model would be necessary in order to investigate further the effect of the mass transfer during the test.

a.

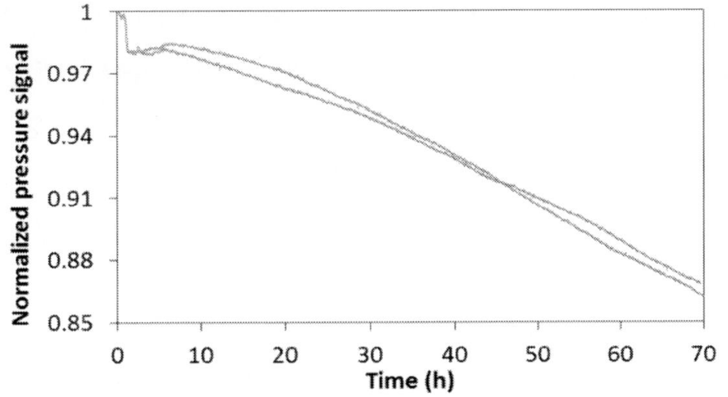

b.

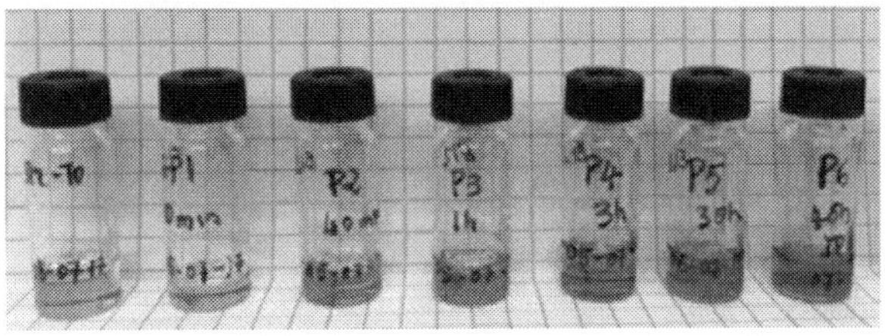

Figure 9: (a) pressure trace recorded during a toluene autoxidation test at 160 °C in the autoclave and (b) illustration of the sample color change during the test.

Major species identified and quantified with GC/MS and FID respectively are presented Figure 10. As expected, three oxygenated species are produced: benzaldehyde, benzyl alcohol and benzoic acid. Two repetitions were performed in order to characterize the robustness of the experimental method. The results demonstrate a good repeatability and emphasize the major differences between these three products. During the first hour, an important oxidation is occurring and this is related to the dramatic pressure decrease observed above. Benzaldehyde remains the most important product in this condition. While benzyl alcohol is produced simultaneously, its concentration remains lower initially and a plateau is then suggested considering the data collected at 120 h. This is of interest compared to the benzoic acid signal which first indicates that its production occurs later around two hours. In addition, at a longer reaction time, its concentration increases to reach almost 1.8 % which is about twice the amount of alcohol concentration.

a.

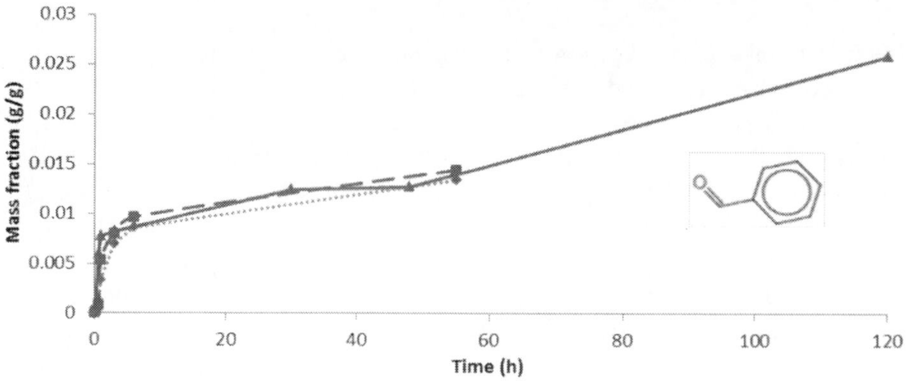

b.

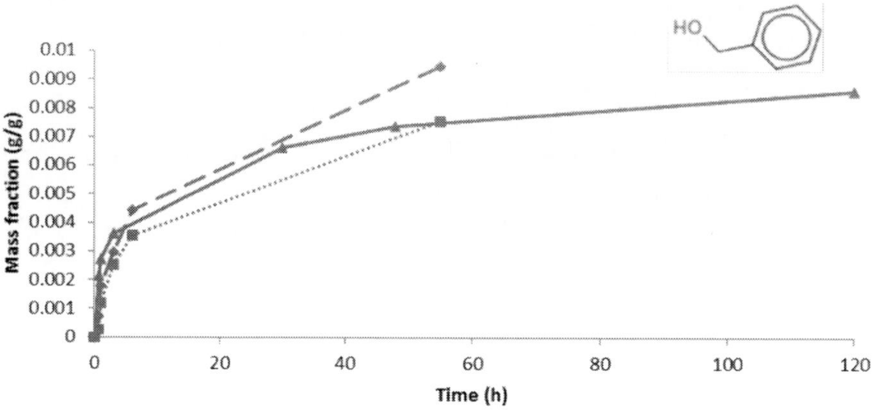

c.

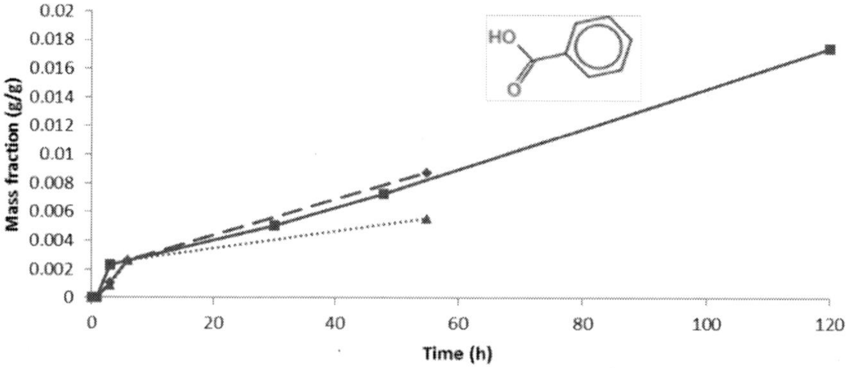

Figure 10: Major species profiles analyzed during the autoxidation of toluene in the autoclave at 160 °C up to 120 h.

In addition to the oxidation product build up, an illustration of the toluene mass fraction during a test is reported in Figure 11. It highlights the fast consumption of toluene at early reaction times followed by a slow consumption. This variation is qualitatively in agreement with the pressure trace. It is interesting to mention a very similar mass fraction at intermediate reaction time around 50 h and at 120 h. While the oxidation products continue to increase, this stabilization suggest a strong interaction with the gas phase.

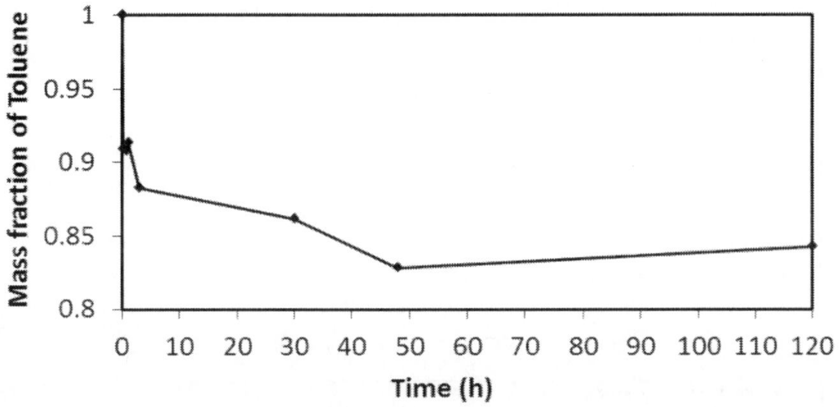

Figure 11: Toluene mass fraction analyzed into the liquid for a test duration of 120 h at 160 °C.

In order to evaluate the detailed kinetic mechanism capabilities regarding the intermediate products formation, a simulation is performed at 160 °C and the comparison between the model and experimental data is illustrated in Figure 12.

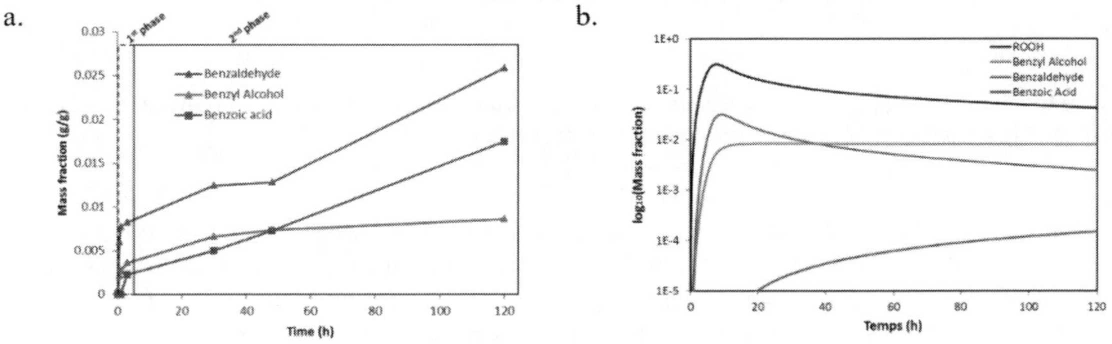

Figure 12: (a) experimental species profiles analyzed (a) and simulation results in the same thermodynamic conditions.

As mentioned earlier, the formation pattern of major autoxidation products can be decomposed into two steps: first an important concentration increase of all species within the first hour and second a slower increasing trend especially for benzyl alcohol which seems to plateau at long reaction time. The modeling results plotted in logarithmic scale, highlight an important contribution of benzylhydroperoxide which is not consistent with the experimental data. Due to a weak oxygen bond energy, this species is expected to decompose within the analysis device. Its decomposition is expected to lead to the formation of benzyl alcohol and benzaldehyde [15] and this could significantly impact the product branching ratio observed experimentally. Further work is currently ongoing to characterize via a quantum chemical approach the expected production branching ratio to evaluate this effect with the detailed kinetic mechanism. Finally, minor species were identified by GC/MS in order to investigate potential deposits precursors that would be of interest for further studies. Figure 13 illustrates the minor species profiles identified in this work.

a.

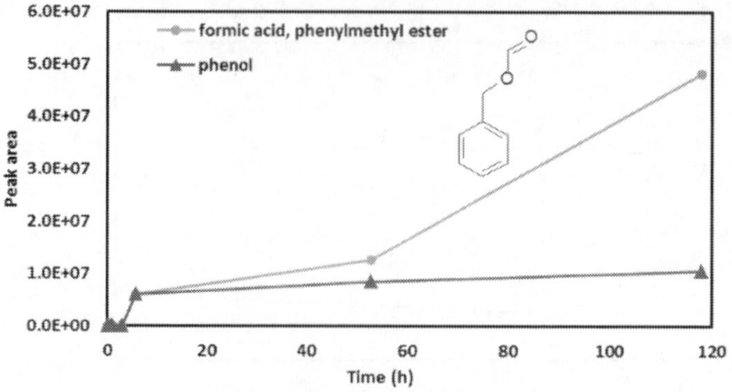

b.

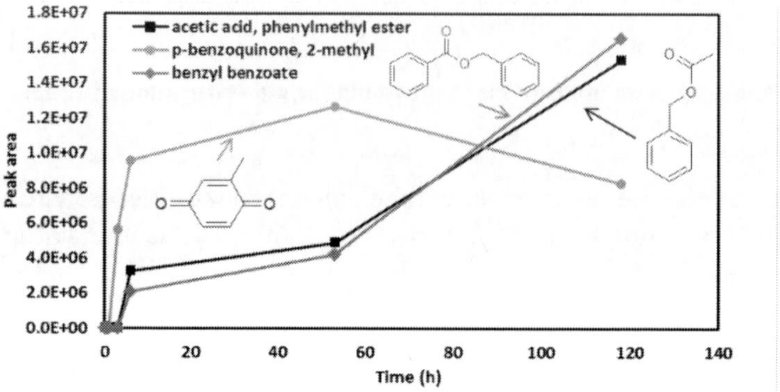

Figure 13: Minor species profiles analyzed by GC/MS from the liquid samples obtained in the autoclave at 160°C and for 120h.Intensities refer to peak area measured.

Interestingly, at early reaction time, a quinone is identified. This class of organic compound is targeted as a potent electrophile and could be involved in catalytic deposit formation for jet fuel [27]. Within five hours, recombination products are observed with the formation of benzyl benzoate but also formic acid and acetic acid phenylmethyl esters. This suggests the formation of low molecular weight acids with a mechanism that has not been elucidated so far.

4.0 CONCLUSIONS AND PERSPECTIVES

This study focuses on the liquid phase oxidation of aromatics as key components of aviation fuels. In a first part, the impact of aromatics on the thermal-oxidation stability of commercial jet fuels is studied using Jet fuel Thermal stability Oxidation Tester (JFTOT) associated with quantitative deposit thickness measurement by ITR. Results obtained from the thermal stability oxidation tests show that the addition of 8 %v/v of a mono aromatic solvent reduces the thermal oxidation stability of the conventional jet A-1 (deposit thickness increases by a factor of 7 at 280 °C). Besides, the addition of 1.6 %v/v of a polyaromatic compound (1-methylnaphthalene) resulted also in a higher deposit formation (deposit thickness increases by 80 % at 280 °C). These results suggest an important role of aromatics in promoting the thermal oxidative degradation of jet fuels.

In order to better understand the liquid phase reactivity of a single-aromatic, the second part of this study proposes a combined experimental and modeling approach to assess the

autoxidation kinetics of toluene as a model compound for aromatics in jet fuels. First, the experimental side addresses the characterization of the temperature dependence for the induction period measured in a PetroOxy apparatus which is a standard test for comparing fuel stability. In parallel, a dedicated setup consisting of a closed batch reactor pressurized with pure oxygen and mixed is used. Time resolved species profiles are obtained combining gas chromatography coupled with mass spectrometry and flame ionization detector.
The modeling of this phenomenon is enabled through the development of a detailed kinetic mechanism first developed automatically with RMG and then improved in terms of thermochemical parameters with the help of quantum chemical calculations. The final mechanism is able to reproduce within an order of magnitude the induction period and also comprises the key pathways for the main autoxidation products experimentally observed: benzaldehyde, benzyl alcohol and benzoic acid. Qualitatively the major trends are well simulated while a major question remain concerning the formation and consumption of an important but unstable intermediate: benzylhydroperoxide. Indeed, the modeling work emphasizes its impact at early reaction times but due to its poor stability it is likely that none of the analytical techniques employed here are adapted to observe this species. Further work is necessary on this aspect to characterize the role of hydroperoxide reactivity in this system.
Finally, the identification of minor species through mass spectrometry highlights the formation of a potent electrophile compound: 2-methyl-p-benzoquinone. Current development are performed to extend the capability of the modeling tool to characterize its effect into the liquid which could be of interest regarding jet fuel deposit formation.

5.0 ACKNOWLEDGMENTS

We would like to thank Karl Chatelain, André Nicolle as well as Anthony Velghe for their contribution to the modeling work and Benjamin Veyrat, Pascal Hayrault for their contribution to the experimental work. Finally, we are grateful to the RMG team at the Massachusetts Institute of Technology for the fruitful discussions regarding the mechanism generation and the thermochemistry update as well as the ORCA developers for providing an early pre-release version of the software.

Part of this work was carried out within a French national research program entitled MOCCASSIN supported by French Directorate-General for Civil Aviation (DGAC).

REFERENCES

[1] ASTM, "ASTM D7566-16, Standard Specification for Aviation Turbine Fuel Containing Synthesized Hydrocarbons," 2016.

[2] Hemighaus, G., Boval, T., Bacha, J., Barnes,, Fred et al., "Aviation Fuels Technical Review," https://www.cgabusinessdesk.com/document/aviation_tech_review.pdf, 2007.

[3] Ben Amara, A., Klopffer, M.-H., Alves-Fortunato, M., and Starck, L., "Composition optimale de carbureacteur stabilite thermique et a l'oxydation ameliorее," INPI 15/58.941, 2015.

[4] Ben Amara, A., Kaoubi, S., and Starck, L., "Impact of Added Cyclic Compounds on The Oxidative And Thermal Stability Of HEFA: Towards A Fit-For-Purpose Formulation," *International Conference on Stability Handling and Use of LIquid Fuels*, 2015.

[5] Ben Amara, A., Kaoubi, S., and Starck, L., "Toward an optimal formulation of alternative jet fuels: Enhanced oxidation and thermal stability by the addition of cyclic molecules," *Fuel* 173:98–105, 2016, doi:10.1016/j.fuel.2016.01.040.

[6] Taylor, W.F., "Jet fuel thermal stability," 1978.

[7] Watkinson, A. and Wilson, D.I., "Chemical reaction fouling: A review," *Exp. Therm Fluid Sci.* 14(4):361–374, 1997, doi:10.1016/S0894-1777(96)00138-0.

[8] Wong, O:, "Design and Development of an Apparatus to Study Aviation Jet Fuel Thermal Stability," Thesis Manuscript, Institute for Aerospace Studies University of Toronto, 2010.

[9] Morris, R.E. and Wagner, R., "Interferometric JFTOT tube deposit measuring device," US Patent No5,293,218, 1994.

[10] Didier Pigeon, "Automated Instrument to Quantify D3241 Absolute Deposit Using Interferometry (Interference Method)," *IASH*, 2013.

[11] ASTM International West Conshohocken, "Standard Test Method for Oxidation Stability of Middle Distillate Fuels—Rapid Small Scale Oxidation Test (RSSOT),," ASTM D7545-14, Rev. 2014.

[12] Laurie Starck, Mickael Sicard, Frederic Ser, and Nicolas Jeuland (eds.), "Potential of alternative Fuels for aircraft: Focus on thermal and oxidation stability, IASH Conference," 2009.

[13] Mickael Sicard, Jean Boulicault, Kevin Coulon, Charlotte Thomasset et al., "Oxidation stability of jet fuel model molecules evaluated by rapid small scale oxidation tests," *IASH, the International Conference on Stability, Handling and Use of Liquid Fuels*.

[14] Araujo, S.V., Rocha, B.S., T. Luna, F., Rola Jr., E.M. et al., "FTIR assessment of the oxidation process of castor oil FAME submitted to PetroOXY and Rancimat methods," *Fuel Processing Technology* 92(5):1152–1155, 2011, doi:10.1016/j.fuproc.2010.12.026.

[15] Chatelain, K., Nicolle, A., Ben Amara, A., Catoire, L. et al., "Wide Range Experimental and Kinetic Modeling Study of Chain Length Impact on n -Alkanes Autoxidation," *Energy Fuels*, 2016, doi:10.1021/acs.energyfuels.5b02470.

[16] Gao, C.W., Allen, J.W., Green, W.H., and West, R.H., "Reaction Mechanism Generator: Automatic construction of chemical kinetic mechanisms," *Computer Physics Communications* 203:212–225, 2016, doi:10.1016/j.cpc.2016.02.013.

[17] Jalan, A., West, R.H., and Green, W.H., "An Extensible Framework for Capturing Solvent Effects in Computer Generated Kinetic Models," *The Journal of Physical Chemistry B* 117(10):2955–2970, 2013, doi:10.1021/jp310824h.

[18] Mintz, C., Burton, K., Acree, W.E., and Abraham, M.H., "Enthalpy of Solvation Correlations For Gaseous Solutes Dissolved in Linear Alkanes (C5 C16) Based on the Abraham Model," *QSAR Comb. Sci.* 27(2):179–186, 2008, doi:10.1002/qsar.200730040.

[19] Goerigk, L. and Grimme, S., "A thorough benchmark of density functional methods for general main group thermochemistry, kinetics, and noncovalent interactions," *Physical chemistry chemical physics : PCCP* 13(14):6670–6688, 2011, doi:10.1039/c0cp02984j.

[20] Marenich, A.V., Cramer, C.J., and Truhlar, D.G., "Universal solvation model based on solute electron density and on a continuum model of the solvent defined by the bulk dielectric constant and atomic surface tensions," *The journal of physical chemistry. B* 113(18):6378–6396, 2009, doi:10.1021/jp810292n.

[21] Ben Amara, A., Nicolle, A., Alves-Fortunato, M., and Jeuland, N., "Toward Predictive Modeling of Petroleum and Biobased Fuel Stability: Kinetics of Methyl Oleate/ n - Dodecane Autoxidation," *Energy Fuels* 27(10):6125–6133, 2013, doi:10.1021/ef401360k.

[22] Matthew J. DeWitt, Zachary West, Steven Zabarnick, Linda Shafer et al., "Effect of Aromatics on the Thermal-Oxidative Stability of Synthetic Paraffinic Kerosene," *Energy & Fuels* 28(6):3696–3703, 2014, doi:10.1021/ef500456e.

[23] Mielczarek, D.C., Matrat, M., Amara, A.B., Bouyou, Y. et al., "Toward the Accurate Prediction of Liquid Phase Oxidation of Aromatics: A Detailed Kinetic Mechanism for Toluene Autoxidation," *Energy & Fuels* 31(11):12893–12913, 2017, doi:10.1021/acs.energyfuels.7b00416.

[24] Denisov, E.T. and Afanas'ev, I.B., "Oxidation and antioxidants in organic chemistry and biology," CRC Press, Boca Raton, FL, ISBN 9780824753566, 2005.

[25] Hoorn, J., Alsters, P.L., and Versteeg, G.F., "A Kinetic Model for Toluene Oxidation Comprising Benzylperoxy Benzoate Ester as Reactive Intermediate in the Formation of Benzaldehyde," *International Journal of Chemical Reactor Engineering* 3(1), 2005, doi:10.2202/1542-6580.1196.

[26] Hermans, I., Peeters, J., Vereecken, L., and Jacobs, P.A., "Mechanism of thermal toluene autoxidation," *Chemphyschem : a European journal of chemical physics and physical chemistry* 8(18):2678–2688, 2007, doi:10.1002/cphc.200700563.

[27] Beaver, B., Gao, L., Burgess-Clifford, C., and Sobkowiak, M., "On the Mechanisms of Formation of Thermal Oxidative Deposits in Jet Fuels. Are Unified Mechanisms Possible for Both Storage and Thermal Oxidative Deposit Formation for Middle Distillate Fuels?," *Energy Fuels* 19(4):1574–1579, 2005, doi:10.1021/ef040090j.

IASH 2017, the 15TH INTERNATIONAL SYMPOSIUM ON STABILITY, HANDLING AND USE OF LIQUID FUELS
Rome, Italy
10-15 September 2017

WALL ROUGHNESS EFFECTS ON DEPOSITION OF THERMALLY STRESSED AVIATION FUEL

Phil Gadsby and Simon G. Blakey[1]

The University of Sheffield, Department of Mechanical Engineering, Sir Fredrick Mappin Building, Mappin Street, Sheffield, S1 3JD, UK, mea08pg@sheffield.ac.uk

Thermal instability in aviation fuels has been quite thoroughly explored over the last 50 years. The problem is somewhat complex, with coupling of chemistry, heat transfer and fluid dynamics. Most efforts have been applied to the chemical kinetics of deposit formation and studies of physical effects such as temperature, flow rate and Reynolds number in a multitude of small to large scale testing devices. Much less attention has been paid to the effects of wall surface roughness. This is surprising, since for turbulent flows, wall roughness enhances heat, momentum and mass transfer between the wall and the free stream, reduces the extent of the laminar portion of the boundary layer, modifies the boundary layer velocity profile and depending on the type of roughness, interferes with turbulence structures responsible for transfer of particles to and from the wall. Furthermore, a rough surface increases the wall surface area, potentially presenting more active sites for metal catalytic activity and increasing fluid residence time in and around roughness troughs.

Two types of roughness were investigated - principally the roughness created from additive manufacturing (an important consideration for engine manufacturers) and drawn microscale tubing. In some cases with high relative roughness (around 4%) and turbulent flow, the deposition rate was doubled with respect to the 'smooth' case. However for other cases, a minimal effect was observed.

IASH 2017, the 15th INTERNATIONAL SYMPOSIUM ON STABILITY, HANDLING AND USE OF LIQUID FUELS
Rome, Italy
10-14 September 2017

GAINING A FUNDAMENTAL UNDERSTANDING OF FUEL PERFORMANCE THROUGH ADVANCED CHEMICAL COMPOSITION MEASUREMENTS

Robert E. Synovec[1], Chris E. Freye[2], and Matthew C. Billingsley[3]

[1] *The University of Washington, Seattle, WA, USA, synovec@chem.washington.edu*
[2] *The University of Washington, Seattle, WA, USA, freyec@uw.edu*
[3] *Air Force Research Laboratory/RQRC, Edwards AFB, CA, USA, matthew.billingsley.1@us.af.mil*

ABSTRACT

Kerosene-based jet and rocket fuels such as Jet A, JP-8, RP-1, and RP-2 are ubiqitous to the aerospace propulsion industry. There is an ongoing need to more fully assess the fuel composition (i.e., specific chemical compounds and compound classes present as a result of refining and feedstock blending) for the purpose of improving quantitative connections between fuel composition, properties, and ultimately system performance. Prediction of fuel performance through modeling puts a greater emphasis on obtaining detailed and accurate fuel property and compositional information. In order to optimally glean chemical composition information from complex fuels, we apply comprehensive two-dimensional gas chromatography with time-of-flight mass spectrometry (GC × GC – TOFMS). By applying multivariate data analysis methods, referred to as chemometrics, we have readily correlated the chemical information for a broad compositional range of distillate or multicomponent aerospace fuels based upon their GC × GC – TOFMS data, to data obtained from the same fuels using ASTM standard measurement methods that infer fuel performance properties (e.g., density, viscosity, and heat of combustion). The chemometric method referred to as partial least squares (PLS) analysis was applied. With an extremely large dataset composed of 72 diverse fuels, reliable PLS models are produced, that relate the chemical information provided by GC × GC – TOFMS measurements of each fuel to the physical properties of each fuel (density, viscosity, and heat of combustion). Only 61 of the 72 fuels were modeled via PLS due to the occurrence of chromatographic overload in 11 "outlier" fuels. These outlier fuels also expressed significant and atypical variation in the chemical compound classes: alkanes (paraffins), cyclic alkanes (cycloparaffins), and aromatics (including aromatics either without or with heteroatoms such as S, N and O), relative to the other 61 fuels.

INTRODUCTION

Chemical composition of kerosene-based fuels is known to affect the fuel properties and performance.[1-4] Traditionally, multicomponent fuels, including middle distillate (kerosene) fractions and blends thereof, are analyzed for known impurities and hydrocarbon classes (e.g. sulfur, olefin, oxygenates, and aromatic compounds).[5] Indeed it is a challenging endeavor to achieve precise control over the chemical composition of multicomponent distillate fuels due to variations in crude oil composition and place of origin, refinery and post-refinery conditions, or even the date and time the material was refined, treated, and formulated to meet the detail specification requirements. Gaining a better understanding between fuel composition and its relationship to performance and physical properties is vital for the design and analysis of energy conversion devices, including those used in aerospace propulsion. Indeed, much work has been performed aimed to establish correlations between the variation in fuel composition and fuel specifications.[6-16] In order to provide a sound connection between fuel composition and performance and physical properties, it is beneficial to evaluate laboratory blends, where the analyst has relative control over chemical composition, as well as to assess the performance of "field" fuels.[6-10, 12-16] However, as the number of fuels analyzed increases, a reliable and straightforward method is needed in order to glean significant information while keeping analysis time short.

Gas chromatography (GC) is a traditional analytical technique that is amenable for the separation and analysis of volatile and semi-volatile mixtures.[17,18] When GC is coupled with mass spectrometry (MS), spectral information can be gathered allowing for further selectivity and the ability to identify chemical compounds. GC-MS has been shown to be a powerful tool for the analysis of kerosene-based fuel.[6-11] Comprehensive two-dimensional (2D) gas chromatography coupled with time-of-flight mass spectrometry (GC × GC – TOFMS) can further improve upon the separation power of one dimensional GC and provide additional insight into complex mixtures of volatile compounds such as those present in kerosene-based fuels.[13-15, 19-21] Traditionally in GC × GC, the first separation dimension uses non-polar stationary phase while the second dimension uses a polar stationary phase. However, a reverse-column configuration (polar first dimension and non-polar second dimension) has been shown to provide better selectivity for petroleum based samples.[13-16] A potential drawback in the application of GC × GC – TOFMS is the sheer magnitude of information that is produced in a single analysis of fuels. This data analysis challenge is compounded by injection replicates for a large set of fuels. Yet, the challenge of obtaining meaningful information can be overcome through application of powerful chemometric software methods which can aid in the interpretation of such complex data sets.

Partial least squares (PLS) analysis is a chemometric method that associates the differences in measurable information for two different data sets. For example, PLS has been used to associate the chemical information obtained from GC × GC – TOFMS and GC × GC – FID to physical properties of kerosene-based fuels.[13-16] Details about PLS can be found elsewhere.[22-25] Briefly, PLS analysis mathematically relates, via linear algebra, two data matrices (X and Y block) through calculation of loadings referred to as latent variables (LVs). Using PLS, in the study herein, models are constructed to account for the variance (ideally, the relevant chemical differences) in both the GC × GC – TOFMS data for a fuel sample set, i.e., the signal intensities (which constitute the X-block) and the respective measured property values, i.e., the ASTM data for the same fuel sample set (which constitute the Y-block). A key output of PLS modeling is a one-to-one correspondence between the *measured* ASTM values relative to the *predicted* ASTM values based upon the relationship to the chemical composition encoded in the GC × GC – TOFMS data.

Herein, we use an expanded fuel set of 72 kerosene-based fuels (i.e., jet and rocket fuels) with varying specifications (e.g. Jet A, JP-5, JP-8, RP-1, RP-2, etc.) in order to gain further insight into the relationship between fuel properties and chemical composition. Table 1 summarizes all the samples. Using PLS, the chemical information obtained from GC × GC – TOFMS is correlated to the measured physical properties of viscosity and heat of combustion, as well as density at 15, 45 and 85 °C, respectively. An important goal is to expand upon the previous work to further gain insight and create robust PLS models with a large, diverse set of fuels.[15,16]

EXPERIMENTAL

ASTM Measured Values

The physical properties for the fuels were obtained in triplicate and the average value was used in the PLS modeling. The complete list of physical and compositional properties measured and methods used to obtain values are the following: viscosity (cSt) at 40 °C according to ASTM D445/D446; heat of combustion (btu/lbm) according to ASTM D4809 using a Parr Instruments Model 6200 calorimeter; and density (g/ml) at 15, 45, and 85 °C according to ASTM D4052 using a Rudolph Research Analytical DDM2911 density meter.

GC × GC Analysis

To investigate the chemical composition of the fuels a GC × GC – TOFMS instrumental platform was used consisting of an Agilent 6890N GC (Agilent Technologies, Palo Alto, CA, USA), a thermal modulator (4D upgrade, LECO, St. Joseph, MI, USA), and a Pegasus III TOFMS (LECO, St. Joseph, MI, USA). Aliquots of the fuel samples were introduced to the GC × GC–TOFMS instrument via a 7683B auto-injector (Agilent Technologies, Palo Alto, CA, USA). The auto-injector was set to inject 1 µl of sample at a 200:1 split at an inlet temperature of 275 °C. Prior to injection, HPLC grade acetone and hexane (Fisher Scientific) were used as solvent rinses. The primary GC × GC column (^{1}D) was a Rxi-17Sil MS: 29.5 m × 250 µm inner diameter (i.d.) × 0.25 µm film thickness, and the secondary GC × GC column (^{2}D) was a Rxi-1MS 1.5 m × 180 µm I.D. × 0.25 µm film thickness. Ultrahigh purity helium (Grade 5, 99.999%, Praxair, Seattle, WA, USA) was used as the carrier gas at a constant flow rate of 2.0 ml/min. The primary ^{1}D oven was held at 40 °C for 1.5 min before being ramped at 5 °C/min to 200 °C where it was held for 1 min. The secondary oven was held at a +12 °C offset relative to the primary oven and the modulator block was held at a +30 °C offset to the primary oven. The modulation period was 3 s (separation run time of the ^{2}D column) with 0.75 s hot and cold pulses for each stage. The transfer line was set to 285 °C and the ion source was 225 °C. Mass channels, *m/z*, 35-334 at unit resolution were collected with an ionization voltage of 70 eV at 100 spectra/s after a 10 s acquisition delay. The 72 samples (Table 1) were analyzed with two GC × GC – TOFMS replicates for each sample.

Sample Number	Sample Name	Type
1	YA2921HW10	RP-2
2	BG1121GP04	RP-1
3	GRC/0-100 HEP	RP-1
4	WC0721HW01	RP-2
5	LB073009-05	RP-1
6	ZI1521HW10	RP-1
7	CG0721HW10	RP-2
8	LB073009-08	RP-1
9	BB0821HW10	RP-2
10	LB080409-05	RP-1
11	ZI2621HW01	RP-2
12	ZJ1321GP01	RP-2
13	RG3021LS06	UL-RP-1
14	RG3021LS05	RP-TS-5
15	POSF 3327	JP-7
16	POSF 4765	JP-900
17	LB073009-02	RP-1
18	112868	RP-1
19	VI2621LS01	RP-1
20	DB131014	RP-1
21	DC310925	RP-1
22	DC310923	RP-1
23	DB131013	RP-1
24	DB131015	RP-1
25	CL031236	RP-1
26	CB1121HW10	RP-2
27	EA130720	RP-1
28	EB220705	RP-1
29	CHC JP-5	JP-5
30	LB080409-01	RP-1
31	LB073009-01	RP-1
32	A0072256	RP-1
33	CL11-3089	D80
34	LB100413-40	RP-1
35	LB073009-03	RP-1
36	A0072256	RP-1
37	SA1421LS03	UL-RP-1
38	ED060739	RP-1
39	CB1121HW10	RP-2
40	LB073009-09	RP-1
41	LB073009-06	RP-1
42	XC2521HW10	RP-1
43	ZK0821HW20	RP-2
44	ZK2121HW10	RP-2
45	CL11-2928	JP-8
46	CL11-2929	IPK
47	CL11-2654	HRJ
48	CL12-3493	Solvent
49	CA2021HW10	RP-2
50	POSF9641	ATJ-2
51	41910	RP-1
52	B01001634-01	RP-1
53	POSF 4751	JP-8
54	POSF 10359	JP-8
55	POSF 10314	JP-8
56	POSF 10312	JP-8
57	PSOF 10316	JP-8
58	POSF 9326	Jet-A
59	POSF 10358	Jet-A
60	POSF 10311	Jet-A
61	POSF 10315	Jet-A
62	POSF 10369	Jet-A
63	POSF 10313	Jet-A
64	POSF 10337	JP-5
65	POSF 10325	Jet-A
66	POSF 10264	JP-8
67	POSF 10289	JP-5
68	POSF 9698	JP-8
69	POSF 3961	Jet-A
70	POSF 3896	JP-8
71	POSF 8535	Biojet
72	POSF 7724	Jet-A

Table 1. Summary of the fuel sample set analyzed by various ASTM methods and also by GC × GC – TOFMS.

Data Analysis

The GC × GC – TOFMS chromatograms were imported into MATLAB 2015b ((MathWorks, Natick, MA) using an in-house converting algorithm. The data was baseline corrected and binned by 4 data points (modulations) on ^{1}D and 15 data points (150 ms) on ^{2}D reducing the sample matrix from (600 ^{1}D data points × 300 ^{2}D data points × 300 *m/z*) to (150 ^{1}D data points × 20 ^{2}D data points × 300 *m/z*). This binning procedure mitigated any minor run-to-run GC retention time misalignment while also reducing computation time.[26] Additionally, the chromatographic data was mean centered. After the preprocessing steps, the chromatographic data was forwarded into the PLS analysis which was performed using PLS Toolbox 8.02 (Eigenvector Research Inc., Wenatchee, WA, USA). The following physical properties were predicted: kinematic viscosity (cSt), and net heat of combustion (btu/lb), and density (g/ml) at 15, 45, and 85 °C. For construction of the PLS models, all underwent Venetian blinds cross validation for testing the overall predictive ability of the models and to determine the correct number of latent variables (LVs). For the Venetian blinds cross validation, a PLS calibration model was prepared with each group of n - 4 fuels, and the given physical property for the remaining 4 fuels were predicted, where n is the total number of fuels being analyzed. For example, with the PLS analysis for 72 fuels, Samples 1,19, 37, 55 were left out of the calibration step but are then predicted, then Samples 2, 20, 38, 56 were left out of the calibration step and are then predicted, and so on in a round robin fashion. If n is not divisible by 4, then the first few groups of 4 excluded from the calibration step will have an additional fuel to make up for the difference (i.e. there will be one or more groups of 5, not to exceed 3 groups). For example, with the PLS analysis for 61 fuels, Samples 1, 16, 31, 46, 61 were left out of the calibration step and then predicted, then Samples 2, 17, 32, 47 were left out of the calibration step and then predicted, and so on. Normally leave-one-out cross validation (LOOCV) is implemented, but with a large dataset, this would result in long computation time and potential overfitting of the data. Linear regression vectors (LRVs), as defined in the PLS toolbox, were investigated to determine the relationship between the chemical composition and physical measurements. Analyte compounds that exhibit positive values in the LRV are correlated with increasing the specified property while negatively correlated analyte compounds exhibit negative values in the LRV and are correlated with decreasing the specified property [13-16, 22-25]. The GC × GC – TOFMS chromatogram replicates were analyzed in separate PLS models and are overlaid on the same graph for ease of visualization. In order to determine the "goodness-of-fit" for the PLS models, the root mean square error of cross validation (RMSECV) was calculated as defined by,

$$RMSECV = \left[\frac{1}{N} * \sum (y_{i,cv} - y_{i,meas})^2\right]^{0.5}$$

where N is the number of fuel samples, $y_{i,cv}$ is the cross validation predicted value of sample (i) and $y_{i,meas}$ is the measured value for the same sample. RMSECV can be normalized (NRMESCV) by dividing the RMESCV by the range of the $y_{,meas}$ values. In addition, principal component analysis (PCA) was performed on the three LRVs from the density measurements for the purpose of interrelating the three separate PLS models at 15, 45 and 85 °C.

RESULTS AND DISCUSSION

Using the GC × GC – TOFMS instrument in reversed column configuration, an excellent two-dimensional (2D) chromatographic separation of the fuels was achieved. Figure 1A shows the 2D separation of a representative fuel, Sample 72: POSF 7724. Figure 1B shows the major compound classes highlighted (if present for a given fuel): paraffins (referred to alkanes herein), cycloparaffins (referred to as cycloalkanes), and aromatics (including aromatics either without or with heteroatoms such as S, N and O). Figure 1C shows the separation of Sample 72 if it had been run using a GC – TOFMS. Figure 1C was prepared by summing the 2D separation provided in Figure 1A onto the first dimension, hence losing the second dimension separation information. Because each GC × GC chromatogram was collected with a multivariate detector (TOFMS), it is possible to select selective *m/z* that are representative of different chemical compound classes that exist in the fuels. To illustrate this point, Figure 2A-F shows several analytical ion chromatograms (AIC) for Sample 72. This figure demonstrates the power of using selected *m/z* (as desired) coupled with the 2D separation advantage to isolate and possibly quantify the various hydrocarbon classes present in these fuels.

The locations of the multitude of analyte compounds in conjunction with the mass spectra are used by the PLS in order to accurately and precisely predict the physical properties. For completion and reference, Figure 3 (with reference to Table 1 numbering) provides the 2D chromatograms (TIC signal) of all 72 fuel samples in a condensed form. Each fuel 2D separation image in Figure 3 can be readily compared to the others in the context of the chemical compound classes in Figure 1B. This comparison provides insight into which compound classes are present (or not) in each fuel. Note that some of the fuels appear to be dramatically different in chemical composition just based on visual comparison of the compound classes present and the relative emphasis of each compound class. Some of the fuels stand out upon visual inspection as being "chemically" very different than a majority of the fuels: 16, 20, 24, 25, 46, 47, 48, 50. Indeed, some of the fuels appear to be comprised of a relatively small number of chemical compounds present at high concentration. We note that several of these fuels possessing atypical composition were included in the sample set for the purpose of evaluating the extent of model validity, and were by no means intended to represent "drop in" replacements for current multicomponent fuels. They may, however, be considered as fuel blend constituents. These fuels as well as a few others will turn out to be troublesome in PLS modeling, but can be handled with statistical rigor.

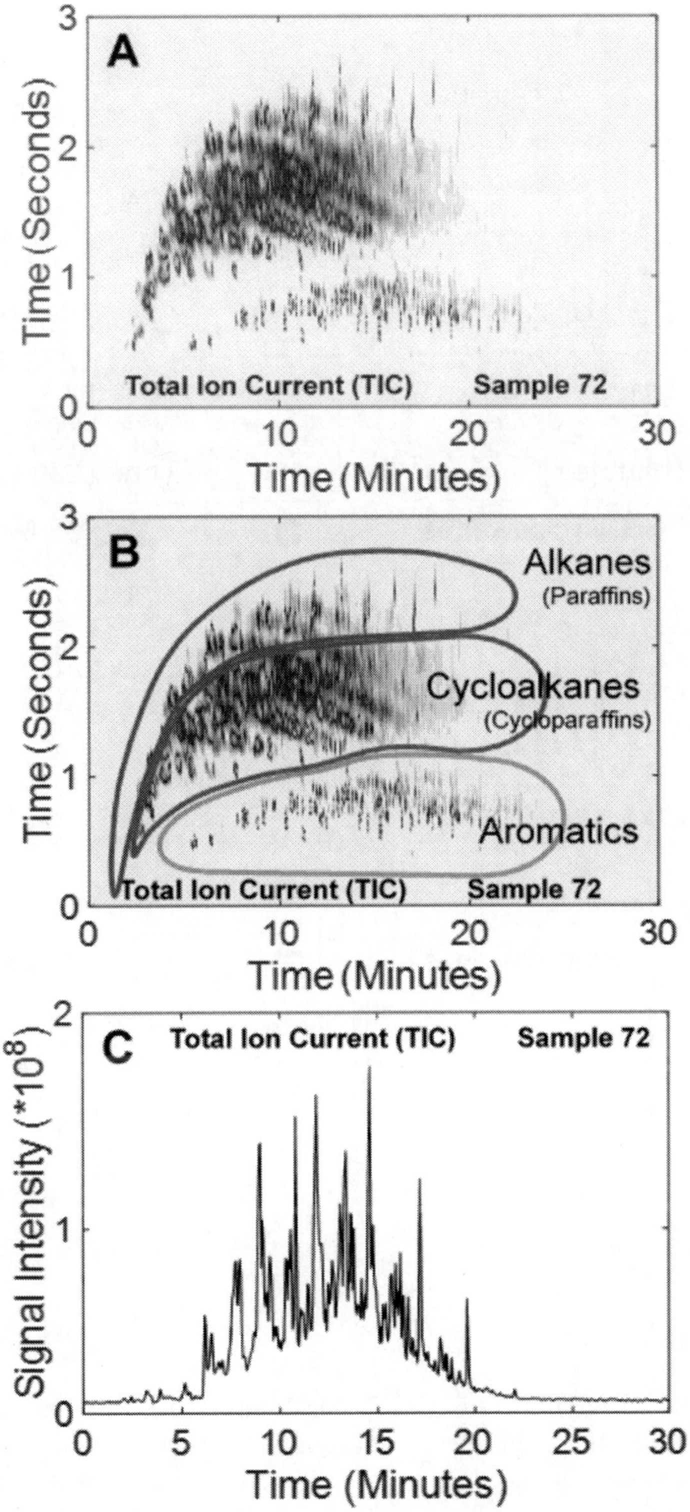

Figure 1. (A) Representative GC × GC – TOFMS chromatogram for the fuel sample set (Sample 72: POSF 7724). (B) The three major chemical classes (alkanes, cycloalkanes, and aromatics) are highlighted for Sample 72. (C) One dimensional GC separation of Sample 72. This figure was prepared by summing the second dimension separations onto the first dimension

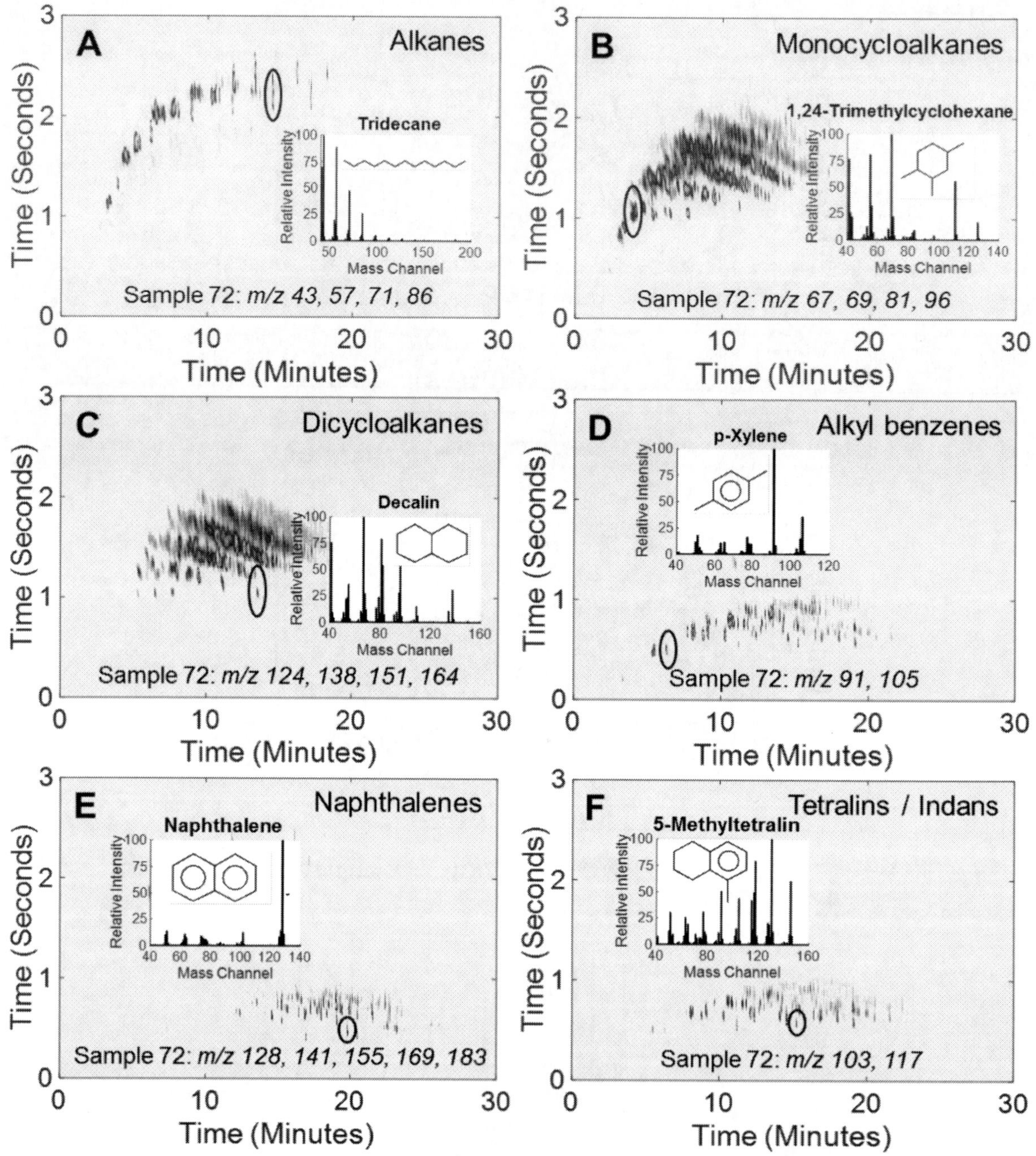

Figure 2. Analytical ion chromatograms (AIC) for Sample 72 showing the ability to use selective *m/z* to identify the chemical compound classes (i.e., monocycloalkanes, dicycloalkanes, alkyl benzenes, etc.). Within each figure is a mass spectrum of an arbitrarily selected analyte compound which has been circled. (A) AIC of Sample 72 for alkanes. (B) AIC of Sample 72 for monocycloalkanes. (C) AIC of Sample 72 for dicycloalkanes. (D) AIC of Sample 72 for alkyl benzenes (E) AIC of Sample 72 for naphthalenes. (F) AIC of Sample 72 for tetralins and indans.

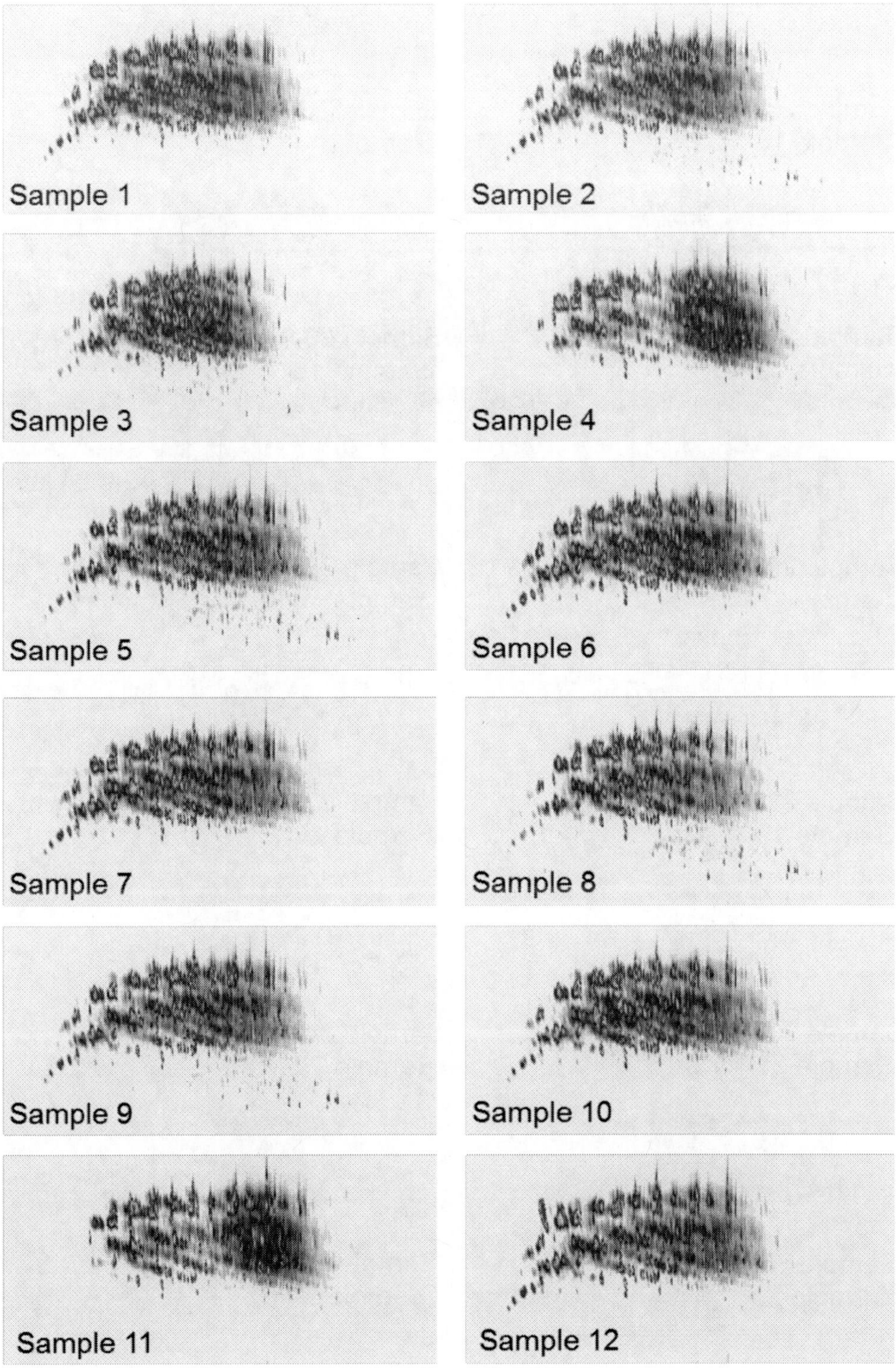
Sample 1
Sample 2
Sample 3
Sample 4
Sample 5
Sample 6
Sample 7
Sample 8
Sample 9
Sample 10
Sample 11
Sample 12

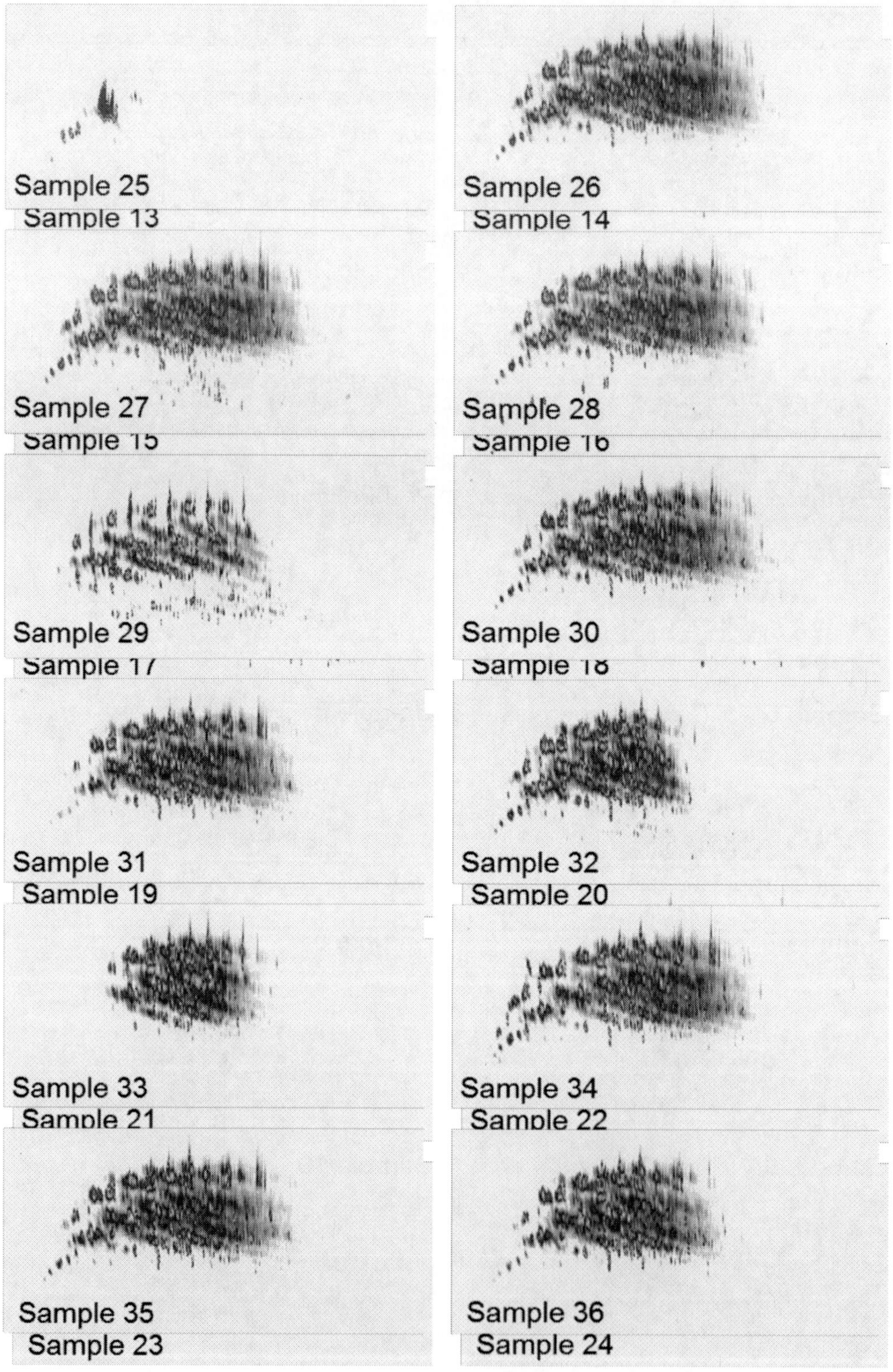
Sample 25
Sample 13
Sample 26
Sample 14
Sample 27
Sample 15
Sample 28
Sample 16
Sample 29
Sample 17
Sample 30
Sample 18
Sample 31
Sample 19
Sample 32
Sample 20
Sample 33
Sample 21
Sample 34
Sample 22
Sample 35
Sample 23
Sample 36
Sample 24

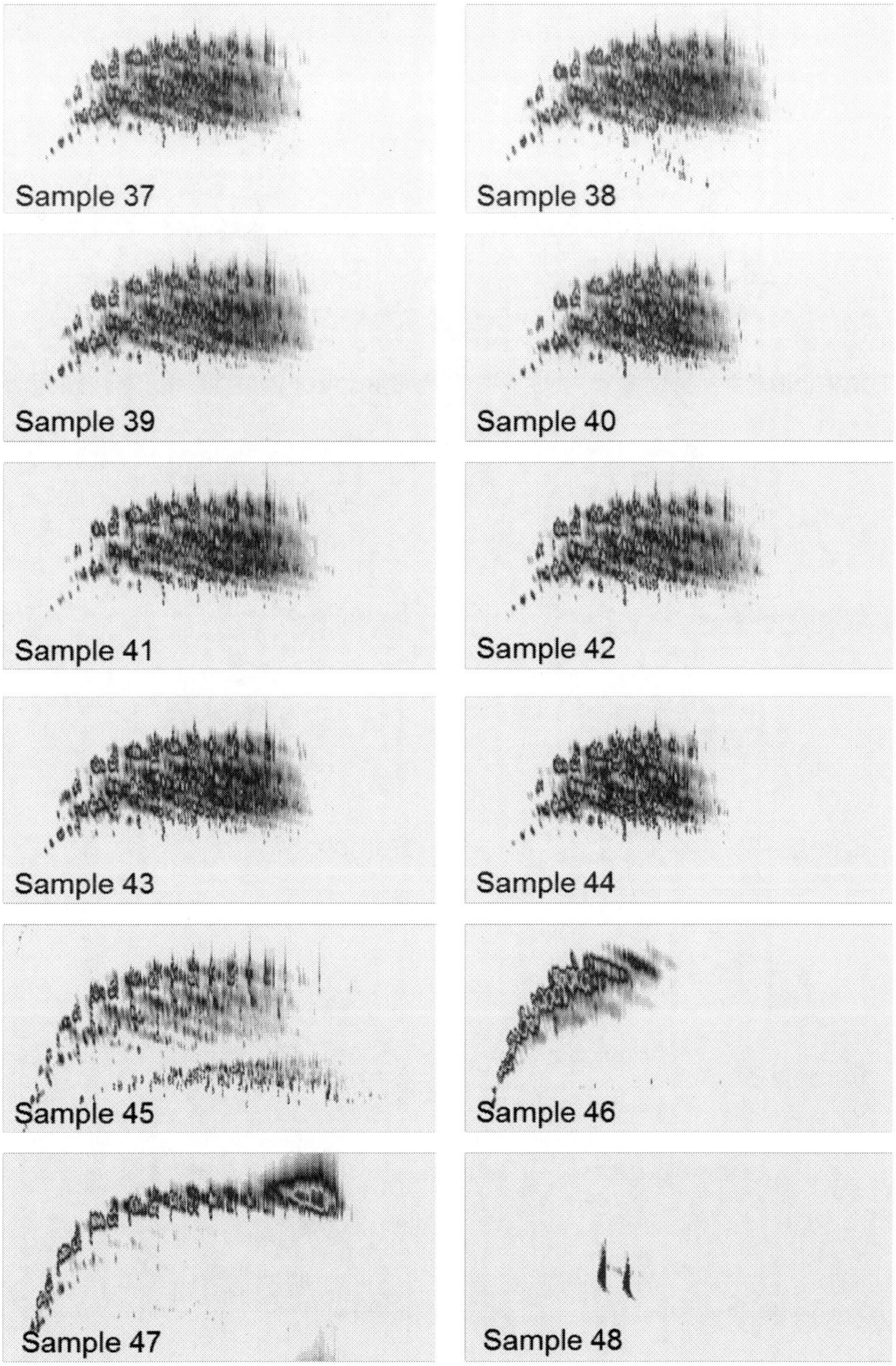
Sample 37
Sample 38
Sample 39
Sample 40
Sample 41
Sample 42
Sample 43
Sample 44
Sample 45
Sample 46
Sample 47
Sample 48

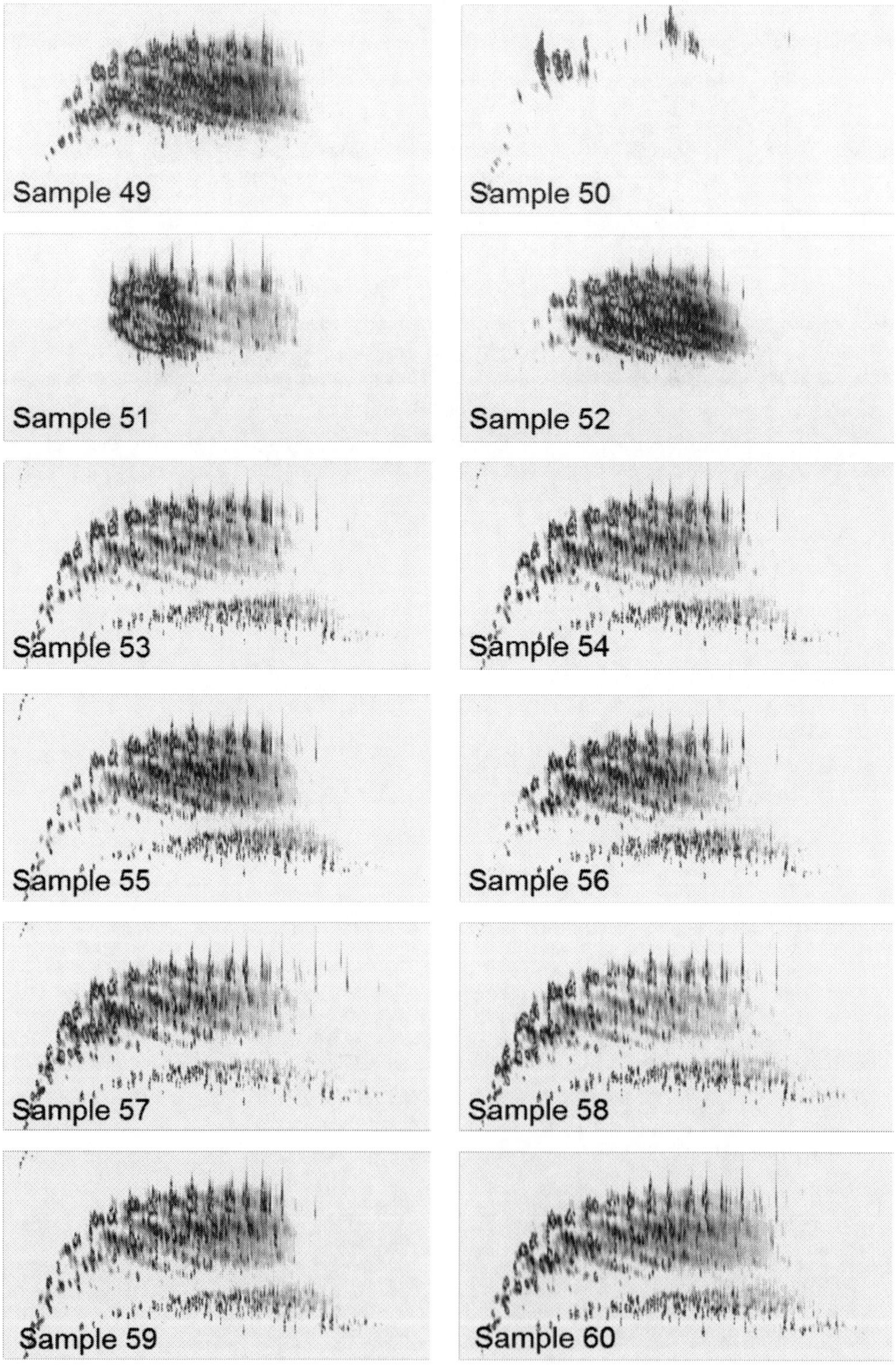
Sample 49
Sample 50
Sample 51
Sample 52
Sample 53
Sample 54
Sample 55
Sample 56
Sample 57
Sample 58
Sample 59
Sample 60

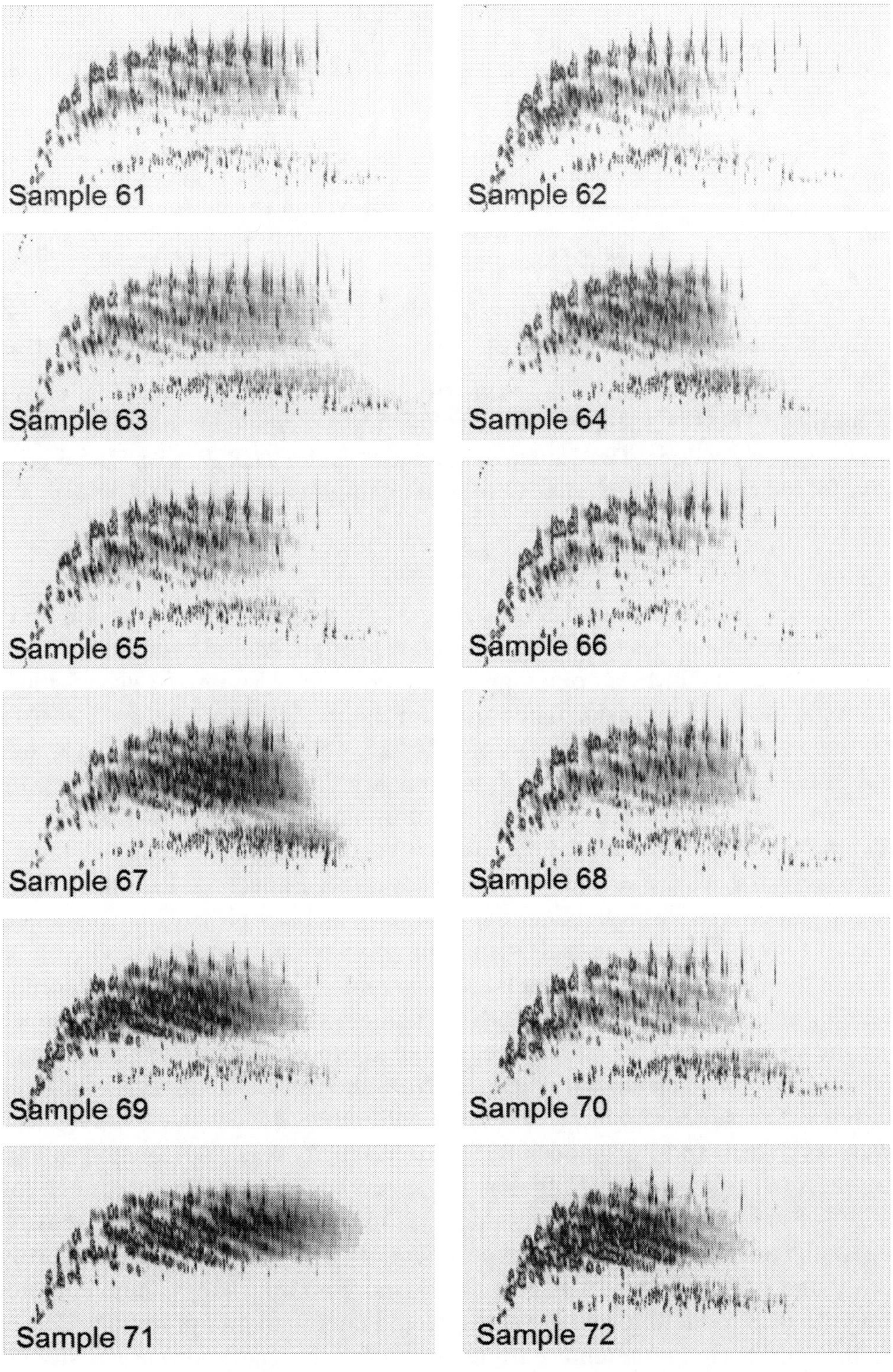

Figure 3. GC × GC – TOFMS chromatograms of all 72 samples. There are 11 samples that are dissimilar in their chemical fingerprint from the majority of the samples.

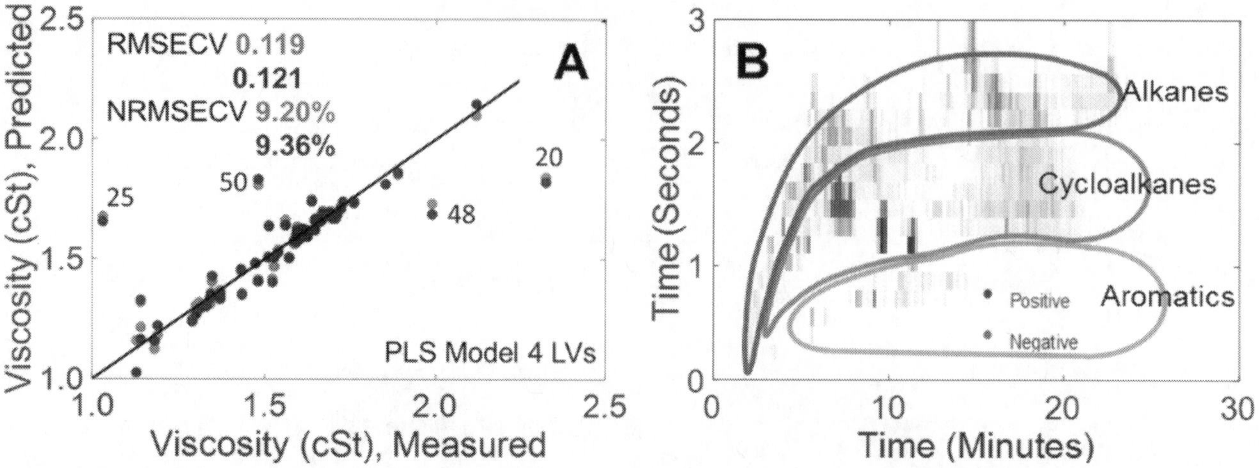

Figure 4. (A) PLS prediction of the viscosity using GC × GC – TOFMS chromatograms of all 72 samples. The black line represents the ideal agreement between the predicted and measured viscosity values. (B) Linear regression vector (LRV) from the PLS prediction of viscosity, folded and presented in a 2D format analogous to a GC × GC chromatogram.

We now turn our attention to the PLS modeling of the physical properties. Using PLS modeling, the measured viscosity values (ASTM D445/D446) for all 72 fuel samples were predicted based upon the GC × GC – TOFMS chromatograms for each fuel. Figure 4A shows the calibration results from the modeled viscosity. The errors for the modeling are below a NRMSECV of 10%. However, there are several samples (Samples 20, 25, 48, and 50) which do not model correctly. Inspection of the LRV in Figure 4B indicate there are several chemical features, three compound regions in particular, ${}^{1}t_{R} \approx 8$ min (large red feature), and 10 and 12 min (large blue features), which adversely influence the model the most. Investigating the Q residuals and Hoteling T^2 (not shown for brevity), there are several fuels that adversely impact the PLS modeling due to large chromatographic differences relative to the "typical" fuel samples. This was mentioned earlier in the context of Figure 3 that some fuels stand out upon visual inspection as being "chemically" very different than a majority of the fuels. For example, Sample 48 only has some minor peaks and two major peaks (chemical compounds) which are grossly overloaded in concentration. Predicting the viscosity as well as the other physical properties will be adversely challenged because this sample is so chemically different from the rest of the fuels. In order to create a model with low Q residuals and low Hoteling T^2, Samples 16, 20, 24, 25, 29, 33, 46, 47, 48, 50, and 51 were excluded, and PLS modeling for the viscosity was re-evaluated in Figure 5 with the remaining fuels (61 of the initial 72 fuels). The errors in the modeling are much lower with NRMSECV ~ 5% in the calibration presented in Figure 5A. The range of viscosity values over which the model accurately represents experimentally measured data is noteworthy. Additionally, the LRV in Figure 5B appear to be more appropriate with no chemical compounds dominating the PLS modeling, and provide a sound chemical interpretation. Compounds with higher boiling points (longer retention times ${}^{1}t_{R}$ on the ^{1}D separation) are positively correlated with increasing viscosity while lower boiling point compounds (shorter ${}^{1}t_{R}$ on the ^{1}D separation) are correlated with lower viscosity.

Likewise, the PLS prediction of the heat of combustion is provided in Figure 6, and was carried out in a similar fashion to the viscosity modeling with the same samples being excluded: 16, 20, 24, 25, 29, 33, 46, 47, 48, 50, and 51. The modeling of the heat of combustion has a slightly higher error than the viscosity and density (see below) with a NRMSECV of ~ 11%. However, the method precision (repeatability) and bias for ASTM D4809 are 0.096 and 0.089

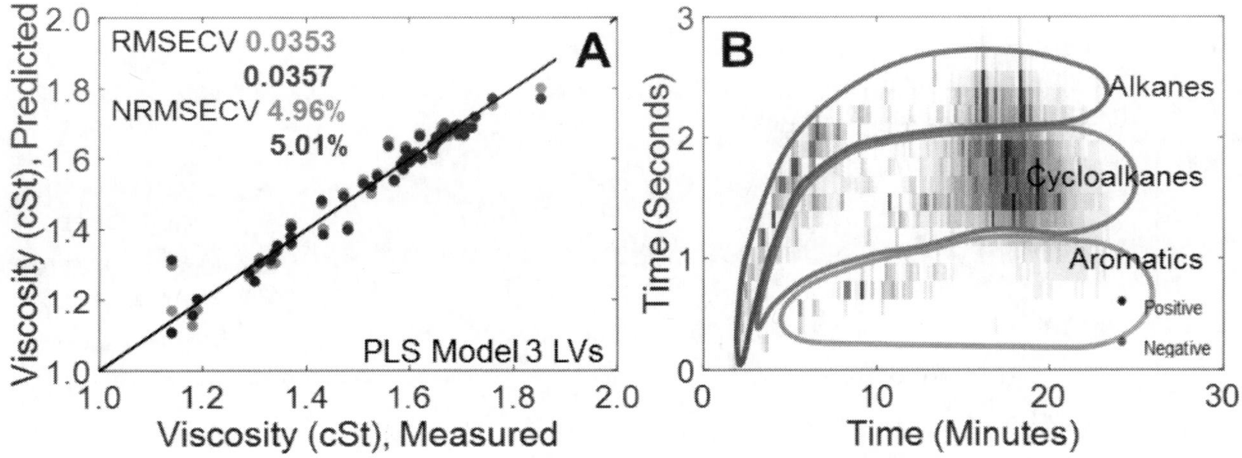

Figure 5. (A) PLS prediction of the viscosity using GC × GC – TOFMS chromatograms using 61 of the initial 72 fuels from Table 1, with Samples 16, 20, 24, 25, 29, 33, 46, 47, 48, 50, and 51 objectively excluded based on statistical tests. The black line represents the ideal agreement between the predicted and measured values. (B) Linear regression vector (LRV) from the PLS prediction of viscosity, folded and presented in a 2D format analogous to a GC × GC chromatogram.

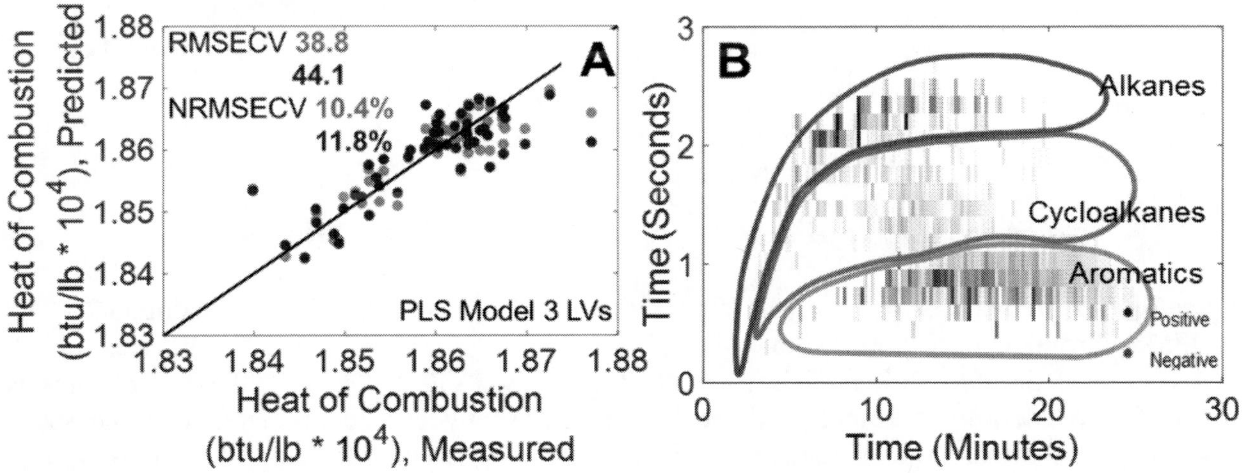

Figure 6. (A) PLS prediction of the heat of combustion using GC × GC – TOFMS chromatograms with Samples 16, 20 ,24, 25, 29, 33, 46, 47, 48, 50, and 51 excluded. The black line represents the ideal agreement between the predicted and measured values. (B) Linear regression vector (LRV) from the PLS prediction of heat of combustion, folded and presented in a 2D format analogous to a GC × GC chromatogram.

MJ/kg (~ 40 btu/lb) respectively, and the model quality based on true values may therefore be better than indicated. Chemical interpretation of the LRV in Figure 6B indicates that the alkanes positively correlate with the heat of combustion while the aromatics negatively correlate with the heat of combustion. The cycloalkanes do not seem to contribute to the heat of combustion based on the PLS modeling. Finally, the PLS modeling of the density at 15 °C, 45 °C, and 85 °C is provided in Figure 7. The density modeling presented in Figure 7A has the lowest error with NRMSECVs below 1%. Initially, the three LRVs in Figures 7B-D appear identical to each other as a function of temperature. In order to investigate this issue further, and to gain insight into the

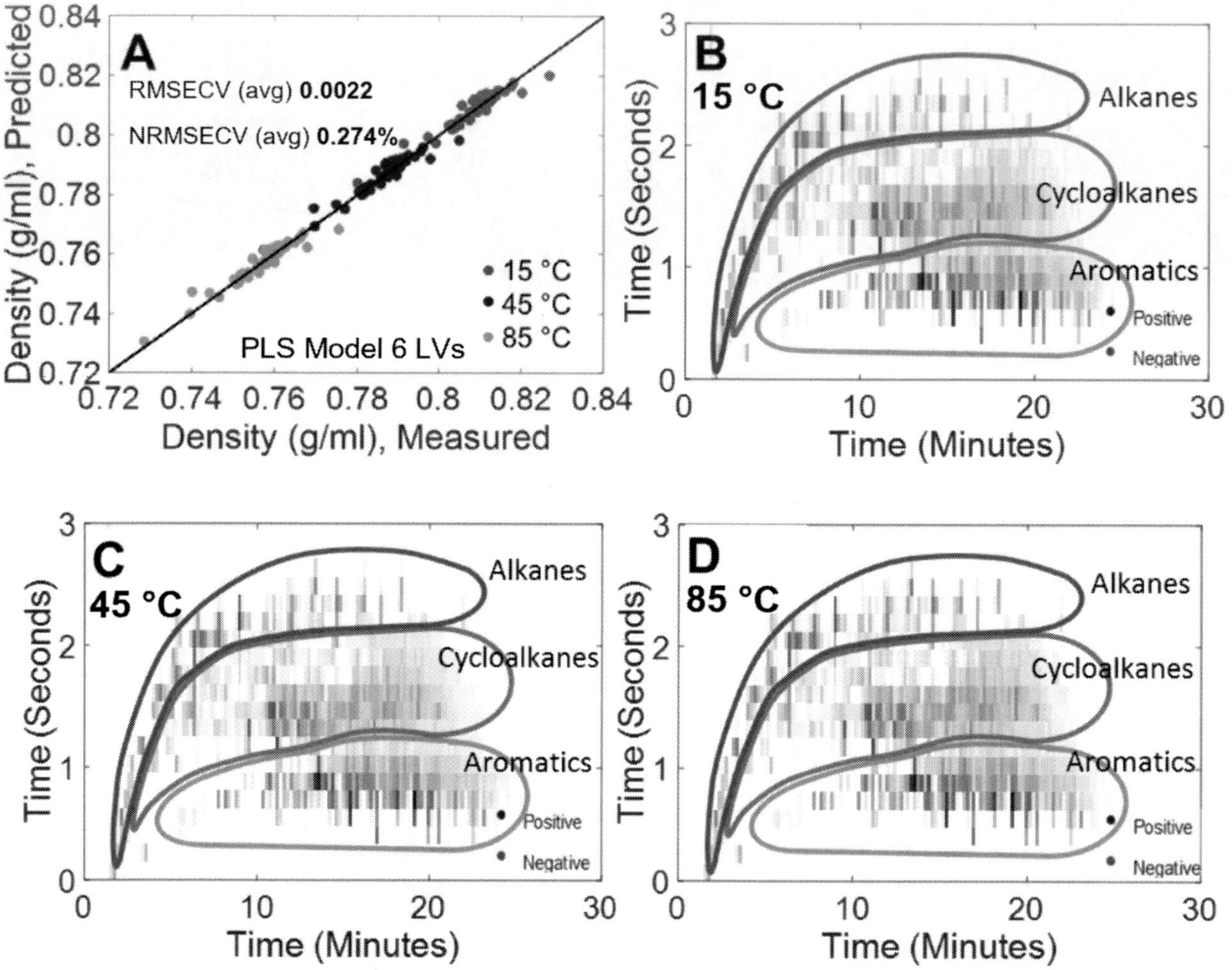

Figure 7. PLS predictions of density at temperatures of 15 °C, 45°C, and 85 °C using GC × GC – TOFMS chromatograms with Samples 16, 20 ,24, 25, 29, 33, 46, 47, 48, 50, and 51 excluded. The average of the two PLS modeling results for the two GC × GC – TOFMS replicates is shown. The black line represents the ideal agreement between the predicted and measured values. (A) PLS prediction of density at 15, 45, and 85 °C. (B) LRV for PLS prediction of density at 15 °C. (C) LRV for PLS prediction of density at 45 °C. (D) LRV for PLS prediction of density at 85 °C.

temperature dependence on density, the LRVs from each PLS modeling of the density (2 replicates × 3 different temperatures) were forwarded to PCA, with results presented in Figure 8. The LRVs separate on PC2 (PC1 not shown for brevity) in the scores plot in Figure 8A, forming a linear trend with respect to temperature. Over a short range of temperatures, density for a bulk sample such as kerosene-based fuel is essentially linear. Investigating the loadings plot for the PCA in Figure 8B reflect what was originally observed in Figure 7, that the alkanes negatively influence density, while the cycloalkanes have a minor positive correlation with density, and aromatics are positively correlated. However, the loadings plot also provides insight into the relationship between density as a function of temperature for each compound.

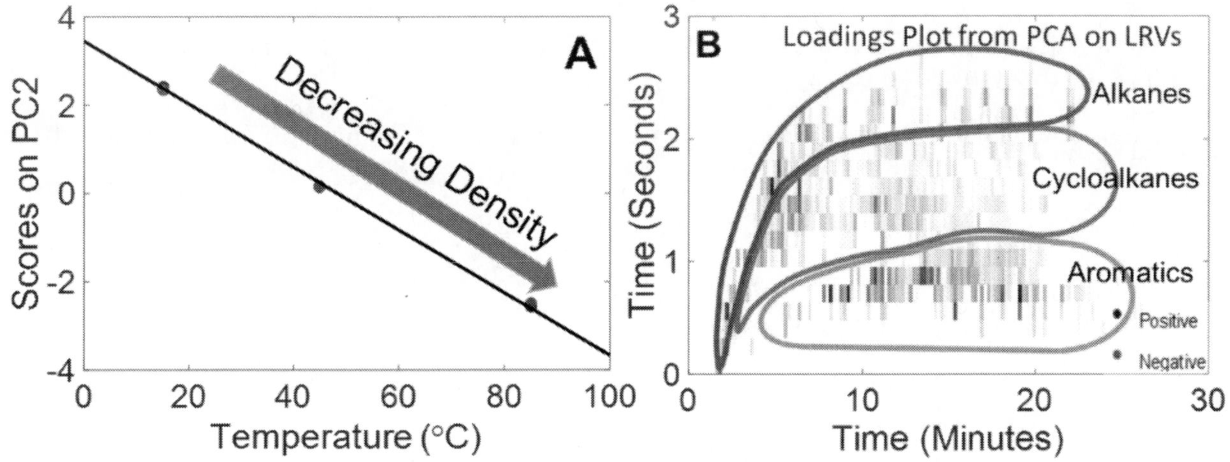

Figure 8. PCA modeling on the LRVs from the PLS prediction of density at 15, 45 and 85 °C, respectively. (A) Scores on PC2 form a linear function with respect to temperature. (B) Loadings plots from the PCA on the LRVs of the PLS prediction of density.

CONCLUSIONS

The overarching goal of this research is to better understand and enhance fuel performance through chemometric modeling, by relating fuel property data (ASTM measurements) to chemical composition data (GC × GC – TOFMS). The predictive PLS modeling approach presented herein may ultimately be used to educate fuel selection. Proof-of-principle of the PLS modeling approach was demonstrated on ASTM measurements of three physical properties: density, viscosity, and heat of combustion. With an extremely large dataset composed of 72 diverse fuels, reliable PLS models were produced, that relate the chemical composition information provided by GC × GC – TOFMS measurements of each fuel to the physical properties of each fuel. Overall success of the approach relied heavily on having a compositionally diverse set of fuels with accurately measured physical properties. Additionally, it should be straight forward to transition this approach to either different fuels (jet fuel, diesel fuel, etc.) and/or different properties or even performance metrics such as thermal stability measurements. Future work should also focus on digging deeper into chemical compound identification that correlates to each physical property.

REFERENCES

1. Cookson, D.; Smith, B. *Energy Fuels* **1990**, *24*, 152-159.
2. Liu, G.; Wang, L; Qu, H.; Shen, H.; Zhang, X.; Zhang, S.; Mi, Z. *Fuel* **2007**, *86*, 2551-2559.
3. Huber, M.L.; Lemmon, E.W.; Bruno, T.J. *Energy Fuels* **2009**, *23*, 5550-5555.
4. DeWitt, M.J.; Edwards, T.; Shafer, L.; Brooks, D.; Striebich, R.; Bagley, S.P.; Wornat, M.J. *Ind. Eng. Chem. Res.* **2011**, *50*, 10434-10451.
5. Lee, I. C.; Ubanyionwu, H. C. *Fuel* **2008**, *87*, 312-318.
6. Billingsley, M.C.; Edwards, J.T.; Shafer, L.M.; Bruno, T.J. 46th American Institute of Aeronautics and Astronautics (AIAA) Joint Propulsion Conference and Exhibit, Nashville, TN, July (2010) 2010-6824.
7. Lovestead, T.M.; Windom, B.C.; Riggs, J.R.; Nickell, C.; Bruno, T.J. *Energy Fuels* **2010**, *24*, 5611-5623.
8. Gough, R.V.; Bruno, T.J. *Energy Fuels*, **2013**, *27*, 294-302.
9. Hsieh, P.Y.; Abel, K.R.; Bruno, T.J. *Energy Fuels* **2013**, *27*, 804-810.
10. Burger, J.L.; Bruno, T.J. *Energy Fuels* **2012**, *26*, 3661-3671.
11. Begue, N.J.; Cramer, J.A.; Von Bargen, C.; Myers, K.M.; Johnson, K.J.; Morris, R.E. *Energy Fuels* **2011**, *25*, 1617- 1623
12. Bruno, T.J.; Ott, L.S.; Lovestead, T.M.; Huber, M.L. *J. Chromatogr. A* **2010**, *1217*, 2703-2715.
13. Billingsley, M.; Keim, N.; Synovec, R.; Hill-Lam, B.; Wilhelm, C. "Progress Toward Quality Assurance Standards for Advanced Hydrocarbon Fuels based on Thermal Performance Testing and Chemometric Modeling," IASH 14th International Symposium on Stability, Handling, and Use of Liquid Fuels, Charleston, SC, **2015**.
14. Billingsley, M.; Keim, N.; Hill-Lam, N.; Synovec, R.; Freye, C. "Hydrocarbon Fuel Thermal Performance Modeling based on Systematic Measurement and Comprehensive Chromatographic Analysis," (AIAA) 2016-4903, 52nd AIAA Joint Propulsion Conference, Salt Lake City, UT, **2016**.
15. Kehimkar, B.; Hoggard, J.C.; Marney, L.C.; Billingsley, M.C.; Fraga, C.G.; Bruno, T.J.; Synovec, R.E. *J. Chromatogr. A* **2014**, *1327*, 132-140.
16. Freye, C.E.; Fitz, B.D.; Billingsley, M.C.; Synovec, R.E. *Talanta* **2016**, *153*, 203-210.
17. Liu, Z.; Phillips, J.B. *J. Chromatogr Sci.* **1991**, *26*, 227-231.
18. Bruckner, C.A.; Prazen, B.J.; Synovec, R.E. *Anal. Chem.* **1998**, *70*, 2796-2804.
19. Johnson, K.J.; Synovec, R.E. *J. Chemom. Intell. Lab. Syst.* **2002**, *60*, 225-237.
20. Fraga, C.G.; Prazen, B.J.; Synovec, R.E. *Anal. Chem.* **2000**, *72*, 4154-4162.
21. Omais, B.; Courtiade, M.; Charon, N.; Thiébaut, D.; Quignard, A.; Hennion, M. C. J. *Chromatogr. A* **2011**, *1218*, 3233–3240.

22. Westad, F.; Afseth, N.K.; Bro, R. *Anal. Chim. Acta* **2007**, *595*, 323-327.
23. Rajalhti, T.; Kyalheim, O.M. *Int. J. Pharm.* **2011**, *417*, 280-290.
24. Gowen, A.A.; Downey, G.; Esquerre, C.; O'Donnell, C.P. *J. Chemom,* **2011**, *25*, 375-381.
25. Geladi, P.; Kowalski, B.R. *Anal. Chim. Acta* **1986**, *185*, 1-17.
26. Marney, L.C.; Siegler, W.C.; Parsons, B.A.; Hoggard, J.C.; Wright, B.W.; Synovec, R.E. *Talanta*, **2013**, *115*, 887-895.

IASH 2017, the 15th INTERNATIONAL SYMPOSIUM ON STABILITY, HANDLING AND USE OF LIQUID FUELS
Rome, Italy
10-14 September 2017

IDENTIFICATION AND QUANTIFICATION OF OXIDATION PRODUCTS OF FUEL COMPONENTS ANALYSED BY MASS SPECTROMETRY

Marcella Frauscher[1,2], Nicole Dörr[1], Charlotte Besser[1], and Alexandra Rögner[3]

[1] *AC2T research GmbH, Viktor-Kaplan-Straße 2C, 2700 Wiener Neustadt, Austria*
Marcella.Frauscher@ac2t.at, Nicole.Doerr@ac2t.at, Charlotte.Besser@ac2t.at
[2] *TU Wien, Institute for Chemical Technologies and Analytics, Getreidemarkt 9/164, 1060 Vienna, Austria, Marcella.Frauscher@ac2t.at*
[3] *OMV Refining & Marketing GmbH, Trabrennstraße 6-8, 1020 Vienna, Austria, Alexandra.Roegner@omv.com*

Oxidative stability is an important issue in respective fuel specifications to support storage stability. However, applied thermo-oxidative test methods provide fragmentary evidence of the fuel´s oxidation stability as either the volatiles (e.g., open set-up EN ISO 12205) or the liquid phase (e.g., rancimat EN 14112) are not considered. Typically, degradation degree is determined by conventional analytical methods such as neutralization number, infrared spectroscopy or determination of water. Thus, chemistries of degradation products remain hidden although the knowledge of the chemical structures and amounts formed from certain product classes may explain specific oxidation behavior and hence support product development.

In order to observe all degradation products – both volatile and liquid – and their respective quantities at different degradation degrees, a custom-design artificial alteration was applied. Model jet fuel mixtures were artificially altered in a closed vessel pressurized with oxygen at 95 °C. Degradation products in samples taken during this procedure were analysed by gas chromatography mass spectrometry(1). Oxidation products were unambiguously assigned to the most relevant chemistries (i.e., alcohols, aldehydes, ketones, and acids) and quantified with different methods such as selected ion monitoring.

Methodology developed enables the qualitative and quantitative determination of the degradation products that dominate fuel degradation. Furthermore, sampling during the stability assessment procedure allows monitoring how oxidative attack of fuel components proceeds over time, hence time-related information on the type and amount of degradation products on the molecular level can be gained. Moreover, variation of alteration parameters (e.g., temperature) as well as addition of commonly occurring impurities (e.g., fatty acid methyl esters) allow the systematic evaluation of such influencing factors on fuel degradation.

(1) Frauscher M., Besser C., Dörr N., Allmaier G. Oxidative degradation products of fuel components identified by mass spectrometry and isotope labelling. 20th Int. Colloquium Tribology: Industrial and Automotive Lubrication, 12-14 January 2016, Stuttgart, Germany

IASH 2017, the 15th INTERNATIONAL SYMPOSIUM ON STABILITY, HANDLING AND USE OF LIQUID FUELS
Rome, Italy
10-14 September 2017

LABORATORY TEST RIG WITH MINIATURE TURBOJET ENGINE AS A TOOL FOR TESTS OF ALTERNATIVE AVIATION FUELS

Jarosław Sarnecki[1], Bartosz Gawron[1], Tomasz Białecki[1]

[1] *Air Force Institute of Technology, 6 Księcia Bolesława Street, 01-494 Warsaw, Poland, jaroslaw.sarnecki@itwl.pl, bartosz.gawron@itwl.pl, tomasz.bialecki@itwl.pl*

Aviation is one of the fastest growing types of transport. Increasing number of aircrafts during recent years is associated with higher intensity of aviation operations carried out. This causes increasing demand for fuel, being power supply for turbine engines, and leads to increase of harmful products which are produced during combustion process. Emitted pollutants, such as harmful exhaust gases, particulate matters and volatile organic components have negative effect on climate, air quality and human health.
In today's research more and more space is devoted to issues related to reduction of harmful emissions from aviation into the environment. Many international research programs are aimed at developing new and greener solutions for aviation industry. One of major considered methods of harmful combustion products reduction is introduction of various components to fossil fuel, including biocomponents and biofuels. However, analysis and measurements on full scale aircraft jet engines are complex and expensive. Moreover, quantities of biofuels especially obtained from experimental installations are limited.
Air Force Institute of Technology, being leading research organization in Poland within fuels for aviation, creates specialized research facilities. In the Department of Fuels and Lubricants laboratory test rig with a miniature turbojet engine (MiniJETRig – Miniature Jet Engine Test Rig) was designed. The test rig has been developed for research aimed at modelling and investigating processes or phenomena occurring in full scale turbine engines. Small volume of fuel is required to carry out tests, which is the main advantage of miniature turbojet engine application.
The paper presents construction of a test rig with brief description of its main components functionalities. Methodology and results, in term of performance and emissions of miniature turbojet engine for few tested fuels (e.g. HEFA fuel) are shown. Measurement results are compared with data for neat Jet A-1 fuel.
The main research applications of MiniJETRig is testing of alternative aviation fuels. In paper, others applications and capabilities of the test rig are shown. For example, the test rig is used for assessment of combustion products impact on living cells.

SCALE DOWN OF DEWAXING PROCESS

Milena Mantarro[1], Viviana Berto[2], Danilo Zufferli[3]

[1] *Chimec S.p.A., Via delle Ande, 19- Rome, Italy, mmantarro@chimec.it*
[2] *Chimec S.p.A., Via delle Ande, 19- Rome, Italy, vberto@chimec.it*
[3] *Chimec S.p.A., Via delle Ande, 19- Rome, Italy dzufferli@chimec.it*

Dewaxing aids is the definition given to the chemicals able to interact with the feed paraffin, modifying the wax crystals during the dewaxing process.

Compared to the baseline without additive, the benefits given by the use of the dewaxing aid are an increase of the throughput (production rate) and dewaxed oil yield, with a consequent positive impact on the refinery's margins.

Test were performed applying an internal procedure based on a laboratory scale simulation of the dewaxing process. The feed was diluted in a solvent mixture and cooled at a fixed rate up to the selected temperature. Then the cooled feed was filtered on a filter resembling the one used in the dewaxing plants. The applied conditions were selected based on those normally applied in the plant.

The performances of the additives were evaluated based on key parameters.

Different applications of dewaxing aids in the plant were allowed to validate the method developed in the laboratory.

IASH 2017, the 15 th INTERNATIONAL SYMPOSIUM ON STABILITY, HANDLING AND USE OF LIQUID FUELS
Rome, Italy
10-14 September 2017

INVESTIGATION OF COMPATIBILITY BETWEEN HFO AND LSFO IN MARITIME INDUSTRY AND THE EFFECT OF STABILIZING ADDITIVES USING FUEL STABILITY ANALYZER (ASTM D 7157)

Milena Mantarro[1], Viviana Berto[2], Antonio Guarascio[3]

[1] *Chimec S.p.A., Via delle Ande, 19- Rome, Italy - mmantarro@chimec.it*
[2] *Chimec S.p.A., Via delle Ande, 19- Rome, Italy - vberto@chimec.it*
[3] *Chimec S.p.A., Via delle Ande, 19- Rome, Italy - aguarascio@chimec.it*

In 2015 IMO regulations set new limits at 0,1% sulfur content of fuel oil in Emission Control Areas (ECA), which presents a significant challenge for the whole maritime industry.

Under the revised MARPOL Annex VI, the global sulphur cap will be reduced from current 3,5% to 0,5%, effective from 1st January 2020. When ships pass through an ECA, they have to switch to an ultra low sulphur fuel oil (LSFO) or other low sulfur fuel, later, when they leave it, they switch back to HFO to avoid extra costs.

Switching from one type of fuel to the other is a delicate operation and it can lead to incompatibility problems. Adding a fuel oil or distillate (Low Sulphur) with a predominance of paraffinic hydrocarbons to a heavy fuel oil (HFO) with a high asphaltene content can promote the precipitation of asphaltenes as sludge, causing the clogging of filters which may cause the engine to shut down.

This paper reports a series of compatibility tests between HFO e LSFO in different ratio. All the tests were performed with a Fuel Stability Analyzer, in compliance with ASTM D 7157 and have shown incompatibility problems between the fuel oils analysed.

The study shows how the fuel oil stability blend can be improved, avoiding the flocculation of asphaltenes, when blended in certain ratios with the use of some stabilizing additives.

IASH 2017, the 15 th INTERNATIONAL SYMPOSIUM ON STABILITY, HANDLING AND USE OF LIQUID FUELS
Rome, Italy
10-14 September 2017

MICROBIOLOGICAL CONTAMINATION OF FUEL – TREATMENT MANAGEMENT FROM STORAGE TANKS TO WASTE WATER SYSTEM AND ON FILLING STATION

Marco Buccolini, Silvia Bozzi , Valter Tandoi[2]

[1] *Chimec S.p.A., Via delle Ande, 19 – Rome, Italy- mbuccolini@chimec.it*
[2] *CNR-IRSA, National Research Council, Water Research Institute, Via Salaria km 29,300, Monterotondo (RM), Italy - tandoi@irsa1.irsa.cnr.it*
[1] *Chimec S.p.A., Via delle Ande, 19 – Rome, Italy - sbozzi@chimec.it*

Since when the presence of Biodiesel in mineral fuel became mandatory, problems related with the presence of microbiological growth and biofouling are becoming more and more evident in storage tanks, in refinery depots and in filling station.
Possible consequences for the filling stations in terms of maintenance and cleaning are quite relevant and can even lead to the stopping of the operation with significant economical loss.
The use of the microbiological treatment set up by Chimec S.p.A., properly managed according to the operative (and environmental) conditions, is the most versatile solution, allowing to get rid of the contamination problems in tanks, refinery and/or filling station without any problems for the waste water biological system (WWTP).
A study together with CNR shows the impact of the use of two different Chimec biocides on a sample of waste water activated sludge, detecting the value of EC 50, applying the standard method ISO 8192: 2007 based on the measurement of oxygen consumption by activated sludge.
The determination of the EC 50 value combined with mass balances in the WWTP, allow to do the biocide treatment in safety conditions, without compromising the WWTP performances.
The study on monitoring the EC 50 values (at 0,5 and 24 hours contact time) with different chemicals have been performed with the target to demonstrate the non toxicity of different biocide chemicals on the activated sludge, the most common wastewater treatment process, utilized: both products belong to the 4th class, with EC 50 > 100 mg/L.

On the other side, to have a complete vision from all the sides of the plant and facilities involved in microbiological contamination, a procedure developed for the cleaning of the tanks of filling station, affected by microbiological contamination, based on the combination of two different chemicals, is reported. This procedure was experimented on true filling station on the Italian distribution net, allowing to clean the fuels from microbiological contamination, avoiding the wasting of fuels and the segregation of the tanks and a loss of economical revenue.
In this paper the modern biomolecular tools utilized for microbial contaminants characterization (Fluorescent in situ Hybridization and PCR) will be shown together how the Chimec know how contributes to manage and control all the production and supply chain involved in microbiological contamination of fuel blended with biofuel form renewable raw material.

IASH 2017, the 15 th INTERNATIONAL SYMPOSIUM ON STABILITY, HANDLING AND USE OF LIQUID FUELS
Rome, Italy
10-14 September 2017

ISSUES WITH NEW RAW MATERIALS AS FEEDSTOCKS IN BIOFUEL

Stefano Cacciatori [1], Silvia Bozzi[2]

[1] *Chimec S.p.A., Via delle Ande, 19- Rome, Italy - scacciatori@chimec.it*
[2] *Chimec S.p.A., Via delle Ande, 19- Rome, Italy - sbozzi@chimec.it*

Critical feedstocks such as UCO (Used Cooking Oils), nowadays are becoming more and more interesting for the production of biodiesel, especially after the issue of the European Directive 2015/1513 which set a cap of 7% of energy in transport in 2020 for the biofuels from food crops.

The UCOs are listed in the part B of the Annex IX, among the raw materials that can count twice their Energetic content, except for different internal rules applied by the Member State. Moreover a biodiesel produced from UCOs can indeed guarantee GHG emission savings (EU amendment 2015/1513) higher than 80%, fully complying with the new European requirements also for new installations.
For this reason there is a growing interest in their use as feedstock for biodiesel to be blended in mineral gasoil.

A critical point of these UCOs is the high variability in quality and characteristics, being related to the composition of the base oil (palm, seeds, tallow…) and to the frying conditions (temperatures, frying cycles, type of food fried, etc.).

The use of this new raw material as blendstocks for mineral diesel often has an impact, in terms of filterability, cold properties, oxidation stability, on the base properties of the mineral fuels and requires the evaluation of new chemical physical properties and new test method.

The aim of this study is to highlight some considerations on the use of this raw material, especially UCOs in the production of biodiesel and on its treatment with specific additives. All the considerations are based on Chimec's experience in this field.

IASH 2017, the 15th INTERNATIONAL SYMPOSIUM ON STABILITY, HANDLING AND USE OF LIQUID FUELS
Rome, Italy
10-14 September 2017

MICROBIAL CONTAMINATION IN ROAD VEHICLE DIESEL-A SURVEY OF GAS STATION UNDERGROUND TANKS IN TWO US STATES.

Joan Kelley[1] and Gerry Herman[1]

[1] *Conidia Bioscience Ltd, Bakeham Lane, Egham, TW20 9TY, UK,* joan@conidia.com

The occurrence of microbial contamination in gas station storage tanks in two states was compared. Florida and New Jersey were selected as regions with contrasting climates and the work was carried out on stations belonging to the same company across both states during a 2-week period in May/June. Testing was carried out on-site using kits compliant with ASTM D8070.
Samples were taken from as close to the tank bottom as the design allowed. Access was gained through the tank fill point and where possible samples were also taken via the Submersible Turbine Pump (STP) access point.
Where contamination was found in the storage tank, nozzle samples were also tested for potential carry over. Positive microbial contamination was defined as that giving results considered 'moderate' or above as indicated by the test kits. A total of 101 tank and nozzle samples were tested from 45 gas stations, 22 in Florida and 23 in New Jersey.
In New Jersey 40% of fill end samples contained a water bottom, 50% in Florida. 51% of fill end samples were contaminated in New Jersey and 50% in Florida. In New Jersey 75% of STP end samples contained a water bottom, 71% in Florida. 63% of STP end samples in New Jersey were contaminated and 71% in Florida. In New Jersey 32% of the nozzle samples tested contained detectable carry over contamination while this rose to 67% in Florida. Overall, in New Jersey 74% of the sites showed contamination and 50% in Florida giving 62% contaminated sites over the 2 States.
Although the percentage of tanks with visible water was approximately the same in both states, significantly more of these sites were contaminated with microorganisms in New Jersey than in Florida. Visible water presence did not always correlate with contamination. Carry thru of contamination into nozzle samples was significant and greater than anticipated at the start of this project.
While trend analysis is normally the preferred method for characterizing microbial loading in fuel tanks. This was a snapshot study carried out in response to repeated reports of blockage and corrosion in 30μ filters. The paper discusses the results in full.

IASH 2017, THE 15th INTERNATIONAL SYMPOSIUM ON STABILITY, HANDLING AND USE OF LIQUID FUELS
Rome, Italy
10-14 September 2017

Comparison of ASTM Method D8073-16 (Small Scale Water Separation Method) with real filter coalescing performance by Single Element Test Stand Water Mapping

Alan J. Fougere[1], Ian Mylrea[2], and Paul Wells[3]

[1] *D-2 Incorporated, 23 Edgerton Drive, North Falmouth, USA, afougere@d-2inc.com*
[2] *Stanhope Seta, London Road, Chertsey, Surry, UK, Ian.Mylrea@stanhope-seta.co.uk*
[3] *ExxonMobil, Paul Wells*

This paper describes the correlation of a new test method, ASTM D8073-16, with water separation characteristics of coalescing filters for aviation fuel. False negative water separation tests in the jet fuel supply chain have significant impacts on aviation fuel distribution, causing disruption and increased cost. By false negative we mean that an aviation fuel which does not meet the required level of result (85, or 70 if containing SDA) and is therefore predicted to have bad water separation characteristic's, actually has good water separation characteristics when passed through the latest design water separator. This can be understood because new 5th edition filter materials outperform materials previously employed for water separation methods, hence test results from existing methods don't always predict filter water separator field performance. The new method uses EI 1581 5th Edition filter material to coalesce the water, which is typical of filter water separators currently employed, and hence is likely to correlate better. A comparison with full scale field filter performance was carried out using the Single Element Test (SET) Water Mapping Protocol previously presented at IASH in 2011 [Wells/Hoskin] to see how the new method compared to existing methodology and how well the new method correlated with the field filter results. A discussion on the results for both weak and strong surfactants and additive blends is provided.

IASH 2017, the 15th INTERNATIONAL SYMPOSIUM ON STABILITY, HANDLING AND USE OF LIQUID FUELS
Rome, Italy
10-14 September 2017

IDENTIFICATION OF OXIDISED COMPOUNDS IN AVIATION FUELS USING NAFION FIBRE SOLID PHASE MICROEXTRACTION

Renée L. Webster[1,2], Jonathan B. Mete[1,2], David J. Evans[1], Paul M. Rawson[1], Philip J. Marriott[2]

[1] *Defence Science and Technology Group, 506 Lorimer St, Fishermans Bend, 3207, VIC, Australia*
[2] *Australian Centre for Research on Separation Science, School of Chemistry, Monash University, Wellington Rd, Clayton, 3800, VIC, Australia*

Thermal stability of aviation fuels is a key consideration facing the operation of high performance aircraft. As modern aircraft engines and fuel systems are required to operate at increasingly high temperatures and employ fuel as a coolant for hydraulic and avionics systems, fuel thermal instability and oxidation reactions are of growing concern. Thermal oxidation occurs when fuel hydrocarbons react with dissolved oxygen, leading to the formation of various oxidised species from hydroperoxides. This is problematic as these oxygenated species are known to attack engine components(1) and form solid deposits leading to blockages and increased frequency of maintenance(2).

Identification of oxygenated compounds is an important aspect of assessing the serviceability of the fuel, and to formulate strategies to mitigate the effects of the oxidised species. However, identification of oxidised species in fuels is complicated by the complex nature and matrix interferences that may arise, and the low concentration of the compounds of interest(3).

Nafion SPME fibres were prepared and applied to the analysis of oxygenated analytes in non-polar aviation fuel matrices. The performance of the Nafion fibres was investigated and exhibited high extraction efficiencies for polar analytes compared with a commercially available polar fibre. Calibration curves were constructed for several oxygenated compounds with linearities up to 0.999 for concentrations of 10-100 mg/L. Oxygen containing compounds were also identified from thermally oxidised aviation fuels using single column and multidimensional gas chromatography with mass spectrometry.

References

1. Fodor, G. E.; Naegeli, D. W.; Kohl, K. B., Peroxide formation in jet fuels. *Energy Fuels* **1988,** *2* (6), 729-734.

2. Heneghan, S. P.; Zabarnick, S., Oxidation of jet fuels and the formation of deposit. *Fuel* **1994,** *73* (1), 35-43.

3. Webster, R. L.; Evans, D. J.; Marriott, P. J., Multidimensional gas chromatographic analysis of low-temperature oxidized jet fuels: formation of alkyldihydrofuranones. *Energy Fuels* **2015,** *29* (4), 2067-2073.

IASH 2017, the 15TH INTERNATIONAL CONFERENCE ON STABILITY, HANDLING AND USE OF LIQUID FUELS
Rome, Italy
10-14 September 2017

DETERMINATION OF IGNITION DELAY AND LEAN BLOWOUT LIMITS OF NARROW DISTILLATION PROFILE ALTERNATIVE FUELS USING A SMALL GAS TURBINE

Ulas Yildirim[a], Glen Rowlinson[b], Sylvester Abanteriba[a] and David J. Evans[b]*

a. School of Aerospace, Mechanical and Manufacturing Engineering, RMIT University, 124a Latrobe St, Melbourne, VIC, 3000, Australia

b. Defence Science and Technology Group, 506 Lorimer St, Fishermans Bend, Victoria, Australia, 3207. David.Evans@dsto.defence.gov.au

Precise determination of Ignition Delay (ID) and Lean Blowout (LBO) characteristics of alternative fuels remains elusive to the fuel community. Currently, test methods exist in the determination of the ID times of fossil and alternative fuels, which require various laboratory apparatus in an attempt to emulate real life scenarios as part of the test methods. Notably, most available data is based on empirical data obtained from fossil fuels. There have been reports that for certain types of alternative fuels these tests fail to provide useful results.

The harshest operating environment for any turbine fuel is within the combustor of a gas turbine. In this experiment, a small gas turbine has been modified and the engine control unit redesigned in order to determine ID and LBO limits of a number of narrow distillation profile alternative fuels.

IASH 2017, the 15th INTERNATIONAL SYMPOSIUM ON STABILITY, HANDLING AND USE OF LIQUID FUELS
Rome, Italy
10-14 September 2017

SINGLE JFTOT TEST FOR UNDERSTANDING THE BREAKPOINT OF A FUEL

David J Abdallah[1], Kenneth E. Thompson[1], Scott K Berkhous[1],

[1]*ExxonMobil Research and Engineering, 600 Billingsport Road, NJ 08066, USA,*
david.j.abdallah@exxonmobil.com

ABSTRACT

The current methodology for determining distillate fuel thermal stability margin involves performing multiple ASTM D3241 measurements at different set point temperatures in order to determine the breakpoint temperature. This process is impractical for routine testing and often can take days to perform. A single test capable of assessing a fuel's thermal stability is described that takes the same amount of time to run as a standard single D3241 test. The enhanced breakpoint methodology is conducted at an elevated temperature and utilizes deposit metrology to evaluate the deposit thickness profile. The tube axial position where deposit reaches a thickness limit or deposit slope limit is correlated to the D3241 Appendix X2 breakpoint temperature.

INTRODUCTION

Jet fuel performance properties are controlled by commercial aviation fuel specifications. Many specification tests produce a value which allows a quantitative understanding of how close a property is to the specification limit. However, some specification tests like the jet fuel thermal oxidative stability test D3241 are pass/fail providing no indication of the fuels margin. A methodology for determining distillate fuel thermal stability margin exists but it is very time consuming and rarely conducted. It involves performing multiple D3241 measurements at different set point temperatures. Quantification of a fuels thermal stability margin can be very important to a refiner or fuel supplier. It allows for thermal stability monitoring over time, control over refinery finishing processes, optimization of blending streams, and quality assurance when fuel is being transported and purchased.

An inherent feature of a D3241 test is a gradient of temperatures across the 60 mm heater tube cylindrical test surface. The gradient is created by cooling of the tube ends through the bus bars while the tube is being electrically heated to the set point temperature of the test. The temperature profile across the tube resembles a Gaussian function curvature. The set point temperature relates to the hottest axial position which is 39 mm from the shoulder on the fuel inlet side of the tube in a standard D3241 test. The tube hotspot is not in the center position due to fuel flow along the axis of the tube. This flow provides additional cooling to the upstream surface of the tube, resulting in a shift of the hotspot. In the current methodology the temperature gradient profile in the 15 mm to 39 mm range is exploited. The proposed thermal stability methodology utilizes the deposit profiles formed in this area from a single test conducted at an elevated temperature. Results were found to correlate to the D3241 Appendix X2 breakpoint.

EXPERIMENTAL

Materials

Jet fuels were obtained from a variety of supply terminals around the world. All fuels in this study exhibited a deposit failure and not a differential pressure failure when conducting the breakpoint testing.

Instrumentation

The Falex 400 jet fuel thermal oxidation apparatus (operating at 500 psi, 3 mL/min, 450 mL, 2.5 hr.) was used in the current evaluation. Alcor aluminum heater tubes were used throughout the testing program. Deposit metrology on the heater tubes was conducted using a Falex 430 Ellipsometer. Ellipsometric measurements comprise 1224 film thickness data points collected in an array of 51 equally spaced points along the axial direction (1 mm increments) of the heater tube at 24 equally spaced circumferential positions (15° increments). The first and last 5 mm of the heater tube surface are not measured.

ASTM D3241 breakpoint (Appendix X2) is defined as the highest temperature in 5 °C intervals where the D3241 test passes. ASTM D3241 Annex 3 ellipsometric measurements were used to determine the pass/fail criteria. A passing test is achieved if the average of the highest deposit thickness values over an area of 2.5 mm^2 (standard spot) is $\leq$ 85 nm.

RESULTS AND DISCUSSION

All D3241 testing was conducted on a single instrument to eliminate variations. It is the experience of the authors that variations exist between the current commercially available instruments. The breakpoint of several fuels was determined using the procedure described in D3241 Appendix X2. A single D3241 test was also performed on the same fuels with a set point test temperature of 350 °C. This temperature was selected as a maximum temperature to remain in the thermal oxidative regime. The total carbon deposition from a single D3241 tests at 350 °C has been used to rank fuel thermal stability in the past.[i] At higher temperatures the chemical degradation mechanism changes from oxidative to pyrolytic and is irrelevant to determining the thermal oxidative stability performance of the fuel. As a result, test temperature was chosen to be 350°C. The 350 °C hotspot temperature creates an approximate temperature gradient profile across the heater tube as shown in Figure 1.

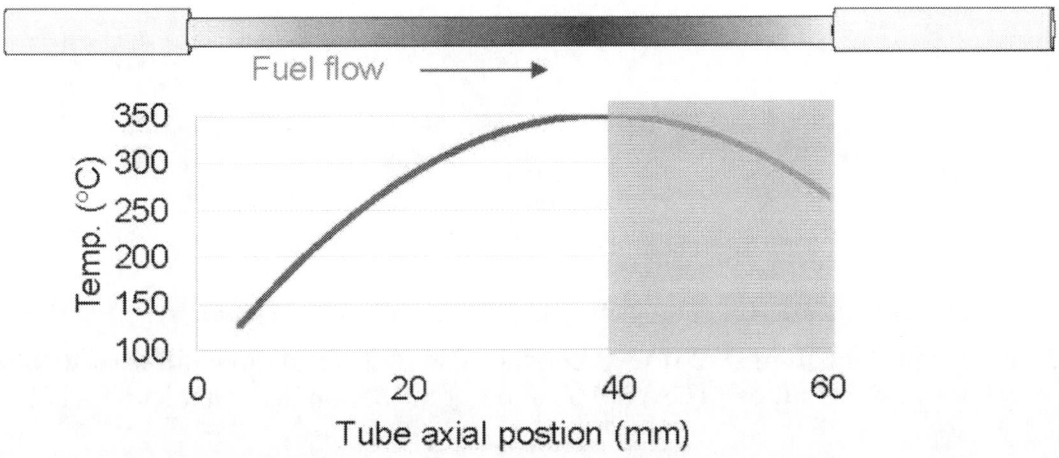

Figure 1. Heater tube temperature profile for D3241 tests using a 350 °C set point temperature.

Heater tube deposits will most likely have a mound-like thickness profile (Gaussian distribution-like curvature) that tracks the surface temperature profile of the tube. Some fuels have been shown to create more complex deposit profiles which can complicate the analysis. Deposit profiles from three D3241 tests conducted at 350 °C using fuels with different breakpoint temperatures are shown in Figure 2. Here, the x-axis represents the axial tube position in mm beginning on the fuel inlet side. The y-axis is the thickness of the deposit in nm. The z-axis represents the circumference of the heater tube as if the tube was cut lengthwise and unrolled. The red arrows highlight the extrapolated axial position where the average annulus deposit thickness is 85 nm. This position was determined by averaging the 24 circumferential ellipsometric measurements at each individual axial position and linearly extrapolating to a 0.1 mm increment between the positions that bound the 85 nm deposit thickness. The axial tube positions where the deposit thickness reaches an average value of 85 nm directly trends with the breakpoint temperature of the fuel.

It is noted that a variety of alternative deposit profile analyses could be used to determine the axial position on the heater tube. For example, a deposit slope limit threshold could be employed where a change in deposit thickness with respect to an incremental change in heater tube axial position is used to designate the axial position on the heater tube. A preliminary evaluation of this technique produced comparable results to that of the 85 nm thickness approach described above.

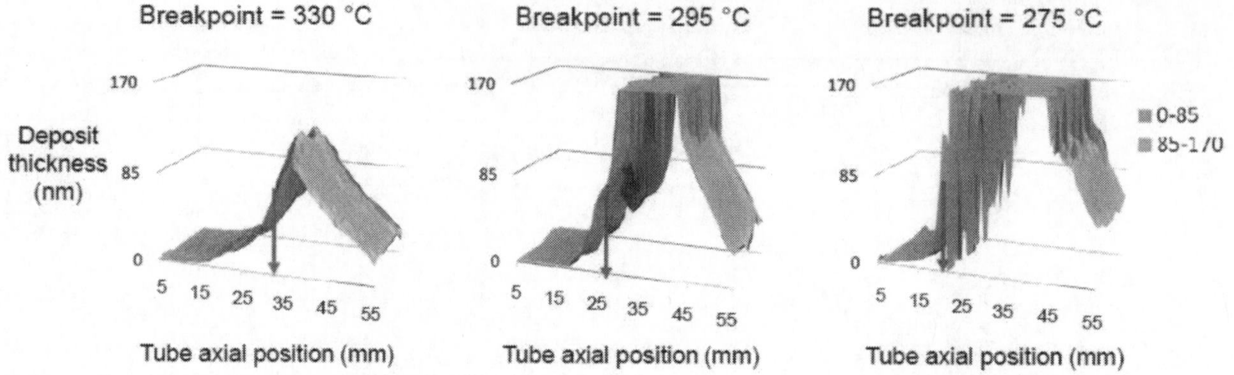

Figure 2. Deposit profiles from D3241 tests conducted at 350 °C on three different fuels with varying breakpoint temperatures. The axial position of the 85 nm deposit thickness is designated by the red arrows.

A comparison of D3241 Appendix X2 breakpoint and the axial position where the 85 nm deposit thickness criteria is achieved in a single 350 °C D3241 test for nine fuels is shown in Figure 3. From this plot the breakpoint relationship can be drawn by plotting a linear regression line through the data. As an example, a fuel with an axial position of 22 mm, corresponding to an average annular deposit thickness of 85 nm, has a correlated fuel breakpoint of 278 °C. This is similar to the fuel with the lowest breakpoint shown in Figure 2. The regression line (Eq 1) reveals an approximately 5 °C increase in breakpoint for each mm increase in axial position along the tube. Just in the way of an example, a potential quantitative specification limit replacement for jet fuel thermal stability would be a minimum 18.5 mm axial position of the 85 nm deposit thickness using this test which would equate to a fuel with at least 260 °C breakpoint.

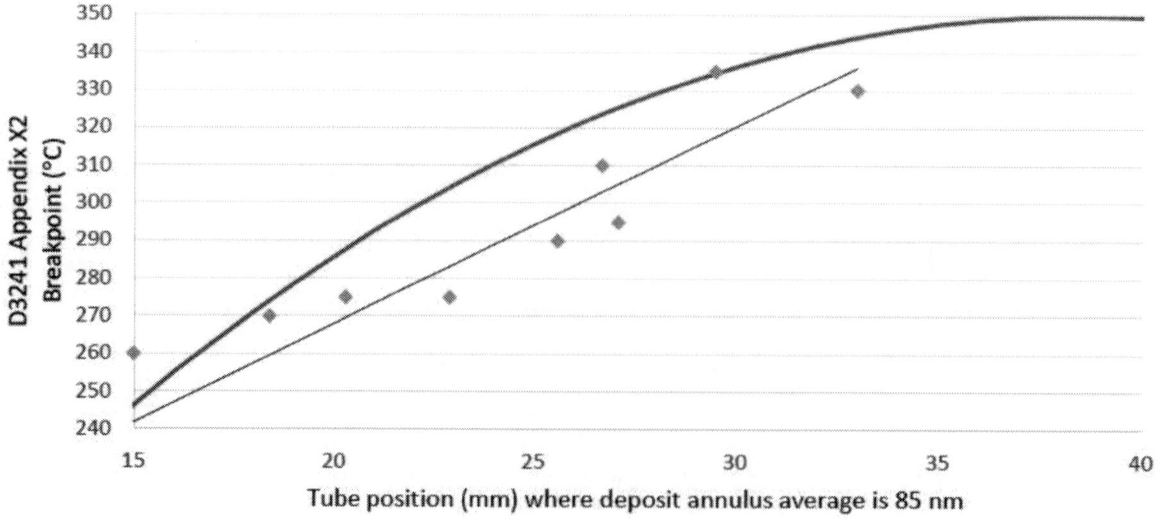

Figure 3. Comparison of the measured D3241 Appendix X2 breakpoint and axial heater tube position of the 85 nm deposit thickness determined from a single 350 °C D3241 test for nine different jet fuels. The red curve highlights the heater tube surface temperature across the axial

position of the tube. The black line is a linear regression showing the agreement between the two measurements. The equation for the linear regression is shown by Equation 1.

Eq 1. Correlated fuel breakpoint = 5.23x + 163.29 (R^2 = 0.8286), where x = the tube position (mm) where deposit annulus average is extrapolated to be 85 nm from a single D3241 tests conducted at 350 °C.

The R^2 value from the linear regression suggests that there is still room for improvement. One potential area of improvement would be to better control the temperature gradient across the heater tube. This could be accomplished by employing a tighter tolerance on the bus bar cooling. It is the experience of the authors that in current commercial instruments, the cooling can drift over the course of the D3241 test, resulting in some variation of the test results.
A second potential area of improvement would be optimizing the test parameters such as the fuel flow rate, set point temperature, and sample volume over the course of the test. Also, utilizing a higher bus bar temperature may decrease the temperature gradient to the hotspot. Currently there is an ASTM effort to improve the precision of the D3241 test. The approach proposed in this paper should be further evaluated once efforts to improve the method precision and equipment are completed.
Another potential area of improvement would be conducting the D3241 Appendix X2 breakpoint at an alternative temperature interval. The current D3241 Appendix X2 methodology calls for 5 °C temperature increments for determination of breakpoint. Once instrumentation improvements have been made, smaller temperature increments could be explored. For example, breakpoint determination using a 1 °C interval could provide a more precise result as compared to the current 5 °C intervals. It is anticipated that better data would improve the correlation.

Although some of the above suggested improvements may improve the agreement between the data, there are some intrinsic chemistry properties that will limit the correlation. Potential properties that could limit the correlation include differences in the high temperature oxidation induction period of the fuels and formation of deposits that do not adhere to the assumed Gaussian-like formation profile.

In addition to the methodology described above for jet fuels, this approach could also be applied to diesel or other distillate fuels where thermal oxidative stability is an important fuel property. With the introduction of high pressure common rail fuel injection systems in modern diesel engines, the diesel industry is currently looking for a suitable thermal oxidative stability test. This methodology may require further improvements to address the automotive industry needs, such as optimization of the instrument test parameters and heater tubes coated with more relevant surface compositions or made from alternative materials, for example stainless steel or other alloys instead of the current aluminum-based tubes.

CONCLUSIONS

This paper has presented a testing methodology with the ability to quantify the thermal oxidative stability of a fuel. The proposed approach brings together a detailed understanding of the D3241 testing equipment and the advantages of deposit metrology to produce a single test capable of quantifying a fuels thermal stability.

The data presented above shows a strong correlation between the axial position on a heater tube, determined by the location of an average annular deposit thickness of 85nm, and the D3241 Appendix X2 breakpoint temperature. Using this approach has advantages of significantly reducing the required fuel volumes, use of test consumables, and time, while approximating the breakpoint temperature of a jet fuel. This approach has far reaching potential in other applications where knowing the thermal oxidative stability of a fuel is critical information.

[i]Kendall, D. R.; Clark R. H.; Stevenson, P. A., The Influence of polar compounds on the stability of jet fuel, Proc. 2nd Int. Conf. Stability and Handling of Liq. Fuels, 1988, p. 694.

IASH 2017, the 15th INTERNATIONAL SYMPOSIUM ON STABILITY, HANDLING AND USE OF LIQUID FUELS
Rome, Italy
10-14 September 2017

PREPARATION, STABILITY AND PROPERTIES OF WATER-EMULSIFIED JET FUEL AT LOW SURFACTANT CONTENT

Moshe Rabaev[1] and Reuven Falkovich[1]

[1]*Israeli Air Force, Fuel and Chemistry Department, POB 02538*

ABSTRACT

Adding water vapor into the combustion process of fuels, while difficult, holds great promise for improving combustion characteristics, reducing unwanted products and providing an "anti-knock" effect. The process of creating emulsions of water in JP-8 fuel for concentrations of water and surfactant lower than 2% has been studied. Surfactant combinations were found which created clear, stable or quasi-stable emulsions with maximal water: surfactant ratios of 0.75-1.9. A critical micelle concentration of 0.10-0.15% was found for one of them. Low-temperature behavior was studied, and a reversible clouding phase transition was observed between -10 and -20°C which is possibly the freezing of "bound" water. In general, easily-measurable clouding point was found to be sensitive to different phase changes in dilute micelles. Two operational properties of the fuel were tested and the emulsions were found to have little to no effect on either foaming or bulk pour point. Effects of both water-soluble (alcohol) and fuel-soluble (ester) co-surfactants were noted. These results demonstrate a promising direction towards improving upon existing jet-fuel and turbine engines.

1. INTRODUCTION

The addition of water or water/alcohol vapor into the combustion process of fuels is a relatively old idea, dating back to even before the advent of internal combustion, and has resurfaced periodically throughout the past decades, especially as an alternative anti-knock and reaction control mechanism to lead-based agents [1]. The general methods of water addition are either direct injection of water vapor into the combustion system, high-shear premixing shortly before combustion, or preparation of a stable emulsified fuel. The proposed and observed effects of these methods with various amounts of added water have been studied extensively both theoretically ([2] and earlier review in [3]) and experimentally – time-resolved combustion observation

and engine-run measurements – mostly on diesel fuels ([4-9] reviewed in [10]) and [11]) and other hydrocarbons [12-13], but also on jet fuel [14]. The strongest effect of water addition to combustion is significant reduction in emitted NO_x compounds, soot and other particulate matter, and hydrocarbons (especially PAH), although there could be a trade-off between these emitted pollutants.
In many of the aforementioned studies, a favorable effect is also observed on combustion profile, fuel efficiency and anti-knock characteristics. These are attributed to an increase in hydroxyl radicals which inhibit radical branching and reduce peak pressure, temperature reduction due to latent heat of vaporization, and, in the case of emulsions, secondary atomization from "micro-explosions" caused by water vaporization.

The first two methods of introducing water into the combustion process, mechanical mixing in the combustion chamber or creating unstable fuel-water premixes, involve high-shear injection of water or steam into the fuel or fuel-air mix, just prior to ignition. These methods have a long commercial record with a variety of patents and implemented designs, but they require a special, dedicated engine, and are therefore largely irrelevant to the purpose of improvement of jet-fuel formulations for existing turbine engines.

The third method is the addition of a water/surfactant mixture to the fuel formulation itself, creating a stable or quasi-stable "emulsified fuel". These emulsion formulations are described in various patents for both automobile gasoline and aviation gasoline and jet-fuel (e.g. [15-17]), with demonstrated improvements in combustion profile and products.

Those formulations, however, and the aforementioned studies, typically include fairly high amounts of water, surfactant and co-surfactant (1, 6 and 4% v/v respectively in the case of Wenzel et al.), although combustion studies show a significant effect on the burning of jet fuel with water concentrations as low as 0.2%.

Jet fuel has, unsurprisingly, very strict demands on conformance to operational and testable properties which are listed in various aviation fuel specifications such as ASTM D1655, DEF STAN 91-91 or their counterparts, and even minute formulation changes undergo a rigorous acceptance process, several aspects of which are detailed in, for example, ASTM D7826. Since it can be expected (and was observed in this study) that high (>5%) concentrations of surfactants and other foreign chemicals have a significant impact on various properties of jet fuel, a path to improving jet-fuel which is currently in use (mainly JP-8 and its counterparts) in existing turbine engines, with minimal structural and procedural adaptations along the fuel supply chain, lies in attempting to introduce "effective" amounts of water (0.1-1%) by using minimal concentrations of surfactants and co-surfactants. The resulting solutions must be stable across a range of operating temperatures and for long storage times, and avoid cross-inhibition of the activity of other additives. Most importantly, the addition of the emulsion must have negligible or manageable effects on the various operational properties of the fuel, especially those most prone to change by addition of surfactants, such as clouding and freezing properties, foaming, distillation curves, and thermal oxidation stability.

While in general the mechanisms underlying micelle creation and thermodynamic stability of microemulsions are mostly well-understood [18], there is still a practical challenge in creating stable or quasi-stable water-in-oil nano- and microemulsions in various configurations, especially near the limit of very low surfactant concentration. Nevertheless, many investigations have been made into the creation, necessary conditions, micelle size distribution and chemical environment of the solubilized water in W/O nano- and microemulsions at high surfactant concentrations, mostly of interest to pharmaceutical and cosmetic applications [19-22], although studies specific to diesel and jet-fuel were also performed, investigating medium-term stability, Ostwald ripening and coalescence, surface effects and other properties [23-26]. Garti et al. have extensively studied the chemical environment of reversed micelles using differential scanning calorimetry [27] and described phenomena which we attempted to identify in this study. These studies usually employ mixtures of nonionic surfactants, with hydrophilic parts composed of sucrose, sorbitol or ethylene glycol groups, and long aliphatic, mostly unbranched, hydrophobic tails. Generally, mixtures composed of surfactants with a high hydrophilic-lipophilic-balance (HLB) and surfactants with low HLB, with an arithmetic average HLB between 5 and 9-10, are described to be useful in creating W/O emulsions. Another aspect of emulsion creation is the shear regime. While microemulsions by definition do not require high-energy methods to create them, low-concentration nanoemulsions have droplet size averages and distributions greatly affected by the shear forces and specific methods involved during mixing. The work in [23-25] give examples of high-energy methods allowing nanoemulsions of water in jet-fuel.

This study aims at the creation of low-concentration stable microemulsions or nanoemulsions with long-term (above 30 days) stability, of water or water/alcohol mix in JP-8 jet fuel, which is the most common aviation fuel used not only in the IAF, but by most military and civilian operators worldwide. Preference is given to the creation of clear or translucent emulsions, which indicate small droplet sizes and increased stability, and low-energy mixing methods, so as not to "hide" mechanisms of emulsion instability.

The obvious operational properties of jet-fuel which can be harmed by emulsification are foaming and low-temperature behavior and stability, and were investigated in this study.

1. MATERIALS AND METHODS

1.1. Materials

JP-8 jet fuel, without any additives (referred to as "pure jet fuel"), was obtained courtesy of the Paz Ashdod Refinery. Regular jet fuel (referred to as "working") consists of a mix of batches of jet fuel with additives from several fuel tanks across the IAF, all of which were approved as conforming to IAF turbine fuel specification, (which is similar to ASTM D1655 except for the lack of DiEGME and ECI) and re-tested for gum content, distillation curves and Millipore filtering. Aromatic content of the jet fuel is unknown, but is less than 25% according to the specification. Different batches were mixed to ensure identical jet-fuel composition. All fuel was filtered through Whatman 41 filter paper (nominal pore size 8μm) prior to use. Non-ionic surfactants Span 80, Span 20, Span 85, Tween 80, Tween 65, Tergitol 15-s-9 and Brij O20 were purchased from Sigma-Aldrich and used as received. All surfactants are soluble in jet fuel at least to some extent. Ethanol, isopropanol and 2-butanol (CP) were purchased from Merck. Butyl Acetate and Cinnamyl Acetate (AR) were obtained courtesy of the Materials and Chemistry Branch of the Ground Forces Technology Division. Type I ultrapure water (resistivity > 18.1MΩ/cm) was obtained using a Merck Direct-Q water purification system from type III ion-replaced water.

1.2. Nomenclature

A surfactant mixture formulation is given a name with three parts - the first lists its components in shortened form, the second lists the composition in terms of weight ratios, and the third part is the calculated arithmetic HLB value [reference]. For example, S8T8-3:1-7 is a mixture of 75% Span 80 and 25% Tween 80 with an HLB value of approximately 7. The contractions of the surfactant names' are S2,S8 and S85 for Span 20, 80 and 85 respectively, T8 and T65 for Tween 80 and 65, Tg for Tergitol 15-s-9 and Bo2 for Brij O20.

1.3. Mixing and testing

Unless specified differently, the preparation of emulsions was done by weighing each surfactant in the formulation and dissolving them in jet fuel. Water was added drop-wise in amounts much higher than expected solubility, and the solution was mixed using a vortex mixer at 40Hz for 120 seconds. The result was most commonly a cloudy or clear homogenous emulsion with a "precipitate" of droplets larger than 5μm collecting on the bottom. In some cases, there would be an additional homogenous aqueous phase visible. The emulsion was left to stand for 4-7 days and the top 10% of the liquid in the test tube (which has usually clarified after that time) was tested for water content.

Water content was analyzed using Karl-Fischer coulometric titration, in a Metrohm 737KF titrator. Start and end drift was set at 10μg/min. The titrator was tested with calibration standards and known-concentration emulsions across a range of concentrations. The presence of the surfactant was not found to affect the measured water content. The accuracy of the titration was determined to be ±50mg/kg for concentrations below 500mg/kg and ±10% of displayed value above 500mg/kg. Further tuning to increase accuracy was deemed unnecessary. In the graphs, water content was reported in mg water per kg of fuel mix. Surfactant concentration was reported in gr surfactant per 100ml fuel mix. However, when calculating the solubilization ratio (conc. water / conc. surfactant) the water content value was multiplied by the fuel specific weight of ~0.8gr/cc.

1.4. Transparency

Each resulting emulsion was given a "clarity rating" by visual inspection of backlight passing through 30mm of fuel. The clarity rating was given from 1 (indistinguishable from pure jet fuel) to 5 (no light is visible through 30mm of emulsion) by comparison to premade solutions with defined clarity ratings.

Visible light absorption was measured using a Bausch&Lomb Spectronic 20 analog spectrophotometer with a quartz cuvette, zeroed for pure jet fuel, at wavelengths of 380 and 500nm. When used, good correlation was found between the measured absorption at 500nm and the visual clarity rating - the following table gives approximate transmission ranges correlating to the different CR's.

Clarity Rating	**Transmission Range [%]**
1	95-100
1.5	88-94
2	75-87
2.5	63-75
3	50-62
3.5-5	<50

1.5. Clouding Point

Measurements were conducted in a Koehler Clouding Point Apparatus designed for measurements according to ASTM D2386. The test cools the sample at 3°C/min and measures the transmission of white light relative to the transmission at room temperature. Instead of reporting on a definitive cloud point as in clean fuel, a full graph of absorbance as a function of temperature was extracted.

1.6. Foaming

Foaming characteristics were tested in a setup designed for measurements according to ASTM D892, which consists of a temperature bath, a 1L graduated cylinder containing 200ml of the tested fluid, and an adjustable air supply through a porous bubbling stone. However, since the procedure is intended for oils of much higher viscosity, two different tests were conducted. In the first one, equivalent to part of ASTM D892, air is bubbled through at a rate of 94mL/min for 5min, so that a stable layer of foam is created, which is measured (in volume) and described. In the second experiment, air is bubbled at the maximum possible rate until the foam fills 500ml, and is then shut off. The time for the disappearance of the foam (opening of foam-less surface) is measured.

2. EXPERIMENTS AND RESULTS

2.1. Tests of different formulations

Nine different 2- and 3-component surfactant formulations were tested as described in 2.3 in concentrations ranging from 0.1% to 1.0% (gr/100ml) to create the appropriate solubility curves as seen in Figure 1 below. The results are summarized in table 1. Solubilization ratio was calculated from the concentration of water divided by the concentration of surfactant.

Table 1: Results of solubilization tests for different formulations

Formulation	HLB	Average Solubilization Ratio	Clarity Rating	Notes
S8T8-3:1-7	6.88	0.82	1-2	Tested multiple times with similar results
S8Tg-2:3-9	9.07	0.03	1-2	
S2T8-4:1-10	10.2	1.95	3-4.5	Water content peaks off at ~9gr/L with ~0.5% surfactant.
S2T8Tg-16:3:1	9.67	0.77	1-1.5	
S8T8Tg-17:2:1	5.55	0.75	2-4.5	
S8T8-9:1-5.5	5.43	0.67	3-4.5	
S8T65-3:1-6	6.00	1.33	-	Water content peaks off at ~6gr/L with ~0.4% surfactant
S8Bo2-3:1-7	7.10	1.59	2-3.5	
S85T8-3:1-7	7.10	0.20	-	

Two formulations with exceptionally high solubility ratios, S2T8-4:1-10 and S8Bo2-3:1-7 (brown and green in the graph), were taken for further testing on stability and the properties of jet fuel. Another formulation taken was S8T8-3:1-7 (dark blue), which exhibited lower water solubility but consistent results, good stability (re-testing all emulsions 60 days after mixing yielded similar water content) and clear solutions. In particular, a further test with a 2.5:1 ratio of S8T8 to water gave a clear, stable emulsion with no precipitate. Two more formulations, S8T65-3:1-6 and S8T8Tg-16:3:1-9.5 (light blue and purple) were also considered but exhibited stability issues early on and were abandoned.

An ionic surfactant (Ammonium Lauryl Amphoacetate) was also tested at similar concentrations but showed no solubilization capacity whatsoever.

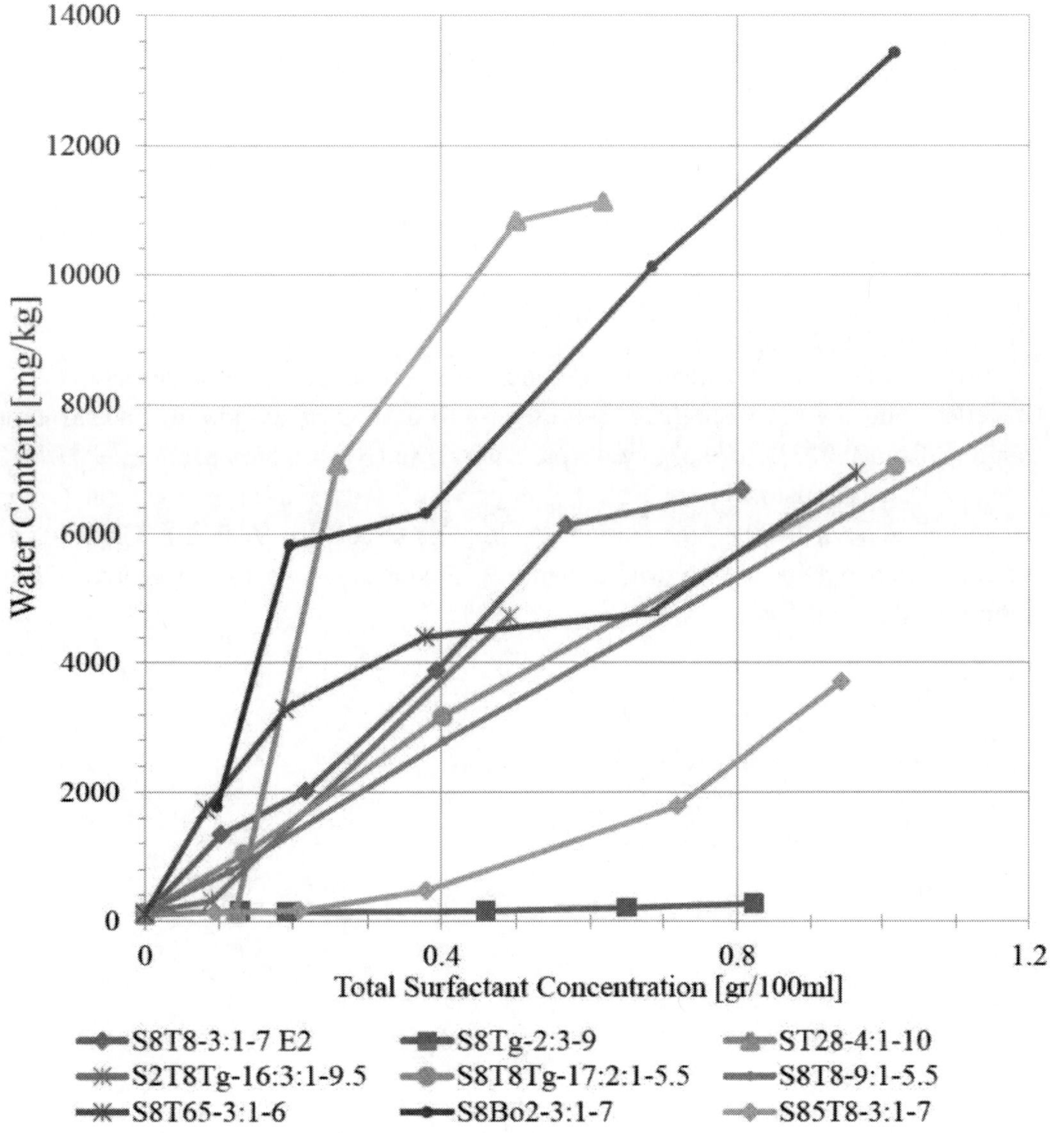

Figure 1: Solubilization Graphs for 9 different formulations tested as described in 2.3 after 7 days settling

2.2. <u>High concentration tests</u>

Four formulations were tested at concentrations ranging from 0.5-5% as described in 2.3, to ascertain the maximum ability to solubilize water. It was found that for surfactant concentrations in the range 1.5-5%, the maximum amount of water one can keep in emulsion does not increase with surfactant concentration, and depends only on the nature of the formulation. The results are summarized in table 2.

Table 2: results of solubilization tests at high concentrations

Formulation	HLB	Maximum water content [gr/L]	Clarity Rating
S8T8-3:1-7	6.99	3.5	1
S2Tg-3:2-10	9.97	8.8	1
S8S2-3:2-6	6.04	7.4	4
Span 20 alone	8.60	5.3	3

2.3. Low concentration tests

In order to examine the critical micelle concentration phenomenon in reversed (W/O) micelles, emulsions were prepared with the S8T8-3:1-7 formulation in concentrations ranging from 0.025% to 0.5%, with pure water and with water containing 20% 2-Butanol. The emulsions were tested 3 days and 24 days after preparation (which should be enough for the precipitation of droplets larger than 600nm). The resulting solubilization graphs are shown in figure 2. Figure 3 shows the respective clarity ratings of the emulsions.

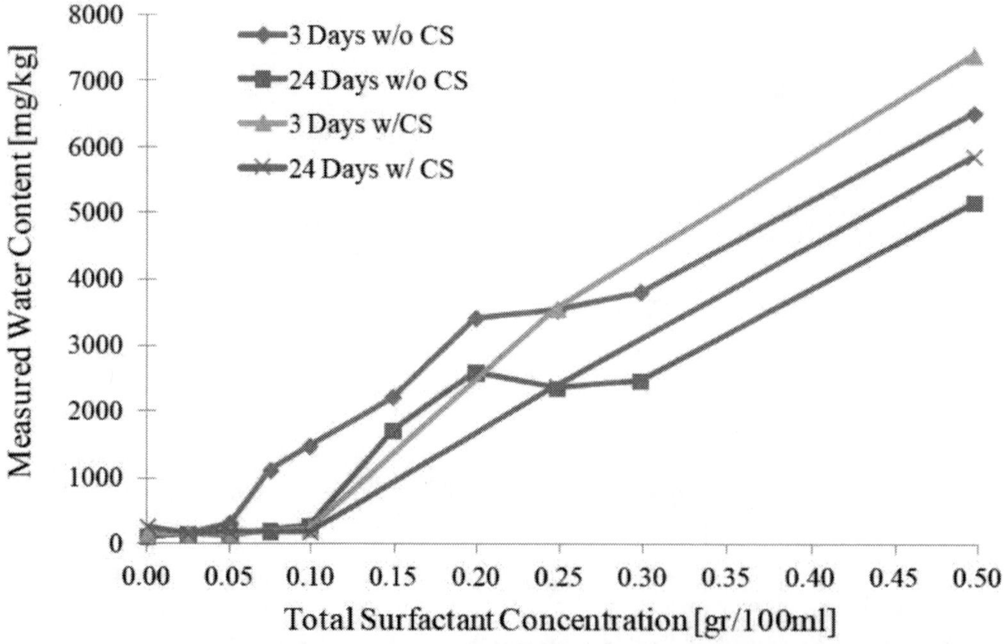

Figure 2: Solubilization graphs for low concentrations of S8T8-3:1-7 for water with and without cosurfactant (20% 2-Butanol in water)

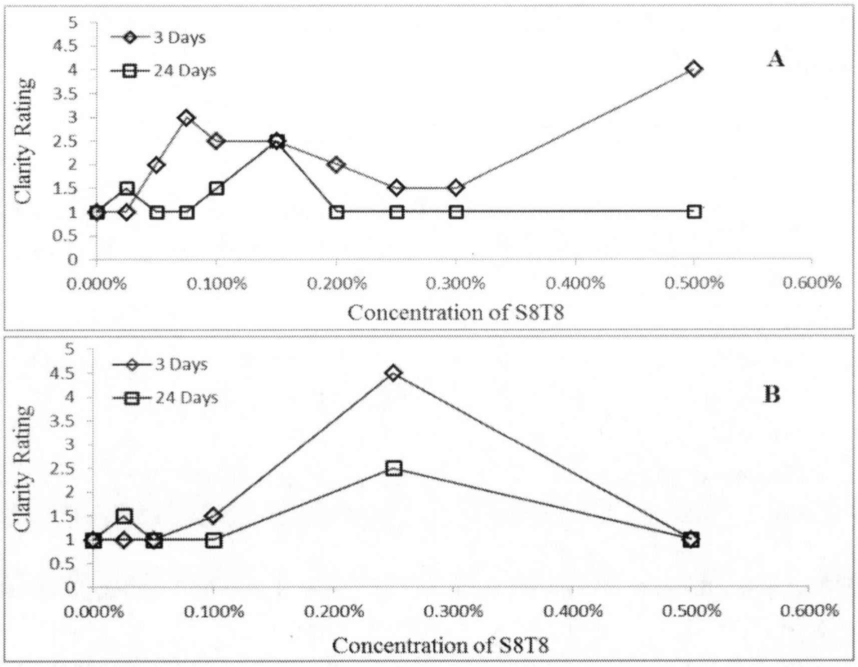

Figure 3: Clarity ratings (1-5) of the low-concentrations emulsions without cosurfactant (a) and with 2-butanol (b) after 3 and 24 days

It can be seen that both with and without the cosurfactant, the S8T8 formulation has a similar critical concentration between 0.10-0.15%, below which no water can be solubilized in a stable emulsion. This is probably the critical micelle concentration, which is independent of alcohol presence, although no examination was made of the micelle structure. Without the co-surfactant, subcritical concentrations (<0.10%) created a temporary, cloudy, unstable emulsion, while with the co-surfactant there was practically no difference between the emulsion after 3 days and after 24 days (further inspection after 60 days revealed no change in clarity ratings).

However, a more general statement about the effect of the co-surfactant on the kinetics of the emulsion settling cannot be made from this test, as the drop in water content from 3 to 24 days is similar in both cases.
In both cases, after settling, the borderline concentration emulsions were cloudy, as opposed to the clear solutions with lower (for lack of emulsified water) and higher (for sufficient surfactant to create nanoscale droplets) concentrations.

2.4. Clouding point

A series of 12 emulsions were prepared for the testing the effect of surfactant type, concentration, and co-surfactant on the low-temperature behavior of the fuel. The emulsions were prepared with surfactant concentrations higher than dictated by the solubility ratio. The components (surfactant and water/water-alcohol mix) were mixed into fuel similarly to 2.3 at 5x the final concentrations, and then diluted appropriately. Most samples (except for #7 and #8) showed a small precipitate of water droplets. Samples 9-12 were left to stand for several days and then filtered through a Whatman №.41 filter paper, while samples 1-8 were filtered immediately after dilution. The samples are detailed in Table 3 below; the listed concentrations are for the emulsions prior to filtering.

Table 3: Samples for the clouding point tests

#	Formulation	Surfactant conc. (gr/100ml)	Water/water-alcohol conc. [gr/100ml]	Added alcohol [%v/v]
1	S8T8-3:1-7	0.75	0.5	-
2	S8T8-3:1-7	0.75	0.5	5% EtOH
3	S8T8-3:1-7	0.75	0.5	20% EtOH
4	S2T8-4:1-10	0.75	0.5	-
5	S8T8-3:1-7	1.0	0.5	-
6	S2T8-4:1-10	1.0	0.5	-
7	S8T8-3:1-7	0.30	0.2	-
8	S8T8-3:1-7	0.15	0.1	-
9	S8T8-3:1-7	0.75	0.625	20% 2-BuOH
10	S8Bo2-3:1-7	0.75	0.5	-
11	S8Bo2-3:1-7	0.75	0.625	20% 2-BuOH
12	S8Bo2-3:1-7	1.5	0.75	-

All samples were tested in a clouding point apparatus as described in 2.5 to yield a graph of absorbance as a function of temperature. The graphs are shown in figures 4-7, grouped for comparisons. The "Fuel Blank" sample was a mix of batches of working jet fuel, similarly filtered. All emulsions demonstrated a clouding effect at a temperature much higher than the fuel freezing point, but below 0°C. This process starts off as the creation of visible "fibers" in the liquid which upon further cooling and mixing cloud the entire emulsion. The temperatures at which it happens seem to bear no correlation to the separate freezing points of the surfactants or the water/water-alcohol mixtures. However, they are in the same range as the freezing of "interfacial" or "bound" water (according to the measurements and distinction from "free" or "bulk" water made by Garti et al. [27]). "Dry" solutions of 0.75% S8T8, S2T8 and S8Bo2 without added water did not exhibit this transition at all, although S2T8 and S8Bo2 did show partial clouding at lower temperatures -20 to -30°C, possibly attributable to partial precipitation of the Span 20 and Brij O20 surfactants.
An additional point of interest is that there is no discernible change in transmission at 0 to -4°C in the emulsions, despite the observation by Garti et al. that even in microemulsions, some of the water freezes at these temperatures. Two possible explanations are (i) ice particles are created at 0°C, but are too small for effective light scattering or (ii) in the case of these low-concentration emulsions, nearly all of the water exists as "interfacial" or "bound" water. A DSC investigation in order to identify the interfacial/free water by their freezing/melting peaks, similarly to [27], detailed in 3.5, was mostly unsuccessful, probably due to the low water concentrations.

Finally, for these emulsions, this process was seen to be reversible: the cloudy emulsion clears up upon mixing and heating back to room temperature. In some emulsions a small amount of water droplets separated after cooling, and disappeared upon vigorous mixing.

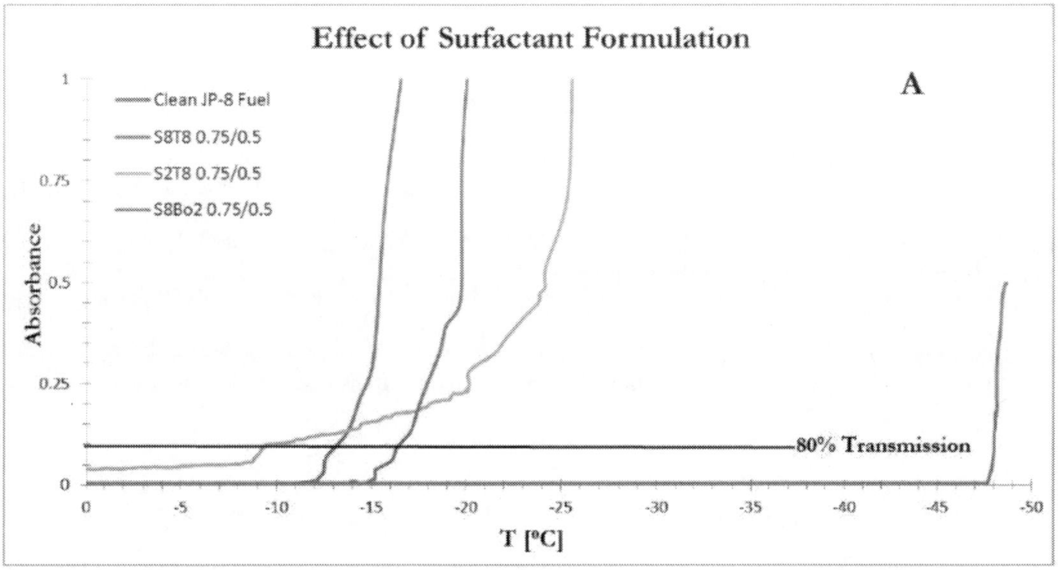

Figure 4a: Temperature-Absorbance graphs of the three formulations with similar initial surfactant (0.75%) and water concentrations (initial 0.5%, measured after filtration 0.2-0.25%). It can be seen that for S8T8 and S8Bo2 the transition is sharp, while for S2T8 it is smoother, and the original room-temperature emulsion is less clear. The black line denotes the 80% transmission point.

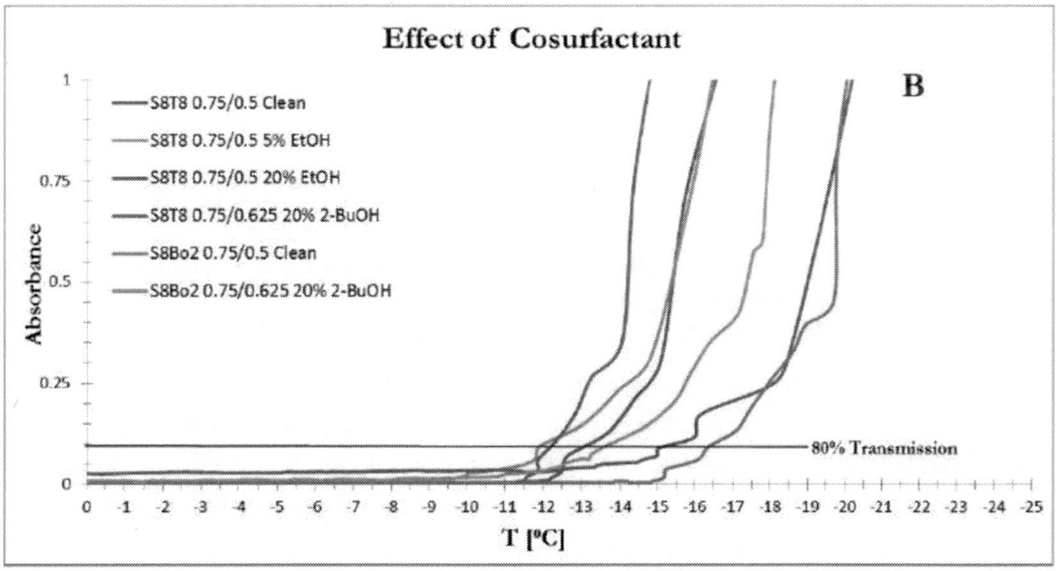

Figure 4b: Temperature-Absorbance graphs of S8T8 and S8Bo2 at similar concentrations, with and without various cosurfactants, as detailed in the label. It can be seen that the cosurfactants have a noticeable but inconsistent effect on the clouding transition.

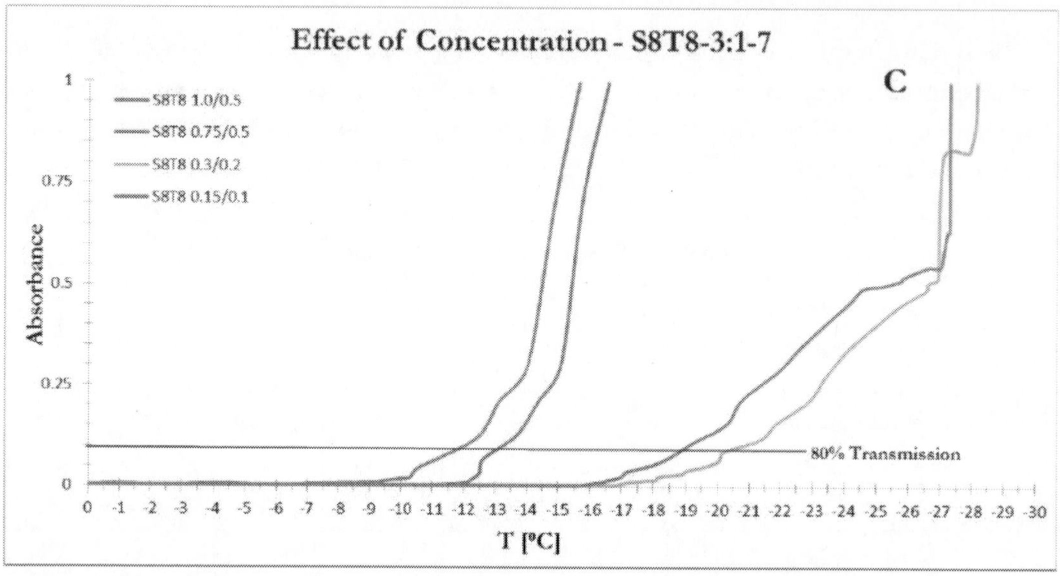

Figure 4c: Temperature-Absorbance graphs of S8T8 emulsions at various surfactant and water concentrations.

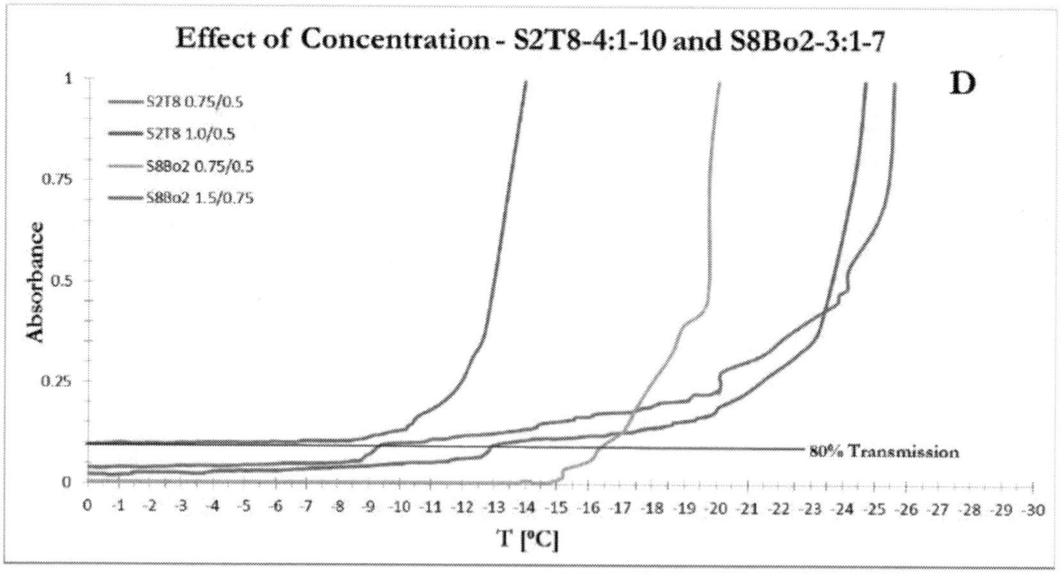

Figure 4d: Temperature-Absorbance graphs of S2T8 and S8Bo2 emulsions at different surfactant and water concentrations.

Table 4 below summarizes the temperatures at which this clouding transition occurs (defined arbitrarily as the temperature at which the transmission drops below 80%), the transmission of 500nm light at room temperature (as described in 2.4) and the measured water content of emulsions 1-8 after filtering, before and after freezing to -50°C.

Table 4: Clouding points, clarity and water content of tested emulsions

#	Composition	Clouding Temperature [°C]	% Transmission @500nm, RT	Water content before freezing [gr/100ml]	Water content after freezing [gr/100ml]
	"Pure" JP-8 Jet Fuel	-48	100		
1	S8T8 0.75/0.5	-14	92	0.190	0.177
2	S8T8 0.75/0.5 5%EtOH	-13	84	0.263	0.282
3	S8T8 0.75/0.5 20%EtOH	-16	90	0.280	0.281
4	S2T8 0.75/0.5	-9	95	0.264	0.322
5	S8T8 1.0/0.5	-12	90	0.274	0.324
6	S2T8 1.0/0.5	-13	75	0.426	0.430
7	S8T8 0.30/0.20	-21	97	0.071	0.110
8	S8T8 0.15/0.10	-19	96	0.047	0.048
9	S8T8 0.75/0.625 20%BuOH	-12	80		
10	S8Bo2 0.75/0.5	-16	80		
11	S8Bo2 0.75/0.625 20%BuOH	-14	80		
12	S8Bo2 1.5/0.75	-9	55		

The water content after filtration drops, in most emulsions, to less than half of original added water. Since presumably the filter passes droplets with a certain cutoff size of several microns, and this cutoff size is larger than scattering size (as can be seen from samples 6 and 12 which are cloudy after filtering), this gives an indicator of droplet distribution.

Under these assumptions, three factors can be seen to increase the amount of small droplets: the addition of alcohol to the water, increase of surfactant concentration and surfactant/water ratio, and formulation itself.

Interestingly, in all samples except one, the water content increased after freezing, in some cases significantly. This could happen if the emulsions, which are not water-saturated, absorbed moisture from the air when cold.

2.5. DSC

Differential Scanning Calorimetry was performed using a TA Q100 calorimeter on several emulsions throughout this study, in the modulated DSC mode. 5-10mg samples in Al pans were cooled under nitrogen from 25°C to -70°C at a rate of 3°C/min, held at -70°C for 5min and then heated back to 25°C at the same rate. The modulation was ±0.79°C every 60s. An exemplary thermogram is shown in figure 5.

The main objective of the DSC analysis, apart from studying the effect of added surfactant on the freezing of bulk jet fuel, was to identify an energetic marker for the clouding transition described in 3.4, and the freezing/melting peaks for "bulk" and "interfacial" water. No such peaks or other markers were found. However, a few of the samples, of water-saturated emulsions, exhibited a clear endotherm at -6 to 0°C during heating, with no corresponding exotherm upon cooling (seen in the thermogram). This indicates a possibility that for these emulsions, freezing induces a gradual and undetectable by DSC precipitation of ice. Upon heating, this ice melts normally at ~0°C.

All samples showed the freezing (exotherm) and melting (endotherm) peaks of the bulk fuel, with no observable difference between clean fuel and the various emulsions. A marked difference between the freezing and melting profiles could be seen in all samples, with a singular freezing peak having a sharp onset at ~-45°C and practically no irreversible component, while the melting process consisted of at least two stages, with an irreversible component accounting for approx. 40% of total heat flow. A similar effect was recorded in the work by Zabarnik and Widmor [28] where it is attributed to supercooling. In addition, nearly all emulsions showed a very small, reversible step transition at temperatures between -25 and -40°C. There is no correlation between the temperatures of this transition and the clouding points observed in 3.4 and no further investigation was made into this phenomenon.

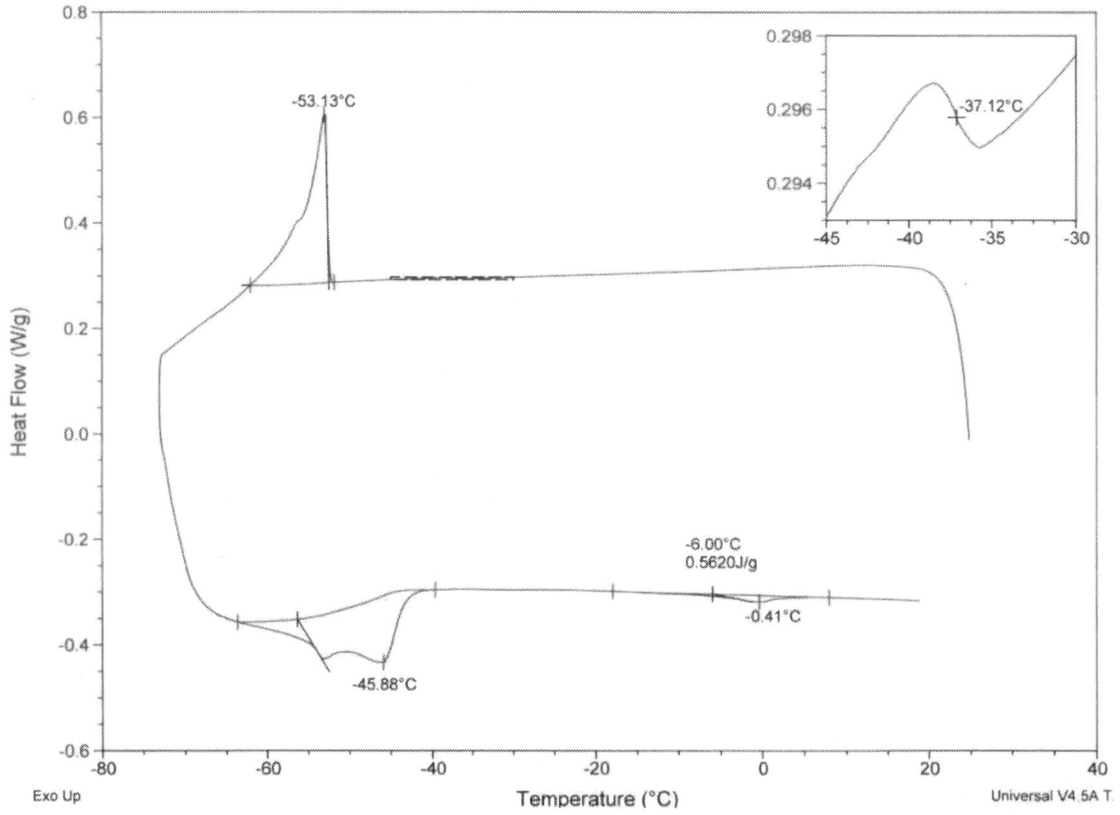

Figure 5: A representative DSC thermogram of an S2T8 water-saturated emulsion (~1% surfactant and 1% water). The top part is from the cooling ramp and the bottom from the heating ramp. Clearly visible is the difference between the freezing exotherm and melting endotherm of the bulk fuel, repeated in all experiments. In addition, there is a melting endotherm around 0°C without a corresponding freezing exotherm. Finally, there is a small phase transition, seen both at cooling and heating, around -25 to -40°C (shown in the inset). This transition appeared in most emulsions and its nature is unknown.

2.6. Foaming

Solutions of various concentrations of S8T8-3:1-7, S2T8-4:1-10 and S2Tg-3:1-10 were prepared with and without water, and tested for two foaming properties (foam volume at 5ml/min flow and 500ml foam disappearance time) as described in 2.6.

It was found that compared to both pure and working jet fuel, S8T8 started showing slightly increased foaming only at concentrations larger than 4%. For S2T8 and S2Tg there was slightly increased foaming at concentrations larger than 1%. Even at surfactant concentrations as high as 10% a stable foam layer could not be created. Finally, the addition of water drastically lowered the foaming of the fuel, presumably by "taking up" surfactant for emulsification leaving less available for foam stabilization.

2.7. Pour Point

Samples no. 1, 3, 4, 5 and 7 from table 3 and pure jet fuel were tested for bulk pour point of the fuel according to ASTM D-97. The test includes gradual cooling of the sample in steps of 3°C, and at each step the sample compartment is automatically lifted and tilted to 45° and then to 90° to observe the flow of the liquid using immersion detectors. If no flow is measured for a specified time period, the temperature is reported as the pour point of the liquid. The test accuracy is ±3°C.

All tested samples gave a pour point of -47°C, except for sample no.4 which gave a pour point of -44°C. Both this test and the DSC measurements indicated that the emulsion has no apparent effect on the bulk freezing of the fuel.

2.8. Ester (fuel-soluble) and alcohol (water-soluble) cosurfactants

A set of 5 emulsions of the S8T8-3:1-7 formulation were prepared with 2 different esteric (fuel-soluble) cosurfactants - butyl acetate and cinnamyl acetate, with and without 2-butanol and tested according to 2.3. All cosurfactants were added to a prepared emulsion and mixed again. Table 5 below details the prepared solutions and resulting solubilization ratios. All concentrations are given in mg/100ml.

Table 5: Solubilization results of esteric cosurfactants

Concentration of S8T8 mix	**Concentration of Bu.Ac.**	**Concentration of Cin. Ac.**	**Conc. of 2-BuOH**	**Solubilization Ratio** vs. surfactant alone	vs. entire mix
540	0	0	0	0.66	
395	161	0	0	0.70	0.50
454	61	0	26	0.70	0.58
391	60	46	0	0.68	0.54
366	61	76	30	0.80	0.55

It can be seen that given a constant amount of S8T8 mix, addition of esters allows more water to be solubilized (solubilization ratio vs. surfactant alone), and addition of 2-butanol helps this process even more. However, the effect of added esters is much weaker than addition of a similar mass of surfactant (solubilization ratio vs. entire

mix). A possible explanation is that the esters act as "fillers" in the micelle structure, but less so than the surfactant molecules themselves.
No difference in clarity or stability was observed between the emulsions of S8T8 alone, S8T8 and the esters and S8T8 + esters + 2-butanol.

The second experiment, which was combined with the testing of new formulations, involved testing an emulsion (according to 2.4) of the new formulation (at a concentration of 0.4gr/100ml) with water alone and with water/alcohol solutions. The results are detailed in table 6 below. Initial testing of clarity rating and water content was done after 4 days. 7 days after that the emulsion was re-tested for water content.

Table 6: Solubilization results with alcohol co-surfactants

Formulation	Dispersed Phase	Clarity Rating	Solubilization Ratio Initial	Solubilization Ratio Change after 7 days [%]
S8T8-3:1-7	Clean	1.5	0.79	0
S8T65-3:1-6	Clean	3	0.30	-8
	10% IPA	4	0.60	-22
S8Bo2-3:1-7	Clean	2	0.34	-2
	5% EtOH	3	0.30	-23
S2T8-4:1-10	Clean	1.5	0.53	0
	5% 2-BuOH	2.5	1.02	0

These results indicate that adding alcohol to the dispersed phase seems to change the droplet distribution to include a larger amount of medium-sized droplets. These droplets initially contribute to the solubilization ratio, doubling it in the case of isopropanol and 2-butanol, but sink within 4 days (for EtOH) to 11 days (for IPA). However, this lies at odds with the results of the low-concentration tests, which indicate that 2-butanol prevents the formation of unstable droplets at sub-critical concentrations of surfactant.
It can be tentatively proposed that 5% 2-butanol allows the smallest droplets (which contribute only slightly to the clouding and take more than 11 days to sink), followed by 10% IPA, and finally 5% EtOH (which sink within 4 days). This is consistent with the predicted changes to water surface tension according to [29].

3. DISCUSSION

3.1. Formulations

Three surfactant formulations, among those tested, were identified as the most promising for future research into applications of solubilizing water in jet fuel.

The first formulation (and one most extensively tested) is S8T8-3:1-7, consisting of Span 80 (75% wt.) and Tween 80 (25% wt.) nonionic sorbitol-ester based surfactants. This formulation gives a maximum solubilization ratio of 0.7-0.8 and at lower water content (surfactant-water ratio 2:1 or lower) creates a completely clear nano- or micro-emulsion, stable for over 60 days without precipitation of water droplets. This formulation has a minimum concentration for water solubilization of approximately 0.10-0.15 gr/100ml. Another advantage is that the two components of the formulation are easily mixable medium viscosity liquids.

The second formulation is S2T8-4:1-10, consisting of Span 20 (80% wt.) and Tween 80 (20% wt.) surfactants. This formulation gives a maximum solubilization ratio of ~1.9 and creates slightly cloudy emulsions. Gravity filtration results in clear or nearly-clear emulsions, stable for over 30 days, with approximately half of original water content.

The third formulation is S8Bo2-3:1-7, consisting of Span 80 (75% wt.) and Brij O20 (25% wt.). This formulation gives a maximum solubilization ratio of ~1.5 and creates cloudy emulsions. A disadvantage of this formulation is that Brij O20 (a waxy solid) has low solubility in jet fuel (requires heating and/or sonication) and precipitation may occur.

3.2. Effects of co-surfactants

An in-depth investigation of the effect of various co-surfactants on the kinetics, structure and properties of the emulsions is yet to be made, but some indications can be gleaned from the separate experiments. There are two theoretical mechanisms of action for the co-surfactants: (i) dissolution in the aqueous phase, lowering the surface tension of the water droplets to a low enough level to sustain a microemulsion, and (ii) mixing into the micelle structure itself, either at the surfactant-water or the surfactant-fuel interface and changing the packing parameter. A third mechanism of lowering the surface tension of the fuel is deemed irrelevant because the fuel-surfactant surface tension is usually negligible compared to the water.

Alcoholic co-surfactants seem to have a strong effect on droplet size distribution and possibly droplet stability. This effect varies greatly with the identity and concentration of the alcohol, and, in emulsion, relies on the lowering of water-surfactant surface tension. Adding alcohol to the water seems to facilitate the breaking up of large droplets into a distribution of small (<200nm) to medium (500-1000nm)-sized droplets. The largest among them settle after a long while, explaining the large drop in water content (this means that alcohol-containing emulsions require a longer time to reach a constant or nearly constant water content). However, the total amount of small water droplets, which defines the water content after settling or filtering, is larger in the presence of alcohol, increasing the solubilization ratio. At surfactant concentrations too low to sustain micelles, an opposite mechanism seems to occur, in which very unstable water droplets form in the fuel-surfactant mix under high shear only without the presence of alcohol.

Ester co-surfactants, butyl and cinnamyl acetate allow increased solubilization of water given the same amount of surfactant, possibly by filling the micelle structure. However, since they are unable to form micelles by themselves and have no amphiphilic structure, per weight the esters are less effective than adding the same

amount of surfactant. The overall effect is slight, and no significant influence was seen on the clarity or stability. Nevertheless, since these small esters are less "damaging" to add to the fuel (being lighter, easier for combustion and less likely to leave residue) than the surfactant itself, they might find use as a low-side-effect method of fine-tuning the emulsion properties.
Finally, it must be noted that all discussion regarding micelle size and structure is only speculative, deriving from theory and observations of clarity, stability and filtering, and not direct observation of the micelles themselves (by microscope) or the true size distribution (by centrifugation or other means), which must be made in the future.

3.3. Properties of the emulsions

A standing question throughout the study is whether the created systems are thermodynamically stable microemulsions. On the one hand, the water solubilization seems to be strongly dependent on mixing and shear conditions. On the other hand, after settling, the top layer of the emulsions is usually clear (or clear after filtration) and stable for a long time with no change in clarity, water content or precipitation.
A probable mechanism is that saturating the surfactant and subjecting it to moderate shear, as was done in most experiments, creates a very wide distribution of micelle sizes, which is strongly dependent on shear conditions, concentrations of water, surfactant and cosurfactant, and other yet-unmeasured parameters, such as temperature and the jet-fuel composition. The droplets which are smaller than a certain cutoff size stay in solution after settling and are stable against coalescing either kinetically or thermodynamically by the low surface tension of the micelles (lowered even further by addition of alcohol cosurfactant). These droplets alone make up the measured water content and are responsible for the solubilization ratio.
All emulsions have, upon cooling, a phase transition occurring between -10 and -20°C, which manifests itself first as white filaments, and finally as clouding of the entire emulsion. Heating the fuel back to room temperature leaves a precipitate of water droplets. If the emulsion is not water-saturated, this is completely reversible, i.e. the droplets disappear upon mixing. This process is specifically a property of the emulsion - solutions of the surfactant alone showed no such transition.
The transition is sharp for S8T8 and S8Bo2 and occurs at -14°C and -16°C for emulsions with the two surfactants respectively, and more spread out for S2T8. Lower concentrations of the emulsion logically lead to a smoother clouding transition. Finally, there is indication that at least some cosurfactants, such as ethanol, lower the temperature at which this clouding occurs.
All these hint at the freezing of "bound" or "interfacial" water - water molecules strongly interacting with the surfactant - as the mechanism responsible for the clouding. On the other hand, no energetic trace of this process was observed by DSC, probably due to the low water concentrations, and the "spreading out" of the transition across a wide temperature range as the interfacial water exists in a variety of chemical environments.
The complete lack of any changes close to 0°C indicates that "free" water at the core

of the micelles does not exist in unsaturated emulsions, which is consistent with the findings of Garti et al.
A complementary description is given by Lam et al. [30] who show that low-temperature precipitation of water droplets (which in our system exist a-priori) out of clean toluene occurs first as supercooled water and only at -30°C- -40°C does ice crystallization occur. This might indicate that the surfactants raise the temperature of crystallization of the supercooled water.
A practical consideration which must be eventually addressed, assuming this clouding is unavoidable, is whether the precipitate may lead to blockage of fuel pumping and filtration systems within the aircraft, or changes in fuel composition as a result of selective water precipitation.

Two operational properties of the bulk fuel were tested - foaming and pour point. It was found that within the concentration limits of this study, there is no detectable influence of the emulsion on either property, although higher concentrations of surfactant may increase foaming.
Some properties specified in ASTM D1655, such as water reaction, microseparometer rating, GUM content or distillation residue will be unavoidably changed. However, these are indicative and not directly operational properties. Relevant operational properties which might be changed and must be tested for are, for example, thermal oxidation stability (JFTOT), corrosion, viscosity at low temperatures, combustion characteristics or flash point. Long-term storage effects must also be recorded. Finally, significant differences in emulsion behavior between "pure" and "working" jet fuel were not observed, hinting that other additives, due to their low (~100ppm total for most additives and <0.25% for DiEGME and FSII) concentrations, have a limited influence on emulsification. However, the reverse effect, of the emulsion on the action of various additives (CI-LI, antioxidants etc.) is likely to be significant and must be tested for.

4. SUMMARY AND FURTHER WORK

This study demonstrates an initial, but promising direction at the challenge of adding effective amounts of water to jet fuel, while at the same time minimizing the total amount of other foreign materials and the influence on operational properties. This is to eventually allow an improvement in the performance of currently used turbine fuels with minimal adaptations and without damage to the existing fuel systems (from supply line to engine).
A large variety of two- and three-component nonionic surfactant formulations was tested, of which three were chosen for further study for their large, consistent solubilization ratios (weight of maximal solubilized water / weight of added surfactant), clear resulting emulsions (with or without filtering) and stability over more than 30 days. At low concentrations of surfactants, the maximum stably solubilized water is about 0.75-1.5 of surfactant concentration and at most ~10gr/L

(1% w/v). However, better emulsion properties (stability toward environmental changes) are achieved with larger surfactant : water ratios (i.e. 2:1).
A critical concentration, below which water solubilization is impossible, was found for one of the formulations to be between 1.0-1.5 gr/L, with the other two formulations expected to have values of the same order. Low-temperature behavior of the emulsions was investigated, and all showed a clouding induced by a phase transition at -10 to -20°C. This process could be the freezing of water molecules which are strongly interacting with the surfactant, although DSC analysis did not show this transition. If so, this means that (i) at low, unsaturated water concentrations, no water exists in a "bulk"/free state (supported by other DSC observations) and (ii) temperature-transmission measurements are a useful substitute for DSC in this aspect for very low water amounts.

Observations of experiments which included alcohol co-surfactants hint at their promise, in addition to being an energetic fuel in their own right, in improving the properties of the emulsion by shifting the droplet size distribution and stabilizing the resulting nano-emulsion. Fuel-soluble co-surfactants, such as small and medium-sized esters, may also find limited use in fine-tuning of the emulsion.

Further investigation into this challenge leads to two additional steps. First, in order to bring the improved fuel closer to practical application, one must subject it to a variety of tests to determine possible effect on operational properties and necessary adaptations. Emulsion-specific properties, such as stability in the long term and against various environmental changes, should be investigated as well. Additionally, various engine run tests must be made to determine whether the added water, at such concentrations, delivers on the promise for more efficient and cleaner combustion.
Separately, an in-depth investigation into micelle structure and size distribution and change (using direct means such as TEM or indirect means such as an optical centrifuge), as well as dependence on untested parameters such as temperature and fuel composition is recommended in order to have a clearer picture of the physics of the system.

The authors would like to thank Ruth Bohmerwald and Mor Ben-Shimol of the Materials Department Central Fuel Laboratory for their vital assistance with the jet fuel testing instruments.

5. REFERENCES

[1] F. Dryer, "Water Addition to Practical Combustion Systems - Concepts and Applications," *Power Systems,* pp. 279-295, 1976.

[2] J. B. Greenberg, Y. Mindelis and R. L. Axelbaum" ,A simple model of a water/fuel spray diffusion flame ",*Fuel* 181 ,pp. 547-556, 2016 .

[3] T. Kadota and H. Yamasaki" ,Recent advances in the combustion of water fuel emulsion ",*Progress in Energy and Combustion Science* 28 ,pp. 385-404, 2002 .

[4] J. W. Park, K. Y. Huh and J. H. Lee" ,Reduction of NOx, smoke and brake specific fuel consumption with optimal injection timing and emulsion ratio of water-emulsified diesel ",*Proceedings of Institution of Mechanical Engineers* ,2015 ,pp. 83-93, 2002 .

[5] O. Armas, R. Ballesteros, F. J. Martos and J. R. Agudelo" ,Characterization of light duty Diesel engine pollutant emissions using water-emulsified fuel ",*Fuel* ,84 ,pp. 1011-1018, 2005 .

[6] T. Yatsufusa, T. Kumura, Y. Nakagawa and Y. Kidoguchi" ,Advantage of Using Water-Emulsified Fuel on Combustion and Emission Characteristics ",*Proceedings of the European Combustion Meeting* .2009 ,

[7] K. Kannan and M. Udayakumar" ,NOx and HC emission control using water emulsified diesel in single cylinder diesel engine ",*ARPN Journal of Engineering and Applied Sciences* ,8 ,pp. 59-62, 2009 .

[8] M. Huo, S. Lin, H. Liu and C. F. Lee" ,Study on the spray and combustion characteristics of water–emulsified diesel ",*Fuel* ,123 ,pp. 218-229, 2014 .

[9] S. Park, S. Woo, H. Kim and K. Lee" ,Effect of Diesel−Water Emulsified Fuel on the NO x and PM Emissions of a Diesel Engine ",*Energy & Fuels* ,30 ,pp. 6070-6079, 2016 .

[10] D. Scarpete" ,DIESEL-WATER EMULSION, AN ALTERNATIVE FUEL TO REDUCE DIESEL ENGINE EMISSONS. A REVIEW ",*Machines, Technologies, Materials* ,7/2013 ,pp. 13-18, 2013 .

[11] H. Shyam Prasad, J. Gonsalvis and V. S. Vijay" ,Effect of Introduction of Water into Combustion Chamber of Diesel Engines - A Review ",*Energy and Power*1)5 ,A), pp. 28-33, 2015 .

[12] C. H. Wang and C. K. Law" ,Microexplosion of Fuel Droplets under High Pressure ", *Combustion and Flame* ,59 'מס ,pp. 53-62, 1985 .

[13] J. P. Wilson ,*Effects of Water Injection and Increased Compression Ratio in a Gasoline Spark Ignition Engine* ,MSc Thesis, University of Idaho, 2011 .

[14] H. Jahani and S. R. Gollahalli" ,Characteristics of Burning Jet A Fuel and Jet A Fuel-Water Emulsion Sprays ",*Combustion and Flame* ,37 ,pp. 145-154, 1980 .

[15] E. C. Wenzel and H. W. Steinmann" ,US Patent 4,083,698: Clear and Stable Liquid Fuel Compositions for Internal Combustion Engines .1978 ",

[16] E. A. Hazbun, S. G. Schon and R. A. Grey ,*US Patent 4,744,796: Microemulsion Fuel System* .1988 ,

[17] H. Rivas et al. ,*U.S. Patent 7,704,288 B2: Water in hydrocarbon emulsion useful as low emission fuel and method for forming same* .2010 ,

[18] E. Ruckenstein "The Origin of Thermodynamic Stability of Microemulsions ",*Chemical Physics Letters* ,57/4 ,pp. 517-521, 1978 .

[19] P. P. Constantinides and S. H. Yiv" ,Particle size determination of phase-inverted water-in-oil microemulsions under different dilution and storage conditions ",*International Journal of Pharmaceutics* ,115 ,pp. 225-234, 1995 .

[20] P. P. Constantinides and J. Scalart" ,Formulation and Physical Characterization of Water-in-Oil Microemulsions Containing Long- Versus Medium-chain Glycerides ", *International Journal of Pharmaceutics* ,158 ,pp. 57-68, 1997 .

[21] P. A. Reynolds, E. P. Gilbert and J. W. White" ,High Internal Phase Water-in-Oil Emulsions and Related Microemulsions Studied by Small Angle Neutron Scattering. 2. The Distribution of Surfactant ",*Journal of Physical Chemistry B* ,105 ,pp. 6925-6932, 2001 .

[22] I. Sole, C. Solans, A. Maestro, C. Gonzales and J. M. Guttierez" ,Study of nano-emulsion formation by dilution of microemulsions ",*Journal of Colloid and Interface Science* ,376 ,pp. 133-139, 2012 .

[23] M. R. Noor El-Din, A. M. Al-Sabagh and R. Miller" ,Preparation of Water-in-Jet Fuel Nano-emulsions Using a High Energy Method ",*International Journal of Green Nanotechnology: Physics and Chemistry* .2010 ,

[24] M. R. Noor El-Din" ,Study on the stability of water-in-kerosene nano-emulsions and their dynamic surface properties ",*Colloids and Surfaces A: Physicochem. Eng. Aspects* , ,390pp. 189-198, 2011 .

[25] A. M. Al-Sabagh, M. M. Emara, M. R. Noor El-Din and W. R. Aly" ,Preparation of Water-in-Diesel Fuel Nanoemulsions Using High-Energy Emulsification Method and a Study of Some of Their Surface Active Properties ",*Journal of Dispersion Science and Technology* ,33 ,pp. 970-976, 2012 .

[26] B. Acharya, P. S. Guru and S. Dash" ,Tween-80–n-Butanol–Diesel–Water Microemulsion System—A Class of Alternative Diesel Fuel ",*Journal of Dispersion Science and Technology* ,35 ,pp. 1492-1591, 2014 .

[27] N. Garti, A. Aserin, I. Tiunova and M. Fanun" ,A DSC Study of Water Behaviour in Water-in-Oil Microemulsions Stabilized by Sucrose Esters and Butanol ",*Colloids and Surfaces A: Physicochem. Eng. Aspects* ,170 ,pp. 1-18, 2000 .

[28] S. Zabarnick and N. Widmor" ,Studies of Jet Fuel Freezing by Differential Scanning Calorimetry ",*Energy & Fuels* ,15 ,pp. 1447-1453, 2001 .

[29] G. Vazquez, E. Alvarez and J. M. Navaza" ,Surface Tension of Alcohol+Water from 20 to 50 C ",*J. Chem. Eng. Data* .1995 ,

[30] J. K.-W. Lam, J. I. Hetherington and M. D. Carpenter" ,Ice growth in aviation jet fuel ", *Fuel* ,113 ,pp. 402-406, 2013 .

IASH 2017, the 15th INTERNATIONAL SYMPOSIUM ON STABILITY, HANDLING AND USE OF LIQUID FUELS
Rome, Italy
10-14 September 2017

A STUDY ON THE STABILITY OF MARINE DISTILLATE FUEL IN THE PRESENCE OF FAME AND HVO

Chrysovalanti E. Tsesmeli[1], George S. Dodos[2] and Fanourios Zannikos[3]

[1]*National Technical University of Athens, Iroon Polytechneiou 9 17580, Athens, Greece, ctsesmeli@cental.ntua.gr*
[2]*National Technical University of Athens, Iroon Polytechneiou 9 17580, Athens, Greece, gdodos@central.ntua.gr*
[3]*National Technical University of Athens, Iroon Polytechneiou 9 17580, Athens, Greece, fzanntua@central.ntua.gr*

The upcoming International Standard EN ISO 8217:2016 refers to the addition of biofuels, such as FAME and HVO, in distillate marine fuels and includes additional specifications for DF grades (DFA, DFZ, DFB) containing FAME up to 7% v/v. Notwithstanding the benefits arisen from biofuels, potential implications regarding the stability of distillate fuels may appeared. Due to its high affinity to water and the tendency to oxidation, FAME is often associated with microbial growth and storage stability issues. Additionally, in spite of the limited available literature data regarding the use of HVO in marine distillate fuels, the latter may also be considered as a renewable substitute. Taking the aforementioned into account, the introduction of bio-derived fuels in marine industry provokes great interest on investigating their impact on marine distillate fuels stability.

The aim of this study was to investigate to what extent the addition of FAME or HVO affects their microbial and oxidation stability of marine distillate fuels and whether they have an impact on marine fuel systems. In specific, FAME and HVO were blended with a distillate fuel used widely in marine industry at the concentration of 7% v/v. Two different arrays of contaminated microcosms along with non-contaminated ones have been set up and stored for a certain period of time under specific conditions in order to examine the behavior of these two biofuels. At predetermined intervals of storage time the microbial proliferation was monitored by the ASTM D7687 method. The oxidation stability was assessed both by the RSSOT (ASTM D7545) and Rancimat (EN 15751) methods which are considered as more suitable for assessing the aging reserve of fuels containing bio-based derivatives. Basic fuel quality parameters were also examined at the end of storage period in order to determine any alterations due to the presence of each biofuel. Moreover, the effect of FAME and HVO in the compatibility with a variety of representative metals was assessed.

IASH 2017, the 15th INTERNATIONAL SYMPOSIUM ON STABILITY, HANDLING AND USE OF LIQUID FUELS
Rome, Italy
10-14 September 2017

Update on EI 1535 Water Mapping Test Method

Keri Petersen[1] and Gary Bessee[2]

[1] *Southwest Research Institute, 6220 Culebra Rd., San Antonio, TX 78238, U.S.A., keri.petersen@swri.org*
[2] *Southwest Research Institute, 6220 Culebra Rd., San Antonio, TX 78238, U.S.A., gary.bessee@swri.org*

After the introduction of the water mapping test protocol to EI 1535 3rd edition in 2015, SwRI is investigating what other factors may impact the results of the test. Since the EI 1581 qualified coalesce/separators are operating outside their designed parameters, a study will be conducted at Southwest Research Institute to evaluate the differences and repeatability in results using filter elements from different manufacturing batches and filter elements from different manufacturers. This study will use an additive with an historical known performance. Two different batches of EI 1581 5th edition coalescers and separators will be ordered from multiple manufacturers. On each test day, elements from each manufacturer will be tested to eliminate variation caused by the fuel. At the end of the test program, these results will be added to SwRI's historical database of results to evaluate the repeatability of results with manufacturer and batch variation, as well as evaluate any notable trends in the results.

IASH 2017, the 15th INTERNATIONAL SYMPOSIUM ON STABILITY, HANDLING AND USE OF LIQUID
Rome, Italy
10-14 September 2017

Phenol in aviation fuel monitoring on IP583 / ASTM D7797 equipment

Paul Spitteler[1]

[1] Stanhope-Seta, London Street, Chertsey, UK, ps@stanhope-seta.co.uk

The primary use for the IP 583 / ASTM D7797 "FAME In Jet Instrument" is to screen aviation fuel for the presence of FAME contamination. Infrared spectra are collected from the sample and the sample stripped from polar components by a sorbent material. In addition to FAME other components like phenols are retained on the sorbent material and therefore show up in the absorbance spectrum constructed from the sample and the processed sample.

Hindered phenols that are generally used as antioxidants are not effectively captured in this system and therefore have no significant impact on the resultant spectra. The phenols that are observed in the spectra are the ones that find their way into the fuel during the refining process. This paper discusses the construction of a Partial Least Square model to determine phenol concentrations in aviation fuel from spectra obtained to determine FAME concentrations. This opens the possibility of developing a system to monitor phenol concentrations which involves no hardware modification and only a software addition to the IP583 / ASTM D7797 test. The spectral region of interest is shared with alcohols and the infrared spectra of phenols vary significantly with the phenol side groups and this poses some challenges to the model. Phenol concentrations in typical aviation fuels can be of the order of several hundred ppm. The reference data used in this study was obtained from a newly developed HPLC extraction method developed by ExxonMobil to distinguish between various phenols and measures their concentrations.

-Paul Spitteler et al., IASH 2013, Phenol in aviation fuel monitoring
on IP583 / ASTM D7797 equipment
-David J. Abdallah et al, IASH 2013, understanding phenol levels in jet fuel

IASH 2017, the 15th INTERNATIONAL SYMPOSIUM ON STABILITY, HANDLING AND USE OF LIQUID FUELS
Rome, Italy
10-14 September 2017

Microbial Contamination Test Methods in Bulk Fuel Storage

By Marlin Vangsness

University of Dayton Research Institute, 300 College Park, Dayton, Ohio 45469

The goal of this DLA Energy funded study is to evaluate low (normal) and elevated levels of microbial contamination in bulk storage of government owned fuel. Fuel was drawn from multiple sampling locations and tested using several off the shelf test kits as well as an AFRL developed procedure. Readings were compiled according to fuel type, test location, and test method. Preliminary observations on the variability in test results for microbial contamination in bulk fuel storage are reported.

IASH 2017, the 15th INTERNATIONAL SYMPOSIUM ON STABILITY, HANDLING AND USE OF LIQUID FUELS
Rome, Italy
10-14 September 2017

Fuel Composition-to-Performance Studies at the United States Air Force Research Laboratory's Fuels and Energy Branch

Edwin Corporan

Air Force Research Lab, Edwin.corporan@us.af.mil

Recent research efforts at the U.S. Air Force Research Laboratory to study the impacts of fuel composition on combustion performance metrics are described. Investigations include: 1) effects of fuel chemical and physical properties on lean blowout, combustion efficiency and ignition at low temperatures in a single nozzle swirl-stabilized combustor, 2) ignition delay studies of relevant jet fuel chemistries in a shock tube, and 3) studies of primary fuel pyrolytic decomposition products and their potential correlations with fuel combustion performance. The fuels evaluated consist of two-component surrogate blends (containing n-, iso-, cyclo-paraffins and/or aromatics), certified synthetic fuel blending formulations, and a conventional jet fuel with average properties. The selected fuels provide a wide range of chemical and physical properties, to help provide insight into their impact on performance. The combustion devices provide data at relevant turbine engine combustor environments and at conditions focusing only fuel chemical kinetics effects (shock tube). Potential correlations of combustor and shock tube performance with fuel physical and chemical properties, fuel reactivity, and primary product selectivities of the thermally cracked fuel are discussed.

IASH 2017, the 15th INTERNATIONAL SYMPOSIUM ON STABILITY, HANDLING, AND USE OF LIQUID FUELS
Rome, Italy
10-14 September 2017

COMBUSTION TESTING OF CANDIDATE ROCKET FUELS: OVERCOMING SPECIFICATION CHALLENGES THROUGH SUBSCALE EVALUATION

Matt Billingsley[1]

[1] *Air Force Research Laboratory/RQ, Edwards AFB, CA, USA, matthew.billingsley.1@us.af.mil*

Aerospace fuels are produced to meet specification limits imposed for the most part by engine operational expectations. Ideally, these metrics are representative of the intended application, guaranteeing fuel quality regardless of production method or source (conventional or alternative). In practice however, complex processes like combustion and regenerative cooling are sensitive to fuel compositional variability not directly controlled by specification tests. For example, chemical variation among several within-specification, highly refined rocket kerosene fuels was recently shown to drastically impact cooling system material compatibility and reliability.[1,2]

From a combustion standpoint, although thermochemical (e.g., heat of combustion and hydrogen content) and physical (e.g., viscosity and density) characteristics are commonplace in fuel specifications, verifying application-specific combustion behavior is not feasible with current test methods. Furthermore, while engine-level evaluation is ideal for confirming fuel fit-for-purpose quality, it is overly expensive and time consuming. In such cases, subscale testing is often employed to evaluate candidate fuels in controlled environments at a low cost, thereby verifying fuel performance and overall fit-for-purpose quality.

This presentation summarizes progress of an experimental campaign in which several candidate kerosene-based rocket fuels were tested in a subscale rocket thrust chamber. Performance characteristics will be reviewed with the purpose of drawing connections between specification properties and system-level behavior. Through this effort, we intend to identify potential deficiencies in the current rocket kerosene specification regarding fuel combustion in future aerospace platforms.

[1] Billingsley, et al., *14th Int'l Symposium on Stability, Handling, and Use of Liquid Fuels*, Charleston, SC, 2015.
[2] Billingsley, et al., (AIAA) 2016-4903, *52nd AIAA Joint Propulsion Conference*, Salt Lake City, UT, 2016.

IASH 2017, the 15th INTERNATIONAL SYMPOSIUM ON STABILITY, HANDLING AND USE OF LIQUID FUELS
Rome, Italy
10-14 September 2017

ASSESSMENT OF THE ROLE OF COPPER CONTAMINATION ON DIESEL FUEL STORAGE STABILITY

Kristina M. Myers[1], Robert E. Morris[1], Thomas N. Loegel[2]

[1] *NOVA Research, Inc., 1900 Elkin Street 230, Alexandria, VA, USA 22308*

[2]*U.S. Naval Research Laboratory, 4555 Overlook Ave. S.W., Washington, D.C., USA 20375.*

It has been shown in the past that high levels of copper are often found in stable F-76 diesel fuels, however there have also been isolated incidents where very unstable F-76 fuels have been found to contain high levels of dissolved copper. Recently, there have been incidents where F-76 diesel fuels from west coast refineries have produced unacceptably high levels of insolubles in storage. In order to help determine whether copper may be either responsible or a contributing factor to poor storage stability, experiments were performed to assess these issues.

A series of storage stability experiments were conducted on various F-76 diesel fuels including low sulfur diesel, ultralow sulfur diesel and high sulfur diesel. Controlled concentrations of dissolved copper were obtained by blending each fuel with the same fuel that had been exposed to copper turnings. Copper concentrations were verified by ICP-AES analysis in accordance with ASTM D7111. Tests were conducted with various copper concentrations with and without a metal deactivator additive. The impact of blending downgraded copper contaminated JP-5 into stored diesel fuel onboard Navy ships was also examined by testing blends of these diesel fuels with copper contaminated JP-5. Copper concentrations in these fuel samples ranged from 65 ppb to about 800 ppb in the neat samples. The samples were stressed according to ASTM D5304, ASTM D4625 and ASTM D7525 in order to gain a broader understanding of what may be occurring in these fuels during storage. During these storage stability tests, total insolubles and hydroperoxides were determined, and GC-MS analyses was conducted to detect any significant chemical changes.

IASH 2017, the 15th INTERNATIONAL SYMPOSIUM ON STABILITY, HANDLING AND USE OF LIQUID FUELS
Rome, Italy
10-14 September 2017

INTRODUCING ASTM METHOD D8071 for PIONA COMPOUND ANALYSIS BY GC-VUV

In March 2017 ASTM published a new test method using new technology to analyze gasoline for paraffins, isoparaffins, olefins (olefins correlate to stability), naphthenes, and aromatics. The method speciates and quantitates benzene, toluene and xylenes. It also quantitates oxygenates – methanol, ethanol, mtbe, etbe, etc The new method is featured in the latest ASTM Standardization news. You may also pull up the D8071 method scope from the ASTM website. Let me know if you need a link to either the scope or the Standardization news. The poster abstract is as follows:

Measurement of bulk composition of hydrocarbon groups and individual compounds in automotive spark-ignition fuels is important for quality control as well as ensuring compliance with various governmental regulations. As such, multiple ASTM methods exist for measuring various aspects of automotive fuel samples. Most of the methods are limited in scope to a subset of hydrocarbon groups or specific compounds of interest, meaning that multiple methods are required for typical production control. During the December 2016 D02 committee meeting ASTM International approved method D8071 for PIONA compound analysis by GC-VUV, known also as the "Standard Test Method for the Determination of Hydrocarbon Group Types and Select Hydrocarbon and Oxygenate Compounds in Automotive Spark-Ignition Engine Fuel using Gas Chromatography with Vacuum Ultraviolet Absorption Spectroscopy Detection (GC-VUV)." The method uses relatively simple instrumentation: a gas chromatograph, a standard 30 m nonpolar column, and a vacuum ultraviolet (VUV) absorption detector. VUV absorbance exhibits sensitivity to all major hydrocarbon classes that occur in automotive spark-ignition fuels, has a high degree of speciation capability, but also exhibits the intuitive property that species belonging to similar hydrocarbon groups have similar absorption characteristics. The setup procedure is straight-forward, with no pre-column tuning or valve timing adjustments. The ability to perform library searches and deconvolution using the inherently three-dimensional dataset means that the new method can use a wide retention index window, so it is more forgiving of variations in retention times than are methods that rely solely on retention time for compound identification or classification. Additionally, since the new method can handle co-elution among various species and hydrocarbon classes, analysis is faster since complete chromatographic separation of all components is not necessary. The new method results in a per-measurement information set that would typically require implementation of multiple ASTM methods, while being inherently more robust and production-worthy than the more comprehensive ASTM automotive fuel methods.

AUTHOR INDEX

International Association for Stability, Handling and Use of Liquid Fuels (IASH)
3525 Piedmont Road
5 Piedmont Center
Suite 300
Atlanta, GA 30305
USA

ISBN 978-1-5108-7066-6
9 781510 870666

CURRAN ASSOCIATES INC.
proceedings.com